微软技术丛书

Windows 程序设计

(第 6 版)

(美) Charles Petzold　著

张大威　汤　铭　段洪秀　译

清华大学出版社

北　京

内 容 简 介

作为 Windows 开发圣经的最新版本，本书的主题是 Windows 8 应用程序开发，全面介绍 Windows 程序设计所涉及的细枝末节，旨在帮助读者从高屋见瓴的角度建立完整的知识体系，为以后的职业生涯奠定良好的基础。全书共两部分 19 章。第 1～12 章着重介绍基础知识。第 13~19 章介绍侧重于 Windows 8 平台开发的触摸、位图、富文本、打印、富文本、GPS/传感器和手写笔等方面。

本书适合任何层次的 Windows 程序员阅读和参考，是帮助他们梳理和建立 Windows 知识体系的理想读物。

北京市版权局著作权合同登记号　图字：01-2007-5032

Authorized translation from the English language edition, entitled **Programming Windows, 6th Edition, Charles Petzold.** published by Pearson Education, Inc, publishing as Microsoft Press, Copyright © 2013 **Charles Petzold**.

图书在版编目(CIP)数据

Windows 程序设计/(美)佩措尔德(Petzold, C.)著；张大成，汤铭，段洪秀译. —6 版. —北京：清华大学出版社，2015(2022.9 重印)

(微软技术丛书)

书名原文：Programming Windows，6th Edition

ISBN 978-7-302-40237-4

Ⅰ. ①W…　Ⅱ. ①佩…　②张…　③汤…　④段…　Ⅲ. ①Windows 操作系统—程序设计　Ⅳ. ①TP316.7

中国版本图书馆 CIP 数据核字(2015)第 101286 号

责任编辑：文开琪
装帧设计：杨玉兰
责任校对：周剑云
责任印制：朱雨萌
出版发行：清华大学出版社

网　址：http://www.tup.com.cn, http://www.wqbook.com
地　址：北京清华大学学研大厦 A 座　　邮　编：100084
社 总 机：010-83470000　　邮　购：010-62786544
投稿与读者服务：010-62776969, c-service@tup.tsinghua.edu.cn
质量反馈：010-62772015, zhiliang@tup.tsinghua.edu.cn

印 装 者：三河市铭诚印务有限公司
经　销：全国新华书店
开　本：185mm×260mm　　印　张：53.25　　字　数：1270 千字
版　次：2015 年 6 月第 1 版　　印　次：2022 年 9 月第 7 次印刷
定　价：128.00 元

产品编号：048961-01

前　言

本书的主题是 Windows 8 应用程序开发。

在阅读本书前，需要一台运行 Windows 8 的计算机并安装 Windows 8 开发工具和软件开发包(SDK)，最简单的办法是下载免费的微软 Visual Studio Express 2012 for Windows 8。下载地址可以从 Windows 8 开发者中心获得：http://msdn.microsoft.com/windows/apps。简体中文版下载地址：http://dev.windows.com/zh-cn/。

要安装 Visual Studio，请单击页面上的“下载工具”连接，然后选择“查找 Visual Studio 的其他版本”。开发者中心主页还提供了注册 Windows 8 开发者账号和向 Windows Store 提交应用程序的相关帮助。

Windows 8 的版本

Windows 8 基本上与 Windows 7 类似，两者都可以运行于同类个人计算机，拥有 32 位和 64 位 Intel x86 微处理器系列的计算机。Windows 8 有一个是标准版本 Windows 8，另外还有一个版本 Windows 8 Pro，功能更多，针对的是技术爱好者和专业人士。

Windows 8 和 Windows 8 Pro 都可以运行两种程序：

- 桌面应用程序
- Windows 8 应用程序，往往也称为“Windows 应用商店应用程序”

桌面应用程序是指传统的 Windows 程序，这些应用程序可以运行于 Windows 7，通过 Windows 应用程序编程接口(Win32 API)与操作系统交互。为运行这些桌面应用程序，Windows 8 提供了人们熟悉的 Windows 桌面界面。

Windows 应用程序打破了传统 Windows 的一贯风格。这种应用程序一般在全屏模式下运行(但两种应用程序可以通过辅屏视图在同一个屏幕上显示)，专门为触摸和平板计算机优化。这些应用程序可以从微软运营的应用商店进行购买和安装。(开发者可以直接通过 Visual Studio 进行部署和测试。)

除了运行在 x86 处理器上的 Windows 8 版本，Windows 8 还有一个版本运行在 ARM 处理器上。这种处理器常见于廉价平板计算机和其他移动设备。此版本的 Windows 8 称为 Windows RT，都已经预装在这些设备上。最早运行 Windows RT 的计算机是微软 Surface。

Windows RT 除了预装的桌面应用程序只能运行 Windows 应用程序。我们不能在 Windows RT 上运行现有的 Windows 7 应用程序，包括 Visual Studio，因而不能在 Windows RT 上开发 Windows 8 应用程序。

Windows 8 用户界面采用了新的设计范式，与 Windows 应用程序的风格协调一致。受到都市标志的启发，这种设计范式通过高反差来突出内容，采用朴素的字体，具有简明的风格和基于磁贴(tile)的界面，并伴随过渡动画效果。

许多开发者最初是通过 Windows Phone 7 认识的 Windows 8 设计范式，可见微软对小尺寸和大尺寸计算机均做了精心设计。过去的几年中，微软试图使传统 Windows 桌面更适合小型设备(如手持计算机和手机)。如今，手机的用户界面设计理念已被带至平板和桌面。

这个全新的设计更注重多点触摸(multitouch)，这种操作方式在很大程度上改变了人机之间的关系。事实上，多点触摸这个词已经有些过时了，因为几乎所有新的触摸设备都能够响应多手指操作。因此直接用“触摸”便足以表达这个意思了。针对 Windows 8 应用程序的部分编程接口用一致的方式融合了触摸、鼠标和触笔输入，因而可以轻松通过这三种输入设备来使用应用程序。

本书要点

本书主要介绍 Windows 应用商店应用程序的开发。有许多介绍 Win32 桌面应用程序开发的图书，其中包括《Windows 程序设计(第 5 版)》。本书偶尔会提到 Win32 API 和桌面应用程序，但会将重点放在 Windows 8 应用程序的开发上。

为了编写这种应用程序，微软引入了一种全新的面向对象 API，称为 Windows 运行时(Windows Runtime)或 WinRT(请勿与运行在 ARM 处理器上的 Windows 8 版本 Windows RT 相混淆)。Windows 运行时内部基于组件对象模型(Component Object Model，COM)，通过/Windows/System32/WinMetadata 目录下扩展名为.winmd 的元数据文件向外暴露接口。从外部看，这套 API 完全是面向对象的。

从应用程序开发者的角度看，Windows 运行时集成了 Silverlight，但 API 内部并不是托管的。对于 Silverlight 开发者而言，最直接的变化或许是命名空间：Silverlight 的命名空间以 System.Windows 开头，新 API 的命名空间将其替换为 Windows.UI.Xaml。

大部分 Windows 8 应用程序不仅由代码构成，还包含标记。这些标记可能是工业标准化的超文本标记语言(HyperText Markup Language，HTML)，也可能是微软的可扩展应用程序标记语言(eXtensible Application Markup Language，XAML)。将代码与标记分离的好处之一是将编码人员与界面设计人员的工作分离。

目前，开发 Windows 8 应用程序的技术主要有 3 种编程语言和标记语言的组合可供选择。

- C++与 XAML
- C#/Visual Basic 与 XAML
- JavaScript 与 HTML 5

对以上三种组合均可以使用 Windows 运行时，但 Windows 运行时为不同语言提供了相应的编程接口。虽然我们不能在同一个项目中混用不同的语言，但可以通过创建一种带有.winmd 扩展名的库(Windows 运行时组件)来实现。这种库可以通过任何 Windows 8 语言访问。

C++程序员可以使用一种 C++的分支 C++组件扩展或 C++/CX。这种扩展使得该语言能够更好地利用 WinRT。C++程序员可以直接访问部分 Win32 和 COM API，也可以访问 DirectX。此外，C++程序可以被编译为本地机器代码。

使用托管语言 C#或 Visual Basic .NET 的程序员很容易上手 WinRT。使用此类语言编写的 Windows 8 应用程序不能像 C++程序那样轻松地访问 Win32、COM 或 DirectX API，但也是可以的。本书第 15 章提供了几个相关示例程序。WinRT 还提供了一个简化的.NET 基础类库，用于完成一些底层任务。

针对 JavaScript，Windows 运行时提供了 Windows Library for JavaScript 或简称 WinJS，使得用 JavaScript 编写的 Windows 8 程序能够调用许多系统级的功能。

经过慎重考虑，我决定在书中主要使用 C#和 XAML 介绍相关技术。托管语言在开发和调试上的优势说服了我，并且我认为 C#与 Windows 运行时最为匹配。希望 C++程序员能够快速熟悉 C#代码，从本书得到更多收获。

肯定还有其他很多 Windows 8 图书介绍如何使用其他语言来编写 Windows 8 应用程序。但我认为，一本书重点介绍一种语言比尝试同时涵盖多种语言更有价值。

不过，鉴于 C++和本地代码在高性能应用中的优势，我还是非常很愿意展开新一轮的讨论。没有哪种工具能够解决任何问题。未来，我会在自己的博客和 *MSDN Magazine* 进一步讨论针对 Windows 8 的 C++和 DirectX 开发。本书配套内容中的示例程序都配有对应的 C++版本。

循序渐进

在阅读本书之前，需要掌握必要的预备知识。首先要熟悉 C#。如果对此知之甚少，建议在阅读本书之前先学习 C#。如果在学习 C#之前有 C 或 C++背景，可以阅读免费的电子书*.NET Book Zero: What the C or C++ Programmer Needs to Know About C# and the .NET Framework*。该书有 PDF 和 XPS 两种格式可供下载，地址为 www.charlespetzold.com/dotnet。

本书假定你了解 XML(可扩展标记语言)的基本语法，因为 XAML 本身也是一种 XML。但本书假定你不熟悉 XAML 和任何基于 XAML 的编程接口。

本书是一本 API 参考书，不是编程工具参考书。本人唯一使用的编程工具是 Microsoft Visual Studio Express 2012 for Windows 8(本书一般简称为 Visual Studio)。

相比程序代码，标记语言一般更受工具的广泛支持。事实上，有些开发者甚至认为 XAML 这样的标记语言完全应该自动生成。Visual Studio 内建可交互的 XAML 设计器，我们可以将控件拖拽至页面。许多开发者逐渐熟悉并喜欢使用 Microsoft Expression Blend 来为他们的应用程序生成复杂的 XAML。Microsoft Expression Blend 包含在前面提到的开发工具和 SDK 中。

这种设计工具非常适合有经验的开发者，我认为新手最好从手工编写 XAML 学起。这也正是本书介绍 XAML 的过程。第 8 章介绍的 XAML Cruncher 工具正好印证了一个观点：虽然它能够在你输入 XAML 的过程中生成相应的对象，但不能为你写 XAML 代码。

也千万不要走向另一个极端。有的开发者非常善于写 XAML，以至于忘记如何在代码中创建和初始化相应的对象！我认为两者都很重要，因此本书将分别介绍如何用代码和标记来完成类似的任务。

在开始写本书之前，我构思了几种介绍 Windows 运行时的方案。一种方案是从底层图形与用户输入开始，演示控件的构建方法，然后介绍现有的控件。

但我最终选择首先介绍对主流开发者最重要的几种技术：在应用程序中组合预定义的控件，然后通过代码和数据将这些控件联系在一起。这就是本书第 I 部分(也就是前 12 章)

的主要内容。

第Ⅱ部分主要介绍底层及较为具体的功能，其中包括触摸、位图、富文本、打印以及屏幕方向和 GPS 传感器。

配套内容

学习新的 API 和学习打篮球与吹奏双簧管非常类似。旁观是很难学会的，必须亲自上手。本书配套的源代码可以通过 Companion Content 链接下载，网址为 https://www.microsoftpressstore.com/ store/programming-windows-9780735671768。

虽然有现成的代码，但最好自己键入代码，效果会更好一些。

计算机配置

为了写这本书，我用的是三星平板电脑 700T 的一个特殊版本。这款平板是 2011 年 9 月的 Microsoft 在 Build 大会上为参会人员提供的。因此，这款平板也有时称为 Build 平板。这款平板计算机配备 1.6 GHz 的 Intel Core i5 处理器，拥有 64 GB 的硬盘，屏幕支持 8 点触摸，分辨率为 1366 × 768 像素(本书绝大部分截图采用的是这个分辨率)，是辅屏视图模式的最低要求。

虽然这个 Build 平板预装的是 Windows 8 的开发者预览版(Developer Preview)，但随着时间的推移，我依次安装了 2012 年 3 月公布的消费者预览版(Consumer Preview，build 8250)、2012 年 6 月的发行预览版(Release Preview，build 8400)，最终安装的是 Windows 8 Pro 的正式发行版。除了测试方向传感器外，我一般都通过底座的 HDMI 接口将这台平板连到 1920×1080 分辨率的外部显示器，并使用外接键盘和鼠标。

微软的 Surface 平板电脑一上市，我就购买拉一台，用于程序测试。为了在 Surface 上部署和调试应用程序，我采用了 Tim Heuer 在博客中介绍的一种技术，网址为 http://timheuer.com/blog/archive/2012/10/26/remote-debugging-windows-store-apps-on-surface-arm-devices.aspx。这种技术在文档中的正式描述为："在远程计算机上从 Visual Studio 运行 Windows 应用商店应用"(http://msdn.microsoft.com/zh-cn/library/hh441469.aspx)。

在测试用到方向传感器的程序来时，采用 Surface 这样的物理平板设备是很有必要的。

大部分情况下，我使用的都是插在底座上的 Build 平板。通过外部的键盘、鼠标和显示器，我可以像以往那样使用微软 Visual Studio 和 Word，而让 Windows 8 程序在带有触摸屏的平板上运行。这样的开发环境我感到非常满意，尤其是和我当年写《Windows 程序设计(第 1 版)》时使用的配置相比。

不过，那已经是 25 年前的事了。

本书的前世今生

本书是《Windows 程序设计》的第 6 版。第 1 版是 1986 年受微软出版社的委托写的。之所以收到这个邀请，是因为我当时在为 *Microsoft Systems Journal(MSDN Magazine* 的前身)

写 Windows 编程系列文章。

那份合同我至今记忆犹新。

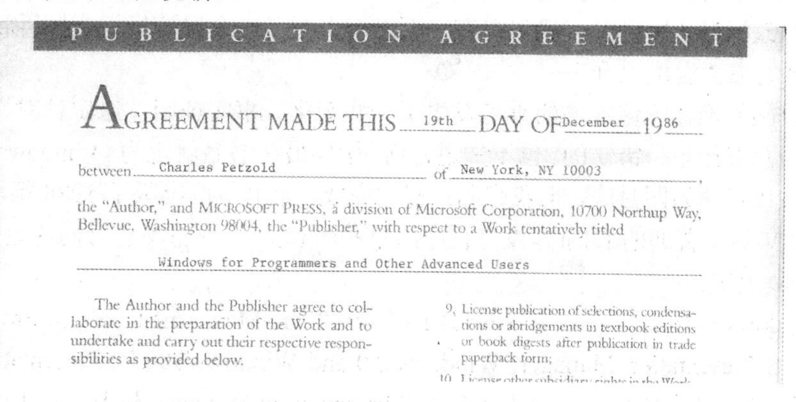

PUBLICATION AGREEMENT

AGREEMENT MADE THIS 19th DAY OF December 1986

between Charles Petzold of New York, NY 10003,

the "Author," and MICROSOFT PRESS, a division of Microsoft Corporation, 10700 Northup Way, Bellevue, Washington 98004, the "Publisher," with respect to a Work tentatively titled

Windows for Programmers and Other Advanced Users

The Author and the Publisher agree to collaborate in the preparation of the Work and to undertake and carry out their respective responsibilities as provided below.

9. License publication of selections, condensations or abridgements in textbook editions or book digests after publication in trade paperback form;

10. License other subsidiary rights in the Work

最值得一提的是它首页下部的一段。

II MANUSCRIPT

The Author agrees to prepare and submit one (1) clean copy and one (1) ASCII readable diskette of the final manuscript of the Work, equivalent to approximately 100,000 words, not later than April 30, 1987, the due date. (A full manuscript page of text consists of approximately 250 words.) The Author's final manuscript shall be in double-spaced typescript or its equivalent, satisfactory to the Publisher in organization, form, content, and style and accompanied by appropriate illustrative material, table of contents, tables, bibliography, and instructional aids ready for reproduction.

typescript(打字稿)这个词意味着，稿件至少要用打字机打印。要求规定了手稿使用双倍行距、固定字宽，每页大概有 250 个字。页数在 400 左右，也就是说微软出版社不希望内容太多。

为了写这本书，我采用了一台 IBM PC/AT 计算机，它采用 8 MHz 的 80286 微处理器、512 KB 的内存和 30 MB 的硬盘。显示器为 16 色的 IBM Enhanced Graphics Adapter，最大分辨率为 640×350。写前面几章时用的是 Windows 1(1985 年 11 月发行的)，后来用上了 Windows 2 的 beta 版本。

当时，编辑和编译 Windows 程序要在 Windows 以外的 MS-DOS 环境下进行。为了编辑源代码，我采用了 WordStar 3.3，也是我写书时所用的工具。MS-DOS 命令行下可以运行微软的 C 编译器，然后在 Windows 上进行运行和测试。为了再次进行编辑和编译，需要退出 Windows，回到 MS-DOS 下。

1987 年有一段时间，我为了写书几乎废寝忘食，越熬越晚，甚至黑白颠倒。当时我家里还没有电视，但会收听地方广播电台 WNYC-FM 不间断播放的古典音乐和美国国家公共电台的节目。在那段时间里，我在 *Morning Edition* 结束后睡觉，刚好在 *All Things Considered* 开始前起床。[①]

① 译注：这是美国国家公共电台最受欢迎的两档新闻广播节目。按美国东部时间，前者一般从早上 5 点持续到 9 点，后者一般从 16 点持续到 18 点。

根据合同约定，我要以磁带和纸质形式将书稿提交给微软出版社。虽然当时我们都有电子邮件，但那时的电子邮件还不支持附件。编辑好的书稿通过邮政包裹发回给我，上面带有许多修改标记和即时贴。我记得有一页上有人画了一个温度标记来倾诉我不断提交的书稿，上面写着“温度上升”！

随着时间的推移，这本书的重点发生了一些变化。事实证明，原定针对程序员和其他高级用户的写书计划是错误的。也不知道是何原因，最终书名就定为《Windows 程序设计》。

合同截止日期为四月底，但我直到八月才完成。本书最终出版于 1988 年，页数达到了 850 页。如果这是普通的书稿(即没有程序清单和图表)，字数将不止合同约定的 10 万字，而会达到 40 万字。

《Windows 程序设计》第 1 版的封面上写着“The Microsoft Guide to Programming for the MS-DOS Presentation Manager: Windows 2.0 and Windows/386”，Presentation Manager[①] 这个名称让我想起针对 Windows 和 OS/2 的 Presentation Manager 和平共存于两个操作系统环境下的那段时光。

《Windows 程序设计》的第 1 版在编程社区中并没有引起太多关注。当 MS-DOS 程序员意识到有必要学习全新的 Windows 时，第 2 版(1990 年，Windows 3)和第 3 版(1992 年，Windows 3.1)也相继出版了。

Windows API 从 16 位升级到 32 位后，《Windows 程序设计》的第 4 版(1996 年，Windows 95)和第 5 版(1998 年，Windows 98)陆续出版。第 5 版目前还在出版发行，根据我收到的读者来信分析，该书在印度和中国最受欢迎。

在前面 5 个版本中，我们用的都是 C 语言。在第 3 版问世后，第 4 版出版之前，我的好友 Jeff Prosise 说他想写《Windows 程序设计与 MFC》，我表示支持。我当时并不太在意 Microsoft Foundation Classes(MFC)，因为我那时认为它只不过是 Windows API 的轻型包装程序，而且我当时也没有深入钻研 C++。

有些程序员希望了解核心机制而不关心程序代码与操作系统之间的细枝末节。随着时间的推移，《Windows 程序设计》在这些程序员当中赢得了广泛的赞誉。

《Windows 程序设计》的早期版本并非如此。当时，为接触到核心需要使用汇编语言将字符直接输出到图形显示内存区域，而只能借助于 MS-DOS 来进行文件 I/O。相对而言，Windows 程序员使用的是高级语言，图形完全没有加速，访问硬件只能通过层层 API 和设备驱动程序来完成。

从 MS-DOS 到 Windows 的变迁意味着用一定的速度和效率来换取某些优势。何种优势？很多经验丰富程序员也说不准。是图形、图片、颜色、漂亮的字体，还是鼠标？这些都不是计算机的全部！一些怀疑论者称其为 WIMP(window-icon-menu-pointer，窗口-图标-菜单-指针)界面。其实，这些特点并不是人们选择这个环境并为其编写代码的动机。

随着时间的推移，高级语言会变成低级语言，多层次接口最终会简化为本地 API(至少在某种语言当中)。如今，有些 C/C++程序员因为效率问题抵触 C#这样的托管语言，而 Windows 也一再成为人们争论的焦点。Windows 8 是 Windows 自 1985 年问世以来革命性地一次升级，但为主流桌面系统引入针对智能电话和平板电脑设计的触摸界面，这种做法

① 译注：Presentation Manager 是 IBM 和微软在 1988 年末引入的一种图形用户界面（GUI）。

备受老版本 Windows 的用户质疑，有的用户甚至会为无法找到熟悉的功能发牢骚。

伴随着 Windows 令人激动和饱受争议的新用户界面，以及现代风格的 API 和编程语言的问世，《Windows 程序设计》也开始掀开新的篇章。

未来计划

在我的编程生涯中，Windows 8 可能会占据一段时间。也就是说我会发表一系列 Windows 8 编程的博文。可以从该网址访问我的博客和订阅 RSS 源：www.charlespetzold.com。

我喜欢解决编程方面的疑难杂症并在博客中与大家分享。如果您有关于 Windows 8 编程方面的疑问希望与我共同探讨并让我尝试解决，可以发送电子邮件给我：cp@charlespetzold.com。

从 MSDN Magazine 的 2013 年 1 月刊开始，我每月为 DirectX Factor 栏目写一篇文章，主要讨论在 Windows 8 和 Windows Phone 8 应用程序中使用 DirectX 的相关议题。MSDN Magazine 可以通过这个网址免费阅读：http://msdn.microsoft.com/magazine。

鸣谢

微软出版社的 Ben Ryan 和 Devon Musgrave 设计了一种出版方式，先将书稿部分章节发表到开发者社区，同时也为最终版本的销售做准备。本书也是以这种方式与大家见面的。

Devon 和我的技术编辑 Marc Young 的部分职责是尽量多发现本书中文字和代码中的错误，以免我陷入尴尬境地。他们确实找到了不少，非常感谢！

感谢 Andrew Whitechapel 对 C++示例代码的反馈。感谢 Brent Rector 通过电子邮件在对某个触摸相关问题为我提供了关键的解决方案，也感谢他为我介绍 IBuffer 的背景知识。感谢 Robert Levy 在触摸技术方面为我提供的反馈。感谢 Prosise 在我遇到疑难问题时给予我关键的支持。感谢 Larry Smith 发现我文字中的诸多错误。感谢 Admiral 鼓励我使 C++开发者也能够从本书中受益。

当然，书中遗留的错误是我的问题。欢迎读者通过本书前面给出的地址将发现的问题发送到出版商，如果是大问题，我也会在 www.charlespetzold.com/pw6 列出。

最后，感谢我的妻子 Deirdre Sinnott 对我的爱与支持。感谢她在生活上做出必要的调整，只为了让我好好写这本书。

Charles Petzold
美国纽约州罗斯科镇与纽约市
2012 年 12 月 31 日

勘误与支持

我们尽力避免本书及其配套内容中出现错误，但错误在所难免。本书出版后报告的错误将在微软出版社的网站上公布。可以通过 https://www.microsoftpressstore.com 搜索本书，然后单击 Errata & Updates 链接。如果发现有未报告的错误，请通过 Submit Errata 链接提交。

如果需要额外的支持，可以发电子邮件到微软出版社的客服，邮箱地址为 mspinput@microsoft.com。

注意，以上地址不提供对微软产品的支持。

意见反馈

在微软出版社，您的满意是对我们最大支持，您的反馈是我们最宝贵的财富。欢迎通过以下网址告诉我们您对本书的看法。

http://aka.ms/tellpress

反馈请尽量简短，我们会阅读您的每一条意见和建议。再次感谢您对我们的支持！

保持联系

让我们保持联系！我们的 Twitter 地址为 http://twitter.com/MicrosoftPress。

目　　录

第I部分　基础知识

第II部分 Windows 8 新特性

第 I 部分　基础知识

第 1 章　标记与代码

自 Brian Kernighan 与 Dennis Ritchie 的经典著作《C 程序设计语言》出版以来，为初学者展示某种“hello, world”程序便成了介绍编程技术前的一种惯例。接下来我们也不妨创建一些针对 Windows 8 的“hello, world”程序。

本书假定读者已安装 Windows 8 及能够创建 Windows 8 应用程序的新版 Microsoft Visual Studio。

下面就让我们在 Windows 8“开始”界面启动 Visual Studio，开始编程吧！

1.1　第一个项目

在 Visual Studio 的首页上，默认选中的是“入门”选项卡。首页的左侧窗格中有个“新建项目”选项。单击它或从“文件”菜单中选择“新建项目”。

“新建项目”对话框打开后，在左侧窗格中选择“模板”，并依次选择 Visual C# | “Windows 应用商店”。在中间区域列出的模板中，选择“空白应用程序”。在对话框底部的“名称”一栏中填写项目名称，这里不妨称其为“Hello”。让“解决方案名称”一栏与之保持一致即可。使用“浏览”按钮可以更改程序创建的位置。下面我们单击“确定”按钮。对于 Visual Studio，本书将会使用“单击”(click)一词；对于所要创建的 Windows 8 应用程序，则会使用“点击”(tap)一词。Visual Studio 多年前就已对触控操作进行了优化。

Visual Studio 会创建一个名为 Hello 的解决方案、一个名为 Hello 的项目以及包含在 Hello 项目中的各种文件。这些文件会显示在 Visual Studio 右侧的解决方案资源管理器中。一般来讲，新建的 Visual Studio 的解决方案会包含一个项目，但也可以包含额外的应用程序项目和类库项目。

如下图所示，当前项目的文件列表中有一个名为 MainPage.xaml 的文件，单击该文件旁边的箭头后会看到子节点是一个名为 MainPage.xaml.cs 的文件。

为查看文件的内容，可以双击文件名，也可以右击文件名并选择“打开”命令。

由于 MainPage.xaml 和 MainPage.xaml.cs 文件都是 MainPage 类定义的一部分，因而在解决方案资源管理器中两者是关联在一起的。对于 Hello 这样简单的程序，MainPage 类包含的便是应用程序的全部可视元素和界面。

MainPage.xaml.cs 这个文件的名称有点特别。其扩展名.cs 代表“C Sharp”。以下代码去掉 MainPage.xaml.cs 中的所有注释，仅保留其主体结构。

```
using System;
using System.Collections.Generic;
using System.IO;
using System.Linq;
using Windows.Foundation;
using Windows.Foundation.Collections;
using Windows.UI.Xaml;
using Windows.UI.Xaml.Controls;
using Windows.UI.Xaml.Controls.Primitives;
using Windows.UI.Xaml.Data;
using Windows.UI.Xaml.Input;
using Windows.UI.Xaml.Media;
using Windows.UI.Xaml.Navigation;

namespace Hello
{
   public sealed partial class MainPage : Page
   {
      public MainPage()
      {
         this.InitializeComponent();
      }

      protected override void OnNavigatedTo(NavigationEventArgs e)
      {
      }
   }
}
```

文件的开始部分通过 using 语句列出了我们可能要用到的命名空间。对于这些命名空间，MainPage.xaml.cs 文件一般不会全部用到，而有时还需额外添加。

按命名空间的前缀，我们可以将这些命名空间分成两大类。

- **System.*** 包含针对 Windows 8 应用程序的.NET 类型
- **Windows.*** 包含 Windows Runtime(或称 WinRT)类型

通过以上 using 语句列表不难看出，带有 Windows.UI.Xaml 前缀的命名空间在 Windows Runtime 中占有重要地位。

在 using 语句下面，MainPage.xaml.cs 文件定义了名为 Hello 的命名空间(与项目名称相同)和名为 MainPage 且派生自 Page 的类。其中，Page 类是由 Windows Runtime 提供的。

Windows 8 API 文档是按命名空间组织的，因此要找到 Page 类需知道其所归属的命名空间。在打开的 MainPage.xaml.cs 源代码中，我们将鼠标悬停在 Page 上面便会发现该类属于 Windows.UI.Xaml.Controls 命名空间。

MainPage 类的构造函数调用了 InitializeComponent 方法(本章稍后会具体介绍)。该类还重写了 OnNavigatedTo 方法。Windows 8 应用程序一般具有类似网页的页面导航结构，因此应用程序中往往包含多个派生自 Page 的类。在导航方面，Page 类定义了几个虚方法，分别是 OnNavigatingFrom、OnNavigatedFrom 和 OnNavigatedTo。其中，OnNavigatedTo 适合用于在页面被激活后执行某些初始化操作。在本书的前几章中，大部分示例都只有一个页面。这里倾向于用“页面”(page)一词，而不使用“窗口”(window)。当然，在应用程序底层，仍存在一个窗口，但其作用远远不及页面。

MainPage 类定义具有一个 partial 关键字①。这个关键字标明该类的定义可以存在于不同的 C#源代码文件中，而实际也如此(稍后便不难发现)。但直观来看，MainPage 类缺失的定义并非来自某个 C#代码文件，而由 MainPage.xaml 文件提供。

```
<Page
    x:Class="Hello.MainPage"
    xmlns="http://schemas.microsoft.com/winfx/2006/xaml/presentation"
    xmlns:x="http://schemas.microsoft.com/winfx/2006/xaml"
    xmlns:local="using:Hello"
    xmlns:d="http://schemas.microsoft.com/expression/blend/2008"
    xmlns:mc="http://schemas.openxmlformats.org/markup-compatibility/2006"
    mc:Ignorable="d">

    <Grid Background="{StaticResource ApplicationPageBackgroundThemeBrush}">

    </Grid>
</Page>
```

组成该文件的标记遵循一种称为“可扩展应用程序标记语言”(eXtensible Application Markup Language，缩写为 XAML，读作“zammel”)的标准。不难发现，XAML 建立在“可扩展标记语言”(eXtensible Markup Language，XML)之上。

一般来讲，我们在 XAML 文件中定义页面的所有可视元素，然后通过 C#文件来处理其余的事务(如执行逻辑计算和响应用户输入)。这个 C#文件通常被称为 XAML 文件对应的“代码隐藏文件”(code-behind file)。

这个 XAML 文件的根节点是 Page。读者可能已经猜到它来自于 Windows Runtime，但请注意 x:Class 特性②。

```
<Page
    x:Class="Hello.MainPage"
```

x:Class 特性只能出现在 XAML 文件的根节点。上述代码我们可以这样理解：在 Hello 命名空间中定义一个名为 MainPage 的类，并使其派生自 Page。没错！这两行代码与代码隐藏文件的类定义表达了相同的意思。

随后的几行代码是 XML 命名空间声明。事实上，这些 URI 并非指向某些网页，而只是公司或组织维护的唯一标识。下面两行是其中比较重要的。

```
xmlns="http://schemas.microsoft.com/winfx/2006/xaml/presentation"
xmlns:x="http://schemas.microsoft.com/winfx/2006/xaml"
```

数字 2006 暗示了 Windows Presentation Foundation 和 XAML 问世的年份。WPF 曾是.NET Framework 3.0 的一部分，之前被称为 WinFX。这就是为什么 winfx 出现在这两个 URI 中的原因。在一定条件下，XAML 文件可以兼容于 WPF、Silverlight、Windows Phone

---

① 译注：具有这个关键字的类被称为“分部类”(partial class)。有关分部类的更多信息，请阅读“C# 编程指南”之“分部类和方法”，网址：http://msdn.microsoft.com/zh-cn/library/wa80x488.aspx。

② 译注：有的文献也将“attribute”一词其翻译为“属性”。由于“property”也被翻译为属性，这里将前者翻译为“特性”以作区分。请读者注意两者形式上的区别。

和 Windows Runtime，前提是 XAML 文件使用的类、属性和功能在上述环境中是共存的。

代码中，第一个命名空间声明没有前缀，用于引用 Windows Runtime 中的公共类、结构和枚举。这些类型包含了 XAML 文件中出现的所有控件和其他元素，也包含本示例中出现的 Page 和 Grid 类。能够使用 XAML 的应用程序不止一种。URI 中的“presentation”代表可视用户界面，因而我们可以根据这一段来区分应用程序的类型。例如，在 Windows Workflow Foundation(WF)中使用的 XAML，其默认命名空间 URI 中会出现“workflow”一词。

第二个命名空间声明将前缀“x”与 XAML 本身固有的元素和特性关联。Windows Runtime 应用程序中只涉及 9 个特性，其中的 x:Class 显然是最常见的。

第三个命名空间也值得一提：

```
xmlns:local="using:Hello"
```

这个声明将 XML 前缀“local”与该应用程序的 Hello 命名空间关联。开发者在应用程序中自行创建的类便可以在 XAML 中用 local 前缀来引用。如果需要引用核心库中的类，还需要定义额外的 XML 命名空间声明来引入这些库的程序集名称和命名空间名称，详情参见后文。

随后的命名空间声明针对的是 Microsoft Expression Blend。Expression Blend 可能会插入一些自用的特殊标记，而这些标记 Visual Studio 编译器应忽略，这就是出现 Ignorable 特性的原因。同样，Ignorable 特性也属于一个命名空间，因而也需要相应的声明。

Page 元素有一个名为 Grid 的子元素。它对应于 Windows.UI.Xaml.Controls 命名空间中定义的 Grid 类。该类很常用，是一种“容器”(container)，因为它能够容纳其他可视对象。由于 Grid 类派生自 Panel 类，因而该类也常被归为“面板”(panel)。在 Windows 8 应用程序的布局方面，派生自 Panel 的类起着至关重要的作用。Visual Studio 为我们创建的 MainPage.xaml 文件中，通过一个预定义的标识符，Grid 被设置了一个背景色(实际为一个 Brush 对象)。第 2 章将具体介绍这种语法。

通常，我们按行和列对 Grid 进行分拆，从而获得单元格(详情参见第 5 章)。这很像是一种改进的 HTML 表格。没有行和列的 Grid 一般被称为“单格 Grid”(single-cell Grid)，它同样非常有用。

我们下面使 Windows Runtime 显示一行文本。为此，要用到 TextBlock 类(也是 Windows.UI.Xaml.Controls 命名空间中定义的)。将 TextBlock 置于单格 Grid 中，然后给它的几个特性赋值，这些特性实际上是 TextBlock 类定义的。

```
项目: Hello | 文件: MainPage.xaml(片段)
<Grid Background="{StaticResource ApplicationPageBackgroundThemeBrush}">
    <TextBlock Text="Hello, Windows 8!"
               FontFamily="Times New Roman"
               FontSize="96"
               FontStyle="Italic"
               Foreground="Yellow"
               HorizontalAlignment="Center"
               VerticalAlignment="Center" />
</Grid>
```

注意　在本书中，有的代码块会带有这样的标题。读者可以根据这个标题在本书的配套资源中找到对应的代码。虽然书中只展示完整文件中的一部分，但通过上下文足可表明其含义。

这些特性的顺序和缩进都无关紧要。为简单起见，可忽略其他特性，只添加 Text。在键入的过程中，会发现 Visual Studio 的“智能感知”(IntelliSense)功能会提示属性的名称和可选的值，我们从中选取即可。在 TextBlock 元素添加完成后，Visual Studio 的设计界面便会展示页面的预览效果。

我们也可以从 Visual Studio 的工具箱中将 TextBlock 直接拖入，然后在表格中设置其属性。但本书并没有这样做，而是指导读者像真正的程序员那样手工输入代码和标记。

按 F5 键或在“调试”菜单中选择“启动调试”，从而编译并运行此程序。即便对于如此简单的程序，最好也通过 Visual Studio 调试器运行。如果一切顺利，你会看到如下图所示的界面。[①]

TextBlock 的 HorizontalAlignment 和 VerticalAlignment 属性设置会使文本居中显示。程序员没必要自己根据显示器和文本的大小计算文本的位置。可以分别尝试将 HorizontalAlignment 设置为 Left 或 Right，将 VerticalAlignment 设置为 Top 或 Bottom，以体验将 TextBlock 置于 Grid 9 个位置的效果。正如第 4 章所要介绍的，Windows Runtime 支持使用像素精确定位可视对象，但通常我们会使用内置的布局功能。

TextBlock 还有 Width 和 Height 属性，但一般无需设置。对于本示例，如果设置了示例中 TextBlock 的这两个属性，可能导致文字被截断或影响文本在页面上的定位。TextBlock 自己知道如何对两者进行设置。

读者可能会在某个能感知方向变化的设备上运行此程序(如在平板电脑上)。果真如此，则会注意到，该页面的内容能够自动适应屏幕方向和横宽比，而无需程序控制。Grid、TextBlock 和 Windows 8 布局系统能够自动完成此项工作。

为终止 Hello 程序，我们可以按快捷键 Shift+F5，也可以从“调试”菜单中选择“停止调试”。该程序不仅被执行，还被部署到 Windows 8 上，并且可以从“开始”屏幕启动。我们创建的这个项目的磁贴(tile)并不美观。该程序的磁贴图片存储在项目的 Assets 目录中，

---

① 译注：如果出现错误，在调试模式下，Visual Studio 一般会直观地报出错误的位置及相关信息。否则，程序可能“闪退”，让开发者不知所措。

可直接替换掉它们。(本书配套示例代码中的磁贴图片已被替换。)我们不必使用 Visual Studio 调试器，从 Windows 8“开始”屏幕可直接运行该程序。

此外，程序也可以在模拟器中调试和运行。我们可以控制模拟器的分辨率、屏幕方向和其他特性。在 Visual Studio 工具栏中，有一个显示为“本地计算机”的下拉列表，将其改为“模拟器”即可。

1.2 图片的使用

传统的“Hello, World”程序是以文本形式显示的，但也不一定拘泥于此。示例项目 HelloImage 通过以下代码从作者的网站上获取一个图片。

项目：HelloImage | 文件：MainPage.xaml(片段)

```
<Grid Background="{StaticResource ApplicationPageBackgroundThemeBrush}">
  <Image Source="http://www.charlespetzold.com/pw6/PetzoldJersey.jpg" />
</Grid>
```

Windows.UI.Xaml.Controls 命名空间下定义的 Image 元素是 Windows Runtime 程序显示位图的标准方式。在默认情况下，图片会被缩放以适应可用空间，但会保持图片的原始长宽比(如下图所示)。

如果页面的大小发生变化，比如改变屏幕方向或切换到辅屏视图(snap view)，图片的大小也将随之改变。

我们可以通过 Image 类定义的 Stretch 属性来修改这个默认的显示行为。该属性的默认值为枚举成员 Stretch.Uniform。下面我们尝试将其设置为 Fill。

```
<Grid Background="{StaticResource ApplicationPageBackgroundThemeBrush}">
  <Image Source="http://www.charlespetzold.com/pw6/PetzoldJersey.jpg"
       Stretch="Fill" />
</Grid>
```

这样，图片的原长宽比被忽略，从而填充整个容器，如下图所示。

将 Stretch 属性设置为 None，则会按原像素值(320×400)显示图片，如下图所示。

与 TextBlock 一样，我们也可以通过 HorizontalAlignment 和 VerticalAlignment 属性控制图片在页面上的位置。

UniformToFill 是 Stretch 属性的第四个选项，可使图片在保持长宽比不变的前提下充满整个容器。但为了做到这一点，只有一个办法，即对图片进行裁剪。至于哪一部分被裁掉，则取决于 HorizontalAlignment 和 VerticalAlignment 属性的设置。

通过 Internet 获取图片依赖于网络连接，并且耗时。为确保图片能够立即显示，可将图片与应用程序本身捆绑。

下面我们通过 Windows 的画图程序创建一个简单位图。运行画图程序，单击“文件”|“属性”，设置图片的大小(如宽 480，高 320)。通过鼠标、手指和笔，我们可以画出自己的问候语。

除了 BMP、JPEG、PNG 和 GIF 这几种常见格式外，Windows Runtime 还支持其他几种。对于上面这张图片，我们不妨将其保存为常见的 PNG 格式，将其命名为 Greeting.png。

下面创建一个新项目，命名为 HelloLocalImage。人们往往愿意将项目用到的图片存储到名为 Images 的目录下。在解决方案资源管理器中，右击项目名称，选择“添加”|“新建文件夹”。如果项目节点在解决方案资源管理器中已被选中，也可以在“项目”菜单中选择“新建文件夹”。将该文件夹命名为 Images。

右击 Images 文件夹，选择“添加”|“现有项”。找到刚才保存的 Greeting.png 文件，然后单击“添加”按钮。在 Greeting.png 文件添加到项目中之后，我们需要确保其作为应用程序的内容存在。在该文件上右击，选择“属性”。在“属性”窗格中，确认“生成操作”的值为“内容”。

引用该图片的 XAML 标记与之前从网上获取图片所使用的标记类似。

```
项目: HelloLocalImage| 文件: MainPage.xaml(片段)
<Grid Background="{StaticResource ApplicationPageBackgroundThemeBrush}">
   <Image Source="Images/Greeting.png"
          Stretch="None" />
</Grid>
```

注意，Source 属性值应包含文件夹和文件的名称。最终效果如下图所示。

有的程序员也会将存储应用程序位图的文件夹命名为 Assets。您或许已经发现，标准的项目模板已将程序的图标放在 Assets 文件夹中。我们同样可以直接利用该文件夹，无需另行创建。

1.3　文字的变形

Grid、TextBlock 和 Image 有时会被看作“控件”。这或许因为人们知道其来自于 Windows.UI.Xaml.Controls 命名空间。但从严格意义上讲，它们并非控件。虽然 Windows Runtime 定义了一个名为 Control 的类，但这 3 个类没有派生自它。下面展示了我们目前遇到的类的层次结构。

```
Object
  DependencyObject
    UIElement
      FrameworkElement
        TextBlock
        Image
        Panel
          Grid
        Control
          UserControl
            Page
```

Page 派生自 Control 类，但 TextBlock 和 Image 并没有，而是派生自 UIElement 的子类 FrameworkElement。正因如此，与普通 XML 文件中的内容一样，我们也将 TextBlock 和 Image 称为“元素”(element)。

元素与控件的区别并不明显，但仍有必要弄清楚。从直观上看，控件由元素构成，其外观可通过模板进行定制。Grid 也是一种元素，但我们一般称其为“面板”(panel)，我们接下来会看到其独特之处。

下面做这样一个试验，在 Hello 项目中，将 Foreground 和所有与字体有关的特性从 TextBlock 元素移至 Page 元素。此时的 MainPage.xaml 文件如下所示。

```
<Page
    x:Class="Hello.MainPage"
    xmlns="http://schemas.microsoft.com/winfx/2006/xaml/presentation"
    xmlns:x="http://schemas.microsoft.com/winfx/2006/xaml"
    xmlns:local="using:Hello"
    xmlns:d="http://schemas.microsoft.com/expression/blend/2008"
    xmlns:mc="http://schemas.openxmlformats.org/markup-compatibility/2006"
    mc:Ignorable="d"
    FontFamily="Times New Roman"
    FontSize="96"
    FontStyle="Italic"
    Foreground="Yellow">

    <Grid Background="{StaticResource ApplicationPageBackgroundThemeBrush}">
        <TextBlock Text="Hello, Windows 8!"
                   HorizontalAlignment="Center"
                   VerticalAlignment="Center" />
    </Grid>
</Page>
```

我们会发现显示效果并未发生改变。这也就是说，针对 Page 元素设置的特性会作用于页面上的所有元素。

下面我们将 TextBlock 的 Foreground 属性设置为 Red，看看会发生什么。

```
<TextBlock Text="Hello, Windows 8!"
           Foreground="Red"
           HorizontalAlignment="Center"
           VerticalAlignment="Center" />
```

Page 的黄色设置被局部改写成了红色。

上面定义的 Page、Grid 和 TextBlock 被称为元素的“可视树”。只不过在 XAML 文件中，这棵树是上下颠倒的，即最上面的 Page 是树的主干，其下面的子孙(Grid 和 TextBlock)构成分支。这让人联想到 Page 定义的字体和 Foreground 属性值会沿着可视树的脉络从父元素传播到子元素。这样看没错，只不过有一点需要注意，即 Grid 并未定义这些属性。此外，在 TextBlock 和 Control 中这些属性的定义也是相互独立的。也就是说，尽管沿途的元素具有不同“基因”，但属性值从 Page 到 TextBlock 的传播仍能得以完成。

如果在文档中查看 Page 或 TextBlock 类的属性，则会发现同一属性具有两个名称不同的定义。例如，TextBlock 具有一个类型为 double 的 FontSize 属性。

```
public double FontSize { set; get; }
```

TextBlack 还有一个类型为 DependencyProperty 的 FontSizeProperty 属性。

```
public static DependencyProperty FontSizeProperty { get; }
```

请注意，FontSizeProperty 属性是一个只读的静态属性。

用于构建 Windows 8 应用程序用户界面的类，有很多都同时具有常规属性和类型为 DependencyProperty 的“依赖属性”(dependency property)。值得一提的是，在上文介绍的类层次结构中，有一个名为 DependencyObject 的类。这两个类型具有一定联系：派生自 DependencyObject 的类通常会声明 DependencyProperty 类型的只读静态属性。DependencyObject 和 DependencyProperty 都定义在 Windows.UI.Xaml 命名空间下，这表明两者对整个 UI 系统起着基础性作用。

依赖属性旨在解决一些与复杂用户界面有关的基础性问题。在 Windows 8 应用程序中，我们能够以多种方式设置属性。例如，正如前文所介绍的，对象的属性可以直接设置，也可以从可视树继承。正如下一章所要介绍的，属性设置可能还来自于样式定义。我们还会看到通过动画设置属性，后面的几章会详细介绍这方面内容。DependencyObject 和 DependencyProperty 类能够为不同属性设置方式设定优先级，从而协调属性在系统中的设置顺序。这里暂不深入讲解该机制，相信读者会在自己定义控件的时候有切身体会。

FontSize 属性背后也伴随着一个名为 FontSizeProperty 的依赖属性。为了简便起见，人们有时会用 FontSize 来指代依赖属性，这并不会产生误解。

UIElement 及其子类定义的许多属性都是依赖属性，但其中只有一部分会沿可视树传播。上文介绍的 Foreground 和与字体有关的属性是可传播的。另外还有一些，本书会在遇到时予以提示。依赖属性具有默认值。对于示例项目 Hello，如果删掉 TextBlock 和 Page 中除 Text 以外的特性，程序会在页面的左上角显示一行 11 像素大小的、采用系统字体的文本。

FontSize 属性以像素为单位，用于设置字体的设计高度。设计高度包含下伸部分(descender)和变音符号(diacritical mark)所占高度。我们一般以磅(point)为单位指定字体大小。一磅等于 1/72 英寸。在实际的设备上，为在像素和磅之间进行换算，我们需要知道显示器的分辨率，即每英寸点数(dots-per-inch，DPI)。在默认条件下，系统假定显示器的分辨率为 96 DPI。字体大小 96 像素等于 72 磅(1 英寸)，那么默认的 11 像素就是 8¼磅。

对于高分辨率的显示器，Windows 会自动调整显示元素的大小与坐标。应用程序可以通过 DisplayProperties 类获得有关信息，该类几乎是 Windows.Graphics.Display 命名空间最重要的类型了。在大多数情况下，将显示器和打印机分辨率假定为 96 DPI 是没有问题的。基于这个假设，我们便可以推算出英寸与像素的常用换算：48(1/2″)，24(1/4″)，12(1/8″)和 6(1/16″)。

如果移除 Foreground 特性，那么系统会以黑底白字显示文本。背景实际并未被设置为黑色，而是由 Grid 引用的预定义的 ApplicationPageBackgroundThemeBrush 标识符所决定的。

示例项目 Hello 还包含一对文件：App.xaml 和 App.xaml.cs。两者共同定义了名为 App 的派生自 Application 的类。虽然每个应用程序可以有多个 Page 类的派生类，但只能有一个 Application 的子类。App 类负责提供设置并管理能够影响整个应用程序的活动。

下面做一个实验：在 App.xaml 文件的根元素中，将 RequestedTheme 特性设置为 Light。

```
<Application
   x:Class="Hello.App"
   xmlns="http://schemas.microsoft.com/winfx/2006/xaml/presentation"
   xmlns:x="http://schemas.microsoft.com/winfx/2006/xaml"
   xmlns:local="using:Hello"
   RequestedTheme="Light">
   ...
</Application>
```

该特性只有 Light 和 Dark 两项设置。此时我们重新编译并运行该程序便会发现背景已经变为浅色。这说明 ApplicationPageBackgroundThemeBrush 标识符引用的颜色发生了变化。如果 Page 或 TextBlock 的 Foreground 未被具体设置，那么文字将显示为黑色。这也就是说，Foreground 属性在不同主题下具有不同的默认值。

书中的大部分示例都默认采用浅色(Light)主题，因为这样印刷效果更好，耗费的墨水也更少。但是要注意，许多小型设备和与日俱增的较大的设备，都配备 OLED(有机发光二极管)技术的显示屏。因而降低其亮度，可以减少电能的消耗。降低电能消耗是深色主题(Dark)被广泛采用的原因之一。

当然，Grid 的 Background 属性和 TextBlock 的 Foreground 属性完全可以自行设置。

```
<Grid Background="Blue">
   <TextBlock Text="Hello, Windows 8!"
              Foreground="Yellow"
              ... />
</Grid>
```

Visual Studio 的“智能感知”功能可以为这类属性提供 140 个标准颜色名称及 Transparent(透明)选项。这些选项都是 Colors 类的静态属性。除了指定名称外，我们还能够以井号打头，通过十六进制的“红绿蓝”(RGB)值来设置颜色，每种色值的范围从 00 到 FF。

```
Foreground="#FF8000"
```

这个颜色具有饱和的红色，半成绿色，没有蓝色。色值的最前面还可以有一个代表 α 通道的可选字节①，用于指定不透明度，其取值范围也从 00 到 FF。下面是一个半透明红色

① 译注：每两个十六进制字符代表一个字节。

的例子。

```
Foreground="#80FF0000"
```

指定 α 值的颜色有时被称为 ARGB 颜色。UIElement 类还定义了一个 Opacity 属性，取值范围介于 0(完全透明)到 1(完全不透明)之间。在 HelloImage 项目中，读者可以尝试将 Grid 的 Background 属性设置为除黑色以外的某种颜色(如 Blue)，然后将 Image 元素的 Opacity 属性设置为 0.5。

字节形式指定的色值与 sRGB(标准 RGB)颜色空间一一对应。该颜色空间的历史可追溯到阴极射线管(CRT)显示器时代。CRT 显示器根据颜色字节控制点亮各像素的电压。非常巧合的是，当时显示器像素亮度的非线性和人类肉眼对亮度感知的非线性叠加后在一定程度上相互抵消，使得这些字节式色值近似表现为线性，或者说准线性。

此外，还有一种名为 scRGB 的颜色空间，通过每种颜色的光线强度表示，每段的取值范围从 0 到 1。下面是中度灰的例子。

```
Foreground="sc# 0.5 0.5 0.5"
```

由于人类肉眼对光强的反应呈对数关系，因而这个灰看起来偏亮，称不上是中度的。

如果需要显示无法通过键盘直接输入的字符，可以使用标准的 XML 转义字符来指定 Unicode 值。例如，如果只有一个美式键盘，但要显示“This costs €55”，可以这样以 Unicode 形式指定其中的欧元符号：

```
<TextBlock Text="This costs &#8364;55" ...
```

或者使用十六进制：

```
<TextBlock Text="This costs &#x20AC;55" ...
```

除了转义字符，还可以将文本直接粘贴到 Visual Studio 编辑器中(正如本章后面的示例所展示的)。

与标准 XML 相同，字符串中可以包含以“&”(and)符号为前缀的转义字符来表示特殊字符。

- &表示&
- '表示单引号
- "表示双引号
- <表示左尖括号(less than，小于号)
- >表示右尖括号(greater than，大于号)

除直接设置 TextBlock 的 Text 属性外，还可以将 TextBlock 标记拆分为开始和结束标记，并将文本以内容形式置于其中。

```
<TextBlock ... >
   Hello, Windows 8!
</TextBlock>
```

正如第 2 章将介绍的，将文本以内容形式置于 TextBlock 内部并不完全等同于设置 Text 属性。事实上，以内容形式设置文本要更为强大。即便不调用特殊功能，这样仍可以添加大量文本，而且不必担心额外的空白和对引号的处理。示例项目 WrappedText 演示了如何通过 TextBlock 的内容来显示整段文本。

```
项目: WrappedText | 文件: MainPage.xaml(片段)
<Grid Background="{StaticResource ApplicationPageBackgroundThemeBrush}">
   <TextBlock FontSize="48"
              TextWrapping="Wrap">
      For a long time I used to go to bed early. Sometimes, when I had put out
      my candle, my eyes would close so quickly that I had not even time to
      say "I'm going to sleep." And half an hour later the thought that it was
      time to go to sleep would awaken me; I would try to put away the book
      which, I imagined, was still in my hands, and to blow out the light; I
      had been thinking all the time, while I was asleep, of what I had just
      been reading, but my thoughts had run into a channel of their own,
      until I myself seemed actually to have become the subject of my book:
      a church, a quartet, the rivalry between François I and Charles V. This
      impression would persist for some moments after I was awake; it did not
      disturb my mind, but it lay like scales upon my eyes and prevented them
      from registering the fact that the candle was no longer burning. Then
      it would begin to seem unintelligible, as the thoughts of a former
      existence must be to a reincarnate spirit; the subject of my book would
      separate itself from me, leaving me free to choose whether I would form
      part of it or no; and at the same time my sight would return and I
      would be astonished to find myself in a state of darkness, pleasant and
      restful enough for the eyes, and even more, perhaps, for my mind, to
      which it appeared incomprehensible, without a cause, a matter dark
      indeed.
   </TextBlock>
</Grid>
```

在解析时，每行尾端的换行符和首部的 8 个空格会被融合成单个空格字符。

请注意 TextWrapping 属性。该属性的默认值为枚举成员 TextWrapping.NoWrap(仅有的另一个成员是 Wrap)。TextAlignment 属性设置为 TextAlignment 枚举的某个成员：Left、Right、Center 或 Justify。其中，Justify 能够在词间插入额外的空白，使每行文本都能够在左右两端对齐。

下图所示程序能够以横屏或竖屏显示。

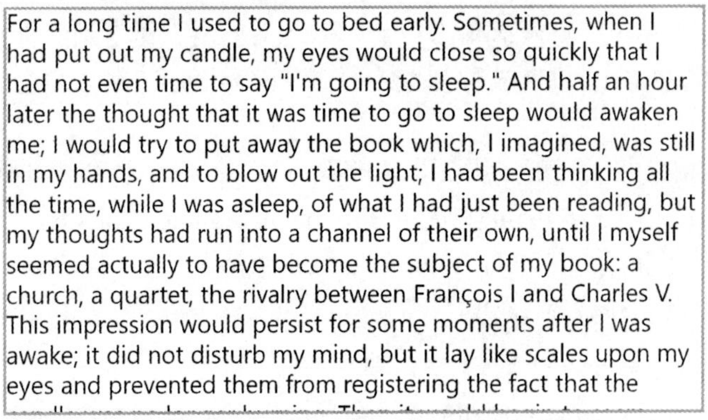

For a long time I used to go to bed early. Sometimes, when I
had put out my candle, my eyes would close so quickly that I
had not even time to say "I'm going to sleep." And half an hour
later the thought that it was time to go to sleep would awaken
me; I would try to put away the book which, I imagined, was still
in my hands, and to blow out the light; I had been thinking all
the time, while I was asleep, of what I had just been reading, but
my thoughts had run into a channel of their own, until I myself
seemed actually to have become the subject of my book: a
church, a quartet, the rivalry between François I and Charles V.
This impression would persist for some moments after I was
awake; it did not disturb my mind, but it lay like scales upon my
eyes and prevented them from registering the fact that the

如果读者的计算机能够响应屏幕方向的变化，则会看到文本格式能够被自动调整。Windows Runtime 会在空格或连字符处断行，但不会在不间断空格( )或不间断连字符(‑)处断行，而且所有软连字符()都会被忽略。

并非所有 XAML 元素都像 TextBlock 一样支持文本内容。例如，Page 和 Grid 就不支持。XAML 的形式并不像 HTML 那样自由，因为 XAML 的语法完全基于底层的类和属性。

但 Grid 支持多个子 TextBlock。示例项目 OverlappedStackedText 在 Grid 中添加了两个颜色和字体大小各不同的两个 TextBlock 元素。

```
项目: OverlappedStackedText | 文件: MainPage.xaml(片段)
<Grid Background="Yellow">
    <TextBlock Text="8"
               FontSize="864"
               FontWeight="Bold"
               Foreground="Red"
               HorizontalAlignment="Center"
               VerticalAlignment="Center" />
    <TextBlock Text="Windows"
               FontSize="192"
               FontStyle="Italic"
               Foreground="Blue"
               HorizontalAlignment="Center"
               VerticalAlignment="Center" />
</Grid>
```

下图所示为该程序的运行效果。

请注意，第二个元素反而显示在第一个元素之上。由于在三维坐标空间中，Z 轴指向外侧，因而这种视觉的分层机制被称为“Z 顺序”(Z order)。第 4 章会介绍如何修改这一行为。

当然，使元素重叠并非处理多块文本的一般方法。第 5 章会介绍如何在 Grid 中定义行和列，但在同一 Grid 单元格中仍可以通过 HorizontalAlignment 和 VerticalAlignment 组织多个元素以防止元素重叠。示例项目 InternationalHelloWorld 在 9 个位置显示了不语言的“hello, world”(来自“谷歌翻译”)。

```
项目: InternationalHelloWorld | 文件: MainPage.xaml(片段)
<Page
    x:Class="InternationalHelloWorld.MainPage"
    ...
    FontSize="40">

    <Grid Background="{StaticResource ApplicationPageBackgroundThemeBrush}">
        <!-- Chinese (simplified) -->
        <TextBlock Text="你好，世界"
                   HorizontalAlignment="Left"
                   VerticalAlignment="Top" />

        <!-- Urdu -->
        <TextBlock Text="هيلو دنيا،"
                   HorizontalAlignment="Center"
                   VerticalAlignment="Top" />
```

```
        <!-- Japanese -->
        <TextBlock Text="こんにちは、世界中のみなさん"
                   HorizontalAlignment="Right"
                   VerticalAlignment="Top" />

        <!-- Hebrew -->
        <TextBlock Text="שלום, עולם"
                   HorizontalAlignment="Left"
                   VerticalAlignment="Center" />

        <!-- Esperanto -->
        <TextBlock Text="Saluton, mondo"
                   HorizontalAlignment="Center"
                   VerticalAlignment="Center" />

        <!-- Arabic -->
        <TextBlock Text="مرحبا، العالم"
                 HorizontalAlignment="Right"
                 VerticalAlignment="Center" />

        <!-- Korean -->
        <TextBlock Text="안녕하세요, 전 세계"
                   HorizontalAlignment="Left"
                   VerticalAlignment="Bottom" />

        <!-- Russian -->
        <TextBlock Text="Здравствуй, мир"
                   HorizontalAlignment="Center"
                   VerticalAlignment="Bottom" />

        <!-- Hindi -->
        <TextBlock Text="नमस्ते दुनिया है,"
                   HorizontalAlignment="Right"
                   VerticalAlignment="Bottom" />
    </Grid>
</Page>
```

需要注意的是，在根元素设置的 FontSize 属性作用于所有 9 个 TextBlock 元素(如下图所示)。属性继承机制是降低 XAML 内容重复的一种有效手段。除此以外，下一章会介绍另外几种方法。

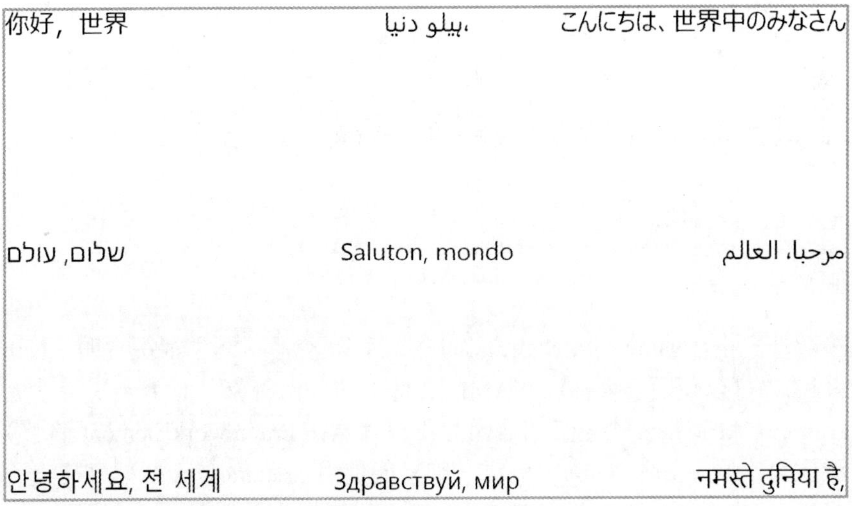

你好，世界　　بيلو دنيا،　　こんにちは、世界中のみなさん

שלום, עולם　　Saluton, mondo　　مرحبا، العالم

안녕하세요, 전 세계　　Здравствуй, мир　　नमस्ते दुनिया है,

1.4 播放媒体文件

截至目前，本章已经展示了文本和位图形式的“hello, world”。示例项目 HelloAudio 展示了如何播放来自网站的音频问候。其中的音频是使用 Windows 8 中的录音机程序录制的，它能够将音频保存为 WMA 格式。该示例的 XAML 文件如下所示。

```
项目: HelloAudio | 文件: MainPage.xaml(片段)
<Grid Background="{StaticResource ApplicationPageBackgroundThemeBrush}">
   <MediaElement Source="http://www.charlespetzold.com/pw6/AudioGreeting.wma" />
</Grid>
```

MediaElement 类派生自 FrameworkElement。虽然没有用于控制音频的用户界面，但该元素为自行开发提供了必要的支持。

我们还可以使用 MediaElement 来播放视频。示例项目 HelloVideo 展示了如何播放来自网站的视频。

```
项目: HelloVideo | 文件: MainPage.xaml(片段)
<Grid Background="{StaticResource ApplicationPageBackgroundThemeBrush}">
   <MediaElement Source="http://www.charlespetzold.com/pw6/VideoGreeting.wmv" />
</Grid>
```

1.5 代码形式的变通

对元素或控件进行初始化，XAML 并非唯一的选择。我们可以利用纯粹的代码完成。从所实现的功能上讲，XAML 所能做的，代码都能做到。代码适合用来创建同种类型的多个对象，因为 XAML 中没有类似 for 循环的机制。

下面我们创建一个名为 HelloCode 的项目，并在 MainPage.xaml 文件中为 Grid 分配一个名称。

```
项目: HelloCode | 文件: MainPage.xaml(片段)
<Grid Name="contentGrid"
      Background="{StaticResource ApplicationPageBackgroundThemeBrush}">

</Grid>
```

为 Grid 设置 Name 属性可以使我们在代码隐藏文件中访问该元素。我们也可以使用 x:Name。

```
<Grid x:Name="contentGrid"
      Background="{StaticResource ApplicationPageBackgroundThemeBrush}">

</Grid>
```

在大多数情况下，Name 与 x:Name 之间并无实际区别。“x”前缀说明 x:Name 特性是 XAML 固有的，可以使用它来标识 XAML 文件中的任何对象。而 Name 特性更为严格：Name 是 FrameworkElement 类定义的，因而我们只能在 FrameworkElement 的子类上使用。对于没有派生自 FrameworkElement 的类，则只能使用 x:Name。一些程序员为了一致性全部采用 x:Name。作者倾向于如果可能就用 Name，其次才用 x:Name。(然而，对于应用程序的程序集中定义的自定义控件，有时必须用 x:Name。)

不论选择 Name 还是 x:Name，元素的命名规则仍要遵循变量的命名规则。例如，名称中不能包含空格，也不能以数字开头。所有名称在同一个 XAML 文件中必须唯一。

我们需要在 MainPage.xaml.cs 文件中额外添加两个 using 语句。

```
项目: HelloCode | 文件: MainPage.xaml.cs(片段)
using Windows.UI;
using Windows.UI.Text;
```

第一个是为了使用 Colors 类，另一个是为了引入 FontStyle 枚举。这两个 using 语句不必手动添加。如果使用 Colors 类或 FontStyle 枚举，Visual Studio 将会在无法解析的标识符下面划上红色波浪线。我们可以在上面右击，然后从快捷菜单中选择“解析”。[①]必要的 using 语句会自动按字符序插入到其他 using 语句中(前提是现有 using 语句是有序的)。

通过右键菜单中的“组织 using”|“移除未使用的 using”可以清理该列表。(示例中的 MainPage.xaml.cs 已经被执行过该操作。)

MainPage 类的构造函数非常适合用来创建 TextBlock，为其设置属性，并将其添加到 Grid 中。

```
项目: HelloCode | 文件: MainPage.xaml.cs(片段)
public MainPage()
{
    this.InitializeComponent();

    TextBlock txtblk = new TextBlock();
    txtblk.Text = "Hello, Windows 8!";
    txtblk.FontFamily = new FontFamily("Times New Roman");
    txtblk.FontSize = 96;
    txtblk.FontStyle = FontStyle.Italic;
    txtblk.Foreground = new SolidColorBrush(Colors.Yellow);
    txtblk.HorizontalAlignment = HorizontalAlignment.Center;
    txtblk.VerticalAlignment = VerticalAlignment.Center;

    contentGrid.Children.Add(txtblk);
}
```

请读者注意，这段代码的最后一行引用了 XAML 文件中名为 contentGrid 的 Grid，就像引用一个普通对象一样。(事实上，这里引用的确实是一个普通的对象，而且所使用的变量还是以字段的形式存在的。)Grid 有一个派生自 Panel 的 Children 属性，类型为 UIElementCollection。该类型是实现 IList<UIElement>和 IEnumerable<UIElement>接口的集合。正因如此，Grid 才能够包含多个子元素，这一点在 XAML 中并未反映出来。

相比在 XAML 中进行定义，代码形式的稍显冗长。这是因为 XAML 解析器在幕后创建了额外的对象并执行了转换。从这段代码可以了解到，我们需要为 FontFamily 属性创建 FontFamily 对象，Foreground 是 Brush 类型，需要创建 Brush 子类(如 SolidColorBrush)的实例。为设置颜色，可以使用 Colors，它包含 141 个类型为 Color 的静态属性。也可以通过 Color.FromArgb 静态方法根据 ARGB 字节来创建 Color 值。

FontStyle、HorizontalAlignment 和 VerticalAlignment 属性都是枚举类型，并且属性与

① 译注：这个示例有些特殊。由于 Control 类恰好有个名为 FontStyle 的属性成员，而 Page 是 Control 类的子类，因而 Visual Studio 会优先认为 FontStyle 是属性名称，而不是枚举类型，也不会在右键菜单中提示“解析”选项。为正确解析该类型，需要手动添加 Windows.UI.Text 命名空间。

类型同名。Text 和 FontSize 都是基本数据类型，分别为字符串和双精度浮点数。

我们可以通过 C# 3.0 中的属性初始化方式来简化以上代码。

```
TextBlock txtblk = new TextBlock
{
    Text = "Hello, Windows 8!",
    FontFamily = new FontFamily("Times New Roman"),
    FontSize = 96,
    FontStyle = FontStyle.Italic,
    Foreground = new SolidColorBrush(Colors.Yellow),
    HorizontalAlignment = HorizontalAlignment.Center,
    VerticalAlignment = VerticalAlignment.Center
};
```

本书广泛地使用了这种风格的代码。(为了不在演示过程中造成不必要的困惑，本书并没有使用包括隐式类型在内的其他 C# 3.0 功能。)不论选择哪种方式，编译并运行 HelloCode 项目都应获得与 XAML 版本相同的结果。两者看起来相同是因为在幕后也是相同的。

除上述代码提供的方法外，也可以在重写的 OnNavigatedTo 方法中创建 TextBlock 并将其添加到 Grid 的 Children 集合中。还可以在构造函数中创建 TextBlock，将其保存在字段中，然后在 OnNavigatedTo 中将其添加到 Grid。

请注意，示例将代码放在了 MainPage 构造函数的 InitializeComponent 方法调用之后。我们可以在 InitializeComponent 方法调用之前创建 TextBlock，但必须在该方法调用之后将 TextBlock 对象添加到 Grid，因为此前 Grid 并不存在。InitializeComponent 方法会在运行时解析 XAML，初始化所有 XAML 对象并将它们添加到一棵可视树上。InitializeComponent 显然是一个非常重要的方法，但让人困惑的是，为何在文档中找不到它。

事情是这样的：Visual Studio 编译应用程序时会生成一些中间文件。使用 Windows 的文件资源管理器导航到示例 HelloCode 的目录，我们来找找这些中间文件。在 obj 目录的子目录中，我们会发现 MainPage.g.cs 和 MainPage.g.i.cs。文件名中的“g”表示该文件是被“生成的”(generated)。这两个文件都定义了 MainPage 类，都带有 partial 关键字并派生自 Page 类。开发者可控的 MainPage.xaml.cs 文件和这两个自动生成的文件构成了最终的 MainPage 类的定义。虽然我们不需要编辑这两个生成的文件，但仍需要对其有一定了解，因为如果在调用 XAML 文件的过程中出现运行时错误，这些自动生成的代码仍会出现在 Visual Studio 中。

MainPage.g.i.cs 文件有两点值得注意。首先，我们可以在该文件中找到 InitializeComponent 方法的定义，该方法会调用静态方法 Application.LoadComponent 来加载 MainPage.xaml 文件。其次是这个分部类包含一个名为 contentGrid 的私有字段，它正是我们在 XAML 中为 Grid 分配的名称。InitializeComponent 方法将该字段赋予了 Application.LoadComponent 创建的具体的 Grid 对象。

contentGrid 字段在整个 MainPage 类中都可以被访问，但在 InitializeComponent 被调用前，该字段为 null。

总的来看，XAML 文件的解析可分为两个阶段。第一阶段发生在编译时，编译器会解析 XAML，从中抽取所有元素的名称(还有一些其他任务)，然后在 obj 目录下生成中间 C# 文件。这些生成的 C#文件会与其他受我们控制的 C#文件一同编译。第二阶段发生在运行时，XAML 文件会再次被解析，初始化所有元素，将这些元素对象组合成可视树，从而获得这些对象的引用。

读者可能想知道作为C#程序入口点的标准Main方法在哪里？在App.g.i.cs中。Visual Studio也会根据App.xaml生成两个文件，App.g.i.cs是其中之一。

作为稍后介绍依赖属性的铺垫，下面我们需要了解一些内幕。

正如前文所介绍的，我们用过的许多属性(如FontFamily和FontSize)都有对应的静态依赖属性(如FontFamilyProperty和FontSizeProperty)。如果将以下普通的属性赋值

```
txtblk.FontStyle = FontStyle.Italic;
```

改成

```
txtblk.SetValue(TextBlock.FontStyleProperty, FontStyle.Italic);
```

似乎让人有些困惑。

后者调用了DependencyObject定义的SetValue方法(TextBlock会继承它)。虽然调用的是TextBlock的方法，但传入的却是TextBlock定义的DependencyProperty类型的TextBlock和我们所要设置的值。虽然上述两种设置FontStyle属性的方法形式不同，但却无实际区别。在TextBlock类的源代码中，FontStyle属性的实现与以下代码非常类似。

```
public FontStyle FontStyle
{
   set
   {
      SetValue(TextBlock.FontStyleProperty, value);
   }
   get
   {
      return (FontStyle)GetValue(TextBlock.FontStyleProperty);
   }
}
```

之所以说“类似”，是因为本人没有Windows Runtime的源代码，并且源代码可能是用C++写的，而不是用C#。但如果FontStyle属性的定义与其他内建的依赖属性一致，那么set和get访问器只不过是使用TextBlock.FontStyleProperty来调用SetValue和GetValue。这里给出的是接近标准的代码。我们自己定义依赖属性时，可以采用下面这种省略部分空白的方式。

```
public FontStyle FontStyle
{
   set { SetValue(TextBlock.FontStyleProperty, value); }
   get { return (FontStyle)GetValue(TextBlock.FontStyleProperty); }
}
```

前文的示例演示了如何通过XAML在Page标签上统一设置Foreground及与字体相关的属性，而不单独设置TextBlock的属性。通过这个示例我们认识到，TextBlock能够继承这些属性。当然，在代码中我们也可以做到。

```
public MainPage()
{
   this.InitializeComponent();
   this.FontFamily = new FontFamily("Times New Roman");
   this.FontSize = 96;
   this.FontStyle = FontStyle.Italic;
   this.Foreground = new SolidColorBrush(Colors.Yellow);

   TextBlock txtblk = new TextBlock();
```

```
    txtblk.Text = "Hello, Windows 8!";
    txtblk.HorizontalAlignment = HorizontalAlignment.Center;
    txtblk.VerticalAlignment = VerticalAlignment.Center;

    contentGrid.Children.Add(txtblk);
}
```

为访问 Page 类的属性和方法，C#并不强制要求使用 this 前缀。只不过在 Visual Studio 编辑器中，键入 this 前缀可以调用“智能感知”功能，提示可选的方法、属性和事件等。

1.6 通过代码显示图片

本章前面的示例项目 HelloImage 和 HelloLocalImage 展示了如何通过 Image 元素来显示图片。在 XAML 中，我们将 Source 属性设置为指向图片的 URI。单从 XAML 文件看，可能会让人认为 Source 属性是字符串类型或 Uri 类型，而实际上要复杂些：Source 属性的类型实际为 ImageSource。该类型的对象封装了 Image 元素所要实际显示的图片。ImageSource 并未定义什么公共成员，该类型本身也不能实例化，但有几个派生于该类的重要类型值得注意。下面是一个相关的类层次结构。

```
Object
  DependencyObject
    ImageSource
      BitmapSource
        BitmapImage
        WriteableBitmap
```

ImageSource 属于 Windows.UI.Xaml.Media 命名空间，但这两个子类都在 Windows.UI.Xaml.Media.Imaging 命名空间下。BitmapSource 也不能被实例化。它定义了一对属性 PixelWidth 和 PixelHeight，以及一个名为 SetSource 的方法，该方法可以从文件或网络流读取位图数据。BitmapImage 继承了这些成员，还定义了 UriSource 属性。

我们可以在代码中使用 BitmapImage 类来显示图片。除了定义了 UriSource 属性外，该类还定义了一个接受 Uri 对象的构造函数。在示例项目 HelloImageCode 中，Grid 被命名为 contentGrid，并在代码隐藏文件中通过 using 关键字引入了 Windows.UI.Xaml.Media.Imaging 命名空间。下面的代码展示了 MainPage 的构造函数。

```
项目：HelloImageCode | 文件：MainPage.xaml.cs(片段)
public MainPage()
{
    this.InitializeComponent();

    Uri uri = new Uri("http://www.charlespetzold.com/pw6/PetzoldJersey.jpg");
    BitmapImage bitmap = new BitmapImage(uri);
    Image image = new Image();
    image.Source = bitmap;
    contentGrid.Children.Add(image);
}
```

为从代码中访问 Grid，并不一定要对其进行命名。Grid 会被设置到 Page 的 Content 属性上，因此以下这行代码：

```
contentGrid.Children.Add(image);
```

可以替换为下面两行：

```
Grid grid = this.Content as Grid;
grid.Children.Add(image);
```

对于这么一个简单的程序，Grid 并非必需。我们可以将 Grid 从可视树上移除，并将 Image 直接设置到 MainPage 的 Content 属性上：

```
this.Content = image;
```

MainPage 类的 Content 属性来自 UserControl 类，类型为 UIElement，因而 MainPage 类只支持一个子元素。一般来讲，MainPage 的子元素是支持多个子元素的 Panel(面板)，但如果只需要一个子元素，则可以直接使用 MainPage 的 Content 属性。

我们可以混合使用 XAML 和代码，即在 XAML 中初始化 Image 元素，而在代码中创建 BitmapImage；也可以在 XAML 中初始化 Image 和 BitmapImage，而在代码中设置 BitmapImage 的 UriSource 属性。示例项目 HelloLocalImageCode 采用了第一种方法，能够显示 Images 目录下的 Greeting.png 文件。XAML 文件声明了 Image 元素，但未使其引用实际的图片文件。

```
项目：HelloLocalImageCode | 文件：MainPage.xaml(片段)
<Grid Background="{StaticResource ApplicationPageBackgroundThemeBrush}">
   <Image Name="image"
          Stretch="None" />
</Grid>
```

在代码隐藏文件中，只需要添加一行代码来设置 Image 的 Source 属性。

```
项目：HelloLocalImageCode | 文件：MainPage.xaml.cs(片段)
public sealed partial class MainPage : Page
{
   public MainPage()
   {
      this.InitializeComponent();
      image.Source = new BitmapImage(new Uri("ms-appx:///Images/Greeting.png"));
   }
}
```

读者可能会注意到代码中引用了图片的特殊 URL。在 XAML 中，其中的前缀是可选的。

选择 XAML 还是代码，有没有什么原则呢？并没有确切答案。除非遇到过度重复，否则本人倾向于在可能的情况下使用 XAML，一般会在遇到三个或更多重复时使用 for，但在将标记转换为代码之前，也常常允许 XAML 中存在较多重复。这种选择在很大程度上取决于对 XAML 简洁程度的要求以及代码维护难易程度的要求。

1.7　纯粹的代码

为了解 Windows Runtime 程序的启动方式，可以读读 OnLaunched 方法重写的源代码，该方法在标准的 App.xaml.cs 文件中。它创建了一个 Frame 对象，通过该对象导航到 MainPage(该页面在这里被实例化)，然后通过静态属性 Window.Current 将这个 Frame 对象设置到当前的 Window 对象上。下面是该过程简化后的代码。

```
var rootFrame = new Frame();
rootFrame.Navigate(typeof(MainPage));
Window.Current.Content = rootFrame;
Window.Current.Activate();
```

Windows 8 应用程序并不强制要求使用 Page 类、Frame 类或任何 XAML 文件。作为本章最后一节，下面我们创建一个名为 StrippedDownHello 的项目。删除 App.xaml、App.xaml.cs、MainPage.xaml、MainPage.xaml.cs 以及整个 Common 文件夹。没错，将这些文件统统删掉。那么项目中便不存在代码和 XAML 文件了，只有程序清单(manifest)、程序集信息以及一些 PNG 文件。

右击项目名称，选择“添加”|“新建”。在“代码”节点中，选择“类”或“代码文件”，并将文件命名为 App.cs。将以下代码添加到新建的文件中。

项目：StrippedDownHello | 文件：App.cs(片段)

```
using Windows.ApplicationModel.Activation;
using Windows.UI;
using Windows.UI.Xaml;
using Windows.UI.Xaml.Controls;
using Windows.UI.Xaml.Media;

namespace StrippedDownHello
{
   public class App : Application
   {
      static void Main(string[] args)
      {
         Application.Start((p) => new App());
      }

      protected override void OnLaunched(LaunchActivatedEventArgs args)
      {
         TextBlock txtblk = new TextBlock
         {
            Text = "Stripped-Down Windows 8",
            FontFamily = new FontFamily("Lucida sans Typewriter"),
            FontSize = 96,
            Foreground = new SolidColorBrush(Colors.Red),
            HorizontalAlignment = HorizontalAlignment.Center,
            VerticalAlignment = VerticalAlignment.Center
         };

         Window.Current.Content = txtblk;
         Window.Current.Activate();
      }
   }
}
```

这就是所需要的全部代码(如果使用 TextBlock 的默认属性，代码显然更少)。静态方法 Main 是程序的入口点。该方法创建了 App 对象，并启动了它。重写的 OnLaunched 方法创建了一个 TextBlock 对象，并将其设置为应用程序的默认窗口。

本书不提倡这样创建 Windows 8 应用程序，这里只是展示这种方式的可行性。

第 2 章　XAML 语法

Windows 8 应用程序由代码和标记构成，两者各司其职。尽管标记不适合处理复杂逻辑和计算，但尽可能多地将程序转换为标记仍是明智之举。标记更易于受编辑工具的支持，能够清晰地表达页面的布局。当然，标记中的一切都是字符串，表达复杂对象不免显得冗长。另外，由于标记没有编程语言所具有的循环处理能力，因而它不擅长表达重复性的内容。

通过 XAML 的语法，其中的一些问题已经得到不同程度的解决。本章将介绍其中较为重要的语法。在开始本话题之前，让我们先来了解一个看似与之完全无关的内容：渐变画笔的定义。

2.1　通过代码定义渐变画笔

Grid 的 Background 属性和 TextBlock 的 Foreground 属性均为 Brush 类型。第 1 章的示例程序将这两个属性均设置为 Brush 的一个子类的实例，该子类为 SolidColorBrush 类。本书演示了如何在代码中创建 SolidColorBrush 类的实例并为其设置 Color 值，这正是 XAML 幕后所做的。

SolidColorBrush 只是四种画笔中的一种。这些画笔的类层次结构如下所示。

```
Object
  DependencyObject
    Brush
      SolidColorBrush
      GradientBrush
        LinearGradientBrush
      TileBrush
        ImageBrush
        WebViewBrush
```

在这些类中，只有 SolidColorBrush、LinearGradientBrush、ImageBrush 和 WebViewBrush 类是可以实例化的。与许多其他图形相关的类一样，大部分画笔都定义在 Windows.UI.Xaml.Media 命名空间下，只有 WebViewBrush 位于 Windows.UI.Xaml.Controls 命名空间。

LinearGradientBrush 能够表现两种或两种以上颜色的渐变效果。下面我们通过该画笔来显示从左到右由蓝色逐渐过渡到红色的文字并将 Grid 的 Background 属性设置为另一种渐变效果。

示例项目 GradientBrushCode 在 XAML 中对 TextBlock 进行了初始化，并对 Grid 和 TextBlock 命了名。

```
项目：GradientBrushCode | 文件：MainPage.xaml(片段)
<Grid Name="contentGrid"
      Background="{StaticResource ApplicationPageBackgroundThemeBrush}">
```

```
    <TextBlock Name="txtblk"
               Text="Hello, Windows 8!"
               FontSize="96"
               FontWeight="Bold"
               HorizontalAlignment="Center"
               VerticalAlignment="Center" />
</Grid>
```

在代码隐藏文件中，MainPage 构造函数创建了两个不同的 LinearGradientBrush 对象，分别设置到 Grid 的 Background 属性和 TextBlock 的 Foreground 属性。

项目：GradientBrushCode | 文件：MainPage.xaml.cs(片段)

```
public MainPage()
{
    this.InitializeComponent();

    // Create the foreground brush for the TextBlock
    LinearGradientBrush foregroundBrush = new LinearGradientBrush();
    foregroundBrush.StartPoint = new Point(0, 0);
    foregroundBrush.EndPoint = new Point(1, 0);

    GradientStop gradientStop = new GradientStop();
    gradientStop.Offset = 0;
    gradientStop.Color = Colors.Blue;
    foregroundBrush.GradientStops.Add(gradientStop);

    gradientStop = new GradientStop();
    gradientStop.Offset = 1;
    gradientStop.Color = Colors.Red;
    foregroundBrush.GradientStops.Add(gradientStop);

    txtblk.Foreground = foregroundBrush;

    // Create the background brush for the Grid
    LinearGradientBrush backgroundBrush = new LinearGradientBrush
    {
          StartPoint = new Point(0, 0),
          EndPoint = new Point(1, 0)
    };
    backgroundBrush.GradientStops.Add(new GradientStop
    {
        Offset = 0,
        Color = Colors.Red
    });
    backgroundBrush.GradientStops.Add(new GradientStop
    {
        Offset = 1,
        Color = Colors.Blue
    });

    contentGrid.Background = backgroundBrush;
}
```

上述代码使用了两种方式对两个画笔进行了初始化，但这两种方式是等价的。LinearGradientBrush 类定义了类型为 Point 的 StartPoint 和 EndPoint 属性。Point 是一个结构类型，定义了表示二维坐标的 X 和 Y 属性。画笔的定位建立在标准窗口坐标系之上，即 X 值随位置从左到右而递增，Y 值随位置自上而下而递增。StartPoint 和 EndPoint 属性所定义的点是相对的点，相对于画笔所针对的对象。坐标(0, 0)和(1, 0)分别代表目标对象的左上角和右上角。画笔沿两点所成线段进行渐变，并且所有的渐变线均与之平行。StartPoint 和 EndPoint 的默认值分别为(0, 0)和(1, 1)，定义了一条从目标对象左上角到右下角的渐变。

LinearGradientBrush 还有一个名为 GradientStops 的属性，它是 GradientStop 对象的集

合。GradientStop 定义了相对于渐变线起点的偏移量(Offset 类型)和该点的颜色(Color 类型)。偏移量的取值范围一般从 0 到 1，但也可以在该范围以外，越过画笔的作用区域。LinearGradientBrush 还定义了两个属性，分别用于控制渐变的计算方式以及最小 Offset 和最大 Offset 之外的效果。

下图为示例程序的运行效果。

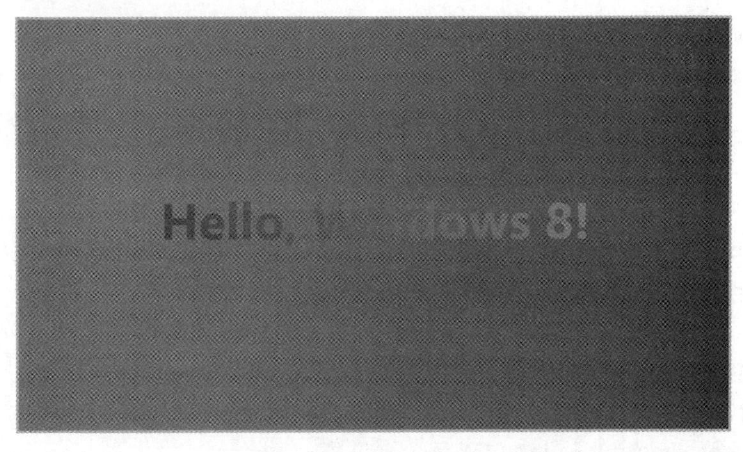

如果在 XAML 中定义这两个画笔，标记的限制便会显现。XAML 允许我们通过指定颜色来定义 SolidColorBrush，但如果通过 Foreground 或 Background 属性来设置渐变效果，又应该如何指定起终点以及两个或更多的偏移量和颜色值？

2.2 属性元素语法

恰好有一种办法。正如前面示例所演示的，如果在 XAML 中使用 SolidColorBrush，一般只需要指定画笔的颜色。

```
<TextBlock Text="Hello, Windows 8!"
           Foreground="Blue"
           FontSize="96" />
```

SolidColorBrush 的实例是在幕后创建的。

有一种语法变形允许我们显式地调用该画笔。首先，我们将 Foreground 属性移除，将 TextBlock 元素分离为开始标签和结束标签。

```
<TextBlock Text="Hello, Windows 8!"
           FontSize="96">

</TextBlock>
```

然后，在这两个标签中插入另一对开始和结束标签。标签名由元素名、英文句点和属性名构成。

```
<TextBlock Text="Hello, Windows 8!"
           FontSize="96">
   <TextBlock.Foreground>

   </TextBlock.Foreground>
</TextBlock>
```

最后，在新建的标签中添加要设置到属性上的对象。

```
<TextBlock Text="Hello, Windows 8!"
           FontSize="96">
   <TextBlock.Foreground>
      <SolidColorBrush Color="Blue" />
   </TextBlock.Foreground>
</TextBlock>
```

这样，Foreground 便被显式设置为 SolidColorBrush 的实例。

这种语法被称为“属性元素语法”(property-element syntax)，是 XAML 的重要特性之一。如果初次接触，这种语法可能让人觉得这是标准 XML 的一种扩展或变形，但实际并非如此。句点是 XML 元素名称中的有效字符。

在上段代码中蕴含三种 XAML 语法。

- TextBlock 和 SolidColorBrush 都是“对象元素”(object element)，因为这些 XML 元素会使对象被创建
- Text、FontSize 和 Color 是“属性特性”，是能够用来指定属性设置的 XML 特性
- TextBlock.Foreground 标记是“属性元素”，是以 XML 元素形式表达的属性

XAML 对属性元素标签有一个限制：起始标签不能包含额外的内容。为属性设置的对象也必须以内容形式置于起始和结束标签之间。

下面这段标记也是通过属性元素标签来设置 SolidColorBrush 的 Color 属性。

```
<TextBlock Text="Hello, Windows 8!"
           FontSize="96">
   <TextBlock.Foreground>
      <SolidColorBrush>
         <SolidColorBrush.Color>
            Blue
         </SolidColorBrush.Color>
      </SolidColorBrush>
   </TextBlock.Foreground>
</TextBlock>
```

我们也可以采用同样方式设置 TextBlock 的另外两个属性。

```
<TextBlock>
   <TextBlock.Text>
      Hello, Windows 8
   </TextBlock.Text>

   <TextBlock.FontSize>
      96
   </TextBlock.FontSize>

   <TextBlock.Foreground>
      <SolidColorBrush>
         <SolidColorBrush.Color>
            Blue
         </SolidColorBrush.Color>
      </SolidColorBrush>
   </TextBlock.Foreground>
</TextBlock>
```

这样做看起来没什么意义。对于这些简单的属性，使用属性特性语法要更简明。相比而言，属性元素语法更适合表达像 LinearGradientBrush 这样较为复杂的对象。下面我们从属性元素标签开始说起。还是沿用上面的例子。

```
<TextBlock Text="Hello, Windows 8!"
           FontSize="96">
   <TextBlock.Foreground>

   </TextBlock.Foreground>
</TextBlock>
```

首先，将 LinearGradientBrush 分为起始标签和结束标签，放置在中间。在起始标签中设置 StartPoint 和 EndPoint 属性。

```
<TextBlock Text="Hello, Windows 8!"
           FontSize="96">
   <TextBlock.Foreground>
      <LinearGradientBrush StartPoint="0 0" EndPoint="1 0">

      </LinearGradientBrush>
   </TextBlock.Foreground>
</TextBlock>
```

请注意，两个类型为 Point 的属性是通过空格分隔两个数字来设置的。

LinearGradientBrush 的 GradientStops 属性是一个 GradientStop 对象的集合，因而我们要通过另外的属性元素来设置该属性。

```
<TextBlock Text="Hello, Windows 8!"
           FontSize="96">
   <TextBlock.Foreground>
      <LinearGradientBrush StartPoint="0 0" EndPoint="1 0">
         <LinearGradientBrush.GradientStops>

         </LinearGradientBrush.GradientStops>
      </LinearGradientBrush>
   </TextBlock.Foreground>
</TextBlock>
```

GradientStops 属性的类型为 GradientStopCollection，也要将该类型的对象添加进来。

```
<TextBlock Text="Hello, Windows 8!"
           FontSize="96">
   <TextBlock.Foreground>
      <LinearGradientBrush StartPoint="0 0" EndPoint="1 0">
         <LinearGradientBrush.GradientStops>
            <GradientStopCollection>

            </GradientStopCollection>
         </LinearGradientBrush.GradientStops>
      </LinearGradientBrush>
   </TextBlock.Foreground>
</TextBlock>
```

最后，在该集合中添加两个 GradientStop 对象。

```
<TextBlock Text="Hello, Windows 8!"
           FontSize="96">
   <TextBlock.Foreground>
      <LinearGradientBrush StartPoint="0 0" EndPoint="1 0">
         <LinearGradientBrush.GradientStops>
            <GradientStopCollection>
               <GradientStop Offset="0" Color="Blue" />
               <GradientStop Offset="1" Color="Red" />
            </GradientStopCollection>
         </LinearGradientBrush.GradientStops>
      </LinearGradientBrush>
   </TextBlock.Foreground>
</TextBlock>
```

这样，我们便使用标记表达了复杂属性的设置。

2.3 内容属性

前一节的示例实例化和初始化 LinearGradientBrush 的语法可能显得有些冗长。如果发现本书前面所示的 XAML 文件都省略了某些属性和元素，就知道这是可行的。下面让我们看一段标记。

```
<Page ... >
   <Grid ... >
      <TextBlock ... />
      <TextBlock ... />
      <TextBlock ... />
   </Grid>
</Page>
```

通过直接使用 C#代码来创建界面，让我们了解到这些 TextBlock 元素会被添加到 Grid 的 Children 集合中，而 Grid 会被设置到 Page 的 Content 属性。那么 Children 和 Content 属性如何体现在标记中？

事实上，这些标记是可以显式添加的。属性元素 Page.Content 和 Grid.Children 允许出现在 XAML 文件中。

```
<Page ... >
   <Page.Content>
      <Grid ... >
         <Grid.Children>
            <TextBlock ... />
            <TextBlock ... />
            <TextBlock ... />
         </Grid.Children>
      </Grid>
   </Page.Content>
</Page>
```

这段标记仍然缺少 Grid.Children 属性所需的 UIElementCollection 对象。我们不能显式地添加该元素，因为只有定义无参公共构造函数的类才能在 XAML 文件中实例化，而 UIElementCollection 类缺少这样的构造函数。

这就带来一个问题：Page.Content 和 Grid.Children 属性元素为何不是 XAML 文件强制要求的？

原因很简单。XAML 中引用的所有的类允许(且只允许)一个属性是“内容”属性。对于这个内容属性，也仅有这个属性，对应的属性元素标签不强制要求。

类中定义的内容属性需要用.NET 特性(attribute)加以修饰。Panel 类(Grid 的父类)的定义利用了一个名为 ContentProperty 的特性①。如果该类是用 C#定义的，那么看起来会像下面这样：

---

① 译注：该类型的完整名称为 ContentPropertyAttribute，派生自 Attribute 类。根据 C#规范，使用特性时，名称中的 Attribute 后缀应被省略，编译器能够自动解析。如果使用完整名称，虽然可以编译通过，但像 StyleCop 这样的代码审查工具会将该问题报告出来。

```
[ContentProperty(Name="Children")]
public class Panel : FrameworkElement
{
    ...
}
```

这个特性的含义很简单。举例来说，如果遇到下面这样的代码：

```
<Grid ... >
   <TextBlock ... />
   <TextBlock ... />
   <TextBlock ... />
</Grid>
```

那么通过 Grid 的 ContentProperty 特性，XAML 解析器便将这些 TextBlock 元素添加至 Children 属性。

类似地，UserControl 类(Page 的父类)将 Content(内容)属性定义为内容属性(听上去让人觉得这是理所当然的)。

```
[ContentProperty(Name="Content")]
public class UserControl : Control
{
    ...
}
```

我们也可以在自定义的类中使用 ContentProperty 特性。为此需要引入 Windows.UI.Xaml.Markup 命名空间。

不幸的是，根据 Windows Runtime 文档(在本书截稿前)，ContentProperty 特性只能修饰类(如 Panel 类文档首页给出的类定义)，而不能直接修饰属性。如果文档在未来发生了变化，则通过示例加以学习并保持良好习惯即可。①

幸运的是，许多内容属性都是类中最常用的属性。例如 LinearGradientBrush 类的内容属性被指定为 GradientStops。虽然 GradientStops 属性的类型为 GradientStopCollection，但 XAML 不强制要求显式设置该集合对象。下面是 LinearGradientBrush 的完整声明。

```
<TextBlock Text="Hello, Windows 8!"
           FontSize="96">
   <TextBlock.Foreground>
      <LinearGradientBrush StartPoint="0 0" EndPoint="1 0">
         <LinearGradientBrush.GradientStops>
            <GradientStopCollection>
               <GradientStop Offset="0" Color="Blue" />
               <GradientStop Offset="1" Color="Red" />
            </GradientStopCollection>
         </LinearGradientBrush.GradientStops>
      </LinearGradientBrush>
   </TextBlock.Foreground>
</TextBlock>
```

其中的 LinearGradientBrush.GradientStops 属性和 GradientStopCollection 标签都是不必要的，因而可以像下面这样简化。

① 译注：至本书中文版截稿前，这种限制同样存在，但也是可以理解的。若一个类允许有两个或多个内容属性，在解析 XAML 时必然会出现歧义，因而只能存在一个内容属性作为缺省条件下的选项。如果 ContentPropertyAttribute 可以修饰属性，那么程序员便可能添加多个来修饰不同属性。但如果只能修饰类，并且只能修饰一次(见该特性类的定义)，便可以避免这种错误的产生，也不必去写代码来检查这种错误。

```
<TextBlock Text="Hello, Windows 8!"
           FontSize="96">
    <TextBlock.Foreground>
        <LinearGradientBrush StartPoint="0 0" EndPoint="1 0">
            <GradientStop Offset="0" Color="Blue" />
            <GradientStop Offset="1" Color="Red" />
        </LinearGradientBrush>
    </TextBlock.Foreground>
</TextBlock>
```

在保持 XML 有效的前提下，标记可以非常简练。

下面我们用纯粹的 XAML 来重写 GradientBrushCode 项目。

项目：GradientBrushMarkup | 文件：MainPage.xaml(片段)

```
<Grid>
    <Grid.Background>
        <LinearGradientBrush StartPoint="0 0" EndPoint="1 0">
            <GradientStop Offset="0" Color="Red" />
            <GradientStop Offset="1" Color="Blue" />
        </LinearGradientBrush>
    </Grid.Background>

    <TextBlock Name="txtblk"
               Text="Hello, Windows 8!"
               FontSize="96"
               FontWeight="Bold"
               HorizontalAlignment="Center"
               VerticalAlignment="Center">
        <TextBlock.Foreground>
            <LinearGradientBrush StartPoint="0 0" EndPoint="1 0">
                <GradientStop Offset="0" Color="Blue" />
                <GradientStop Offset="1" Color="Red" />
            </LinearGradientBrush>
        </TextBlock.Foreground>
    </TextBlock>
</Grid>
```

即便使用属性元素语法，其可读性也要胜过对应的代码。代码方式可以清晰地展示可视树的构建过程，而标记则直观地展现了可视树的结构。

有一点值得注意。假设要为一个带有多个子元素的 Grid 定义一个属性元素。

```
<Grid>
    <Grid.Background>
        <SolidColorBrush Color="Blue" />
    </Grid.Background>

    <TextBlock Text="one" />
    <TextBlock Text="two" />
    <TextBlock Text="three" />
</Grid>
```

可以将属性元素置于底端。

```
<Grid>
    <TextBlock Text="one" />
    <TextBlock Text="two" />
    <TextBlock Text="three" />

    <Grid.Background>
        <SolidColorBrush Color="Blue" />
    </Grid.Background>
</Grid>
```

但不可以将属性元素混在内容中间。

```
<!-- This doesn't work! -->
<Grid>
   <TextBlock Text="one" />

   <Grid.Background>
      <SolidColorBrush Color="Blue" />
   </Grid.Background>

   <TextBlock Text="two" />
   <TextBlock Text="three" />
</Grid>
```

为什么这样不行呢？我们将 Children 属性变成属性元素，答案便不言而喻了。

```
<!-- This doesn't work! -->
<Grid>
   <Grid.Children>
      <TextBlock Text="one" />
   </Grid.Children>

   <Grid.Background>
      <SolidColorBrush Color="Blue" />
   </Grid.Background>

   <Grid.Children>
      <TextBlock Text="two" />
      <TextBlock Text="three" />
   </Grid.Children>
</Grid>
```

Children 属性被定义了两次，这显然是不合法的。

2.4　TextBlock 的内容属性

第 1 章的示例程序 WrappedText 展示了如何以内容形式指定 TextBlock 的文本。然而 TextBlock 的内容属性并非 Text 属性，而是一个名为 Inlines 的属性，其类型为 InlineCollection。该类型可以包含若干 Inline 对象，确切地讲是 Inline 子类的实例。Inline 类及其子类定义于 Windows.UI.Xaml.Documents 命名空间下。相关类的层次结构如下所示。

```
Object
  DependencyObject
    TextElement
      Block
        Paragraph
      Inline
        InlineUIContainer
        LineBreak
        Run(用于定义 Text 属性)
        Span(用于定义 Inlines 属性)
          Bold
          Italic
          Underline
```

这些类使我们能够在单个 TextBlock 中指定格式化的文本。TextElement 定义了 Foreground 和所有与字体相关的属性：FontFamily、FontSize、FontStyle、FontWeight(用于设置字体粗细)、FontStretch(能够为支持的字体设置伸缩)和 CharacterSpacing。这些属性被所有子类继承。

Block 和 Paragraph 类主要供 RichTextBlock 类使用，后者是 TextBlock 的升级版本。第 16 章将详细介绍有关 RichTextBlock 类的内容，而本章将重点介绍 Inline 的派生类。

Run 元素是唯一定义了 Text 属性的类，而该属性恰好也是内容属性。InlineCollection 中的文本内容都会被转换为 Run 对象，除非这些内容已经是 Run 对象。我们也可以显式地用 Run 对象来指定文本字符串的各种字体属性。

与 TextBlock 一样，Span 类也定义了 Inlines 属性，因而 Span 及其子类可以嵌套。Span 的 3 个子类均是某种快捷方式。例如，Bold 类等价于将 Span 的 FontWeight 特性设置为 Bold。

下面我们来分析一个具体例子。这段标记定义了一个 TextBlock 元素，在 Inlines 集合中添加几个嵌套的快捷方式类。

```
<TextBlock>
   Text in <Bold>bold</Bold> and <Italic>italic</Italic> and
   <Bold><Italic>bold italic</Italic></Bold>
</TextBlock>
```

在解析这段标记时，每段零散的文本都会被转换为一个 Run 对象。那么这个 TextBlock 的 Inlines 集合便有 6 个对象，它们的类型依次为 Run、Bold、Run、Italic、Run 和 Bold。第一个 Bold 和第一个 Italic 对象的 Inlines 集合都包含一个子 Run 对象。第二个 Bold 对象的 Inlines 包含一个 Italic 对象，而这个 Italic 对象的 Inlines 集合又包含一个 Run 对象。

这个 TextBlock、Bold 和 Italic 联用的示例告诉我们，XAML 语法是以类和属性为基础的。如果 Bold 类不具有 Inlines 集合属性，那么也就不可能在 Bold 中嵌入 Italic 标签。

下面这个示例定义了一个略微有些复杂的 TextBlock 元素，展示了更多格式化功能。

```
项目: TextFormatting | 文件: MainPage.xaml(片段)
<Grid Background="{StaticResource ApplicationPageBackgroundThemeBrush}">
   <TextBlock Width="400"
              FontSize="24"
              TextWrapping="Wrap"
              HorizontalAlignment="Center"
              VerticalAlignment="Center">
      Here is text in a
      <Run FontFamily="Times New Roman">Times New Roman</Run> font,
      as well as text in a
      <Run FontSize="36">36-pixel</Run> height.
      <LineBreak />
      <LineBreak />
      Here is some <Bold>bold</Bold> and here is some
      <Italic>italic</Italic> and here is some
      <Underline>underline</Underline> and here is some
      <Bold><Italic><Underline>bold italic underline and
      <Span FontSize="36">bigger and
      <Span Foreground="Red">Red</Span> as well</Span>
      </Underline></Italic></Bold>.
   </TextBlock>
</Grid>
```

这个 TextBlock 的宽为 400 像素，因而该元素不会被胀得过宽。正如下图中第一段文本中展示的，Run 元素可用于对文本进行格式化。但如果要使用嵌套的格式(并与几个快捷方式类联用)，则需要使用 Span 类及其子类。

Here is text in a Times New Roman font, as well as text in a 36-pixel height.

Here is some **bold** and here is some *italic* and here is some underline and here is some ***bold italic underline*** ***and bigger and Red as well.***

不难看出，LineBreak 元素可以实现在任意处断行。理论上，InlineUIContainer 类允许在文本中嵌入任意 UIElement(如 Image 元素)，但实际只有 RichTextBlock 支持 InlineUIContainer 类，而 TextBlock 不支持。

2.5 画笔和其他资源的共享

假设页面上有多个 TextBlock，但希望它们共享同一个画笔。如果使用的是 SolidColorBrush，那么重复标记并无大碍。但如果是 LinearGradientBrush，重复的标记便会显得冗长。定义 LinearGradientBrush 至少需要定义 6 个标签，重复这样的标记让人感到非常乏味，尤其是要修改代码的时候。

Windows Runtime 有一种叫“XAML 资源”的功能，允许我们在多个元素之间共享对象。共享画笔是 XAML 资源的应用之一，而应用最多的是用来定义和共享样式。

XAML 资源一般都存在 ResourceDictionary 类的对象中，后者是键和值均为 object 类型的字典。但键的实际类型一般是字符串。FrameworkElement 和 Application 都定义了一个名为 Resources 的属性，其类型为 ResourceDictionary。

示例项目 SharedBrush 展示了在一个页面中的多个元素间共享一个 LinearGradientBrush(和其他几个对象)的常规方法。XAML 文件的顶部定义了一个 Resources 属性元素，其中包含页面要用到的资源。

```
项目：SharedBrush | 文件：MainPage.xaml(片段)
<Page ... >

   <Page.Resources>
      <x:String x:Key="appName">Shared Brush App</x:String>

      <LinearGradientBrush x:Key="rainbowBrush">
         <GradientStop Offset="0" Color="Red" />
         <GradientStop Offset="0.17" Color="Orange" />
         <GradientStop Offset="0.33" Color="Yellow" />
         <GradientStop Offset="0.5" Color="Green" />
         <GradientStop Offset="0.67" Color="Blue" />
         <GradientStop Offset="0.83" Color="Indigo" />
         <GradientStop Offset="1" Color="Violet" />
      </LinearGradientBrush>
```

```
        <FontFamily x:Key="fontFamily">Times New Roman</FontFamily>

        <x:Double x:Key="fontSize">96</x:Double>
    </Page.Resources>
    ...
</Page>
```

XAML 文件接近顶部用来定义资源的部分叫“资源区段”(resource section)。对于上面这段标记，Resources 字典中初始化了 4 个不同类型：String、LinearGradientBrush、FontFamily 和 Double。请注意，String 和 Double 标签有“x”前缀。这两个类型是.NET 框架提供的基本数据类型，而非 Windows Runtime 提供，因而它们不在默认的 XAML 命名空间中。类似的类型还包括 x:Boolean 和 x:Int32。

还要注意的是，这些资源中的对象都有 x:Key 特性。x:Key 特性只在 Resources 字典中有效。顾名思义，x:Key 特性代表资源在字典中的键。

在 XAML 文件的主体中，引用资源时使用这个键，但要用到一种叫“XAML 标记扩展”的特殊标记。

XAML 标记扩展有几种，它们最显著的特点是都带有大括号。用于引用资源的标记扩展由关键字 StaticResource 和被引用资源的键构成。事实上，我们已经见过无数次 StaticResource 标记扩展了，为默认的 Grid 提供背景画笔用的就是这种语法。这个示例 XAML 文件的主体部分通过 StaticResource 标记扩展获取了定义在 Resources 字典中的资源。

```
项目: SharedBrush | 文件: MainPage.xaml(片段)
<Page ... >
    ...
    <Grid Background="{StaticResource ApplicationPageBackgroundThemeBrush}">
        <TextBlock Text="{StaticResource appName}"
                   FontSize="48"
                   HorizontalAlignment="Center"
                   VerticalAlignment="Center" />

        <TextBlock Text="Top Text"
                   Foreground="{StaticResource rainbowBrush}"
                   FontFamily="{StaticResource fontFamily}"
                   FontSize="{StaticResource fontSize}"
                   HorizontalAlignment="Center"
                   VerticalAlignment="Top" />

        <TextBlock Text="Left Text"
                   Foreground="{StaticResource rainbowBrush}"
                   FontFamily="{StaticResource fontFamily}"
                   FontSize="{StaticResource fontSize}"
                   HorizontalAlignment="Left"
                   VerticalAlignment="Center" />

        <TextBlock Text="Right Text"
                   Foreground="{StaticResource rainbowBrush}"
                   FontFamily="{StaticResource fontFamily}"
                   FontSize="{StaticResource fontSize}"
                   HorizontalAlignment="Right"
                   VerticalAlignment="Center" />

        <TextBlock Text="Bottom Text"
                   Foreground="{StaticResource rainbowBrush}"
                   FontFamily="{StaticResource fontFamily}"
                   FontSize="{StaticResource fontSize}"
                   HorizontalAlignment="Center"
                   VerticalAlignment="Bottom" />
    </Grid>
</Page>
```

该示例程序的运行效果如下图所示。

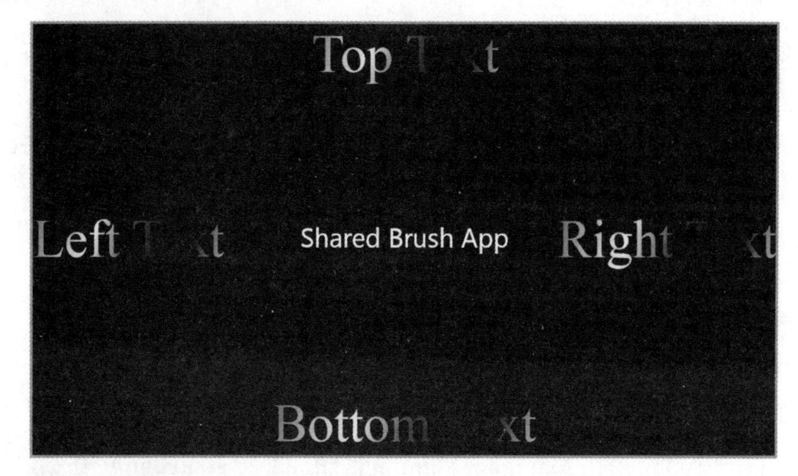

这段标记有几点需要说明。

4 个 TextBlock 元素各自都引用了 3 个资源，只是为了演示标记扩展的使用，但这种做法有悖于一种更高效的技术，即“样式”。有关样式的详细内容本章稍后会做介绍。

从语法上讲，在 XAML 文件中使用资源之前必须定义它们。这也就是为什么 Resources 字典一般都出现在 XAML 文件顶部且定义在根元素上的原因。

FrameworkElement 的派生类都支持 Resources 字典，因而可以在可视树的分支上定义资源。单个 Resources 字典内的键必须唯一，但不同字典之间键可以相同。在遇到 StaticResource 标记扩展时，XAML 解析器会搜索可视树，并选择最先遇到的。Resources 的值可以在分支处被重写。

如果 XAML 解析器无法在可视树中找到匹配的键，则会在 Application 对象的 Resources 字典中查找。App.xaml 文件中也可以定义资源，但这些资源由整个应用程序共享。为在多个应用程序之间共享资源，可以创建单独的 XAML 文件，定义一个名为 ResourceDictionary 的根元素，并添加要共享的资源。然后将这个 XAML 文件添加到要引用这些资源的项目中，并在该项目的 App.xaml 文件中导入该文件的内容，这样便可以在该项目中使用这些资源了。

在针对 Windows 8 应用程序的项目模板中，Visual Studio 碰巧提供了一个例子。Common 文件夹包含一个名为 StandardStyles.xaml 的文件，该文件根元素的类型就是 ResourceDictionary。

```
<ResourceDictionary
   xmlns="http://schemas.microsoft.com/winfx/2006/xaml/presentation"
   xmlns:x="http://schemas.microsoft.com/winfx/2006/xaml">

   ...

</ResourceDictionary>
```

该文件被默认的 App.xaml 文件引用。事实上，默认的 App.xaml 文件也恰好展示了引用资源集合的方法。

```
<Application
   x:Class="App1.App"
   xmlns="http://schemas.microsoft.com/winfx/2006/xaml/presentation"
   xmlns:x="http://schemas.microsoft.com/winfx/2006/xaml"
   xmlns:local="using:App1">
```

```
    <Application.Resources>
        <ResourceDictionary>
            <ResourceDictionary.MergedDictionaries>
                <ResourceDictionary Source="Common/StandardStyles.xaml"/>
            </ResourceDictionary.MergedDictionaries>

        </ResourceDictionary>
    </Application.Resources>
</Application>
```

为添加资源，可以在 MergedDictionaries 集合中插入 ResourceDictionary 标签，也可以直接在 App 对象的 Resources 字典中添加资源。

下面这段代码演示了如何通过代码访问 Resources 字典。在 InitializeComponent 方法调用的后面，我们可以通过索引器从字典获取资源。

```
FontFamily fntfam = this.Resources["fontFamily"] as FontFamily;
```

可以做这样一个试验：在 MainPage.xaml 文件中注释掉 fontFamily 资源。在 MainPage 的构造函数 InitializeComponent 方法调用的前面，将该资源添加到字典中。

```
this.Resources.Add("fontFamily", new FontFamily("Times New Roman"));
```

在 XAML 文件被 InitializeComponent 解析时，该资源便可以在 XAML 文件中被引用了。

ResourceDictionary 类未定义在可视树父对象中搜索字典的公共方法。如果要在代码中搜索资源，可以使用 FrameworkElement 定义的 Parent 属性和 Windows.UI.Xaml.Media 命名空间定义的 VisualTreeHelper 类来搜索可视树。应用程序的 Application 对象可以通过静态属性 Application.Current 获得。

MSDN 文档似乎并未介绍预定义的资源(如 Grid 引用的 ApplicationPageBackgroundThemeBrush)，但我们可以在以下文件中找到相关的值(该文件定义了 Default、Light 和 High Contrast 这 3 个主题，都具有预定义资源)。

```
C:\Program Files (x86)\Windows Kits\8.0\Include\winrt\xaml\design\themeresources.xaml
```

有两个重要的预定义资源，第一个资源的标识符为 ApplicationPageBackgroundThemeBrush，另一个是紧跟其后的 ApplicationForegroundThemeBrush。ApplicationForegroundThemeBrush 在浅色主题中呈现黑色，在深色主题中呈现白色。如果希望使用一个与背景有适当反差的颜色(稍后会介绍)，则可以使用它。如果需要一种与背景和前景都有反差的高亮颜色，可以使用 UISettings 对象的 UIElementColor 方法并传入枚举成员 Highlight，该方法会返回所需要的 Color 对象①。

2.6 资源是共享的

资源对象真的在引用它们的对象间得到共享吗？对于每次引用，StaticResource 难道不会创建新的实例吗？

① 译注：原书为 SolidColorBrush 对象，可实际为 Color 对象。这里加以更正。想了解该方法的更多信息，可以访问以下网址：http://msdn.microsoft.com/zh-cn/library/windows/apps/br229470。

为解除这个疑惑，在 SharedBrush.xaml.cs 文件中，InitializeComponent 方法调用的后面插入以下代码。

```
TextBlock txtblk = (this.Content as Grid).Children[1] as TextBlock;
LinearGradientBrush brush = txtblk.Foreground as LinearGradientBrush;
brush.StartPoint = new Point(0, 1);
brush.EndPoint = new Point(0, 0);
```

这段代码通过 Grid 的 Children 集合引用了的第二个 TextBlock 的 LinearGradientBrush 对象，修改了 StartPoint 和 EndPoint 属性。请注意，所有引用 LinearGradientBrush 的 TextBlock 元素均受到了影响，如下图所示。

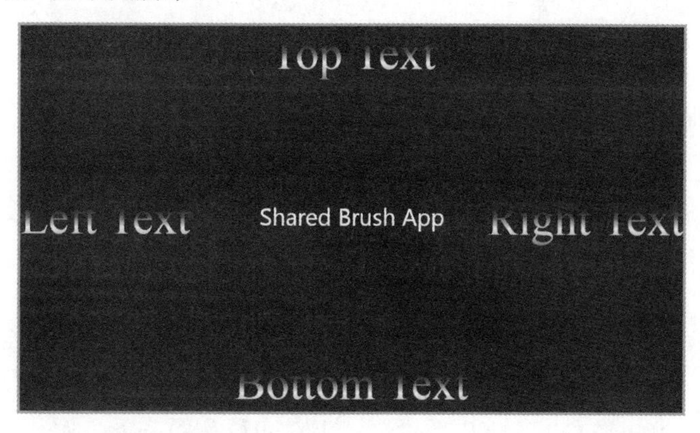

由此可以得出结论：资源是共享的。

另外，未被引用的资源同样会被实例化。如果有兴趣，也可以加以验证。

2.7 探究矢量图形

前面介绍过，在 Windows 8 应用程序中显示文本和图片要分别创建 TextBlock 和 Image 对象，并将其附加到可视树上。其中并未涉及“绘制”的概念，至少在应用程序层面没有。在 Windows Runtime 内部，TextBlock 和 Image 元素都能够呈现(render)自身。

类似地，如果要显示某些矢量图形(如直线、曲线和填充的区域)，我们并不调用 DrawLine 和 DrawBezier 这样的方法。事实上，这些方法在 Windows Runtime 中也并不存在。DirectX 提供了类似的方法，可以在 Windows 8 应用程序中使用，但如果使用 Windows Runtime，则需要通过创建 Line、Polyline、Polygon 和 Path 对象来实现。这些类派生自 Shape 类(派生自 FrameworkElement)，定义于 Windows.UI.Xaml.Shapes 命名空间。该命名空间定义的类型一般统称为“图形库”(Shapes library)。

Polyline 和 Path 类是图形库的主要成员。Polyline 用于呈现一系列相连的直线，但其真正的强大之处在于绘制复杂曲线。在绘制曲线时，要使每段直线尽量短并提供足够的量。通过 Polyline 添加数以千计的直线不成问题，它擅长于此。

下面我们用 Polyline 来画一条“阿基米德螺旋线”(Archimedean spiral)。示例程序 Spiral 的 XAML 文件初始化了一个 Polyline 对象，但并未添加绘制该图形的点。

```
项目: Spiral | 文件: MainPage.xaml(片段)
<Grid Background="{StaticResource ApplicationPageBackgroundThemeBrush}">
    <Polyline Name="polyline"
```

```
            Stroke="{StaticResource ApplicationForegroundThemeBrush}"
            StrokeThickness="3"
            HorizontalAlignment="Center"
            VerticalAlignment="Center" />
</Grid>
```

Stroke 属性(继承于 Shape)包含用于绘制直线的画笔。该画笔一般为 SolidColorBrush 类型，但并非必须如此(正如稍后要介绍的)。这段标记通过 StaticResource 指定了预定义的资源。该资源能够在深色主题下提供白色画笔，在浅色主题下提供黑色画笔。StrokeThickness(继承于 Shape)用于指定直线的宽度，单位为像素。此外，这段标记还包含我们之前用过的 HorizontalAlignment 和 VerticalAlignment。

为一个矢量图形指定 HorizontalAlignment 和 VerticalAlignment 似乎让人有些困惑。这里有必要解释一下。

二维矢量图要用到笛卡儿坐标系和(*X*, *Y*)形式的坐标点。其中，*X* 为横轴坐标，*Y* 为纵轴坐标。Windows Runtime 中的矢量图所采用的定位方式与窗口环境密切相关。*X* 的值随位置右移而递增(合乎数学惯例)，但 *Y* 值随位置下移而递增(与数学惯例相反)。

X 和 *Y* 的值一般是正数，原点(0, 0)位于图的左上角。

负数坐标用于表示纵轴以左和横轴以上的点。在计算布局时，Windows Runtime 会忽略矢量图对象的负数坐标。举例来说，假设要绘制一个折线，一条线段的 *X* 坐标范围从-100 到 300，*Y* 坐标为 400，另一条线段 X 坐标为 300，*Y* 坐标范围从-200 到 400。这意味着这个折线理论上具有 400 像素宽，600 像素高。但在布局和对齐屏幕后，这个折线看起来只有 300 像素宽，400 像素高。

为使得矢量图在 Windows Runtime 中的布局系统中均以可预见的方式处理，规定左上角为(0, 0)点。在布局层面上讲，可见点的 *X* 坐标最大正值为元素的宽，*Y* 坐标最大正值为元素的高。

为指定坐标点，Windows.Foundation 命名空间提供了 Point 结构。该结构具有两个 double 类型的属性 *X* 和 *Y*。此外，Windows.UI.Xaml.Media 命名空间还提供了 PointCollection 类型来作为 Point 对象的集合。

Polyline 本身只定义了一个 PointCollection 类型的属性 Points。在 XAML 中，可以将点的集合赋给 Points 属性，但通过某种算法计算得到的点，则要通过代码添加。在示例程序中，MainPage 类[①]的构造函数通过一个 for 循环将角度从 0 递增到 3600，刚好绘 10 圈。

```
项目: Spiral | 文件: MainPage.xaml.cs(片段)
public MainPage()
{
    this.InitializeComponent();

    for (int angle = 0; angle < 3600; angle++)
    {
        double radians = Math.PI * angle / 180;
        double radius = angle / 10;
        double x = 360 + radius * Math.Sin(radians);
        double y = 360 + radius * Math.Cos(radians);
        polyline.Points.Add(new Point(x, y));
    }
}
```

① 译注：原文为 Spiral，但示例中不存在这个类。根据下面的示例可知，作者要表达的是 MainPage。

循环体的第一步将角度转换为弧度并赋给变量 radians，以便供下面.NET 的三角函数使用。变量 radius(半径)根据角度计算得到，范围从 0 到 360，也就是说，最大半径为 360。静态方法 Math.Sin 和 Math.Cos 返回的值乘以半径得到横纵坐标，范围均在－360 到 360(像素)之间。

我们需要平移此图形，以使所有像素点相对于左上角都是正值。为此，要在这两个乘积上各加上 360。这样，螺旋线的中央便是(360, 360)，并且各方向的边界不超 360 像素。

循环体的最后一步创建了 Point 结构的实例，并将其添加到 Polyline 的 Points 集合中。程序运行的效果如下图所示。

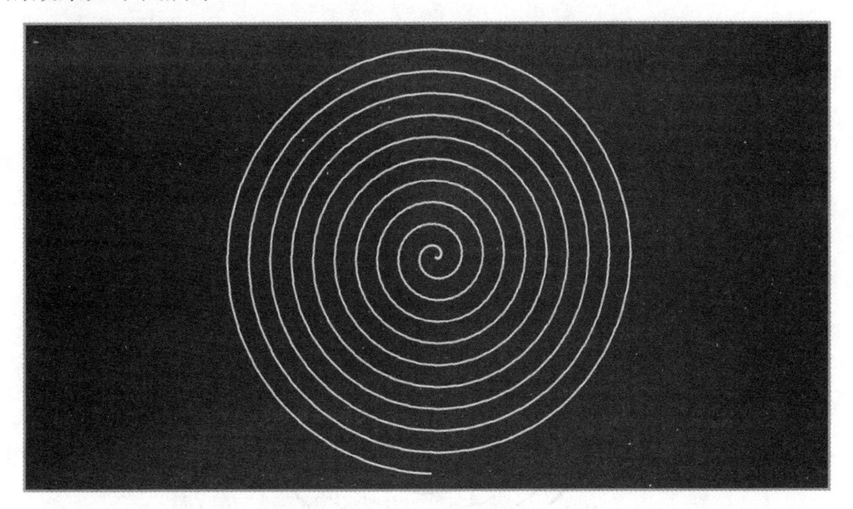

若没有 HorizontalAlignment 和 VerticalAlignment 设置，此图形则与页面的左上角对齐。如果在计算时不调整螺旋线中心的位置，那么它的中心则会在页面的左上角，并且有四分之三的部分都不可见。如果将 HorizontalAlignment 和 VerticalAlignment 设置为 Center，但不调整螺旋线的中心，那么图形则会偏向左上方。

这个螺旋线几乎充满了屏幕，但这只是因为本人屏幕的高度为 768 像素。如果不论屏幕大小都使螺旋线充满整个屏幕，该怎么做?

一个方案是直接将屏幕的分辨率代入螺旋线坐标的计算中。我们会在第 3 章中了解到如何实现。

另一个方案要用到 Shape 类定义的一个名为 Stretch 的属性。该属性与 Image 的 Stretch 属性用法完全一致。Polyline 的 Stretch 属性默认为枚举成员 Stretch.None(不拉伸)。但我们可以将其设置为 Uniform，以便在保持长宽比的前提下使图形充满容器。

示例项目 StretchedSpiral 演示了这个方案。此外，XAML 文件将线调得更宽。

```
项目: StretchedSpiral | 文件: MainPage.xaml(片段)
<Grid Background="{StaticResource ApplicationPageBackgroundThemeBrush}">
    <Polyline Name="polyline"
              Stroke="{StaticResource ApplicationForegroundThemeBrush}"
              StrokeThickness="6"
              Stretch="Uniform" />
</Grid>
```

在代码隐藏文件中，螺旋线坐标的计算可以使用任意半径。示例中螺旋线的最大半径为 1000。

```
项目: StretchedSpiral | 文件: MainPage.xaml.cs(片段)
public MainPage()
{
    this.InitializeComponent();

    for (int angle = 0; angle < 3600; angle++)
    {
        double radians = Math.PI * angle / 180;
        double radius = angle / 3.6;
        double x = 1000 + radius * Math.Sin(radians);
        double y = 1000 - radius * Math.Cos(radians);
        polyline.Points.Add(new Point(x, y));
    }
}
```

这个示例在计算纵坐标 y 时将原来的加号改成了减号，因而螺旋线终点的位置从下端变成了上端(如下图所示)。此外，这里还使用 ApplicationForegroundThemeBrush 资源，以便在切换到浅色主题时方便地修改 Stroke 颜色。

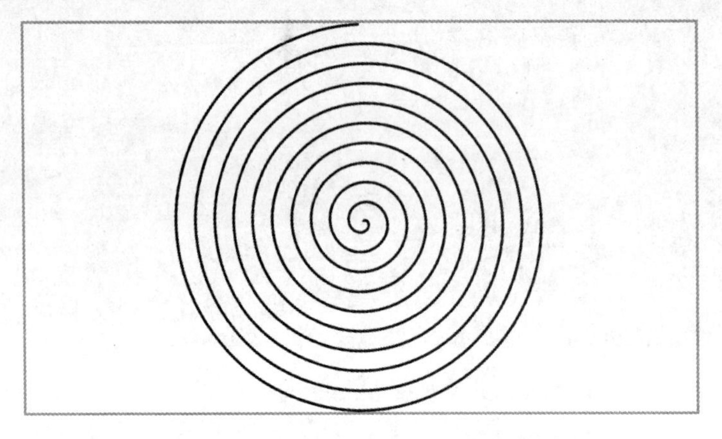

如果将 Stretch 属性设置为 Fill，圆螺旋线将变成椭圆螺旋线。

读者可能还记得，不论目标元素的大小 LinearGradientBrush 画笔都能够自动适应。该画笔也能够适应矢量图。下面让我们了解一下 ImageBrush 元素，它是一种基于位图的画笔。

示例项目 ImageBrushedSpiral 的代码隐藏类与示例项目 StretchedSpiral 中的一样，但 XAML 文件将线宽增大许多并利用了 ImageBrush。

```
项目: ImageBrushedSpiral | 文件: MainPage.xaml(片段)
<Grid Background="{StaticResource ApplicationPageBackgroundThemeBrush}">
    <Polyline Name="polyline"
              StrokeThickness="25"
              Stretch="Uniform">
        <Polyline.Stroke>
            <ImageBrush ImageSource="http://www.charlespetzold.com/pw6/PetzoldJersey.jpg"
                        Stretch="UniformToFill"
                        AlignmentY="Top" />
        </Polyline.Stroke>
    </Polyline>
</Grid>
```

与 Image 的 Source 属性一样，ImageBrush 的 ImageSource 属性的类型也为 ImageSource。在 XAML 中，我们只需要为该画笔设置一个 URL。ImageBrush 有自己的 Stretch 属性，默认值为 Fill。也就是说，位图会被拉伸以填充整个区域，而不会保持图片的长宽比。对于示例所用到的图片，这种设置会让其中的人物可能看起来胖一些，因而将其设置为 UniformToFill。该设置会保持图片的长宽比并使图片填充整个区域。与此同时，图片的一部分会被裁掉。AlignmentX 和 AlignmentY 可用于控制图片与图形的相对位置，从而决定哪一部分被裁掉。如下图所示，这个示例倾向于将图片的下部裁掉，只露出人物的头部[①]。

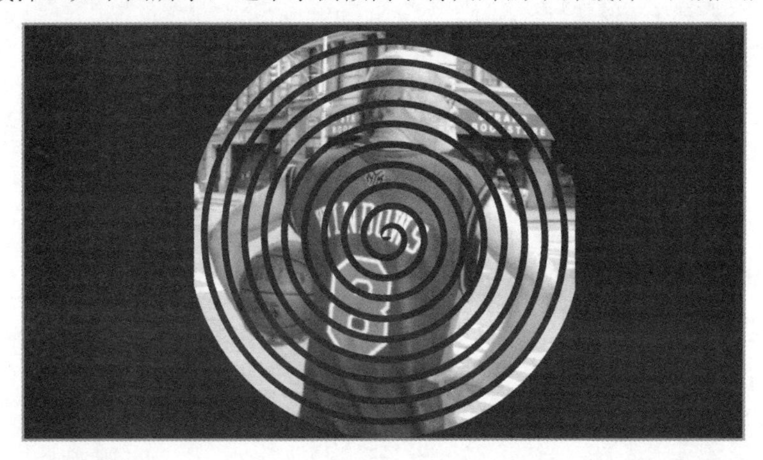

请注意，图片是按螺旋线的几何线对齐的，而并未按 25 像素宽的线条对齐。这使得上、左、右的边缘貌似被削掉了。这个问题可以通过 ImageBrush 的 Transform 属性解决，但这不在本章的讨论范围内。

ImageBrush 类派生自 TileBrush。这个继承关系意味着可以使位图在水平和竖直方向铺开，但 Windows Runtime 不支持这一功能。

任何可通过函数描述的曲线，均可以通过 Polyline 来呈现。但如果要绘制复杂的弧线(椭圆的弧长)、三次贝塞尔曲线(标准形式)或者二次贝塞尔曲线(只有一个可控点)，则不需要使用 Polyline。这些类型的曲线直接受 Path 元素支持。

Path 本身只定义了一个叫 Data 的属性，其类型为 Windows.UI.Xaml.Media 命名空间下的 Geometry 类。在 Windows Runtime 中，Geometry 和相关的类代表纯粹的解析几何图形。使用 Geometry 对象可以通过坐标点定义直线和曲线，Path 会通过选定的画笔和指定的粗细将线条呈现出来。

在 Geometry 派生类中，最强大且最灵活的是 PathGeometry 类。PathGeometry 的内容属性为 Figures，即 PathFigure 对象的集合。每个 PathFigure 代表一系列相连的直线和曲线。PathFigure 的内容属性是 Segments，即 PathSegment 对象的集合。PathSegment 的子类包括 LineSegment、PolylineSegment、BezierSegment、PolyBezierSegment、QuadraticBezierSegment、PolyQuadraticBezierSegment 和 ArcSegment。

下面，我们使用 Path 和 PathGeometry 来绘制一个 HELLO。

```
项目：HelloVectorGraphics | 文件：MainPage.xaml(片段)
<Grid Background="{StaticResource ApplicationPageBackgroundThemeBrush}">
    <Path Stroke="Red"
          StrokeThickness="12"
          StrokeLineJoin="Round"
          HorizontalAlignment="Center"
          VerticalAlignment="Center">
        <Path.Data>
            <PathGeometry>
                <!-- H -->
```

① 译注：图中人物正是作者本人。

```
            <PathFigure StartPoint="0 0">
               <LineSegment Point="0 100" />
            </PathFigure>
            <PathFigure StartPoint="0 50">
               <LineSegment Point="50 50" />
            </PathFigure>
            <PathFigure StartPoint="50 0">
               <LineSegment Point="50 100" />
            </PathFigure>

            <!-- E -->
            <PathFigure StartPoint="125 0">
               <BezierSegment Point1="60 -10" Point2="60 60" Point3="125 50" />
               <BezierSegment Point1="60 40" Point2="60 110" Point3="125 100" />
            </PathFigure>

            <!-- L -->
            <PathFigure StartPoint="150 0">
               <LineSegment Point="150 100" />
               <LineSegment Point="200 100" />
            </PathFigure>

            <!-- L -->
            <PathFigure StartPoint="225 0">
               <LineSegment Point="225 100" />
               <LineSegment Point="275 100" />
            </PathFigure>

            <!-- O -->
            <PathFigure StartPoint="300 50">
               <ArcSegment Size="25 50" Point="300 49.9" IsLargeArc="True" />
            </PathFigure>
         </PathGeometry>
      </Path.Data>
   </Path>
</Grid>
```

每个字符由一个或多个 PathFigure 对象构成。PathFigure 要求指定相连线段的起点。PathSegment 的派生类都会从该起点出发。例如，为绘制字母 E，BezierSegment 要通过两个点来控制开始和结束。前一个 BezierSegment 的终点为后一个 BezierSegment 的起点。(在使用 ArcSegment 时，弧的终点不能与起点相同，否则图形不会被绘制。这就是为什么示例中设置了 0.1 像素的间隔。避免出现问题的另一个办法是使用两个 ArcSegment，各用于绘制两半椭圆弧。)

事实上，使用一对贝塞尔曲线并非绘制大写字母 E 的最佳方式(如下图所示)。

若将 Path 的 Stretch 属性设置为 Fill，则会得到一个充满整个屏幕的“hello”图形(如下图所示)。

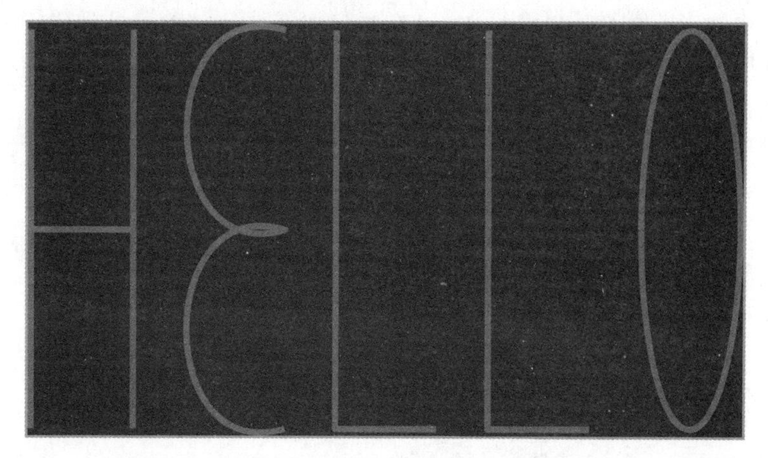

当然，也可以在代码中组合使用 PathFigure 和 PathSegment，但我们不妨了解一种在 XAML 中定义图形的简便方法。有一种“路径标记语法”(Path Markup Syntax)，每段指令由单字母开始，在必要时提供坐标点、尺寸和布尔值[①]。这种语法能够极大地减少标记量。示例项目 HelloVectorGraphicsPath 使用这种语法绘制了与 HelloVectorGraphics 相同的图形。

```
项目: HelloVectorGraphicsPath | 文件: MainPage.xaml(片段)
<Grid Background="{StaticResource ApplicationPageBackgroundThemeBrush}">
   <Path Stroke="Red"
         StrokeThickness="12"
         StrokeLineJoin="Round"
         HorizontalAlignment="Center"
         VerticalAlignment="Center"
         Data="M 0 0 L 0 100 M 0 50 L 50 50 M 50 0 L 50 100
               M 125 0 C 60 -10, 60 60, 125 50, 60 40, 60 110, 125 100
               M 150 0 L 150 100, 200 100
               M 225 0 L 225 100, 275 100
               M 300 50 A 25 50 0 1 0 300 49.9" />
</Grid>
```

Data 属性被设置了一个很长的字符串，这里将它分成了 5 行，对应 5 个字母图形。M 是“移动”命令，后面跟坐标点的 X 和 Y 值。这个点作为画笔的起点。L 用于绘制线段(更准确地说是多段线)，后面跟一个或多个点。C 是三次贝塞尔曲线命令，后面依次跟两个控制点和一个终点。A 是椭圆弧线命令，是这里面最复杂的。前两个数字分别代表椭圆的 X 轴半径和 Y 轴半径，第三个数字用于控制椭圆旋转的度数。后面两个数字分别用于控制 isLargeArc 标识和是否按照正角方向绘制弧线。最后的两个数字用于指定终点。这里有一个 Z 命令[②]未被用到，但它比较常用。该命令能够绘制一条返回起点的直线，将图形闭合。

使用一系列“路径标记语法”来定义复杂几何图形，是只能在 XAML 中完成的任务之一。Windows Runtime 未公开直接提供这种功能的类。只有 XAML 解析器可以在内部使用。为在代码中将“路径标记语言”字符串转换为 Geometry 对象，需要通过某种方式在代码中将 XAML 转换为对象。

① 译注：这里的布尔值要用数字来表示，即 0 代表 false，1 代表 true。

② 译注：Z 指令也叫“关闭指令”。

系统中恰好有这种转换工具，即 Windows.UI.Xaml.MarkupXamlReader 命名空间下的 XamlReader.Load 静态方法。传入 XAML 字符串，该方法会对树进行初始化和组装，最后输出根元素的实例。虽然 XamlReader.Load 有些限制(例如不能解析外部代码中的事件处理程序)，但是它仍是非常强大的工具。第 8 章有一个叫 XamlCruncher 的示例项目，它允许我们方便地进行有关 XAML 的试验。

这个示例项目展示了如何在代码中使用 Path 和“路径标记语法”。

项目：PathMarkupSyntaxCode | 文件：MainPage.xaml(片段)

```
using Windows.UI;                   // for Colors
using Windows.UI.Xaml;
using Windows.UI.Xaml.Controls;
using Windows.UI.Xaml.Markup;       // for XamlReader
using Windows.UI.Xaml.Media;
using Windows.UI.Xaml.Shapes;       // for Path

namespace PathMarkupSyntaxCode
{
    public sealed partial class MainPage : Page
    {
        public MainPage()
        {
            this.InitializeComponent();

            Path path = new Path
            {
                Stroke = new SolidColorBrush(Colors.Red),
                StrokeThickness = 12,
                StrokeLineJoin = PenLineJoin.Round,
                HorizontalAlignment = HorizontalAlignment.Center,
                VerticalAlignment =  VerticalAlignment.Center,
                Data = PathMarkupToGeometry(
                    "M 0 0 L 0 100 M 0 50 L 50 50 M 50 0 L 50 100 " +
                    "M 125 0 C 60 -10, 60 60, 125 50, 60 40, 60 110, 125 100 " +
                    "M 150 0 L 150 100, 200 100 " +
                    "M 225 0 L 225 100, 275 100 " +
                    "M 300 50 A 25 50 0 1 0 300 49.9")
            };

            (this.Content as Grid).Children.Add(path);
        }

        Geometry PathMarkupToGeometry(string pathMarkup)
        {
            string xaml =
                "<Path " +
                "xmlns='http://schemas.microsoft.com/winfx/2006/xaml/presentation'>" +
                "<Path.Data>" + pathMarkup + "</Path.Data></Path>";

            Path path = XamlReader.Load(xaml) as Path;

            // Detach the PathGeometry from the Path
            Geometry geometry = path.Data;
            path.Data = null;
            return geometry;
        }
    }
}
```

在代码中使用 Path 类时有一点需要特别注意：Visual Studio 生成的 MainPage.xaml.cs 文件并未包含相关代码来引入 Path 所在的 Windows.UI.Xaml.Shapes 命名空间，但该文件引入了 System.IO 命名空间。虽然后者也定义了 Path 类，但用于编辑文件和目录路径。

代码底部的方法是最关键的。该方法生成了一小段 XAML。这段 XAML 将 Path 作为根元素，通过属性元素语法包裹了“路径标记语法”字符串。注意，这段 XAML 中必须包含引用标准 XML 命名空间的声明。如果 XamlReader.Load 未遇到错误，它将返回一个 Path 对象，其 Data 属性会被设置为一个 PathGeometry 对象。在与当前 Path 取消关联之前，PathGeometry 对象不能关联另一个 Path 对象。为取消关联，可以将 Path 的 Data 设置为 null。

2.8 通过 Viewbox 实现拉伸

Image 类和 Shape 类都定义了 Stretch 属性，能够根据容器的大小拉伸位图和矢量图形，但并非 FrameworkElement 的所有派生类都支持这样一个属性。到底为何要通过 Viewbox 实现对 TextBlock 之类已经支持 Stretch 属性的元素进行拉伸呢？

有时我们需要以特殊方式进行拉伸。假设要显示一系列带有文本标题的对象。每个标题受限于特定的矩形区域，同时不同的对象和标题组合需要看起来一致。文本的长度是可变的(例如来自用户的输入)。如果文本字数过多，则希望它缩短一些，以便其能够适应所在的矩形区域。虽然可以在代码隐藏类中计算适当的 FontSize，但最好使 TextBlock 自动调整大小以适应特定空间。

Viewbox 正擅长于此。该元素有一个 Child 属性，类型为 UIElement。Viewbox 能够将子元素拉伸到自身大小。与 Image 和 Shape 一样，Viewbox 也定义了 Stretch 属性。该属性的默认值为 Uniform(与 Image.Stretch 属性的默认值一样)。示例程序将 Stretch 设置为 Fill，从而忽略 TextBlock 的长宽比，使其充满整个屏幕。

```
项目: TextStretch | 文件: MainPage.xaml(片段)
<Grid Background="{StaticResource ApplicationPageBackgroundThemeBrush}">
   <Viewbox Stretch="Fill">
      <TextBlock Text="Stretch Windows 8!" />
   </Viewbox>
</Grid>
```

TextBlock 会计算其自身包含“变音符”和“下伸部”的高度，即便两者未出现在这段文本中。这就是字母没有完全扩展至窗口高度的原因。

当然，这段文本原本的长宽比随着缩放发生了变化，如下图所示。

与 Image 和 Shape 不同的是，Viewbox 定义了 StretchDirection 属性，可选的值包括 UpOnly、DownOnly 和 Both(默认值)。Viewbox 通过改设置来确定是只增大子元素、只缩

小或者视情况而定。

下面将对示例项目 HelloVectorGraphics 进行修改，使每个字母呈现不同颜色。我们无法通过单个 Path 的 Stretch 属性使所有字母都符合窗口的高度，因为每个字母的高度不同。为此，需要将单个 Path 拆分成 5 个不同的 Path 元素。

下面，我们将 5 个 Path 元素置于一个 Grid 中，将这个 Grid 放入 Viewbox。

项目：VectorGraphicsStretch | 文件：MainPage.xaml(片段)

```
<Grid Background="{StaticResource ApplicationPageBackgroundThemeBrush}">
    <Viewbox Stretch="Fill">
        <Grid Margin="6 6 0 0">
            <!-- H -->
            <Path Stroke="Red"
                  StrokeThickness="12"
                  StrokeLineJoin="Round"
                  Data="M 0 0 L 0 100 M 0 50 L 50 50 M 50 0 L 50 100" />

            <!-- E -->
            <Path Stroke="#C00040"
                  StrokeThickness="12"
                  StrokeLineJoin="Round"
                  Data="M 125 0 C 60 -10, 60 60, 125 50, 60 40, 60 110, 125 100" />

            <!-- L -->
            <Path Stroke="#800080"
                  StrokeThickness="12"
                  StrokeLineJoin="Round"
                  Data="M 150 0 L 150 100, 200 100" />

            <!-- L -->
            <Path Stroke="#4000C0"
                  StrokeThickness="12"
                  StrokeLineJoin="Round"
                  Data="M 225 0 L 225 100, 275 100" />

            <!-- O -->
            <Path Stroke="Blue"
                  StrokeThickness="12"
                  StrokeLineJoin="Round"
                  Data="M 300 50 A 25 50 0 1 0 300 49.9" />
        </Grid>
    </Viewbox>
</Grid>
```

这样，不同矢量图形便会被看作一个整体进行缩放(如下图所示)。

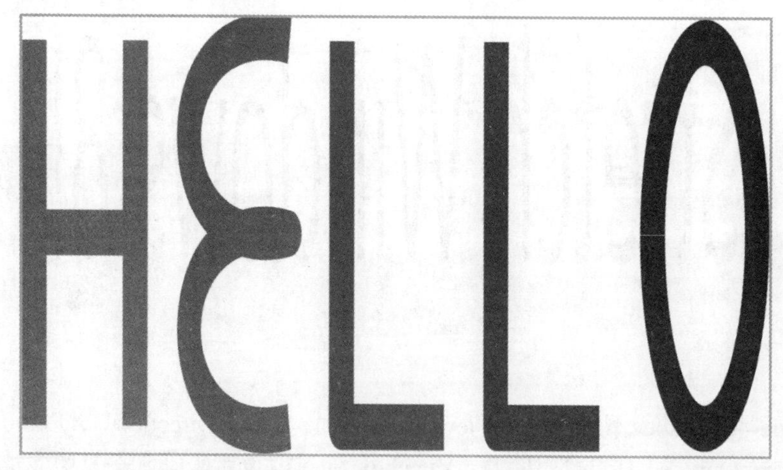

有一点需要注意，通过 Viewbox 实现的拉伸会改变笔画(stroke)的宽度，而通过 Path 自身的 Stretch 实现的拉伸不会。

2.9　样　　式

本章介绍了如何将画笔定义为资源，从而在不同元素间共享。资源最重要的作用是定义样式，而样式表现为 Style 的实例。简单地讲，样式是可以在多个元素间共享的属性集合。样式不仅能减少重复的标记，还能够实现属性设置的统筹管理。

在本节讨论过后，对于 Common 文件夹下 Visual Studio 生成的 StandardStyles.xaml 文件，您将能够读懂大部分内容。其中的 ControlTemplate 将在第 11 章中介绍。

示例项目 SharedBrushWithStyle 与之前的示例项目 SharedBrush 类似，只不过前者通过 Style 合并了集合属性。示例样式位于 Resources 区段的底部。

```
项目: SharedBrushWithStyle | 文件: MainPage.xaml(片段)
<Page.Resources>
   <x:String x:Key="appName">Shared Brush with Style</x:String>

   <LinearGradientBrush x:Key="rainbowBrush">
      <GradientStop Offset="0" Color="Red" />
      <GradientStop Offset="0.17" Color="Orange" />
      <GradientStop Offset="0.33" Color="Yellow" />
      <GradientStop Offset="0.5" Color="Green" />
      <GradientStop Offset="0.67" Color="Blue" />
      <GradientStop Offset="0.83" Color="Indigo" />
      <GradientStop Offset="1" Color="Violet" />
   </LinearGradientBrush>

   <Style x:Key="rainbowStyle" TargetType="TextBlock">
      <Setter Property="FontFamily" Value="Times New Roman" />
      <Setter Property="FontSize" Value="96" />
      <Setter Property="Foreground" Value="{StaticResource rainbowBrush}" />
   </Style>
</Page.Resources>
```

与普通的资源一样，Style 的开始标签也包含 x:Key 特性。此外，Style 还要求指定 TargetType 特性，其值必须为 FrameworkElement 或其子类。样式只能作用于 FrameworkElement 的派生类。

这个 Style 的主体当中包含一组 Setter 元素，并为每个 Setter 指定 Property 和 Value 特性。其中的一个 Setter 通过 StaticResource 将 Value 特性设置为之前定义的 LinearGradientBrush。需要注意的是，为使这种引用能够生效，在 XAML 文件中，Style 必须在被引用的画笔之后定义，但样式可以在可视树分支的某个“资源区段”中定义。

元素需要通过自身的 Style 属性来引用样式。与引用资源一样，引用样式也需要使用 StaticResource 标记扩展。

```
项目: SharedBrushWithStyle | 文件: MainPage.xaml(片段)
<Grid Background="{StaticResource ApplicationPageBackgroundThemeBrush}">
   <TextBlock Text="{StaticResource appName}"
              FontSize="48"
              HorizontalAlignment="Center"
              VerticalAlignment="Center" />

   <TextBlock Text="Top Text"
              Style="{StaticResource rainbowStyle}"
```

```
                HorizontalAlignment="Center"
                VerticalAlignment="Top" />

     <TextBlock Text="Left Text"
                Style="{StaticResource rainbowStyle}"
                HorizontalAlignment="Left"
                VerticalAlignment="Center" />

     <TextBlock Text="Right Text"
                Style="{StaticResource rainbowStyle}"
                HorizontalAlignment="Right"
                VerticalAlignment="Center" />

     <TextBlock Text="Bottom Text"
                Style="{StaticResource rainbowStyle}"
                HorizontalAlignment="Center"
                VerticalAlignment="Bottom" />
</Grid>
```

这个示例的界面与之前 SharedBrush 示例的界面是一样的。

通过 Style 来定义 LinearGradientBrush 还有另外一种方式。为通过较复杂标记来定义某个对象，我们往往会在元素上使用“属性元素语法”。与之类似，我们也可以对 Setter 类的 Value 使用这种语法。

```
<Style x:Key="rainbowStyle" TargetType="TextBlock">
   <Setter Property="FontFamily" Value="Times New Roman" />
   <Setter Property="FontSize" Value="96" />
   <Setter Property="Foreground">
      <Setter.Value>
         <LinearGradientBrush>
            <GradientStop Offset="0" Color="Red" />
            <GradientStop Offset="0.17" Color="Orange" />
            <GradientStop Offset="0.33" Color="Yellow" />
            <GradientStop Offset="0.5" Color="Green" />
            <GradientStop Offset="0.67" Color="Blue" />
            <GradientStop Offset="0.83" Color="Indigo" />
            <GradientStop Offset="1" Color="Violet" />
         </LinearGradientBrush>
      </Setter.Value>
   </Setter>
</Style>
```

乍看起来在样式中定义画笔有些奇怪，但这是很普遍的做法。LinearGradientBrush 自身并未设置 x:Key 特性，而只有 Resources 集合中的顶层 Style 元素具有该特性。

在代码中定义 Style 可以这样做。

```
Style style = new Style(typeof(TextBlock));
style.Setters.Add(new Setter(TextBlock.FontSizeProperty, 96));
style.Setters.Add(new Setter(TextBlock.FontFamilyProperty,
                             new FontFamily("Times New Roman")));
```

将这个样式添加到 Page 的 Resources 集合需要在 InitializeComponent 被调用前完成。这样，XAML 文件中定义的 TextBlock 元素才能使用它。另外，也可以将 Style 对象直接设置到 TextBlock 的 Style 属性上。但这种做法并不常见，因为在代码中我们可以通过其他方式为相同的属性赋值，例如通过 for 或 foreach 循环。

请注意上述示例代码中 Setter 构造函数的第一个参数。该参数的类型为 DependencyProperty。样式的目标类型或其父类会定义一些 DependencyProperty 类型的静态属性，我们应通过这些静态属性来指定该参数。依赖属性可以在不创建类实例的前提下指定类的属性，这个示例恰好展示了这样设计的好处。

这段代码还暗示了一点，即 Style 所针对的属性只能为依赖属性。正如前文提到的，依赖属性能够沿着可视树传播。举例来说，假设程序中有以下标记。

```
<TextBlock Text="Top Text"
           Style="{StaticResource rainbowStyle}"
           FontSize="24"
           HorizontalAlignment="Center"
           VerticalAlignment="Top" />
```

Style 定义了 FontSize 值，但 FontSize 属性又在 TextBlock 本地被重新设置。正如人们希望的，本地设置优先于样式设置，并且本地设置和样式设置优先于沿着可视树传播的 FontSize 值。

一旦 Style 对象被设置到某元素的 Style 属性上，便不能被更改。此时，我们可以在这个元素上设置一个不同的 Style 对象，可以修改样式引用的对象(如画笔)的属性，但不能设置或移除 Setter 对象，也不能修改 Value 属性。

样式可以通过 Style 类的 BasedOn 属性从其他样式继承属性设置。为此，一般使用 StaticResource 标记扩展来引用之前定义的 Style 资源。

```
<Style x:Key="baseTextBlockStyle" TargetType="TextBlock">
   <Setter Property="FontFamily" Value="Times New Roman" />
   <Setter Property="FontSize" Value="24" />
</Style>

<Style x:Key="gradientStyle" TargetType="TextBlock"
       BasedOn="{StaticResource baseTextBlockStyle}">
   <Setter Property="FontSize" Value="96" />
   <Setter Property="Foreground">
      <Setter.Value>
         <LinearGradientBrush>
            <GradientStop Offset="0" Color="Red" />
            <GradientStop Offset="1" Color="Blue" />
         </LinearGradientBrush>
      </Setter.Value>
   </Setter>
</Style>
```

名称为 gradientStyle 的 Style 基于之前定义的 baseTextBlockStyle 样式。前者继承了后者的 FontFamily 设置，改写了 FontSize 设置并引入了 Foreground 设置。

再看一个例子。

```
<Style x:Key="centeredStyle" TargetType="FrameworkElement">
   <Setter Property="HorizontalAlignment" Value="Center" />
   <Setter Property="VerticalAlignment" Value="Center" />
</Style>

<Style x:Key="rainbowStyle" TargetType="TextBlock"
       BasedOn="{StaticResource centeredStyle}">
   <Setter Property="FontSize" Value="96" />
   <Setter Property="Foreground">
      <Setter.Value>
         <LinearGradientBrush>
            <GradientStop Offset="0" Color="Red" />
            <GradientStop Offset="1" Color="Blue" />
         </LinearGradientBrush>
      </Setter.Value>
   </Setter>
</Style>
```

在这个例子中，第一个 Style 的 TargetType(目标类型)为 FrameworkElement，这说明该

样式只能包含 FrameworkElement 本身定义和它继承的属性。该样式可以应用到 TextBlock 上，因为 TextBlock 派生自 FrameworkElement。第二个 Style 基于 centeredStyle 样式，目标类型为 TextBlock，也就是说它可以包含 TextBlock 特有的属性。TargetType 所指定的类型必须与 BasedOn 的样式所针对的类型相同或是 BasedOn 的样式所针对的类型的子类。

尽管前文说资源都要有键，但实际上 Style 是个例外。没有 x:Key 的 Style 叫“隐式样式”(implicit style)。示例项目 ImplicitStyle 的“资源区段”演示了这种样式的使用。

```
项目: ImplicitStyle | 文件: MainPage.xaml(片段)
<Page.Resources>
    <x:String x:Key="appName">Implicit Style App</x:String>

    <Style TargetType="TextBlock">
        <Setter Property="FontFamily" Value="Times New Roman" />
        <Setter Property="FontSize" Value="96" />
        <Setter Property="Foreground">
            <Setter.Value>
                <LinearGradientBrush>
                    <GradientStop Offset="0" Color="Red" />
                    <GradientStop Offset="0.17" Color="Orange" />
                    <GradientStop Offset="0.33" Color="Yellow" />
                    <GradientStop Offset="0.5" Color="Green" />
                    <GradientStop Offset="0.67" Color="Blue" />
                    <GradientStop Offset="0.83" Color="Indigo" />
                    <GradientStop Offset="1" Color="Violet" />
                </LinearGradientBrush>
            </Setter.Value>
        </Setter>
    </Style>
</Page.Resources>
```

事实上，Windows RT 幕后会为这个属性创建一个键。该键是类型为 RuntimeType 的对象(非公共类)。对于这个例子，所生成的 RuntimeType 对象用于指代样式所针对的 TextBlock 类型。

隐式样式非常有用。可视树上任何未设置 Style 属性的 TextBlock，都将获得这个隐式样式。如果页面中已经添加了许多 TextBlock 元素，但之后决定统一其样式，使用隐式样式最方便。请注意，在这个示例中，TextBlock 元素都没有设置 Style 属性。

```
项目: ImplicitStyle | 文件: MainPage.xaml(片段)
<Grid Background="{StaticResource ApplicationPageBackgroundThemeBrush}">

    <TextBlock Text="{StaticResource appName}"
               FontFamily="Portable User Interface"
               FontSize="48"
               Foreground="{StaticResource ApplicationForegroundThemeBrush}"
               HorizontalAlignment="Center"
               VerticalAlignment="Center" />

    <TextBlock Text="Top Text"
               HorizontalAlignment="Center"
               VerticalAlignment="Top" />

    <TextBlock Text="Left Text"
               HorizontalAlignment="Left"
               VerticalAlignment="Center" />

    <TextBlock Text="Right Text"
               HorizontalAlignment="Right"
               VerticalAlignment="Center" />

    <TextBlock Text="Bottom Text"
```

```
            HorizontalAlignment="Center"
            VerticalAlignment="Bottom" />
</Grid>
```

这里将隐式样式应用到了页面上的大多数 TextBlock 元素，但如果不希望应用到第一个元素上(显示在中央的文本块)，该怎么办？如果不希望页面上的某个元素受隐式样式影响，可以赋予其显式样式，可以通过本地设置覆盖 Style 对象中的属性设置，也可以将该元素的 Style 属性设置为 null。在这个示例中，我们修改第一个 TextBlock 的 FontFamily 和 FontSize 属性的默认值，并使用预定义的资源设置 Foreground，这些都会覆盖隐式样式对该元素的设置。

开发者创建的样式不能继承隐式样式，但隐式样式可以继承非隐式样式。为此，提供 TargetType 和 BasedOn 特性，但不添加 x:Key 即可。

虽然隐式样式非常强大，但凡事必有利弊。在大型应用程序中，样式定义随处可见，而可视树可能由多个 XAML 文件共同构建。如果样式隐式地应用到某个元素上，则很难确定该样式到底在哪里定义的。

现在，我们可以开始使用(或阅读一下)StandardStyles.xaml 文件中针对 TextBlock 的样式。这些样式包括 BasicTextStyle、BaselineTextStyle、HeaderTextStyle、SubheaderTextStyle、TitleTextStyle、ItemTextStyle、BodyTextStyle、CaptionTextStyle、PageHeaderTextStyle、PageSubheaderTextStyle 和 SnappedPageHeaderTextStyle。显然，这些样式也是以文本的形式存在的。

2.10 初探数据绑定

使用数据绑定(data binding)是另一种在 XAML 文件中共享对象的方法。简单地讲，数据绑定是一种建立在两属性间的连接。数据绑定主要用于在页面可视元素与数据源之间建立连接，并且是实现“模型-视图-视图模型”(Model-View-ViewModel，MVVM)模式的主要手段(详情参见第 6 章)。MVVM 的绑定目标是视图中的可视元素，绑定源是视图模型中的属性。在定义用于显示数据对象的模板时，数据绑定是一个关键步骤(第 11 章将具体介绍)。

我们可以使用数据绑定来连接两个元素的属性。与 StaticResource 一样，Binding 也是以标记扩展形式表达的，要在大括号中间声明。但比 StaticResource 更进一步，Binding 可以通过属性元素语法来表达。

下面的代码来自示例项目 SharedBrushWithBinding 的“资源区段”。

```
项目: SharedBrushWithBinding | 文件: MainPage.xaml(片段)
<Page.Resources>
    <x:String x:Key="appName">Shared Brush with Binding</x:String>

    <Style TargetType="TextBlock">
        <Setter Property="FontFamily" Value="Times New Roman" />
        <Setter Property="FontSize" Value="96" />
    </Style>
</Page.Resources>
```

TextBlock 的隐式样式不再拥有 Foreground 属性。LinearGradientBrush 被定义在第一个使用该画笔的 TextBlock 元素上。后面的 3 个 TextBlock 元素通过绑定引用该画笔。

项目：SharedBrushWithBinding | 文件：MainPage.xaml(片段)

```
<Grid Background="{StaticResource ApplicationPageBackgroundThemeBrush}">
    <TextBlock Text="{StaticResource appName}"
               FontFamily="Portable User Interface"
               FontSize="48"
               HorizontalAlignment="Center"
               VerticalAlignment="Center" />

    <TextBlock Name="topTextBlock"
               Text="Top Text"
               HorizontalAlignment="Center"
               VerticalAlignment="Top">
        <TextBlock.Foreground>
            <LinearGradientBrush>
                <GradientStop Offset="0" Color="Red" />
                <GradientStop Offset="0.17" Color="Orange" />
                <GradientStop Offset="0.33" Color="Yellow" />
                <GradientStop Offset="0.5" Color="Green" />
                <GradientStop Offset="0.67" Color="Blue" />
                <GradientStop Offset="0.83" Color="Indigo" />
                <GradientStop Offset="1" Color="Violet" />
            </LinearGradientBrush>
        </TextBlock.Foreground>
    </TextBlock>

    <TextBlock Text="Left Text"
               HorizontalAlignment="Left"
               VerticalAlignment="Center"
               Foreground="{Binding ElementName=topTextBlock, Path=Foreground}" />

    <TextBlock Text="Right Text"
               HorizontalAlignment="Right"
               VerticalAlignment="Center"
               Foreground="{Binding ElementName=topTextBlock, Path=Foreground}" />

    <TextBlock Text="Bottom Text"
               HorizontalAlignment="Center"
               VerticalAlignment="Bottom">
        <TextBlock.Foreground>
            <Binding ElementName="topTextBlock" Path="Foreground" />
        </TextBlock.Foreground>
    </TextBlock>
</Grid>
```

数据绑定需要有“源”和“目标”。目标总是绑定设置所在的属性，而源则是绑定所引用的属性。在这个例子中，名为topTextBlock的TextBlock是绑定源，而后面三个TextBlock的 Foreground 则是绑定目标。前两个目标使用的是以 XAML 标记扩展方式表达的 Binding 对象。

```
Foreground="{Binding ElementName=topTextBlock, Path=Foreground}"
```

XAML 标记扩展必须在大括号中声明。Binding 标记扩展要求设置两个属性值。不同属性间用英文逗号分隔。ElementName 属性用于指定源元素(值来源于该元素)的名称，Path 属性用于指定源属性的名称。

在键入 Binding 标记扩展时，人们可能要习惯性地对属性值加引号。这是错误的。引号不应出现在绑定表达式中。

示例的最后一个 TextBlock 展示了以“属性元素语法”形式表达的 Binding 对象。

```
<TextBlock.Foreground>
    <Binding ElementName="topTextBlock" Path="Foreground" />
</TextBlock.Foreground>
```

这种语法要求在 ElementName 和 Path 属性的值上加引号。

我们也可以在代码中创建 Binding 对象，通过 FrameworkElement 的 SetBinding 方法将该对象设置到目标属性上。在这个过程中会发现，绑定目标必须是依赖属性。

Binding 类的 Path 属性之所以叫做 Path，是因为实际的值可能是多个由句点分隔的名称。还是以本节的示例项目为例，用下面这行标记来替换其中的某个 Text 设置。

```
Text="{Binding ElementName=topTextBlock, Path=FontFamily.Source}"
```

Path 值的第一部分表示我们要从 FontFamily 属性获取内容。该属性返回的是 FontFamily 的对象。该对象有一个 Source 属性，返回的是字体的名称。这个 TextBlock 将显示“Times New Roman”。(C++程序目前还不支持这种复合的、索引形式的路径。)

可以在示例项目的任意 TextBlock 上尝试以下设置。

```
Text="{Binding RelativeSource={RelativeSource Self}, Path=FontSize}
```

这是一种 Binding 标记扩展内部的标记扩展，即 RelativeSource 标记扩展。可以用这种语法使元素能够引用自身的属性。

熟悉 StaticResource、Binding 和 RelativeSource 后，便能够理解 Windows Runtime 支持的大部分标记扩展。第 11 章会介绍另一种标记扩展，即 TemplateBinding 标记扩展。

其余的标记扩展并不十分常用，但有时也不可或缺。假设已为 Grid 定义好了一种设置 Background 属性的隐式样式，但希望其中一个 Grid 的 Background 属性为 null。那么在标记中如何指定 null 呢？可以像下面这样做。

```
Background="{x:Null}"
```

再举一个例子，假设已经定义了一种隐式样式，但不希望其中的某个元素以任何方式受该样式影响，则可以像下面这样做。

```
Style="{x:Null}"
```

至此，本书已经介绍了 XAML 文件中针对 Windows Runtime 可能出现的大部分带有“x”前缀的数据类型、元素、特性和标记扩展。其中的数据类型包括 x:Boolean、x:Double、x:Int32 和 x:String，特性包括 x:Class、x:Name 和 x:Key，标记扩展包括 x:Null。还有一个指令之前未提到过，即 x:Uid，该指令用于引用国际化资源，值为应用程序全局唯一的字符串。

第 3 章　基本事件的处理

前面两章演示了如何在 XAML 与代码中实例化和初始化元素及其他对象。XAML 一般用来定义最初的页面布局和元素的外观，然后通过代码在运行时修改元素的属性。

之前还介绍过，在 XAML 中为元素分配 Name 或 x:Name 后，Page 类中便会生成对应的字段。这样，代码隐藏文件便能够通过该字段访问 XAML 中的元素。在代码与 XAML 交互方面，这是两种主要方式之一，另一种是通过事件(event)。事件是一种对象间的通信机制。事件被一个对象“引发”(或者说“触发”)，被其他订阅该事件的若干对象“处理”。在 Windows Runtime 中，事件主要用来通知来自触摸、鼠标屏、手写笔或键盘的用户输入。

初始化完成之后，Windows Runtime 程序便会驻留于内存并等待所关注的事件。几乎所有的子程序都在事件发生后执行，所以本书下面介绍的内容绝大部分都与事件处理相关。

3.1 Tapped 事件

UIElement 类定义了所有用户输入事件，可分成以下几类。

- 8 个涵盖触摸、鼠标和手写笔的输入的事件，这些事件的名称以 Pointer 为前缀。
- 5 个汇集了来自多点触摸的输入的事件，这些事件的名称都以 Manipulation 为前缀。
- 2 个响应键盘输入的事件，这两个事件的名称以 Key 为前缀。
- 4 个高级事件，分别为 Tapped、DoubleTapped、RightTapped 和 Holding。

RightTapped 事件并非由右手手指触摸引发，而用于响应鼠标右键单击。可以通过手指点击并保持不动，然后抬手来模拟触摸板的右击，注意，该操作也会引发 Holding 事件。应处理哪个事件则要具体问题具体分析。

第 13 章会介绍与触摸、鼠标和手写笔相关的事件。UIElement 还定义了以下两组与用户输入有关的事件。

- 接受键盘输入的元素所具有的 GotFocus 和 LostFocus 事件
- 与拖放有关的 DragEnter、DragOver、DragLeave 和 Drop 事件

我们先从 Tapped 这个比较有代表性且较为简单的事件说起。派生自 UIElement 的元素都有 Tapped 事件，能够在用户触摸、鼠标单击、手写笔点按时引发。为引发 Tapped 事件，手指、鼠标或手写笔按下后不能有太大移位，并且迅速抬起。

所有用户输入事件的模式都类似。UIElement 定义的 Tapped 用 C#语法可以这样表达。

```
public event TappedEventHandler Tapped;
```

TappedEventHandler 定义于 Windows.UI.Xaml.Input 命名空间，是一种有事件处理程序签名的委托类型。

```
public delegate void TappedEventHandler(object sender, TappedRoutedEventArgs e);
```

这个委托签名的第一个参数指向引发事件的对象(一般为 UIElement 的实例)，第二个参数是 Tapped 事件特有的。[①]

示例项目 TapTextBlock 的 XAML 文件定义了一个带有 Name 特性和 Tapped 事件处理程序的 TextBlock。

```
<Grid Background="{StaticResource ApplicationPageBackgroundThemeBrush}">
   <TextBlock Name="txtblk"
              Text="Tap Text!"
              FontSize="96"
              HorizontalAlignment="Center"
              VerticalAlignment="Center"
              Tapped="txtblk_Tapped_1" />
</Grid>
```

在 XAML 中键入 TextBlock 的特性时，“智能感知”功能会提示可用的属性和事件。两者可以通过名称左端的小图标加以区别：属性用“扳手”图标表示，事件用“闪电”图标表示。还有一些大括号图标，第 4 章会介绍。如果需要，“智能感知”功能还可以自动生成事件处理程序的名称。从 XAML 语法无法区分特性和事件。

实际的事件处理程序要在代码隐藏文件中实现。除了可以生成处理程序的名称外，Visual Studio 还可以生成空白事件处理程序。下面的空白事件处理程序是在 MainPage.xaml.cs 文件中生成的。

```
private void txtblk_Tapped_1(object sender, TappedRoutedEventArgs e)
{

}
```

该方法会在 TextBlock 被用户点击后引发。对于后面示例程序中的事件处理程序，本人会根据个人偏好修改其名称，即移除 private 关键字(因为这是默认的)，去掉名称中的下划线，添加单词 On(如 OnTextBlockTapped)，并将参数 e 重命名为 args。在代码中重新对方法名命名后，可以单击全局重命名小图标，XAML 文件中的方法引用能够随之变化。[②]

这个示例程序可以在 TextBlock 被单击后随机生成一种颜色。为此，MainPage 类中定义了一个 Random 对象和一个表示红绿蓝三色的字节数组。

```
项目：TapTextBlock | 文件：MainPage.xaml.cs(片段)
public sealed partial class MainPage : Page
{
   Random rand = new Random();
   byte[] rgb = new byte[3];

   public MainPage()
   {
      this.InitializeComponent();
   }

   private void txtblk_Tapped_1(object sender, TappedRoutedEventArgs e)
   {
      rand.NextBytes(rgb);
```

① 译注：这是事件委托的标准形式。虽然编译器允许事件使用任意签名的委托类型，但根据.NET 平台规范，事件的处理程序应带有 sender 和 e 这两个参数，前者类型为 object，后者是派生自 EventArgs 的类。

② 译注：使用“重命名”对话框(按功能键 F2)进行重命名也是较常用的方式。在反复编辑名称时，“重命名”图标可能会由于反复修改而消失，而“重命名”对话框可以确保重命名操作能够被执行(是否彻底是另一回事)。此外，通过“重命名”对话框可以预览受影响的所有文件及位置。编辑名称后，按 Enter 键关闭对话框。可以自始至终不使用鼠标操作，因而同样高效。

```
        Color clr = Color.FromArgb(255, rgb[0], rgb[1], rgb[2]);
        txtblk.Foreground = new SolidColorBrush(clr);
    }
}
```

这个示例移除了默认添加的 OnNavigatedTo 方法，因为这里并未用到。Tapped 事件处理程序通过 Random 的 NextBytes 方法生成了 3 个随机的字节，传入静态方法 Color.FromArgb 便可获得 Color 值。最后，这个事件处理程序将 TextBlock 的 Foreground 属性设置为基于此 Color 值的 SolidColorBrush。

运行这个示例程序，并用手指、鼠标或手写笔点击屏幕中央的 TextBlock，文字会呈现随机的颜色。点击 TextBlock 以外的屏幕区域则没有任何变化。使用鼠标或手写笔也不必点击到字母的笔画上。点击笔画和笔画间，事件都能够被引发。好像这个 TextBlock 具有一个不可见的背景，其高度包含字体变音符号和下伸部所占高度。事实的确如此。

Visual Studio 生成的 MainPage.g.cs 文件中会包含一个名为 Connect 的方法。该方法将事件处理程序与 TextBlock 的 Tapped 事件关联。我们也可以自己订阅事件。将 MainPage.xaml 文件中的 Tapped 处理程序设置去掉，并在代码隐藏文件的构造函数中订阅事件(如下所示)。

```
public MainPage()
{
    this.InitializeComponent();
    txtblk.Tapped += txtblk_Tapped_1;
}
```

实际效果无差别。

为使 TextBlock 的 Tapped 事件工作，需要设置几个属性。IsHitTestVisible 和 IsTapEnabled 属性必须为 true(默认值)。Visibility 属性必须为 Visibility.Visible(默认值)。如果将 Visibility 属性设置为 Visibility.Collapsed，那么 TextBlock 将不可见且无法响应用户输入。

事件处理程序 txtblk_Tapped_1 的第一个参数是引发事件的元素(对照本示例来说，这个元素就是 TextBlock)。第二个参数用于提供当前事件的信息，其中包括点击发生的坐标点以及指针设备类型(手指、鼠标或手写笔)。第 13 章会具体介绍有关内容。

3.2 路由事件的处理

Tapped 事件处理程序的第一个参数 sender 传入的是引发该事件的元素，因此不必为访问该元素而为其设置名称。这里，我们可以将参数 sender 转换为 TextBlock 类型的对象。这样可以使多个元素共享同一个事件处理程序(正如在示例项目 RoutedEvents0 中所展示的)。

路由事件是 Windows Runtime 的重要功能。本章的示例中有一系列演示路由事件的项目。RoutedEvents0 项目并未特别突出路由事件，因而其名称后缀为 0。该示例创建了一个 Tapped 处理程序，其签名和名称都根据本人偏好做了调整。

```
项目: RoutedEvents0 | 文件: MainPage.xaml.cs(片段)
public sealed partial class MainPage : Page
{
    Random rand = new Random();
    byte[] rgb = new byte[3];
```

```
    public MainPage()
    {
        this.InitializeComponent();
    }

    void OnTextBlockTapped(object sender, TappedRoutedEventArgs args)
    {
        TextBlock txtblk = sender as TextBlock;
        rand.NextBytes(rgb);
        Color clr = Color.FromArgb(255, rgb[0], rgb[1], rgb[2]);
        txtblk.Foreground = new SolidColorBrush(clr);
    }
}
```

请注意，事件处理程序的第一行将 sender 参数转换成了 TextBlock。

由于代码隐藏文件中已包含了事件处理程序，所以在 XAML 文件中添加事件时，Visual Studio 的“智能感知”功能会提示该处理程序的名称。因此，为示例中的 9 个 TextBlock 元素添加事件轻而易举。

项目：RoutedEvents0 | 文件：MainPage.xaml(片段)

```
<Page
    x:Class="RoutedEvents0.MainPage"
    ...
    FontSize="48">

    <Grid Background="{StaticResource ApplicationPageBackgroundThemeBrush}">
        <TextBlock Text="Left / Top"
                   HorizontalAlignment="Left"
                   VerticalAlignment="Top"
                   Tapped="OnTextBlockTapped" />

        ...

        <TextBlock Text="Right / Bottom"
                   HorizontalAlignment="Right"
                   VerticalAlignment="Bottom"
                   Tapped="OnTextBlockTapped" />
    </Grid>
</Page>
```

为解释这段标记，这里没必要罗列所有元素。由于在 Page 元素上设置了 FontSize，所以全部 TextBlock 元素都将继承该属性。运行程序，并点击任意元素。我们会发现每个元素的颜色能够独立变化(如下图所示)。

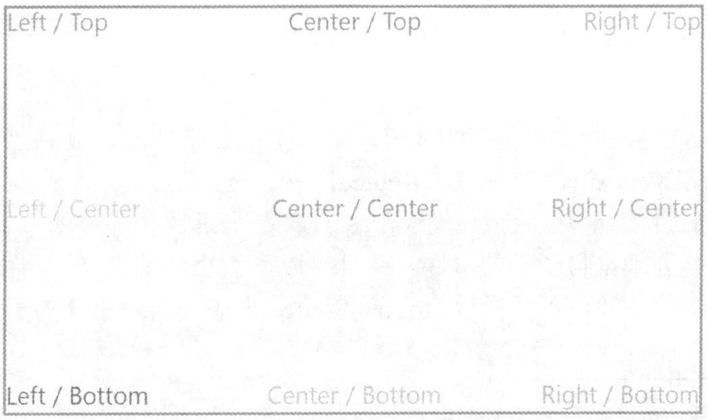

Left / Top　Center / Top　Right / Top
Left / Center　Center / Center　Right / Center
Left / Bottom　Center / Bottom　Right / Bottom

单击空白区域则不会产生任何效果。

在 XAML 文件中为 9 个元素设置事件处理程序或许让人觉得枯燥乏味。下面这个示例程序恰好解决了这个问题。RoutedEvents1 程序利用了“路由输入处理”(routed input handling)，输入事件(如 Tapped)被引发后会沿着可视树向上传播。这个示例并没有单独为每个 TextBlock 元素设置 Tapped 处理程序，而是在某元素的父元素(如 Grid)上设置。下面这段代码来自 RoutedEvents1 的 XAML 文件。

```
项目：RoutedEvents1 | 文件：MainPage.xaml(片段)
<Grid Background="{StaticResource ApplicationPageBackgroundThemeBrush}"
      Tapped="OnGridTapped">

    <TextBlock Text="Left / Top"
               HorizontalAlignment="Left"
               VerticalAlignment="Top" />

    ...

    <TextBlock Text="Right / Bottom"
               HorizontalAlignment="Right"
               VerticalAlignment="Bottom" />
</Grid>
```

除了将每个 TextBlock 的 Tapped 处理程序设置转移到 Grid 上，处理程序的名称也被相应地修改。

事件处理程序的内容也要做相应修改。上一个示例的 Tapped 处理程序将 sender 参数转换成了 TextBlock。之所以可以这样进行类型转换，是因为该处理程序只作用于 TextBlock 类型的元素。如果将该处理程序设置到 Grid 上，那么它的参数 sender 的类型便是 Grid。那么问题来了，如何区分哪个 TextBlock 被点击了呢？

这并不困难，TappedRoutedEventArgs 类(事件处理程序第二个参数的类型)有一个名为 OriginalSource 的属性，用于获取当前事件的来源。对于这个示例，OriginalSource 属性可能返回 TextBlock(文本被点击)，也可能返回 Grid(空白区域被点击)，因而要在转换前做类型检查。

```
项目：RoutedEvents1 | 文件：MainPage.xaml.cs(片段)
void OnGridTapped(object sender, TappedRoutedEventArgs args)
{
    if (args.OriginalSource is TextBlock)
    {
        TextBlock txtblk = args.OriginalSource as TextBlock;
        rand.NextBytes(rgb);
        Color clr = Color.FromArgb(255, rgb[0], rgb[1], rgb[2]);
        txtblk.Foreground = new SolidColorBrush(clr);
    }
}
```

若先做类型转换[①]，再判断结果是否为非空，依此来确定事件源，效率会略微高一些。

TappedRoutedEventArgs 派生自 RoutedEventArgs 类[②]。后者只定义一个名为 OriginalSource 的属性。显然，OriginalSource 属性是处理路由事件的重要工具。该属性允许可视树中的元素处理来自子孙的事件，并且可以得知事件的来源。路由事件使父元素能够了解子元素的情况。我们可以通过 OriginalSource 来识别引发事件的子元素。

---

① 译注：在 C#中用关键字 as 进行类型转换(正如示例代码所做的)。

② 译注：这里所说的 RoutedEventArgs 来自 Windows.UI.Xaml 命名空间。System.Windows 命名空间也有一个同名类。

另外，这个示例的Tapped处理程序也可以从Grid转移到MainPage上。通过MainPage，我们可以采用一种特殊方式处理事件。之前提到过，UIElement类定义了所有用户输入事件。这些事件被所有派生类继承，但Control类添加了自己的事件接口，包括一组与这些输入事件对应的可视方法(visual method)。例如，对于UIElement定义的Tapped事件，Control类定义了名为OnTapped的可视方法。这些可视方法的名称都以On为前缀，后面跟事件名，因此这些方法也被称为"On方法"。Page类间接通过Control类派生自UserControl，所以Page和MainPage都继承了这些方法。

下面这段代码来自示例项目RoutedEvents2的XAML文件。该文件未引用任何事件处理程序。

```
项目: RoutedEvents2 | 文件: MainPage.xaml(片段)
<Page
    x:Class="RoutedEvents2.MainPage"
    xmlns="http://schemas.microsoft.com/winfx/2006/xaml/presentation"
    xmlns:x="http://schemas.microsoft.com/winfx/2006/xaml"
    xmlns:local="using:RoutedEvents2"
    xmlns:d="http://schemas.microsoft.com/expression/blend/2008"
    xmlns:mc="http://schemas.openxmlformats.org/markup-compatibility/2006"
    mc:Ignorable="d"
    FontSize="48">

    <Grid Background="{StaticResource ApplicationPageBackgroundThemeBrush}">
        <TextBlock Text="Left / Top"
                   HorizontalAlignment="Left"
                   VerticalAlignment="Top" />

        ...

        <TextBlock Text="Right / Bottom"
                   HorizontalAlignment="Right"
                   VerticalAlignment="Bottom" />
    </Grid>
</Page>
```

对应的代码隐藏文件重写了OnTapped方法。

```
项目: RoutedEvents2 | 文件: MainPage.xaml.cs(片段)
protected override void OnTapped(TappedRoutedEventArgs args)
{
    if (args.OriginalSource is TextBlock)
    {
        TextBlock txtblk = args.OriginalSource as TextBlock;
        rand.NextBytes(rgb);
        Color clr = Color.FromArgb(255, rgb[0], rgb[1], rgb[2]);
        txtblk.Foreground = new SolidColorBrush(clr);
    }
    base.OnTapped(args);
}
```

在Visual Studio中，如果要重写OnTapped这样的可视方法，只需要键入关键字override并按空格，Visual Studio会显示父类定义的所有可视方法。选择所要重写的方法后，Visual Studio会创建一个方法存根，该存根只调用基类被重写的方法。虽然重写基类方法并不一定要调用基类的对应方法，但包含这样的代码是一种良好的习惯。是在最初、最后、中间某处调用，还是根本不调用，取决于被重写的方法的设计。

On方法基本上与事件处理程序是等价的。前者没有sender参数，因为它是多余的，sender与this都指向处理该事件的Page实例。[①]

① 译注：前提是被处理的事件和对应的可视方法均属于同一个类，并且在该类或其派生类中处理它们。

示例项目 RoutedEvents3 会在元素被点击时为 Grid 随机分配一种颜色。XAML 文件与前一个示例一样，但 OnTapped 方法有了些改变。

项目：RoutedEvents3 | 文件：MainPage.xaml.cs(片段)
```
protected override void OnTapped(TappedRoutedEventArgs args)
{
    rand.NextBytes(rgb);
    Color clr = Color.FromArgb(255, rgb[0], rgb[1], rgb[2]);
    SolidColorBrush brush = new SolidColorBrush(clr);

    if (args.OriginalSource is TextBlock)
        (args.OriginalSource as TextBlock).Foreground = brush;

    else if (args.OriginalSource is Grid)
        (args.OriginalSource as Grid).Background = brush;

    base.OnTapped(args);
}
```

运行程序。单击 TextBlock 元素，它自身的颜色会改变。单击空白区域，Grid 的颜色会改变。

假设出于某种原因，我们需要像最初那样显式地为每个 TextBlock 元素定义事件处理程序来改变文本的颜色，同时保留更改背景颜色的 OnTapped 方法重写。示例项目 RoutedEvents4 的 XAML 文件恢复 TextBlock 元素对 Tapped 事件处理程序的引用，并为 Grid 分配了一个名称。

项目：RoutedEvents4 | 文件：MainPage.xaml(片段)
```
<Grid Name="contentGrid"
      Background="{StaticResource ApplicationPageBackgroundThemeBrush}">

    <TextBlock Text="Left / Top"
               HorizontalAlignment="Left"
               VerticalAlignment="Top"
               Tapped="OnTextBlockTapped" />

    ...

    <TextBlock Text="Right / Bottom"
               HorizontalAlignment="Right"
               VerticalAlignment="Bottom"
               Tapped="OnTextBlockTapped" />
</Grid>
```

设置 TextBlock 和 Grid 颜色的代码得到了分离，因而不必使用 if-else 块。TextBlock 元素的 Tapped 处理程序可毫无顾虑地转换 sender 的类型，而 OnTapped 方法重写可通过名称来访问 Grid。

项目：RoutedEvents4 | 文件：MainPage.xaml.cs(片段)
```
public sealed partial class MainPage : Page
{
    Random rand = new Random();
    byte[] rgb = new byte[3];

    public MainPage()
    {
        this.InitializeComponent();
    }

    void OnTextBlockTapped(object sender, TappedRoutedEventArgs args)
    {
        TextBlock txtblk = sender as TextBlock;
```

```
        txtblk.Foreground = GetRandomBrush();
    }

    protected override void OnTapped(TappedRoutedEventArgs args)
    {
        contentGrid.Background = GetRandomBrush();
        base.OnTapped(args);
    }

    Brush GetRandomBrush()
    {
        rand.NextBytes(rgb);
        Color clr = Color.FromArgb(255, rgb[0], rgb[1], rgb[2]);
        return new SolidColorBrush(clr);
    }
}
```

然而，这样的代码可能也不是我们所期望的。点击 TextBlock 时，随着事件沿可视树向上路由，OnTapped 方法重写也会对其进行处理，因而不仅 TextBlock 会变色，Grid 也会变色！若这不是你所期望的行为，可以利用 TappedRoutedEventArgs 对象的属性来防止这种情况的发生。如果 OnTextBlockTapped 处理程序将该对象的 Handled 属性设为 true，事件则不会继续被可视树的上层元素处理。

示例项目 RoutedEvents5 演示了这种做法。除了事件处理方法 OnTextBlockTapped 外，其余与示例项目 RoutedEvents4 相同。

```
项目: RoutedEvents5 | 文件: MainPage.xaml.cs(片段)
void OnTextBlockTapped(object sender, TappedRoutedEventArgs args)
{
    TextBlock txtblk = sender as TextBlock;
    txtblk.Foreground = GetRandomBrush();
    args.Handled = true;
}
```

3.3　重写 Handled 设置

正如上一节所介绍的，如果处理元素的事件(如 Tapped)并将事件参数的 Handled 属性设置为 true，则可以中断事件路由。由于事件对可视树上层的元素不可见，因此它不会被继续处理。

在某些情况下，这不是人们期望的。例如，虽然在事件处理程序中将 Handled 设置为 true，但还希望事件对可视树中的上层元素可见。一种方案是修改代码，但可能无法实现。例如，元素在动态链接库中定义，并且没有源代码。

示例项目 RoutedEvents6 中的 XAML 与 RoutedEvents5 是一样的。所有 TextBlock 都设置了 Tapped 事件的处理程序。Tapped 处理程序将 Handled 属性设置为 true。代码隐藏类还定义了一个名为 OnPageTapped 的处理程序，设置了 Grid 的背景颜色。

```
项目: RoutedEvents6 | 文件: MainPage.xaml.cs(片段)
public sealed partial class MainPage : Page
{
    Random rand = new Random();
    byte[] rgb = new byte[3];

    public MainPage()
    {
```

```
        this.InitializeComponent();

        this.AddHandler(UIElement.TappedEvent,
                        new TappedEventHandler(OnPageTapped),
                        true);
    }

    void OnTextBlockTapped(object sender, TappedRoutedEventArgs args)
    {
        TextBlock txtblk = sender as TextBlock;
        txtblk.Foreground = GetRandomBrush();
        args.Handled = true;
    }

    void OnPageTapped(object sender, TappedRoutedEventArgs args)
    {
        contentGrid.Background = GetRandomBrush();
    }

    Brush GetRandomBrush()
    {
        rand.NextBytes(rgb);
        Color clr = Color.FromArgb(255, rgb[0], rgb[1], rgb[2]);
        return new SolidColorBrush(clr);
    }
}
```

请注意，构造函数用了一种特殊的方法订阅 Page 的 Tapped 事件。我们一般用下面这种方式来附加事件的处理程序：

```
this.Tapped += OnPageTapped;
```

倘若这样，OnPageTapped 处理程序不会在 TextBlock 收到 Tapped 事件时被调用，因为 TextBlock 的处理程序将 Handled 设置成了 true。这个示例使用一个名为 AddHandler 的方法来附加处理程序。

```
this.AddHandler(UIElement.TappedEvent,
                new TappedEventHandler(OnPageTapped),
                true);
```

AddHandler 方法是 UIElement 类定义的。UIElement 类还定义了静态属性 UIElement.TappedEvent，此属性是 Routed Event 类型的。

FontSize 属性伴随有一个类型为 DependencyProperty、名称为 FontSizeProperty 的属性。与之类似，Tapped 路由事件也有一个对应的静态属性 TappedEvent，其类型为 RoutedEvent。RoutedEvent 本身未定义公共成员，它的存在只是为了在不创建元素实例的前提下订阅事件。

AddHandler 方法能够将处理程序附加到事件上。该方法第二个参数的类型为 object，需要创建引用事件处理程序的委托对象。最关键的是最后一个参数：如果它为 true，事件处理程序将依然能够收到 Handled 为 true 的路由事件。

AddHandler 方法虽不常用，但我们应该了解它的存在，以备不时之需。

3.4　输入、对齐与背景

RoutedEvents 系列示例项目还有最后一个，展示了几个与输入事件有关的重要概念。

示例项目 RoutedEvents7 的 XAML 文件只有一个 TextBlock，且未定义事件处理程序。

项目：RoutedEvents7 | 文件：MainPage.xaml（片段）
```
<Page ...
   FontSize="48">

   <Grid Background="{StaticResource ApplicationPageBackgroundThemeBrush}">
      <TextBlock Text="Hello, Windows 8!"
                 Foreground="Red" />
   </Grid>
</Page>
```

由于没有设置 HorizontalAlignment 和 VerticalAlignment，所以 TextBlock 会出现在 Grid 的左上角。

与示例项目 RoutedEvents3 一样，RoutedEvents7 中源自 TextBlock 和 Grid 的事件也是在代码隐藏文件中被分别处理的。

项目：RoutedEvents7 | 文件：MainPage.xaml.cs（片段）
```
public sealed partial class MainPage : Page
{
   Random rand = new Random();
   byte[] rgb = new byte[3];

   public MainPage()
   {
      this.InitializeComponent();
   }

   protected override void OnTapped(TappedRoutedEventArgs args)
   {
      rand.NextBytes(rgb);
      Color clr = Color.FromArgb(255, rgb[0], rgb[1], rgb[2]);
      SolidColorBrush brush = new SolidColorBrush(clr);

      if (args.OriginalSource is TextBlock)
         (args.OriginalSource as TextBlock).Foreground = brush;

      else if (args.OriginalSource is Grid)
         (args.OriginalSource as Grid).Background = brush;

      base.OnTapped(args);
   }
}
```

该程序的运行效果如下图所示。

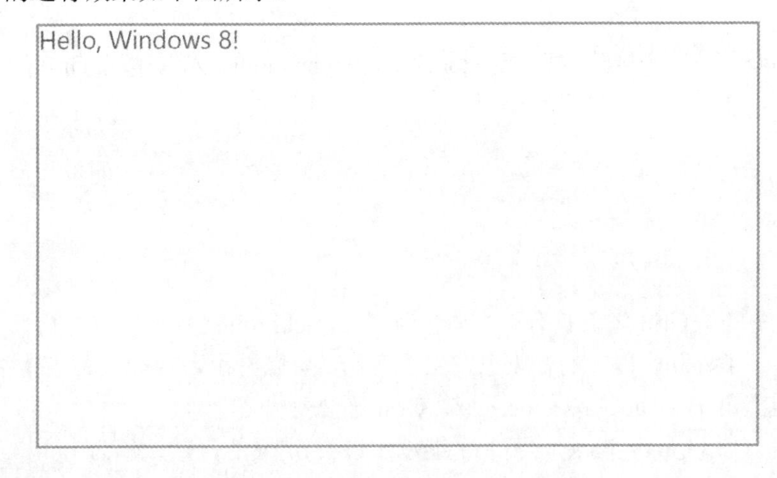

点击文本后，它会像之前一样随机变换一种颜色，但点击文本以外的区域，它也会变色，而 Grid 不变色。这看起来就像 TextBlock 占据了整个页面而捕获了所有 Tapped 事件一样。

事实的确如此。TextBlock 的 HorizontalAlignment 和 VerticalAlignment 属性具有默认值。通过这个示例，直觉告诉我们它们的默认值分别为 Left 和 Top，但实际上均为 Stretch。这意味着 TextBlock 会拉伸到与父元素(Grid)一样大小。这很难察觉，因为字体大小为 48 像素，没有一同被拉伸。另外，虽然充满整个页面，但 TextBlock 的背景是透明的。

Windows Runtime 中所有元素的 HorizontalAlignment 和 VerticalAlignment 属性默认值均为 Stretch。这是在使用 Windows Runtime 布局系统时需要特别注意的地方。第 4 章会具体介绍有关内容。

下面我们修改一下 HorizontalAlignment 和 VerticalAlignment 属性的值。

```
<Grid Background="{StaticResource ApplicationPageBackgroundThemeBrush}">
   <TextBlock Text="Hello, Windows 8!"
              HorizontalAlignment="Left"
              VerticalAlignment="Top"
              Foreground="Red" />
</Grid>
```

如下所示，样一来，TextBlock 只占据页面左上角的一片小区域，并且单击 TextBlock 之外的空白区域，Grid 会相应地变色。

将 HorizontalAlignment 属性替换为 TextAlignment。

```
<Grid Background="{StaticResource ApplicationPageBackgroundThemeBrush}">
   <TextBlock Text="Hello, Windows 8!"
              TextAlignment="Left"
              VerticalAlignment="Top"
              Foreground="Red" />
</Grid>
```

程序外观不会发生变化。文本依然位于页面的左上角，但单击 TextBlock 右侧的空白区域，TextBlock 会变色，而 Grid 不变色。由于具有一个默认值为 Stretch 的 HorizontalAlignment 属性，因而 TextBlock 占据了屏幕的整个宽度。在 TextBlock 所占据的这个宽度内，文本是左对齐的。

这个试验告诉我们，虽然 HorizontalAlignment 和 TextAlignment 的视觉效果一样，但还是有区别的。

下面我们做另外一个试验，即恢复 HorizontalAlignment 的设置，移除 Grid 的 Background 属性。

```
<Grid>
   <TextBlock Text="Hello, Windows 8!"
              HorizontalAlignment="Left"
              VerticalAlignment="Top"
              Foreground="Red" />
</Grid>
```

在浅色主题下，Grid 之前具有一个白色背景。Background 属性被移除后，页面的背景变成了黑色。除了外观，程序的行为也发生了变化，可以看出，TextBlock 依然能够在被点击后变色，但点击 TextBlock 以外的区域，Grid 不会变色。

Background 属性是由 Panel 类(Grid 继承于该类)定义的，默认值为 null。如果背景为

null，Grid 则不会捕获点击事件，而是将其忽略。

为保持透明效果，但又不使其忽略点击事件，一种办法是修改可视外观，显式地将 Grid 的 Background 属性设置为 Transparent。

```
<Grid Background="Transparent">
   <TextBlock Text="Hello, Windows 8!"
              HorizontalAlignment="Left"
              VerticalAlignment="Top"
              Foreground="Red" />
</Grid>
```

虽然最初的效果并未发生变化，但 Tapped 事件发生时 OriginalSource 已经能够返回 Grid 对象了。

这几个试验印证了一点，即表象往往是有欺骗性的。HorizontalAlignment 和 VerticalAlignment 为默认值的元素与使用 Left 和 Top 值的元素，从表面上看并无差别，但前者会填充整个容器并遮挡下层的元素。Background 属性值为 null 或 Transparent，Panel 外观也无差别，但前者会导致触摸事件被忽略。

这两个问题很有可能出现在未来的开发中，所导致的错误不易被发现。即便有多年 XAML 布局系统使用经验，遇到这样的错误仍然会让你晕头转向。

当然，这只是本人的经验之谈。

3.5　大小与方向的变化

很多年前，当 Windows 还是早期版本时，有关该系统的编程资料非常难找。直到 1986 年 12 月第一篇有关 Windows 编程的文章在 *Microsoft Systems Journal*(*MSDN Magazine* 的前身)上发表，情况才开始有所改观。该文章介绍了一个叫 WHATSIZE 的程序(确实都是大写字母)。该程序的功能非常简单，只显示程序窗口的大小，能够在窗口大小变化时更新显示的值。

显然，那时的 WHATSIZE 程序还是用 Windows API 编写的，它在遇到 WM_PAINT 消息时重绘界面。对于当时的 Windows API，程序窗口的任何内容变为无效，该消息都会被引发，要求程序进行重绘。程序可以定义自己的窗口，只要大小发生变化，窗口视图则变为无效。

Windows Runtime 没有 WM_PAINT 消息这样的机制，整个图形编程方法已发生了翻天覆地的变化。早期的 Windows 采用“直接模式”(direct mode)图形系统。应用程序可以直接修改显存。当然，这是通过一个软件层(Graphics Device Interface)和设备驱动程序完成的，但有一些绘图函数确实是直接修改显存。

如今的 Windows Runtime 已大为不同。公共编程接口中并没有绘制的概念。Windows 8 应用程序需要创建元素(派生自 FrameworkElement 的类)，并将其添加到应用程序的可视树中。这些元素负责呈现自身。例如，如果 Windows 8 应用程序要显示文本，它并不“绘制”文本，而是创建 TextBlock 对象；如果要显示位图，则创建 Image 元素；如果要绘制线段、贝塞尔曲线或椭圆，则创建 Polyline 或 Path 元素。

Windows Runtime 采用的是一种“保留模式”(retained mode)图形系统。应用程序与显示设备之间有一个复合层次。在这一层上，所有生成的输出会在呈现给用户前被组合。这种保留模式最显著的优势或许在于其能够避免闪烁，这一点可以在下面的示例中加以体会。

虽然 Windows Runtime 与之前版本 Windows 所采用的图形系统不同，但 Windows 8 应用程序与传统应用程序也有一定的相通之处。一旦程序加载到内存并开始运行，就会驻留于内存等待某些通知。通知可能是事件，也可能是回调。大部分用于通知用户输入，但也有一些表示其他活动的发生。OnNavigatedTo 方法是回调方式的典型应用。对于单页面程序，构造函数返回后，该方法会立即被调用。

为实现 WHATSIZE 程序的功能，Windows 8 应用程序可能要用到一个名为 SizeChanged 的事件。示例程序 WhatSize(Windows 8 版本)的 XAML 文件如下所示。请注意，订阅 SizeChanged 事件处理程序的是根元素。

```
项目: WhatSize | 文件: MainPage.xaml(片段)
<Page
   x:Class="WhatSize.MainPage"
   ...
   FontSize="36"
   SizeChanged="OnPageSizeChanged">

   <Grid Background="{StaticResource ApplicationPageBackgroundThemeBrush}">
      <TextBlock HorizontalAlignment="Center"
                 VerticalAlignment="Top">
         &#x21A4; <Run x:Name="widthText" /> pixels &#x21A6;
      </TextBlock>

      <TextBlock HorizontalAlignment="Center"
                 VerticalAlignment="Center"
                 TextAlignment="Center">
         &#x21A5;
         <LineBreak />
         <Run x:Name="heightText" /> pixels
         <LineBreak />
         &#x21A7;
      </TextBlock>
   </Grid>
</Page>
```

这段 XAML 标记定义了两个 TextBlock 元素，两个 Run 对象周围各有两个箭头。(稍后会展示显示效果。)将第二个 TextBlock 的三个属性均设置为 Center 似乎有些极端，但这样做是必要的。前两个值为 Center 的属性能够将 TextBlock 置于页面中央，而将 TextAlignment 设置为 Center 可使箭头相对文本处于中间位置。两个 Run 元素都具有 x:Name 特性，这样便可以在 SizeChanged 事件处理程序中设置该元素的 Text 属性。

```
项目: WhatSize | 文件: MainPage.xaml.cs(片段)
public sealed partial class MainPage : Page
{
   public MainPage()
   {
      this.InitializeComponent();
   }

   void OnPageSizeChanged(object sender, SizeChangedEventArgs args)
   {
      widthText.Text = args.NewSize.Width.ToString();
      heightText.Text = args.NewSize.Height.ToString();
   }
}
```

事件的参数恰好提供了我们需要的页面尺寸值，其类型为 Size 结构。我们只需要将 Width 和 Height 属性转换为字符串并分别将该字符串设置到两个 Run 元素的 Text 属性上即可(见下图)。

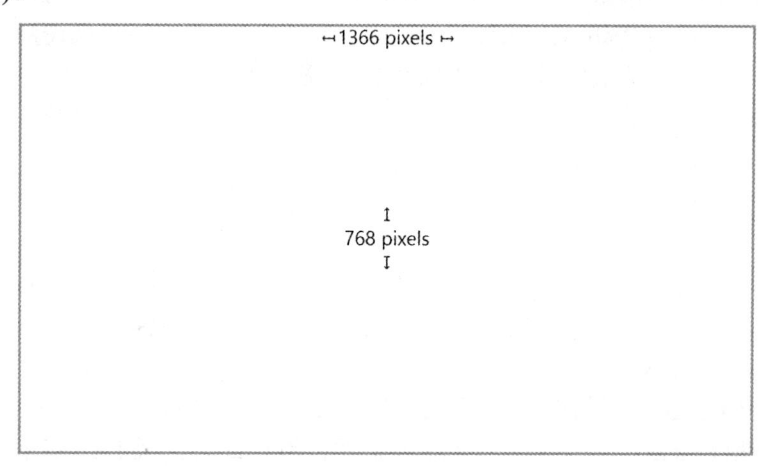

如果在能够响应方向变化的设备上运行此程序，可以旋转屏幕并观察数值变化。也可以通过滑动手势将该程序切换到辅屏视图，调整此程序和另一个程序之间分隔的位置，然后观察数值变化。

SizeChanged 事件处理程序不必在 XAML 中订阅，在 Page 的构造函数中也可以订阅。

```
this.SizeChanged += OnPageSizeChanged;
```

SizeChanged 是 FrameworkElement 类定义的，由所有子类继承。尽管 SizeChangedEventArgs 派生自 RoutedEventArgs，但该事件并非路由事件。之所以这么说有三个原因：①该事件参数的 OriginalSource 属性总是为 null；②不存在 SizeChangedEvent 属性，③子元素会在该事件中提供自身宽度。我们可以订阅任何元素的 SizeChanged 事件。一般来讲，该事件的引发顺序是沿着可视树自上而下。以这个示例来说，MainPage 优先接到该事件，然后才依次是 Grid 和 TextBlock。

如果要在 SizeChanged 事件上下文之外获得元素的实际尺寸，可以使用 FrameworkElement 定义的 ActualWidth 和 ActualHeight 属性。事实上，与使用 SizeChanged 处理程序提供的尺寸相比，使用这两个属性更简便。

```
void OnPageSizeChanged(object sender, SizeChangedEventArgs args)
{
   widthText.Text = this.ActualWidth.ToString();
   heightText.Text = this.ActualHeight.ToString();
}
```

我们可能不希望使用 Width 和 Height 属性。虽然这两个属性也是 FrameworkElement 定义的，但两者的默认值均为 NaN(Not a Number)①。程序可以显式地为 Width 和 Height 赋值(正如第 2 章的示例项目 TextFormatting 所做的)，但通常应保持这两个属性的默认值。它们并不适合用来确定元素的实际尺寸。FrameworkElement 还定义了默认值为 NaN 的

① 译注：NaN 是 double 结构的一个特殊值，可以通过常量字段 double.NaN 获取这个值。若对 double 执行了未定义的操作(例如 x/0.0d)，则会产生这个值。

MinWidth、MaxWidth、MinHeight 和 MaxHeight 属性，但这些属性在实际应用当中并不常用。

如果在页面的构造函数中访问 ActualWidth 和 ActualHeight 属性，它们均会返回 0。InitializeComponent 刚被调用后，虽然已经构建了可视树，但此时的可视树尚未经过布局处理。在构造函数执行完毕后，以下页面事件会被依次引发。

- OnNavigatedTo
- SizeChanged
- LayoutUpdated
- Loaded

在这些事件被引发后，如果页面的大小变化，则会引发 SizeChanged 事件和 LayoutUpdated 事件。如果元素被添加到可视树、从中被移除，或者某元素的变化导致布局的改变，LayoutUpdated 事件也会被引发。

在布局初始化之后，所有可视树中元素尺寸为非零。如果需要此时执行某些初始化操作，则可以使用 Loaded 事件。我们可能经常需要订阅 Page 派生类的 Loaded 事件。Loaded 事件一般只会在 Page 对象生命周期中执行一次。之所以说“一般”是因为，如果 Page 对象脱离父元素(Frame)，并重新被附加，Loaded 事件也会被引发。这种情况一般不会发生，除非开发者有意为之。此外，如果页面从可视树脱离，系统会通过 Unloaded 事件进行通知。

所有 FrameworkElement 的派生类都有 Loaded 事件。在可视树的构建过程中，可视树子元素会先于父元素引发 Loaded 事件，最终止于 Page 的派生类。当 Page 对象的 Loaded 事件被引发时，则可以认为所有子对象的 Loaded 事件均被引发，并且所有尺寸已调整完成。

在 Page 类中处理 Loaded 事件十分普遍，有的开发者甚至在构造函数中通过匿名处理程序来执行 Loaded 操作。

```
public MainPage()
{
this.InitializeComponent();

   Loaded += (sender, args) =>
      {
         ...
      };
}
```

Windows 8 应用程序有时需要获得屏幕方向改变的通知。第 1 章的 InternationalHelloWorld 程序在横屏模式下显示正确，但在竖屏模式下文本可能会出现重叠。示例项目 InternationalHelloWorld 的代码隐藏文件会在竖屏模式下将页面的 FontSize 属性修改为 24，从而解决了这个问题。

```
项目: ScalableInternationalHelloWorld | 文件: MainPage.xaml.cs(片段)
public sealed partial class MainPage : Page
{
   public MainPage()
   {
      this.InitializeComponent();
      SetFont();
      DisplayProperties.OrientationChanged += OnDisplayPropertiesOrientationChanged;
   }

   void OnDisplayPropertiesOrientationChanged(object sender)
   {
```

```
        SetFont();
    }

    void SetFont()
    {
        bool isLandscape =
            DisplayProperties.CurrentOrientation == DisplayOrientations.Landscape ||
            DisplayProperties.CurrentOrientation == DisplayOrientations.LandscapeFlipped;

        this.FontSize = isLandscape ? 40 : 24;
    }
}
```

DisplayProperties 类和 DisplayOrientations 枚举定义于 Windows.Graphics.Display 命名空间。DisplayProperties.OrientationChanged 是静态事件。如果该事件被引发，可以通过静态属性 DisplayProperties.CurrentOrientation 来获取当前的屏幕方向。

使用 Windows.UI.ViewManagement 命名空间下 ApplicationView 类的 ViewStateChanged 事件，我们可以获得包括当前是否处于辅屏视图在内的更多信息。第 12 章会详细介绍有关内容。

3.6　尝试绑定到 Run 元素

第 2 章介绍了数据绑定。数据绑定可以连接两个属性。如果源属性发生变化，目标属性也会随之变化。这种机制可以减少对事件的依赖。

可否重写示例项目 WhatSize，使用数据绑定来替代 SizeChanged 处理程序？这值得一试。

在 WhatSize 项目中，从 MainPage.xaml.cs 文件中移除 OnPageSizeChanged 处理程序(如果不想过多地修改此文件，可以将其注释掉)。移除 MainPage.xaml 文件根标签的 SizeChanged 特性，并将 MainPage 命名为 page。在两个 Run 对象上通过 Binding 标记扩展来引用页面的 ActualWidth 和 ActualHeight 属性。

```
<Page ...
    Name="page"
    FontSize="36">

    <Grid Background="{StaticResource ApplicationPageBackgroundThemeBrush}">
        <TextBlock HorizontalAlignment="Center"
                   VerticalAlignment="Top">
            &#x21A4;
            <Run Text="{Binding ElementName=page, Path=ActualWidth}"/>
            pixels &#x21A6;
        </TextBlock>

        <TextBlock HorizontalAlignment="Center"
                   VerticalAlignment="Center"
                   TextAlignment="Center">
            &#x21A5;
            <LineBreak />
            <Run Text="{Binding ElementName=page, Path=ActualHeight}"/> pixels
            <LineBreak />
            &#x21A7;
        </TextBlock>
    </Grid>
</Page>
```

这个程序能够编译通过，并且不会出现运行时异常。只不过它有一个问题：两个数字均为 0。

这不免让人困惑，尤其是我们将相同的绑定设置从 Run 转移到 TextBlock 的 Text 属性上时。

```
<Page ...
   Name="page"
   FontSize="36">

   <Grid Background="{StaticResource ApplicationPageBackgroundThemeBrush}">
      <TextBlock HorizontalAlignment="Center"
                 VerticalAlignment="Top"
                 Text="{Binding ElementName=page, Path=ActualWidth}" />
      <TextBlock HorizontalAlignment="Center"
                 VerticalAlignment="Center"
                 TextAlignment="Center"
                 Text="{Binding ElementName=page, Path=ActualHeight}" />
   </Grid>
</Page>
```

效果如下图所示。

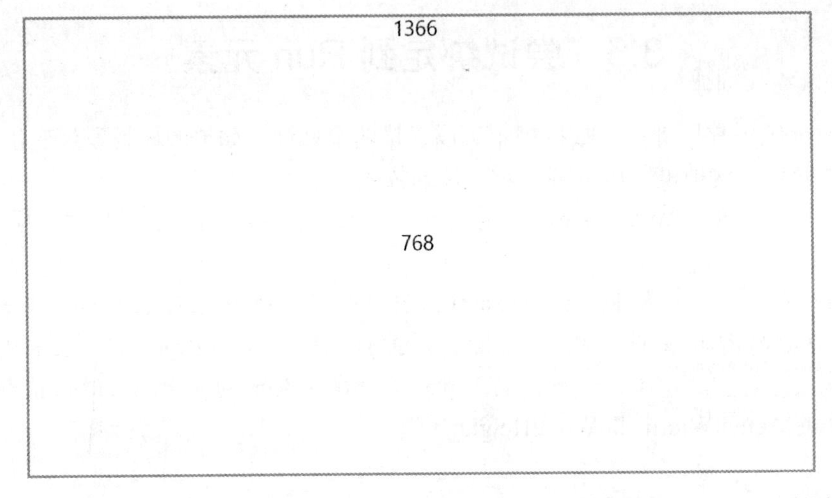

这段标记运行后，最初的状态是正确的。对于本书截稿前的 Windows 8 版本，改变屏幕方向或页面大小，这两个数字均不会被更新，但它们应该被更新。[①]从理论上讲，数据绑定只有在源属性发生改变时才会通知目标属性，因而程序的源代码中不必使用事件处理程序或任何更新机制，这正是数据绑定的优势所在。

为何数据绑定只对 TextBlock 的 Text 属性有效，而对 Run 的 Text 属性不起作用呢？

这是因为数据绑定的目标必须是依赖属性。在代码中通过 SetBinding 方法来定义数据绑定时便会发现这一点。两个元素的 Text 属性不同：TextBlock 的 Text 属性伴随有相应的 TextProperty 依赖属性，而 Run 的 Text 属性没有。XAML 解析器应禁止在 Run 的 Text 属性上设置绑定，但它并没有这么做。

在这个试验中，放弃绑定到 Run 元素转而绑定 TextBlock，致使文本周围的箭头被舍

① 译注：虽然数据绑定能够在源属性变化时进行通知，但要求源对象主动发起(与界面有关的要求在主线程上执行)。虽然 ActualWidth 和 ActualHeight 属性的值发生变化，但没有进行通知。文档笼统地解释道：“它们是在运行时以异步方式计算得到的。”不论怎样，在实际项目中，暂时都不要绑定这两个属性，除非这一限制在未来可以解决。

弃。第 4 章将介绍如何通过 StackPanel 使箭头回归。第 16 章会介绍如何通过 RichTextBlock 来实现这种界面。

3.7　计时器与动画

Windows 8 应用程序有时需要以某个固定时间间隔接收事件。例如，时钟程序可能需要每秒钟更新一次界面。为此，最好使用 DispatcherTimer 类，设置其时间间隔并订阅它的 Tick 事件。

时钟示例程序的 XAML 文件非常简单，只是显示一个字体较大的 TextBlock(如下所示)。

项目：DigitalClock | 文件：MainPage.xaml(片段)

```
<Grid Background="{StaticResource ApplicationPageBackgroundThemeBrush}">
    <TextBlock Name="txtblk"
               FontFamily="Lucida Console"
               FontSize="120"
               HorizontalAlignment="Center"
               VerticalAlignment="Center" />
</Grid>
```

代码隐藏文件创建了一个间隔为 1 秒的 DispatcherTimer，并在 Tick 事件处理程序中设置了 TextBlock 的 Text 属性。

项目：DigitalClock | 文件：MainPage.xaml.cs(片段)

```
public sealed partial class MainPage : Page
{
    public MainPage()
    {
        this.InitializeComponent();

        DispatcherTimer timer = new DispatcherTimer();
        timer.Interval = TimeSpan.FromSeconds(1);
        timer.Tick += OnTimerTick;
        timer.Start();
    }

    void OnTimerTick(object sender, object e)
    {
        txtblk.Text = DateTime.Now.ToString("h:mm:ss tt");
    }
}
```

该示例程序的运行效果如下图所示。

8:47:03 PM

调用 Tick 事件处理程序的线程与用户界面的执行线程是同一个。如果该线程比较繁忙，Tick 处理程序就不会被调用并中断该线程的工作。在这种情况下，Tick 事件可能会变得不规则，甚至可能跳过几个周期。在多页面应用程序中，最好在 OnNavigatedTo 方法重写中启动计时器，并在 OnNavigatedFrom 中终止它，以防它在页面不可见的情况下做“无用功”。

从这个示例可以看出 Windows 桌面应用程序与 Windows 8 应用程序在显示方面的差异。这两种应用程序都可以通过计时器来实现时钟应用，但 Windows 8 时钟应用并非通过定期使窗口内容失效的方法来绘制或重绘界面，而仅仅是修改现有元素的属性来改变其视觉外观。

DispatcherTimer 的间隔时间可以设置得再短一些，但最好不要使 Tick 处理程序的调用过于频繁，甚至超过显示器的刷新频率(大概为 60 Hz，或者说间隔不要低于 17 毫秒)。当然，更新界面的频率高于显示器的刷新频率也没有意义。以适当的频率进行更新可以使动画更为平滑。如果要实现动画，则不应使用 DispatcherTimer，而应使用静态事件 CompositionTarget.Rendering。该事件刚好会在界面被刷新前调用。

除了 CompositionTarget.Rendering 事件，Windows Runtime 还提供了很多与动画有关的类型。这些类型允许我们在 XAML 或代码中定义动画。它们提供了很多选项，并且其中的一些是在后台线程上执行的。

在第 9 章专门介绍动画类之前，CompositionTarget.Rendering 是最适合实现动画的工具。这种方式也称为“手动”(manual)动画，因为程序需要根据逝去的时间自行计算动画的参数。

配套资源中有一个 ExpandingText 示例项目。该项目在 CompositionTarget.Rendering 事件处理程序中修改 TextBlock 的 FontSize，使文本变大或变小。初始化这个 TextBlock 的 XAML 如下所示。

项目：ExpandingText | 文件：MainPage.xaml(片段)

```
<Grid Background="{StaticResource ApplicationPageBackgroundThemeBrush}">
   <TextBlock Name="txtblk"
              Text="Hello, Windows 8!"
              HorizontalAlignment="Center"
              VerticalAlignment="Center" />
</Grid>
```

在该项目的代码隐藏文件中，构造函数订阅了 CompositionTarget.Rendering 事件。事件处理程序第二个参数的类型为 object，但实际的类型为 RenderingEventArgs。这个参数类型有一个类型为 TimeSpan 的 RenderingTime 属性，可提供自程序启动以来已经流逝的时间。

项目：ExpandingText | 文件：MainPage.xaml.cs(片段)

```
public sealed partial class MainPage : Page
{
   public MainPage()
   {
      this.InitializeComponent();
      CompositionTarget.Rendering += OnCompositionTargetRendering;
   }

   void OnCompositionTargetRendering(object sender, object args)
   {
      RenderingEventArgs renderArgs = args as RenderingEventArgs;
      double t = (0.25 * renderArgs.RenderingTime.TotalSeconds) % 1;
      double scale = t < 0.5 ? 2 * t : 2 - 2 * t;
```

```
        txtblk.FontSize = 1 + scale * 143;
    }
}
```

这段代码具有一定的代表性。在 4 秒的周期内，变量 t 在 0 到 1 之间变化，变量 scale 则从 0 变到 1 再到 0，从而使 FontSize 从 1 变到 144 再到 1。(这段代码不会使 FontSize 的值变为 0，否则会有异常抛出。)程序刚开始运行时会让人感觉有些卡顿，这是因为要对不同大小的字体进行像素化。但在程序步入正轨后，动画会变平滑，且无闪烁。

颜色也可以随动画变化，这里提供两种方法。虽然后一种方法优于前一种，但作为对比，下面先来看看第一种方法。示例项目 ManualBrushAnimation 的 XAML 文件如下所示。

```
项目: ManualBrushAnimation | 文件: MainPage.xaml(片段)
<Grid Name="contentGrid">
    <TextBlock Name="txtblk"
               Text="Hello, Windows 8!"
               FontFamily="Times New Roman"
               FontSize="96"
               FontWeight="Bold"
               HorizontalAlignment="Center"
               VerticalAlignment="Center" />
</Grid>
```

Grid 和 TextBlock 均未定义画笔。根据变化的颜色创建画笔的工作是由 CompositionTarget.Rendering 事件处理程序完成的。

```
项目: ManualBrushAnimation | 文件: MainPage.xaml.cs(片段)
public sealed partial class MainPage : Page
{
    public MainPage()
    {
        this.InitializeComponent();
        CompositionTarget.Rendering += OnCompositionTargetRendering;
    }

    void OnCompositionTargetRendering(object sender, object args)
    {
        RenderingEventArgs renderingArgs = args as RenderingEventArgs;
        double t = (0.25 * renderingArgs.RenderingTime.TotalSeconds) % 1;
        t = t < 0.5 ? 2 * t : 2 - 2 * t;

        // Background
        byte gray = (byte)(255 * t);
        Color clr = Color.FromArgb(255, gray, gray, gray);
        contentGrid.Background = new SolidColorBrush(clr);

        // Foreground
        gray = (byte)(255 - gray);
        clr = Color.FromArgb(255, gray, gray, gray);
        txtblk.Foreground = new SolidColorBrush(clr);
    }
}
```

随着 Grid 背景颜色由黑变白再变黑，TextBlock 字体颜色则从白到黑再到白，周而复始。

虽然效果不错，但每一帧(大概 60 次每秒)都会创建两个 SolidColorBrush 对象，然后它们又被迅速销毁，其实这是没有必要的。更好的方式是在 XAML 文件中创建两个 SolidColorBrush 对象。

```
项目: ManualColorAnimation | 文件: MainPage.xaml(片段)
<Grid>
    <Grid.Background>
```

```
        <SolidColorBrush x:Name="gridBrush" />
    </Grid.Background>

    <TextBlock Text="Hello, Windows 8!"
               FontFamily="Times New Roman"
               FontSize="96"
               FontWeight="Bold"
               HorizontalAlignment="Center"
               VerticalAlignment="Center">
        <TextBlock.Foreground>
            <SolidColorBrush x:Name="txtblkBrush" />
        </TextBlock.Foreground>
    </TextBlock>
</Grid>
```

对于这个示例，这两个 SolidColorBrush 对象的生命周期与程序的生命周期相当。为其命名后，我们便可以在 CompositionTarget.Rendering 处理程序中访问它们。

项目: ManualColorAnimation | 文件: MainPage.xaml.cs(片段)

```
void OnCompositionTargetRendering(object sender, object args)
{
    RenderingEventArgs renderingArgs = args as RenderingEventArgs;
    double t = (0.25 * renderingArgs.RenderingTime.TotalSeconds) % 1;
    t = t < 0.5 ? 2 * t : 2 - 2 * t;

    // Background
    byte gray = (byte)(255 * t);
    gridBrush.Color = Color.FromArgb(255, gray, gray, gray);

    // Foreground
    gray = (byte)(255 - gray);
    txtblkBrush.Color = Color.FromArgb(255, gray, gray, gray);
}
}
```

运行之初可能没有什么不同，因为每一帧都有两个 Color 对象被创建和销毁。这里不应称其为“对象”，因为 Color 不是类，而是结构。称之为 Color 值更为贴切。Color 值存在栈(stack)上，不需要像堆(heap)那样开辟内存空间。

要避免在堆上频繁开辟空间。每秒 60 次的速度在堆上创建对象会涉及较大的性能开销。这个示例的改进之处在于 SolidColorBrush 对象会一直保持在 Windows Runtime 组合系统中。该程序有效地利用了组合层并只修改画笔的属性，因此显示效果有所改善。

该程序还展示了依赖属性的独特之处，即依赖属性能够以一种较为结构化的方式响应变化。正如本书后面要介绍的，Windows Runtime 内建的动画工具只能用依赖属性进行操纵。基于 CompositionTarget.Rendering 的手动动画也有类似的限制。幸运的是，TextBlock 的 Foreground 属性和 Grid 的 Background 属性均是类型为 Brush 的依赖属性，并且 SolidColorBrush 的 Color 属性也是依赖属性。

事实上，我们每当遇到依赖属性时，都应该思考“如何在动画中使用它”。举例来说，GradientStop 类的 Offset 属性是依赖属性，因而可以通过它来实现某种特殊动画效果。

示例项目 RainbowEight 的 XAML 文件如下所示。

项目: RainbowEight | 文件: MainPage.xaml(片段)

```
<Grid Background="{StaticResource ApplicationPageBackgroundThemeBrush}">
    <TextBlock Name="txtblk"
               Text="8"
               FontFamily="CooperBlack"
               FontSize="1"
               HorizontalAlignment="Center">
```

```
      <TextBlock.Foreground>
         <LinearGradientBrush x:Name="gradientBrush">
            <GradientStop Offset="0.00" Color="Red" />
            <GradientStop Offset="0.14" Color="Orange" />
            <GradientStop Offset="0.28" Color="Yellow" />
            <GradientStop Offset="0.43" Color="Green" />
            <GradientStop Offset="0.57" Color="Blue" />
            <GradientStop Offset="0.71" Color="Indigo" />
            <GradientStop Offset="0.86" Color="Violet" />
            <GradientStop Offset="1.00" Color="Red" />
            <GradientStop Offset="1.14" Color="Orange" />
            <GradientStop Offset="1.28" Color="Yellow" />
            <GradientStop Offset="1.43" Color="Green" />
            <GradientStop Offset="1.57" Color="Blue" />
            <GradientStop Offset="1.71" Color="Indigo" />
            <GradientStop Offset="1.86" Color="Violet" />
            <GradientStop Offset="2.00" Color="Red" />
         </LinearGradientBrush>
      </TextBlock.Foreground>
   </TextBlock>
</Grid>
```

Offset 值大于 1 的 GradientStop 对象是不可见的。TextBlock 元素也不易被察觉，因为它的 FontSize 为 1。在 Loaded 事件的处理程序中，Page 类获得了这个 TextBlock 元素的 ActualHeight，将其保存到一个字段中，并随后订阅 CompositionTarget.Rendering 事件。

项目：RainbowEight | 文件：MainPage.xaml.cs（片段）

```
public sealed partial class MainPage : Page
{
   double txtblkBaseSize;  // ie, for 1-pixel FontSize

   public MainPage()
   {
      this.InitializeComponent();
      Loaded += OnPageLoaded;
   }

   void OnPageLoaded(object sender, RoutedEventArgs args)
   {
      txtblkBaseSize = txtblk.ActualHeight;
      CompositionTarget.Rendering += OnCompositionTargetRendering;
   }

   void OnCompositionTargetRendering(object sender, object args)
   {
      // Set FontSize as large as it can be
      txtblk.FontSize = this.ActualHeight / txtblkBaseSize;

      // Calculate t from 0 to 1 repetitively
      RenderingEventArgs renderingArgs = args as RenderingEventArgs;
      double t = (0.25 * renderingArgs.RenderingTime.TotalSeconds) % 1;

      // Loop through GradientStop objects
      for (int index = 0; index < gradientBrush.GradientStops.Count; index++)
         gradientBrush.GradientStops[index].Offset = index / 7.0 - t;
   }
}
```

在 CompositionTarget.Rendering 事件的处理程序中，TextBlock 的 FontSize 属性根据 Page 的 ActualHeight 属性计算得到，这个行为非常类似于一种手动的 Viewbox。字体不会与页面同高，因为 TextBlock 的 ActualHeight 包含下伸部和变音符号的高度。字体被尽可能地放大，并且能够随屏幕方向的变化而变化。

此外，CompositionTarget.Rendering 事件处理程序会不断地修改 LinearGradientBrush

中所有 GradientStop 对象的 Offset 属性，因此字符会呈现下图所示的彩虹动画效果(可惜书中无法展现，不过可以访问作者网站查看)。

有人可能会质疑：以显示器的帧速修改 TextBlock 的 FontSize 属性，这样做是否有必要？订阅 Page 的 SizeChanged 事件并在该事件的处理程序中修改这个属性，这样是否更为可取？

示例的做法可能显得有些低效。但依赖属性有一种特性：只有属性值正发生变化时对象才会进行通知。如果属性被设为原值，则不会有任何影响，不信你可以尝试通过订阅 TextBlock 本身的 SizeChanged 事件来加以验证。

第 4 章　基于 Panel 的布局

Windows Runtime 程序一般包含一个或多个派生自 Page 的类。每个页面对应一个由若干元素组成的可视树，元素是分层组织的。Page 对象的 Content 属性只支持单个元素，该元素一般是 Panel 派生类的实例。Panel 类定义了名为 Children 的属性，属性类型为 UIElementCollection(UIElement 派生类和其他面板实例的集合)。

Panel 及其派生类是 Windows Runtime 动态布局系统的核心。它能够在页面大小或方向发生改变时重新组织其子元素，使其填充可用空间。不同类型的 Panel 组织其子元素的方式有所不同。例如，Grid 按网格方式组织子元素。StackPanel 能够在水平或竖直方向依次排列子元素。VariableSizedWrapGrid 不仅可以在水平或竖直方向上排列子元素，还可以在需要时添加行或列(类似于 Windows 8“开始”界面)。Canvas 允许通过像素坐标来指定元素位置。

为平衡父元素与子元素需求的矛盾，布局系统变得越来越复杂。局部来看，布局系统是“子元素驱动的”(child-driven)，即每个子元素决定自身尺寸，父元素需要为其提供足够的空间。但布局系统也是“父元素驱动的”(parent-driven)，在这种情况下，父元素的尺寸受限，无法为子元素提供超过自身大小的空间，从而限制子元素的尺寸。

网页中也有类似的概念。例如，具有宽度的 HTML 页面是父驱动的，因为它受显示器或浏览器窗口宽度的制约。然而，页面的高度是子驱动的，因为页面的高度取决于内容。如果内容超过浏览器窗口的高度，则会显示滚动条。

再举一个 Windows 8“开始”界面的例子。在垂直方向上，应用程序磁贴的数目是父元素驱动的，因为其受制于屏幕的高度。水平方向是子驱动的，因为如果磁贴在水平方向超出屏幕显示范围，则需要滑动滚动条才能查看超出的部分。

4.1　Border 元素

HorizontalAlignment 和 VerticalAlignment 是两个与布局有关的最重要的属性，由 FrameworkElement 类定义，可选的值来自同名的枚举类型：HorizontalAlignment 和 VerticalAlignment。

正如第 3 章介绍的，HorizontalAlignment 和 VerticalAlignment 的默认值并非 Left 和 Top，而分别是 HorizontalAlignment.Stretch 和 VerticalAlignment.Stretch。Stretch 设置说明布局是父驱动的，即元素自动拉伸到父元素大小，但这一点可能不是特别直观。上一章有一个示例，TextBlock 被拉伸到父元素的大小，并捕获了父元素的所有 Tapped 事件。

只要元素的 HorizontalAlignment 或 VerticalAlignment 属性的值不是默认的 Stretch，布局便会切换至子驱动模式，也就是说，具有该设置的元素会根据内容设置自身宽度或高度。

在页面上处理父元素和子元素关系时，HorizontalAlignment 和 VerticalAlignment 的重要性就很明显。假设要为 TextBlock 显示一个边框，但会发现它没有与边框有关的属性。

Windows.UI.Xaml.Controls 命名空间有一个名为 Border 的元素，它有 Child 属性。这样，便可以将 TextBlock 置于 Border 内，再将这个 Border 置于 Grid 内(如下所示)。

项目：NaiveBorderedText | 文件：MainPage.xaml(片段)

```
<Page ...>

    <Grid Background="{StaticResource ApplicationPageBackgroundThemeBrush}">

        <Border BorderBrush="Red"
                BorderThickness="12"
                CornerRadius="24"
                Background="Yellow">

            <TextBlock Text="Hello Windows 8!"
                       FontSize="96"
                       Foreground="Blue"
                       HorizontalAlignment="Center"
                       VerticalAlignment="Center" />
        </Border>

    </Grid>
</Page>
```

Border 元素定义的 BorderThickness 属性允许为矩形的四边设置不同的粗细。如果指定四个值，则设置顺序依次为：左、上、右、下。如果只指定两个值，则第一个值作用于左右两边，第二个值作用于上下两边。CornerRadius 属性用于指定四个圆角的半径。

请注意 TextBlock 的 HorizontalAlignment 和 VerticalAlignment 属性设置。虽然标记看起来合理，但实际效果可能并非所期待的(参见下图)。

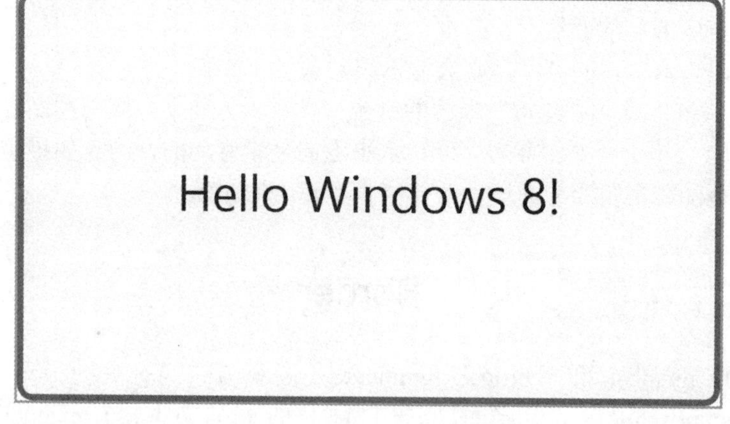

由于 Border 派生自 FrameworkElement，它也具有 HorizontalAlignment 和 VerticalAlignment 属性，其默认值均为 Stretch。这使得 Border 会被拉伸到父元素的尺寸大小。为获得所期望的效果，需要将 HorizontalAlignment 和 VerticalAlignment 设置从 TextBlock 转移到 Border 上。

项目：BetterBorderedText | 文件：MainPage.xaml(片段)

```
<Grid Background="{StaticResource ApplicationPageBackgroundThemeBrush}">

    <Border BorderBrush="Red"
            BorderThickness="12"
            CornerRadius="24"
            Background="Yellow"
            HorizontalAlignment="Center"
            VerticalAlignment="Center">
```

```
        <TextBlock Text="Hello Windows 8!"
                   FontSize="96"
                   Foreground="Blue"
                   Margin="24" />
    </Border>

</Grid>
```

这段代码还通过 Margin 属性为 TextBlock 添加了四分之一英寸的边距。这使得 Border 比文本每边大四分之一英寸(参见下图)。

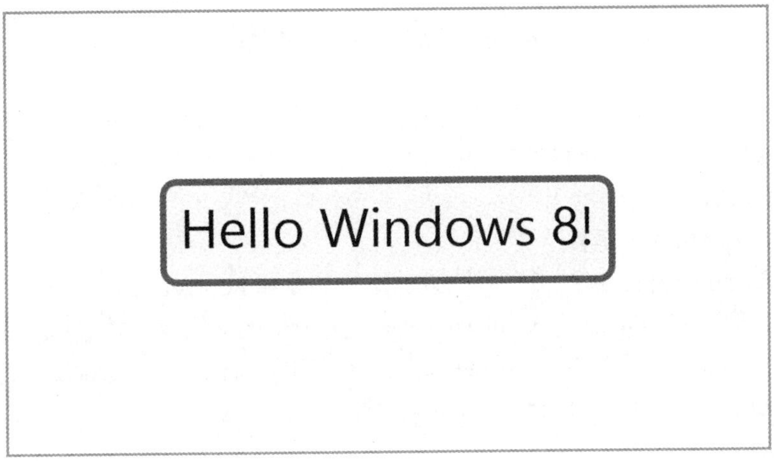

Margin 属性是 FrameworkElement 类定义的，因而被所有元素继承。该属性的类型为 Thickness(与 BorderThickness 属性的类型相同)，即带有 Left、Top、Right 和 Bottom 属性的结构。Margin 用于限定元素周围空白区域的大小，因而元素之间不必彼此紧贴。该属性在实践当中较为常用。与 BorderThickness 类似，Margin 可以包含四个不同的值。在 XAML 中，这四个值依次用于表示左、上、右、下的边距。如果只指定两个值，则第一个值表示左右边距，第二个值表示上下边距。

此外，Border 还定义了名为 Padding 的属性。与 Margin 不同的是，Padding 表示元素内部的空间。下面让我们移除 TextBlock 的 Margin 属性，并为 Border 添加 Padding 属性。

```
<Border BorderBrush="Red"
        BorderThickness="12"
        CornerRadius="24"
        Background="Yellow"
        HorizontalAlignment="Center"
        VerticalAlignment="Center"
        Padding="24" >

    <TextBlock Text="Hello Windows 8!"
               FontSize="96"
               Foreground="Blue"/>
</Border>
```

效果是相同的。不论哪种情况，TextBlock 上的任何 HorizontalAlignment 或 VerticalAlignment 设置均无关紧要了。

从布局的角度看，Margin 定义的空间会被计入元素尺寸，但此空间不受元素控制。例如，元素不能控制边距部分的背景颜色。这部分颜色取决于父元素的设置。再如，设置边距的元素无法获取边距区域的用户输入。如果点击该区域，父元素就会收到 Tapped 事件。

Padding 属性的类型也为 Thickness，但只有几个类定义了 Padding 属性，其中包括 Control、Border、TextBlock、RichTextBlock 和 RichTextBlockOverflow。Padding 属性用于定义元素内部的边距区域。该区域会被看作元素的一部分，支持包括获取用户输入在内的所有特性。

如果希望 TextBlock 能够响应点击事件，并且不局限于文本部分，包含文本周围 100 像素范围内，则需要将 TextBlock 的 Padding 属性设置为 100，而不应设置 Margin 属性。

4.2 矩形与椭圆

第 2 章介绍过，Windows.UI.Xaml.Shapes 命名空间包含用于呈现矢量图形的类。可绘制的矢量图形包括线段、曲线和填充的区域。Shape 类本身派生自 FrameworkElement，定义了许多属性，如 Stroke(用于指定绘制线段或曲线的画笔)、StrokeThickness 和 Fill(用于指定用于呈现闭合区域的画笔)。

派生自 Shape 的类有 6 个。其中的 Line、Polyline 和 Polygon 可以根据坐标点绘制直线段，Path 可以通过 Windows.UI.Xaml.Media 中的类来呈现直线段、弧线和贝塞尔曲线。

另外两个 Shape 的派生类为 Rectangle(矩形)和 Ellipse(椭圆)。虽然名称很简单，但需要注意的是，这两种元素不能用坐标点来定义图形。举例来说，下面这段 XAML 呈现了一个椭圆。

```
项目：SimpleEllipse | 文件：MainPage.xaml(片段)
<Grid Background="{StaticResource ApplicationPageBackgroundThemeBrush}">
   <Ellipse Stroke="Red"
            StrokeThickness="24"
            Fill="Blue" />
</Grid>
```

请注意，这个椭圆是填充整个容器的(参见下图)。

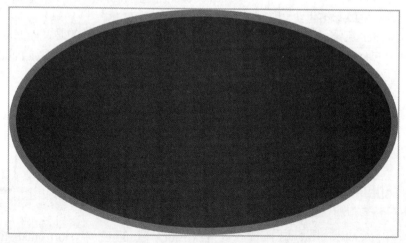

与其他 FrameworkElement 的派生类一样，Ellipse 也具有默认值为 Stretch 的 HorizontalAlignment 和 VerticalAlignment 属性。不同的是，Ellipse 不会有意避免某些设置导致的不良后果。

若将 Ellipse 的 HorizontalAlignment 或 VerticalAlignment 属性设置为非默认值会怎样？

试一下！结果是椭圆会收缩，直至完全消失。事实上，我们很难想象它会有什么其他合理行为。如果不希望 Ellipse 或 Rectangle 元素填充整个容器，则只能显式地设置它们的 Height 和 Width 属性。

Shape 类还定义了 Stretch 属性(请不要与 HorizontalAlignment 和 VerticalAlignment 属性的默认值 Stretch 相混淆)。该属性与 Image 和 Viewbox 定义的 Stretch 属性类似。例如，在示例项目 SimpleEllipse 中，若将 Stretch 属性设置为 Uniform，则椭圆的长轴和短轴会变为等长，进而形成一个正圆。将 Stretch 属性设置为 UniformToFill 也会使其变成正圆，但其直径取决于容器边长的最大值，而这会使这个圆被截断，如下图所示。

HorizontalAlignment 和 VerticalAlignment 属性可以用来控制哪部分被截断。

与 Ellipse 非常类似，Rectangle 也具有 Border 的一些性质。下表列出了两元素相关的属性。

Border	Rectangle
BorderBrush	Stroke
BorderThickness	StrokeThickness
Background	Fill
CornerRadius	RadiusX / RadiusY

Border 与 Rectangle 最大的区别是 Border 具有 Child 属性，而 Rectangle 没有。

4.3　StackPanel

Panel 及其派生类是 Windows Runtime 布局系统的核心。Panel 本身只定义了少数几个属性，其中最重要的是 Children。只有 Panel 及其派生类允许包含多个子元素。

Panel 及相关类的继承关系如下所示。

Object
　DependencyObject
　　UIElement
　　　FrameworkElement

Panel
  Canvas
  Grid
  StackPanel
  VariableSizedWrapGrid

除以上 4 个派生类之外，另外还有几个不允许在 ItemsControl 上下文之外使用的。第 11 章会具体介绍这些类。第 5 章会介绍 Grid 类，而本章将重点放在其余 3 个 Panel 派生类上。

这几个标准面板元素中，StackPanel 无疑是最好用的。顾名思义，它会横向或纵向排列其子元素(在默认情况下纵向排列)。如果纵向排列，子元素可以具有任意高度，但 StackPanel 只提供其所需要的空间。示例项目 SimpleVerticalStack 展示了这一特性。

项目：SimpleVerticalStack | 文件：MainPage.xaml(片段)

```
<Grid Background="{StaticResource ApplicationPageBackgroundThemeBrush}">
   <StackPanel>
      <TextBlock Text="Right-Aligned Text"
                 FontSize="48"
                 HorizontalAlignment="Right" />

      <Image Source="http://www.charlespetzold.com/pw6/PetzoldJersey.jpg"
             Stretch="None" />

      <TextBlock Text="Figure 1. Petzold heading to the basketball court"
                 FontSize="24"
                 HorizontalAlignment="Center" />

      <Ellipse Stroke="Red"
               StrokeThickness="12"
               Fill="Blue" />

      <TextBlock Text="Left-Aligned Text"
                 FontSize="36"
                 HorizontalAlignment="Left" />
   </StackPanel>
</Grid>
```

XAML 中 StackPanel 子元素的顺序与子元素显示顺序相同(参见下图)。

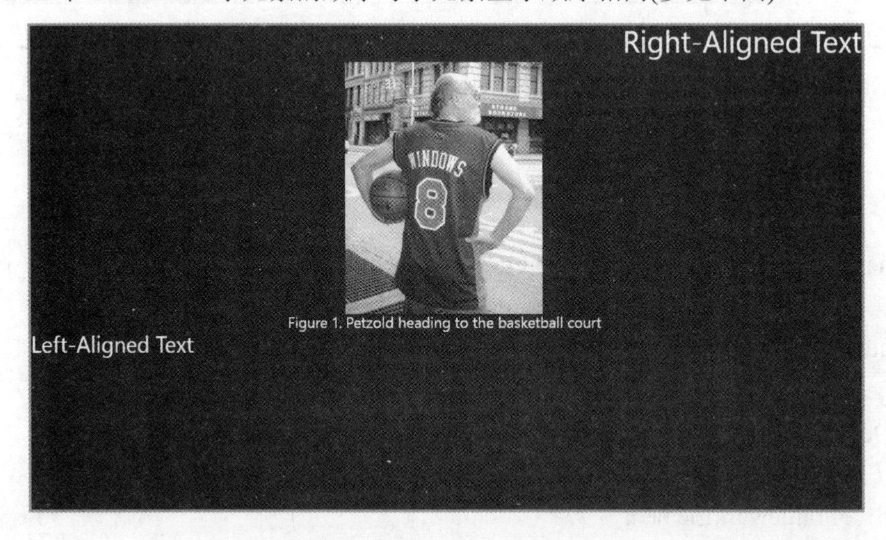

请注意，这里将 StackPanel 作为 Grid 的子元素。Panel 可以嵌套，而且这种情况十分常见。对于这个示例，我们可以用 StackPanel 来替换 Grid，并为其设置相同的 Background 属性值。

StackPanel 中的元素只占据它所需要的高度，但会保持与面板同宽(正如示例中分别右对齐和左对齐的两个 TextBlock 所展示的)。对于纵向排列的 StackPanel，子元素的任何 VerticalAlignment 设置均会被忽略。

Image 的 Stretch 属性被设置为 None，即按原尺寸显示位图。如果使用默认值 Uniform，那么 Image 会被拉伸至 StackPanel 的宽度(与 Page 同宽)，图片按比例放大。这样，图片会将下面的元素挤压到屏幕下部，甚至挤出显示区域。

这段 XAML 还定义了一个 Ellipse。它为何没有被显示出来？与 StackPanel 的其他子元素一样，Ellipse 会在垂直方向上得到所需的空间，但实际上它一点也不需要，因此收缩至消失。若希望这个 Ellipse 可见(比如下图)，需要将 Height 属性设置为一个非零值(如 48)。

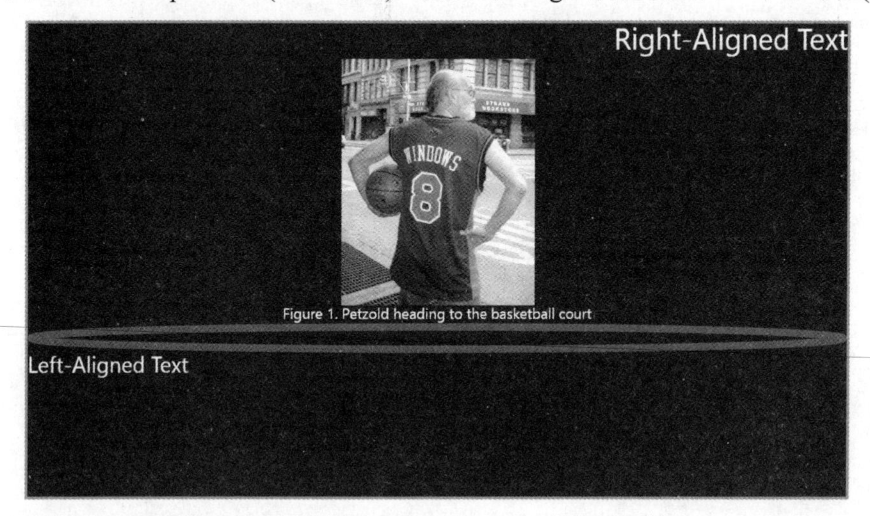

若将 Ellipse 的 Stretch 属性设置为 Uniform，该元素就会变成一个正圆，但不会变得很大。

这个 StackPanel 占据了整个屏幕空间。为什么这么说呢？在使用面板时，要知道它占据了多大空间，为其设置一个特别的 Background 即可。例如：

```
<StackPanel Background="Blue">
```

与 FrameworkElement 的派生类一样，StackPanel 也具有 HorizontalAlignment 和 VerticalAlignment 属性。如果被设置为非默认值，StackPanel 则会尽可能地压缩其子元素，而且变化非常剧烈。下图展示了一个 Background 为 Blue 的 StackPanel，它的 HorizontalAlignment 和 VerticalAlignment 均被设置为 Center。

这种情况下，StackPanel 的宽度取决于最宽子元素的宽度。对于本示例，StackPanel 无疑与图片下方的标题同宽。

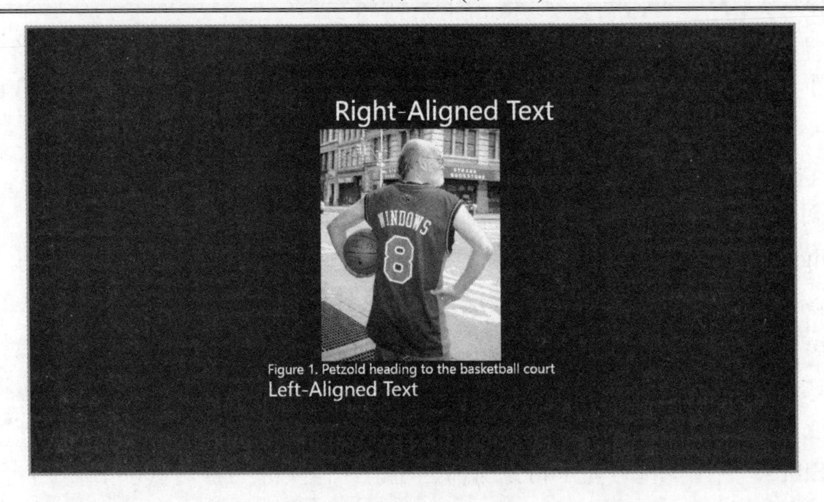

4.4 横向的 StackPanel

如果将 StackPanel 的 Orientation 属性设置为 Horizontal，元素就会横向依次排列。示例项目 SimpleHorizontalStack 展示了这一行为。

项目：SimpleHorizontalStack | 文件：MainPage.xaml（片段）

```
<Grid Background="{StaticResource ApplicationPageBackgroundThemeBrush}">

    <StackPanel Orientation="Horizontal"
                VerticalAlignment="Center"
                HorizontalAlignment="Center">

        <TextBlock Text="Rectangle: "
                   VerticalAlignment="Center" />

        <Rectangle Stroke="Blue"
                   Fill="Red"
                   Width="72"
                   Height="72"
                   Margin="12 0"
                   VerticalAlignment="Center" />

        <TextBlock Text="Ellipse: "
                   VerticalAlignment="Center" />

        <Ellipse Stroke="Red"
                 Fill="Blue"
                 Width="72"
                 Height="72"
                 Margin="12 0"
                 VerticalAlignment="Center" />

        <TextBlock Text="Petzold: "
                VerticalAlignment="Center" />

        <Image Source="http://www.charlespetzold.com/pw6/PetzoldJersey.jpg"
               Stretch="Uniform"
               Width="72"
               Margin="12 0"
               VerticalAlignment="Center" />

    </StackPanel>
</Grid>
```

该程序的运行效果如下图所示。

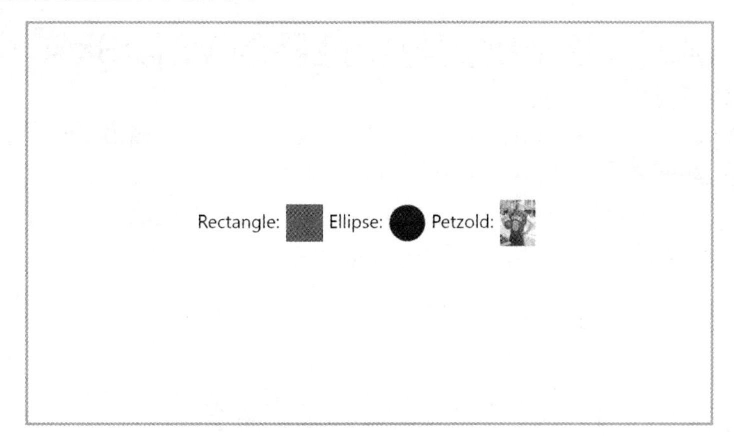

标记中的诸多对齐设置可能让人感觉有些多余。移除所有 VerticalAlignment 和 HorizontalAlignment 设置之后，得到下图所示的效果。

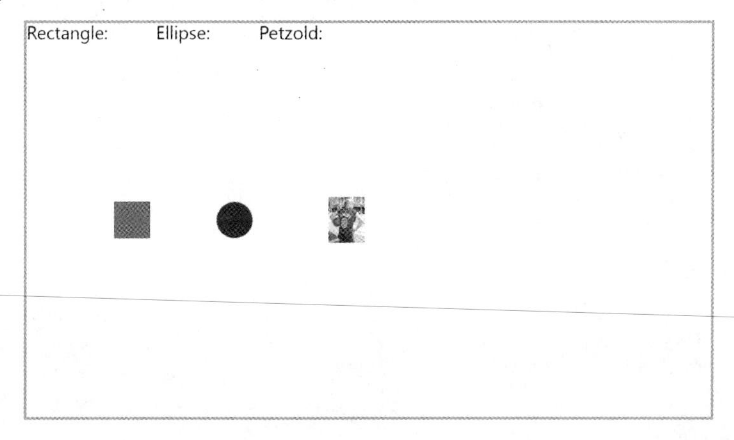

此时，StackPanel占据了整个页面。子元素与StackPanel同高。TextBlock会顶端对齐，其他元素则在竖直方向上居中。将Panel的HorizontalAlignment和VerticalAlignment设置为Center会使得Panel被收紧并被移至屏幕中央(如下图所示)。

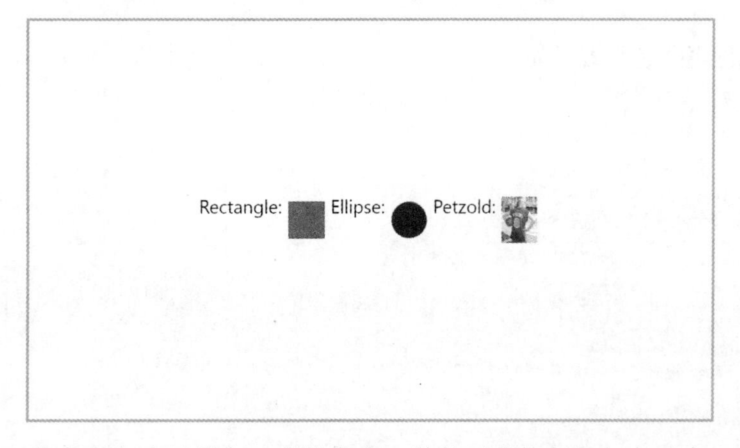

StackPanel 的高度取决于最高元素的高度，其他元素均保持与之同高。为使所有元素相对彼此居中对齐，最简单的办法是将它们的 VerticalAlignment 设置为 Center。

4.5 基于绑定与转换器的 WhatSize

第 3 章介绍了一个叫 WhatSize 的程序，它没能利用绑定，因为 Run 类的 Text 属性不是依赖属性，前面曾说过，只有依赖属性支持数据绑定。

幸运的是，对于包含单行文本的 Run 对象，我们可以通过在横向的 StackPanel 中使用多个 TextBlock 来模拟。示例项目 WhatSizeWithBindings 的标记如下所示。

```
项目: WhatSizeWithBindings | 文件: MainPage.xaml(片段)
<Page
    x:Class="WhatSizeWithBindings.MainPage"
    ...
    FontSize="36"
    Name="page">

    <Grid Background="{StaticResource ApplicationPageBackgroundThemeBrush}">
        <StackPanel Orientation="Horizontal"
                    HorizontalAlignment="Center"
                    VerticalAlignment="Top">
            <TextBlock Text="&#x21A4; " />
            <TextBlock Text="{Binding ElementName=page, Path=ActualWidth}" />
            <TextBlock Text=" pixels &#x21A6;" />
        </StackPanel>

        <StackPanel HorizontalAlignment="Center"
                    VerticalAlignment="Center">
            <TextBlock Text="&#x21A5;" TextAlignment="Center" />

            <StackPanel Orientation="Horizontal"
                        HorizontalAlignment="Center">
                <TextBlock Text="{Binding ElementName=page, Path=ActualHeight}" />
                <TextBlock Text=" pixels" />
            </StackPanel>

            <TextBlock Text="&#x21A7;" TextAlignment="Center" />
        </StackPanel>
    </Grid>
</Page>
```

请注意，根元素被分配了一个名称 page。该名称用在了获取 ActualWidth 和 ActualHeight 属性的数据绑定标记扩展中。与前一个版本的不同在于这个程序未在代码隐藏文件中订阅任何事件。该程序的运行效果如下图所示。

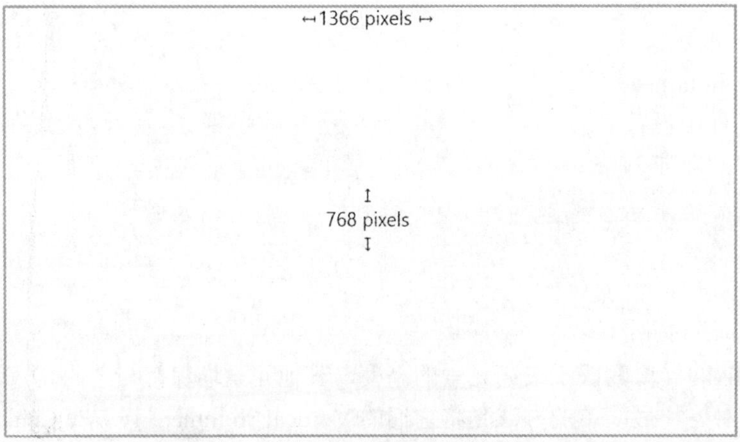

↤1366 pixels ↦

↥
768 pixels
↧

虽然程序的初始状态是正确的，但本书截稿前的 Windows 8 版本不能在屏幕方向变化或切换至辅屏视图时更新界面上的尺寸值。

这两个数据绑定能够自动将 double 值转换为字符串对象。那么转换逻辑可否控制？例如，希望结果保留指定位数的小数。或者，就这个示例程序而言，希望宽度值中包含千分位符(如 1,366)。

我们可以自定义数据转换逻辑，并将其指定给 Binding 对象。Binding 类有一个名为 Converter 的属性，类型为 IValueConverter。该接口类型具有两个方法：Convert 和 ConvertBack。Convert 用于将绑定源转换至绑定目标，ConvertBack 用于在双向绑定中将绑定目标转换至绑定源。

为创建自定义的转换器，需要创建一个派生自 IValueConverter 的类，并实现该接口的两个方法。下面两个方法只是返回原值，而不包含任何其他逻辑。

```
public class NothingConverter : IValueConverter
{
    public object Convert(object value, Type targetType, object parameter, string language)
    {
        return value;
    }
    public object ConvertBack(object value, Type targetType, object parameter, string language)
    {
        return value;
    }
}
```

如果只希望在单向数据绑定中使用转换器，则不需要实现 ConvertBack 方法。Convert 方法的参数 value 来自绑定源(对于示例程序 WhatSize 来说，value 的实际类型为 double)。TargetType 是目标类型(在 WhatSize 中为 string 类型)。

对于 WhatSize，为将浮点数转换为字符串，要求包含逗号分隔符且不带小数点，则可以如下实现 Convert 方法。

```
public object Convert(object value, Type targetType, object parameter, string language)
{
    return ((double)value).ToString("N0");
}
```

转换器一般具有一定的通用性。例如，如果传入的 value 的类型实现了 IFormattable 接口(如包括 double 在内的数字类型和 DateTime)，最好能加以处理。IFormattable 接口定义了带有两个参数的 ToString 方法：第一个参数为格式字符串，另一个参数是实现 IFormatProvider 的对象(一般为 CultureInfo 对象)。

除了 value 和 targetType，Convert 方法还有 parameter 和 language 参数。这两个参数一般在 XAML 文件中设置。这样一来，可以利用传给 Convert 方法的 parameter 参数来指定 ToString 的格式，而通过 language 参数来创建 CultureInfo 对象。下面的代码是一种实现。

```
项目: WhatSizeWithBindingConverter | 文件: FormattedStringConverter.cs
using System;
using System.Globalization;
using Windows.UI.Xaml.Data;

namespace WhatSizeWithBindingConverter
{
    public class FormattedStringConverter : IValueConverter
    {
        public object Convert(object value, Type targetType, object parameter, string language)
```

```
        {
            if (value is IFormattable &&
                parameter is string &&
                !String.IsNullOrEmpty(parameter as string) &&
                targetType == typeof(string))
            {
                if (String.IsNullOrEmpty(language))
                    return (value as IFormattable).ToString(parameter as string, null);

                return (value as IFormattable).ToString(parameter as string,
                                                new CultureInfo(language));
            }

            return value;
        }
        public object ConvertBack(object value, Type targetType, object parameter, string language)
        {
            return value;
        }
    }
}
```

Convert 方法只在满足一定条件的情况下才会调用 ToString 方法。如果条件不满足，则直接返回传入的 value 参数。

在 XAML 文件中，绑定转换器一般能够以资源的形式定义，进而可由多个绑定共享。

```
项目: WhatSizeWithBindingConverter | 文件: MainPage.xaml(片段)
<Page
    x:Class="WhatSizeWithBindingConverter.MainPage"
    ...
    FontSize="36"
    Name="page">

    <Page.Resources>
        <local:FormattedStringConverter x:Key="stringConverter" />
    </Page.Resources>

    <Grid Background="{StaticResource ApplicationPageBackgroundThemeBrush}">
        <StackPanel Orientation="Horizontal"
                    HorizontalAlignment="Center"
                    VerticalAlignment="Top">
            <TextBlock Text="&#x21A4; " />
            <TextBlock Text="{Binding ElementName=page,
                                      Path=ActualWidth,
                                      Converter={StaticResource stringConverter},
                                      ConverterParameter=N0}" />
            <TextBlock Text=" pixels &#x21A6;" />
        </StackPanel>

        <StackPanel HorizontalAlignment="Center"
                    VerticalAlignment="Center">
            <TextBlock Text="&#x21A5;" TextAlignment="Center" />

            <StackPanel Orientation="Horizontal"
                        HorizontalAlignment="Center">
                <TextBlock Text="{Binding ElementName=page,
                                          Path=ActualHeight,
                                          Converter={StaticResource stringConverter},
                                          ConverterParameter=N0}" />
                <TextBlock Text=" pixels" />
            </StackPanel>

            <TextBlock Text="&#x21A7;" TextAlignment="Center" />
        </StackPanel>
    </Grid>
</Page>
```

请注意 Binding 语法。为了使其更易读(且易于印刷)，这里将它拆分成了 4 行。Binding 标记扩展中包含一个 StaticResource 标记扩展，用来引用转换器资源。两个标记扩展均未使用引号。

经过格式化，呈现的数字得到了一定的改善(参见下图)。

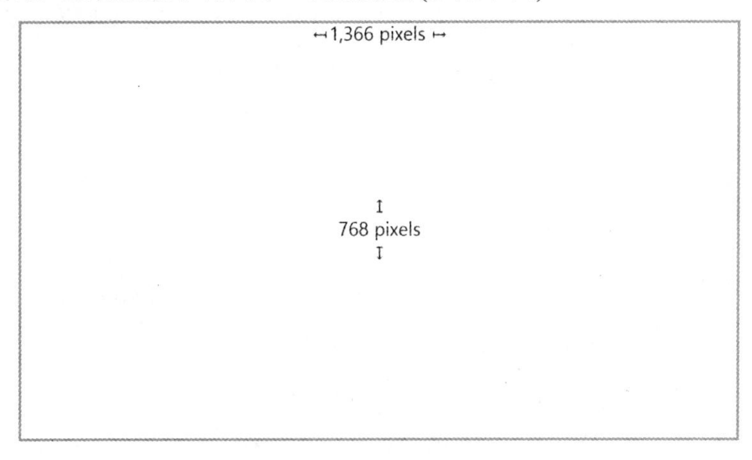

4.6　ScrollViewer 方案

如果元素过多，StackPanel 无法显示该怎么办？在实践当中，这种情况十分常见。解决方案一般是将带有较多元素的 StackPanel 放入 ScrollViewer。

ScrollViewer 有一个名为 Content 的属性，可设置一个屏幕空间无法容纳的元素(如一个较大的 Image)。ScrollViewer 为使用鼠标的用户提供了滚动条。用户也可以通过手势来移动其中的内容。在默认情况下，ScrollViewer 还支持通过两个手指对内容进行拉伸和缩放。若要禁用该功能，可将 ZoomMode 属性设置为 Disabled。

我们经常需要 ScrollViewer 纵向滚动(如配合纵向的 StackPanel)。ScrollViewer 的 VerticalScrollBarVisibility 属性的默认值是枚举成员 ScrollBarVisibility.Visible。此设置并不意味着滚动条总是可见的。对于使用鼠标的用户，只有在鼠标移动到 ScrollViewer 右侧的滚动条才会显示，而鼠标移走则会消失。如果用手指滑动，则会显示一个非常细的滚动条。

与纵向滚动不同，用于设置横向滚动的 HorizontalScrollBarVisibility 属性的默认值为 Disabled。为启用横向滚动，需要修改该属性。该属性还有另外两个值：Hidden(只支持手指操作而不支持鼠标)和 Auto(如果内容超出屏幕显示范围，行为同 Visible；如果不超出屏幕，则同 Disabled)。

示例项目 StackPanelWithScrolling 的 XAML 文件中，ScrollViewer 内部定义了一个 StackPanel。FontSize 属性是在根标签中设置的，所以作用于整个页面。

```
项目：StackPanelWithScrolling | 文件：MainPage.xaml(片段)
<Page
    x:Class="StackPanelWithScrolling.MainPage"
    ...
    FontSize="26">

    <Grid Background="{StaticResource ApplicationPageBackgroundThemeBrush}">
        <ScrollViewer>
```

```
            <StackPanel Name="stackPanel" />
        </ScrollViewer>
    </Grid>
</Page>
```

代码隐藏类只需要生成足够多的子项目，从而使 StackPanel 无法在一屏中完全呈现。那么怎么获得如此多的项目呢？Colors 类定义了 141 个静态属性，类型为 Color，可以通过反射获取全部属性值。

```
项目: StackPanelWithScrolling | 文件: MainPage.xaml.cs(片段)
public sealed partial class MainPage : Page
{
    public MainPage()
    {
        this.InitializeComponent();

        IEnumerable<PropertyInfo> properties =
                        typeof(Colors).GetTypeInfo().DeclaredProperties;

        foreach (PropertyInfo property in properties)
        {
            Color clr = (Color)property.GetValue(null);
            TextBlock txtblk = new TextBlock();
            txtblk.Text = String.Format("{0} \x2014 {1:X2}-{2:X2}-{3:X2}-{4:X2}",
                                property.Name, clr.A, clr.R, clr.G, clr.B);
            stackPanel.Children.Add(txtblk);
        }
    }
}
```

Windows 8 的反射与.NET 反射稍有不同。为从 Type 对象获取信息，需要调用一个 Windows 8 特有的扩展方法 GetTypeInfo。该方法会返回 TypeInfo 对象，能够提供的信息比 Type 更多。在这个程序中，TypeInfo 的 DeclaredProperties 属性能够以 PropertyInfo 对象的形式返回 Colors 类的所有属性信息。Colors 中的所有属性均为静态的。通过调用 PropertyInfo 对象的 GetValue 方法[①]，并传入 null，便可以得到这些属性的值。TextBlock 对象用于显示颜色的名称、英文破折号(Unicode 值为 0x2014)及十六进制颜色字节值。该程序的运行效果如下图所示。

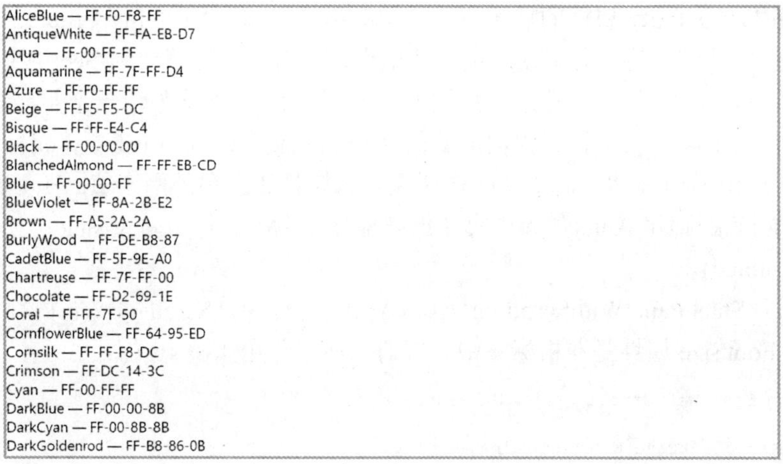

AliceBlue — FF-F0-F8-FF
AntiqueWhite — FF-FA-EB-D7
Aqua — FF-00-FF-FF
Aquamarine — FF-7F-FF-D4
Azure — FF-F0-FF-FF
Beige — FF-F5-F5-DC
Bisque — FF-FF-E4-C4
Black — FF-00-00-00
BlanchedAlmond — FF-FF-EB-CD
Blue — FF-00-00-FF
BlueViolet — FF-8A-2B-E2
Brown — FF-A5-2A-2A
BurlyWood — FF-DE-B8-87
CadetBlue — FF-5F-9E-A0
Chartreuse — FF-7F-FF-00
Chocolate — FF-D2-69-1E
Coral — FF-FF-7F-50
CornflowerBlue — FF-64-95-ED
Cornsilk — FF-FF-F8-DC
Crimson — FF-DC-14-3C
Cyan — FF-00-FF-FF
DarkBlue — FF-00-00-8B
DarkCyan — FF-00-8B-8B
DarkGoldenrod — FF-B8-86-0B

① 译注：该示例调用的 PropertyInfo.GetValue 方法重载接受一个 Object 类型的参数，能够返回该对象的属性值。返回哪个属性的值取决于传入的 PropertyInfo 对象。如果传入 null，则返回该静态属性的值。

当然，可以使用手指或鼠标进行滚动。

为简化反射的使用，C++版本的示例中使用了本人通过 C#实现的 ReflectionHelper 类库。这个类库也会被本章或其他章节的示例所引用。稍后再对这个类库做进一步阐述。

用手指滑动屏幕会发现 ScrollViewer 能够流畅地响应手指的运动，并具有减速和回弹效果。人们可能希望 ScrollViewer 能够满足所有对滚动的需求，但会发现许多元素类型已内建了 ScrollViewer，从而本身支持滚动(其中包括第 11 章要介绍的 ListBox 和 GridView)。如果 Windows 8 “开始” 界面也使用了 ScrollViewer，这并不让人感到意外。

对于这个示例，除了名称和色值外，如果能让用户看到实际的颜色岂不是更好？本章后面有一个示例正是这样做的。

本书到目前为止展示过几个类的继承层次结构。若在 Windows 8 的文档中查找这些类的层次结构，可能会发现文档中只提供了祖先类的层次结构，而没有派生类。那么本书展示的类层次结构是如何得出的呢？本书的配套资源中有一个名为 DependencyObjectClassHierarchy 的示例程序，它可以通过 ScrollViewer 和 StackPanel 来显示 DependencyObject 的所有派生类。

```
highlightBrush =
new SolidColorBrush(new UISettings().UIElementColor(UIElementType.Highlight));
```

这个示例与上一个示例的 XAML 文件类似，只不过这个示例的字号要小一些。

```
项目: DependencyObjectClassHierarchy | 文件: MainPage.xaml(片段)
<Page
   x:Class="DependencyObjectClassHierarchy.MainPage"
   ...
   FontSize="{StaticResource ControlContentThemeFontSize}">

   <Grid Background="{StaticResource ApplicationPageBackgroundThemeBrush}">
      <ScrollViewer>
         <StackPanel Name="stackPanel" />
      </ScrollViewer>
   </Grid>
</Page>
```

该程序能够构建一个类继承关系树。每个节点代表一个类，直接子节点为它的直接子类。为表示节点，以及节点与子节点间的关系，示例中有这样一个文件。

```
项目: DependencyObjectClassHierarchy | 文件: ClassAndSubclasses.cs
using System;
using System.Collections.Generic;

namespace DependencyObjectClassHierarchy
{
   class ClassAndSubclasses
   {
      public ClassAndSubclasses(Type parent)
      {
         this.Type = parent;
         this.Subclasses = new List<ClassAndSubclasses>();
      }

      public Type Type { protected set; get; }
      public List<ClassAndSubclasses> Subclasses { protected set; get; }
   }
}
```

我们可以通过反射获取类定义的所有属性，也可以通过反射来获取程序集中定义的所有公共类。这些类可通过 Assembly 对象的 ExportedTypes 属性获得。从概念上讲，整个 Windows Runtime 只与单个程序集关联，为获取该程序集，只需要一个类型，通过该类型获得 TypeInfo 对象，从该对象的 Assembly 属性获得 Assembly 对象。

项目：DependencyObjectClassHierarchy | 文件：MainPage.xaml.cs(片段)

```
public sealed partial class MainPage : Page
{
    Type rootType = typeof(DependencyObject);
    TypeInfo rootTypeInfo = typeof(DependencyObject).GetTypeInfo();
    List<Type> classes = new List<Type>();
    Brush highlightBrush;

    public MainPage()
    {
        this.InitializeComponent();
        highlightBrush = new SolidColorBrush(new UISettings().UIElementColor(UIElementType.Highlight));

        // Accumulate all the classes that derive from DependencyObject
        AddToClassList(typeof(Windows.UI.Xaml.DependencyObject));

        // Sort them alphabetically by name
        classes.Sort((t1, t2) =>
            {
                return String.Compare(t1.GetTypeInfo().Name, t2.GetTypeInfo().Name);
            });

        // Put all these sorted classes into a tree structure
        ClassAndSubclasses rootClass = new ClassAndSubclasses(rootType);
        AddToTree(rootClass, classes);

        // Display the tree using TextBlock's added to StackPanel
        Display(rootClass, 0);
    }

    void AddToClassList(Type sampleType)
    {
        Assembly assembly = sampleType.GetTypeInfo().Assembly;

        foreach (Type type in assembly.ExportedTypes)
        {
            TypeInfo typeInfo = type.GetTypeInfo();

            if (typeInfo.IsPublic && rootTypeInfo.IsAssignableFrom(typeInfo))
                classes.Add(type);
        }
    }

    void AddToTree(ClassAndSubclasses parentClass, List<Type> classes)
    {
        foreach (Type type in classes)
        {
            Type baseType = type.GetTypeInfo().BaseType;

            if (baseType == parentClass.Type)
            {
                ClassAndSubclasses subClass = new ClassAndSubclasses(type);
                parentClass.Subclasses.Add(subClass);
                AddToTree(subClass, classes);
            }
        }
    }

    void Display(ClassAndSubclasses parentClass, int indent)
    {
        TypeInfo typeInfo = parentClass.Type.GetTypeInfo();
```

```
        // Create TextBlock with type name
        TextBlock txtblk = new TextBlock();
        txtblk.Inlines.Add(new Run { Text = new string(' ', 8 * indent) });
        txtblk.Inlines.Add(new Run { Text = typeInfo.Name });

        // Indicate if the class is sealed
        if (typeInfo.IsSealed)
           txtblk.Inlines.Add(new Run
              {
                 Text = " (sealed)",
                 Foreground = highlightBrush
              });

        // Indicate if the class can't be instantiated
        IEnumerable<ConstructorInfo> constructorInfos = typeInfo.DeclaredConstructors;
        int publicConstructorCount = 0;

        foreach (ConstructorInfo constructorInfo in constructorInfos)
           if (constructorInfo.IsPublic)
              publicConstructorCount += 1;

        if (publicConstructorCount == 0)
           txtblk.Inlines.Add(new Run
           {
              Text = " (non-instantiable)",
              Foreground = highlightBrush
           });

        // Add to the StackPanel
        stackPanel.Children.Add(txtblk);

        // Call this method recursively for all subclasses
        foreach (ClassAndSubclasses subclass in parentClass.Subclasses)
           Display(subclass, indent + 1);
    }
}
```

这段代码通过向 Inlines 集合中添加 Run 对象来构造显示每个类的 TextBlock。在某些情况下，需要为类层次结构显示额外的信息。该程序会检查类是否被标记为 sealed 及是否可被实例化。对于 Windows Presentation Foundation 和 Silverlight，不能被实例化的类一般会被定义为 abstract(抽象的)；而在 Windows Runtime 中，这种类的构造函数是受保护的。

下图展示了一段 Panel 派生类的类层次结构。

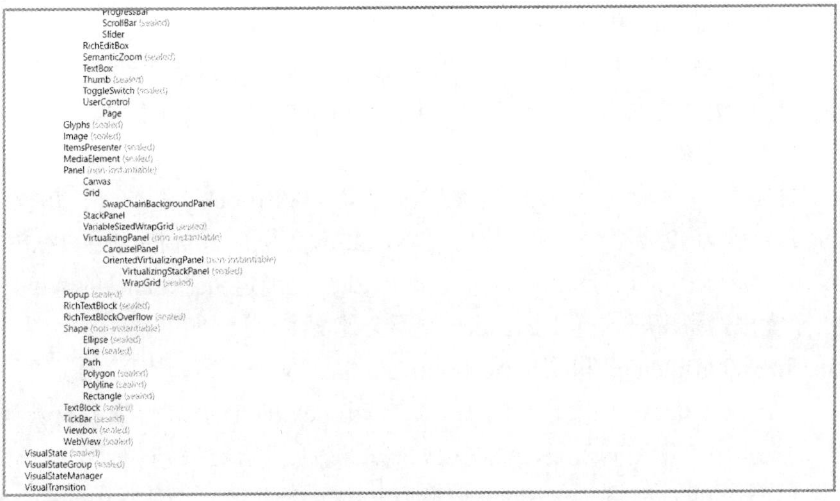

```
                        ProgressBar
                        ScrollBar (sealed)
                        Slider
                RichEditBox
                SemanticZoom (sealed)
                TextBox
                Thumb (sealed)
                ToggleSwitch (sealed)
                UserControl
                        Page
        Glyphs (sealed)
        Image (sealed)
        ItemsPresenter (sealed)
        MediaElement (sealed)
        Panel (non-instantiable)
                Canvas
                Grid
                        SwapChainBackgroundPanel
                StackPanel
                VariableSizedWrapGrid (sealed)
                VirtualizingPanel (non-instantiable)
                        CarouselPanel
                        OrientedVirtualizingPanel (non-instantiable)
                                VirtualizingStackPanel (sealed)
                                WrapGrid (sealed)
        Popup (sealed)
        RichTextBlock (sealed)
        RichTextBlockOverflow (sealed)
        Shape (non-instantiable)
                Ellipse (sealed)
                Line (sealed)
                Path
                Polygon (sealed)
                Polyline (sealed)
                Rectangle (sealed)
        TextBlock (sealed)
        TickBar (sealed)
        Viewbox (sealed)
        WebView (sealed)
VisualState (sealed)
VisualStateGroup (sealed)
VisualStateManager
VisualTransition
```

4.7　布局中的怪异现象

理解布局机制是优秀 Windows Runtime 开发者的必备技能。而为理解布局机制，最好的方法是自己编写一个 Panel 的派生类。第 11 章会展示相关的例子，但亲自动手会让人受益良多。

假设有一个 StackPanel，在这个 StackPanel 中添加一个 ScrollViewer，然后在这个 ScrollViewer 中添加另一个 StackPanel。为得知这样做到底会发生什么，可以如下修改 StackPanelWithScrolling 项目的 XAML 文件。

```
<Grid Background="{StaticResource ApplicationPageBackgroundThemeBrush}">
   <StackPanel>
      <ScrollViewer>
         <StackPanel Name="stackPanel" />
      </ScrollViewer>
   </StackPanel>
</Grid>
```

运行程序后会发现，这样做是行不通的，无法滚动，原因何在呢？

StackPanel 和 ScrollViewer 计算自身高度的方式不同，并发生了冲突。StackPanel 根据所有子元素的高度之和计算自身高度。在纵向模式下(默认的)，StackPanel 完全是子驱动的。为计算总高度，它会先为子元素提供无穷的(infinite)高度。(在编写自定义的 Panel 时，会发现这样说并不模糊或抽象。它确实用了 Double.PositiveInfinity 值。)子元素会根据自身的自然大小计算高度。StackPanel 最终将这些高度累加得到自身高度。

ScrollViewer 的高度是父驱动的，即它的高度是其父元素决定的。在前面的示例中，其高度与 Grid、Page 和窗口的高度相同。ScrollViewer 之所以允许用户滚动其内容，是因为它知道自身高度与子元素(一般为 StackPanel)的差距。

这里将纵向滚动的 ScrollViewer 作为 StackPanel 的子元素。为确定 ScrollViewer 的尺寸，StackPanel 先提供一个无限的高度。那么 ScrollViewer 如何知道自身应该多高呢？ScrollViewer 的高度此时变成了子驱动的，而非父驱动的，它的高度就是子元素 StackPanel 的高度，也就是 StackPanel 中所有子元素高度之和。

从 ScrollViewer 的角度看，其高度与内容相同意味着它不需要滚动。

换言之，纵向滚动的 ScrollViewer 被置于纵向排列的 StackPanel 中，从而失去滚动功能，这是完全合乎逻辑的。

布局方面还有一种较为常见的怪异现象：为 TextBlock 设置一长串文本，并将 TextWrapping 设置为 Wrap。在大多数情况下，文本会换行。若将这个 TextBlock 置于 Orientation 属性为 Horizontal 的 StackPanel 元素中，那么为确定 StackPanel 的宽度，StackPanel 需要提供一个无限的宽度，这样 TextBlock 便不会使文本换行。

在 WhatSizeWithBindings 和 WhatSizeWithBindingConverter 程序中，横向的 StackPanel 有效地串联多个 TextBlock 元素。其中的一个 TextBlock 的 Text 属性设置有数据绑定。若希望在换行的文本中实现相同效果，该怎么做呢？我们无法通过横向的 StackPanel 实现，因为文本不会换行；也不能将 TextBlock 替换为 Run 元素来实现，因为 Run 的 Text 属性不是依赖属性。一种解决方案是在代码中生成文本，另一种解决方案是使用 RichTextBlock(详情参见第 16 章)。

横向的 StackPanel 无法使作为子元素的 TextBlock 的文本换行，但纵向的 StackPanel 可以。纵向的 StackPanel 的宽度有限，适合容纳需要换行的 TextBlock 元素(正如稍后要介绍的)。

4.8　编写一个简单的电子书应用

对于包含在纵向 StackPanel 元素中的 TextBlock，我们可以将 TextWrapping 属性设置为 Wrap。该选项使 TextBlock 可以包含较长段落的文本，而不仅仅用来显示一两个词。Image 元素也可以包含在 StackPanel 中。这些为编写一个简易的电子书应用创造了条件。

在著名的 Project Gutenberg 网站[①]上，可以找到比阿特丽克斯・波特(Beatrix Potter)的经典童话故事书《小猫汤姆的故事》(*The Tale of Tom Kitten*，http://www.gutenberg.org/ebooks/14837)，因此可将示例项目命名为 TheTaleOfTomKitten。在项目中创建一个名为 Images 的文件夹。通过下载 Project Gutenberg 提供的本书 HTML 版本，可以轻松以 JPEG 格式获得所有插图。插图名称的格式为 tomxx.jpg，其中 xx 代表不同插图在原书中出现的页码。在 Visual Studio 项目中，将 28 个插图文件添加到 Images 文件夹中。

这个示例剩下的工作与 MainPage.xaml 文件有关。这本童话书的每一段落对应一个 TextBlock，而 Images 文件夹中的每个 JPEG 文件对应一个 Image 元素。

我们可以从 Project Gutenberg 网站提供的 HTML 文件找到文字与图片位置关系的一些线索。通过原版《小猫汤姆的故事》的 PDF(http://archive.org/details/ taleoftomkitten00pottuoft)可以得知作者对此书图片与文本的组织方式，有两种模式。

(1)　文字出现在偶数页的左侧，插图在奇数页的右侧。

(2)　文本出现在奇数页的右侧，插图在偶数页的左侧。

为保持与原书一致的格式，需要针对每种情况修改文本与图片的顺序。因此，在 XAML 文件中会出现一些书页布局的交替变化。

由于要定义如此多的 TextBlock 和 Image 元素，所以样式显然是必不可少的。

项目：TheTaleOfTomKitten | 文件：MainPage.xaml(片段)

```
<Page.Resources>
   <Style x:Key="commonTextStyle" TargetType="TextBlock">
      <Setter Property="FontFamily" Value="Century Schoolbook" />
      <Setter Property="FontSize" Value="36" />
      <Setter Property="Foreground" Value="Black" />
      <Setter Property="Margin" Value="0 12" />
   </Style>

   <Style x:Key="paragraphTextStyle" TargetType="TextBlock"
         BasedOn="{StaticResource commonTextStyle}">
      <Setter Property="TextWrapping" Value="Wrap" />
   </Style>
```

① 译注：这是一个电子书网站，免费提供数以万计的、高质量的电子书。

```
    <Style x:Key="frontMatterTextStyle" TargetType="TextBlock"
           BasedOn="{StaticResource commonTextStyle}">
        <Setter Property="TextAlignment" Value="Center" />
    </Style>

    <Style x:Key="imageStyle" TargetType="Image">
        <Setter Property="Stretch" Value="None" />
        <Setter Property="HorizontalAlignment" Value="Center" />
    </Style>
</Page.Resources>
```

Margin 设置为段落之间提供了间空。TextBlock 元素引用 paragraphTextStyle(正文段落的样式)或 frontMatterTextStyle(书中前面的标题及其他信息的样式)。可以为 Image 元素设置隐式样式，只需要移除样式的 x:Key 特性及 Image 元素的 Style 特性即可。

用来显示前言的 TextBlock 具有自己的 FontSize 设置。由于此书是白纸黑字印刷，因而这里将 TextBlock 的 Foreground 属性硬编码为黑色，将 Grid 的 Background 属性硬编码为白色。为限制每行的长度，这里将 StackPanel 的 MaxWidth 设置为 640，并将 StackPanel 限定在 ScrollViewer 中间。下面的标记片段展示了 TextBlock 和 Image 元素的关系。

```
项目: TheTaleOfTomKitten | 文件: MainPage.xaml(片段)
<Grid Background="White">
    <ScrollViewer>
        <StackPanel MaxWidth="640"
                    HorizontalAlignment="Center">
            ...
            <!-- pg. 38 -->
            <TextBlock Style="{StaticResource paragraphTextStyle}">
                  Mittens laughed so that she fell off the
                wall. Moppet and Tom descended after her; the pinafores
                and all the rest of Tom's clothes came off on the way down.
            </TextBlock>

            <TextBlock Style="{StaticResource paragraphTextStyle}">
                  “Come! Mr. Drake Puddle-Duck,” said Moppet
                — “Come and help us to dress him! Come and button up Tom!”
            </TextBlock>

            <Image Source="Images/tom39.jpg" Style="{StaticResource imageStyle}" />

            <!-- pg. 41 -->
            <TextBlock Style="{StaticResource paragraphTextStyle}">
                  Mr. Drake Puddle-Duck advanced in a slow
                sideways manner, and picked up the various articles.
            </TextBlock>

            <Image Source="Images/tom40.jpg" Style="{StaticResource imageStyle}" />
            ...
        </StackPanel>
    </ScrollViewer>
</Grid>
```

如下图所示，每段开头的两个 字符为空格，用于提供首行缩进，TextBlock 类并未提供该功能，但 RichTextBlock 提供了(详见第 16 章)。

这个电子书程序支持横向显示模式和纵向显示模式。

4.9　StackPanel 子项的定制

前面展示了一个程序，能够显示 Windows Runtime 中 141 种预定义的颜色。该程序能够显示颜色的名称和 RGB 值。下面我们来实现一个名为 ColorList1 的示例。先来看一下该程序完成后的效果(参见下图)，这样更能够做到有的放矢。

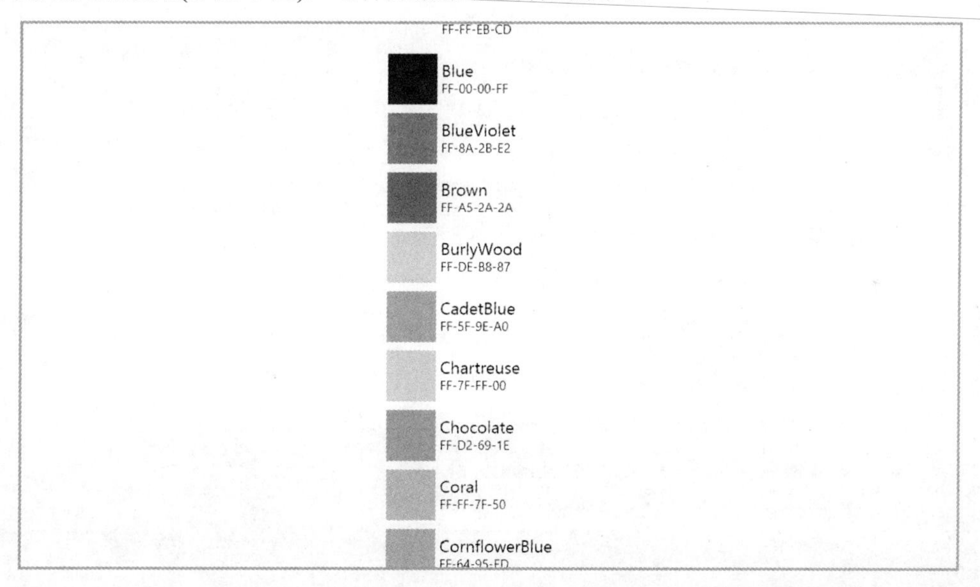

此程序总共包含 283 个 StackPanel 元素。141 种颜色，每种颜色需要两个 StackPanel 元素：一个纵向的 StackPanel，用于容纳一对 TextBlock 元素；一个横向的 StackPanel，用于容纳一个 Rectangle 和一个纵向的 StackPanel。在 ScrollViewer 中，这些横向的 StackPanel 作为主 StackPanel 的子元素。主 StackPanel 通过 XAML 文件创建。

```
项目: ColorList1 | 文件: MainPage.xaml(片段)
<Grid Background="{StaticResource ApplicationPageBackgroundThemeBrush}">
   <ScrollViewer>
      <StackPanel Name="stackPanel"
                  HorizontalAlignment="Center" />
   </ScrollViewer>
</Grid>
```

虽然 StackPanel 在 ScrollViewer 中是中间对齐的(与最宽的子元素同宽)，但 ScrollViewer 仍然占据整个页宽。这样，可见的滚动条或滑块均会出现在页面的最右端。另外，我们也可以将 HorizontalAlignment 设置移至 ScrollViewer。在这种情况下，内容仍会居中，但 ScrollViewer 与 StackPanel 同宽。

代码隐藏类的构造函数会在迭代 Colors 类的静态属性时为每种颜色创建嵌套的 StackPanel 元素。

```
项目: ColorList1 | 文件: MainPage.xaml.cs(片段)
public sealed partial class MainPage : Page
{
   public MainPage()
   {
      this.InitializeComponent();

      IEnumerable<PropertyInfo> properties = typeof(Colors).GetTypeInfo().DeclaredProperties;

      foreach (PropertyInfo property in properties)
      {
         Color clr = (Color)property.GetValue(null);

         StackPanel vertStackPanel = new StackPanel
         {
            VerticalAlignment = VerticalAlignment.Center
         };

         TextBlock txtblkName = new TextBlock
         {
            Text = property.Name,
            FontSize = 24
         };
         vertStackPanel.Children.Add(txtblkName);

         TextBlock txtblkRgb = new TextBlock
         {
            Text = String.Format("{0:X2}-{1:X2}-{2:X2}-{3:X2}",
                                 clr.A, clr.R, clr.G, clr.B),
            FontSize = 18
         };
         vertStackPanel.Children.Add(txtblkRgb);

         StackPanel horzStackPanel = new StackPanel
         {
            Orientation = Orientation.Horizontal
         };

         Rectangle rectangle = new Rectangle
         {
            Width = 72,
            Height = 72,
            Fill = new SolidColorBrush(clr),
            Margin = new Thickness(6)
         };
         horzStackPanel.Children.Add(rectangle);
         horzStackPanel.Children.Add(vertStackPanel);
         stackPanel.Children.Add(horzStackPanel);
```

```
        }
    }
}
```

这段代码虽然没什么错误，但可能有无数种更好的实现方式。所谓“更好”，并非更快或更高效，而是更简洁、更优雅，最重要的是更易于维护和修改。

这个示例并未结束，下一节将进一步优化这个程序。第 11 章会介绍一种绝佳的方案，到那时这个程序才算完成。

4.10　UserControl 的定制

带有颜色的列表项(嵌套的 StackPanel、TextBlock 和 Rectangle)是示例项目 ColorList1 的亮点。这原本看上去难以实现(我们不能在 MainPage.xaml 文件中实现这种效果)，因为无法通过 XAML 来“生成”列表项的实例，而复制粘贴这 141 个列表项可能是最糟糕的选择。

下面我们来创建 ColorList2，以便了解另一种实现这种界面的常见方法。创建 ColorList2 项目后，在解决方案资源管理器中右击项目名称，选择“添加”|“新建项”。在“添加新项”对话框中，选择“用户控件”，将其命名为“ColorItem.xaml”。Visual Studio 会创建一对文件，分别为 ColorItem.xaml 和对应的代码隐藏文件 ColorItem.xaml.cs。

在 Visual Studio 创建的 ColorItem.xaml.cs 文件中，ColorList2 命名空间下定义了一个名为 ColorItem 的类，该类派生自 UserControl。

```
namespace ColorList2
{
    public sealed partial class ColorItem : UserControl
    {
        public ColorItem(string name, Color clr)
        {
            this.InitializeComponent();
        }
    }
}
```

Visual Studio 创建的 ColorItem.xaml 文件以 XAML 的形式定义了以下内容。

```
<UserControl
    x:Class="ColorList2.ColorItem"
    xmlns="http://schemas.microsoft.com/winfx/2006/xaml/presentation"
    xmlns:x="http://schemas.microsoft.com/winfx/2006/xaml"
    xmlns:local="using:ColorList2"
    xmlns:d="http://schemas.microsoft.com/expression/blend/2008"
    xmlns:mc="http://schemas.openxmlformats.org/markup-compatibility/2006"
    mc:Ignorable="d"
    d:DesignHeight="300"
    d:DesignWidth="400">
    <Grid>
    </Grid>
</UserControl>
```

您或许在前文中读到过 UserControl，因为 Page 派生自 UserControl 类。名称中的 user 所指的并非应用程序的最终用户，而是程序员。派生 UserControl 类是创建自定义控件最为简单的一种方法，因为我们可以在 XAML 文件中定义控件的可视元素。UserControl 定义了一个名为 Content 的属性。由于该属性是内容属性，UserControl 标签内添加的所有元素都会被设置到 Content 属性上。

ColorItem.xaml 文件中的 d:DesignHeight 和 d:DesignWidth 属性是供 Microsoft Expression Blend 使用的，我们可以忽略。控件的实际大小取决于其内容。

下面，我们在 ColorItem.xaml 文件中定义颜色列表项的可视元素。

```
项目: ColorList2 | 文件: ColorItem.xaml(片段)
<UserControl
   x:Class="ColorList2.ColorItem" ...>

   <Grid>
      <StackPanel Orientation="Horizontal">
         <Rectangle Name="rectangle"
                    Width="72"
                    Height="72"
                    Margin="6" />

         <StackPanel VerticalAlignment="Center">

            <TextBlock Name="txtblkName"
                       FontSize="24" />

            <TextBlock Name="txtblkRgb"
                       FontSize="18" />
         </StackPanel>
      </StackPanel>
   </Grid>
</UserControl>
```

列表项中元素的层次结构与 ColorList1 中的一致，不同的是 Rectangle 和两个 TextBlock 元素都具有名称，因此可以在代码隐藏文件中访问它们。

```
项目: ColorList2 | 文件: ColorItem.xaml.cs(片段)
public sealed partial class ColorItem : UserControl
{
   public ColorItem(string name, Color clr)
   {
      this.InitializeComponent();

      rectangle.Fill = new SolidColorBrush(clr);
      txtblkName.Text = name;
      txtblkRgb.Text = String.Format("{0:X2}-{1:X2}-{2:X2}-{3:X2}",
                                     clr.A, clr.R, clr.G, clr.B);
   }
}
```

这个类的构造函数被重新定义，接受颜色名称和 Color 值为参数。构造函数通过这两个参数来设置 Rectangle 和两个 TextBlock 元素的属性。

需要说明的是，在 UserControl 的派生类中使用带参数的构造函数是不被推荐的做法，最好通过定义属性来设置参数。这里之所以没这么做，是因为这些属性必须是依赖属性，就演示 UserCongtrol 而言，这样做可能有些喧宾夺主。

由于没有无参构造函数，ColorItem 类不能在 XAML 中被实例化。但对于这个示例是没问题的，因为该类不会在 XAML 中被实例化。与 ColorList1 相比，ColorList2 的 MainPage.xaml 文件并未发生变化，但代码隐藏文件有所不同。

```
项目: ColorList2 | 文件: MainPage.xaml.cs(片段)
public sealed partial class MainPage : Page
{
   public MainPage()
   {
      this.InitializeComponent();
```

```
            IEnumerable<PropertyInfo> properties = typeof(Colors).GetTypeInfo().DeclaredProperties;

            foreach (PropertyInfo property in properties)
            {
                Color clr = (Color)property.GetValue(null);
                ColorItem clrItem = new ColorItem(property.Name, clr);
                stackPanel.Children.Add(clrItem);
            }
        }
    }
```

每个 ColorItem 通过颜色名称和 Color 值实例化，然后被添加至 StackPanel。

4.11　Windows Runtime 类库的创建

作为 ColorList 系列的下一个程序，这次我们在类库中实现 ColorItem 类，使其能够共享给其他项目。

Visual Studio 创建的解决方案一般只有一个项目，但在现有的应用程序项目中添加类库的做法也十分常见。可以在同一解决方案中创建一个应用程序项目来测试类库中的代码。将必要的代码置于类库中，好让其他应用程序可以在需要时引用它。

下面我们来创建一个名为 ColorList3 的程序。在解决方案资源管理器中，右击解决方案名称，选择“添加”|“新建项目”。(或者在“文件”菜单中选择“新建项目”。)在“新建项目”对话框的左侧窗格中展开 Visual C# |“Windows 8 应用商店”节点。在右侧的项目模板列表中选择“类库”。

一般来讲，类库名一般由句点分隔的多级名称组成。项目创建后，类库名会被用作项目的默认命名空间。类库名一般由公司名(或对等的名称)开头。对于这个示例，可将类库命名为 Petzold.Windows8.Controls。本人不太喜欢多级程序集名称，因此去掉了其中的句点，变成 PetzoldWindow8Controls。在应用程序创建后，将命名空间修改为 Petzold.Windows8.Controls。

Visual Studio 会自动为新类库创建一个名为 Class1.cs 的文件。我们可以删掉它。右击项目名称，选择“添加”|“新建”。在“新建项目”对话框中，选择“用户控件”。将其命名为 ColorItem。在这个示例中，我们稍微美化一下 ColorItem 的外观。

```
解决方案: ColorList3 | 项目: PetzoldWindows8Controls | 文件: ColorItem.xaml(片段)
<UserControl ...>
    <Grid>
        <Border BorderBrush="{StaticResource ApplicationForegroundThemeBrush}"
                BorderThickness="1"
                Width="336"
                Margin="6">
            <StackPanel Orientation="Horizontal">
                <Rectangle Name="rectangle"
                           Width="72"
                           Height="72"
                           Margin="6" />

                <StackPanel VerticalAlignment="Center">

                    <TextBlock Name="txtblkName"
                               FontSize="24" />

                    <TextBlock Name="txtblkRgb"
```

```
                        FontSize="18" />
            </StackPanel>
        </StackPanel>
    </Border>
  </Grid>
</UserControl>
```

该控件具有一个 Border，并显式地设置了 Width 属性和 Margin 属性。控件宽度是根据最长的颜色名(LightGoldenrodYellow)确定的。BorderBrush 被设置为一个预定义的画笔。该画笔在浅色主题下为黑色，在深色主题下为白色。主题是应用程序决定的(而非类库决定的，因为类库没有用于设置主题的 App 类)，因而此画笔的颜色由引用这个 ColorItem 的应用程序的主题决定。

创建一个名为 ColorList3 的应用程序项目。虽然该项目与类库项目位于同一解决方案内，但应用程序项目仍然需要显式地引用这个类库。在 ColorList3 项目的“引用”节点上右击，选择“添加引用”。在“引用管理器”对话框的左侧窗格中选择“解决方案”(表示所要引用的程序集位于同一解决方案内)，在右侧窗格中选择 PetzoldWindows8Controls，然后单击“确定”按钮。

将两个项目置于同一解决方案内有一个好处：如果 PetzoldWindows8Controls 库被更新，那么在生成 ColorList3 项目时，Visual Studio 也会自动生成 PetzoldWindows8Controls。

ColorList3 中的 MainPage.xaml 文件与前两个示例相同。代码隐藏类需要引入类库的命名空间，其他代码与 ColorList2 一致。

项目：ColorList3 | 文件：MainPage.xaml.cs

```
using System.Collections.Generic;
using System.Reflection;
using Windows.UI;
using Windows.UI.Xaml.Controls;
using Petzold.Windows8.Controls;

namespace ColorList3
{
    public sealed partial class MainPage : Page
    {
        public MainPage()
        {
            this.InitializeComponent();

            IEnumerable<PropertyInfo> properties =
                        typeof(Colors).GetTypeInfo().DeclaredProperties;

            foreach (PropertyInfo property in properties)
            {
                Color clr = (Color)property.GetValue(null);
                ColorItem clrItem = new ColorItem(property.Name, clr);
                stackPanel.Children.Add(clrItem);
            }
        }
    }
}
```

程序的运行效果如下图所示。

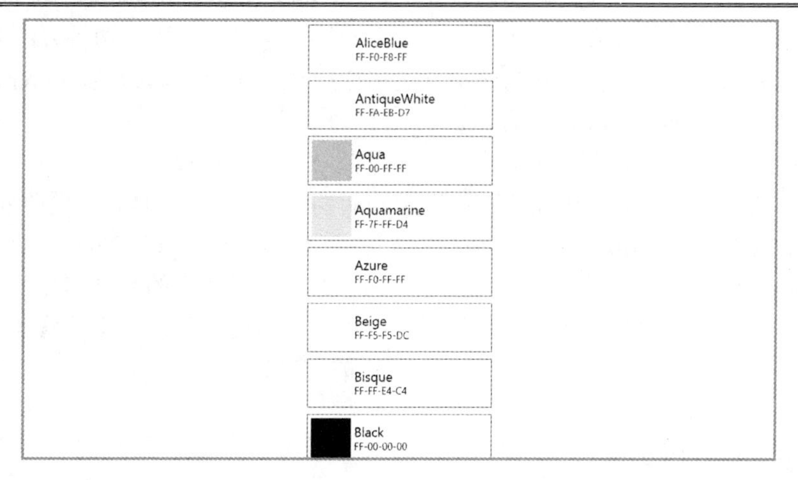

创建 Petzold.Windows8.Controls 类库时，我们从“添加新项目”对话框中选择的是“类库”。此外，我们也可以选择“Windows 运行时组件”(Windows Runtime Component)。对于 ColorList3 这个示例，选择哪种库并无差别，程序都能够正常运行。事实上，甚至可以在项目名称上右击，选择“属性”，在“应用程序”选项卡中，将“输出类型”由“类库”改为“Windows 运行时组件”。

两种项目类型的差别在于：通过“类库”创建的库只能通过 C#和 Visual Basic 来访问。但除了这两种应用程序，“Windows 运行时组件”还可以通过 C++和 JavaScript 来访问。“Windows 运行时组件”支持针对 Windows 8 应用程序的语言互操作性(language interoperability)。

凡事有利必有弊。“Windows 运行时组件”有一些“类库”没有的限制。例如，公共类必须是封闭的(sealed)。如果从 ColorItem 控件的定义中移除 sealed 关键字，那么该类将不能作为“Windows 运行时组件”的一部分。另一个主要限制与结构类型有关，即不能暴露任何非字段形式的公共成员。另外，组件 API 使用的数据类型也限制在 Windows Runtime 数据类型中。①

本书配套资源中，StackPanelWithScrolling 的 C++版本使用了一个用 C#编写的“Windows 运行时组件”ReflectionHelper，用于降低在 C++程序中使用反射的难度。第 15 章会介绍一种反向的实现，即通过用 C++编写的“Windows 运行时组件”使 C#程序能够访问 DirectX 类。

4.12 换行的替代方案

下面我们在另一个项目中引用 PetzoldWindows8Controls 库。引用该项目有 3 种方法。

方法 1：在库项目所在的解决方案中添加新的应用程序项目。对于本示例，可以在 ColorList3 解决方案中添加。这是最简单的方法。同一解决方案中的两个应用程序可能具有某种联系。

① 译注：“Windows 运行时组件”具有诸多限制，有关详情可阅读 MSDN 中标题为“用 C# 和 Visual Basic 创建 Windows 运行时组件”的文章，网址：http://msdn.microsoft.com/zh-cn/library/br230301。

这里并不打算采用这种方法，而是选择以下两种方法之一。下面两种方法需要创建新的解决方案和应用程序。我们将新的解决方案和应用程序项目命名为 ColorWrap，通过该项目来引用 PetzoldWindows8Controls 库。

方法 2：在 ColorWrap 项目中右击“引用”节点，然后选择“添加引用”。在“引用管理器”中，选择左侧窗格中的“浏览”节点，并单击右下角的“浏览”按钮。找到 PetzoldWindows8Controls.dll 文件所在目录(位于 ColorList3 解决方案 PetzoldWindows8Controls 项目的 bin/Debug 目录)，然后选择所需的 DLL 文件。

采用这种方式的前提是，库项目的代码已完成，不需要任何进一步的修改。这样引用的是 DLL，而非带有源代码的项目。在本书截稿前，这样做最大的弊端在于，Windows 8 不兼容包含 XAML 文件的库。

那么只有最后一种方法了。

方法 3：在 ColorWrap 解决方案中，右击解决方案名，选择“添加”|“现有项目”。在“添加现有项目”对话框中，找到 PetzoldWindows8Controls.csproj 文件。这是 ColorList3 解决方案中的项目文件，由 Visual Studio 维护。选择该文件。这个库项目不会被复制，目标解决方案只包含这个库项目的引用。Visual Studio 会决定在何种情况下重新生成这个库。

虽然 PetzoldWindows8Controls 项目成了 ColorWrap 解决方案的一部分，但 ColorWrap 应用程序项目仍然需要显式地引用这个库项目。在 ColorWrap 项目的“引用”节点上右击。从解决方案中选择这个库，正如在示例 ColorList3 中所做的那样。

分别加载 ColorList3 和 ColorWrap 解决方案的两个 Visual Studio 实例可能同时运行，并都允许对 PetzoldWindows8Controls 库进行修改。只要在修改过后保存或编译即可。如果同一文件在两个 Visual Studio 实例中均被打开，并通过其中的一个实例对该文件进行修改并保存，那么另一个实例则会在被激活后通知有更改发生。

经过一番准备后，下面我们将注意力转移到 ColorWrap 程序上。该程序将通过 VariableSizedWrapGrid 面板来显示各种颜色。尽管它取名“可变尺寸”，但它要求所有子元素具有相同尺寸，因而本示例显式地设置了 ColorItem 中 Border 元素的 Width 属性。

与 StackPanel 类似，VariableSizedWrapGrid 也具有 VariableSizedWrapGrid 属性，默认值为 Vertical。Children 集合中开始的几个子元素会显示为一列。与 StackPanel 不同的是，VariableSizedWrapGrid 支持若干列，与 Windows 8“开始”界面很像。这种方式要求 VariableSizedWrapGrid 可在水平方向上滚动，因而需要对 ScrollViewer 属性做相应设置。ColorWrap 项目的 XAML 文件如下所示。

```
项目: ColorWrap | 文件: MainPage.xaml(片段)
<Grid Background="{StaticResource ApplicationPageBackgroundThemeBrush}">
    <ScrollViewer HorizontalScrollBarVisibility="Visible"
                  VerticalScrollBarVisibility="Disabled">
        <VariableSizedWrapGrid Name="wrapPanel" />
    </ScrollViewer>
</Grid>
```

代码隐藏文件与上一个示例类似，只不过这里要将子元素添加至 wrapPanel。

```
项目: ColorWrap | 文件: MainPage.xaml.cs(片段)
public sealed partial class MainPage : Page
{
    public MainPage()
    {
```

```
        this.InitializeComponent();

        IEnumerable<PropertyInfo> properties = typeof(Colors).GetTypeInfo().DeclaredProperties;

        foreach (PropertyInfo property in properties)
        {
            Color clr = (Color)property.GetValue(null);
            ColorItem clrItem = new ColorItem(property.Name, clr);
            wrapPanel.Children.Add(clrItem);
        }
    }
}
```

程序的运行效果如下图所示。

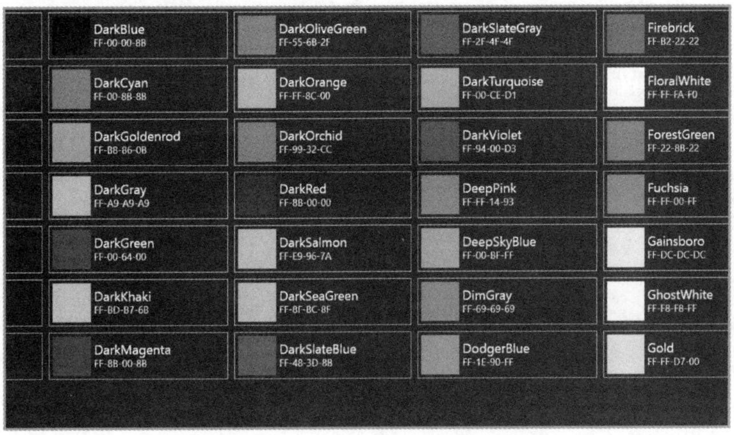

面板可水平滚动。

4.13　Canvas 与附加属性

本章要讨论的最后一个 Panel 派生类是 Canvas。从某种程度上讲，Canvas 是最“传统”的面板类型，因为它根据具体的像素坐标来定位元素。

子元素的哪个属性用来指定自己相对于 Canvas 的位置？在 UIElement 和 FrameworkElement 定义的诸多属性中，没有名为 Location、Position、X 或 Y 的属性。只有用于绘制矢量图的元素具有指定坐标位置的属性，其他元素没有。在 Windows Runtime 中，这样的属性没有太大意义，因为 Grid、StackPanel、WrapPanel 等布局元素不需要它们。我们通常不使用像素坐标来定位元素，但 Canvas 是个例外。

正因如此，Canvas 定义了用于相对于自身定位元素的属性。这些属性被称为“附加属性”(attached property)。附加属性由一个类(如 Canvas)定义，可设置到其他类的实例上(如 Canvas 的子元素)。设置附加属性的对象并不需要读取属性值，也不需要知道属性的来源。

下面让我们看看它是如何工作的。示例项目 TextOnCanvas 的 XAML 文件在默认的 Grid 中定义了一个 Canvas(也可以将 Grid 替换成 Canvas)。这个 Canvas 包含 3 个 TextBlock 元素。

项目：TextOnCanvas | 文件：MainPage.xaml(片段)

```
<Page
    x:Class="TextOnCanvas.MainPage"
    ...
    FontSize="48">
```

```
    <Grid Background="{StaticResource ApplicationPageBackgroundThemeBrush}">
        <Canvas>
            <TextBlock Text="Text on Canvas at (0, 0)"
                       Canvas.Left="0"
                       Canvas.Top="0" />

            <TextBlock Text="Text on Canvas at (200, 100)"
                       Canvas.Left="200"
                       Canvas.Top="100" />

            <TextBlock Text="Text on Canvas at (400, 200)"
                       Canvas.Left="400"
                       Canvas.Top="200" />
        </Canvas>
    </Grid>
</Page>
```

该程序的运行效果如下图所示(略显简陋)。

Text on Canvas at (0, 0)

Text on Canvas at (200, 100)

Text on Canvas at (400, 200)

请注意标记中的特殊语法。

```
<TextBlock Text="Text on Canvas at (200, 100)"
           Canvas.Left="200"
           Canvas.Top="100" />
```

从名称上看，Canvas.Left 和 Canvas.Top 特性是 Ganvas 类定义的，但被设置到了 Canvas 的子元素上以便对子元素进行定位。这种通过类名和属性名指定的 XAML 特性就是“附加属性”。

有趣的是，Canvas 实际并未定义任何名为 Left 和 Top 的属性。但它有定义名称与之相近的属性。

为了进一步理解附加属性的工作方式，可以看看在代码中如何设置它们。示例项目 TapAndShowPoint 的 XAML 文件在默认创建的 Grid 中定义了一个带有名称的 Canvas。

```
项目: TapAndShowPoint | 文件: MainPage.xaml(片段)
<Grid Background="{StaticResource ApplicationPageBackgroundThemeBrush}">
    <Canvas Name="canvas" />
</Grid>
```

其他的任务由代码隐藏文件完成(重写 OnTapped 方法)。该方法会创建点(实际为 Ellipse 元素)和 TextBlock，并将两者添加至 Canvas 被点击处。

```
项目: TapAndShowPoint | 文件: MainPage.xaml.cs(片段)
public sealed partial class MainPage : Page
{
    public MainPage()
    {
        this.InitializeComponent();
```

```
    }

    protected override void OnTapped(TappedRoutedEventArgs args)
    {
        Point pt = args.GetPosition(this);

        // Create dot
        Ellipse ellipse = new Ellipse
        {
            Width = 3,
            Height = 3,
            Fill = this.Foreground
        };

        Canvas.SetLeft(ellipse, pt.X);
        Canvas.SetTop(ellipse, pt.Y);
        canvas.Children.Add(ellipse);

        // Create text
        TextBlock txtblk = new TextBlock
        {
            Text = String.Format("({0})", pt),
            FontSize = 24,
        };

        Canvas.SetLeft(txtblk, pt.X);
        Canvas.SetTop(txtblk, pt.Y);
        canvas.Children.Add(txtblk);

        args.Handled = true;
        base.OnTapped(args);
    }
}
```

当屏幕被点击时，点和文本会出现在被点击处(如下图所示)。

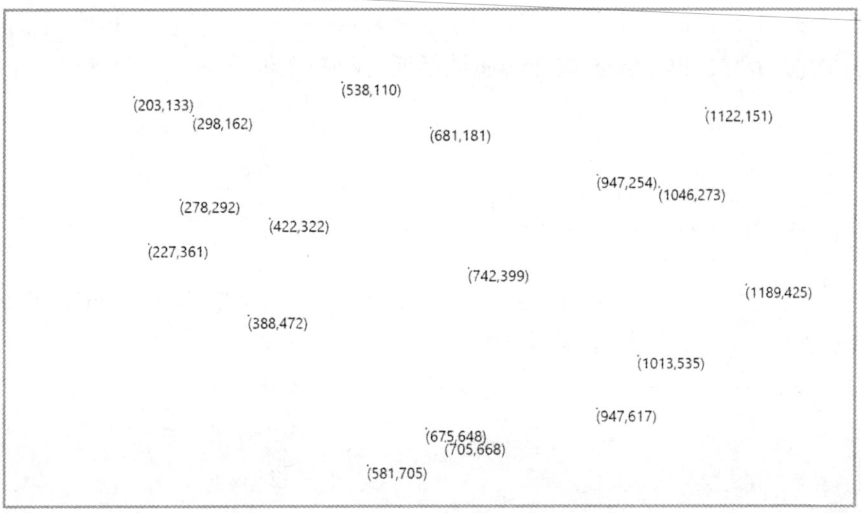

在代码中，点的位置是像下面这样指定的。在设置位置后，这个点会被添加到 Canvas 的 Children 集合中。

```
Canvas.SetLeft(ellipse, pt.X);
Canvas.SetTop(ellipse, pt.Y);
canvas.Children.Add(ellipse);
```

两个步骤的顺序无关紧要：可以先将元素添加到 Canvas，然后设置其位置。

Canvas.SetLeft 和 Canvas.SetTop 静态方法与 XAML 中的 Canvas.Left 和 Canvas.Top 特性的作用是相同的，都是指定子元素位置的坐标点。(示例中设置点坐标的方法有个小问题，这个问题随着 Ellipse 变大而越发明显。在调用 Canvas.SetLeft 和 Canvas.SetTop 时，应使用点的中心来表示被点击的位置，而不应使用 Ellipse 的左上角。若希望将 Ellipse 的中心坐标设置为变量 pt，可以将 pt.X 减去宽度除以 2 得到横坐标，将 pt.Y 减去高度除以 2 得到纵坐标。)

前面提到过，Canvas 并未具体定义 Left 和 Top 属性，而是定义了 SetLeft 和 SetTop 静态方法和类型为 DependencyProperty 的静态属性。如果 Canvas 类是用 C#编写的，这两个返回 DependencyProperty 对象的属性应该像这样定义。

```
public static DependencyProperty LeftProperty { get; }
public static DependencyProperty TopProperty { get; }
```

正如后文要介绍的，这些特殊类型的依赖属性可以设置到 Canvas 之外的元素上。

有一点值得注意。示例项目 TapAndShowPoint 是像下面这样调用 Canvas.SetLeft 和 Canvas.SetTop 静态方法的。

```
Canvas.SetLeft(ellipse, pt.X);
Canvas.SetTop(ellipse, pt.Y);
```

除此以外，还有一种方法(也是合法有效的，并且完全等价)。那就是像下面这样调用子元素的 SetValue，并引用 Canvas 定义的静态属性。

```
ellipse.SetValue(Canvas.LeftProperty, pt.X);
ellipse.SetValue(Canvas.TopProperty, pt.Y);
```

这种方式与调用 Canvas.SetLeft 和 Canvas.SetTop 是完全等价的，选择哪种取决于开发者的个人偏好。

前文介绍过 SetValue 方法。该方法是 DependencyObject 定义的，Windows Runtime 中的许多类都继承于它。FontSize 这样的属性是通过将静态依赖属性传给 SetValue 方法实现的。

```
public double FontSize
{
   set { SetValue(FontSizeProperty, value); }
   get { return (double)GetValue(FontSizeProperty); }
}
```

事实上，虽然本人并未读过 Canvas 类内部代码，但可以确定的是，Canvas 中的静态方法 SetLeft 和 SetTop 与以下代码是等价的。

```
public static void SetLeft(DependencyObject element, double value)
{
   element.SetValue(LeftProperty, value);
}
public static void SetTop(DependencyObject element, double value)
{
   element.SetValue(TopProperty, value);
}
```

这两个方法清楚地展示了一点：依赖属性实际被设置到子元素上，而非 Canvas 本身。

Canvas 还定义了 GetLeft 和 GetTop 方法，方式与上述附加属性相同。

```
public static double GetLeft(DependencyObject element)
{
   return (double)element.GetValue(LeftProperty);
```

```
}
public static double GetTop(DependencyObject element)
{
   return (double)element.GetValue(TopProperty);
}
```

Canvas 类内部通过这两个方法来获取每个子元素的 Left 和 Top 属性设置，从而在布局过程中将子元素呈现在指定位置。

静态方法 SetLeft、SetTop、GetLeft 和 GetTop 的实现暗示了依赖属性系统用到了某种字典。SetValue 方法允许 Canvas.LeftProperty 这样的附加属性存储在某个元素中，而该元素不必预先知道该属性的存在，也无需了解它存在的目的。Canvas 可以通过获取这些属性的值来决定子元素相对于自身的位置。

4.14　Z-Index

Canvas 还有一个附加属性，可以在 XAML 中通过 Canvas.ZIndex 特性设置。属性名 ZIndex 中的“Z”代表三维坐标系统的一个维度，方向垂直于屏幕向外，指向用户。

如果元素发生重叠，则一般会按照元素出现在可视树中的顺序显示，即 Children 集合前面的元素会被后面的元素遮挡。看一下这段标记。

```
<Grid>
   <TextBlock Text="Blue Text" Foreground="Blue" FontSize="96" />
   <TextBlock Text="Red Text" Foreground="Red" FontSize="96" />
</Grid>
```

红色的文本会遮挡蓝色的文本。

我们可以通过附加属性 Canvas.ZIndex 来重写此行为。有趣的是，它对所有面板元素均有效，而不仅仅针对 Canvas。为使颜色文本显示在红色文本之上，可以这样做。

```
<Grid>
   <TextBlock Text="Blue Text" Foreground="Blue" FontSize="96" Canvas.ZIndex="1" />
   <TextBlock Text="Red Text" Foreground="Red" FontSize="96" Canvas.ZIndex="0" />
</Grid>
```

4.15　使用 Canvas 的注意事项

本章前面有关布局的介绍有很多 Canvas 都不适用。Canvas 的布局是子驱动的。Canvas 会为子元素分配逻辑上无穷的空间。这意味着，每个子元素会以其自然尺寸呈现，这个尺寸也就是元素所占据的空间。Canvas 子元素的 HorizontalAlignment 和 VerticalAlignment 设置会被忽略。类似地，如果将 Image 作为 Canvas 的子元素，那么 Image 的 Stretch 属性也会被忽略，因为 Image 总会以图片的原像素尺寸显示。如果不显式地对 Rectangle 和 Ellipse 设置宽度和高度，那么它们将收缩至消失。

虽然 HorizontalAlignment 和 VerticalAlignment 对 Canvas 的子元素无效，但对 Canvas 本身是有效的。对于其他面板，如果将这两个对齐属性设置为 Stretch 以外的值，这些面板则会尽可能地收缩并紧紧包裹子元素，但 Canvas 不同。如果将 Canvas 的 HorizontalAlignment 和 VerticalAlignment 设置为 Stretch 以外的值，那么不论子元素多大，Canvas 都会收缩至消失。

虽然 Canvas 会收缩至零尺寸，但子元素的显示并不受影响。从概念上讲，Canvas 更像是参照点，而不是容器，Canvas 子元素的大小会在布局系统中被忽略。

我们可以对 Canvas 的这种特性加以利用。例如，在一个较小的 Grid 中显示一个较大的 TextBlock。

```
<Grid Width="200" Height="100">
   <TextBlock Text="Text in a Small Grid" FontSize="144" />
</Grid>
```

TextBlock 超出 Grid 之外的部分会被截掉。当然，可以增大 Grid，但某些情况下 Grid 的大小受制约(如受子元素的影响)。如何在这种情况下仍然使这个 TextBlock 与其他元素对齐，而不被 Grid 截断呢？

在这种情况下，最好的办法是在 Grid 中添加一个 Canvas，然后将 TextBlock 置于 Canvas 中。

```
<Grid Width="200" Height="100">
   <Canvas>
      <TextBlock Text="Text in a Small Grid" FontSize="144" />
   </Canvas>
</Grid>
```

虽然 Canvas 会被 Grid 截断，但作为 Canvas 子元素的 TextBlock 不会。这个 TextBlock 会像人们期望那样显示(仍与 Grid 的左上角对齐)，不被截断，但会延伸到常规布局之外。

在类似的场景中，这种方法简单有效。

第 5 章 控件与交互

本书第 1 章介绍了 FrameworkElement 的派生类和 Control 的派生类之间的区别。为避免混淆，我们一般将 FrameworkElement 的派生类称为“元素”(如 TextBlock 和 Image)，但这里有必要对其做进一步解释。

本章的标题暗示元素主要用于显示，而控件注重交互，但实际并非一成不变。UIElement 类定义了所有用户输入事件，涉及来自触摸屏、鼠标、触控笔和键盘的输入。这说明元素和控件都支持丰富的交互方式。

元素本身在布局、样式和数据绑定方面并不见长。为此，FrameworkElement 类定义了包括 Width、Height、HorizontalAlignment、VerticalAlignment 和 Margin 在内的许多布局属性，还定义了用于指定样式的 Style 属性和实现数据绑定的 SetBinding 方法。

5.1 Control 的特别之处

从视觉上和功能上讲，FrameworkElement 的派生类是基本元素(类似原子)，而 Control 的派生类是由这些基本元素组合而成的产物(类似分子)。例如，Button 实际由 Border 和 TextBlock 构成(在多数情况下如此)。Slider 由一对带有 Thumb 控件的 Rectangle 元素构成，而 Thumb 本身也是通过 Rectangle 实现的 Control。可视内容由文本、位图和矢量图形组合而成的元素大多直接或间接地派生自 Control。

由于 Control 的派生类需要由其他元素进行组合，所以 Template 成为 Control 定义最重要的属性之一。正如第 11 章所要演示的，该属性允许我们通过自定义的可视树来重新定义控件的外观。重新定义 Button 的外观是有意义的。例如，为将 Button 放在应用栏(app bar)中，我们希望该控件是圆的，而不是矩形的。相对而言，重新定义 TextBlock 或 Image 的外观毫无意义，因为除了添加额外的文本或图片外，其他没有什么可做的。如果要为 TextBlock 或 Image 增添些什么，则需要定义 Control，因为这需要用基本元素构建可视树来实现。

虽然可以从 FrameworkElement 派生来创建自定义控件，但之后会立刻发现我们对这个派生类无能为力，甚至无法为其添加可视元素。但若从 Control 类派生，则可以使用 XAML 来定义可视树，从而为自定义控件建立默认外观。

Control 类定义了许多本身并不需要的属性。这些属性是为其派生类提供的。TextBlock 用到了其中的 CharacterSpacing、FontFamily、FontSize、FontStretch、FontStyle、FontWeight 和 Foreground 属性，而 Border 用到了其中的 Background、BorderBrush、BorderThickness 和 Padding 属性。并非 Control 的每个派生类都要用到文本或边框，但如果在创建新控件或新模板时需要，则可以方便地利用相关的属性。Control 类还提供了两个属性来定义控件外观，分别是 HorizontalContentAlignment 和 VerticalContentAlignment。

Control 的派生类往往还会定义额外的属性和事件。控件通常会处理来自指针设备、鼠标、触控笔和键盘的用户输入，并将其转换为更高级的事件。例如，ButtonBase 类(所有按钮的基类)定义了 Click 事件，Slider 定义了用于通知其 Value 属性发生改变的 ValueChanged 事件，TextBox 定义了用于通知其 Text 属性发生改变的 TextChanged 事件。

在实践当中，Control 的派生类更多用来与用户交互，正如本章标题所要表达的。为了更方便地获取用户输入，Control 提供了受保护的虚方法，这些虚方法与 UIElement 定义的用户输入事件一一对应。例如，对应于 UIElement 定义的 Tapped 事件，Control 定义了受保护的虚方法 OnTapped。Control 还定义了 IsEnabled 属性，可以在控件不适用的情况下停止接收用户输入，该属性在改变时会引发 IsEnabledChanged 事件，它是 Control 定义的唯一的公共事件。

Windows 8 中也有“输入焦点”(input focus)的概念。在控件获得输入焦点后，用户希望该控件能够收到大多数键盘事件。(有些键盘事件与输入焦点无关，如 Windows 键。)为此，Control 定义了 Focus 方法以及虚方法 OnGotFocus 和 OnLostFocus。

有一个与输入焦点相关的功能，即通过 Tab 键在不同控件间切换焦点。Control 通过 IsTabStop、TabIndex 和 TabNavigation 属性使该功能成为可能。

Windows.UI.Xaml.Controls 命名空间中有许多 Control 的派生类，Windows.UI.Xaml.Controls.Primitives 命名空间中也有几个。后一个命名空间中的控件用于构建其他控件，但这只是一种建议，并非约束。

Control 的大多数派生类都直接派生自 Control，但有 4 个较为重要的 Control 派生类建立了自己的类别。这些类的关系如下所示。

```
Object
  DependencyObject
    UIElement
      FrameworkElement
        Control
          ContentControl
          ItemsControl
          RangeBase
          UserControl
```

很多重要的类(如 Button、ScrollViewer 和 AppBar)派生自 ContentControl，但 ContentControl 类只不过定义了一个名为 Content 的属性，其类型为 object。例如，若要改变 Button 外观，可以通过 Content 属性进行设置，一般可将其设置为文本或图片，但也可以设置为包含其他内容的面板。

注意，ContentControl 的 Content 属性的类型为 object，而非 UIElement。这样设计不无道理。我们几乎可以将任何类型的对象作为 Button 的内容，甚至可以通过模板(可视树的形式)来告知 Button 如何显示其内容。我们对于 Button 不常用到模板，而对于 ItemsControl 的派生类却经常用到。第 11 章会介绍如何定义内容模板。

ItemsControl 的派生类用于显示项目集合，其中包括常见的 ListBox 和 ComboBox 以及 Windows 8 新控件 FlipView、GridView 和 ListView。第 11 章会一并介绍。

创建自定义控件的方式有很多种。最简单的是定义控件的 Style，但较复杂的外观需要用到模板。在某些情况下，我们可以从现有控件派生新控件，并为其添加额外的功能；如果 ContentControl 或 ItemsControl 提供了所需的功能，也可以考虑从这两个类派生。

创建自定义控件最常见的方法是从 UserControl 派生。虽然这种方法不适合实现作为商业产品的自定义控件库，但在应用程序中使用这种技术却不失为一种好的选择。

5.2　用于设置范围的 Slider 控件

前文展示的 Control 相关类的继承层次结构中还有一个重要的类没有介绍，它就是 RangeBase。该类有 3 个派生类：ProgressBar、ScrollBar 和 Slider。

这 3 个类中哪个最为独特？显然是 ProgressBar。ProgressBar 只用到了 RangeBase 的几个属性：Minimum、Maximum、SmallChange、LargeChange 和 Value。RangeBase 的 Value 属性的类型为 double，取值范围通过 Minimum(最小值)和 Maximum(最大值)属性来设置。ScrollBar 和 Slider 的 Value 属性会在用户操纵控件时发生变化，而 ProgressBar 的 Value 属性需要程序设置，从而提示用户耗时操作的进度。

ProgressBar 有一种非确定性模式。在该模式下，ProgressBar 会表现为一串依次穿越屏幕的点。ProgressRing 与之类似，只不过点会被限制在圆环中。

在 Windows 约 25 个年头的演变中，ScrollBar 一度在控件库中占据重要地位，如今却只能在 ScrollViewer 控件中见到了。单独实例化 Windows Runtime 版本的 ScrollBar 界面上不会显示任何内容。我们必须为其提供模板。与 RangeBase 一样，ScrollBar 也定义在 Windows.UI.Xaml.Controls.Primitives 命名空间中，表示该控件并不是供应用程序开发者平时使用的。

对于几乎所有选择数值范围的需求，ScrollBar 均已被 Slider 取代。由于具有支持触摸屏的界面，Slider 变得更好用。Slider 在默认配置下是没有箭头的。在触摸 Slider 或滑动手指/鼠标到某一点时，该控件会计算出该点所对应的值。

以编程方式或通过用户操作都可以修改 Slider 的 Value 属性。要想在 Value 属性变化时获得通知，可以订阅 ValueChanged 事件(正如示例项目 SliderEvents 所展示的)。

```
项目: SliderEvents | 文件: MainPage.xaml(片段)
<Grid Background="{StaticResource ApplicationPageBackgroundThemeBrush}">
    <StackPanel>
        <Slider ValueChanged="OnSliderValueChanged" />

        <TextBlock HorizontalAlignment="Center"
                   FontSize="48" />

        <Slider ValueChanged="OnSliderValueChanged" />

        <TextBlock HorizontalAlignment="Center"
                   FontSize="48" />
    </StackPanel>
</Grid>
```

这里的两个 Slider 控件共用一个事件处理程序。这个示例的功能很简单，即将每个 Slider 的当前状态显示在各自下面的 TextBlock 中。若不为这些控件分配名称，实现起来就可能有些难度。为找到目标元素，事件处理程序有两个假设：一是 Slider 的父元素是 Panel；二是在同一个 Panel 中，Slider 的下一个子元素为 TextBlock。

```
项目: SliderEvents | 文件: MainPage.xaml.cs(片段)
void OnSliderValueChanged(object sender, RangeBaseValueChangedEventArgs args)
{
    Slider slider = sender as Slider;
    Panel parentPanel = slider.Parent as Panel;
    int childIndex = parentPanel.Children.IndexOf(slider);
    TextBlock txtblk = parentPanel.Children[childIndex + 1] as TextBlock;
    txtblk.Text = args.NewValue.ToString();
}
```

这里用到了多种访问可视树中元素的方法，这可能让人觉得有些复杂。最后一步，TextBlock 的 Text 属性被设置为事件参数的 NewValue 值(需要转换为字符串)。这一步也可以使用 Slider 的 Value 属性。

```
txtblk.Text = slider.Value.ToString();
```

虽然 RangeBaseValueChangedEventArgs 派生自 RoutedEvent，但 ValueChanged 并不是路由事件。该事件不会沿可视树从分支传播到主干。参数 sender 引用的总是 Slider 对象，而该事件的 OriginalSource 属性总是 null。

运行该程序后会发现，TextBlock 元素最初什么都不显示。在 Value 属性从 0(默认值)变为其他值之前，ValueChanged 事件是不会被引发的。

在触摸或用鼠标单击 Slider 的某个位置时，Value 属性会直接变为该位置所对应的值。可以用手指或鼠标指针拖动滑块来改变 Slider 的值。在操作 Slider 的过程中，我们会看到被选定的值在 0 到 100 之间变化(如下图所示)。

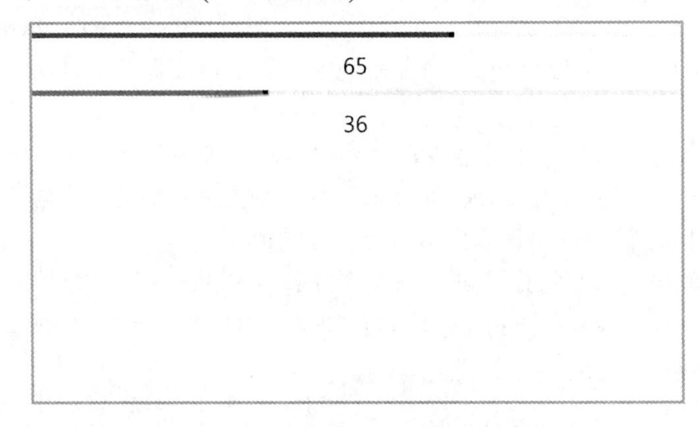

这个范围是 Minimum 和 Maximum 属性的默认值决定的，最小值和最大值分别为 0 和 100。虽然 Value 属性为 double 类型，但由于 StepFrequency 属性的默认值为 1，因而最终的 Value 总是整数。

默认情况下，Slider 是横向显示的。我们可以通过 Orientation 属性将其更改为纵向的。Slider 的宽度不能通过设置属性来更改，而只能使用模板来重新定义。该控件的布局宽度要比可视宽度大一些。从布局角度讲，横向 Slider 的默认高度为 60 像素，纵向 Slider 的默认宽度为 45 像素；从使用角度讲，这种尺寸设计足以消除触摸误差所带来的影响。

程序运行时按 Tab 键可以更改键盘的输入焦点，在两个 Slider 间切换，然后可以使用方向键修改选定的值。按 Home 和 End 键可以分别将值调至最小和最大。

下面要介绍的示例项目 SliderBindings 是上一示例的另一种形式，即所有更新逻辑从代码隐藏文件被转移至 XAML。StackPanel 中有 3 个 Slider 控件，每个 Slider 下面都有一个

TextBlock 元素。针对 TextBlock 的隐式样式可以避免过多的标记重复。

```
项目: SliderBindings | 文件: MainPage.xaml(片段)
<Grid Background="{StaticResource ApplicationPageBackgroundThemeBrush}">
   <Grid.Resources>
      <Style TargetType="TextBlock">
         <Setter Property="FontSize" Value="48" />
         <Setter Property="HorizontalAlignment" Value="Center" />
      </Style>
   </Grid.Resources>

   <StackPanel>
      <Slider Name="slider1" />

      <TextBlock Text="{Binding ElementName=slider1, Path=Value}" />

      <Slider Name="slider2"
              IsDirectionReversed="True"
              StepFrequency="0.01" />

      <TextBlock Text="{Binding ElementName=slider2, Path=Value}" />

      <Slider Name="slider3"
              Minimum="-1"
              Maximum="1"
              StepFrequency="0.01"
              SmallChange="0.01"
              LargeChange="0.1" />

      <TextBlock Text="{Binding ElementName=slider3, Path=Value}" />
   </StackPanel>
</Grid>
```

数据绑定主动获取源属性的初始值，而不会等待某些事件(如 ValueChanged)。在用户操作 Slider 的过程中，数据绑定能够跟踪值的变化(参见下图)。

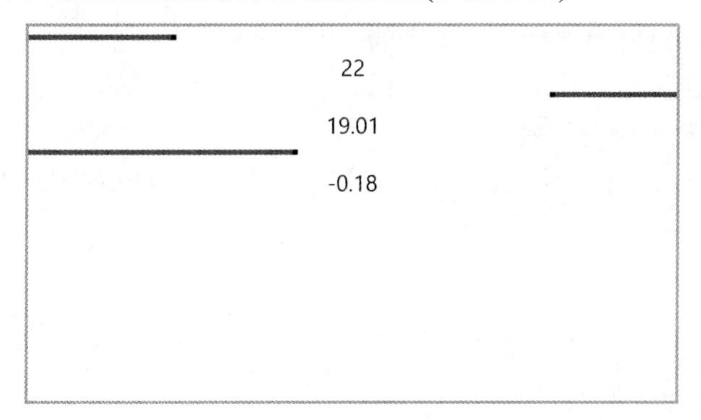

第二个 Slider 的 StepFrequency 属性被设置为 0.01，IsDirectionReversed 属性被设置为 true，因而滑块在最右端时 Slider 的值为最小(零)。实践当中，横向 Slider 的 IsDirectionReversed 很少被设置为 true，该设置更适合纵向的 Slider。默认情况下，纵向 Slider 在滑块位于最下端时值为最小，但在某些情况下我们希望值为最大。

若用键盘方向键操作第二个 Slider，我们会发现其步长为 1，而非 StepFrequency 所设置的 0.01。键盘操作的步长由 SmallChange 属性控制，其默认值为 1。

第三个 Slider 的范围从-1 到 1。该 Slider 显示之初，滑块位于正中央，默认值为 0。StepFrequency 和 SmallChange 属性均为 0.01，LargeChange 属性为 0.1，但目前尚未发现通过鼠标或键盘以 LargeChange 设置触发跳跃的方法。

Slider 还定义了 TickFrequency 和 TickPlacement 属性用于显示刻度。如果设置 Slider 的 Background 和 Foreground 属性，Slider 会将 Foreground 用作滑道最小值一侧的颜色，将 Background 用作滑道最大值一侧的颜色，但在鼠标悬停或操作过程中，Slider 会呈现默认颜色。

在本书创建更多Slider之前，有必要优化一下其布局。下面让我们进一步认识一下Grid。

5.3 关于 Grid

您或许已对 Grid 司空见惯了，因为本书几乎每个示例都有它的身影。虽然如此，但不对它的介绍并不深入。本书后面的许多示例不会使用单格模式的 Grid，而会为其添加行和列。

Grid 与 HTML 中的 table 有许多共通之处，但不尽相同。对于 Grid，我们不能单独定义每个单元格的边框和边距。Grid 只用于布局。任何显示效果都取决于其父元素或子元素。例如，Grid 可以包含在 Border 中，而 Border 也可以作为 Grid 单元格的内容。

Grid 的行数和列数必须显式指定，无法通过子元素的数量自动确定。Grid 的每个子元素可以属于指定的某个单元格(行和列的交汇处)，也可以跨越多行和多列。

虽然我们能够以编程方式在运行时修改行数和列数，但在实践中很少这样做。在大多数情况下，我们都是在 XAML 文件中定义固定的行数和列数。定义行和列需要将 RowDefinition 和 ColumnDefinition 对象分别添加到 Grid 的 RowDefinitions 和 ColumnDefinitions 集合中。

行和列的宽度通过以卜 3 种方式指定：

- 以像素值的形式显式指定行和列的宽度
- Auto，由子元素的尺寸决定
- 星号(*)，按比例分配剩余空间

在 XAML 中，可以使用“属性元素语法”来填充 RowDefinitions 和 ColumnDefinitions 集合(如下所示)。

```
<Grid>
   <Grid.RowDefinitions>
      <RowDefinition Height="Auto" />
      <RowDefinition Height="55" />
      <RowDefinition Height="*" />
   </Grid.RowDefinitions>

   <Grid.ColumnDefinitions>
      <ColumnDefinition Width="Auto" />
      <ColumnDefinition Width="10*" />
      <ColumnDefinition Width="20*" />
      <ColumnDefinition Width="Auto" />
   </Grid.ColumnDefinitions>

   <!-- Children go here -->

</Grid>
```

请注意，Grid 集合属性的名称 RowDefinitions 和 ColumnDefinitions 为复数，而对象的类型名 RowDefinition 和 ColumnDefinition 为单数。若不设置 RowDefinitions，Grid 则为单

行；若不设置 ColumnDefinitions，则为单列。

上述标记通过 3 种方式定义了 3 行 4 列的 Grid。纯数字表示按像素设置宽度(或高度)。这种将行高或列宽设置为具体数值的方式不如另外两种方式常用。

Auto 设置会使 Grid 根据子元素决定具体尺寸。计算得出的行高(或列宽)取决于该行(列)中最高(宽)的子元素。

与 HTML 类似，星号会使 Grid 自动分配剩余可用空间。对于前文定义的 Grid，第三行的高度是通过将 Grid 的总高度减去第一行和第二行的高度计算得出。第二列和第三列的宽度为 Grid 的总宽度减去第一列和第四列的宽度。星号前面的数字为权值。也就是说，第三列的宽度是第二列宽度的两倍。

需要注意的是，仅当 Grid 为父驱动时星号才适用。举例来说，若将上述 Grid 作为纵向 StackPanel 的子元素，StackPanel 就会为 Grid 分配逻辑上无限的高度。那么 Grid 如何为中间一行分配无限的高度呢？这是不会发生的。在这种情况下，星号会按 Auto 设置进行处理。

类似地，若将 Grid 作为 Canvas 的子元素，且不显式设置 Grid 的 Height 和 Width 属性，那么所有星号都会按 Auto 设置进行处理。将 HorizontalAlignment 和 VerticalAlignment 属性改为非默认值时，这种情况也会发生。对于前文展示的 Grid，由于子元素的原因，第二列可能比第三列还要宽。

然而，如果没有 RowDefinition 对象使用星号设置，Grid 的高度就由子元素驱动。我们可以将这样的 Grid 置于 Canvas 或纵向的 StackPanel 中，也可以修改 VerticalAlignment，这样做都不会发生莫名其妙的问题。

RowDefinition 的 Height 属性和 ColumnDefinition 的 Width 属性的类型均为 GridLength。该类型为 Windows.UI.Xaml 命名空间中定义的结构。我们可以通过它在代码中通过 Auto 或星号来指定尺寸。RowDefinition 还定义了 MinHeight 和 MaxHeight 属性，ColumnDefinition 也相应地定义了 MinWidth 和 MaxWidth 属性。这些属性的类型均为 double，用于以像素为单位设置最小或最大尺寸。通过 RowDefinition 的 ActualHeight 属性和 ColumnDefinition 的 ActualWidth 属性可以获得行和列的实际尺寸。

Grid 还定义了 4 个可以在子元素上设置的附加属性：Grid.Row 和 Grid.Column 的默认值为 0，Grid.RowSpan 和 Grid.ColumnSpan 的默认值为 1。我们可以通过前两个属性来指定子元素所从属的单元格以及子元素所跨越的行列数。一个单元格可以容纳多个元素。

Grid 可以在单元格中嵌套其他 Grid 和面板，但嵌套面板可能会降低布局效果。如果深层嵌套的元素在动画过程中改变自身尺寸，或者子元素被频繁添加至 Children 集合，或者从中移除，则要特别注意。应尽量避免以显示器刷新频率来不断计算页面布局！

已发表的第一篇有关 Windows 编程的文章介绍了一个名为 WHATSIZE 的程序。本书第 3 章展示了一个 Windows 8 版本的 WHATSIZE。第 3 篇介绍 Windows 编程的文章发表于 1987 年 5 月的 Microsoft Systems Journal。该文章展示了一个名为 COLORSCR(color scroll)的程序。该程序在 Windows 2 beta 版中的运行效果如下图所示。

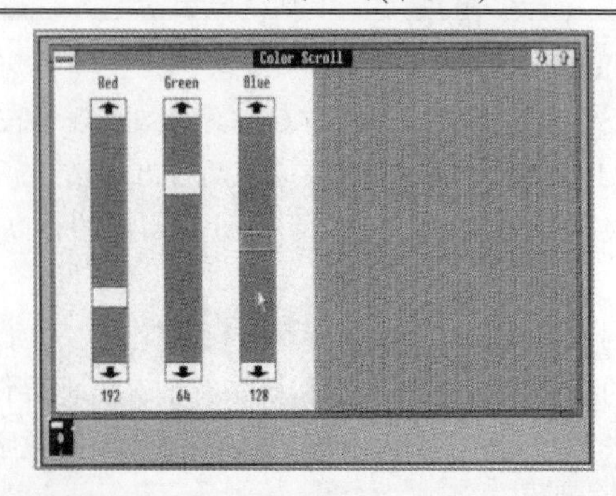

三个滚动条分别对应红绿蓝三色，程序界面右侧是这三种颜色混合后的颜色。当时的显示设备无法显示全彩色。对于设备无法呈现的颜色，则会采用抖动(dithering)方式显示。三种色值会分别显示在三个滚动条的下面。此程序采用了动态布局方式(计算量较大)，滚动条的宽度会随窗口尺寸的改变而改变。

Grid 非常适合用来实现这种布局。示例项目 SimpleColorScroll 中有 6 个 TextBlock 和 3 个 Slider。为便于样式设置，XAML 文件定义了两种隐式样式。

```
项目: SimpleColorScroll | 文件: MainPage.xaml(片段)
<Page.Resources>
    <Style TargetType="TextBlock">
        <Setter Property="Text" Value="00" />
        <Setter Property="FontSize" Value="24" />
        <Setter Property="HorizontalAlignment" Value="Center" />
        <Setter Property="Margin" Value="0 12" />
    </Style>

    <Style TargetType="Slider">
        <Setter Property="Orientation" Value="Vertical" />
        <Setter Property="IsDirectionReversed" Value="True" />
        <Setter Property="Maximum" Value="255" />
        <Setter Property="HorizontalAlignment" Value="Center" />
    </Style>
</Page.Resources>
```

色值最好以十六进制显示，因而 TextBlock 的 Style 将 Text 属性初始化为“00”，对应 Slider 在最小值位置的十六进制值。

Grid 具有三行四列。这三行分别用于容纳 Slider 和上下两个 TextBlock。请注意，左侧连续的三列均是一倍宽，而第四列为三倍宽。

```
项目: SimpleColorScroll | 文件: MainPage.xaml(片段)
<Grid Background="{StaticResource ApplicationPageBackgroundThemeBrush}">
    <Grid.ColumnDefinitions>
        <ColumnDefinition Width="*" />
        <ColumnDefinition Width="*" />
        <ColumnDefinition Width="*" />
        <ColumnDefinition Width="3*" />
    </Grid.ColumnDefinitions>

    <Grid.RowDefinitions>
        <RowDefinition Height="Auto" />
        <RowDefinition Height="*" />
        <RowDefinition Height="Auto" />
    </Grid.RowDefinitions>

    ...

</Grid>
```

这个 XAML 文件其余的标记为 Grid 实例化 10 个子元素并且都设置了附加属性 Grid.Row 和 Grid.Column(虽然设置为 0 是不必要的)。在指定 Grid 子元素特性时，本人倾向于将能够快速识别元素的特性(如 Name 或 Text)放在首位，后面跟附加属性。

项目：SimpleColorScroll | 文件：MainPage.xaml(片段)

```
<Grid Background="{StaticResource ApplicationPageBackgroundThemeBrush}">

    ...

    <!-- Red -->
    <TextBlock Text="Red"
               Grid.Column="0"
               Grid.Row="0"
               Foreground="Red" />

    <Slider Name="redSlider"
            Grid.Column="0"
            Grid.Row="1"
            Foreground="Red"
            ValueChanged="OnSliderValueChanged" />

    <TextBlock Name="redValue"
               Grid.Column="0"
               Grid.Row="2"
               Foreground="Red" />

    <!-- Green -->
    <TextBlock Text="Green"
               Grid.Column="1"
               Grid.Row="0"
               Foreground="Green" />

    <Slider Name="greenSlider"
            Grid.Column="1"
            Grid.Row="1"
            Foreground="Green"
            ValueChanged="OnSliderValueChanged" />

    <TextBlock Name="greenValue"
               Grid.Column="1"
               Grid.Row="2"
               Foreground="Green" />

    <!-- Blue -->
    <TextBlock Text="Blue"
               Grid.Column="2"
               Grid.Row="0"
               Foreground="Blue" />

    <Slider Name="blueSlider"
            Grid.Column="2"
            Grid.Row="1"
            Foreground="Blue"
            ValueChanged="OnSliderValueChanged" />

    <TextBlock Name="blueValue"
               Grid.Column="2"
               Grid.Row="2"
               Foreground="Blue" />

    <!-- Result -->
    <Rectangle Grid.Column="3"
               Grid.Row="0"
               Grid.RowSpan="3">
        <Rectangle.Fill>
            <SolidColorBrush x:Name="brushResult"
```

```
                    Color="Black" />
    </Rectangle.Fill>
  </Rectangle>
</Grid>
```

每组 TextBlock 和 Slider 元素的 Foreground 属性被设置为各自所代表的颜色。

Grid 底部的 Rectangle 有一个附加属性 Grid.RowSpan，值为 3 表示该元素需要跨三行。SolidColorBrush 被设置为 Black，该颜色正是三个 Slider 初始值所对应的颜色。除了在 XAML 文件中进行初始化，也可使用代码隐藏类的构造函数(或 Loaded 事件)来完成此任务。

三个 Slider 控件共用代码隐藏文件中的同一个 ValueChanged 事件处理程序。

```
项目：SimpleColorScroll | 文件：MainPage.xaml.cs(片段)
public sealed partial class MainPage : Page
{
    public MainPage()
    {
        this.InitializeComponent();
    }

    void OnSliderValueChanged(object sender, RangeBaseValueChangedEventArgs args)
    {
        byte r = (byte)redSlider.Value;
        byte g = (byte)greenSlider.Value;
        byte b = (byte)blueSlider.Value;

        redValue.Text = r.ToString("X2");
        greenValue.Text = g.ToString("X2");
        blueValue.Text = b.ToString("X2");

        brushResult.Color = Color.FromArgb(255, r, g, b);
    }
}
```

这个程序本可以通过对 sender 参数进行类型转换来获得实际引发事件的 Slider 控件，并通过 RangeBaseValueChangedEventArgs 对象获取新选定的值。但不论哪个 Slider 的值发生变化，此处理程序都需要根据三个选定的值来重新创建 Color 值。这段代码唯一显得重复的操作是在任意 Slider 发生变化时都要设置 3 个文本。但为解决这个问题，需要找到引发事件的 Slider 所对应的 TextBlock。就单纯演示 Slider 控件而言，这样做似乎有些喧宾夺主。

如图所示，此程序可合成 16 777 216 种颜色。

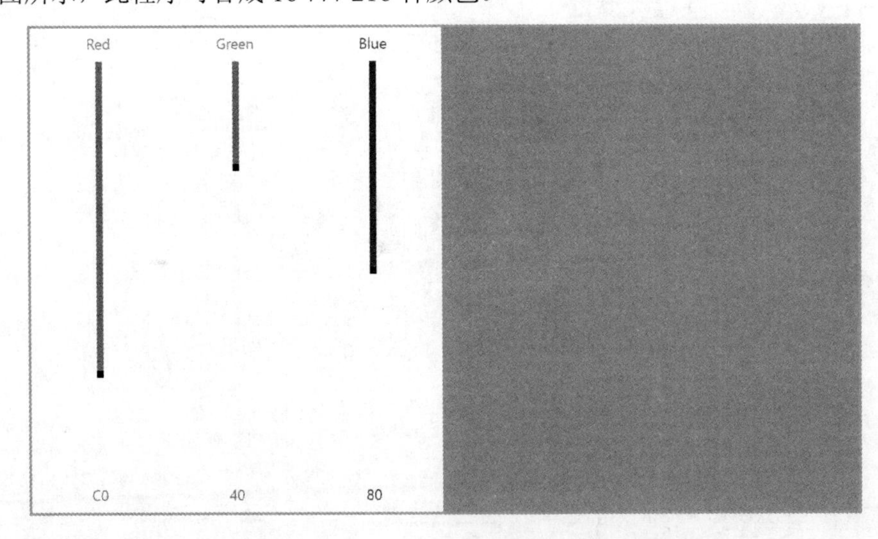

5.4 屏幕方向与比例

如果在平板电脑上运行 SimpleColorScroll 并转动屏幕使其切换至纵向视图，布局则会显得拥挤不堪。而在横向视图下将程序切换至辅屏视图会导致文本标签的重叠。为解决此类问题，我们可以在代码隐藏文件中添加一些逻辑，以便根据显示方向和比例来调整布局。

对于这个特定的程序，为方便布局的调整，可以将原来的 Grid 一分为二，使两者能够嵌套。内层的 Grid 有三行三列，用于容纳 TextBlock 元素和 Slider 控件。外层 Grid 包含两个子元素，分别为 Grid 和 Rectangle。在横向视图下，外层 Grid 只有两列；而在纵向视图下，它只有两行。

示例项目 OrientableColorScroll 延续了 SimpleColorScroll 项目在 XAML 中的 Style 定义。外层 Grid 的标记如下所示。

项目：OrientableColorScroll | 文件：MainPage.xaml(片段)

```
<Grid Background="{StaticResource ApplicationPageBackgroundThemeBrush}"
      SizeChanged="OnGridSizeChanged">

    <Grid.ColumnDefinitions>
        <ColumnDefinition Width="*" />
        <ColumnDefinition x:Name="secondColDef" Width="*" />
    </Grid.ColumnDefinitions>

    <Grid.RowDefinitions>
        <RowDefinition Height="*" />
        <RowDefinition x:Name="secondRowDef" Height="0" />
    </Grid.RowDefinitions>

    <Grid Grid.Row="0"
          Grid.Column="0">

    ...

    </Grid>

    <!-- Result -->
    <Rectangle Name="rectangleResult"
               Grid.Column="1"
               Grid.Row="0">
        <Rectangle.Fill>
            <SolidColorBrush x:Name="brushResult"
                             Color="Black" />
        </Rectangle.Fill>
    </Rectangle>
</Grid>
```

外层 Grid 的 RowDefinitions 和 ColumnDefinitions 有两种初始化方式：两行一列和一行两列。每个集合的第二个元素均被命名，以便在代码中访问它们。这个程序假定在初始状态下为横向视图，因而第二行的高度被设置为 0。

内层的 Grid(包含 TextBlock 元素和 Slider 控件)总是位于第一列或第一行。

```
<Grid Grid.Row="0"
      Grid.Column="0">

...

</Grid>
```

在 Grid 上设置 Grid.Row 和 Grid.Column 特性可能让人觉得有些困惑。该设置并非针对当前 Grid，而是用于设置其在父 Grid 中所处的行和列。这两个附加属性的默认值均为 0，因而实际并不需要显式地设置这两个特性。

在初始状态下，Rectangle 元素位于第一行第二列。

```
<Rectangle Name="rectangleResult"
           Grid.Column="1"
           Grid.Row="0">
    ...
</Rectangle>
```

在本示例的第一个版本中，Rectangle 被命了名，因而这两个附加属性可以在代码隐藏文件中修改。该操作可以在外层 Grid 的 SizeChanged 事件处理程序中完成。

项目：OrientableColorScroll | 文件：MainPage.xaml.cs(片段)

```
void OnGridSizeChanged(object sender, SizeChangedEventArgs args)
{
    // Landscape mode
    if (args.NewSize.Width > args.NewSize.Height)
    {
        secondColDef.Width = new GridLength(1, GridUnitType.Star);
        secondRowDef.Height = new GridLength(0);

        Grid.SetColumn(rectangleResult, 1);
        Grid.SetRow(rectangleResult, 0);
    }
    // Portrait mode
    else
    {
        secondColDef.Width = new GridLength(0);
        secondRowDef.Height = new GridLength(1, GridUnitType.Star);

        Grid.SetColumn(rectangleResult, 0);
        Grid.SetRow(rectangleResult, 1);
    }
}
```

这段代码修改了外层 Grid 的 RowDefinition 和 ColumnDefinition，另外还有 Rectangle 元素所处的单元格。这样，Rectangle 元素便能够在横向视图下被置于第一行第二列，而在纵向视图下被置于第二行第一列。

该程序的运行效果如下图所示。

第 12 章会进一步讨论针对辅屏视图布局的调整。

5.5　Slider 与格式化字符串转换器

对于前文介绍的两个 ColorScroll 程序，底部的 TextBlock 标签能够以十六进制显示 Slider 的值。我们不仅可以在代码隐藏文件中设置这些标签的值，还可以通过数据绑定将 Slider 的值传给 TextBlock。后一种方法只需要写一个转换器，将 double 类型转换为双字符的十六进制字符串。

第 4 章在示例项目 WhatSizeWithBindingConverter 中介绍了一个名为 FormattedStringConverter 的类，但此转换器在这里并不适用。读者不妨试一下，但最终会发现格式字符串“X2”只能在整数类型上使用，而 Slider 的 Value 属性为 double。

有时我们可以编写一些简单实用的绑定转换器来达到一举多得的效果(正如下一节要介绍的)。

5.6　工具提示与转换

在 ColorScroll 程序中操纵 Slider 控件时，您或许会发现 Slider 会给出一个能够显示当前值的工具提示(tooltip)。这个功能虽然非常不错，但它显示的是十进制数，与标签的十六进制数在形式上不一致。

如果觉得 Slider 同时显示十进制和十六进制值无伤大雅，则可直接跳过本节。倘若希望工具提示与标签的内容相同(十六进制值)，则可利用 Slider 定义的 ThumbToolTipValueConverter 属性。我们可以通过该属性指定用于格式化文本的对象。该对象的类型必须实现 IValueConverter 接口，该接口也是实现绑定转换器所要实现的接口。

需要注意的是，设置到 ThumbToolTipValueConverter 属性的转换器不能像数据绑定转换器那样复杂，因为无法指定转换参数。好处是，这种转换器只针对特定情况，实现起来较为简单。

示例项目 ColorScrollWithValueConverter 定义了转换器，专门用于将 double 值转换为双字符十六进制字符串。这个类十分简单，甚至类名和实现代码都差不多长。

项目：ColorScrollWithValueConverter | 文件：DoubleToStringHexByteConverter.cs

```
using System;
using Windows.UI.Xaml.Data;

namespace ColorScrollWithValueConverter
{
   public class DoubleToStringHexByteConverter : IValueConverter
   {
      public object Convert(object value, Type targetType, object parameter, string language)
      {
         return ((int)(double)value).ToString("X2");
      }
      public object ConvertBack(object value, Type targetType, object parameter, string language)
      {
         return value;
      }
   }
}
```

这个转换器不仅适用于 Slider 的工具提示，也适用于用来显示 Slider 值的 TextBlock。下面这个版本的 ColorScroll 程序展示了该转换器的使用。(为了使程序简单明了，该示例并未调整显示比例。)转换器是在 XAML 文件的 Resources 区段中实例化的。

```
项目: ColorScrollWithValueConverter | 文件: MainPage.xaml(片段)
<Page.Resources>
    <local:DoubleToStringHexByteConverter x:Key="hexConverter" />
    ...
</Page.Resources>
```

下面这段标记展示了第一组 TextBlock 和 Slider。Slider 通过简单的 StaticResource 来引用 hexConverter 资源，而 TextBlock 则通过 Binding 引用该资源(为了便于阅读，绑定代码被拆分为三行)。

```
项目: ColorScrollWithValueConverter | 文件: MainPage.xaml(片段)
<!-- Red -->
<TextBlock Text="Red"
           Grid.Column="0"
           Grid.Row="0"
           Foreground="Red" />

<Slider Name="redSlider"
        Grid.Column="0"
        Grid.Row="1"
        ThumbToolTipValueConverter="{StaticResource hexConverter}"
        Foreground="Red"
        ValueChanged="OnSliderValueChanged" />

<TextBlock Text="{Binding ElementName=redSlider,
                          Path=Value,
                          Converter={StaticResource hexConverter}}"
           Grid.Column="0"
           Grid.Row="2"
           Foreground="Red" />
```

由于 ValueChanged 事件处理程序不再需要更新 TextBlock 标签，这里已将相关代码移除，只保留计算合成颜色的部分。

```
项目: ColorScrollWithValueConverter | 文件: MainPage.xaml.cs(片段)
void OnSliderValueChanged(object sender, RangeBaseValueChangedEventArgs args)
{
    byte r = (byte)redSlider.Value;
    byte g = (byte)greenSlider.Value;
    byte b = (byte)blueSlider.Value;

    brushResult.Color = Color.FromArgb(255, r, g, b);
}
```

我们也可以将每个 Slider 标签中的 ThumbToolTipValueConverter 设置转移至针对 Slider 的样式。

```
<Style TargetType="Slider">
    <Setter Property="Orientation" Value="Vertical" />
    <Setter Property="IsDirectionReversed" Value="True" />
    <Setter Property="Maximum" Value="255" />
    <Setter Property="HorizontalAlignment" Value="Center" />
    <Setter Property="ThumbToolTipValueConverter" Value="{StaticResource hexConverter}" />
</Style>
```

可否进一步利用数据绑定来彻底剔除 ValueChanged 事件处理程序呢？若能够将几个 Slider 绑定到 Color 的相应属性上，这便是可行的。

```
<!-- Doesn't work! -->
<Rectangle Grid.Column="3"
           Grid.Row="0"
           Grid.RowSpan="3">
   <Rectangle.Fill>
      <SolidColorBrush>
         <SolidColorBrush.Color>
            <Color A="255"
                   R="{Binding ElementName=redSlider, Path=Value}"
                   G="{Binding ElementName=greenSlider, Path=Value}"
                   B="{Binding ElementName=blueSlider, Path=Value}" />
         </SolidColorBrush.Color>
      </SolidColorBrush>
   </Rectangle.Fill>
</Rectangle>
```

这段标记最大的问题在于，绑定目标必须是依赖属性，但 Color 的这些属性不是。依赖属性只能在 DependencyObject 的派生类中实现，但 Color 根本就不是类，而是结构。

SolidColorBrush 的 Color 属性是依赖属性，可以作为绑定目标。然而，对于这个程序，Color 属性需要通过三个值计算得出，而 Windows Runtime 不支持多绑定源的数据绑定。

一种解决方案是创建一个能够根据红绿蓝色值创建 Color 对象的类。第 6 章会具体介绍这种方法。

5.7　用 Slider 绘制草图

本人不打算在这里展示下面这个示例的截图。这个示例项目名为 SliderSketch，它用 Slider 实现了一个大约 50 年前就已经问世的程序。用户需要分别通过横向和纵向的 Slider 来操控一个概念上的笔尖来逐步延伸一条连续的多段线。之所以不展示它的屏幕截图，是因为此程序甚是难用，本人实在难以绘制出一幅像样的图案。

此程序的 XAML 文件定义了一个两行两列的 Grid。屏幕的绝大部分区域被其中的一个较大的单元格所占据。该单元格包含一个 Border 和一个 Polyline。纵向的 Slider 位于最左侧，横向的 Slider 位于最底端。左下角的单元格为空。

项目: SliderSketch | 文件: MainPage.xaml(片段)

```
<Grid Background="{StaticResource ApplicationPageBackgroundThemeBrush}">
   <Grid.RowDefinitions>
      <RowDefinition Height="*" />
      <RowDefinition Height="Auto" />
   </Grid.RowDefinitions>

   <Grid.ColumnDefinitions>
      <ColumnDefinition Width="Auto" />
      <ColumnDefinition Width="*" />
   </Grid.ColumnDefinitions>

   <Slider Name="ySlider"
           Grid.Row="0"
           Grid.Column="0"
           Orientation="Vertical"
           IsDirectionReversed="True"
           Margin="0 18"
           ValueChanged="OnSliderValueChanged" />

   <Slider Name="xSlider"
```

```
            Grid.Row="1"
            Grid.Column="1"
            Margin="18 0"
            ValueChanged="OnSliderValueChanged" />

    <Border Grid.Row="0"
            Grid.Column="1"
            BorderBrush="{StaticResource ApplicationForegroundThemeBrush}"
            BorderThickness="3 0 0 3"
            Background="#C0C0C0"
            Padding="24"
            SizeChanged="OnBorderSizeChanged">

        <Polyline Name="polyline"
                  Stroke="#404040"
                  StrokeThickness="3"
                  Points="0 0" />
    </Border>
</Grid>
```

将 Grid 最外侧行和列的宽度设置为 Auto 而通过星号使内部区域占据绝大部分空间，这是较为常见的做法。这样，边缘的内容会自动贴靠。虽然 Windows 8 没有 DockPanel，但可以通过 Grid 来模拟。

Slider 控件的 Margin 属性是根据经验来设置的。为使程序正确工作，Slider 值的范围应与代表笔尖的像素的最大和最小坐标值一致，Slider 滑块的位置应尽量保持与该像素对齐。每当显示区域的大小发生改变，Slider 的 Minimum 和 Maximum 值就都需要重新计算。

```
项目: SliderSketch | 文件: MainPage.xaml.cs(片段)
public sealed partial class MainPage : Page
{
    public MainPage()
    {
        this.InitializeComponent();
    }

    void OnBorderSizeChanged(object sender, SizeChangedEventArgs args)
    {
        Border border = sender as Border;
        xSlider.Maximum = args.NewSize.Width - border.Padding.Left
                                             - border.Padding.Right
                                             - polyline.StrokeThickness;

        ySlider.Maximum = args.NewSize.Height - border.Padding.Top
                                              - border.Padding.Bottom
                                              - polyline.StrokeThickness;
    }

    void OnSliderValueChanged(object sender, RangeBaseValueChangedEventArgs args)
    {
        polyline.Points.Add(new Point(xSlider.Value, ySlider.Value));
    }
}
```

实现“绘制”功能的方法实际只有一行，位于这段代码底部。作用是将新建的 Point 添加到 Polyline。

请不要尝试通过翻转或摇晃平板电脑来重置图画，因为这个功能尚未实现。

5.8　按钮的几种变体

Windows Runtime 支持多种按钮，它们都派生自 ButtonBase 类。

```
Object
  DependencyObject
    UIElement
      FrameworkElement
        Control
          ContentControl
            ButtonBase
              Button
              HyperlinkButton
              RepeatButton
              ToggleButton
                CheckBox
                RadioButton
```

示例项目 ButtonVarieties 展示了这些按钮的默认外观。

项目：ButtonVarieties | 文件：MainPage.xaml(片段)

```
<Grid Background="{StaticResource ApplicationPageBackgroundThemeBrush}">
   <StackPanel>
      <Button Content="Just a plain old Button" />
      <HyperlinkButton Content="HyperlinkButton" />
      <RepeatButton Content="RepeatButton" />
      <ToggleButton Content="ToggleButton" />
      <CheckBox Content="CheckBox" />

      <RadioButton Content="RadioButton #1" />
      <RadioButton>RadioButton #2</RadioButton>
      <RadioButton>
         <RadioButton.Content>
            RadioButton #3
         </RadioButton.Content>
      </RadioButton>
      <RadioButton>
         <RadioButton.Content>
            <TextBlock Text="RadioButton #4" />
         </RadioButton.Content>
      </RadioButton>

      <ToggleSwitch />
   </StackPanel>
</Grid>
```

此程序创建了 4 个 RadioButton 的实例(参见下图)。虽然不同实例设置 Content 属性的方式不同，但是它们彼此之间是等价的。

如果觉得这些按钮的外观不合适，可以通过 ControlTemplate 彻底改变(详情参见第 11 章)。

与所有 FrameworkElement 派生类一样，这些按钮的 HorizontalAlignment 和 VerticalAlignment 属性的默认值均为 Stretch。但在按钮加载之初，HorizontalAlignment 属性为 Left，VerticalAlignment 属性为 Center，Padding 属性也不为 0。虽然 Margin 属性值为 0，但 Border 外侧仍有较窄的边距。

ButtonBase 定义了 Click 事件，能在手指、鼠标、触控笔按控件并释放时引发。该行为可以通过 ClickMode 属性更改。另外，程序可以通过一种命令接口在按钮被单击时进行通知，详情参见第 6 章。

Button 是传统的按钮。HyperlinkButton 与 Button 极为相似，只不过 HyperlinkButton 的外观是由另一种模板定义的。RepeatButton 能够在按钮被按下并保持一段后引发一系列 Click 事件，通常用于实现 ScrollBar 的重复行为。

单击 ToggleButton 能够使其在“开”和“关”两种状态之间进行切换。前面的屏幕截图展示了该控件“开”的状态。CheckBox 未定义任何公共成员，它只不过继承了 ToggleButton 的所有功能，并通过模板改了外观。

ToggleButton 通过 IsChecked 属性来表明其当前状态。该控件还定义了 Checked 和 Unchecked 事件来通知其状态的变化。这两个事件一般都需要订阅，但可以共用同一个处理程序。

ToggleButton 的 IsChecked 属性并不是 bool 类型，而是 Nullable<bool>类型。这说明该属性可以返回 null。开关按钮的这种“中间”状态可能会让人不解。不妨通过一个例子来说明：字处理程序通常会有用于设置“加粗”(Bold)的 CheckBox。如果被选中的文本已被加粗，该复选框则被选中。如果选中的文本未被加粗，该复选框则不被选中。如果被选中的文字有加粗的和未加粗的，那么复选框则处于一种中间状态。为启用这种状态，要将 IsThreeState 设置为 true。如果需要，还可以订阅 Indeterminate 事件。处于这种中间状态的 ToggleButton 外观会略有不同，按钮会显示一个小方块，而不是对号。

说到这里，您或许会注意到另一个实现开关的控件 ToggleSwitch。它是专为 Windows 8 应用程序设计的。虽然 ToggleSwitch 并不是派生自 ButtonBase，但示例中还是将其放在最后一并列出。正如示例程序所展示的，它的默认标签为“关闭”(Off)和“启用”(On)，但可以修改。该控件还有一个标题，第 8 章会进行讲解。

RadioButton 是 ToggleButton 的一种特殊形式，用于在一组选项中选择某一个。该控件的名称源于老式汽车收音机，这种收音机上有用于切换预设电台的按钮：一个按钮被按下后，之前被按下的按钮则会弹起。类似地，当某个 RadioButton 控件被选中，其他 RadioButton 按钮则会取消选中，前提是这些按钮为同一面板的子元素。(需要注意的是，如果将某个 RadioButton 置于 Border 中，那么它将与其他 RadioButton 脱离关系。如果需要为 RadioButton 添加 Border，就应该使用模板。)如果需要将同一面板中的多个 RadioButton 分成若干不相关的组，则可以使用 GroupName 属性。

Control 类定义了 Foreground 属性、字体相关属性、与 Border 相关的属性以及控制按钮外观的属性。例如，我们可以这样初始化一个 Button。

```
<Button Content="Not just a plain old Button anymore"
        Background="Yellow"
        BorderBrush="Red"
        BorderThickness="12"
        Foreground="Blue"
        FontSize="48"
        FontStyle="Italic" />
```

效果如下图所示。

虽然设置了这些属性，但有些视觉效果仍然由模板控制。例如，当鼠标悬停于按钮上面或按下它，黄色背景会立即变为标准颜色。此外，虽然我们可以修改 Border 的颜色和粗细，但无法使其具有圆角。

ButtonBase 派生自 ContentControl。后者定义了一个名为 Content 的属性。虽然 Content 属性一般用于设置文本，但也可以被设置为 Image 或面板。这使派生的控件更为强大。例如，我们可以为 Button 设置一个图片和一个图片标题。

```
<Button>
    <StackPanel>
        <Image Source="http://www.charlespetzold.com/pw6/PetzoldJersey.jpg"
               Width="100" />
```

```
        <TextBlock Text="Figure 1"
                   HorizontalAlignment="Center" />
    </StackPanel>
</Button>
```

Content 属性几乎可以被设置为任何对象，而通过模板可以定义该对象的外观。第 11 章会具体介绍有关内容。

下面我们一起制作一个简单的电话拨号盘。拨号按键通过 Button 控件实现，而电话号码通过 TextBlock 显示。

在下面这个 XAML 文件中，拨号键盘位于 HorizontalAlignment 和 VerticalAlignment 均为 Center 的 Grid 中，因而拨号键盘显示于屏幕正中央。先不论拨号键盘和按钮内容有多大，但至少 12 个按钮的尺寸是相同的。示例中采用了两种方法来设置按钮的宽度和高度。容纳拨号键盘的 Grid 的宽度为 288(大约为 3 英寸)。这个宽度值是具体的，因为用户可能按下多位号码，不应使 Grid 适应一个过大的 TextBlock。每个 Button 的高度通过隐式样式来指定。

```
项目: SimpleKeypad | 文件: MainPage.xaml(片段)
<Grid Background="{StaticResource ApplicationPageBackgroundThemeBrush}">

    <Grid HorizontalAlignment="Center"
          VerticalAlignment="Center"
          Width="288">

        <Grid.Resources>
            <Style TargetType="Button">
                <Setter Property="ClickMode" Value="Press" />
                <Setter Property="HorizontalAlignment" Value="Stretch" />
                <Setter Property="Height" Value="72" />
                <Setter Property="FontSize" Value="36" />
            </Style>
        </Grid.Resources>

        <Grid.RowDefinitions>
            <RowDefinition Height="Auto" />
            <RowDefinition Height="Auto" />
            <RowDefinition Height="Auto" />
            <RowDefinition Height="Auto" />
            <RowDefinition Height="Auto" />
        </Grid.RowDefinitions>

        <Grid.ColumnDefinitions>
            <ColumnDefinition Width="*" />
            <ColumnDefinition Width="*" />
            <ColumnDefinition Width="*" />
        </Grid.ColumnDefinitions>

        <Grid Grid.Row="0" Grid.Column="0" Grid.ColumnSpan="3">
            <Grid.ColumnDefinitions>
                <ColumnDefinition Width="*" />
                <ColumnDefinition Width="Auto" />
            </Grid.ColumnDefinitions>

            <Border Grid.Column="0"
                    HorizontalAlignment="Left">

                <TextBlock Name="resultText"
                           HorizontalAlignment="Right"
                           VerticalAlignment="Center"
                           FontSize="24" />
            </Border>
```

```
            <Button Name="deleteButton"
                    Content="&#x21E6;"
                    Grid.Column="1"
                    IsEnabled="False"
                    FontFamily="Segoe Symbol"
                    HorizontalAlignment="Left"
                    Padding="0"
                    BorderThickness="0"
                    Click="OnDeleteButtonClick" />
        </Grid>

        <Button Content="1"
                Grid.Row="1" Grid.Column="0"
                Click="OnCharButtonClick" />

        <Button Content="2"
                Grid.Row="1" Grid.Column="1"
                Click="OnCharButtonClick" />

        <Button Content="3"
                Grid.Row="1" Grid.Column="2"
                Click="OnCharButtonClick" />

        <Button Content="4"
                Grid.Row="2" Grid.Column="0"
                Click="OnCharButtonClick" />

        <Button Content="5"
                Grid.Row="2" Grid.Column="1"
                Click="OnCharButtonClick" />

        <Button Content="6"
                Grid.Row="2" Grid.Column="2"
                Click="OnCharButtonClick" />

        <Button Content="7"
                Grid.Row="3" Grid.Column="0"
                Click="OnCharButtonClick" />

        <Button Content="8"
                Grid.Row="3" Grid.Column="1"
                Click="OnCharButtonClick" />

        <Button Content="9"
                Grid.Row="3" Grid.Column="2"
                Click="OnCharButtonClick" />

        <Button Content="*"
                Grid.Row="4" Grid.Column="0"
                Click="OnCharButtonClick" />

        <Button Content="0"
                Grid.Row="4" Grid.Column="1"
                Click="OnCharButtonClick" />

        <Button Content="#"
                Grid.Row="4" Grid.Column="2"
                Click="OnCharButtonClick" />
    </Grid>
</Grid>
```

号码显示区域位于第一行。此行需要包含一个 TextBlock 来显示已输入的号码和一个删除按钮，因而 Grid 的第一行嵌套了另一个 Grid 来显示这两个元素。删除按钮重写了隐式样式的许多设置。请注意，删除按钮初始状态下是被禁用的，只有拨号后才会被启用。

TextBlock 幕后的逻辑较复杂。在正常输入状态下，它是左对齐的，但如果显示的号码过长，应截断 TextBlock 左侧字符，而非右侧的。为解决这个问题，可以将此 TextBlock 置于 Border 中。

```
<Border Grid.Column="0"
        HorizontalAlignment="Left">

    <TextBlock Name="resultText"
               HorizontalAlignment="Right"
               VerticalAlignment="Center"
               FontSize="24" />
</Border>
```

Border 对 TextBlock 的宽度做了限制：后者的宽度不能超过外层 Grid 与删除按钮的宽度之差。在这个区域内，Border 是左对齐的。由于 TextBlock 与 Border 等宽，尽管对齐方式不同，但 TextBlock 仍位于左端。在键入较多号码后，TextBlock 的宽度会超过 Border 的宽度。此时，值为 Right 的 HorizontalAlignment 设置便开始起作用，即将 TextBlock 的左侧超出的部分遮盖。

此程序除第一行以外，都相对简单。隐式样式在很大程度上简化了 10 个数字和 2 个符号按钮的 XAML 标记。

代码隐藏文件包含删除按钮和拨号按钮的 Click 事件处理程序。其中，12 个拨号按钮共用同一处理程序。

```
项目：SimpleKeypad | 文件：MainPage.xaml.cs(片段)
public sealed partial class MainPage : Page
{
    string inputString = "";
    char[] specialChars = { '*', '#' };

    public MainPage()
    {
        this.InitializeComponent();
    }

    void OnCharButtonClick(object sender, RoutedEventArgs args)
    {
        Button btn = sender as Button;
        inputString += btn.Content as string;
        FormatText();
    }

    void OnDeleteButtonClick(object sender, RoutedEventArgs args)
    {
        inputString = inputString.Substring(0, inputString.Length - 1);
        FormatText();
    }

    void FormatText()
    {
        bool hasNonNumbers = inputString.IndexOfAny(specialChars) != -1;

        if (hasNonNumbers || inputString.Length < 4 || inputString.Length > 10)
            resultText.Text = inputString;

        else if (inputString.Length < 8)
            resultText.Text = String.Format("{0}-{1}", inputString.Substring(0, 3),
                                                        inputString.Substring(3));
        else
```

```
            resultText.Text = String.Format("({0}) {1}-{2}", inputString.Substring(0, 3),
                                                              inputString.Substring(3, 3),
                                                              inputString.Substring(6));
        deleteButton.IsEnabled = inputString.Length > 0;
    }
}
```

删除按钮的处理程序能够移除 inputString 字段中的一个字符，另一个处理程序能够向该字段添加一个字符。这两个处理程序最后会调用 FormatText 方法，将字符串以电话号码格式显示。该方法的最后，在输入字符串含有字符的情况下会启用删除按钮。效果如下图所示。

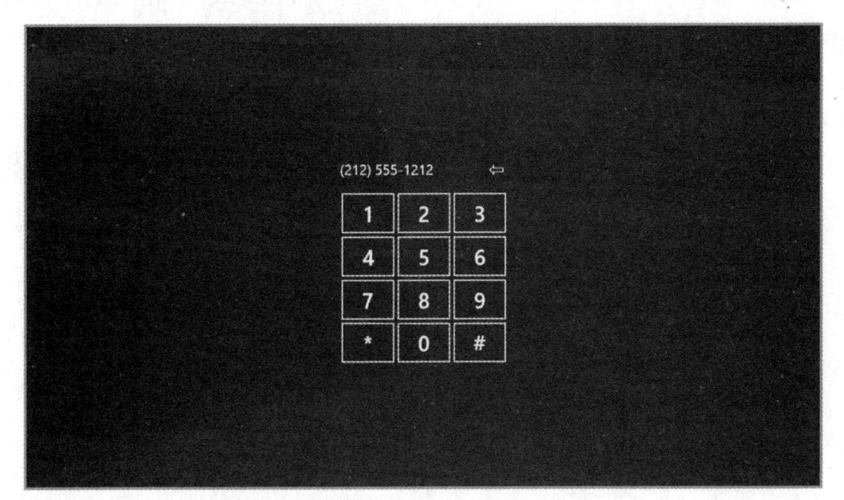

OnCharButtonClick 事件处理程序通过被按下按钮的 Content 属性来决定要追加到输入字符串的字符。按钮的 Content 属性值与按钮的功能之间这种关联并非总是存在的。在多个按钮共用同一个事件处理程序的情况下，该处理程序需要获得更多有关按钮的信息。FrameworkElement 定义了一个 Tag 属性，此时刚好可以用。我们可以在 XAML 文件中为 Tag 设置一个用于标识所属元素的字符串或对象，并在事件处理程序中读取该属性。本章稍后介绍 RadioButton 时会提供相关演示。

5.9　依赖属性的定义

假设某应用程序要求所有 Button 控件通过渐变画笔来显示其文本。当然，我们可以分别为每个 Button 的 Foreground 属性设置 LinearGradientBrush，但最终的标记会有些长。也可以将 Style 的 Foreground 属性设置为 LinearGradientBrush，但不同的 Button 将共用渐变效果完全相同的 LinearGradientBrush。如果要求更加灵活些呢？

现在要求创建一种 Button，开发者可以通过名为 Color1 和 Color2 的属性来设置渐变色的两个颜色。我们可以从 Button 派生一个类，在该类的构造函数中创建 LinearGradientBrush 对象，然后定义 Color1 和 Color2 属性来控制画笔的颜色。

Color1 和 Color2 属性可否为带有 set 和 get 访问器的普通.NET 属性？是的，这样做是可以的。然而，定义这样的属性会限制该控件的应用范围。这样的属性不能作为样式、绑定和动画的目标属性。唯有依赖属性能够兼顾。

依赖属性要比普通属性复杂些，但掌握如何定义依赖属性是开发者应具备的一项重要技能。创建一个项目，添加一个新项，在列表中选择“类”并将其命名为GradientButton。在类文件中，将其定义为公共类并使其继承于Button。

```
public class GradientButton : Button
{

}
```

下面将完善该类的定义，另外还需要添加若干using指令。

添加两个类型为Color的属性，分别命名为Color1和Color2。相应地，我们需要定义类型为DependencyProperty、名称分别为Color1Property和Color2Property的依赖属性。

```
public static DependencyProperty Color1Property { private set; get; }
public static DependencyProperty Color2Property { private set; get; }
```

DependencyProperty对象可以在静态构造函数中创建。DependencyProperty类定义了一个名为Register的静态方法，专门用来创建DependencyProperty对象。

```
static GradientButton()
{
    Color1Property =
        DependencyProperty.Register("Color1",
            typeof(Color),
            typeof(GradientButton),
            new PropertyMetadata(Colors.White, OnColorChanged));

    Color2Property =
        DependencyProperty.Register("Color2",
            typeof(Color),
            typeof(GradientButton),
            new PropertyMetadata(Colors.Black, OnColorChanged));
}
```

还有一个稍有不同的静态方法DependencyProperty.RegisterAttached，用于注册附加属性。

DependencyProperty.Register的第一个参数是属性的名称。XAML解析器可能会用到这个值。第二个参数是属性的类型。第三个参数是注册当前依赖属性的类的类型。

第四个参数需要传入类型为PropertyMetadata的对象。该类型的构造函数有两个。第一个用于指定属性的默认值，第二个用于指定属性发生更改时要调用的方法。如果设置到依赖属性的新值与原值相同，则该方法不会被调用。

PropertyMetadata构造函数的第一个参数默认值的类型必须匹配Register第二个参数指定的类型，否则会产生运行时异常。这并非听上去那么无关紧要。例如，程序员往往为double类型的属性设置默认值0。在编译时，0会被认为是整型，因而在运行时会出现类型不匹配，进而产生异常。如果定义double类型的依赖属性，可以将默认值设置为0.0，这样编译器便会将该参数识别为正确的类型。

除了使用静态构造函数，还可以定义私有静态字段。先在字段上直接初始化DependencyProperty对象，然后通过公共静态属性将这些对象暴露出来。

```
static readonly DependencyProperty color1Property =
    DependencyProperty.Register("Color1",
        typeof(Color),
        typeof(GradientButton),
        new PropertyMetadata(Colors.White, OnColorChanged));
```

```
static readonly DependencyProperty color2Property =
   DependencyProperty.Register("Color2",
      typeof(Color),
      typeof(GradientButton),
      new PropertyMetadata(Colors.Black, OnColorChanged));

public static DependencyProperty Color1Property
{
   get { return color1Property; }
}

public static DependencyProperty Color2Property
{
   get { return color2Property; }
}
```

显式的静态构造函数并不强制要求。我们可以沿用 WPF 或 Silverlight 的风格，直接暴露公共的静态字段，而不定义公共静态属性。就本示例而言，我们可以定义两个分别名为 Color1Property 和 Color2Property 的字段。

```
public static readonly DependencyProperty Color1Property =
   DependencyProperty.Register("Color1",
      typeof(Color),
      typeof(GradientButton),
      new PropertyMetadata(Colors.White, OnColorChanged));

public static readonly DependencyProperty Color2Property =
   DependencyProperty.Register("Color2",
      typeof(Color),
      typeof(GradientButton),
      new PropertyMetadata(Colors.Black, OnColorChanged));
```

虽然 Windows 8 支持这种方式，但本人不倾向于这么做，因为标准 Windows Runtime 控件都是通过静态公共属性暴露的 DependencyProperty 对象，而不是通过字段。

不论通过静态属性还是静态字段来暴露 DependencyProperty 对象，我们都需要为 GradientButton 定义两个分别名为 Color1 和 Color2 的常规.NET 属性。这些属性的形式非常统一。

```
public Color Color1
{
   set { SetValue(Color1Property, value); }
   get { return (Color)GetValue(Color1Property); }
}

public Color Color2
{
   set { SetValue(Color2Property, value); }
   get { return (Color)GetValue(Color2Property); }
}
```

访问器 set 要调用 SetValue 方法(继承于 DependencyObject 类)并传入依赖属性对象，访问器 get 要调用 GetValue 并将结果转换为当前属性的类型。若不希望该属性被外界修改，可以用 protected 或 private 来修饰访问器 set。

我们需要将 GradientButton 控件的 Foreground 属性设置为 LinearGradientBrush 对象。Color1 和 Color2 属性分别用于设置两个 GradientStop 对象的颜色。这两个 GradientStop 对象是以字段的形式定义的。

```
GradientStop gradientStop1, gradientStop2;
```

我们可以在类的常规实例构造函数中创建这两个对象及画笔对象 LinearGradientBrush，最后将该画笔对象设置到 Foreground 属性上。

```
public GradientButton()
{
    gradientStop1 = new GradientStop
    {
        Offset = 0,
        Color = this.Color1
    };

    gradientStop2 = new GradientStop
    {
        Offset = 1,
        Color = this.Color2
    };

    LinearGradientBrush brush = new LinearGradientBrush();
    brush.GradientStops.Add(gradientStop1);
    brush.GradientStops.Add(gradientStop2);

    this.Foreground = brush;
}
```

请注意这段代码通过 Color1 和 Color2 属性来初始化 GradientStop 对象的过程。LinearGradientBrush 就这样通过两个依赖属性获得了默认颜色。

前文定义这两个依赖属性时提到了一个名为 OnColorChanged 的方法。该方法会在 Color1 或 Color2 属性发生变化时被调用。由于这个属性变更通知方法是在静态构造函数中引用的，那么该方法必须也是静态的。

```
static void OnColorChanged(DependencyObject obj, DependencyPropertyChangedEventArgs args)
{

}
```

定义 GradientButton 类是为了在应用程序中多次复用，而现在却要在类的实例中定义一个 Color1 或 Color2 属性发生更改时调用的静态方法。这似乎让人难以捉摸。我们如何才能得知该方法作用于哪个实例？

其实也容易，使用第一个参数即可。OnColorChanged 方法的第一个参数总是会传入相关属性被修改的 GradientButton 对象。我们可以放心地将该对象转换为 GradientButton，然后访问 GradientButton 的实例字段和属性。

本人倾向于在这个静态方法中调用同名的实例方法，并传入第二个参数。

```
static void OnColorChanged(DependencyObject obj, DependencyPropertyChangedEventArgs args)
{
    (obj as GradientButton).OnColorChanged(args);
}

void OnColorChanged(DependencyPropertyChangedEventArgs args)
{

}
```

我们可以通过第二个方法来访问该类的实例字段和属性。

DependencyPropertyChangedEventArgs 对象为我们提供了一些有价值的信息。它的 Property 属性的类型为 DependencyProperty，能够指明源对象的哪个属性被更改。就本例而言，Property 属性可能为 Color1Property 或 Color2Property。DependencyPropertyChangedEventArgs 对象还有名为 OldValue 和 NewValue 的属性，类型均为 object。

GradientButton 类中的这个属性变更处理程序可以根据 NewValue 来设置相应 GradientStop 对象的 Color 属性。

```
void OnColorChanged(DependencyPropertyChangedEventArgs args)
{
   if (args.Property == Color1Property)
      gradientStop1.Color = (Color)args.NewValue;

   if (args.Property == Color2Property)
      gradientStop2.Color = (Color)args.NewValue;
}
```

上述内容是 GradientButton 所需的全部代码。剩下的只需要将这些代码合理地组织在 GradientButton 类中。本人习惯于将所有字段置于最顶端，随后依次放置静态构造函数、静态属性、实例构造函数、实例属性，最后是所有方法。下面是示例项目 DependencyProperties 中 GradientButton 类的完整代码。

项目：DependencyProperties | 文件：GradientButton.cs

```
using Windows.UI;
using Windows.UI.Xaml;
using Windows.UI.Xaml.Controls;
using Windows.UI.Xaml.Media;

namespace DependencyProperties
{
   public class GradientButton : Button
   {
      GradientStop gradientStop1, gradientStop2;

      static GradientButton()
      {
         Color1Property =
            DependencyProperty.Register("Color1",
               typeof(Color),
               typeof(GradientButton),
               new PropertyMetadata(Colors.White, OnColorChanged));

         Color2Property =
            DependencyProperty.Register("Color2",
               typeof(Color),
               typeof(GradientButton),
               new PropertyMetadata(Colors.Black, OnColorChanged));
      }

      public static DependencyProperty Color1Property { private set; get; }

      public static DependencyProperty Color2Property { private set; get; }

      public GradientButton()
      {
         gradientStop1 = new GradientStop
         {
            Offset = 0,
            Color = this.Color1
         };

         gradientStop2 = new GradientStop
         {
            Offset = 1,
            Color = this.Color2
         };
```

```
        LinearGradientBrush brush = new LinearGradientBrush();
        brush.GradientStops.Add(gradientStop1);
        brush.GradientStops.Add(gradientStop2);

        this.Foreground = brush;
    }

    public Color Color1
    {
        set { SetValue(Color1Property, value); }
        get { return (Color)GetValue(Color1Property); }
    }

    public Color Color2
    {
        set { SetValue(Color2Property, value); }
        get { return (Color)GetValue(Color2Property); }
    }

    static void OnColorChanged(DependencyObject obj,
                               DependencyPropertyChangedEventArgs args)
    {
        (obj as GradientButton).OnColorChanged(args);
    }

    void OnColorChanged(DependencyPropertyChangedEventArgs args)
    {
        if (args.Property == Color1Property)
            gradientStop1.Color = (Color)args.NewValue;

        gradientStop1.Color = this.Color1;

        if (args.Property == Color2Property)
            gradientStop2.Color = (Color)args.NewValue;
    }
  }
}
```

属性更改处理程序有很多种写法。若为不同属性指定单独的处理程序，则不需要检查事件参数的 Property 属性。

另一种写法是直接访问类的实例属性，而不使用 NewValue 属性。例如：

```
gradientStop1.Color = this.Color1;
```

属性更改处理程序被调用时，Color1 属性已被设置为新值。

Color1 和 Color2 属性的值实际存储在哪里？本人猜测是某种字典，或许是某些优化的机制(但愿如此)，但都无法通过 API 直接访问。这些属性的状态由操作系统管理，我们只能通过 SetValue 和 GetValue 方法来访问它们的值。

该示例项目的 XAML 文件定义了两个样式。第一个样式通过 Setter 元素设置 Color1 和 Color2 属性，作用于 GradientButton 的两个实例。任何引用 GradientButton 的 XAML 文件需要预先通过 XML 命名空间 local 来引入 GradientButton 所在命名空间 DependencyProperties。请注意，两个 Style 的 TargetType 和按钮的实例化标记都使用了前缀 local。

```
项目: DependencyProperties | 文件: MainPage.xaml(片段)
<Page ...
    xmlns:local="using:DependencyProperties"
    ... >
```

```
    <Page.Resources>
        <Style x:Key="baseButtonStyle" TargetType="local:GradientButton">
            <Setter Property="FontSize" Value="48" />
            <Setter Property="HorizontalAlignment" Value="Center" />
            <Setter Property="Margin" Value="0 12" />
        </Style>

        <Style x:Key="blueRedButtonStyle"
               TargetType="local:GradientButton"
               BasedOn="{StaticResource baseButtonStyle}">
            <Setter Property="Color1" Value="Blue" />
            <Setter Property="Color2" Value="Red" />
        </Style>
    </Page.Resources>

    <Grid Background="{StaticResource ApplicationPageBackgroundThemeBrush}">
        <StackPanel>
            <local:GradientButton Content="GradientButton #1"
                                  Style="{StaticResource baseButtonStyle}" />

            <local:GradientButton Content="GradientButton #2"
                                  Style="{StaticResource blueRedButtonStyle}" />

            <local:GradientButton Content="GradientButton #3"
                                  Style="{StaticResource baseButtonStyle}"
                                  Color1="Aqua"
                                  Color2="Lime" />
        </StackPanel>
    </Grid>
</Page>
```

从下图可知，在这三个按钮中，第一个采用的是 Color1 和 Color2 的默认设置，第二个采用的是 Style 中定义的设置，第三个采用的是局部设置。

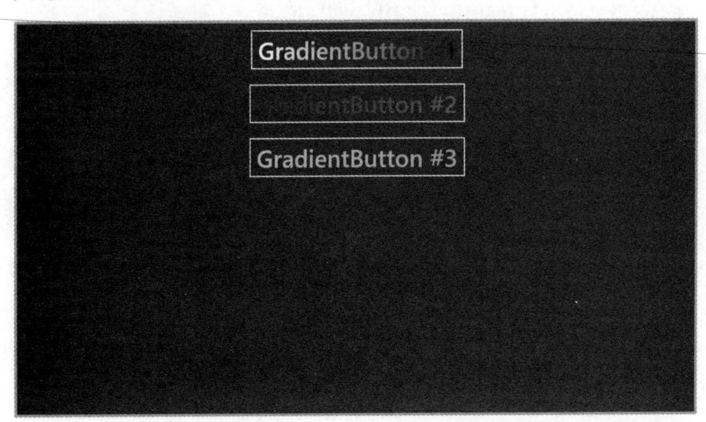

下面将介绍另一种创建 GradientButton 的方法，即在 XAML 中定义 LinearGradientBrush，并避免使用属性变更处理程序。下面我们看看具体如何实现。

在一个单独的项目中，添加新项。为创建 GradientButton 类，我们这次不选择“类”模板，而选择“用户控件”，将其命名为 GradientButton。这样，我们便会得到一对文件：GradientButton.xaml 和 GradientButton.xaml.cs。这个 GradientButton 类派生自 UserControl。GradientButton.xaml.cs 文件中的类是下面这样定义的。

```
public sealed partial class GradientButton : UserControl
{
    public GradientButton()
    {
        this.InitializeComponent();
```

```
   }
}
```

我们将父类由 UserControl 改为 Button。

```
public sealed partial class GradientButton : Button
{
   public GradientButton()
   {
      this.InitializeComponent();
   }
}
```

这个类的主体部分与之前的 GradientButton 类非常类似，但这里的实例构造函数只调用 InitializeComponent 方法，而不做其他操作。另外，这个 GradientButton 控件也不定义属性变更处理程序。示例项目 DependencyPropertiesWithBindings 展示了这个类。

项目：DependencyPropertiesWithBindings | 文件：GradientButton.xaml.cs(片段)

```
public sealed partial class GradientButton : Button
{
   static GradientButton()
   {
      Color1Property =
         DependencyProperty.Register("Color1",
            typeof(Color),
            typeof(GradientButton),
            new PropertyMetadata(Colors.White));

      Color2Property =
         DependencyProperty.Register("Color2",
            typeof(Color),
            typeof(GradientButton),
            new PropertyMetadata(Colors.Black));
   }

   public static DependencyProperty Color1Property { private set; get; }

   public static DependencyProperty Color2Property { private set; get; }

   public GradientButton()
   {
      this.InitializeComponent();
   }

   public Color Color1
   {
      set { SetValue(Color1Property, value); }
      get { return (Color)GetValue(Color1Property); }
   }

   public Color Color2
   {
      set { SetValue(Color2Property, value); }
      get { return (Color)GetValue(Color2Property); }
   }
}
```

GradientButton.xaml 文件最初被创建时，根节点声明这个类派生自 UserControl。

```
<UserControl
   x:Class="DependencyPropertiesWithBindings.GradientButton" ... >
   ...
</UserControl>
```

我们需要像下面这样将这个父类改为 Button。

```
<Button
  x:Class="DependencyPropertiesWithBindings.GradientButton" ... >
  ...
</Button>
```

一般情况下，XAML 的根标签中的内容会被设置到 Content 属性上。本示例不希望设置 Button 的 Content 属性，而要将 GradientButton 的 Forground 属性设置为 LinearGradientBrush。为此我们要用到属性元素标签 Button.Foreground。XAML 文件的完整内容如下所示。

项目：DependencyPropertiesWithBindings | 文件：GradientButton.xaml

```
<Button
  x:Class="DependencyPropertiesWithBindings.GradientButton"
  xmlns="http://schemas.microsoft.com/winfx/2006/xaml/presentation"
  xmlns:x="http://schemas.microsoft.com/winfx/2006/xaml"
  Name="root">

  <Button.Foreground>
    <LinearGradientBrush>
      <GradientStop Offset="0"
                    Color="{Binding ElementName=root,
                             Path=Color1}" />
      <GradientStop Offset="1"
                    Color="{Binding ElementName=root,
                             Path=Color2}" />
    </LinearGradientBrush>
  </Button.Foreground>
</Button>
```

请注意 GradientStop 对象的 Color 属性是如何被设置的：为了让两个数据绑定能够引用自定义的依赖属性，这里将根元素命名为 root，使其可以作为数据绑定的源对象。

此项目的 MainPage.xaml 文件和最终结果与上一个示例相同，这里不再赘述。

5.10　RadioButton

用户可以通过一组 RadioButton 控件在若干互斥的选项中选取一项。从程序的角度看，同一组 RadioButton 控件的不同实例最好与枚举成员一一对应，通过 RadioButton 对象来获取枚举值。这样，同一组的单选按钮便可以共用一个事件处理程序。

为此，我们可以利用 Tag 属性。Tag 属性可以被设置为任何一个能够标识当前控件的对象。假设要写一个程序来测试 Shape 定义的 StrokeStartLineCap、StrokeEndLineCap 和 StrokeLineJoin 属性。在呈现较粗的线时，这三个属性可以控制线的端部及对接处的形状。StrokeStartLineCap 和 StrokeEndLineCap 属性需要被设置为枚举类型 PenLineCap 的成员，而 StrokeLineJoin 属性需要被设置为枚举类型 PenLineJoin 的成员。

例如，枚举类型 PenLineJoin 有一个名为 Bevel 的成员。那么我们便可以这样定义代表该选项的 RadioButton。

```
<RadioButton Content="Bevel join"
             Tag="Bevel"
             ... />
```

问题在于 Bevel 会被 XAML 解析器解析为字符串，因而在代码隐藏文件的处理程序中，我们需要通过 switch 和 case 语句来区分不同字符串，或者使用 Enum.TryParse 将字符串转换为 PenLineJoin 枚举值。

为避免这种转换，可以将 Tag 属性以属性元素的形式显式地为其指定 PenLineJoin 类型的值。

```
<RadioButton Content="Bevel join"
          ... >
   <RadioButton.Tag>
      <PenLineJoin>Bevel</PenLineJoin>
   </RadioButton.Tag>
</RadioButton>
```

当然，这样的标记略显冗长。尽管如此，示例项目 LineCapsAndJoins 仍然采用了这种方法。该项目的 XAML 文件定义了 3 组 RadioButton 控件，分别用来设置 3 个 Shape 属性。每组设置包含三四个单选按钮控件，与枚举成员一一对应。

项目：LineCapsAndJoins | 文件：MainPage.xaml(片段)

```
<Grid Background="{StaticResource ApplicationPageBackgroundThemeBrush}">
   <Grid.RowDefinitions>
      <RowDefinition Height="*" />
      <RowDefinition Height="Auto" />
   </Grid.RowDefinitions>

   <Grid.ColumnDefinitions>
      <ColumnDefinition Width="Auto" />
      <ColumnDefinition Width="*" />
      <ColumnDefinition Width="Auto" />
   </Grid.ColumnDefinitions>

   <StackPanel Name="startLineCapPanel"
               Grid.Row="0" Grid.Column="0"
               Margin="24">

      <RadioButton Content="Flat start"
                   Checked="OnStartLineCapRadioButtonChecked">
         <RadioButton.Tag>
            <PenLineCap>Flat</PenLineCap>
         </RadioButton.Tag>
      </RadioButton>

      <RadioButton Content="Round start"
                   Checked="OnStartLineCapRadioButtonChecked">
         <RadioButton.Tag>
            <PenLineCap>Round</PenLineCap>
         </RadioButton.Tag>
      </RadioButton>

      <RadioButton Content="Square start"
                   Checked="OnStartLineCapRadioButtonChecked">
         <RadioButton.Tag>
            <PenLineCap>Square</PenLineCap>
         </RadioButton.Tag>
      </RadioButton>

      <RadioButton Content="Triangle start"
                   Checked="OnStartLineCapRadioButtonChecked">
         <RadioButton.Tag>
            <PenLineCap>Triangle</PenLineCap>
         </RadioButton.Tag>
      </RadioButton>
   </StackPanel>

   <StackPanel Name="endLineCapPanel"
               Grid.Row="0" Grid.Column="2"
               Margin="24">
```

```
        <RadioButton Content="Flat end"
                     Checked="OnEndLineCapRadioButtonChecked">
            <RadioButton.Tag>
                <PenLineCap>Flat</PenLineCap>
            </RadioButton.Tag>
        </RadioButton>

        <RadioButton Content="Round end"
                     Checked="OnEndLineCapRadioButtonChecked">
            <RadioButton.Tag>
                <PenLineCap>Round</PenLineCap>
            </RadioButton.Tag>
        </RadioButton>

        <RadioButton Content="Square end"
                     Checked="OnEndLineCapRadioButtonChecked">
            <RadioButton.Tag>
                <PenLineCap>Square</PenLineCap>
            </RadioButton.Tag>
        </RadioButton>

        <RadioButton Content="Triangle End"
                     Checked="OnEndLineCapRadioButtonChecked">
            <RadioButton.Tag>
                <PenLineCap>Triangle</PenLineCap>
            </RadioButton.Tag>
        </RadioButton>
    </StackPanel>

    <StackPanel Name="lineJoinPanel"
                Grid.Row="1" Grid.Column="1"
                HorizontalAlignment="Center"
                Margin="24">

        <RadioButton Content="Bevel join"
                     Checked="OnLineJoinRadioButtonChecked">
            <RadioButton.Tag>
                <PenLineJoin>Bevel</PenLineJoin>
            </RadioButton.Tag>
        </RadioButton>

        <RadioButton Content="Miter join"
                     Checked="OnLineJoinRadioButtonChecked">
            <RadioButton.Tag>
                <PenLineJoin>Miter</PenLineJoin>
            </RadioButton.Tag>
        </RadioButton>

        <RadioButton Content="Round join"
                     Checked="OnLineJoinRadioButtonChecked">
            <RadioButton.Tag>
                <PenLineJoin>Round</PenLineJoin>
            </RadioButton.Tag>
        </RadioButton>
    </StackPanel>

    <Polyline Name="polyline"
              Grid.Row="0"
              Grid.Column="1"
              Points="0 0, 500 1000, 1000 0"
              Stroke="{StaticResource ApplicationForegroundThemeBrush}"
              StrokeThickness="100"
              Stretch="Fill"
              Margin="24" />
</Grid>
```

每组 RadioButton 控件被置于相应的 StackPanel 中。属于同一个 StackPanel 的单选按钮控件共用一个 Checked 事件处理程序。

这段标记未使任何 RadioButton 初始处于选中状态。这个操作由代码隐藏类的构造函数定义的 Loaded 处理程序完成。(若在构造函数中直接进行初始化，而不使用 Loaded 处理程序，则只有设置线对接处样式的那组 RadioButton 控件会被正确初始化，而其他两组设置不会，这甚是怪异。)

这段标记最后定义了一个较粗的 Polyline，用于展示 StrokeStartLineCap、StrokeEndLineCap 和 StrokeLineJoin 属性变化的效果。属性设置实际发生在代码隐藏文件的 Checked 事件处理程序中。

项目: LineCapsAndJoins | 文件: MainPage.xaml.cs(片段)

```
public sealed partial class MainPage : Page
{
   public MainPage()
   {
      this.InitializeComponent();

      Loaded += (sender, args) =>
         {
            foreach (UIElement child in startLineCapPanel.Children)
               (child as RadioButton).IsChecked =
                  (PenLineCap)(child as RadioButton).Tag == polyline.StrokeStartLineCap;

            foreach (UIElement child in endLineCapPanel.Children)
               (child as RadioButton).IsChecked =
                  (PenLineCap)(child as RadioButton).Tag == polyline.StrokeEndLineCap;

            foreach (UIElement child in lineJoinPanel.Children)
               (child as RadioButton).IsChecked =
                  (PenLineJoin)(child as RadioButton).Tag == polyline.StrokeLineJoin;
         };
   }

   void OnStartLineCapRadioButtonChecked(object sender, RoutedEventArgs args)
   {
      polyline.StrokeStartLineCap = (PenLineCap)(sender as RadioButton).Tag;
   }

   void OnEndLineCapRadioButtonChecked(object sender, RoutedEventArgs args)
   {
      polyline.StrokeEndLineCap = (PenLineCap)(sender as RadioButton).Tag;
   }

   void OnLineJoinRadioButtonChecked(object sender, RoutedEventArgs args)
   {
      polyline.StrokeLineJoin = (PenLineJoin)(sender as RadioButton).Tag;
   }
}
```

Loaded 处理程序对每组 RadioButton 控件进行迭代，如果当前 RadioButton 的 Tag 值匹配 Polyline 的对应属性，则将这个 RadioButton 的 IsChecked 属性设置为 true。后续 RadioButton 的选中状态取决于用户的操作。Checked 事件处理程序只需要根据被选中的 RadioButton 的 Tag 属性来修改 Polyline 的相应属性即可。此程序的运行效果如下图所示。

虽然这段标记显式地将 Tag 属性设置为 PenLineCap 或 PenLineJoin 枚举的成员，但 XAML 解析器实际上会为 Tag 赋予一个与枚举成员对应的整数值。我们可以轻而易举地将这个整数转换为对应的枚举成员，但 Tag 属性存储的绝非枚举成员本身。

通过定义自定义控件，示例项目 LineCapsAndJoins 的大部分标记都可以省略。这些自定义控件的标签(tag)属性不必是依赖属性，使用具有特定类型的普通.NET 属性即可。

示例项目 LineCapsAndJoinsWithCustomClass 展示了这种实现方式。这个派生自 RadioButton 的控件专门用来设置 PenLineCap 值。

项目：LineCapsAndJoinsWithCustomClass | 文件：LineCapRadioButton.cs

```
using Windows.UI.Xaml.Controls;
using Windows.UI.Xaml.Media;

namespace LineCapsAndJoinsWithCustomClass
{
    public class LineCapRadioButton : RadioButton
    {
        public PenLineCap LineCapTag { set; get; }
    }
}
```

类似地，用于设置 PenLineJoin 的控件如下所示。

项目：LineCapsAndJoinsWithCustomClass | 文件：LineJoinRadioButton.cs

```
using Windows.UI.Xaml.Controls;
using Windows.UI.Xaml.Media;

namespace LineCapsAndJoinsWithCustomClass
{
    public class LineJoinRadioButton : RadioButton
    {
        public PenLineJoin LineJoinTag { set; get; }
    }
}
```

这里将通过一段 XAML 来演示(上一示例中三组 RadioButton 控件的)属性元素语法是如何被省略的。

项目：LineCapsAndJoinsWithCustomClass | 文件：MainPage.xaml(片段)

```
<StackPanel Name="lineJoinPanel"
            Grid.Row="1" Grid.Column="1"
```

```
            HorizontalAlignment="Center"
            Margin="24">

    <local:LineJoinRadioButton Content="Bevel join"
                               LineJoinTag="Bevel"
                               Checked="OnLineJoinRadioButtonChecked" />

    <local:LineJoinRadioButton Content="Miter join"
                               LineJoinTag="Miter"
                               Checked="OnLineJoinRadioButtonChecked" />

    <local:LineJoinRadioButton Content="Round join"
                               LineJoinTag="Round"
                               Checked="OnLineJoinRadioButtonChecked" />
</StackPanel>
```

在 Visual Studio 键入此元素时，“智能感知”功能能够正确地将 LineCapTag 和 LineJoinTag 属性识别为相应的枚举类型，并提示可选的枚举成员，这个必须得赞一个！

使用自定义的 RadioButton 派生类后，变化基本上集中在 XAML 文件。代码隐藏文件除了省略了部分类型转换，其他大致相同。

```
项目: LineCapsAndJoinsWithCustomClass | 文件: MainPage.xaml.cs(片段)
public sealed partial class MainPage : Page
{
    public MainPage()
    {
        this.InitializeComponent();

        Loaded += (sender, args) =>
            {
                foreach (UIElement child in startLineCapPanel.Children)
                    (child as LineCapRadioButton).IsChecked =
                        (child as LineCapRadioButton).LineCapTag == polyline.StrokeStartLineCap;

                foreach (UIElement child in endLineCapPanel.Children)
                    (child as LineCapRadioButton).IsChecked =
                        (child as LineCapRadioButton).LineCapTag == polyline.StrokeEndLineCap;

                foreach (UIElement child in lineJoinPanel.Children)
                    (child as LineJoinRadioButton).IsChecked =
                        (child as LineJoinRadioButton).LineJoinTag == polyline.StrokeLineJoin;
            };
    }

    void OnStartLineCapRadioButtonChecked(object sender, RoutedEventArgs args)
    {
        polyline.StrokeStartLineCap = (sender as LineCapRadioButton).LineCapTag;
    }

    void OnEndLineCapRadioButtonChecked(object sender, RoutedEventArgs args)
    {
        polyline.StrokeEndLineCap = (sender as LineCapRadioButton).LineCapTag;
    }

    void OnLineJoinRadioButtonChecked(object sender, RoutedEventArgs args)
    {
        polyline.StrokeLineJoin = (sender as LineJoinRadioButton).LineJoinTag;
    }
}
```

5.11　键盘输入与 TextBox

触摸键盘允许用户通过点击屏幕来输入文本，这使得 Windows 8 应用程序的键盘输入变得复杂化。虽然触摸键盘对平板电脑和没有物理键盘的设备来说必不可少，但人们仍有可能为其添加物理键盘。

如此一来，使触摸键盘能够适时地弹出和消失就显得至关重要。为此，许多控件(包括自定义控件)并不自动接收键盘输入。若果真如此，系统则要在控件得到输入焦点时调用触摸键盘。从另一个角度讲，如果创建自定义控件并订阅 KeyUp 和 KeyDown 事件处理程序(或重写 OnKeyUp 和 OnKeyDown 方法)，则会发现它们不会被调用。我们需要写必要的代码来使控件获得输入焦点。

如果只希望从物理键盘获得键盘输入，而不关心触摸键盘(或许只是为了测试)，可以采用一种非常简单的方法。首先，在页面的构造函数中获取应用程序的 CoreWindow 对象。

```
CoreWindow coreWindow = Window.Current.CoreWindow;
```

CoreWindow 类位于 Windows.UI.Core 命名空间。我们可以订阅该对象的 KeyDown 和 KeyUp 事件(能够提示哪个键被按下)以及 CharacterReceived(能够将按键转换为字符)。

如果所创建的自定义控件需要从物理键盘和触摸键盘获取键盘输入，实现起来就要稍微复杂些。我们需要从 FrameworkElementAutomationPeer 类(实现了 ITextProvider 和 IValueProvider 接口)派生一个类，并通过重写自定义控件的 OnCreateAutomationPeer 方法来返回这个派生类的对象。

显然，实际做起来并不那么容易。第 16 章会展示具体步骤。

如果某个应用程序只需要文本输入，那么直接利用以下任何一个控件最为经济。

- TextBox 支持以统一的字体输入单行或多行文本。该控件在多行模式下与 Windows 中传统的“记事本”程序非常相似
- RichEditBox 支持格式化的文本，与 Windows 中传统的“写字板”程序非常相似
- PasswordBox 支持单行的密码输入

下面将重点介绍 TextBox，后面的章节也有一些示例。第 16 章会着重介绍 RichTextBox。

TextBox 定义了一个 Text 属性，我们可以通过它来设置 TextBox 的文本，也可以通过它获取 TextBox 的当前文本。SelectedText 属性能够返回被选中的文本(如果有)，SelectionStart 和 SelectionLength 属性能够返回被选中文本的起始位置和长度。如果 SelectionLength 为 0，那么 SelectionStart 返回的则是光标的位置。将 IsReadOnly 属性设置为 true 可以避免文本内容被修改，但允许用户将被选中的文本复制到“剪贴板”。“剪切”、“复制”和“粘贴”操作可以通过上下文菜单完成。TextBox 还定义了 TextChanged 事件和 SelectionChanged 事件。

TextBox 在默认情况下只允许输入单行文本。有两个属性可以修改这个行为。

- TextBox 默认忽略 Enter(回车)符。若将 AcceptsReturn 设置为 true，那么 TextBox 会在用户按下 Enter 键后另起一行。
- TextWrapping 属性的默认值为 NoWrap。若将该属性设置为 Wrap，那么超出一行

的文本会自动换行。

这两个属性可以单独设置。不论通过哪种方式换行，换行后 TextBox 的高度都会增长。TextBox 内建了 ScrollViewer。如果不希望 TextBox 无限增长，可以设置 MaxLength 属性。

触摸键盘并非只有一种。它们有的适合输入数字，有的适合输入电子邮件地址，而有的适合输入 URI。我们可以通过 TextBox 的 InputScope 属性来指定触摸键盘的类型。

示例项目 TextBoxInputScopes 展示了不同的键盘布局、多行 TextBox 的不同模式。PasswordBox 也在此一并展示。

项目：TextBoxInputScopes | 文件：MainPage.xaml(片段)

```
<Page ...>
   <Page.Resources>
      <Style TargetType="TextBlock">
         <Setter Property="FontSize" Value="24" />
         <Setter Property="VerticalAlignment" Value="Center" />
         <Setter Property="Margin" Value="6" />
      </Style>

      <Style TargetType="TextBox">
         <Setter Property="Width" Value="320" />
         <Setter Property="VerticalAlignment" Value="Center" />
         <Setter Property="Margin" Value="0 6" />
      </Style>
   </Page.Resources>

   <Grid Background="{StaticResource ApplicationPageBackgroundThemeBrush}">
      <Grid HorizontalAlignment="Center">
         <Grid.RowDefinitions>
            <RowDefinition Height="Auto" />
            <RowDefinition Height="Auto" />
            <RowDefinition Height="Auto" />
            <RowDefinition Height="Auto" />
            <RowDefinition Height="Auto" />
            <RowDefinition Height="Auto" />
            <RowDefinition Height="Auto" />
            <RowDefinition Height="Auto" />
            <RowDefinition Height="Auto" />
            <RowDefinition Height="Auto" />
         </Grid.RowDefinitions>

         <Grid.ColumnDefinitions>
            <ColumnDefinition Width="Auto" />
            <ColumnDefinition Width="Auto" />
         </Grid.ColumnDefinitions>

         <!-- Multiline with Return, no wrapping -->
         <TextBlock Text="Multiline (accepts Return, no wrap):"
                    Grid.Row="0" Grid.Column="0" />

         <TextBox AcceptsReturn="True"
                 Grid.Row="0" Grid.Column="1" />

         <!-- Multiline with no Return, wrapping -->
         <TextBlock Text="Multiline (ignores Return, wraps):"
                    Grid.Row="1" Grid.Column="0" />

         <TextBox TextWrapping="Wrap"
                  Grid.Row="1" Grid.Column="1" />

         <!-- Multiline with Return and wrapping -->
         <TextBlock Text="Multiline (accepts Return, wraps):"
                    Grid.Row="2" Grid.Column="0" />

         <TextBox AcceptsReturn="True"
```

```
                TextWrapping="Wrap"
                Grid.Row="2" Grid.Column="1" />

        <!-- Default input scope -->
        <TextBlock Text="Default input scope:"
                   Grid.Row="3" Grid.Column="0" />

        <TextBox Grid.Row="3" Grid.Column="1"
                 InputScope="Default" />

        <!-- Email address input scope -->
        <TextBlock Text="Email address input scope:"
                   Grid.Row="4" Grid.Column="0" />

        <TextBox Grid.Row="4" Grid.Column="1"
                 InputScope="EmailSmtpAddress" />

        <!-- Number input scope -->
        <TextBlock Text="Number input scope:"
                   Grid.Row="5" Grid.Column="0" />

        <TextBox Grid.Row="5" Grid.Column="1"
                 InputScope="Number" />

        <!-- Search input scope -->
        <TextBlock Text="Search input scope:"
                   Grid.Row="6" Grid.Column="0" />

        <TextBox Grid.Row="6" Grid.Column="1"
                 InputScope="Search" />

        <!-- Telephone number input scope -->
        <TextBlock Text="Telephone number input scope:"
                   Grid.Row="7" Grid.Column="0" />

        <TextBox Grid.Row="7" Grid.Column="1"
                 InputScope="TelephoneNumber" />

        <!-- URL input scope -->
        <TextBlock Text="URL input scope:"
                   Grid.Row="8" Grid.Column="0" />

        <TextBox Grid.Row="8" Grid.Column="1"
                 InputScope="Url" />

        <!-- PasswordBox -->
        <TextBlock Text="PasswordBox:"
                   Grid.Row="9" Grid.Column="0" />

        <PasswordBox Grid.Row="9" Grid.Column="1" />
      </Grid>
    </Grid>
</Page>
```

在选择多行模式或设置 InputScope 时，不妨通过这个程序来做试验。

5.12　触摸与 Thumb

第 13 章将讨论触摸输入以及如何通过触摸输入来操纵屏幕上的对象。相对而言，Thumb 是一个较为简单的控件，提供了基本的触摸功能。Thumb 定义于 Windows.UI.Xaml.Controls.Primitives 命名空间，主要用作 Slider 和 Scrollbar 的构建块。第 8 章将进一步介绍一种自定义的网格分隔控件。

Thumb 控件能够根据鼠标、触控笔或触摸相对自身的滑动生成 3 种事件：DragStarted、DragDelta 和 DragCompleted。当手指点击 Thumb 控件或者鼠标单击该控件时，DragStarted 事件会被引发。随后，DragDelta 事件会被接连引发，从而提示手指或鼠标的移动轨迹。我们可以通过这两个事件来移动 Thumb(和其他控件)，如果将 Canvas 作为容器，则实现起来最为方便。DragCompleted 事件可提示手指抬起或鼠标按钮被释放。

示例程序 AlphabetBlocks 定义了一些标有字母、数字和符号的按钮。这些按钮环绕在屏幕四周。单击任意一个按钮，屏幕上便会显示一个可用手指或鼠标拖动的字母块。人们可能希望快速滑动手指使字母块在屏幕上自由滑动，但这种现象不会发生。Thumb 不支持触摸惯性。为了获得惯性效果，需要用到以 Manipulation 开头的触摸事件。

字母块本身是 UserControl 的派生类，名为 Block。XAML 文件中定义了边长为 144 像素的正方形区域，包含图片、Thumb、矢量图和 TextBlock。

项目：AlphabetBlocks | 文件：Block.xaml

```
<UserControl
    x:Class="AlphabetBlocks.Block"
    xmlns="http://schemas.microsoft.com/winfx/2006/xaml/presentation"
    xmlns:x="http://schemas.microsoft.com/winfx/2006/xaml"
    xmlns:local="using:AlphabetBlocks"
    Width="144"
    Height="144"
    Name="root">

    <Grid>
        <Thumb DragStarted="OnThumbDragStarted"
               DragDelta="OnThumbDragDelta"
               Margin="18 18 6 6" />

        <!-- Left -->
        <Polygon Points="0 6, 12 18, 12 138, 0 126"
                 Fill="#E0C080" />

        <!-- Top -->
        <Polygon Points="6 0, 18 12, 138 12, 126 0"
                 Fill="#F0D090" />

        <!-- Edge -->
        <Polygon Points="6 0, 18 12, 12 18, 0 6"
                 Fill="#E8C888" />

        <Border BorderBrush="{Binding ElementName=root, Path=Foreground}"
                BorderThickness="12"
                Background="#FFE0A0"
                CornerRadius="6"
                Margin="12 12 0 0"
                IsHitTestVisible="False" />

        <TextBlock FontFamily="Courier New"
                   FontSize="156"
                   FontWeight="Bold"
                   Text="{Binding ElementName=root, Path=Text}"
                   HorizontalAlignment="Center"
                   VerticalAlignment="Center"
                   Margin="12 18 0 0"
                   IsHitTestVisible="False" />
    </Grid>
</UserControl>
```

图形 Polygon 与 Polyline 类似，只不过前者能够自动闭合图形，然后使用 Fill 属性指定的画笔来填充内部区域。

此程序订阅了 Thumb 的 DragStarted 和 DragDelta 事件。这个 Thumb 之上有两个元素(Border 和 TextBlock)，在视觉上遮住了 Thumb。但这两个元素的 IsHitTestVisible 属性被设置为 false，因此二者不会拦截到达 Thumb 的触摸输入。

Border 的 BorderBrush 属性被绑定至根元素的 BorderBrush 属性。您或许还记得，BorderBrush 属性是 Control 类定义的，也被 UserControl 继承，能够传播至整个可视树。TextBlock 的 Foreground 属性能够自动获取这个画笔。TextBlock 的 Text 属性与这个用户控件的 Text 属性绑定。UserControl 并没有定义 Text 属性，这说明该属性可能是另行添加的。

代码隐藏文件证实了这一假设。这个 Text 属性是以依赖属性的形式定义的。

项目：AlphabetBlocks | 文件：Block.xaml.cs

```
using Windows.UI.Xaml;
using Windows.UI.Xaml.Controls;
using Windows.UI.Xaml.Controls.Primitives;

namespace AlphabetBlocks
{
   public sealed partial class Block : UserControl
   {
      static int zindex;

      static Block()
      {
         TextProperty = DependencyProperty.Register("Text",
            typeof(string),
            typeof(Block),
            new PropertyMetadata("?"));
      }

      public static DependencyProperty TextProperty { private set; get; }

      public static int ZIndex
      {
         get { return ++zindex; }
      }

      public Block()
      {
         this.InitializeComponent();
      }

      public string Text
      {
         set { SetValue(TextProperty, value); }
         get { return (string)GetValue(TextProperty); }
      }

      void OnThumbDragStarted(object sender, DragStartedEventArgs args)
      {
         Canvas.SetZIndex(this, ZIndex);
      }

      void OnThumbDragDelta(object sender, DragDeltaEventArgs args)
      {
         Canvas.SetLeft(this, Canvas.GetLeft(this) + args.HorizontalChange);
         Canvas.SetTop(this, Canvas.GetTop(this) + args.VerticalChange);
      }
   }
}
```

Block 类还定义了静态属性 ZIndex，这里需要特别说明一下。当用户单击程序中的按钮后，Block 对象会被创建并添加至 Canvas。按 Block 在集合中出现的顺序，后创建的 Block

会显示在先创建的 Block 之上。而实际上，我们希望在手指点击 Block 之后，它能够显示在屏幕的最上层。也就是说，它的 z-index 值应大于其他 Block。

静态的 ZIndex 属性能够实现这一行为。这个值每次返回前都会自增。DragStarted 事件被引发说明用户点击了某个 Block。Canvas.SetZIndex 方法会为当前 Block 分配一个高于其他 Block 的 z-index 值。在 ZIndex 属性达到最大值后，程序会出错。但实际上这种情况难以发生。(Windows Runtime 所支持的最大 z-index 值为 1 000 000。也就是说，如果每秒钟移动一次字母块，那么程序运行到第 12 天才会抛出异常。)

Thumb 的 DragDelta 事件会通过 HorizontalChange 和 VerticalChange 属性通知程序触摸点或鼠标指针相对该控件移动的距离。我们可以直接使用这两个属性来累加附加属性 Canvas.Left 和 Canvas.Top。

本示例的 MainPage.xaml 文件颇为简单，基本上只是在屏幕中央显示程序的名称。

项目: AlphabetBlocks | 文件: MainPage.xaml(片段)

```
<Grid Background="{StaticResource ApplicationPageBackgroundThemeBrush}"
      SizeChanged="OnGridSizeChanged">

    <TextBlock Text="Alphabet Blocks"
               FontStyle="Italic"
               FontWeight="Bold"
               FontSize="96"
               TextWrapping="Wrap"
               HorizontalAlignment="Center"
               VerticalAlignment="Center"
               TextAlignment="Center"
               Opacity="0.1" />

    <Canvas Name="buttonCanvas" />
    <Canvas Name="blockcanvas" />
</Grid>
```

请注意，这里订阅了 Grid 的 SizeChanged 事件。当页面尺寸变化时，相应的处理程序能够重新创建所有 Button 对象，使其环绕在屏幕四周。代码隐藏类的大部分代码都是为了实现这个效果。

项目: AlphabetBlocks | 文件: MainPage.xaml.cs(片段)

```
public sealed partial class MainPage : Page
{
    const double BUTTON_SIZE = 60;
    const double BUTTON_FONT = 18;
    string blockChars = "ABCDEFGHIJKLMNOPQRSTUVWXYZ0123456789!?-+*/%=";
    Color[] colors = { Colors.Red, Colors.Green, Colors.Orange, Colors.Blue, Colors.Purple };
    Random rand = new Random();

    public MainPage()
    {
        this.InitializeComponent();
    }

    void OnGridSizeChanged(object sender, SizeChangedEventArgs args)
    {
        buttonCanvas.Children.Clear();

        double widthFraction = args.NewSize.Width /
                    (args.NewSize.Width + args.NewSize.Height);
        int horzCount = (int)(widthFraction * blockChars.Length / 2);
        int vertCount = (int)(blockChars.Length / 2 - horzCount);
        int index = 0;

        double slotWidth = (args.NewSize.Width - BUTTON_SIZE) / horzCount;
```

```
        double slotHeight = (args.NewSize.Height - BUTTON_SIZE) / vertCount + 1;

        // Across top
        for (int i = 0; i < horzCount; i++)
        {
            Button button = MakeButton(index++);
            Canvas.SetLeft(button, i * slotWidth);
            Canvas.SetTop(button, 0);
            buttonCanvas.Children.Add(button);
        }

        // Down right side
        for (int i = 0; i < vertCount; i++)
        {
            Button button = MakeButton(index++);
            Canvas.SetLeft(button, this.ActualWidth - BUTTON_SIZE);
            Canvas.SetTop(button, i * slotHeight);
            buttonCanvas.Children.Add(button);
        }

        // Across bottom from right
        for (int i = 0; i < horzCount; i++)
        {
            Button button = MakeButton(index++);
            Canvas.SetLeft(button, this.ActualWidth - i * slotWidth - BUTTON_SIZE);
            Canvas.SetTop(button, this.ActualHeight - BUTTON_SIZE);
            buttonCanvas.Children.Add(button);
        }

        // Up left side
        for (int i = 0; i < vertCount; i++)
        {
            Button button = MakeButton(index++);
            Canvas.SetLeft(button, 0);
            Canvas.SetTop(button, this.ActualHeight - i * slotHeight - BUTTON_SIZE);
            buttonCanvas.Children.Add(button);
        }
    }

    Button MakeButton(int index)
    {
        Button button = new Button
        {
            Content = blockChars[index].ToString(),
            Width = BUTTON_SIZE,
            Height = BUTTON_SIZE,
            FontSize = BUTTON_FONT,
            Tag = new SolidColorBrush(colors[index % colors.Length]),
        };
        button.Click += OnButtonClick;
        return button;
    }

    void OnButtonClick(object sender, RoutedEventArgs e)
    {
        Button button = sender as Button;

        Block block = new Block
        {
            Text = button.Content as string,
            Foreground = button.Tag as Brush
        };
        Canvas.SetLeft(block, this.ActualWidth / 2 - 144 * rand.NextDouble());
        Canvas.SetTop(block, this.ActualHeight / 2 - 144 * rand.NextDouble());
        Canvas.SetZIndex(block, Block.ZIndex);
        blockcanvas.Children.Add(block);
    }
}
```

Block 对象是在 Button 的 Click 事件处理程序中创建的。该处理程序会将其随机置于屏幕上的某个位置。读者可以尝试通过摆放字母块得到一个特别的 Hello Windows 8 程序(如下图所示)。

第 6 章　WinRT 与 MVVM

“概念隔离”是软件开发最重要的原则之一。大型应用程序最好分层进行开发、调试和维护。在交互性较强的图形环境中，内容的表示与内容本身的分离是最显著的。表示层是程序显示控件(和其他图形)及与用户交互的部分。表示层下面是业务逻辑与数据提供程序。

为帮助开发者认识和实现概念隔离，一些架构模式应运而生。在基于 XAML 的编程环境中，“模型-视图-视图模型”(Model-View-ViewModel，简称 MVVM)颇受欢迎。在 MVVM 模式下，XAML 可用来实现表示层，然后通过数据绑定和命令将表示层与底层业务逻辑关联起来。

与本书类似，很多书籍一般都通过一些小程序来演示某种功能或概念。为适应某种架构模式，小程序往往会变得臃肿。MVVM 对小程序来说不但大材小用，还可能容易让人混淆。

数据绑定和命令是 Windows Runtime 的重要组成部分。理解这两个概念有助于理解 MVVM 架构的实现方法。

6.1　MVVM 简介

顾名思义，采用“模型-视图-视图模型”(Model-View-ViewModel，简称 MVVM)模式的应用程序一般分为三层。

- “模型”是处理数据和原始内容的层次。该层次一般用来获取和维护来自文件或 Web 服务的数据。
- “视图”是控件和图形的表示层，一般通过 XAML 实现。
- “视图模型”位于“模型”和“视图”之间。一般情况下，该层次负责使来自“模型”的数据或内容更易于在“视图”中呈现。

“模型”层一般不可或缺，但本章的示例程序不涉及这部分。

如果三个层次间的交互是通过过程式的方法调用实现的，那么调用层次可能是这样的：

```
"视图" → "视图模型" → "模型"
```

除了事件，反方向的调用是违背设计原则的。“模型”可以定义供“视图模型”订阅的事件，“视图模型”可以定义供“视图”订阅的事件。事件允许“视图模型”通知“视图”数据被更新，而“视图”可以通过“视图模型”来获取最新的数据。

“视图”和“视图模型”之间一般是通过数据绑定和命令来交互的。也就是说，大部分(甚至全部)方法调用和事件处理实际是在幕后进行的。数据绑定和命令可以实现以下三种交互场景。

- “视图”将用户输入传给“视图模型”。

- “视图模型”在数据被更新后通知“视图”。
- “视图”从“视图模型”获取用于呈现的最新数据。

简化代码隐藏文件的代码(至少在页面和窗口一级)是 MVVM 的目标之一。最理想的情况是“视图”与“视图模型”完全通过 XAML 文件定义的数据绑定来建立关联。

6.2 数据绑定通知

第 5 章介绍过这样的数据绑定。

```
<TextBlock Text="{Binding ElementName=slider, Path=Value}" />
```

这个绑定发生在两个 FrameworkElement 派生类的对象之间，绑定目标是这个 TextBlock 的 Text 属性，绑定源是 Slider 对象的 Value 属性，对象名为 slider。这里的绑定目标和源均是以依赖属性方式实现的属性。绑定目标必须是依赖属性，但绑定源不必是(正如接下来要展示的)。

TextBlock 显示的文本随 Slider 的 Value 属性变化而变化。这是如何实现的？若绑定源是依赖属性，幕后的机制则取决于 Windows Runtime 的实现。毋庸置疑，此时会引发某个事件。Slider 具有能够在 Value 属性变化时进行通知的事件，而 Binding 对象会订阅该事件。在事件发生时，Binding 对象会将最新的 double 值转换为 string，并将结果设置到 TextBlock 的 Text 属性。如果知道 Slider 拥有公共的 ValueChanged 事件，并且该事件能够在 Value 属性变化时被引发，那么这个过程就不难理解了。

对于“视图模型”，数据绑定的一端发生了变化：绑定的目标仍旧是 XAML 文件中的元素，但绑定源变成“视图模型”的属性。这就是“视图模型”和“视图”(XAML 文件)之间来回传输数据的基本方式。

绑定源不必是依赖属性，但为了使数据绑定能够正常工作，绑定源必须实现某种通知机制，以便在属性变化时通知 Binding 对象。这种通知并不是凭空发生的，必须通过事件来实现。

作为绑定源的“视图模型”一般需要实现 System.ComponentModel 命名空间中的 INotifyPropertyChanged 接口。该接口的定义非常简单。

```
public interface INotifyPropertyChanged
{
   event PropertyChangedEventHandler PropertyChanged;
}
```

PropertyChangedEventHandler 委托在签名中使用了 PropertyChangedEventArgs 类。该类定义了一个名为 PropertyName 的属性，类型为 string。实现 INotifyPropertyChanged 的类需要在属性值变化时引发该接口的 PropertyChanged 事件。

下面是一个实现 INotifyPropertyChanged 的类。TotalScore 属性会在它的值发生变化时引发 PropertyChanged 事件。

```
public class SimpleViewModel : INotifyPropertyChanged
{
   double totalScore;

   public event PropertyChangedEventHandler PropertyChanged;
```

```
        public double TotalScore
        {
            set
            {
                if (totalScore != value)
                {
                    totalScore = value;

                    if (PropertyChanged != null)
                        PropertyChanged(this, new PropertyChangedEventArgs("TotalScore"));
                }
            }
            get
            {
                return totalScore;
            }
        }
    }
```

TotalScore 属性后面对应一个名为 totalScore 的字段。这个属性会将传入 set 访问器的值与 totalScore 字段进行比较。如有不同，则引发 PropertyChanged 事件。请不要图一时省事而省略这个步骤！这个事件名为 PropertyChanged，而非 PropertySetAndPerhapsChangedOrMaybeNot[①]。

虽然系统允许实现 INotifyPropertyChanged 接口的类总是不引发 PropertyChanged 事件，但不应这样做。

若类中存在较多属性，则有必要定义一个受保护、名为 OnPropertyChanged 的方法，并通过该方法来引发事件。此过程还可以自动执行，正如稍后要介绍的。

在设计“视图”和“视图模型”时，应考虑将控件看作数据类型的可视化表示。“视图”中的控件通过数据绑定与“视图模型”的属性进行关联。例如，将 Slider 看作 double 的可视化表示，将 TextBox 与 string 对应，将 CheckBox 和 ToggleSwitch 与 bool 对应，将一组 RadioButton 控件与特定的枚举类型对应。[②]

6.3 ColorScroll 的视图模型

第 5 章介绍的 ColorScroll 程序演示了如何通过数据绑定将 Slider 的 Value 属性更新到 TextBlock。但通过定义数据绑定来根据三个 Slider 的值来更新颜色则遇到些困难。这可以办到吗？

我们可以通过一个单独的类根据 Red、Green 和 Blue 属性来创建 Color 对象。这三个属性中的任意一个发生改变都重新计算 Color 属性。在 XAML 文件中，将三个 Slider 控件分别与 Red、Green 和 Blue 属性绑定，将 SolidColorBrush 与 Color 属性绑定。即便没有明说这个类是“视图模型”，但它的确承担了这个角色。

在示例项目中，RgbViewModel 类实现了 INotifyPropertyChanged 接口，能够在 Red、Green、Blue 和 Color 属性改变时引发 PropertyChanged 事件。

---

① 译注：这个名称的事件并不存在，而只是为了提醒：此事件引发时属性值已被设置并且发生了改变，不能模棱两可。

② 译注：我们可以将“视图模型”看作“视图”的抽象。也就是将“视图模型”看作可以通过代码操纵的“视图”。“视图”的数据和表现的行为在“视图模型”中都能找到对应的成员。

```
项目: ColorScrollWithViewModel | 文件: RgbViewModel.cs
using System.ComponentModel;      // for INotifyPropertyChanged
using Windows.UI;                 // for Color

namespace ColorScrollWithViewModel
{
    public class RgbViewModel : INotifyPropertyChanged
    {
        double red, green, blue;
        Color color = Color.FromArgb(255, 0, 0, 0);

        public event PropertyChangedEventHandler PropertyChanged;

        public double Red
        {
            set
            {
                if (red != value)
                {
                    red = value;
                    OnPropertyChanged("Red");
                    Calculate();
                }
            }
            get
            {
                return red;
            }
        }

        public double Green
        {
            set
            {
                if (green != value)
                {
                    green = value;
                    OnPropertyChanged("Green");
                    Calculate();
                }
            }
            get
            {
                return green;
            }
        }

        public double Blue
        {
            set
            {
                if (blue != value)
                {
                    blue = value;
                    OnPropertyChanged("Blue");
                    Calculate();
                }
            }
            get
            {
                return blue;
            }
        }

        public Color Color
        {
            protected set
```

```
            {
               if (color != value)
               {
                  color = value;
                  OnPropertyChanged("Color");
               }
            }
            get
            {
               return color;
            }
        }

        void Calculate()
        {
            this.Color = Color.FromArgb(255, (byte)this.Red, (byte)this.Green, (byte)this.Blue);
        }

        protected void OnPropertyChanged(string propertyName)
        {
            if (PropertyChanged != null)
               PropertyChanged(this, new PropertyChangedEventArgs(propertyName));
        }
    }
}
```

这个类最后的 OnPropertyChanged 方法接受属性的名称并进而引发 PropertyChanged 事件。

为了使用数据绑定，Red、Green 和 Blue 属性的类型被定义为 double。这三个属性可以算作“视图模型”的输入。由于输入很可能来自 Slider，因而选用 double 类型可以更好地实现对接。

Red、Green 和 Blue 属性的 set 访问器会在值发生变化后引发 PropertyChanged 事件并调用 Calculate 方法。Calculate 方法会重新设置 Color 属性，而这会再一次引发 PropertyChanged 事件来通知 Color 属性发生了变化。Color 属性本身有一个受保护的 set 访问器，这表明该类并非反过来根据 Color 值来计算 Red、Green 和 Blue 属性。(稍后会继续讨论这个问题。)

RgbViewModel 类是示例项目 ColorScrollWithViewModel 的一部分。该类在 MainPage.xaml 文件的 Resources 区段被实例化。

```
项目: ColorScrollWithViewModel | 文件: MainPage.xaml(片段)
<Page.Resources>
   <local:RgbViewModel x:Key="rgbViewModel" />
   ...
</Page.Resources>
```

请注意，这里使用 local 来指定命名空间。

使 XAML 文件能够访问“视图模型”对象有两种简单方法。其中之一是将该对象作为资源。正如第 2 章介绍的，在 Resources 区段引用的类只会被实例化一次，供所有 StaticResource 引用共享。多个绑定引用一个对象的情况，都可以如法炮制。

三个 Slider 控件大致相同，这里展示其中一个即可。

```
项目: ColorScrollWithViewModel | 文件: MainPage.xaml(片段)
<!-- Red -->
<TextBlock Text="Red"
           Grid.Column="0"
           Grid.Row="0"
           Foreground="Red" />
```

```
<Slider Grid.Column="0"
        Grid.Row="1"
        Value="{Binding Source={StaticResource rgbViewModel},
                        Path=Red,
                        Mode=TwoWay}"
        Foreground="Red" />

<TextBlock Text="{Binding Source={StaticResource rgbViewModel},
                          Path=Red,
                          Converter={StaticResource hexConverter}}"
           Grid.Column="0"
           Grid.Row="2"
           Foreground="Red" />
```

请注意，Slider 元素不再需要 Name 特性，因为 XAML 中没有其他元素引用它，代码隐藏文件中也不需要引用它。上一个版本中的 ValueChanged 事件处理程序也不再需要了。代码隐藏文件只包含对 InitializeComponent 方法的调用。

请特别注意一下 Slider 上绑定的定义。

```
<Slider ...
      Value="{Binding Source={StaticResource rgbViewModel},
                      Path=Red,
                      Mode=TwoWay}" ... />
```

这个绑定定义略长，因而把它拆分成三行。这里不需要指定 ElementName，因为绑定没有引用 XAML 文件中的其他元素。它引用的是以 XAML 资源形式实例化的对象，因而必须通过 Source 和 StaticResource 语法来进行引用。这个绑定的目标为 Slider 的 Value 属性，源为 RgbViewModel 实例的 Red 属性。

数据如何传至“视图模型”？难道 Slider 不需要向 RgbViewModel 传值？

需要，但是 RgbViewModel 必须作为绑定源，而不能是目标。这个“视图模型”不能作为绑定目标是因为它没有依赖属性。尽管表面上 Slider 的 Value 属性是绑定目标，但实际上我们希望 Slider 将值传给 Red 属性。为此，Binding 的 Mode 属性被设置为 TwoWay。此设置意味着以下两点。

- 源属性被更新促使目标属性随之更新(数据绑定的一般情况)。
- 目标值被更新促使源属性随之更新(这里所讨论的重点)。

Mode 属性的默认设置为 OneWay。还有一个选项为 OneTime，意为仅在绑定建立之初将绑定源更新到绑定目标。在 OneTime 模式下，如果源属性发生后续的更改，则不会进行目标的更新。若绑定源没有通知机制，则可以使用 OneTime 模式。

还需要注意，TextBlock 在这个示例中也被绑定至 RgbViewModel 对象。

```
<TextBlock Text="{Binding Source={StaticResource rgbViewModel},
                          Path=Red,
                          Converter={StaticResource hexConverter}}" ... />
```

此元素的绑定可以像之前的示例项目一样直接引用 Slider，但引用 RgbViewModel 更为可取。默认的 OneWay 模式在这里适用是因为只需要将值从源传给目标。

SolidColorBrush 的 Color 属性的绑定也可以采用 OneWay 模式。

项目：ColorScrollWithViewModel | 文件：MainPage.xaml(片段)

```
<Rectangle Grid.Column="3"
           Grid.Row="0"
           Grid.RowSpan="3">
    <Rectangle.Fill>
```

```
        <SolidColorBrush Color="{Binding Source={StaticResource rgbViewModel},
                                         Path=Color}" />
    </Rectangle.Fill>
</Rectangle>
```

SolidColorBrush 不再需要 x:Name 特性，因为我们不需要在代码隐藏文件中引用该画笔。

当然，相比从代码隐藏文件移除的 ValueChanged 事件处理程序，RgbViewModel 类中的代码要更多一些。正如前文所提到的，MVVM 对小程序来说有些臃肿。对于大型应用程序，起初为获得更清晰的架构的确要付出一定的代价，但将表示与业务逻辑分离是长远之举。

RgbViewModel 类中，Color 属性的 set 访问器被定义为受保护的，因而只能在类内部访问它。有人或许会问，这样做有必要吗？或许可以这样设计：外界对 Color 属性进行修改后使 Red、Green 和 Blue 属性被重新计算。

```
public Color Color
{
    set
    {
        if (color != value)
        {
            color = value;
            OnPropertyChanged("Color");
            this.Red = color.R;
            this.Green = color.G;
            this.Blue = color.B;
        }
    }
    get
    {
        return color;
    }
}
```

乍一看这是自找麻烦，因为这会导致属性和 OnPropertyChanged 方法不断递归。但这种情况实际不会发生，如果属性值未发生改变，set 访问器不会进行任何操作，因而这样做是安全的。

但这样做还是有缺陷的。举例来说，若 Color 属性当前为 RGB 值(0, 0, 0)，但被设置为(255, 128, 0)。当 Red 属性被设置为 255，PropertyChanged 事件会被引发，Color(和 color 字段)被设置为(255, 0, 0)，最终 Green 和 Blue 会被设置为 0。

与其试图防止重入(re-entry)的发生，不如通过逻辑上的修改来实现最初的目标。下面这个版本可以工作，即便它会使 PropertyChanged 事件“泛起一些涟漪”。

```
public Color Color
{
    set
    {
        if (color != value)
        {
            color = value;
            OnPropertyChanged("Color");
            this.Red = value.R;
            this.Green = value.G;
            this.Blue = value.B;
        }
    }
    get
```

```
    {
        return color;
    }
}
```

本章将在此程序的下一个版本中将 Color 属性的 set 访问器定义为公共的。

6.4 精简的语法

通过前文展示的 RgbViewModel，可能让人觉得实现 INotifyPropertyChanged 是一件繁琐的事。的确如此。为简化这一过程，Visual Studio 会在“网格应用程序”和“拆分视图应用程序”的 Common 文件夹中创建 BindableBase 类。(请勿将 BindableBase 类与 Binding 的父类 BindableBase 类混淆。)

Visual Studio 不会为“空白应用程序”创建 BindableBase 类。我们不妨读一下这个类的代码，看看从中能够学到什么。

BindableBase 类所处命名空间的名称由项目名、句点和 Common 组成。下面是去掉注释和部分特性(attribute)后的代码。

```
public abstract class BindableBase : INotifyPropertyChanged
{
    public event PropertyChangedEventHandler PropertyChanged;

    protected bool SetProperty<T>(ref T storage, T value,
        [CallerMemberName] String propertyName = null)
    {
        if (object.Equals(storage, value)) return false;

        storage = value;
        this.OnPropertyChanged(propertyName);
        return true;
    }

    protected void OnPropertyChanged([CallerMemberName] string propertyName = null)
    {
        var eventHandler = this.PropertyChanged;
        if (eventHandler != null)
        {
            eventHandler(this, new PropertyChangedEventArgs(propertyName));
        }
    }
}
```

派生于 BindableBase 的类可以在属性的 set 访问器中调用 SetProperty 方法。SetProperty 方法的签名虽然看起来有些复杂，使用起来却很简单。例如，对于名为 Red、类型为 double 的属性，可以像下面这样定义其背后的字段。

```
double red;
```

在 set 访问器中调用 SetProperty 的方法如下所示。

```
SetProperty<double>(ref red, value, "Red");
```

请注意 BindableBase 中 CallerMemberName 的使用。此特性(attribute)是.NET 4.5 引入的，C# 5.0 可以通过它来获取主调属性或方法的信息。由于利用了此特性，所以调用 SetProperty 一般不必指定最后一个参数。如果在 Red 属性的 set 访问器中像下面这样调用

SetProperty，属性名就会被自动设置。

```
SetProperty<double>(ref red, value);
```

如果属性值实际被更改，SetProperty 会返回 true。若希望针对新值进一步执行某种逻辑，则可以利用这个返回值。接下来要介绍的示例项目名为 ColorScrollWithDataContext。此项目也包含一个 RgbViewModel 类，但它借用了 BindableBase 中的一些代码。在这个项目中，Color 属性的 set 访问器被声明为公共成员。

项目：ColorScrollWithDataContext | 文件：RgbViewModel.cs

```
using System.ComponentModel;
using System.Runtime.CompilerServices;
using Windows.UI;

namespace ColorScrollWithDataContext
{
   public class RgbViewModel : INotifyPropertyChanged
   {
      double red, green, blue;
      Color color = Color.FromArgb(255, 0, 0, 0);

      public event PropertyChangedEventHandler PropertyChanged;

      public double Red
      {
         set
         {
            if (SetProperty<double>(ref red, value, "Red"))
               Calculate();
         }
         get
         {
            return red;
         }
      }

      public double Green
      {
         set
         {
            if (SetProperty<double>(ref green, value))
               Calculate();
         }
         get
         {
            return green;
         }
      }

      public double Blue
      {
         set
         {
            if (SetProperty<double>(ref blue, value))
               Calculate();
         }
         get
         {
            return blue;
         }
      }

      public Color Color
      {
         set
```

```
            {
                if (SetProperty<Color>(ref color, value))
                {
                    this.Red = value.R;
                    this.Green = value.G;
                    this.Blue = value.B;
                }
            }
            get
            {
                return color;
            }
        }

        void Calculate()
        {
            this.Color = Color.FromArgb(255, (byte)this.Red, (byte)this.Green, (byte)this.Blue);
        }

        protected bool SetProperty<T>(ref T storage, T value,
                                [CallerMemberName] string propertyName = null)
        {
            if (object.Equals(storage, value))
                return false;

            storage = value;
            OnPropertyChanged(propertyName);
            return true;
        }

        protected void OnPropertyChanged(string propertyName)
        {
            if (PropertyChanged != null)
                PropertyChanged(this, new PropertyChangedEventArgs(propertyName));
        }
    }
}
```

这种实现 INotifyPropertyChanged 的方法更简洁。下一节要介绍的 ColorScrollWithDataContext 项目会采用这个版本的 RgbViewModel。

6.5 DataContext 属性

到目前为止，本书介绍了三个为绑定指定源对象的选项：ElementName、RelativeSource 和 Source。ElementName 适合引用 XAML 内部的元素，RelativeSource 使绑定能够引用目标对象的属性。(第 11 章会介绍 RelativeSource 另一种更重要的作用。)第三个选项 Source 一般与 StaticResource 联用来访问 Resources 集合中的对象。

还有一种指定绑定源的方法：如果 ElementName、RelativeSource 和 Source 均为 null，Binding 对象会选择目标对象的 DataContext 属性。

DataContext 属性是 FrameworkElement 定义的。它有一个有趣(且重要)的性质，即其值能够传播至整个可视树。像这样传播的属性并不多，其中包括 Foreground 和所有与字体相关的属性，但鲜有其他属性。DataContext 是一个特例。代码隐藏类的构造函数可以用来实例化视图模型，并将该实例设置到页面的 DataContext 属性上。示例项目 ColorScrollWithDataContext 的 MainPage.xaml.cs 文件正是这样做的。

```
项目: ColorScrollWithDataContext | 文件: MainPage.xaml.cs(片段)
public MainPage()
{
    this.InitializeComponent();
    this.DataContext = new RgbViewModel();

    // Initialize to highlight color
    (this.DataContext as RgbViewModel).Color =
            new UISettings().UIElementColor(UIElementType.Highlight);
}
```

选择在代码中实例化视图模型的原因可能有很多。例如，视图模型有一个含参构造函数。这样的对象 XAML 是无法实例化的。

为了测试 Color 属性，这里将其设置为系统的高亮颜色。

使用 DataContext 的好处之一是数据绑定语法得到了简化。由于不需要再设置 Source，因而绑定标记可以像下面这样写。

```
<Slider ... Value="{Binding Path=Red, Mode=TwoWay}" ... />
```

如果 Path 设置是绑定标记的第一个属性，不可以省略 Path=。

```
<Slider ... Value="{Binding Red, Mode=TwoWay}" ... />
```

这样，Binding 语法便得到进一步简化。

不论选择哪种绑定源，Path=都可以省略，但前提是 Path 是绑定语法中的第一个属性。本人倾向于将 Source 或 ElementName 作为第一个属性，而在使用 DataContext 时才省略 Path=。

下面这段 XAML 标记展示了新语法的使用方法。由于它们变得非常简单，因而不再需要拆成多行展示。

```
项目: ColorScrollWithDataContext | 文件: MainPage.xaml(片段)
<!-- Red -->
<TextBlock Text="Red"
           Grid.Column="0"
           Grid.Row="0"
           Foreground="Red" />

<Slider Grid.Column="0"
        Grid.Row="1"
        Value="{Binding Red, Mode=TwoWay}"
        Foreground="Red" />

<TextBlock Text="{Binding Red, Converter={StaticResource hexConverter}}"
           Grid.Column="0"
           Grid.Row="2"
           Foreground="Red" />
...
<!-- Result -->
<Rectangle Grid.Column="3"
           Grid.Row="0"
           Grid.RowSpan="3">
    <Rectangle.Fill>
        <SolidColorBrush Color="{Binding Color}" />
    </Rectangle.Fill>
</Rectangle>
```

两种方法可以配合使用。例如，可以在 XAML 文件的 Resources 集合中实例化“视图模型”。

```
<Page.Resources>
   ...
   <local:RgbViewModel x:Key="rgbViewModel" />
   ...
</Page.Resources>
```

同时，在可视树开始的某处设置 DataContext 属性：

```
<Grid ... DataContext="{StaticResource rgbViewModel}" ... >
```

或

```
<Grid ... DataContext="{Binding Source={StaticResource rgbViewModel}}" ... >
```

如果希望将 DataContext 设置为“视图模型”的某个属性，而非“视图模型”本身，则需要采用第二种方式。第 11 章在讨论集合时会展示更多示例。

6.6 绑定与 TextBox

将底层业务逻辑隔离，好处之一是能够在不使用“视图模型”的情况下构建用户界面。假设需要一个类似 ColorScroll 的颜色选择程序，但每种颜色是在 TextBox 中输入的。虽然这样的程序难以使用，但的确是可以实现的。

示例项目 ColorTextBoxes 采用了 ColorScrollWithDataContext 项目的 RgbViewModel 类。两个项目代码隐藏类的构造函数也是一致的。

项目：ColorTextBoxes | 文件：MainPage.xaml.cs(片段)

```
public MainPage()
{
   this.InitializeComponent();
   this.DataContext = new RgbViewModel();

   // Initialize to highlight color
   (this.DataContext as RgbViewModel).Color =
          new UISettings().UIElementColor(UIElementType.Highlight);
}
```

XAML 文件实例化了三个 TextBox 控件，并将其分别与 RgbViewModel 的 Red、Green 和 Blue 属性进行了绑定。

项目：ColorTextBoxes | 文件：MainPage.xaml(片段)

```
<Page ...>

   <Page.Resources>
      <Style TargetType="TextBlock">
         <Setter Property="FontSize" Value="24" />
         <Setter Property="Margin" Value="24 0 0 0" />
         <Setter Property="VerticalAlignment" Value="Center" />
      </Style>

      <Style TargetType="TextBox">
         <Setter Property="Margin" Value="24 48 96 48" />
         <Setter Property="VerticalAlignment" Value="Center" />
      </Style>
   </Page.Resources>

   <Grid Background="{StaticResource ApplicationPageBackgroundThemeBrush}">
      <Grid.ColumnDefinitions>
         <ColumnDefinition Width="*" />
         <ColumnDefinition Width="*" />
      </Grid.ColumnDefinitions>
```

```
        <Grid Grid.Column="0">
           <Grid.RowDefinitions>
              <RowDefinition Height="Auto" />
              <RowDefinition Height="Auto" />
              <RowDefinition Height="Auto" />
           </Grid.RowDefinitions>

           <Grid.ColumnDefinitions>
              <ColumnDefinition Width="Auto" />
              <ColumnDefinition Width="*" />
           </Grid.ColumnDefinitions>

           <TextBlock Text="Red: "
                      Grid.Row="0"
                      Grid.Column="0" />

           <TextBox Text="{Binding Red, Mode=TwoWay}"
                    Grid.Row="0"
                    Grid.Column="1" />

           <TextBlock Text="Green: "
                      Grid.Row="1"
                      Grid.Column="0" />

           <TextBox Text="{Binding Green, Mode=TwoWay}"
                    Grid.Row="1"
                    Grid.Column="1" />

           <TextBlock Text="Blue: "
                      Grid.Row="2"
                      Grid.Column="0" />

           <TextBox Text="{Binding Blue, Mode=TwoWay}"
                     Grid.Row="2"
                     Grid.Column="1" />
        </Grid>

        <!-- Result -->
        <Rectangle Grid.Column="1">
           <Rectangle.Fill>
              <SolidColorBrush Color="{Binding Color}" />
           </Rectangle.Fill>
        </Rectangle>
     </Grid>
  </Page>
```

程序运行后，每个 TextBox 控件将会呈现具体色值(参见下图)。所有必要的数据类型转换发生在幕后。

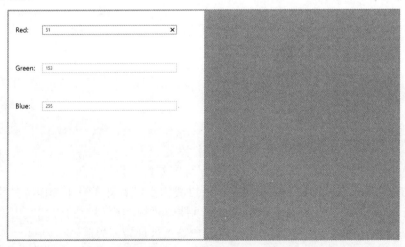

点击其中的某个 TextBox 并填入其他数值。什么都不会发生。单击另一个 TextBox，或按 Tab 键将输入焦点切换至下一个 TextBox。这时，最初在 TextBox 中输入的数字才被接受并用于更新屏幕右侧的颜色。

我们在测试此程序的过程中会发现 Windows Runtime 默许字符串中出现字母和符号，而不会抛出任何异常，并且在 TextBox 中输入的新值只有在该控件失去焦点后才生效。

这两个行为是有意而为之的。举例来说，假设有一个绑定到 TextBox 的“视图模型”要用到一个“模型”。这个“模型”通过网络连接来更新数据库。用户向 TextBox 键入文本的过程中(可能会出现某些错误并退格)，我们显然不希望每次按键都得进行网络请求。正因如此，仅当 TextBox 失去焦点时，程序才认为 TextBox 的输入已完成，并可以进行后续处理。

不幸的是，目前还没有修改这一行为的选项，也没有任何方法可以在数据绑定过程中进行数据验证。如果 TextBox 的绑定行为是不可接受的，并且希望通过自定义的控件来绕过 TextBox 的这种行为，最现实的选择就是弃用数据绑定，而用 TextChanged 事件处理程序。

示例项目 ColorTextBoxesWithEvents 展示了这种方法。此项目沿用了之前的 RgbViewModel 类。XAML 文件中 TextBox 控件被命名并订阅了 TextChanged 事件处理程序，其余与前一个示例一致。

项目：ColorTextBoxesWithEvents | 文件：MainPage.xaml(片段)

```
<TextBlock Text="Red: "
           Grid.Row="0"
           Grid.Column="0" />

<TextBox Name="redTextBox"
         Grid.Row="0"
         Grid.Column="1"
         Text="0"
         TextChanged="OnTextBoxTextChanged" />

<TextBlock Text="Green: "
           Grid.Row="1"
           Grid.Column="0" />

<TextBox Name="greenTextBox"
         Grid.Row="1"
         Grid.Column="1"
         Text="0"
         TextChanged="OnTextBoxTextChanged" />

<TextBlock Text="Blue: "
           Grid.Row="2"
           Grid.Column="0" />

<TextBox Name="blueTextBox"
         Grid.Row="2"
         Grid.Column="1"
         Text="0"
         TextChanged="OnTextBoxTextChanged" />
```

Rectangle 仍然像前一个示例那样采用数据绑定方式来获得更新。

由于这里替换了之前的双向数据绑定，所以我们不仅需要订阅 TextBox 的事件，也需要订阅 RgbViewModel 的 PropertyChanged 事件。“视图模型”的属性发生变化时更新 TextBox 的逻辑非常简单，但这里增加了对用户输入进行验证的逻辑。

项目：ColorTextBoxesWithEvents | 文件：MainPage.xaml.cs(片段)

```
public sealed partial class MainPage : Page
{
    RgbViewModel rgbViewModel;
    Brush textBoxTextBrush;
    Brush textBoxErrorBrush = new SolidColorBrush(Colors.Red);

    public MainPage()
    {
        this.InitializeComponent();

        // Get TextBox brush
        textBoxTextBrush = this.Resources["TextBoxForegroundThemeBrush"] as SolidColorBrush;

        // Create RgbViewModel and save as field
        rgbViewModel = new RgbViewModel();
        rgbViewModel.PropertyChanged += OnRgbViewModelPropertyChanged;
        this.DataContext = rgbViewModel;

        // Initialize to highlight color
        rgbViewModel.Color = new UISettings().UIElementColor(UIElementType.Highlight);
    }

    void OnRgbViewModelPropertyChanged(object sender, PropertyChangedEventArgs args)
    {
        switch (args.PropertyName)
        {
            case "Red":
                redTextBox.Text = rgbViewModel.Red.ToString("F0");
                break;

            case "Green":
                greenTextBox.Text = rgbViewModel.Green.ToString("F0");
                break;

            case "Blue":
                blueTextBox.Text = rgbViewModel.Blue.ToString("F0");
                break;
        }
    }

    void OnTextBoxTextChanged(object sender, TextChangedEventArgs args)
    {
        byte value;

        if (sender == redTextBox && Validate(redTextBox, out value))
            rgbViewModel.Red = value;

        if (sender == greenTextBox && Validate(greenTextBox, out value))
            rgbViewModel.Green = value;

        if (sender == blueTextBox && Validate(blueTextBox, out value))
            rgbViewModel.Blue = value;
    }

    bool Validate(TextBox txtbox, out byte value)
    {
        bool valid = byte.TryParse(txtbox.Text, out value);
        txtbox.Foreground = valid ? textBoxTextBrush : textBoxErrorBrush;
        return valid;
    }
}
```

Validate 方法通过标准的 TryParse 方法将文本转换为 byte 值。如果成功，“视图模型”的相应属性会被更新。否则，文本呈现红色以表明存在错误。

如果数字以空格或零开头，这段代码会出现一个小问题。如果在第一个 TextBox 中键

入 **0**。这是一个有效的 byte，因而 RgbViewModel 的 Red 属性会被更新为这个值，进而触发 PropertyChanged 方法，最终 TextBox 的 Text 属性会被设置为“0”。到目前为止一切正常。再键入 **5**。这时 TextBox 中显示“05”。TryParse 方法会认为这是一个有效的 byte 字符串，所以 Red 属性被更新为 5。PropertyChanged 处理程序将 TextBox 的 Text 设置为“5”，从而覆盖原来的“05”。而且光标会跳到 5 的前面，而不会保持在它的后面。

为解决这个问题，最好的办法是在 TextChanged 处理程序设置“视图模型”的属性期间忽略来自“视图模型”的 PropertyChanged 事件。使用一个标志即可做到。

```
bool blockViewModelUpdates;

...

void OnRgbViewModelPropertyChanged(object sender, PropertyChangedEventArgs args)
{
    if (blockViewModelUpdates)
        return;
    ...
}

void OnTextBoxTextChanged(object sender, TextChangedEventArgs args)
{
    blockViewModelUpdates = true;
    ...
    blockViewModelUpdates = false;
}
```

有人或许希望在 TextBox 失去输入焦点后清理其中显示的文本。

在某些情况下，数据验证更适合在“视图模型”的上下文中进行，而不在“视图”层面进行。

6.7 按钮与 MVVM

介绍到这里，为在一定程度上削弱对代码隐藏文件的依赖，读者可能认为 MVVM 只对能够生成值的控件有效，而对按钮则无用武之地。Button 能够引发 Click 事件，但该事件必须在代码隐藏文件中处理。如果“视图模型”要针对按钮实现某种逻辑(这非常可能)，可以通过 Click 事件处理程序调用“视图模型”。这样做在架构上是合理的，但实施起来却较为繁琐。

幸运的是，有一种替代 Click 事件的机制可以很好地配合 MVVM 使用，我们一般称其为“命令接口”。ButtonBase 定义了名为 Command(类型为 ICommand)和 CommandParameter(类型为 object)的属性，能够使 Button 通过数据绑定调用“视图模型”。Command 和 CommandParameter 均为依赖属性，因而可以作为绑定目标。Command 属性总是作为数据绑定的目标，而 CommandParameter 是可选的。后者可在多个按钮绑定到同一个 Command 对象时对按钮加以区分，作用类似于 Tag 属性。

假设有一个计算器程序，计算器的引擎通过“视图模型”实现，并将它设置到 DataContext 属性上。“+”(加法)命令的按钮可以在 XAML 中这样实例化。

```
<Button Content="+"
        Command="{Binding CalculateCommand}"
        CommandParameter="add" />
```

“视图模型”有一个名为 CalculateCommand、类型为 ICommand 的属性，可以像下面这样定义。

```
public ICommand CalculateCommand { protected set; get; }
```

“视图模型”必须初始化 CalculateCommand 属性，将其设置为实现 ICommand 接口的类的实例。这个接口的定义如下所示。

```
public interface ICommand
{
   void Execute(object param);
   bool CanExecute(object param);
   event EventHandler<object> CanExecuteChanged;
}
```

当加法按钮被单击后，CalculateCommand 属性所引用的对象的 Execute 方法便会被调用，通过参数传入字符串 add。这就是 Button 调用“视图模型”(更确切地说，应该是包含 Execute 方法的类)的基本过程。

ICommand 接口的另外两个成员都包含“can execute”字样。从字面上理解，它与命令在特定情况下的有效性有关。如果命令当前是无效的(如当前没输入数字，计算器无法进行加法运算)，那么 Button 应被禁用。

就本示例而言，幕后的逻辑是这样的：XAML 在运行时被解析和加载后，Button 的 Command 属性会被绑定到 CalculateCommand 对象。Button 会订阅该对象的 CanExecuteChanged 事件，调用 CanExecute 方法并通过参数传入 add。如果 CanExecute 返回 false，Button 则会自动变为禁用状态。当 CanExecuteChanged 事件再次被引发，Button 还会调用 CanExecute，如此往复。

为在“视图模型”中使用命令，必须提供实现 ICommand 接口的类。然而，这个类很有可能需要访问“视图模型”中的属性，也可能有反方向的访问。

那么有人可能会问：这两个类可否合二为一？

从理论上讲，这是可行的，但前提是同一页面的所有按钮都可以共用同一对 Execute 和 CanExecute 方法。这要求每个按钮都必须拥有唯一的 CommandParameter，以便这两个方法能够区分不同按钮。下面让我们了解一下在“视图模型”中使用命令的一般方法。

6.8　DelegateCommand 类

第 5 章介绍了一个名为 SimpleKeypad 的程序，它能够收集按键序列并生成格式化的文本。下面让我们用新的方法来重写这个程序。“视图模型”除了实现 INotifyPropertyChanged 接口外，还要处理来自拨号盘按钮的命令，通过命令来替代 Click 事件处理程序。

问题在于，为使“视图模型”处理按钮命令，它必须拥有一个或多个 ICommand 类型的属性。这需要一个或多个实现 ICommand 接口的类。实现 ICommand 的类必须包含 Execute 和 CanExecute 方法以及 CanExecuteChanged 事件。这些方法的内部必然要与“视图模型”的某些部分交互。

解决方案之一是在“视图模型”中用不同名称定义所有 Execute 和 CanExecute 方法。然后，我们通过一个实现了 ICommand 的类来调用“视图模型”中具体的 Execute 和

CanExecute 方法。

这个类名为 DelegateCommand。通过查阅资料会发现，这样的类有多种实现，其中包括微软 Prism 框架中的(此框架旨在帮助开发者在 Windows Presentation Foundation(WPF)和 Silverlight 中应用 MVVM)。下面要介绍本人的实现。

实现 ICommand 接口意味着 DelegateCommand 类包含 Execute 和 CanExecute 方法和 CanExecuteChanged 事件，但 DelegateCommand 仍需要另一个方法来引发此事件。我们不妨称这个额外的方法为 RaiseCanExecuteChanged。为达到这个目的，我们首先定义一个派生于 ICommand 的接口，并添加这个方法。

```
项目：KeypadWithViewModel | 文件：IDelegateCommand.cs
using System.Windows.Input;

namespace KeypadWithViewModel
{
   public interface IDelegateCommand : ICommand
   {
      void RaiseCanExecuteChanged();
   }
}
```

DelegateCommand 类实现了 IDelegateCommand 接口，并利用了 System 空间中的几个简单(而实用)的泛型委托。这些预定义的委托的名称为 Action 和 Func，拥有 1 到 16 个类型参数。Func 委托可以返回指定类型的对象，而 Action 委托无返回值。在示例程序中，Action<object>委托代表接受单个 object 参数、返回 void 的方法(匹配 Execute 方法的签名)。Func<object, bool>委托代表接受 object 参数、返回 bool 的方法(匹配 CanExecute 方法的签名)。DelegateCommand 定义了这两个委托类型的字段，用于存储对应签名的方法。

```
项目：KeypadWithViewModel | 文件：DelegateCommand.cs
using System;

namespace KeypadWithViewModel
{
   public class DelegateCommand : IDelegateCommand
   {
      Action<object> execute;
      Func<object, bool> canExecute;

      // Event required by ICommand
      public event EventHandler CanExecuteChanged;

      // Two constructors
      public DelegateCommand(Action<object> execute, Func<object, bool> canExecute)
      {
         this.execute = execute;
         this.canExecute = canExecute;
      }
      public DelegateCommand(Action<object> execute)
      {
         this.execute = execute;
         this.canExecute = this.AlwaysCanExecute;
      }

      // Methods required by ICommand
      public void Execute(object param)
      {
         execute(param);
      }
      public bool CanExecute(object param)
```

```
        {
            return canExecute(param);
        }

        // Method required by IDelegateCommand
        public void RaiseCanExecuteChanged()
        {
            if (CanExecuteChanged != null)
                CanExecuteChanged(this, EventArgs.Empty);
        }

        // Default CanExecute method
        bool AlwaysCanExecute(object param)
        {
            return true;
        }
    }
}
```

这个类实现了 Execute 和 CanExecute 方法，但这些方法只是调用字段形式的委托。这两个字段通过类构造函数的参数设置。

假设计算器的“视图模型”有一个用于执行计算的命令，则可以这样定义一个 CalculateCommand 属性。

```
public IDelegateCommand CalculateCommand { protected set; get; }
```

“视图模型”还需要可以定义 ExecuteCalculate 和 CanExecuteCalculate 方法。

```
void ExecuteCalculate(object param)
{
    ...
}

bool CanExecuteCalculate(object param)
{
    ...
}
```

“视图模型”类的构造函数可通过这两个方法来实例化 DelegateCommand，并将结果赋给 CalculateCommand 属性。

```
this.CalculateCommand = new DelegateCommand(ExecuteCalculate, CanExecuteCalculate);
```

了解实现命令的基本过程之后，下面让我们看看拨号盘的“视图模型”。针对输入到拨号盘和由拨号盘显示的文本，“视图模型”分别定义名为 InputString 的属性和提供格式化文本的 DisplayText 属性。

“视图模型”还定义了名为 AddCharacterCommand(用于输入所有数字及符号)和 DeleteCharacterCommand 的属性，类型均为 IDelegateCommand。这两个属性会被初始化为 DelegateCommand 对象。实例化 DelegateCommand 对象时会传入 ExecuteAddCharacter、ExecuteDeleteCharacter 和 CanExecuteDeleteCharacter 委托。由于所有数字和符号按键在任何情况下都是有效的，因而不需要定义 CanExecuteAddCharacter 方法。

```
项目: KeypadWithViewModel | 文件: KeypadViewModel.cs
using System;
using System.ComponentModel;
using System.Runtime.CompilerServices;

namespace KeypadWithViewModel
{
```

```
public class KeypadViewModel : INotifyPropertyChanged
{
    string inputString = "";
    string displayText = "";
    char[] specialChars = { '*', '#' };

    public event PropertyChangedEventHandler PropertyChanged;

    // Constructor
    public KeypadViewModel()
    {
        this.AddCharacterCommand = new DelegateCommand(ExecuteAddCharacter);
        this.DeleteCharacterCommand =
            new DelegateCommand(ExecuteDeleteCharacter, CanExecuteDeleteCharacter);
    }

    // Public properties
    public string InputString
    {
        protected set
        {
            bool previousCanExecuteDeleteChar = this.CanExecuteDeleteCharacter(null);

            if (this.SetProperty<string>(ref inputString, value))
            {
                this.DisplayText = FormatText(inputString);

                if (previousCanExecuteDeleteChar != this.CanExecuteDeleteCharacter(null))
                    this.DeleteCharacterCommand.RaiseCanExecuteChanged();
            }
        }

        get { return inputString; }
    }

    public string DisplayText
    {
        protected set { this.SetProperty<string>(ref displayText, value); }
        get { return displayText; }
    }

    // ICommand implementations
    public IDelegateCommand AddCharacterCommand { protected set; get; }

    public IDelegateCommand DeleteCharacterCommand { protected set; get; }

    // Execute and CanExecute methods
    void ExecuteAddCharacter(object param)
    {
        this.InputString += param as string;
    }

    void ExecuteDeleteCharacter(object param)
    {
        this.InputString = this.InputString.Substring(0, this.InputString.Length - 1);
    }

    bool CanExecuteDeleteCharacter(object param)
    {
        return this.InputString.Length > 0;
    }

    // Private method called from InputString
    string FormatText(string str)
    {
        bool hasNonNumbers = str.IndexOfAny(specialChars) != -1;
        string formatted = str;
```

```
            if (hasNonNumbers || str.Length < 4 || str.Length > 10)
            {
            }
            else if (str.Length < 8)
            {
                formatted = String.Format("{0}-{1}", str.Substring(0, 3),
                                          str.Substring(3));
            }
            else
            {
                formatted = String.Format("({0}) {1}-{2}", str.Substring(0, 3),
                                              str.Substring(3, 3),
                                              str.Substring(6));
            }
            return formatted;
        }

        protected bool SetProperty<T>(ref T storage, T value,
                          [CallerMemberName] string propertyName = null)
        {
            if (object.Equals(storage, value))
                return false;

            storage = value;
            OnPropertyChanged(propertyName);
            return true;
        }

        protected void OnPropertyChanged(string propertyName)
        {
            if (PropertyChanged != null)
                PropertyChanged(this, new PropertyChangedEventArgs(propertyName));
        }
    }
}
```

ExecuteAddCharacter 方法传入的参数是用户键入的字符。这就是多个按钮能够共用同一个命令对象的原因。

CanExecuteDeleteCharacter 只在有可删除的字符的情况下才返回 true。如果没有字符，删除按钮应被禁用。这个方法在绑定刚建立时被首次调用，后续的调用发生在 CanExecuteChanged 事件被引发时。引发此事件的逻辑位于 InputString 属性的 set 访问器(比较输入字符串在修改前后 CanExecuteDeleteCharacter 的返回值)。

XAML 文件将这个“视图模型”以资源的形式实例化，然后将其指定给 Grid 的 DataContext 属性。请注意 13 个 Button 控件上的 Command 绑定的简化语法以及数字和符号键对 CommandParameter 属性的使用。

项目：KeypadWithViewModel | 文件：MainPage.xaml(片段)

```
<Page ...>

    <Page.Resources>
        <local:KeypadViewModel x:Key="viewModel" />
    </Page.Resources>

    <Grid Background="{StaticResource ApplicationPageBackgroundThemeBrush}"
          DataContext="{StaticResource viewModel}">

        <Grid HorizontalAlignment="Center"
              VerticalAlignment="Center"
              Width="288">

            <Grid.Resources>
                <Style TargetType="Button">
```

```
            <Setter Property="ClickMode" Value="Press" />
            <Setter Property="HorizontalAlignment" Value="Stretch" />
            <Setter Property="Height" Value="72" />
            <Setter Property="FontSize" Value="36" />
         </Style>
      </Grid.Resources>

      <Grid.RowDefinitions>
         <RowDefinition Height="Auto" />
         <RowDefinition Height="Auto" />
         <RowDefinition Height="Auto" />
         <RowDefinition Height="Auto" />
         <RowDefinition Height="Auto" />
      </Grid.RowDefinitions>

      <Grid.ColumnDefinitions>
         <ColumnDefinition Width="*" />
         <ColumnDefinition Width="*" />
         <ColumnDefinition Width="*" />
      </Grid.ColumnDefinitions>

      <Grid Grid.Row="0" Grid.Column="0" Grid.ColumnSpan="3">
         <Grid.ColumnDefinitions>
            <ColumnDefinition Width="*" />
            <ColumnDefinition Width="Auto" />
         </Grid.ColumnDefinitions>

         <Border Grid.Column="0"
               HorizontalAlignment="Left">

            <TextBlock Text="{Binding DisplayText}"
                       HorizontalAlignment="Right"
                       VerticalAlignment="Center"
                       FontSize="24" />
         </Border>

         <Button Content="&#x21E6;"
                 Command="{Binding DeleteCharacterCommand}"
                 Grid.Column="1"
                 FontFamily="Segoe Symbol"
                 HorizontalAlignment="Left"
                 Padding="0"
                 BorderThickness="0" />
      </Grid>

      <Button Content="1"
              Command="{Binding AddCharacterCommand}"
              CommandParameter="1"
              Grid.Row="1" Grid.Column="0" />

      ...

      <Button Content="#"
              Command="{Binding AddCharacterCommand}"
              CommandParameter="#"
              Grid.Row="4" Grid.Column="2" />
   </Grid>
  </Grid>
</Page>
```

这个项目的代码隐藏文件得到了极大的简化，只包含对 InitializeComponent 的调用。至此，大功告成！

第7章 异 步

现代应用程序不提倡频繁使用消息框(message box)。但要以最直接的方式获得重要信息或得到“是”、“否”或“取消”之类的答复，使用消息框肯定更让人感到得心应手。

Windows Runtime 支持通过 MessageDialog 类实现的消息框。这种消息框最多支持三个按钮，但按钮标签可以任意修改。这个类没有 Show 方法，取而代之的是 ShowAsync 方法。

后缀 Async 是“asynchronous”(异步的)的缩写。这个后缀在 Windows Runtime 中有非凡的含义。它不仅充当方法名称的一部分，而且改变了异步编程的方式，还改变了现代操作系统(如 Windows 8)的编程思想。

7.1 线程与用户界面

与 Windows 早期版本上的应用程序类似，Windows 8 应用程序也是一种状态机(state machine)。程序初始化完毕后，它会驻留于内存等待某些事件的发生。事件一般用来通知用户交互的发生，有时也用来通知系统级别的变化(如屏幕方向的变换)。

程序应尽快处理事件，在完成处理后将控制权交还给操作系统，并继续等待其他事件。如果程序无法快速完成事件处理，则会处于无法响应状态，进而惹恼用户。我们应使用户界面专用的线程尽量处于空闲状态，不应使其承担繁重的处理任务。

如果 Windows Runtime 中的某个方法需要耗费较长时间，该怎么办？这个问题是否应该由应用开发者来处理，使其在单独的工作线程上运行？

不，应用开发者没必要这样做。在设计 Windows Runtime 时，微软的开发者已筛选出要耗时至少 50 毫秒才将控制权交给应用程序的方法。Windows Runtime 中大约有 10%～15%这样的方法。这些方法被设计成异步的，以便在单独的工作线程中处理耗时的任务。这些方法在调用后能够立即返回，并在完成后通知应用程序。

我们往往在文件 I/O 或访问 Internet 时会用到异步方法。在调用 Windows 8 中的对话框(如 MessageDialog 和本章要介绍的文件选择对话框)时，也会用到异步方法。Windows Runtime 中所有异步方法的名称均带有 Async 后缀，并采用了相同的定义模式。幸运的是，借助于强大的.NET 类库和改进后的 C#编程语言，异步方法用起来并不困难。

异步编程在未来若干年是一项需要我们掌握的重要技术。对于过去的消费类计算机，所有线程都在同一个处理器上执行。操作系统负责对不同线程进行切换，使其表面上看是并行的。近些年的计算机往往配备多个处理器(一般是具有多核配置的单一芯片)。这种硬件允许在不同处理器上运行不同线程。

某些繁重的计算任务(如数组的处理)可以借助于多核处理器实现并行处理。为支持异步和并行处理，.NET 增加了一种叫“基于任务的异步模式”(Task-based Asynchronous Pattern，TAP)的支持，围绕 System.Threading.Tasks 命名空间中的 Task 类实现。Windows Runtime 应用程序可以通过 C#和 Visual Basic 来使用.NET 的这部分功能，这比直接使用

Windows Runtime 本身对异步的支持要容易得多。

7.2 MessageDialog 的使用

下面让我们通过 MessageDialog 来初步认识一下异步函数。MessageDialog 的构造函数接受一个内容字符串和一个(可选的)标题字符串。默认情况下只显示一个标有“关闭”的按钮。这种默认的消息框可用于为用户提供某些重要信息。我们可以通过 UICommand 对象最多定义三个自定义按钮。示例项目 HowToAsync1 给了这样的一个例子。

```
MessageDialog msgdlg = new MessageDialog("Choose a color", "How To Async #1");
msgdlg.Commands.Add(new UICommand("Red", null, Colors.Red));
msgdlg.Commands.Add(new UICommand("Green", null, Colors.Green));
msgdlg.Commands.Add(new UICommand("Blue", null, Colors.Blue));
```

UICommand 构造函数的第一个参数用于指定按钮上显示的文本，第三个参数为 object 类型，用于指定按钮的 ID，可以传入任何能够标识按钮的对象。这里选择使用标签所对应的 Color 值来标识按钮。第二个参数稍后介绍。

UICommand 类实现了 IUICommand 接口。MessageDialog 通过 IUICommand 来通知程序某个按钮被用户按下。

ShowAsync 方法的调用按异步方式处理。此方法没有参数，并能够立即返回。消息框会在工作线程[①]中执行。此方法可以像下面这样调用。

```
IAsyncOperation<IUICommand> asyncOp = msgdlg.ShowAsync();
```

ShowAsync 会返回一个实现泛型接口 IAsyncOperation 的对象。泛型参数为 IUICommand 接口类型。这也就是说，MessageDialog 间接地返回了 IUICommand 类型的对象。该对象在消息框的按钮被按下并且 MessageDialog 消失前是不会返回的，但在调用 ShowAsync 方法时 MessageDialog 尚未显示。正因如此，IAsyncOperation 也被看作一种“前景”(future)或“许诺”(promise)。

IAsyncInfo 接口定义了 Cancel 和 Close 方法，还定义了 Id、Status 和 ErrorCode 属性。IAsyncOperation<T>接口派生自 IAsyncInfo 接口，并定义 Completed 属性，属性类型为 AsyncOperationCompletedHandler<T>委托。

我们需要在代码中将 Completed 属性设置为某个回调方法。虽然 Completed 是属性，但它很像事件，能够在某些事项发生后通知程序。(不同在于，事件可以有多个处理程序，而属性只能有一个。)该属性的赋值方法如下所示。

```
asyncOp.Completed = OnMessageDialogShowAsyncCompleted;
```

若程序中的某个方法调用了 ShowAsync 并将 Completed 属性设置为某个处理程序，那么这个处理程序将稍后执行。调用 ShowAsync 方法的代码将控制器交给操作系统后 MessageDialog 才会显示。

MessageDialog 在专门的线程上运行。虽然程序的用户界面在 MessageDialog 显示后被禁用，但程序的用户界面线程未被阻塞，仍然可以继续工作。

---

① 译注：“工作线程”是相对于“UI 线程”而言的。

读者或许会注意到下图中标记为 Red 的按钮与其他按钮颜色的不同。这种样式代表该按钮是默认按钮，用户按下 Enter 键便可触发。我们可以通过 MessageDialog 的 DefaultCommandIndex 属性来指定哪个按钮为默认的，还可以通过 CancelCommandIndex 属性来设置用户按下 Esc 键所触发的按钮。

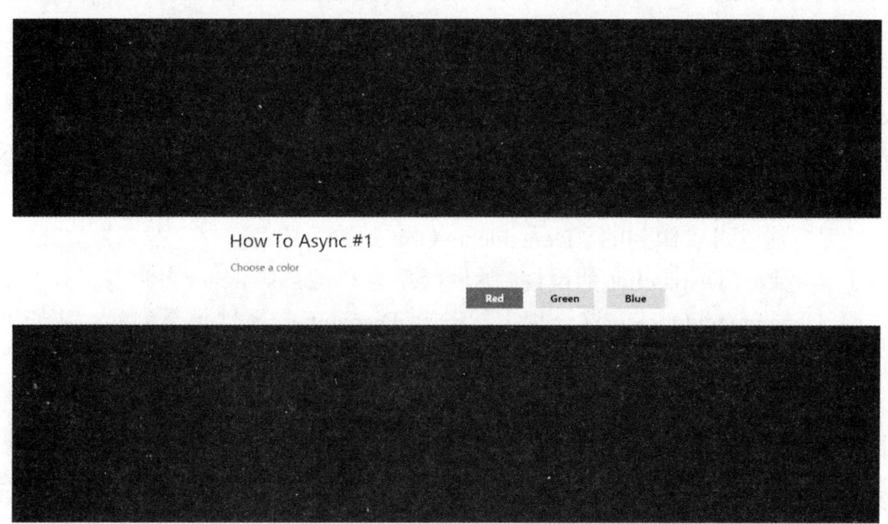

用户按下某个按钮后，消息框消失，Completed 回调方法会被调用。这个回调的第一个参数传入的与 ShowAsync 返回的是同一个对象。只不过这里变量的名称发生了一些变化(asyncInfo)，因为它可以为我们提供额外的信息。

```
void OnMessageDialogShowAsyncCompleted(IAsyncOperation<IUICommand> asyncInfo,
                                       AsyncStatus asyncStatus)
{
   // Get the Color value
   IUICommand command = asyncInfo.GetResults();
   clr = (Color)command.Id;
   ...
}
```

IAsyncOperation 接口有一个名为 Status 的属性，属性类型为 AsyncStatus 的枚举。这个枚举的成员包括 Started、Completed、Canceled 和 Error。这个属性的值通过第二个参数传入 Completed 处理程序。如果发生错误(一般与 MessageDialog 类无关，而由文件 I/O 或 Internet 访问引起)，IAsyncOperation 的 ErrorCode 属性会被设置为一个 Exception 类型的对象。

调用 GetResults 前应该检查状态是否为 Completed。GetResults 方法会返回类型取决于 IAsyncOperation 的泛型参数。对于本示例，该方法会返回代表被按下按钮的 IUICommand 对象。我们可以通过该对象的 Id 属性获得一个标识对象，该对象是之前通过 UICommand 构造函数的第三个参数传入的。本示例将其转换成 Color 值。

现在，程序或许可以通过这个 Color 值来设置 Grid 的背景颜色。

```
contentGrid.Background = new SolidColorBrush(clr);
```

别急，还不行！

当程序调用 ShowAsync 时，MessageDialog 会创建单独的工作线程来显示对话框和按钮。当用户按下按钮后，我们指定的 Completed 处理程序会被调用。处理程序的调用也发生在这个工作线程上，但这个线程不允许访问用户界面的对象！

不论哪个窗口，应用程序都只有一个线程来处理用户输入和呈现用于交互的控件和图形，这个线程被称为“UI 线程”。对 Windows 应用程序来说，UI 线程非常重要且特殊，因为与用户的所有交互都发生在此线程上。但只有运行在这个线程上的代码才能够访问构成用户界面的元素和控件。

这个限制可以如此描述：DependencyObject 不是线程安全(thread safe)的。任何继承于 DependencyObject 的类所产生的对象只能由创建该对象的线程访问。

就本示例而言，Color 值可以在工作线程上创建，因为 Color 是一个结构，并不是派生自 DependencyObject。但使这个 Color 值作用于用户界面的代码必须运行在 UI 线程上。

幸运的是，这是可以做到的。DependencyObject 类不是线程安全的，为弥补这一缺憾，该类暴露了一个名为 Dispatcher 的属性，返回类型为 CoreDispatcher 的对象。对于不能从其他线程访问 DependencyObject 这一规定，访问 Dispatcher 属性是个例外。我们可以通过 CoreDispatcher 的 HasThreadAccess 属性来获悉当前 DependencyObject 是否可以在当前线程上访问。不管能或不能，需要在创建该对象的线程上执行的代码都可以被放入一个队列中等待执行。

通过 CoreDispatcher 定义的 RunAsync 方法，我们可以将代码放入队列中来使其在 UI 线程上执行。这个方法也是一个异步方法，可以传入需要在 UI 线程上运行的方法。

```
void OnMessageDialogShowAsyncCompleted(IAsyncOperation<IUICommand> asyncInfo,
                                       AsyncStatus asyncStatus)
{
    ...
    // Use a Dispatcher to run in the UI thread
    this.Dispatcher.RunAsync(CoreDispatcherPriority.Normal, OnDispatcherRunAsyncCallback);
}

void OnDispatcherRunAsyncCallback()
{
    // Set the background brush
    contentGrid.Background = new SolidColorBrush(clr);
}
```

来自 Dispatcher 属性的 CoreDispatcher 对象通常不被保存在变量中。就本示例而言，RunAsync 方法是直接在 Dispatcher 属性上调用的。传入 RunAsync 方法的回调可以安全地访问用户界面。但需要注意的是，我们不能通过参数向这个方法传入任何信息，OnMessageDialogShowAsyncCompleted 须先将 Color 值保存到字段中。

从哪个对象获得 CoreDispatcher 对象都无所谓，因为所有用户界面对象都在同一个 UI 线程上创建，这些 UI 对象的工作方式也是一致的。

CoreDispatcher 的 RunAsync 方法会返回 IAsyncAction 类型的对象(前一段代码并没有展示)。

```
IAsyncAction asyncAction = this.Dispatcher.RunAsync(CoreDispatcherPriority.Normal,
                                                    OnDispatcherRunAsyncCallback);
```

IAsyncAction 非常类似于 MessageDialog 的 ShowAsync 方法返回的 IAsyncOperation 对象。两者均实现了 IAsyncInfo 接口。最大的区别在于 IAsyncOperation 针对的是有返回值的异步方法(因而需要提供泛型参数)，而 IAsyncAction 针对的是无返回值的异步方法。

下面是相关接口的层次结构。

IAsyncInfo
  IAsyncAction
  IAsyncActionWithProgress<TProgress>
  IAsyncOperation<TResult>
  IAsyncOperationWithProgress<TResult, TProgress>

其中有两个接口定义了在执行异步任务时报告进度的方法。

对于 CoreDispatcher 对象的 RunAsync 方法返回的 IAsyncAction 对象，我们可以将其 Completed 属性设置为某个处理程序来访问用户界面。

```
void OnMessageDialogShowAsyncCompleted(IAsyncOperation<IUICommand> asyncInfo,
                                        AsyncStatus asyncStatus)
{
   ...
   IAsyncAction asyncAction = this.Dispatcher.RunAsync(CoreDispatcherPriority.Normal,
                                             OnDispatcherRunAsyncCallback);

   asyncAction.Completed = OnDispatcherRunAsyncCompleted;
}

void OnDispatcherRunAsyncCompleted(IAsyncAction asyncInfo, AsyncStatus asyncStatus)
{
   contentGrid.Background = new SolidColorBrush(clr);
}
```

这个 Completed 处理程序运行在 UI 线程。OnDispatcherRunAsyncCallback 的存在无实际意义，指定这个方法是因为 RunAsync 的第二个参数不能为 null。

下面是 HowToAsync1 项目 XAML 文件的主要内容。这段 XAML 定义了一个专门用于调用 MessageDialog 的按钮。

项目：HowToAsync1 | 文件：MainPage.xaml(片段)

```
<Grid Name="contentGrid"
      Background="{StaticResource ApplicationPageBackgroundThemeBrush}">

   <Button Content="Show me a MessageDialog!"
           HorizontalAlignment="Center"
           VerticalAlignment="Center"
           Click="OnButtonClick" />

</Grid>
```

代码隐藏类到目前为止还没有什么特别之处。

项目：HowToAsync1 | 文件：MainPage.xaml.cs(片段)

```
public sealed partial class MainPage : Page
{
   Color clr;

   public MainPage()
   {
      this.InitializeComponent();
   }

   void OnButtonClick(object sender, RoutedEventArgs args)
   {
      MessageDialog msgdlg = new MessageDialog("Choose a color", "How To Async #1");
      msgdlg.Commands.Add(new UICommand("Red", null, Colors.Red));
      msgdlg.Commands.Add(new UICommand("Green", null, Colors.Green));
      msgdlg.Commands.Add(new UICommand("Blue", null, Colors.Blue));
```

```
        // Show the MessageDialog with a Completed handler
        IAsyncOperation<IUICommand> asyncOp = msgdlg.ShowAsync();
        asyncOp.Completed = OnMessageDialogShowAsyncCompleted;
    }

    void OnMessageDialogShowAsyncCompleted(IAsyncOperation<IUICommand> asyncInfo,
                                           AsyncStatus asyncStatus)
    {
        // Get the Color value
        IUICommand command = asyncInfo.GetResults();
        clr = (Color)command.Id;

        // Use a Dispatcher to run in the UI thread
        IAsyncAction asyncAction = this.Dispatcher.RunAsync(CoreDispatcherPriority.Normal,
                                                            OnDispatcherRunAsyncCallback);
    }

    void OnDispatcherRunAsyncCallback()
    {
        // Set the background brush
        contentGrid.Background = new SolidColorBrush(clr);
    }
}
```

UICommand 构造函数的第二个可选参数为 UICommandInvokedHandler 类型，可以指定下面这样签名的方法作为回调方法。

```
void OnMessageDialogCommand(IUICommand command)
{
    ...
}
```

这个回调在 UI 线程执行。因此，它或许是一种获取按钮命令的简便方法。

7.3 Lambda 函数形式的回调

为了更好地实现回调方法，C# 3.0 引入了“匿名方法”(anonymous method)，也被称为 Lambda 函数或 Lambda 表达式。示例项目 HowToAsync1 中的所有回调逻辑均能够移入 Click 处理程序，以 Lambda 函数的形式实现，并且 Color 值不再需要通过字段传递。这正如示例项目 HowToAsync2 所展示的。

```
项目: HowToAsync2 | 文件: MainPage.xaml.cs(片段)
void OnButtonClick(object sender, RoutedEventArgs args)
{
    MessageDialog msgdlg = new MessageDialog("Choose a color", "How To Async #2");
    msgdlg.Commands.Add(new UICommand("Red", null, Colors.Red));
    msgdlg.Commands.Add(new UICommand("Green", null, Colors.Green));
    msgdlg.Commands.Add(new UICommand("Blue", null, Colors.Blue));

    // Show the MessageDialog with a Completed handler
    IAsyncOperation<IUICommand> asyncOp = msgdlg.ShowAsync();
    asyncOp.Completed = (asyncInfo, asyncStatus) =>
        {
            // Get the Color value
            IUICommand command = asyncInfo.GetResults();
            Color clr = (Color)command.Id;

            // Use a Dispatcher to run in the UI thread
            IAsyncAction asyncAction = this.Dispatcher.RunAsync(CoreDispatcherPriority.Normal,
                                                                () =>
                {
```

```
                // Set the background brush
                contentGrid.Background = new SolidColorBrush(clr);
            });
        };
}
```

虽然所有逻辑均被移入 Click 处理程序，但显然这段代码并非一次全部执行。MessageDialog 的 Completed 处理程序仅在消息框消失后执行，而 CoreDispatcher 的回调仅在 UI 线程空闲时执行。

嵌套两个 Lambda 函数或许还说得过去，但复杂的嵌套不免让人感到困惑。例如，文件 I/O 往往涉及一系列的操作，其中的一些可能是异步的，而选择使用嵌套的 Lambda 函数会导致代码结构混乱不堪。虽然 Lambda 函数用起来方便，但往往也会削弱程序的可读性。某些情况下，为了完成简单的 return 语句或异常处理，使用 Lambda 函数可能会遇到麻烦。

我们需要其他解决方案。幸运的是，的确有一种。

7.4　神奇的 await 运算符

与回调形式不同，C# 5.0 的 await 关键字可以像普通方法一样来执行异步操作。这是之前用于获取 IAsyncOperation 对象的代码。

```
IAsyncOperation<IUICommand> asyncOp = msgdlg.ShowAsync();
```

上一个示例通过回调方法来获得代表被按下按钮的 IUICommand 对象。await 运算符可以直接从 IAsyncOperation 对象提取 IUICommand 对象。

```
IUICommand command = await asyncOp;
```

通常以上两语句是合二为一的，正如示例项目 HowToAsync3 程序所展示的。此程序与之前两个是等效的。

项目：HowToAsync3 | 文件：MainPage.xaml.cs（片段）

```
async void OnButtonClick(object sender, RoutedEventArgs args)
{
    MessageDialog msgdlg = new MessageDialog("Choose a color", "How To Async #3");
    msgdlg.Commands.Add(new UICommand("Red", null, Colors.Red));
    msgdlg.Commands.Add(new UICommand("Green", null, Colors.Green));
    msgdlg.Commands.Add(new UICommand("Blue", null, Colors.Blue));

    // Show the MessageDialog
    IUICommand command = await msgdlg.ShowAsync();

    // Get the Color value
    Color clr = (Color)command.Id;

    // Set the background brush
    contentGrid.Background = new SolidColorBrush(clr);
}
```

代码很漂亮，不是吗？

await 关键字是 C#运算符，嵌在复杂代码中也是合乎语法的。例如，可以将上段代码的后三条语句合并为一条。

```
contentGrid.Background = new SolidColorBrush((Color)(await msgdlg.ShowAsync()).Id);
```

这里要特别强调一下：HowToAsync3 在功能上与前两个程序完全等同。但

HowToAsync3 的语法更为简洁，这都要归功于 await 运算符。await 运算符似乎消除了所有凌乱的回调，并直接返回 IUICommand。这就像魔术一样，但幕后的实现则被隐藏了。C#编译器能够识别 ShowAsync 方法的模式，并生成回调和对 GetResults 的调用。

从本质上讲，await 运算符会对它要调用的方法进行分解，将其变成一个状态机。OnButtonClick 方法最初会按常规方式执行，直到通过 await 调用 ShowAsync。尽管关键字 await 字面上有等待之意，但它实际并不会等待操作执行完毕。相反，OnButtonClick 方法会在此时暂停。控制权会被交还给 Windows。程序用户界面的其他代码得以执行，包括 MessageDialog 本身。当 MessageDialog 被关闭，结果返回，UI 线程空闲， IUICommand 对象被返回，OnButtonClick 方法才会继续执行。若再次遇到 await 运算符，依此类推。

在 await 出现之前，本人认为在 C#中执行异步操作会破坏语言的命令式结构(imperative structure)。await 运算符使命令式结构得以回归，使异步调用看上去是普通方法的顺序调用。但在 await 带来了便利的同时，也应时刻牢记一点：出现 await 的方法在幕后会被拆成多段回调代码。

某些情况下，这种拆分也会造成一些问题。在 Windows 调用程序中的方法时，若方法将控制权交还给操作系统，Windows 便会认为它执行完毕。如果方法中使用了 await 运算符，情况则不同了。对于含有 await 的方法，await 运算符后面的代码执行之前，控制权会被交还给 Windows。

为使 Windows 知悉使用 await 运算符的方法尚未执行完毕，需要用到一个“延期”(deferral)对象。在本章介绍 Application 类的 Suspending 事件时会具体讲解这个工作机制。

await 运算符也有一些限制。例如，它不能出现在异常处理程序的 catch 或 finally 子句中，但它可以出现在 try 子句中。我们可以在 try 子句中捕获异步方法抛出的异常，或决定异步操作是否被取消(稍后会介绍)。

包含 await 运算符的方法必须被标记为 async，正如这个 Click 处理程序所展示的。

```
async void OnButtonClick(object sender, RoutedEventArgs args)
{
   // ... code with await operators
}
```

这个 async 关键字没有太多作用。由于 C#的早期版本未将 await 当作关键字，程序员可以将其作为变量名、属性名或其他名称。C# 5.0 引入的 await 关键字会破坏这种代码，但将 await 限制在 async 关键字修饰的方法中则可避免这个问题。async 修饰符并不影响方法的签名，上述方法仍是有效的 Click 处理程序。但我们不能对入口方法使用 async(因而也不能使用 await)，具体来说就是 Main 函数和类的构造函数。

如果需要在页面初始化过程中调用异步方法，就要将这些方法调用置于 Loaded 事件处理程序中，并将这个处理程序标记为 async。

```
public MainPage()
{
   this.InitializeComponent();
   ...
   Loaded += OnLoaded;
}

async void OnLoaded(object sender, RoutedEventArgs arg)
```

```
{
   ...
}
```

如果希望以匿名方法的形式来定义 Loaded 处理程序，则可以像下面这样做。

```
public MainPage()
{
   this.InitializeComponent();
   ...
   Loaded += async (sender, args) =>
      {
         ...
      };
}
```

注意到参数列表前的 async 关键字了吗？

7.5　异步操作的撤销

并非所有异步操作都可以像调用 MessageDialog 的 ShowAsync 方法一样简单明了。异步操作有三个特性会使其变复杂。

- **撤销**　由于用户可能有意终止或由于其他原因，许多异步操作需要实现撤消。
- **进度**　有些异步操作能够报告耗时操作的进展情况。用户往往希望通过 ProgressBar 或文本形式看到进度报告。
- **错误**　异步操作执行期间可能会出现错误(例如，尝试打开不存在的文件)。

下面先来说说撤销的问题。撤销消息框(即在用户按下按钮之前使其从屏幕消失)非常普遍，但或许在特定场景下讨论更有意义。

IAsyncInfo 接口(Windows Runtime 中 4 个标准异步接口所继承的接口)定义了一个名为 Cancel 的方法，用于撤销操作。正如前文所介绍的，IAsyncInfo 接口还定义了一个名为 Status 的属性，类型为 AsyncStatus 枚举。这个枚举有 4 个成员：Started、Completed、Canceled 和 Error。该接口的另一个属性为 Exception 类型的 ErrorCode 属性。

如果以回调形式执行异步操作，需要在回调方法的起始处检查 Status 属性，确保在调用 GetResults 方法前该属性的值为 Completed，而非 Canceled 或 Error。

await 应在 try 块中使用。如果异步操作中途被撤销，它会抛出 TaskCanceledException 类型的异常。若异步操作执行期间发生真正的错误，所抛出的异常则包含错误信息。

示例项目 HowToAsync3 是这样调用 MessageDialog 的 ShowAsync 方法的。

```
IUICommand command = await msgdlg.ShowAsync();
```

为探究幕后的 IAsyncOperation 对象，可以将这条语句拆成两条。

```
IAsyncOperation<IUICommand> asyncOp = msgdlg.ShowAsync();
IUICommand command = await asyncOp;
```

两种写法效果上并无差别。这意味着我们可以将 asyncOp 对象以字段形式保存，这样类内部的其他方法便可以调用该对象的 Cancel 方法。

下面让我们通过计时器来触发 MessageDialog 的撤销。示例项目 HowToCancelAsync 会在 MessageDialog 显示后启动一个 5 秒的 DispatcherTimer。如果 MessageDialog 在 5 秒内

未被用户主动撤销，计时器的 Tick 事件处理程序会调用保存在字段中的 IAsyncOperation 对象的 Cancel 方法。

项目：HowToCancelAsync | 文件：MainPage.xaml.cs(片段)

```
public sealed partial class MainPage : Page
{
    IAsyncOperation<IUICommand> asyncOp;

    public MainPage()
    {
        this.InitializeComponent();
    }

    async void OnButtonClick(object sender, RoutedEventArgs args)
    {
        MessageDialog msgdlg = new MessageDialog("Choose a color", "How To Cancel Async");
        msgdlg.Commands.Add(new UICommand("Red", null, Colors.Red));
        msgdlg.Commands.Add(new UICommand("Green", null, Colors.Green));
        msgdlg.Commands.Add(new UICommand("Blue", null, Colors.Blue));

        // Start a five-second timer
        DispatcherTimer timer = new DispatcherTimer();
        timer.Interval = TimeSpan.FromSeconds(5);
        timer.Tick += OnTimerTick;
        timer.Start();

        // Show the MessageDialog
        asyncOp = msgdlg.ShowAsync();
        IUICommand command = null;

        try
        {
            command = await asyncOp;
        }
        catch (Exception)
        {
            // The exception in this case will be TaskCanceledException
        }

        // Stop the timer
        timer.Stop();

        // If the operation was cancelled, exit the method
        if (command == null)
            return;

        // Get the Color value and set the background brush
        Color clr = (Color)command.Id;
        contentGrid.Background = new SolidColorBrush(clr);
    }

    void OnTimerTick(object sender, object args)
    {
        // Cancel the asynchronous operation
        asyncOp.Cancel();
    }
}
```

这个可撤销版本的逻辑显然要更复杂些，但这段代码并不比平时使用 try-catch 块的代码复杂，并且仍保持命令式结构。这里要再次提醒一下：await 运算符之前的代码会先执行。MessageDialog 消失后，try 块中的代码会继续执行。不论是否存在异常，程序都会检查 try 块中赋值的 command 变量是否为 null。

7.6　File I/O 的处理

熟悉.NET 的开发者应该对实现文件 I/O 的 System.IO 命名空间不陌生。其中的一些经验在 Windows 8 中仍然适用，但 Windows 8 的 System.IO 命名空间更精简。以 Windows.Storage 为前缀的几个命名空间包含 Windows Runtime 对文件 I/O 的大部分支持。这必然涌现许多新的文件 I/O 类和概念。文件与流的整套接口得到改进，访问磁盘的所有方法也均被改为异步形式。

Windows 8 应用程序有三种进行文件 I/O 的方法。下面三节将逐一介绍。

应用程序本地存储

如果应用程序需要保存对其他应用程序或用户无关的信息，最好将其存储在应用程序的本地存储中。本地存储有时也称为“独立存储”，是应用程序在硬盘上的专属区域，但应用程序不需要关心它的实际位置。如果应用程序从系统中被移除，这片区域也会被自动释放。

访问独立存储需要用到 Windows.Storage 命名空间的 ApplicationData 类。应用程序当前可用的 Windows.Storage 对象可通过 Current 属性获得。

```
ApplicationData appData = ApplicationData.Current;
```

为方便使用，ApplicationData 类定义了几个属性。

LocalSettings 和 RoamingSettings 属性允许我们访问 ApplicationDataContainer 对象。ApplicationDataContainer 对象以字典形式存储应用程序设置。这些应用程序设置仅限于 Windows Runtime 的基本数据类型(数值与字符串)。

LocalFolder、RoamingFolder 和 TemporaryFolder 属性能够返回 StorageFolder 类型的对象。StorageFolder 类是 Windows.Storage 命名空间的重要组成部分。StorageFolder 类代表目录，对本地存储来说它代表应用程序专属的目录。StorageFolder 类包含用于创建子目录及创建和访问文件的方法。文件由 StorageFile 对象表示，可以打开，并返回可供读/写的流。

文件选择器

FileOpenPicker、FileSavePicker 和 FolderPicker 是 Windows.Storage.Pickers 命名空间中三个较为重要的类，代表供 Windows 8 应用程序使用的三种标准对话框，可以用来打开和保存标准数据文件夹中的文件(标准数据文件夹包括“文档库”、“音乐库”和“图片库”)。

与 MessageDialog 类似，FileOpenPicker 和 FileSavePicker 具有用于显示对话框的异步方法，能够返回 StorageFile 类型的对象，而 FolderPicker 能够返回 StorageFolder 类型的对象。

用户能够通过这些选择器浏览文件系统，并授权应用程序对其进行访问，因而这些选择器非常灵活。应用程序通常只接受特定类型的文件。以 FileOpenPicker 为例，在使用这个选择器时，应通过 FileTypeFilter 属性至少为其指定一种文件类型(如“.txt”)。文件类型描述不能包含通配符。

虽然 FileOpenPicker 能够显示各种类型的文件，但它只会显示应用程序通过 FileTypeFilter 属性指定的文件类型，而无法显示所有类型的文件。

批量访问

应用程序可以通过 Windows.Storage.BulkAccess 命名空间中定义的 FileInformation 和 FolderInformation 类来直接访问文件系统。这两个类允许应用程序以非常灵活的方式查询和操纵文件夹中的子文件夹和文件。

这种文件访问方式不受用户干预，因而应用程序需要事先声明这个需求。使用批量访问的应用程序需要通过 package.appxmanifest 声明应用程序所要访问的存储区域。在 Visual Studio 中，我们可以通过对话框编辑 package.appxmanifest 文件。为访问相应的存储区，需要在对话框的“功能”选项卡中选中对应的“文档库”、“音乐库”和“视频库”。该选项卡列出的功能选项用于定义应用程序的权限。若选择“文档库”，则必须在“声明”选项卡中添加“文件类型关联”，应用程序所针对的所有文件类型必须显式地声明。用户所能看到的文件也仅限于所声明的这些文件类型。

第 14 章介绍 PhotoScatter 程序时会演示如何通过这些类来访问“图片库”。本章会将重点放在另外两种文件 I/O 方式上。

7.7 文件选择器和文件 I/O

为进一步认识 FileOpenPicker 和 FileSavePicker 类，我们来编写一个类似于 Windows“记事本”的小程序 PrimitivePad。这个程序带有几个命令按钮和一个较大的 TextBox。命令按钮一般是通过应用栏(application bar)实现的，但本书将应用栏的相关话题留到第 8 章进行讨论。

PrimitivePad 有两个用于文件 I/O 的按钮：Open 和 Save As。在实际的应用程序中，可能还有 New 和 Save 按钮，并在用户按下 New 或 Open 后提示是保存还是放弃当前文件的更改。下一章会具体描述这个逻辑。

PrimitivePad 还有一个用于更改换行模式的按钮，该模式的状态会以程序设置的形式保存。此程序的 XAML 文件如下所示。

项目: PrimitivePad | 文件: MainPage.xaml(片段)

```
<Page ... >
    <Page.Resources>
        <Style x:Key="buttonStyle" TargetType="ButtonBase">
            <Setter Property="HorizontalAlignment" Value="Center" />
            <Setter Property="Margin" Value="0 12" />
        </Style>
    </Page.Resources>

    <Grid Background="{StaticResource ApplicationPageBackgroundThemeBrush}">
        <Grid.RowDefinitions>
```

```
            <RowDefinition Height="Auto" />
            <RowDefinition Height="*" />
        </Grid.RowDefinitions>

        <Grid.ColumnDefinitions>
            <ColumnDefinition Width="*" />
            <ColumnDefinition Width="*" />
            <ColumnDefinition Width="*" />
        </Grid.ColumnDefinitions>

        <Button Content="Open..."
                Grid.Row="0"
                Grid.Column="0"
                Style="{StaticResource buttonStyle}"
                Click="OnFileOpenButtonClick" />

        <Button Content="Save As..."
                Grid.Row="0"
                Grid.Column="1"
                Style="{StaticResource buttonStyle}"
                Click="OnFileSaveAsButtonClick" />

        <ToggleButton Name="wrapButton"
                      Content="No Wrap"
                      Grid.Row="0"
                      Grid.Column="2"
                      Style="{StaticResource buttonStyle}"
                      Checked="OnWrapButtonChecked"
                      Unchecked="OnWrapButtonChecked" />

        <TextBox Name="txtbox"
                 Grid.Row="1"
                 Grid.Column="0"
                 Grid.ColumnSpan="3"
                 FontSize="24"
                 AcceptsReturn="True" />
    </Grid>
</Page>
```

如下图所示，这三个按钮分列于屏幕顶端，我们可以在大的这个 TextBox 中写一首诗。

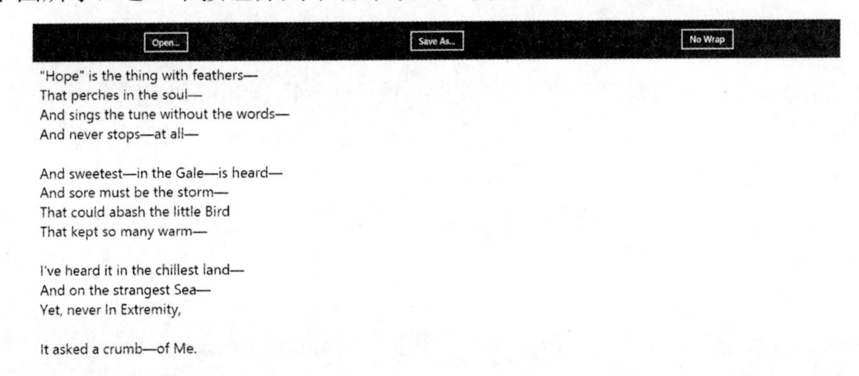

FileOpenPicker 和 FileSavePicker 类调用的对话框会完全遮住应用程序原本的界面，并在对话框被关闭后控制权才会被交给应用程序。如果这个行为不满足需求，则要采用批量访问的方式自行查询文件目录。

这两个类都能够返回 StorageFile 类型的对象给应用程序。(FileOpenPicker 支持多选，因而一次可以返回多个 StorageFile 对象。)StorageFile 类位于 Windows.Storage 命名空间中，表示未打开的文件。StorageFile 对象的 Open 方法能够返回流对象。这个流对象实现了 Windows.Storage.Streams 命名空间下定义的 IInputStream 或 IRandomAccessStream 接口。我们可以将 DataReader 和 DataWriter 对象附加到这个流对象上来分别进行读操作和写操作。System.IO 命名空间中定义的扩展方法可以将 Windows Runtime 流对象转换为我们熟悉的.NET 对象(如 StreamReader 和 StreamWriter)来对文件进行读和写。这样我们便可以重用基于.NET 流的现有代码，也可以通过.NET 流对象对 XML 文件进行读和写。

在 FileTypeFilter 属性中指定一个合法的字符串(如“.txt”)是使用 FileOpenPicker 唯一的前提条件。设置该属性后，调用 PickSingleFileAsync 方法便可以打开标准的文件选择器。用户可以从中选择现有的文件，然后“打开”，或者直接“取消”。如果使用 await 来调用该方法，程序会直接返回代表用户选择的文件的 StorageFile 对象。下面的代码展示了 Open 按钮的 Click 事件处理程序。

```
项目: PrimitivePad | 文件: MainPage.xaml.cs(片段)
async void OnFileOpenButtonClick(object sender, RoutedEventArgs args)
{
    FileOpenPicker picker = new FileOpenPicker();
    picker.FileTypeFilter.Add(".txt");
    StorageFile storageFile = await picker.PickSingleFileAsync();

    // If user presses Cancel, result is null
    if (storageFile == null)
        return;

    using (IRandomAccessStream stream = await storageFile.OpenReadAsync())
    {
        using (DataReader dataReader = new DataReader(stream))
        {
            uint length = (uint)stream.Size;
            await dataReader.LoadAsync(length);
            txtbox.Text = dataReader.ReadString(length);
        }
    }
}
```

PickSingleFileAsync 方法实际返回的是 IAsyncOperation<StorageFile>对象。该方法是少数几个泛型参数所代表的对象可能为 null 的方法。当用户按下文件打开选择器的“取消”按钮后，则会返回 null 值。此时便无需执行后续的操作了。

为打开 StorageFile 对象以便读取文件内容，可以调用该对象的 OpenReadAsync 方法。由于 OpenReadAsync 方法要访问磁盘，因而该方法也是一个异步的。这个方法实际会返回 IAsyncOperation<IRandomAccessStreamWithContentType> 类型的对象。IRandomAccessStreamWithContentType 接口实现了 IRandomAccessStream 接口，因而这段代码使用的是这个较短的接口名称。由于 IRandomAccessStream 实现了 IDisposable 接口，因而最好将 IRandomAccessStream 对象置于 using 块中，以便使其自动被释放。

DataReader 类也实现了 IDisposable 接口。这个类为 Windows Runtime 基本类型提供了若干“读”方法(如 ReadString)。这些“读”方法是同步方法，因为它们不涉及磁盘访问，而只是从存储于内存的内部缓存(IBuffer 类型)读取数据并将其转换为具体的数据类型。实际访问磁盘文件的方法为 LoadAsync 方法。在这些“读”方法被调用前，LoadAsync 方法会将一定量的字节从文件加载到缓存。大文件最好分段加载。DataReader 的

UnconsumedBufferLength 属性正用于支持这种加载方式。

如果不使用 await 运算符，则需要为这三个异步方法分别指定回调方法。为在 UI 线程中设置 TextBox 的 Text 属性，还需要定义一个回调方法。

文件保存的逻辑与读取类似(如下所示)。

```
项目: PrimitivePad | 文件: MainPage.xaml.cs(片段)
async void OnFileSaveAsButtonClick(object sender, RoutedEventArgs args)
{
    FileSavePicker picker = new FileSavePicker();
    picker.DefaultFileExtension = ".txt";
    picker.FileTypeChoices.Add("Text", new List<string> { ".txt" });
    StorageFile storageFile = await picker.PickSaveFileAsync();

    // If user presses Cancel, result is null
    if (storageFile == null)
        return;

    using (IRandomAccessStream stream = await storageFile.OpenAsync(FileAccessMode.ReadWrite))
    {
        using (DataWriter dataWriter = new DataWriter(stream))
        {
            dataWriter.WriteString(txtbox.Text);
            await dataWriter.StoreAsync();
        }
    }
}
```

DataWriter 的 StoreAsync 方法会返回实现 IAsyncOperation<uint>接口的对象。uint 值表示写入文件的字节数。StoreAsync 是这段程序调用的最后的方法。有人或许会问，为何这个方法还要使用 await 运算符。一般来说，在不关心返回值时，调用异步方法可以不使用 await。但应牢记于心的是：在异步方法运行过程中主调方法会继续执行。如果主调方法在继续执行后隐式地期待异步方法完成，则会出现问题。就本示例而言，之所以没有省略 await，是因为 using 块会隐式地关闭 DataWriter 和 IRandomAccessStream 对象，而这不应发生在 StoreAsync 执行完毕之前。

示例程序 PrimitivePad 中还有一个 ToggleButton 控件，允许用户选择文本是否换行。这正是 MainPage 代码隐藏文件最后一部分代码所实现的。

```
项目: PrimitivePad | 文件: MainPage.xaml.cs(片段)
public sealed partial class MainPage : Page
{
    ApplicationDataContainer appData = ApplicationData.Current.LocalSettings;

    public MainPage()
    {
        this.InitializeComponent();

        Loaded += (sender, args) =>
            {
                if (appData.Values.ContainsKey("TextWrapping"))
                    txtbox.TextWrapping = (TextWrapping)appData.Values["TextWrapping"];

                wrapButton.IsChecked = txtbox.TextWrapping == TextWrapping.Wrap;
                wrapButton.Content = (bool)wrapButton.IsChecked ? "Wrap" : "No Wrap";

                txtbox.Focus(FocusState.Programmatic);
            };
    }
    ...
    void OnWrapButtonChecked(object sender, RoutedEventArgs args)
```

```
        {
            txtbox.TextWrapping = (bool)wrapButton.IsChecked ? TextWrapping.Wrap :
                                                               TextWrapping.NoWrap;
            wrapButton.Content = (bool)wrapButton.IsChecked ? "Wrap" : "No Wrap";
            appData.Values["TextWrapping"] = (int)txtbox.TextWrapping;
        }
    }
```

这个程序有一个引用 ApplicationDataContainer 对象的字段。该对象的 Values 属性是一个字典，应用程序可以通过它来保存基本数据类型的设置。在 Loaded 事件处理程序中，如果这个字典包含 TextWrapping 设置，则根据它来初始化 TextBox 的属性，ToggleButton 也会被相应地初始化。

若 ToggleButton 的状态发生变化，处理程序会设置 TextBox 的 TextWrapping 属性，并将新设置保存在字典中。

这是保存应用程序设置的途径之一。本章稍后还会介绍一种通过 Application 类的 Suspending 属性保存应用程序设置的方法。如果希望探寻这些设置(和其他本地存储)在硬盘上的位置，可以先通过 Visual Studio 打开 Package.appxmanifest 文件，在“打包”选项卡中查看应用程序的“包名称”。(也可以以 XML 格式查看该文件 Identity 元素的 Name 特性。)这个名称是能够标识应用程序的 GUID。程序设置和本地数据可以在以下目录中找到。

```
C:\Users\[user-name]\AppData\Local\Packages\[app-guid]
```

7.8 异常处理

我们可以有意使 PrimitivePad 程序崩溃。例如，按下 PrimitivePad 程序的 Open 按钮，在选择器中选中某个文件，通过 Windows 资源管理器(或其他程序)删除该文件，然后按选择器的“打开”按钮。PrimitivePad 在尝试打开不存在的文件时会引发异常。

为了捕获这个错误，可以将检查变量 storageFile 是否为 null 之后的代码放在 try 块中。但应注意，不能在 catch 块中通过 MessageDialog 来提示用户有错误出现，因为 catch 块中不允许出现 await 运算符。下面给出一种较为合理的异常处理方法。

```
async void OnFileOpenButtonClick(object sender, RoutedEventArgs args)
{
    ...
    Exception exception = null;

    try
    {
        using (IRandomAccessStream stream = await storageFile.OpenReadAsync())
        {
            using (DataReader dataReader = new DataReader(stream))
            {
                uint length = (uint)stream.Size;
                await dataReader.LoadAsync(length);
                txtbox.Text = dataReader.ReadString(length);
            }
        }
    }
    catch (Exception exc)
    {
        exception = exc;
    }
```

```
    if (exception != null)
    {
        MessageDialog msgdlg = new MessageDialog(exception.Message,
                                                 "File Read Error");
        await msgdlg.ShowAsync();
    }
}
```

最后面的 if 语句通过检查 exception 变量是否为 null 值来判断是否存在异常。如果存在异常，MessageDialog 就会显示错误消息。

7.9　多个异步调用的合并

我们可以将所有的文件打开和保存逻辑提取成独立的方法，然后在按钮的 Click 事件处理程序中调用这些方法。这样，如果程序需要在多处使用选择器来打开或保存文件，便可以复用这些方法了。

为了达到这个目的，可以通过一个名为 LoadFile 的方法来显示 FileOpenPicker，读取文本文件的内容，然后返回一个字符串。这样一来 OnFileOpenButtonClick 方法便可以得到简化。

```
void OnFileOpenButtonClick(object sender, RoutedEventArgs args)
{
    txtbox.Text = LoadFile();
}
```

但事实上，我们不能这样简化该方法。LoadFile 方法不能返回 string，因为在所有异步操作完成之前是得不到 string 的。这里再次提醒读者，await 运算符会使编译器生成回调方法，就好像我们有意为之一样。可以尝试编写一个 LoadFile 方法来显式地调用回调方法，并从这个 LoadFile 方法返回 string，但最终会发现这是做不到的。

我们需要在调用 LoadFile 时使用 await 运算符，将该方法重命名为 LoadFileAsync。

```
async void OnFileOpenButtonClick(object sender, RoutedEventArgs args)
{
    txtbox.Text = await LoadFileAsync();
}
```

这样我们便可以不在 LoadFileAsync 中进行异常处理，而在主调程序中进行(例如 OnFileOpenButtonClick 处理程序)。

但真正的问题在于：LoadFileAsync 到底应该返回什么类型？根据 Windows Runtime 中实现的异步方法猜测，它或许应该返回 IAsyncOperation<string>类型。但实际上是做不到的。关键问题在于 Windows Runtime 未提供实现该接口的公共类。

C#程序员习惯使用.NET 中的类来实现异步操作。也就是说，LoadFileAsync 方法的最佳返回类型为 Task<string>。因而该方法可以像下面这样定义。

```
async Task<string> LoadFileAsync()
{
    FileOpenPicker picker = new FileOpenPicker();
    picker.FileTypeFilter.Add(".txt");
    StorageFile storageFile = await picker.PickSingleFileAsync();

    // If user presses Cancel, result is null
    if (storageFile == null)
```

```
        return null;

    using (IRandomAccessStream stream = await storageFile.OpenReadAsync())
    {
        using (DataReader dataReader = new DataReader(stream))
        {
            uint length = (uint)stream.Size;
            await dataReader.LoadAsync(length);
            return dataReader.ReadString(length);
        }
    }
}
```

虽然这个方法被命名为 LoadFileAsync，但其中的代码都运行在 UI 线程上。由于该方法幕后有一部分仍运行于另外的线程上，因而该方法仍被认为是异步方法。应注意的是，如果用户按下文件选择器的“取消”按钮，该方法会返回 null。TextBox 的 Text 属性不能设置为 null，因而 Click 处理程序必须对这种情况进行处理。

```
async void OnFileOpenButtonClick(object sender, RoutedEventArgs args)
{
    string text = await LoadFileAsync();

    if (text != null)
        txtbox.Text = text;
}
```

那么 Task 是什么呢？Task 是定义于 System.Threading.Tasks 命名空间下的类，是.NET 实现异步和并行处理的核心。它有泛型和非泛型的版本。对于无返回值的异步方法，可以使用非泛型的版本。下面这个包含所有保存逻辑的方法展示了 Task 非泛型版本的使用。

```
async void OnFileSaveAsButtonClick(object sender, RoutedEventArgs args)
{
    await SaveFileAsync(txtbox.Text);
}

async Task SaveFileAsync(string text)
{
    FileSavePicker picker = new FileSavePicker();
    picker.DefaultFileExtension = ".txt";
    picker.FileTypeChoices.Add("Text", new List<string> { ".txt" });
    StorageFile storageFile = await picker.PickSaveFileAsync();

    // If user presses Cancel, result is null
    if (storageFile == null)
        return;

    using (IRandomAccessStream stream = await storageFile.OpenAsync(FileAccessMode.ReadWrite))
    {
        using (DataWriter dataWriter = new DataWriter(stream))
        {
            dataWriter.WriteString(text);
            await dataWriter.StoreAsync();
        }
    }
}
```

.NET 和 Windows Runtime 对异步处理的支持是一致的，两者相关的类型也是可以相互转换的。Task 有一个名为 AsAsyncAction 的扩展方法，能够返回 IAsyncAction；Task<T> 有一个名为 AsAsyncOperation<T>的扩展方法，能够返回 IAsyncOperation<T>。相应地，IAsyncAction 有一个名为 AsTask 的方法，能够返回 Task；IAsyncOperation<T>有一个名为 AsTask<T>的方法，能够返回 Task<T>。

然而，相比 Windows Runtime 中的 Task 对异步的支持，.NET 中的 Task 更为强大，能够管理并行处理，还能够等待一组任务。用一本书来介绍 Task 也不为过，但这不是本书的重点。本章的重点是如何通过 Task 来处理耗时的任务。

7.10　高效的文件 I/O

虽然程序员应该熟练使用 DataReader 和 DataWriter 来进行文件 I/O，但大部分文件 I/O 操作都可以通过一些更简便的方法来完成。Windows.Storage 命名空间的 FileIO 和 PathIO 类提供了一些方法。有了这些方法，单次调用便可以实现对文件的读和写。

对文本文件来说，FileIO.ReadLinesAsync 方法可以读取文件内容并返回 string 对象的 IList(每个 string 对应文本中的一行)，而 FileIO.ReadTextAsync 能够通过单个 string 对象返回整个文件内容。就示例项目 PrimitivePad 而言，OnFileOpenButtonClick 中的两个嵌套的 using 语句可以替换为下面这行代码。

```
txtbox.Text = await FileIO.ReadTextAsync(storageFile);
```

类似地，该项目的文件保存逻辑可以替换为下面这行代码。

```
await FileIO.WriteTextAsync(storageFile, txtbox.Text, UnicodeEncoding.Utf8);
```

对于二进制文件，我们可以使用 ReadBufferAsync 和 WriteBufferAsync 方法。这两个方法利用了 IBuffer 类型的对象。IBuffer 对象不过是系统内存中的字节序列。所有 IBuffer 的引用都会被跟踪，这样 Windows 便在不需要它们的时候将其从内存中移除。

IBuffer 对象在 C#程序中无法直接访问，但可以间接获得。为创建二进制文件，可以创建 DataWriter 对象，向其中写入内容，然后保存 DataWriter 创建的 IBuffer 对象。

```
DataWriter dataWriter = new DataWriter();
// ... write to dataWriter
await FileIO.WriteBufferAsync(storageFile, dataWriter.DetachBuffer());
```

为读取二进制文件，可以先通过读取文件的方法来获得 IBuffer 对象，然后通过该对象来创建 DataReader。

```
IBuffer buffer = await FileIO.ReadBufferAsync(storageFile);
DataReader dataReader = DataReader.FromBuffer(buffer);
// ... read from dataReader
```

如果引入 System.Runtime.InteropServices.WindowsRuntime 命名空间，则可以将 IBuffer 对象转换为.NET 的 Stream 对象，然后通过 Stream 对象来创建 System.IO 命名空间定义的其他对象：其中包括 BinaryReader、BinaryWriter、StreamReader 和 StreamWriter。我们还可以将 IBuffer 转换为字节数组。

PathIO 类与 FileIO 相似，但前者的静态方法接受的不是 StorageFile 对象，而是 URI 字符串。为访问应用程序本身的内容文件，其 URI 一般要以“ms-appx:///”开头；为访问应用程序存储中的文件，其 URI 要以“ms-appdata:///”开头(稍后会进行演示)。

HttpClient 类用于通过网络上传和下载文件。如果不需要特别的灵活性，可以考虑使用简单易用的 RandomAccessStreamReference 类。

```
Uri uri = new Uri("http://...");
RandomAccessStreamReference streamRef = RandomAccessStreamReference.CreateFromUri(uri);

using (IRandomAccessStream stream = await streamRef.OpenReadAsync())
{
    ...
}
```

调用 IRandomAccessStream 对象的 ReadAsync 可以将文件内容读取到 IBuffer，然后将 IBuffer 传至 DataReader.FromBuffer 静态方法。

7.11 应用程序的生命周期

PrimitivePad 程序存在一个不易被察觉的问题，尚待解决。我们都知道，如果运行常规的 Windows 桌面“记事本”程序，键入一些文本后尝试关闭该程序(可以按右上角的“关闭”按钮，按组合键 Alt + F4，从“文件”菜单中选择“退出”，或者关闭 Windows)，会有一个消息框显示“是否将更改保存到……？”可以选择“保存”、“不保存”或“取消”。

我们对这个过程应该不陌生，但这对于 Windows 8 应用程序来说不是一个好的解决方案。如今，人们未必会坐在办公桌前，打开计算机，做一些工作，然后将计算机关闭。人们更倾向于从提包中拿出平板电脑，放在咖啡桌上，解锁屏幕，用一会，然后将其放回包中。用户可能按下开关按钮将平板电脑置于休眠模式，或使其自动进入休眠模式。

人们是否愿意在计算机休眠前被 Windows 应用程序询问是否保存数据？显然答案是否定的。将计算机置于休眠(或将视线从屏幕上移开)说明用户暂时不想与计算机进行交互。

但问题在于，如果放在咖啡桌上的平板电脑电量过低，甚至连休眠状态也无以为继，并即将关机，这时该怎么办？实践当中无法提示用户。

在我们所使用的计算机上，Windows 也有可能需要腾出一些内存。一种方式是终止久置不用的应用程序。同样，用户不会希望被通知这一切的发生。

出于上述原因的考虑，优秀的 Windows 8 应用程序会保存信息，不论被终止与否都能提供连贯流畅的用户体验。如果应用程序包含由于丢失而会让用户感到懊悔的数据，那么在程序被终止并再次运行后，这些数据应能够重新呈现。(当然，不同应用程序的数据，重要性也不同。例如，如果计算器程序丢弃一些数据可能并无大碍，但电子表格丢失数据则让人无法接受。)

这应该不难做到，对吗？作为程序员，我们一般认为应用程序终止前有一个事件会被引发。该事件可以用来将未保存的数据保存在应用程序的存储中，以便在程序下次运行时将这些数据恢复。

问题在于没有这样的事件。

但有一个事件可通知应用程序即将挂起(suspend)。应用程序在被彻底终止之前，总会先进入挂起状态(除非应用程序以非正常方式终止，如崩溃)，但挂起并不意味着程序即将终止。还有一个事件可能在程序挂起后通知其即将被恢复(resume)。

程序在后台不运行时会被挂起(如切换至 Windows“开始”屏幕或在屏幕左边缘滑动手指将另一个程序切换至前台)。按组合键 Alt+F4 终止应用程序或使计算机进入休眠状态，程序也会先被挂起。不论何种原因引发挂起，在程序被真正挂起前有 10 秒左右的延迟，以防止中断 Windows(和应用程序)正在进行的工作。

程序挂起之后可能被恢复，也可能被终止。程序挂起后可能历经较长一段时间才会恢复或终止。由于没有通知程序即将被终止的事件，应用程序必须通过挂起来保存未来进行恢复所必需的数据。程序有可能挂起还未完成就被终止。(现在或许可以在一定程度上理解为什么没有专门用于通知程序终止的事件：如果程序已挂起，为引发终止事件，Windows 要先恢复应用程序。)

为此，Application 类定义了两个事件：Suspending 和 Resuming。Suspending 事件的重要性远远高于 Resuming 事件。应用程序通过 Suspending 事件将未保存的状态保存在应用程序的本地存储中。应用程序不需要在 Resuming 事件中恢复这些数据，Windows 会自动完成这项工作。但应用程序需要在下次运行时加载数据。

应用程序可以选择在 Suspending 事件发生时执行其他任务(如尝试释放较大的、可以重新创建的资源，从而减少其对内存的占用)，而在 Resuming 事件发生时撤销所执行的任务或执行某些刷新工作(如将来自 Web 源的数据刷新到界面)。

应用程序通过 Visual Studio 调试器运行时挂起和恢复的方式与独立运行时所采用的方式有所不同。独立运行的应用程序在后台会挂起，而通过 Visual Studio 调试器运行的应用程序不会。

另一点不同在于：如果程序异常终止，程序不会在终止之前被挂起。未处理异常或使用 Visual Studio 的“停止调试”功能(在实践当中可以留意一下)都会导致非正常终止。

然而，如果通过组合键 Alt + F4 终止 Visual Studio 调试器运行的应用程序，程序会收到 Suspending 事件，然后终止，这个过程大概耗有 10 秒。在这段时间内，Visual Studio 仍然会认为程序正在运行。

为解决调试的不便，Visual Studio 提供了名为“调试位置”的工具栏，可以用来手动执行“挂起”(Suspend)、“继续”(Resume)和“挂起并关闭”(Suspend And Shutdown)命令。当程序通过调试器运行时，这些命令在开发与挂起和恢复相关的代码时极为有用。

通过 Visual Studio 运行的程序不容易观察到常规的 Suspending 和 Resuming 事件，因而这里提供一个实验程序，将这些事件的日志记录在应用程序的本地存储中。建议脱离 Visual Studio 调试器运行此程序。

示例项目 SuspendResumeLog 的 XAML 中仅包含一个只读的 TextBox。

```
项目: SuspendResumeLog | 文件: MainPage.xaml(片段)
<Grid Background="{StaticResource ApplicationPageBackgroundThemeBrush}">
    <TextBox Name="txtbox"
             AcceptsReturn="True"
             IsReadOnly="True" />
</Grid>
```

对应的代码隐藏文件订阅了 3 个事件：MainPage 的 Loaded 事件(仅在程序启动时执行，并且只执行一次)以及当前 Application 对象的 Suspending 和 Resuming 事件。所有事件都会记录到应用程序本地存储的 logfile.txt 文件中。

```
项目: SuspendResumeLog | 文件: MainPage.xaml.cs(片段)
public sealed partial class MainPage : Page
{
    StorageFile logfile;

    public MainPage()
```

```
    {
        this.InitializeComponent();

        Loaded += OnLoaded;
        Application.Current.Suspending += OnAppSuspending;
        Application.Current.Resuming += OnAppResuming;
    }

    async void OnLoaded(object sender, RoutedEventArgs args)
    {
        // Create or obtain the log file
        StorageFolder localFolder = ApplicationData.Current.LocalFolder;
        logfile = await localFolder.CreateFileAsync("logfile.txt",
                                CreationCollisionOption.OpenIfExists);

        // Load the file and display it
        txtbox.Text = await FileIO.ReadTextAsync(logfile);

        // Log the launch
        txtbox.Text += String.Format("Launching at {0}\r\n", DateTime.Now.ToString());
        await FileIO.WriteTextAsync(logfile, txtbox.Text);
    }

    async void OnAppSuspending(object sender, SuspendingEventArgs args)
    {
        SuspendingDeferral deferral = args.SuspendingOperation.GetDeferral();

        // Log the suspension
        txtbox.Text += String.Format("Suspending at {0}\r\n", DateTime.Now.ToString());
        await FileIO.WriteTextAsync(logfile, txtbox.Text);

        deferral.Complete();
    }

    async void OnAppResuming(object sender, object args)
    {
        // Log the resumption
        txtbox.Text += String.Format("Resuming at {0}\r\n", DateTime.Now.ToString());
        await FileIO.WriteTextAsync(logfile, txtbox.Text);
    }
}
```

Loaded 事件被引发后，程序会获取代表当前应用程序本地存储的 StorageFolder 对象，然后创建名为 logfile.txt 的文件。由于将枚举值 CreationCollisionOption.OpenIfExists 传入 CreateFileAsync 方法，在文件已存在的情况下(如程序第二次运行及后续的运行)该方法与 GetFileAsync 是等效的。

OpenIfExists 枚举成员的名称并不十分准确。通过该枚举访问文件时文件并没有打开供读写，因而将该枚举命名为 GetIfExists 更为贴切。若所要获取的文件不存在，CreateFileAsync 方法则会创建长度为零的文件，并获取对该文件的引用。FileIO.ReadTextAsync 和 FileIO.WriteTextAsync 方法会真正打开文件，并在分别完成读和写之后将文件关闭。

请注意 Suspending 事件处理程序中的 SuspendingDeferral 对象。如果没有此对象，Windows 会认为 Suspending 处理程序在调用 WriteTextAsync 时便已完成，因为调用该方法时该处理程序首次退出。

一般而言，如果程序在本地存储中维护数据，只需要在 Loaded 事件(或其他初始化事件)被引发时加载数据，并在 Suspending 事件发生时保存数据。SuspendResumeLog 程序在 Loaded 事件和 Resuming 事件发生时都保存了数据。该程序是为在 Visual Studio 调试器以外的环境下运行而设计的。之所以随时保存数据是考虑到程序可能通过 Visual Studio 调试器运行且被“停止调试”终止。否则，若这样终止程序，Suspending 处理程序是不会被调用的，进而造成数据丢失。

若在 Visual Studio 调试器环境中测试程序的数据保存与恢复，最好养成使用“挂起并关闭”命令来终止程序的习惯，而不要使用“停止调试”。

上面使用 FileIO.ReadTextAsync 的那段代码可以替换为下面一行。

```
txtbox.Text = await PathIO.ReadTextAsync("ms-appdata:///local/logfile.txt");
```

使用 FileIO.WriteTextAsync 的那段代码可以替换为下面一行。

```
await PathIO.WriteTextAsync("ms-appdata:///local/logfile.txt", txtbox.Text);
```

前缀 ms-appdata 表示应用程序的独立存储。在程序中表现为文件目录。根据目录所属类别的不同，其名称可能为 local、roaming 和 temp。即便使用这种 URI 来读写文件内容，仍需要通过 StorageFolder 的方法来创建 StorageFile 对象。

在正常情况下，更新日志文件的代码需要向现有文件追加文本。FileIO 和 PathIO 都有追加文本的方法，但 SuspendResumeLog 程序并未采用。除了该程序所采用的方式外，可以将同样的文本同时追加到 TextBox 和日志文件，也可以在每次追加后将文件重新加载到 TextBox。

SuspendResumeLog 程序仅包含一个 TextBox，并将内容存储在应用程序的独立存储中。示例项目 QuickNotes 与 SuspendResumeLog 类似。QuickNotes 允许用户在 TextBox 中输入内容，并能够自动保存，以便下次打开程序时呈现之前的内容。这个示例程序的 XAML 文件如下所示。

```
项目: QuickNotes | 文件: MainPage.xaml(片段)
<Grid Background="{StaticResource ApplicationPageBackgroundThemeBrush}">
   <TextBox Name="txtbox"
            AcceptsReturn="True"
            TextWrapping="Wrap" />
</Grid>
```

对应的代码隐藏文件通过 FileIO.ReadTextAsync 来读取文件内容(可以借助于现有的 StorageFile 对象)，通过 PathIO.WriteTextAsync 来向文件写入内容。

```
项目: QuickNotes | 文件: MainPage.xaml.cs(片段)
public sealed partial class MainPage : Page
{
   public MainPage()
   {
      this.InitializeComponent();
      Loaded += OnLoaded;
      Application.Current.Suspending += OnAppSuspending;
   }

   async void OnLoaded(object sender, RoutedEventArgs args)
   {
      StorageFolder localFolder = ApplicationData.Current.LocalFolder;
      StorageFile storageFile = await localFolder.CreateFileAsync("QuickNotes.txt",
                                        CreationCollisionOption.OpenIfExists);
      txtbox.Text = await FileIO.ReadTextAsync(storageFile);
      txtbox.SelectionStart = txtbox.Text.Length;
      txtbox.Focus(FocusState.Programmatic);
   }
```

```
    async void OnAppSuspending(object sender, SuspendingEventArgs args)
    {
        SuspendingDeferral deferral = args.SuspendingOperation.GetDeferral();
        await PathIO.WriteTextAsync("ms-appdata:///local/QuickNotes.txt", txtbox.Text);
        deferral.Complete();
    }
}
```

7.12 自定义的异步方法

前面演示了如何编写后缀为 Async 的异步方法,如何在这些方法中调用其他异步方法。这些异步方法本身的代码可以运行在 UI 线程上,而它们所调用的其他异步方法则在不同线程上运行。

应用程序有时需要执行一些耗时的任务,可能阻塞 UI 线程。若能将每个耗时的任务拆分成若干小块,则可以使用 DispatcherTimer 类或 CompositionTarget.Rendering 事件来分别运行。虽然事件处理程序运行在 UI 线程上,但任务本身会通过某种方式运行在其他线程上,从而使用户界面保持可响应状态。

我们可以自己想办法让任务运行在工作线程上。为此,可以利用 Windows.System.Threading 命名空间下的 ThreadPool 类,也可以利用功能更为强大的 Task(稍后演示)。

Task.Run 方法最简单的重载只有一个 Action 类型的参数(代表没有参数且没有返回值的委托),能够利用线程池中的线程来运行通过参数指定的方法。我们一般通过 Lambda 函数来指定该参数。

假设有一个可能需要运行很久的方法(可能有几个参数)。

```
void BigJob(object arg1, object arg2)
{
    // ... heavy processing job
}
```

该方法不应直接运行在 UI 线程上,但可以将该方法置于 Lambda 函数体中,将 Lambda 函数传入 Task.Run,并通过 await 运算符等待其运行完毕。

```
await Task.Run(()⇨BigJob("abc", 555));
```

由于 Task.Run 在工作线程上运行 BigJob 方法,因而该方法不能包含任何访问用户界面对象的代码。(如果确实包含访问用户界面的代码,则需要通过 CoreDispatcher 的 RunAsync 方法运行。如果需要在 RunAsync 调用过程中通过 await 等待 BigJob 运行完毕,则必须用 async 来修饰 BigJob 方法并使该方法返回 Task 对象。)

再举一个耗时方法的例子,但这一次方法有返回值。

```
double CalculateMagicNumber(string str, double x)
{
    double magicNumber = 0;

    // ... big job

    return magicNumber;
}
```

显然，我们不希望在 UI 线程上调用这个方法，但可以通过 Task.Run 来运行它。

```
double magicNum = await Task.Run(() =>
   {
      return CalculateMagicNumber("abc", 5);
   });
```

这段代码将 CalculateMagicNumber 方法的调用放在 Lambda 函数的主体中，并将 Lambda 函数传入 Task.Run。CalculateMagicNumber 返回类型是 double，因而 Task.Run 返回类型为 Task<double>。应注意的是，await 运算符返回的是 CalculateMagicNumber 计算得到的 double 值。

我们也可以像下面这样定义一个异步方法 CalculateMagicNumberAsync。

```
Task<double> CalculateMagicNumberAsync(string str, double x)
{
   return Task.Run(() =>
   {
      return CalculateMagicNumber(str, x);
   });
}
```

因而我们便可以像下面这样在 UI 线程上调用此方法。

```
double magicNum = await CalculateMagicNumberAsync("xyz", 333);
```

我们也可以将上述方法合并为一个。

```
Task<double> CalculateMagicNumberAsync(string str, double x)
{
   return Task.Run(() =>
      {
         double magicNumber = 0;

         // ... big job in non-UI thread

         return magicNumber;
      });
}
```

如果计算过程中需要调用其他异步方法，可以使用 await 来调用，并使用 async 来修饰 Lambda 函数。

```
Task<double> CalculateMagicNumberAsync(string str, double x)
{
   return Task.Run(async () =>
      {
         double magicNumber = 0;

         // ... big job with await' s

         return magicNumber;
      });
}
```

如果涉及取消和进度报告功能，这种单一方法的编码形式可以降低实现难度。

有的异步方法可能包含某种循环。

```
Task<double> CalculateMagicNumberAsync(string str, double x)
{
   return Task.Run(async () =>
   {
      double magicNumber = 0;
```

```
        for (int i = 0; i < 100; i++)
        {
            // ... big job with await's
        }
        return magicNumber;
    });
}
```

循环非常适合用来实现撤销检查和进度报告功能，但应谨慎。我们不希望以每秒上千次(可能过快)或每五秒一次(可能过慢)的频率来进行撤销检查或进度报告。通常一秒一次或一秒几次为宜。对于要进行上千次或上百万次的循环，需要通过某种逻辑来检查撤销操作或报告进度(例如在循环变量可以被 100 整除时进行)。

为使此方法能够被撤销，可以添加一个 CancellationToken 类型的参数，并在适当位置调用该参数的 ThrowIfCancellationRequested 方法。

```
Task<double> CalculateMagicNumberAsync(string str, double x,
                                       CancellationToken cancellationToken)
{
    return Task.Run(async () =>
    {
        double magicNumber = 0;
        for (int i = 0; i < 100; i++)
        {
            cancellationToken.ThrowIfCancellationRequested();

            // ... big job with await's
        }
        return magicNumber;
    }, cancellationToken);
}
```

请注意，cancellationToken 参数作为第二个参数传至 Task.Run。这样，如果任务在开始时就被撤销，则任务不会被执行。

这样一来，我们需要在调用 CalculateMagicNumberAsync 方法时通过最后一个参数传入 CancellationToken 对象。为共享 CancellationToken 对象，需要将其声明为字段。

```
CancellationTokenSource cts;
```

事实上，这个变量必须声明为字段，因为触发撤销的方法要访问它(这很有可能是用户的操作)。

```
void OnCancelButtonClick(object sender, RoutedEventArgs args)
{
    cts.Cancel();
}
```

调用 CalculateMagicNumberAsync 之前，必须先创建 CancellationTokenSource 对象，将被调用的方法置于 try 块中，并将该对象的 Token 属性传入被调用的方法。

```
cts = new CancellationTokenSource();
double magicNum = 0;

try
{
    magicNum = await CalculateMagicNumberAsync("xyz", 333, cts.Token);
}
catch (OperationCanceledException)
{
    // ... cancellation logic
}
```

```
catch (Exception exc)
{
   // ... other exceptions logic
}
```

一旦 CancellationTokenSource 的 Cancel 方法被调用，在执行到 CancellationToken 对象的 ThrowIfCancellationRequested 方法时便会有异常抛出。该异常的类型为 OperationCanceledException，会被调用异步方法的代码接收。其他可能的异常也应一并通过 catch 块捕获。

若希望异步方法能够报告进度，可以为该方法添加另一个专用的参数，它的类型为 IProgress<T>。该泛型接口的类型参数 T 为进度值的类型，一般为 double 类型。取值范围是从 0 到 1，还是 0 到 100，这完全取决于开发者。如果是后一种情形，T 可以是 int。有甚者将 T 指定为 bool 类型，用 true 表示任务完成。

报告进度可以在一个适当的位置进行(如在进行撤销检查之前)。

```
Task<double> CalculateMagicNumberAsync(string str, double x,
                                       CancellationToken cancellationToken,
                                       IProgress<double> progress)
{
   return Task.Run(async () =>
   {
      double magicNumber = 0;

      for (int i = 0; i < 100; i++)
      {
         cancellationToken.ThrowIfCancellationRequested();
         progress.Report((double)i);

         // ... big job with await's
      }
      return magicNumber;
   }, cancellationToken);
}
```

这段代码只是将范围在 0 到 100 的循环变量转换为 double 类型来表示进度(这样可以直接设置 ProgressBar 的 Value 属性)。在某些情况下，可能需要有意地在方法开始处报告最小进度值，而在方法最后报告最大进度值。

我们还需要一个显示进度的方法，该方法应接受进度值类型的参数。

```
void ProgressCallback(double progress)
{
   progressBar.Value = progress;
}
```

这个方法应在 UI 线程上调用。

在调用 CalculateMagicNumberAsync 时(就像在前面 try 块中那样调用)，需要通过刚刚定义的回调方法创建一个 Progress 类型的对象，并将该对象作为最后一个参数传入 CalculateMagicNumberAsync 方法。

显示进度的回调不一定要单独定义，也可以使用 Lambda 表达式。

```
magicNum = await CalculateMagicNumberAsync("xyz", 333, cts.Token,
         new Progress<double>((percent) => progressBar.Value = percent));
```

下面让我们来看一个更为完整的例子。

演示异步操作最困难的地方在于如何找到一个适当的例子，执行需要耗费一定时间而又简单易懂。这里要有意编写一些效率较低的代码，一是为肉眼能够观察到 ProgressBar 的变化，二是为单击 Cancel 按钮留出时间。

示例项目 WordFreq 能够读取文本文件并计算词频。这里用的文本文件是来自著名的 Project Gutenberg 网站的纯文本电子书，即赫尔曼·梅尔维尔(Herman Melville)的《白鲸》(*Moby-Dick*)，该程序要统计每个词(如“whale”)在书中出现的次数。事实上，WordFreq 程序硬编码了《白鲸》这部小说。

GetWordFrequenciesAsync 方法有一个.NET Stream 类型的参数，因为该方法需要通过.NET 中的 StreamReader 来逐行读取电子书文件。该方法还接受 CancellationToken 和 IProgress 类型的参数。

此方法返回的结果有些细碎，因而这里通过.NET 中的 Dictionary 对象来聚合文本文件中的单词。该方法最后通过 LINQ 中的 OrderByDescending 方法按字典中“值”的大小对字典项进行排序，即词频高的单词排在最前面。结果是以下类型的集合。

```
KeyValuePair<string, int>
```

OrderByDescending 实际返回的集合为泛型类型 IOrderedEnumerable 的对象。

```
IOrderedEnumerable<KeyValuePair<string, int>>
```

也就是说 GetWordFrequenciesAsync 方法返回的是下面这个类型。

```
Task<IOrderedEnumerable<KeyValuePair<string, int>>>
```

这个方法的完整代码如下所示。

项目：WordFreq | 文件：MainPage.xaml.cs(片段)

```
Task<IOrderedEnumerable<KeyValuePair<string, int>>> GetWordFrequenciesAsync(Stream stream,
                                            CancellationToken cancellationToken,
                                            IProgress<double> progress)
{
    return Task.Run(async () =>
        {
            Dictionary<string, int> dictionary = new Dictionary<string, int>();

            using (StreamReader streamReader = new StreamReader(stream))
            {
                // Read the first line
                string line = await streamReader.ReadLineAsync();

                while (line != null)
                {
                    cancellationToken.ThrowIfCancellationRequested();
                    progress.Report(100.0 * stream.Position / stream.Length);

                    string[] words = line.Split(' ', ',', '.', ';', ':');

                    foreach (string word in words)
                    {
                        string charWord = word.ToLower();

                        while (charWord.Length > 0 && !Char.IsLetter(charWord[0]))
                            charWord = charWord.Substring(1);

                        while (charWord.Length > 0 &&
                                    !Char.IsLetter(charWord[charWord.Length - 1]))
                            charWord = charWord.Substring(0, charWord.Length - 1);

                        if (charWord.Length == 0)
                            continue;
```

```
                if (dictionary.ContainsKey(charWord))
                    dictionary[charWord] += 1;
                else
                    dictionary.Add(charWord, 1);
            }
            line = await streamReader.ReadLineAsync();
        }
    }

    // Return the dictionary sorted by Value (the word count)
    return dictionary.OrderByDescending(i => i.Value);
}, cancellationToken);
}
```

注意，传入Task.Run的方法体通过await运算符来调用StreamReader的ReadLineAsync方法。该方法读取文件中的每一行前都会检查CancellationToken，并根据Stream对象读取文件的多少来报告进度(百分比形式)。Project Gutenberg的《白鲸》电子版本大约有两万两千行。为了使代码简单明了，这里没有降低对这两个方法的调用频率，因而显得过于频繁。若要减少两者的调用次数可以跟踪文件读取的行数。

此方法本身没有进行异常处理。如果StreamReader的构造函数或ReadLineAsync抛出异常，则要由主调方法处理。

WordFreq程序的XAML定义了Start和Cancel按钮(后者最初被禁用)、用于报告进度的ProgressBar、用于报告错误的TextBlock以及ScrollViewer中用于展示词频结果的StackPanel。

项目: WordFreq | 文件: MainPage.xaml(片段)

```
<Grid Background="{StaticResource ApplicationPageBackgroundThemeBrush}">
    <Grid HorizontalAlignment="Center">
        <Grid.RowDefinitions>
            <RowDefinition Height="Auto" />
            <RowDefinition Height="Auto" />
            <RowDefinition Height="Auto" />
            <RowDefinition Height="*" />
        </Grid.RowDefinitions>

        <Grid.ColumnDefinitions>
            <ColumnDefinition Width="*" />
            <ColumnDefinition Width="*" />
        </Grid.ColumnDefinitions>

        <Button Name="startButton"
                Content="Start"
                Grid.Row="0" Grid.Column="0"
                HorizontalAlignment="Center"
                Margin="24 12"
                Click="OnStartButtonClick" />

        <Button Name="cancelButton"
                Content="Cancel"
                Grid.Row="0" Grid.Column="1"
                IsEnabled="false"
                HorizontalAlignment="Center"
                Margin="24 12"
                Click="OnCancelButtonClick" />

        <ProgressBar Name="progressBar"
                     Grid.Row="1" Grid.Column="0" Grid.ColumnSpan="2"
                     Margin="24" />

        <TextBlock Name="errorText"
```

```
              Grid.Row="2" Grid.Column="0" Grid.ColumnSpan="2"
              FontSize="24"
              TextWrapping="Wrap" />

        <ScrollViewer Grid.Row="3" Grid.Column="0" Grid.ColumnSpan="2">
            <StackPanel Name="stackPanel" />
        </ScrollViewer>
    </Grid>
</Grid>
```

除了 GetWordFrequenciesAsync 方法外，代码隐藏文件中还包含几个实现操作撤销和进度报告的方法。

项目：WordFreq | 文件：MainPage.xaml.cs(片段)

```
public sealed partial class MainPage : Page
{
    // Project Gutenberg ebook of Herman Melville's "Moby-Dick"
    Uri uri = new Uri("http://www.gutenberg.org/ebooks/2701.txt.utf-8");
    CancellationTokenSource cts;

    public MainPage()
    {
        this.InitializeComponent();
    }

    async void OnStartButtonClick(object sender, RoutedEventArgs args)
    {
        ...
    }

    void OnCancelButtonClick(object sender, RoutedEventArgs args)
    {
        cts.Cancel();
    }

    void ProgressCallback(double progress)
    {
        progressBar.Value = progress;
    }

    Task<IOrderedEnumerable<KeyValuePair<string, int>>> GetWordFrequenciesAsync(Stream stream,
                                                    CancellationToken cancellationToken,
                                                    IProgress<double> progress)
    {
        ...
    }
}
```

这里要特别说明一下 Start 按钮的 Click 处理程序。该处理程序在程序运行过程中可能被执行多次，但它是不可重入的。也就是说，在它执行完成之前，不能执行下一次。这个处理程序的主要作用是操纵界面元素，包括初始化 StackPanel，初始化 ProgressBar，以及启用和禁用按钮。所有文件访问及对 GetWordFrequenciesAsync 的调用，都是在 try 块中进行的。

项目：WordFreq | 文件：MainPage.xaml.cs(片段)

```
async void OnStartButtonClick(object sender, RoutedEventArgs args)
{
    stackPanel.Children.Clear();
    progressBar.Value = 0;
    errorText.Text = "";
    startButton.IsEnabled = false;
    IOrderedEnumerable<KeyValuePair<string, int>> wordList = null;

    try
```

```
        {
            RandomAccessStreamReference streamRef = RandomAccessStreamReference.CreateFromUri(uri);

            using (IRandomAccessStream raStream = await streamRef.OpenReadAsync())
            {
                using (Stream stream = raStream.AsStream())
                {
                    cancelButton.IsEnabled = true;
                    cts = new CancellationTokenSource();

                    wordList = await GetWordFrequenciesAsync(stream, cts.Token,
                                                    new Progress<double>(ProgressCallback));

                    cancelButton.IsEnabled = false;
                }
            }
        }
        catch (OperationCanceledException)
        {
            progressBar.Value = 0;
            cancelButton.IsEnabled = false;
            startButton.IsEnabled = true;
            return;
        }
        catch (Exception exc)
        {
            progressBar.Value = 0;
            cancelButton.IsEnabled = false;
            startButton.IsEnabled = true;
            errorText.Text = "Error: " + exc.Message;
            return;
        }

        // Transfer the list of word and counts to the StackPanel
        foreach (KeyValuePair<string, int> word in wordList)
        {
            if (word.Value > 1)
            {
                TextBlock txtblk = new TextBlock
                {
                    FontSize = 24,
                    Text = word.Key + " \x2014 " + word.Value.ToString()
                };
                stackPanel.Children.Add(txtblk);
            }

            await Task.Yield();
        }

        startButton.IsEnabled = true;
    }
```

这里遇到一个问题：当异步方法返回后，Click 处理程序需要将返回的键值对填入 StackPanel。这项工作是在处理程序尾部的 foreach 块中完成的。foreach 循环会频繁地操纵用户界面上的对象(创建 TextBlock 并将其添加至 StackPanel)，并且该操作不能在 UI 线程以外的线程上执行。即便对《白鲸》的词频结果进行筛选限制(正如示例程序所做的)，要显示的词条也将近一万个。这样的循环可能会造成用户界面被阻塞，进而无法响应用户输入，甚至还会导致词条在较长一段时间内无法显示到界面上。

有一个语句可以提供一种解决方案(可能不是最好的)。

```
await Task.Yield();
```

通过 await 调用此方法为 UI 线程上的其他代码的执行留出了机会。其他代码执行完毕后，控制权会重新回到这里。就本示例而言，所谓的“其他代码”包括 StackPanel 类中实现对作为子元素的 TextBlock 进行布局的代码以及允许用户滚动 ScrollViewer 中的 StackPanel 的代码。

若不调用 Task.Yield，则无法在 ProgressBar 达到最大值后的 5 秒内显示词条列表。反复调用 Task.Yield 会在很大程度上减慢循环的速度。虽然循环要花更长时间才能完成(通过进度变化和 Start 按钮重新恢复可用的时间可以感受得到)，但却可以立即让用户看到效果。此外，用户还能够在循环完成之前滚动词条列表。最终我们可以看到 whale 一次出现了 963 次(见下图)。

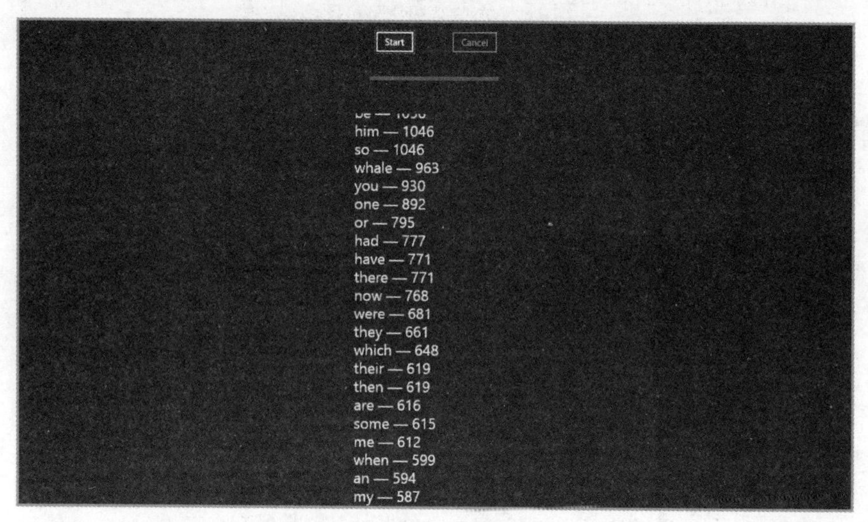

还有一种解决方案，甚至不用 StackPanel。正如第 11 章要介绍的，有专门用于显示列表项的控件。这些控件利用 VirtualizingStackPanel 面板，仅加载用户滑动出来的列表项。

虽然 Windows 8、.NET 和 C#的改进使异步方法的使用变得相当简便，但我们仍需注意细节并进行测试。举例来说，GetWordFrequenciesAsync 在本人的计算机上要花三四秒的时间才能完成。然而，如果将撤销检查和进度报告功能去掉，该方法在一秒内就能完成。不知读者怎么看，但本人对一秒内能完成的异步方法实现撤销检查和进度报告的做法表示怀疑。

权衡这些问题并不容易，之所以这样是因为一个矛盾的存在：人们希望计算机完成尽可能多的任务，但希望它看上去好像是闲着的。我们应使 Windows 8 应用程序看上去能够承担繁重的任务而不会被卡住，这仍然是一项挑战。

第 8 章　应用栏和弹出式窗口

组合第 4 章以及第 5 章所提到的元素、控件和面板，就可以在页面上构造出完整的用户界面。但是许多程序都喜欢把大多数命令和程序选项保持隐藏状态，而在用户明确需要时才出现。

过去的 Windows 应用程序一般使用菜单和对话框来合并命令和选项。顶级菜单总是可见，而实际命令通常在下拉子菜单上。一些菜单命令会启用对话框呈现一组相关程序选项。

而 Windows 8 强调应用内容，而不是强调好看。在许多情况下，以前应用菜单上的程序选项会移到应用栏，而应用栏通常是隐藏状态，但如果用户在屏幕顶部或底部边缘滑动手指或者移动鼠标指针到应用栏位置，就会启用应用栏。应用栏是 ContentControl 的派生类，称为 AppBar，本章会展示如何使用。

此外，Windows 8 程序可以用简单的 PopupMenu 类型对象来显示命令列表 (通常用作快捷菜单)，或者通过 Popup 元素向用户提供更广泛的控件集合。本章会展示如何使用这两种弹出式窗口。

本章结尾会给出本书目前最全的应用 XamlCruncher，该程序可以用 XAML 进行交互实验。

8.1　实施快捷菜单

快捷菜单是点击鼠标右键或者用“按住-保持-松开”手势而被启用的菜单。在屏幕被触碰的位置弹出快捷菜单，而如果选择其中的一个命令，快捷菜单一般会随之消失。快捷菜单通常与特定控件或单个控件的特定区域相关，这就是“快捷”名称的由来。

TextBox 控件包含一个快捷菜单。运行本书中使用 TextBox 的任何程序就能看到。在其中输入并选中一些文本，再右键单击控件或执行“按住-保持-松开”手势，就会出现一个菜单。取决于所选内容和剪贴板状态，菜单最多包含下图所示的五项命令。

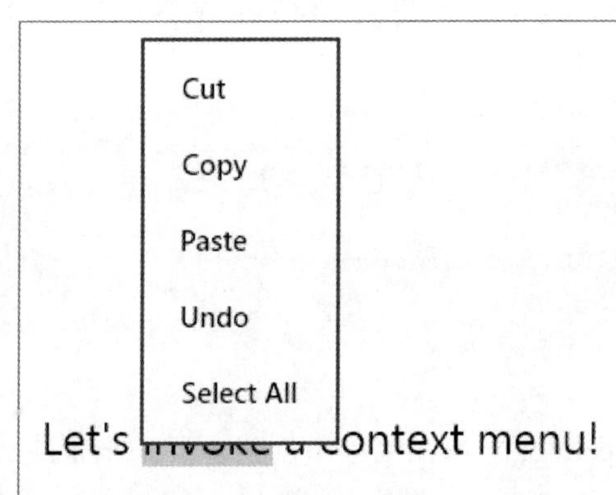

为获得定制快捷菜单，需要创建 PopupMenu 类型的对象。Windows.UI.Popups 命名空间、MessageDialog(第 7 章提到过)和 UICommand 共同定义了该类，可用来在 MessageDialog 和 PopupMenu 中具体指定命令。

PopupMenu 派生自 Object，因此不大可能会在 XAML 文件可视树中进行实例化。相反，你可能想在启用的时候完全用代码来构造 PopupMenu 对象，而最有可能的就是响应 RightTapped 事件的时候。

以下 XMAL 文件中包含位于页面中央的 TextBlock 以及 RightTapped 事件处理程序：

项目：SimpleContextMenu | 文件：MainPage.xaml(片段)

```
<Grid Background="{StaticResource ApplicationPageBackgroundThemeBrush}">
    <TextBlock Name="textBlock"
               FontSize="24"
               HorizontalAlignment="Center"
               VerticalAlignment="Center"
               TextAlignment="Center"
               RightTapped="OnTextBlockRightTapped">
        Simple Context Menu
        <LineBreak />
        <LineBreak />
        (right-click or press-and-hold-and-release to invoke)
    </TextBlock>
</Grid>
```

使用 MessageDialog，通过 UICommand 实例来显示菜单上要出现的命令。调用 ShowAsync 来显示：

项目：SimpleContextMenu | 文件：MainPage.xaml.cs(片段)

```
public sealed partial class MainPage : Page
{
    public MainPage()
    {
        this.InitializeComponent();
    }

    async void OnTextBlockRightTapped(object sender, RightTappedRoutedEventArgs args)
    {
        PopupMenu popupMenu = new PopupMenu();
        popupMenu.Commands.Add(new UICommand("Larger Font", OnFontSizeChanged, 1.2));
        popupMenu.Commands.Add(new UICommand("Smaller Font", OnFontSizeChanged, 1 / 1.2));
        popupMenu.Commands.Add(new UICommandSeparator());
        popupMenu.Commands.Add(new UICommand("Red", OnColorChanged, Colors.Red));
        popupMenu.Commands.Add(new UICommand("Green", OnColorChanged, Colors.Green));
        popupMenu.Commands.Add(new UICommand("Blue", OnColorChanged, Colors.Blue));

        await popupMenu.ShowAsync(args.GetPosition(this));
    }

    void OnFontSizeChanged(IUICommand command)
    {
        textBlock.FontSize *= (double)command.Id;
    }

    void OnColorChanged(IUICommand command)
    {
        textBlock.Foreground = new SolidColorBrush((Color)command.Id);
    }
}
```

注意，UICommandSeparator 对象在菜单中创建了一条水平线。

通过 MessageDialog，ShowAsync 调用返回 IAsyncOperation < IUICommand >类型的对

象，从而可以获取用户所选命令。而我选择为 UICommand 构造函数中的命令指定两个定制处理程序，我还使用该构造函数的第三个变元来指定值，以帮助处理程序尽可能简单的处理命令。

ShowAsyn 方法需要 Point 值来确定在哪里显示菜单。这个点应该和应用窗口相对，通常就是和页面相对。菜单一般以此点水平居中而垂直居上。在触摸界面上这么做是有道理的：你不会希望用户的手盖住菜单！

右击 Simple 的‘S’顶部，菜单会如下图所示。

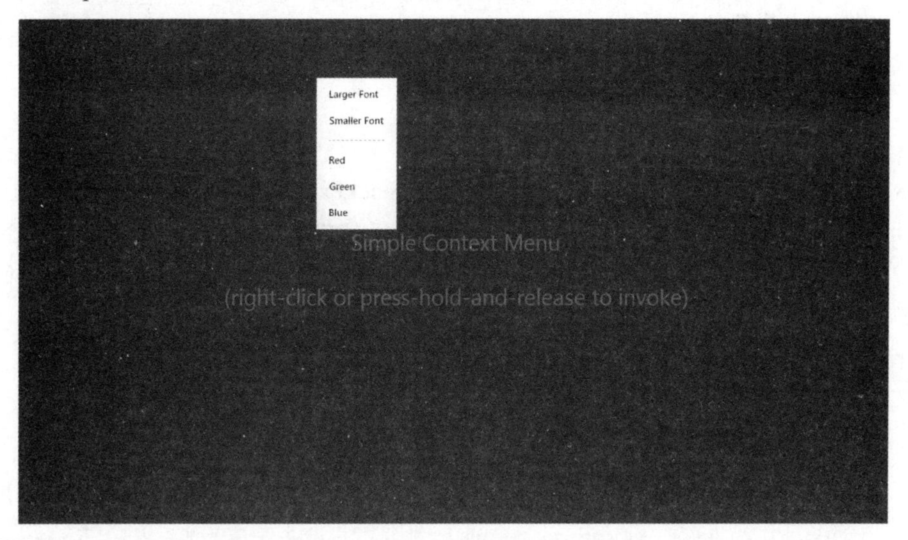

当然，如果点击 TextBlock 之外的地方，什么也不会出现。

如果所指定的点太靠近窗口左边缘、顶部或者右边缘，菜单位置会自动转移，菜单就不会被截断。无论 RequestedTheme 是什么值，白色背景上的菜单总是显示为黑色文字。

菜单提供了键盘接口，但也没什么特别：可以用方向键在条目上移动选项，并按 Enter 键选中。选择命令之后，轻点或者点击菜单之外的任何地方或者按键盘上除 Enter 之外的任意键，菜单都会消失。如果选择处理 ShowAsync 返回的 IUICommand 对象，而如果没有选择命令就释放菜单，该对象会为空。

PopupMenu 的全部功能基本就这么多。ShowForSelectionAsync 是另一种启用菜单的唯一方法。这种方法需要 Rect 值和 Placement 可选枚举值，包括 Default、Above、Below、Left 和 Right。这只是偏好位置：可以选择实际位置，以便整个菜单出现在程序窗口内。

如果 PopupMenu 处于禁用状态，则不能显示其中任何命令。如果目前不适用特定命令，就不要将它包括在菜单中！

也不能用选中标记来显示被选中的命令项目。如果想显示的不只是简单命令，就需要从 PopupMenu 升级到 Popup。

8.2　Popup 对话框

Popup 类(派生自 FrameworkElement)是 Windows Runtime 最接近于传统对话框的类。Popup 有 UIElement 类型的 Child 属性，可以设置为包含很多控件的 Panel，或者设置为 Panel

子类的 Border。

SimpleContextDialog 和之前项目的功能一样，XAML 文件也非常相似：

```
项目：SimpleContextDialog | 文件：MainPage.xaml(片段)
<Grid Background="{StaticResource ApplicationPageBackgroundThemeBrush}">
    <TextBlock Name="textBlock"
               FontSize="24"
               HorizontalAlignment="Center"
               VerticalAlignment="Center"
               TextAlignment="Center"
               RightTapped="OnTextBlockRightTapped">
        Simple Context Dialog
        <LineBreak />
        <LineBreak />
        (right-click or press-hold-and-release to invoke)
    </TextBlock>
</Grid>
```

TextBlock 中的 RightTapped 事件处理程序在 StackPanel 里集合了两个 Button 控件和三个 RadioButton，StackPanel 由设置为 Popup 的 Child 属性的子 Border 构成。这部分代码比较多，而代码较少的部分是 Button 控件的 Click 处理程序和 RadionButton 控件的 Checked 处理程序。

```
项目：SimpleContextDialog | 文件：MainPage.xaml.cs(片段)
public sealed partial class MainPage : Page
{
    public MainPage()
    {
        this.InitializeComponent();
    }

    void OnTextBlockRightTapped(object sender, RightTappedRoutedEventArgs args)
    {
        StackPanel stackPanel = new StackPanel();

        // Create two Button controls and add to StackPanel
        Button btn1 = new Button
        {
            Content = "Larger font",
            Tag = 1.2,
            HorizontalAlignment = HorizontalAlignment.Center,
            Margin = new Thickness(12)
        };
        btn1.Click += OnButtonClick;
        stackPanel.Children.Add(btn1);

        Button btn2 = new Button
        {
            Content = "Smaller font",
            Tag = 1 / 1.2,
            HorizontalAlignment = HorizontalAlignment.Center,
            Margin = new Thickness(12)
        };
        btn2.Click += OnButtonClick;
        stackPanel.Children.Add(btn2);

        // Create three RadioButton controls and add to StackPanel
        string[] names = { "Red", "Green", "Blue" };
        Color[] colors = { Colors.Red, Colors.Green, Colors.Blue };

        for (int i = 0; i < names.Length; i++)
        {
            RadioButton radioButton = new RadioButton
            {
```

```
                Content = names[i],
                Foreground = new SolidColorBrush(colors[i]),
                IsChecked = (textBlock.Foreground as SolidColorBrush).Color == colors[i],
                Margin = new Thickness(12)
            };
            radioButton.Checked += OnRadioButtonChecked;
            stackPanel.Children.Add(radioButton);
        }

        // Create a Border for the StackPanel
        Border border = new Border
        {
            Child = stackPanel,
            Background =
                this.Resources["ApplicationPageBackgroundThemeBrush"] as SolidColorBrush,
            BorderBrush = this.Resources["ApplicationForegroundThemeBrush"] as SolidColorBrush,
            BorderThickness = new Thickness(1),
            Padding = new Thickness(24),
        };

        // Create the Popup object
        Popup popup = new Popup
        {
            Child = border,
            IsLightDismissEnabled = true
        };

        // Adjust location based on content size
        border.Loaded += (loadedSender, loadedArgs) =>
            {
                Point point = args.GetPosition(this);
                point.X -= border.ActualWidth / 2;
                point.Y -= border.ActualHeight;
                // Leave at least a quarter inch margin
                popup.HorizontalOffset =
                    Math.Min(this.ActualWidth - border.ActualWidth - 24,
                        Math.Max(24, point.X));

                popup.VerticalOffset =
                    Math.Min(this.ActualHeight - border.ActualHeight - 24,
                        Math.Max(24, point.Y));

                // Set keyboard focus to first element
                btn1.Focus(FocusState.Programmatic);
            };

        // Open the popup
        popup.IsOpen = true;
    }

    void OnButtonClick(object sender, RoutedEventArgs args)
    {
        textBlock.FontSize *= (double)(sender as Button).Tag;
    }

    void OnRadioButtonChecked(object sender, RoutedEventArgs args)
    {
        textBlock.Foreground = (sender as RadioButton).Foreground;
    }
}
```

为了定位 Popup，就需要设置相对于程序窗口的 HorizontalOffset 和 VerticalOffset 属性值。然而，如果不知道 Popup 中有多少内容，就无法智能设置这些属性，而且一般要先显示 Popup，才知道有多少内容。出于此原因，以上代码设置了 Popup 内容元素 Border 的 Loaded 处理程序。Popup 定位在右触点正上方(很像 PopupMenu)，但也允许 Popup 和程序窗口之间至少有 24 个像素留白。

RightTapped 处理程序运行结果把 Popup 的 IsOpen 属性设置为 true，Popup 则显示在屏幕上。正常情况下，用户仍然可以与程序页面的其他部分进行交互。但要注意，Popup 的 IsLightDismissEnabled 属性设置为 true。通过点击或者轻击 Popup 以外的区域，或者按 Esc 键，就会释放 Popup。如果没有设置这个属性，就会显示多个对话框副本，很可能响应子控件事件的时候，程序需要设置 IsOpen 属性为 false 来移除 Popup。如果需要此信息用于初始化或者清屏，Popup 还定义了 Opened 和 Closed 属性。

点击右括号上方，字体已放大，如下图所示。

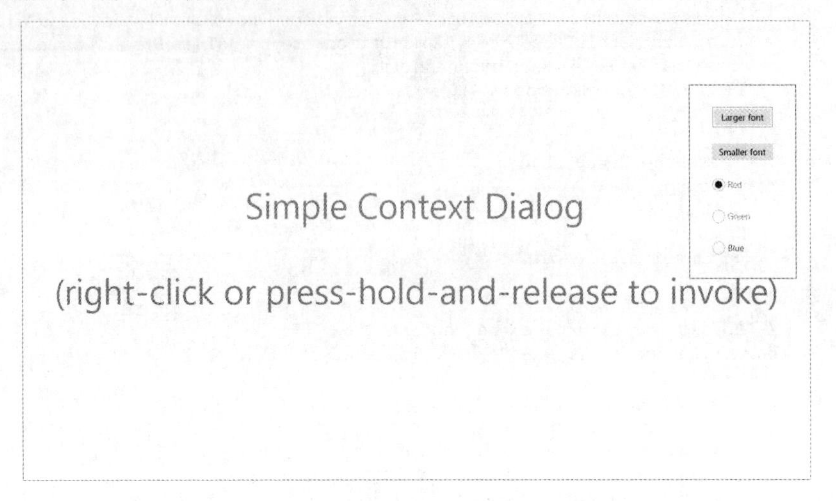

可以使用 Tab 键来浏览条目。默认情况下，这些对话框和应用程序有相同颜色主题，因此，使用像我写的 Border 能帮助在页面上显示对话框。

该对话框没有 OK 和 Cancel 按钮。而我用了以上方法来实施对话框，点击按钮会立即改变下面的显示，点击或轻击对话框之外的地方会关闭 Popup。如果对话框比较复杂，则需要一个按钮来恢复默认值。

当然，在代码中定义 Popup 的全部内容是一件麻烦事。更常见的做法是专门为对话框中定义 UserControl 控件，再实例 Popup 子类。然而，需要为该 UserControl 提供某种方式把用户选择传给程序，最佳方法是直接或通过视图模式来绑定对话框和应用。本章稍后会提到这两种方法的例子。

8.3 应 用 栏

Windows 8 应用栏想通过类似传统菜单或工具栏的方式来实施程序命令和选项。应用栏是称为 AppBar 的类，如果用户手指滑过屏幕顶部或底部，就会启用应用栏。应用栏可以出现在页面顶部、底部或者两者兼而有之。如果选择了命令，应用栏通常会消失，但也不是必须消失。

Page 类定义了两个属性，TopAppBar 和 BottomAppBar，在 XAML 中一般设置为 AppBar 标记。AppBar 派生自 ContentControl，通常会把 Content 属性设置为包含显示在应用栏中控件的面板。AppBar 没有固定高度，其高度基于其承载的控件。

当然，探索 Windows 8 自带的一些标准应用，是熟悉实际程序使用应用栏的最佳方法。大多数时候，应用栏包含一行圆形 Button 控件，但 Windows 8 版本的 IE 浏览器中应用栏则说明应用栏可以包含多种控件。在 IE 浏览器中，底部应用栏包含可以输入 URL 的 Textbox，顶部应用栏显示已访问网页集合。

以下程序提供了两个非常规的应用栏：

```
项目: UnconventionalAppBar | 文件: MainPage.xaml(片段)
<Page ... >
   <Grid Background="LightGray">
      <TextBlock Name="textBlock"
                 Text="Unconventional App Bar"
                 HorizontalAlignment="Center"
                 VerticalAlignment="Center"
                 FontSize="{Binding ElementName=slider, Path=Value}" />
    </Grid>

    <Page.TopAppBar>
       <AppBar Name="topAppBar">
           <Slider Name="slider"
                   Minimum="8"
                   Maximum="196"
                   Value="24" />
       </AppBar>
</Page.TopAppBar>

    <Page.BottomAppBar>
       <AppBar Name="bottomAppBar">
           <StackPanel Orientation="Horizontal"
                       HorizontalAlignment="Right">

              <Button Content="Red"
                      Foreground="Red"
                      Margin="24 12"
                      Click="OnAppBarButtonClick" />

              <Button Content="Green"
                      Foreground="Green"
                      Margin="24 12"
                      Click="OnAppBarButtonClick" />

              <Button Content="Blue"
                      Foreground="Blue"
                      Margin="24 12"
                      Click="OnAppBarButtonClick" />
           </StackPanel>
        </AppBar>
      </Page.BottomAppBar>
</Page>
```

屏幕中心位置有一个 TextBlock，但其 FontSize 受限于 Slider，而后者是顶部 AppBar 中的唯一内容。第二个 AppBar 设置为 BottomAppBar 属性，包含有三个 Button 控件的水平 StackPanel，这些 Button 控制在代码隐藏中共享一个 Click 处理程序：

```
项目: UnconventionalAppBar | 文件: MainPage.xaml.cs(片段)
void OnAppBarButtonClick(object sender, RoutedEventArgs args)
{
   textBlock.Foreground = (sender as Button).Foreground;
}
```

一般情况下，处理应用栏比处理 Popup 或者 PopupMenu 简单得多，因为 AppBar 是页面可视树的一部分，简化了绑定设置和事件处理程序。应用栏通常不可见，除非用户手指滑过屏幕顶部或者底部，这是控件在 AppBar 上和在界面上的唯一区别。此时，用户可以和控件进行交互，如下图所示。

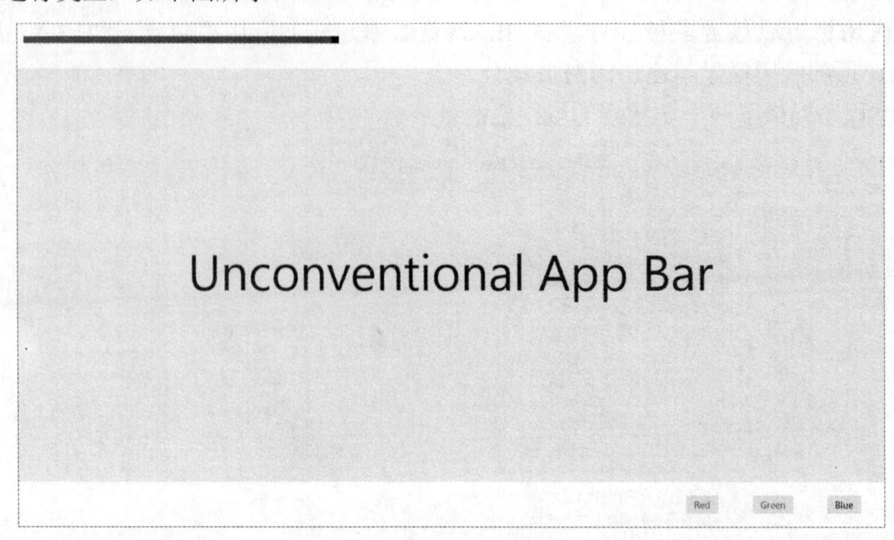

应用栏和控件的着色取决于应用的 RequestedTheme，在本程序中设置为 Light。主 Grid 配了 LightGray 背景，以对比那些颜色。使用应用栏的大多数程序好像都采用 Dark 主题。

如果点击或者点按应用栏之外的任何地方，或者按 Esc 键，应用栏会自动消失，并回到隐藏状态。如果不喜欢用这种方式关闭应用栏，可以把 AppBar 的 IsSticky 属性设置为 true。此时，为了去除应用栏，用户需要使用另一个手指滑过，或者需要把隐藏代码文件中一个或者两个 AppBar 对象的 IsOpen 属性设置为 false。

有些情况下，程序可能要用代码关闭应用。例如，在本特定程序中，为了改变文本颜色，用户需要滑动手指，以显示出应用栏，按一下按钮，再滑动另一个手指或按应用栏以外的地方以关闭应用栏。按压按钮的时候，可以选择用代码关闭应用栏：

```
void OnAppBarButtonClick(object sender, RoutedEventArgs args)
{
   textBlock.Foreground = (sender as Button).Foreground;
   topAppBar.IsOpen = false;
   bottomAppBar.IsOpen = false;
}
```

这种方式很常见，如果厌倦了点击按钮来关闭应用程序栏，就需要这样做。然而，有时用户会发现设置一行几个选项而不需要重新启用应用程序栏非常方便。这就要根据需要来判断了。

程序第一次运行的时候，一些应用会要求用户与应用栏进行交互。在这种情况下,可以把 IsOpen 属性初始化为 true。就像 Popup 一样，AppBar 初始化或者清空时，会执行 Opened 和 Closed 事件。

8.4　应用栏按钮样式

许多 Windows 8 应用只有底部应用栏，里面包含一行圆形 Button 控件。这些按钮通常是在圆圈中放一个符号，再加上简短的文本命令。

圆形按钮基于 Standard.xaml 所定义的 Style，键名为 AppBarButtonStyle。Standard.xaml 文件位于 Visual Studio 所创建的 C#、Visual Basic 或者 C++ Windows 8 项目的 Common 文件夹中。该文件包含在 App.xaml 文件的 Resource 部分，任何 Windows 8 应用都可以使用。

AppBarButtonStyle Style 定义包含 ControlTemplate 长代码，定义了圆形按钮的视觉效果。看完第 11 章后，你可能想仔细研究 ControlTemplate。而同时，即使完全不了解该模板，也可以使用该 Style。

AppBarButtonStyle 包含一个 Setter 对象，可以把 FontFamily 设置为 Segoe UI Symbol，按钮上的符号会使用这些字体的字符。

顾名思义，Segoe UI Symbol 是符号字体。然而，它又不像那些用很多符号代替常见字母和数字的老式符号字体。该字体仍然有可以用于常规目的之常规字符。但通过定义字符编码范围从 0xE000 到 0xf8ff 为“私人使用区”，Unicode 标准允许这种字体定制字体。也就是说，这些字符编码为特定字型。Segoe UI Symbol 不使用定制字体填满整个区域，但 0xE100 到 0xE1F4 的范围是图形文字集合，能代表很多常见计算机任务，因此适用于应用栏按钮。

例如，如果想显示一个按钮，按钮上有一个小房子和单词“Home”，用以下方法可以把这种按钮放到应用栏上。

```
<Page.BottomAppBar>
   <AppBar>
      <StackPanel Orientation="Horizontal">
         <Button Style="{StaticResource AppBarButtonStyle}"
                 Content="&#xE10F;"
                 AutomationProperties.Name="Home"
                 Click="OnButtonClick" />
         ...
      </StackPanel>
   </AppBar>
</Page.BottomAppBar>
```

前面提到过 Content 和 Click 属性。AutomationProperties 类是附加属性集合，而 Name 就是其中之一。这些属性通常允许识别用户界面元素，以达到测试目的，也允许辅助技术访问，比如屏幕阅读器。AppBarButtonStyle 所定义的 ControlTemplate 引用 AutomationProperties.Name 属性，以在按钮下方显示文本字符串。该特别按钮在暗色主题的显示效果，如下图所示。

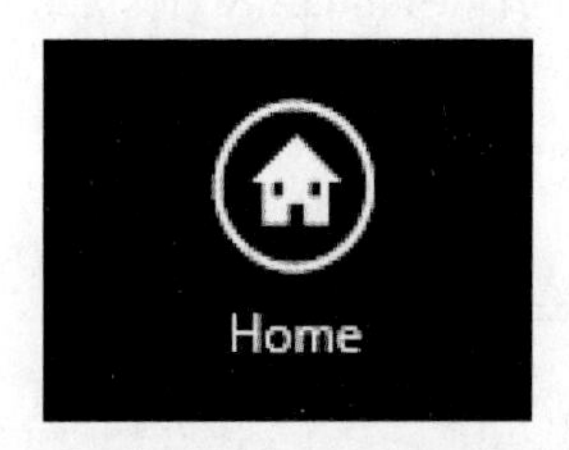

StandardStyles.xaml 文件还为许多(但非全部)Segoe UI Symbol 字符编码定义了基于 AppBarButtonStyle 的单独样式，范围从 0xE100 到 0xE1E9。例如，以下是 HomeAppBarButtonStyle 的 Style 定义：

```
<Style x:Key="HomeAppBarButtonStyle" TargetType="ButtonBase"
                          BasedOn="{StaticResource AppBarButtonStyle}">
   <Setter Property="AutomationProperties.AutomationId" Value="HomeAppBarButton"/>
   <Setter Property="AutomationProperties.Name" Value="Home"/>
```

```
    <Setter Property="Content" Value="&#xE10F;"/>
</Style>
```

显然，这些样式非常方便，因为已经有人匹配好了符号、名称以及建议功能。然而，在标准 StandardStyles.xaml 文件中，这些样式都加了注释，需要删除注释才能使用。在 XAML 文件中如何引用该 Style，如下所示。

```
<Button Style="{StaticResource HomeAppBarButtonStyle}"
        Click="OnButtonClick" />
```

如果想的话，也可以指定自己的文本。

```
<Button Style="{StaticResource HomeAppBarButtonStyle}"
        AutomationProperties.Name="Head on Home"
        Click="OnButtonClick" />
```

如果能用符号和文本标签来获取 StandardStyles.xaml 所定义应用栏按钮样式的完整列表，不是很好吗？LookAtAppBarButtonStyles 程序就提供了这项功能。其 XAML 文件包含一个准备填充的 ScrollViewer 和 StackPanel，以及有两个标准 RadioButton 控件的应用栏。

```
项目: LookAtAppBarButtonStyles | 文件: MainPage.xaml(片段)
<Page ... >
    <Grid Background="{StaticResource ApplicationPageBackgroundThemeBrush}">
        <ScrollViewer FontSize="20">
            <StackPanel Name="stackPanel" />
        </ScrollViewer>
    </Grid>

    <Page.BottomAppBar>
        <AppBar>
            <StackPanel Orientation="Horizontal">
                <RadioButton Name="symbolSortRadio"
                             Content="Sort by symbol"
                             Checked="OnRadioButtonChecked" />

                <RadioButton Name="textSortRadio"
                             Content="Sort by text"
                             Checked="OnRadioButtonChecked" />
            </StackPanel>
        </AppBar>
    </Page.BottomAppBar>
</Page>
```

在 Loaded 事件处理程序中，代码隐藏文件通过引用与当前应用实例相关 Resources 集合的 MergedDictionaries 属性，访问由 StandardStyles.xaml 提供的 ResourceDictionary。代码用键名“AppBarButtonStyle”定位 Style，并保存所有 Style 实例，BasedOn 属性同时和 class 的中的 Style 相等，class 是内部类集合。

```
项目: LookAtAppBarButtonStyles | 文件: MainPage.xaml.cs(片段)
public sealed partial class MainPage : Page
{
    class Item
    {
        public string Key;
        public char Symbol;
        public string Text;
    }

    List<Item> appbarStyles = new List<Item>();
    FontFamily segoeSymbolFont = new FontFamily("Segoe UI Symbol");

    public MainPage()
```

```
    {
      this.InitializeComponent();
      Loaded += OnLoaded;
    }

void OnLoaded(object sender, RoutedEventArgs args)
{
    // Basically gets StandardStyles.xaml
    ResourceDictionary dictionary = Application.Current.Resources.MergedDictionaries[0];
    Style baseStyle = dictionary["AppBarButtonStyle"] as Style;

    // Find all styles based on AppBarButtonStyle
    foreach (object key in dictionary.Keys)
    {
        Style style = dictionary[key] as Style;

        if (style != null && style.BasedOn == baseStyle)
        {
            Item item = new Item
            {
                Key = key as string
            };

            foreach (Setter setter in style.Setters)
            {
                if (setter.Property.Equals(AutomationProperties.NameProperty))
                    item.Text = setter.Value as string;

                if (setter.Property.Equals(ButtonBase.ContentProperty))
                    item.Symbol = (setter.Value as string)[0];
            }

            appbarStyles.Add(item);
        }
    }

    // Display items by checking RadioButton
    symbolSortRadio.IsChecked = true;
  }
  ...
}
```

Loaded 事件检查应用中的两个 RadioButton 控件之一，并得出结果。这会调用 RadioButton 的 Checked 处理程序，并以两种不同方式中的一种排列样式集合：

项目: LookAtAppBarButtonStyles | 文件: MainPage.xaml.cs(片段)

```
void OnRadioButtonChecked(object sender, RoutedEventArgs args)
{
    if (sender == symbolSortRadio)
    {
      // Sort by symbol
      appbarStyles.Sort((item1, item2) =>
      {
         return item1.Symbol.CompareTo(item2.Symbol);
      });
    }
    else

    {
      // Sort by text
      appbarStyles.Sort((item1, item2) =>
      {
         return item1.Text.CompareTo(item2.Text);
      });
    }

    // Close app bar and display the items
```

```
        this.BottomAppBar.IsOpen = false;
        DisplayList();
    }
```

Checked 处理程序会调用 DisplayList，DisplayList 为每一项创建文本。(注意，每一行第一个 TextBlock 的 FontFamily 使用 Segoe UI Symbol 字体)。每一项都会增加到 ScrollViewer 的 StackPanel 中。

项目: LookAtAppBarButtonStyles | 文件: MainPage.xaml.cs(片段)

```
void DisplayList()
{
    // Clear the StackPanel
    stackPanel.Children.Clear();

    // Loop through the styles
    foreach (Item item in appbarStyles)
    {

        // A StackPanel for each item
        StackPanel itemPanel = new StackPanel
        {
            Orientation = Orientation.Horizontal,
            Margin = new Thickness(0, 6, 0, 6)
        };

        // The symbol itself
        TextBlock textBlock = new TextBlock
        {
            Text = item.Symbol.ToString(),
            FontFamily = segoeSymbolFont,
            Margin = new Thickness(24, 0, 24, 0)
        };
        itemPanel.Children.Add(textBlock);

        // The Unicode identifier
        textBlock = new TextBlock
        {
            Text = "0x" + ((int)item.Symbol).ToString("X4"),
            Width = 96
        };
        itemPanel.Children.Add(textBlock);

        // The text for the button
        textBlock = new TextBlock
        {
            Text = "\"" + item.Text + "\"",
            Width = 240,
        };
        itemPanel.Children.Add(textBlock);

        // The key name
        textBlock = new TextBlock
        {
            Text = item.Key
        };
        itemPanel.Children.Add(textBlock);

        stackPanel.Children.Add(itemPanel);
    }
}
```

列表节选如下图所示。

0xE144	"Keyboard"	KeyboardAppBarButtonStyle
0xE145	"Dock Left"	DockLeftAppBarButtonStyle
0xE146	"Dock Right"	DockRightAppBarButtonStyle
0xE147	"Dock Bottom"	DockBottomAppBarButtonStyle
0xE148	"Remote"	RemoteAppBarButtonStyle
0xE149	"Sync"	SyncAppBarButtonStyle
0xE14A	"Rotate"	RotateAppBarButtonStyle
0xE14B	"Shuffle"	ShuffleAppBarButtonStyle
0xE14C	"List"	ListAppBarButtonStyle
0xE14D	"Shop"	ShopAppBarButtonStyle
0xE14E	"Select All"	SelectAllAppBarButtonStyle
0xE14F	"Orientation"	OrientationAppBarButtonStyle
0xE150	"Import"	ImportAppBarButtonStyle
0xE151	"Import All"	ImportAllAppBarButtonStyle
0xE155	"Browse Photos"	BrowsePhotosAppBarButtonStyle
0xE156	"Webcam"	WebcamAppBarButtonStyle
0xE158	"Pictures"	PicturesAppBarButtonStyle
0xE159	"Save Local"	SaveLocalAppBarButtonStyle
0xE15A	"Caption"	CaptionAppBarButtonStyle
0xE15B	"Stop"	StopAppBarButtonStyle

不要受限于以上列表条目。可以在应用栏按钮中使用 Segoe UI Symbol 中的任何字符，也可以指定不同字体。

8.5　深入 Segoe UI Symbol 字体

Segoe UI Symbol 字体还支持字符编码从 0x2600 到 0x26FF 的字符，Unicode 标准将这些字符分类为“杂项符号”。其中一些字符也适用于应用栏按钮。

Segoe UI Symbol 字体也超越了 16 位编码范围，包含映射编码 0x1F300 到 0x1F5FF 表情字符的图形符号。这些图标字符起源于日本，但微软 Windows Phone 和苹果 iPhone 也都在使用。

Segoe UI Symbol 字体还支持常见表情符号，编码范围从 0x1F600 到 0x1F64F，其中包括九个猫表情符号和非礼勿视、非礼勿听以及非礼勿言三只猴子符号。

Segoe UI Symbol 字体还支持从 0x1F680 到 0x1F6C5 的运输和地图符号。

为了帮助你(和我)为应用栏选择更多符号，我写了一个叫 SegoeSymbols 的程序，可以显示 Segoe UI Symbol 字体中编码从 0 到 0x1FFFF 的所有字符。

正如你所知道的，Unicode 刚开始的时候采用 16 位字符，编码范围从 0x0000 到 0xFFFF。而到了大于 65536 个编码点不够用的时候，Unicode 合并了从 0x10000 到 0x10FFFF 的字符编码，字符数量增加到了 110 万以上。经过扩展，Unicode 还包括一套系统，用一对 16 位值来展示更多字符。

使用单个 32 位编码来表示 Unicode 字符，被称为 32 位 Unicode 转换格式或者 UTF-32。但这有点用词不当，因为 UTF-32 不存在转换：32 位数字编码和象形符号之间是一对一的映射关系。

UTF-32 极其罕见，而且大多数人也根本没有把 Unicode 当成 32 位字符编码，因为在事实上，Unicode 的 32 位部分是附加在 16 位编码上的。

与之对应，大多数现代编程语言和操作系统都支持 UTF-16。Windows Runtime 中的 Char 结构基本上是 16 位整数，这也是 C#中 char 数据类型的基础。为了表示范围从 0x10000 到 0x10FFFF 的更多字符，UTF-16 使用了两个 16 位字符序列。这些被称为“代理项”(surrogate)，Unicode 留出一个特殊 16 位编码范围供其使用。主代理项范围从 0xD800 到 0xDBFF，而后续代理项范围从 0xDC00 到 0xDFFF。也就是 1024 个可能的主代理项，1024 个可能的后续代理项。这样就足够满足从 0x10000 到 0 x10FFFF 的共 1 048 576 个编码。(稍后有实际算法)

使用拉丁字母的语言文本主要限定于 0x0020 到和 0x007E 之间的 ASCII 字符编码，因此，大多数网页和其他文件都使用 UTF-8 系统存储文本以节约大量空间。UTF-8 直接编码 7 位字符，不过对于其他 Unicode 字符，会使用一到三位额外字节。

我写的 SegoeSymbols 项目主要是为了检查应用栏中可能的有用符号，所以程序只执行到 0x1FFFF 的字符编码。其 XAML 文件包含一个简单标题，一个行列等待显示 256 个字符块的 Grid 以及一个 Slider。

```
项目: SegoeSymbols | 文件: MainPage.xaml(片段)
<Page ... >

    <Page.Resources>
        <local:DoubleToStringHexByteConverter x:Key="hexByteConverter" />
    </Page.Resources>

    <Grid Background="{StaticResource ApplicationPageBackgroundThemeBrush}">
        <Grid.RowDefinitions>
            <RowDefinition Height="Auto" />
            <RowDefinition Height="*" />
            <RowDefinition Height="Auto" />
        </Grid.RowDefinitions>

        <TextBlock Name="titleText"
                   Grid.Row="0"
                   Text="Segoe UI Symbol"
                   HorizontalAlignment="Center"
                   Style="{StaticResource HeaderTextStyle}" />

        <Grid Name="characterGrid"
              Grid.Row="1"
              HorizontalAlignment="Center"
              VerticalAlignment="Center" />

        <Slider Grid.Row="2"
                Orientation="Horizontal"
                Margin="24 0"
                Minimum="0"
                Maximum="511 "
                SmallChange="1"
                LargeChange="16"
                ThumbToolTipValueConverter="{StaticResource hexByteConverter}"
                ValueChanged="OnSliderValueChanged" />
    </Grid>
</Page>
```

注意，Slider 最大值(Maximum)为 511，我想显示被 256 除的最大字符(0x1FFFF)。

Resource 部分引用的 DoubleToStringHexByteConverter 类和以前见过的类似，但它还显示了和屏幕视觉效果一致的两条下划线。

```
项目: SegoeSymbols | 文件: DoubleToStringHexByteConverter.cs(片段)
public class DoubleToStringHexByteConverter : IValueConverter
{
    public object Convert(object value, Type targetType, object parameter, string language)
    {
        return ((int)(double)value).ToString("X2") + "__";
    }
    public object ConvertBack(object value, Type targetType, object parameter, string language)
    {
        return value;
    }
}
```

每个 Slider 值对应显示一个 16×16 数组的 256 个字符。用于构造显示 256 个字符 Grid 的代码相当麻烦，因为我认为字符的所有行列之间都应该有分隔线，而这些分隔线在 Grid 里应该也有各自的行列。

```
项目: SegoeSymbols | 文件: MainPage.xaml.cs(片段)
public sealed partial class MainPage : Page
{
    const int CellSize = 36;
    const int LineLength = (CellSize + 1) * 16 + 18;
    FontFamily symbolFont = new FontFamily("Segoe UI Symbol");

    TextBlock[] txtblkColumnHeads = new TextBlock[16];
    TextBlock[,] txtblkCharacters = new TextBlock[16, 16];

    public MainPage()
    {
        this.InitializeComponent();

        for (int row = 0; row < 34; row++)
        {

            RowDefinition rowdef = new RowDefinition();

            if (row == 0 || row % 2 == 1)
                rowdef.Height = GridLength.Auto;
            else
                rowdef.Height = new GridLength(CellSize, GridUnitType.Pixel);

            characterGrid.RowDefinitions.Add(rowdef);

            if (row != 0 && row % 2 == 0)
            {
                TextBlock txtblk = new TextBlock
                {
                    Text = (row / 2 - 1).ToString("X1"),
                    VerticalAlignment = VerticalAlignment.Center
                };
                Grid.SetRow(txtblk, row);
                Grid.SetColumn(txtblk, 0);
                characterGrid.Children.Add(txtblk);
            }

            if (row % 2 == 1)
            {
```

```
            Rectangle rectangle = new Rectangle
            {
                Stroke = this.Foreground,
                StrokeThickness = row == 1 || row == 33 ? 1.5 : 0.5,
                Height = 1
            };
            Grid.SetRow(rectangle, row);
            Grid.SetColumn(rectangle, 0);
            Grid.SetColumnSpan(rectangle, 34);
            characterGrid.Children.Add(rectangle);
        }
    }

    for (int col = 0; col < 34; col++)
    {
        ColumnDefinition coldef = new ColumnDefinition();
        if (col == 0 || col % 2 == 1)
            coldef.Width = GridLength.Auto;
        else
            coldef.Width = new GridLength(CellSize);

        characterGrid.ColumnDefinitions.Add(coldef);

        if (col != 0 && col % 2 == 0)
        {
            TextBlock txtblk = new TextBlock
            {
                Text = "00" + (col / 2 - 1).ToString("X1") + "_",
                HorizontalAlignment = HorizontalAlignment.Center
            };
            Grid.SetRow(txtblk, 0);
            Grid.SetColumn(txtblk, col);
            characterGrid.Children.Add(txtblk);
            txtblkColumnHeads[col / 2 - 1] = txtblk;
        }

        if (col % 2 == 1)
        {
            Rectangle rectangle = new Rectangle
            {
                Stroke = this.Foreground,
                StrokeThickness = col == 1 || col == 33 ? 1.5 : 0.5,
                Width = 1
            };
            Grid.SetRow(rectangle, 0);
            Grid.SetColumn(rectangle, col);
            Grid.SetRowSpan(rectangle, 34);
            characterGrid.Children.Add(rectangle);
        }
    }

for (int col = 0; col < 16; col++)
    for (int row = 0; row < 16; row++)
    {
        TextBlock txtblk = new TextBlock
        {
            Text = ((char)(16 * col + row)).ToString(),
            FontFamily = symbolFont,
            FontSize = 24,
            HorizontalAlignment = HorizontalAlignment.Center,
            VerticalAlignment = VerticalAlignment.Center
```

```
                };
                Grid.SetRow(txtblk, 2 * row + 2);
                Grid.SetColumn(txtblk, 2 * col + 2);
                characterGrid.Children.Add(txtblk);
                txtblkCharacters[col, row] = txtblk;
            }
        }
        ...
    }
```

Slider 的 ValueChanged 处理程序功能相对容易，把正确文本插入到已存在的 TextBlock 元素，但是处理大于 0xFFFF 的字符编码也会有麻烦。

项目：SegoeSymbols | 文件：MainPage.xaml.cs(片段)

```
void OnSliderValueChanged(object sender, RangeBaseValueChangedEventArgs args)
{
    int baseCode = 256 * (int)args.NewValue;

    for (int col = 0; col < 16; col++)
    {
        txtblkColumnHeads[col].Text = (baseCode / 16 + col).ToString("X3") + "_";

        for (int row = 0; row < 16; row++)
        {
            int code = baseCode + 16 * col + row;
            string strChar = null;

            if (code <= 0x0FFFF)
            {
                strChar = ((char)code).ToString();
            }
            else
            {
                code -= 0x10000;
                int lead = 0xD800 + code / 1024;
                int trail = 0xDC00 + code % 1024;
                strChar = ((char)lead).ToString() + (char)trail;
            }
            txtblkCharacters[col, row].Text = strChar;
        }
    }
}
```

代码结尾的四行语句表明了数学运算方法，该方法把 0x10000 和 010FFFF 之间的一个 Unicode 字符编码分解成两个 10 位值，以构建主代理项和后续代理项，两者共同组成字符串以定义字符。

如果不想搞明白具体过程，可以用以下语句取代这四行：

```
strChar = Char.ConvertFromUtf32(code);
```

如果 code 不大于 0xFFFF，Char.ConvertFromUtf32 就返回由一个字符组成的字符串；如果 code 大于 0xFFFF，则该字符串含有两个字符。向该方法传递代理编码(从 0xD800 到 0xDFFF)会报异常。

构造应用栏按钮最有趣的区域开始于 0x2600(杂项符号区)，0xE100(Seqoe UI　Symbol 使用的私人使用区)和 0x1F300(图形文字、表情符号、交通和地图符号)。下图是图形文字的首屏截图。

Segoe UI Symbol

可以在 XAML 中指定超出 0xFFFF 的字符。

```
<TextBlock FontFamily="Segoe UI Symbol"
           FontSize="24"
           Text="&#x1F3B7;" />
```

上面指定了萨克斯管符号。Visual Studio 有时会发出警告，但程序还是能编译并运行。

以下应用栏按钮显然是用于音乐应用。

```
<Button Style="{StaticResource AppBarButtonStyle}"
        Content="&#x1F3B7;"
        AutomationProperties.Name="Saxophone"
Click="OnMusicButtonClick" />

<Button Style="{StaticResource AppBarButtonStyle}"
        Content="&#x1F3B8;"
        AutomationProperties.Name="Guitar"
        Click="OnMusicButtonClick" />

<Button Style="{StaticResource AppBarButtonStyle}"
        Content="&#x1F3B9;"
        AutomationProperties.Name="Piano"
        Click="OnMusicButtonClick" />

<Button Style="{StaticResource AppBarButtonStyle}"
        Content="&#x1F3BA;"
        AutomationProperties.Name="Trumpet"
        Click="OnMusicButtonClick" />

<Button Style="{StaticResource AppBarButtonStyle}"
        Content="&#x1F3BB;"
        AutomationProperties.Name="Violin"
        Click="OnMusicButtonClick" />
```

下图是界面。

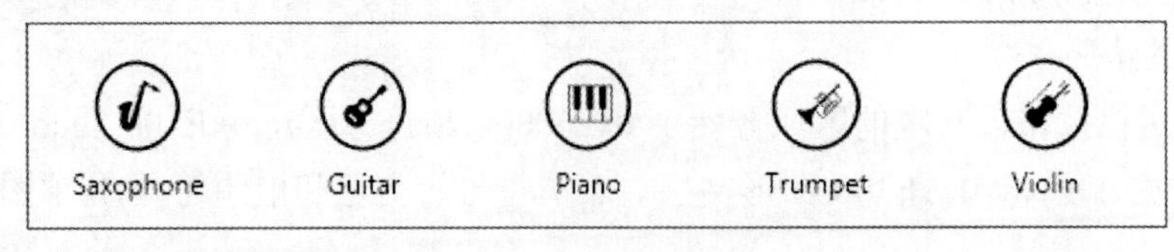

8.6　应用栏 CheckBox 和 RadioButton

在好多 Windows 8 标准应用中，在应用栏里都可以看到类似于 CheckBox 或 RadioButton 的圆形按钮。在日历应用中，日、周、月按钮就像一组 RadioButton 控件，而地图应用的显示交通按钮则像 CheckBox。

AppBarButtonStyle 包含 ButtonBase 的 TargetType，也就是说，可以用来设计 CheckBox 或 RadioButton 样式。然而，在我看到的 StandardStyles.xaml 版本中，AppBarButtonStyle 的 ControlTemplate 引用了 BackgroundCheckedGlyph，但模板中并没有定义。如果 CheckBox 或 RadioButton 使用这些样式时发生错误，去掉注释引用了 BackgroundCheckedGlyph 的 ObjectAnimationUsingKeyFrames 对象即可。

以下是我写的 TextFormattingAppBar 代码，页面中心有一个 TextBlock，应用栏包含三个 CheckBox 控件和三个 RadioButton 控件，所有样式都基于 AppBarButtonStyle。

项目：TextFormattingAppBar | 文件：MainPage.xaml(片段)

```
<Page ... >
    <Grid Background="LightGray">
        <TextBlock Name="textBlock"
                   FontFamily="Times New Roman"
                   FontSize="96"
                   HorizontalAlignment="Center"
                   VerticalAlignment="Center">
            <Run>Text Formatting AppBar</Run>
        </TextBlock>
    </Grid>

    <Page.BottomAppBar>
        <AppBar>
            <StackPanel Orientation="Horizontal">
                <CheckBox Style="{StaticResource BoldAppBarButtonStyle}"
                          Checked="OnBoldAppBarCheckBoxChecked"
                          Unchecked="OnBoldAppBarCheckBoxChecked" />

                <CheckBox Style="{StaticResource ItalicAppBarButtonStyle}"
                          Checked="OnItalicAppBarCheckBoxChecked"
                          Unchecked="OnItalicAppBarCheckBoxChecked" />

                <CheckBox Style="{StaticResource UnderlineAppBarButtonStyle}"
                Checked="OnUnderlineAppBarCheckBoxChecked"
                Unchecked="OnUnderlineAppBarCheckBoxChecked" />

                <Polyline Points="0 12, 0 48"
                          Stroke="{StaticResource ApplicationForegroundThemeBrush}"
                          VerticalAlignment="Top" />

                <RadioButton Name="redRadioButton"
                             Style="{StaticResource FontColorAppBarButtonStyle}"
                             Foreground="Red"
                             AutomationProperties.Name="Red"
                             Checked="OnFontColorAppBarRadioButtonChecked" />

                <RadioButton Style="{StaticResource FontColorAppBarButtonStyle}"
                             Foreground="Green"
                             AutomationProperties.Name="Green"
                             Checked="OnFontColorAppBarRadioButtonChecked" />

                <RadioButton Style="{StaticResource FontColorAppBarButtonStyle}"
                             Foreground="Blue"
```

```
                                    AutomationProperties.Name="Blue"
                                    Checked="OnFontColorAppBarRadioButtonChecked" />
            </StackPanel>
        </AppBar>
    </Page.BottomAppBar>
</Page>
```

如下图所示，如果选中某个按钮，会显示成相反颜色反转，但 RadioButton 显示颜色的效果没有那么好。

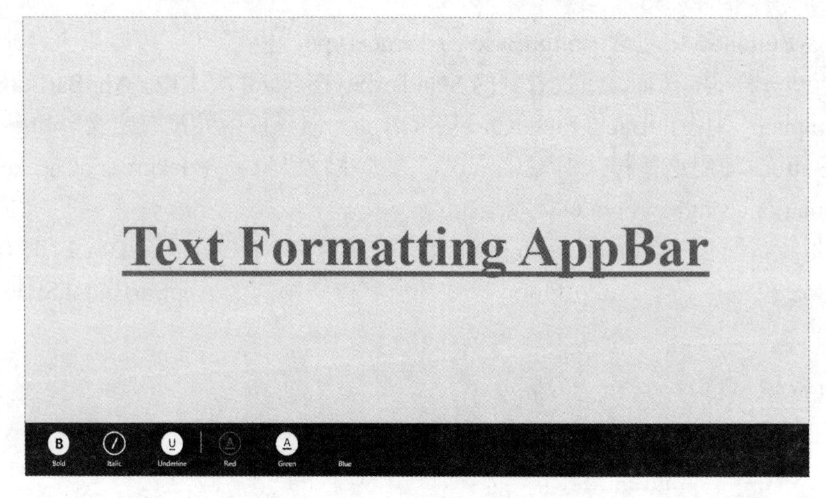

如果想另外设置所选 CheckBox 或者 RadioButton 的颜色，更改 AppBarButtonStyle 即可。

隐藏代码文件和你期望的一样，只不过实现下划线选项的代码格外乱。

项目: TextFormattingAppBar | 文件: MainPage.xaml(片段)

```
public sealed partial class MainPage : Page
{
    public MainPage()
    {
        this.InitializeComponent();
    }

    void OnBoldAppBarCheckBoxChecked(object sender, RoutedEventArgs args)
    {
        CheckBox chkbox = sender as CheckBox;
        textBlock.FontWeight = (bool)chkbox.IsChecked ? FontWeights.Bold: FontWeights.Normal;
    }

    void OnItalicAppBarCheckBoxChecked(object sender, RoutedEventArgs args)
    {
        CheckBox chkbox = sender as CheckBox;
        textBlock.FontStyle = (bool)chkbox.IsChecked ? FontStyle.Italic : FontStyle.Normal;
    }

    void OnUnderlineAppBarCheckBoxChecked(object sender, RoutedEventArgs args)
    {
        CheckBox chkbox = sender as CheckBox;
        Inline inline = textBlock.Inlines[0];

    if  ((bool)chkbox.IsChecked && !(inline is Underline))
    {
        Underline underline = new Underline();
        textBlock.Inlines[0] = underline;
        underline.Inlines.Add(inline);
    }
```

```
        else if  (!(bool)chkbox.IsChecked && inline is Underline)
        {
            Underline underline = inline as Underline;
            Run run = underline.Inlines[0] as Run;
            underline.Inlines.Clear();
            textBlock.Inlines[0] = run;
        }
    }

    void OnFontColorAppBarRadioButtonChecked(object sender, RoutedEventArgs args)
    {
        textBlock.Foreground = (sender as RadioButton).Foreground;
    }
}
```

设计 CheckBox 或 RadioButton 样式并不是实现这种功能的唯一方法。天气应用阐述了模仿 CheckBox 的另一种方法。轻击，Button 的标签“Change to Celsius”会变成“Change to Fahrenheit”。

应用栏上的按钮还会启用 PopupMenu 或者 Popup。例如，按下 Windows 8 IE 浏览器的扳手图标按钮，会出现一个小弹窗，里面至少包含两项命令：Find on page 和 View on the desktop。也可以在地图里按下地图样式按钮，会看到两个互斥选择 Road View 和 Aerial View，其中一个有选中标记，表明是当前选项。还可以在相机应用里按下相机选项命令，会看到一个弹窗，里面有组合框、切换开关和一个 More 链接，展开后会显示一个更大的弹出对话框。

在应用栏中使用 PopupMenu 和 Popup，和通过右键启用非常类似，只需要对其智能定位即可，稍后会进行演示。

应用运行的时候，如果用户手指滑到屏幕右边，就会看到标准列表：Search、Share、Start、Devices 和 Settings。第 17 章会演示应用如何接入这些功能。特别值得一提的是，Settings 按钮经常启用选项列表，包括 About、Help 以及 Settings 等选项。然而，一些应用会在应用栏里包括 Options 项，而应用栏也包含 Settings 项。事实上，StandardStyles.xaml 包含的 SettingsAppBarButtonStyle 显示为齿轮图标及单词“Setting”。如何在这些项目目中分离程序功能取决于你，但一般而言，列表访问频率比 Settings 高的项目，应用栏上的 Option 按钮会用于访问频率比 Settings 高的项目，而 Settings 按钮会用于访问频率比 Setting 功能高的项目。

8.7　记事本应用栏

第 7 章的 PrimitivePad 程序在界面顶部有三个按钮，分别标记为 Open、Save As，还有一个在 Wrap 和 No Wrap 之间交替显示的 ToggleButton。我们把它们转换为应用栏按钮，同时文字换行选项改为 Popup，并增加按钮，以增加和减小字体大小。但我不想把文件 I/O 逻辑搞得太复杂。

以下是 AppBarPad 的 MainPage.xaml 文件。

```
项目: AppBarPad | 文件: MainPage.xaml(片段)
<Page ... >
    <Grid Background="{StaticResource ApplicationPageBackgroundThemeBrush}">
        <TextBox Name="txtbox"
                 IsEnabled="False"
```

```
            FontSize="24"
            AcceptsReturn="True" />
    </Grid>

        <Page.BottomAppBar>
            <AppBar>
                <Grid>
                  <StackPanel Orientation="Horizontal"
                              HorizontalAlignment="Left">

                    <Button Style="{StaticResource FontIncreaseAppBarButtonStyle}"
                            Click="OnFontIncreaseAppBarButtonClick" />

                    <Button Style="{StaticResource FontDecreaseAppBarButtonStyle}"
                            Click="OnFontDecreaseAppBarButtonClick" />

                    <Button Style="{StaticResource SettingsAppBarButtonStyle}"
                            AutomationProperties.Name="Wrap Option"
                            Click="OnWrapOptionAppBarButtonClick" />
                 </StackPanel>

                 <StackPanel Orientation="Horizontal"
                             HorizontalAlignment="Right">

                     <Button Style="{StaticResource OpenFileAppBarButtonStyle}"
                             Click="OnOpenAppBarButtonClick" />

                     <Button Style="{StaticResource SaveAppBarButtonStyle}"
                             AutomationProperties.Name="Save As"
                             Click="OnSaveAsAppBarButtonClick" />
                   </StackPanel>
                </Grid>
            </AppBar>
        </Page.BottomAppBar>
</Page>
```

一般而言，一些按钮放在应用程序栏左边，一些放在右边。手持平板电脑时，这样做更方便使用(相较于按钮放在中间)。XAML 有几种方法把按钮分到左边和右边。也许最简单的方法是在单格 Grid 里放置两个水平 StackPanel 元素，并保持右对齐和左对齐。

建议在最右侧放一个 New(或者 Add)按钮，虽然本程序没有 New 按钮，但是与文件相关的其他按钮也应该出现在右边，因为和 New 相关。我提供了一个 Save As 按钮，以取代采用 SaveAppBarButtonStyle 样式的 Save 按钮。

程序选项都在左边：增加和减小字体大小按钮，还有另一个(使用通用 SettingsAppBarButtonStyle)用于文字换行设置的按钮。用户手指滑过屏幕顶部或底部时，就会看到如下图所示的内容。

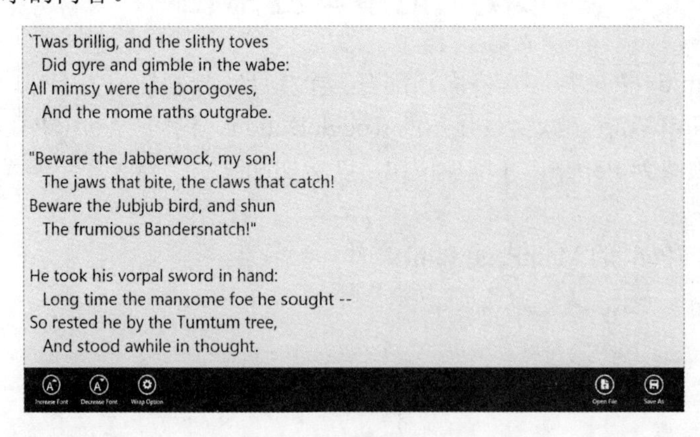

程序响应 Application 所定义的 Suspending 事件的时候，会保存用户设置(和 TextBox 内容)。Loaded 处理程序会载入这些设置。为方便起见，可以在 MainPage 构造函数中把两者定义为匿名方法。以下简单处理程序用于字体大小增减按钮：

项目：AppBarPad | 文件：MainPage.xaml.cs(片段)

```
public sealed partial class MainPage : Page
{
    public MainPage()
    {
        this.InitializeComponent();

        // Get local settings object
        ApplicationDataContainer appData = ApplicationData.Current.LocalSettings;

        Loaded += async (sender, args) =>
            {
                // Load TextBox settings
                if (appData.Values.ContainsKey("TextWrapping"))
                txtbox.TextWrapping = (TextWrapping)appData.Values["TextWrapping"];

                if (appData.Values.ContainsKey("FontSize"))
                txtbox.FontSize = (double)appData.Values["FontSize"];

                // Load TextBox content
                StorageFolder localFolder = ApplicationData.Current.LocalFolder;
                StorageFile storageFile = await localFolder.CreateFileAsync("AppBarPad.txt",
                                                    CreationCollisionOption.OpenIfExists);
                txtbox.Text = await FileIO.ReadTextAsync(storageFile);

                // Enable the TextBox and give it input focus
                txtbox.IsEnabled = true;
                txtbox.Focus(FocusState.Programmatic);
            };

        Application.Current.Suspending += async (sender, args) =>
        {
            // Save TextBox settings
            appData.Values["TextWrapping"] = (int)txtbox.TextWrapping;
            appData.Values["FontSize"] = txtbox.FontSize;

            // Save TextBox content
            SuspendingDeferral deferral = args.SuspendingOperation.GetDeferral();
            await PathIO.WriteTextAsync("ms-appdata:///local/AppBarPad.txt", txtbox.Text);
                                    deferral.Complete();
        };
}

    void OnFontIncreaseAppBarButtonClick(object sender, RoutedEventArgs args)
    {
        ChangeFontSize(1.1);
    }

    void OnFontDecreaseAppBarButtonClick(object sender, RoutedEventArgs args)
    {
        ChangeFontSize(1/1.1);
    }

    void ChangeFontSize(double multiplier)
    {
        txtbox.FontSize *= multiplier;
    }
    ...
}
```

单击 Wrap Options 按钮，程序会显示一个包含 Wrap 和 No Wrap 的小对话框。我把对话框的布局定义为名为 WrapOptionsDialog 的 UserControl。该 XAML 文件展示了 RadioButton 控件的两个选项。

```
项目: AppBarPad | 文件: WrapOptionsDialog.xaml(片段)
<UserControl ... >

    <Grid Background="{StaticResource ApplicationPageBackgroundThemeBrush}">
        <StackPanel Name="stackPanel"
                    Margin="24">
            <RadioButton Content="Wrap"
                         Checked="OnRadioButtonChecked">
                <RadioButton.Tag>
                    <TextWrapping>Wrap</TextWrapping>
                </RadioButton.Tag>
            </RadioButton>

            <RadioButton Content="No wrap"
                         Checked="OnRadioButtonChecked">
                <RadioButton.Tag>
                    <TextWrapping>NoWrap</TextWrapping>
                </RadioButton.Tag>
            </RadioButton>
        </StackPanel>
    </Grid>
</UserControl>
```

你会发现 Grid 使用了标准背景刷。需要有某种刷子，否则背景会是透明的。该程序中保留了暗色主题，因此，对话框会有白色前景和黑色背景，并和 TextBox 形成对比。

对话框的隐藏代码文件定义了一个 TextWrapping 类型的 TextWrapping 从属属性。设置此属性时，属性更改处理程序会检查 RadioButton，如果用户选择 RadioButton，就设置属性。

```
项目: AppBarPad | 文件: WrapOptionsDialog.xaml.cs(片段)
public sealed partial class WrapOptionsDialog : UserControl
{
    static WrapOptionsDialog()
    {
        TextWrappingProperty = DependencyProperty.Register("TextWrapping",
            typeof(TextWrapping),
            typeof(WrapOptionsDialog),
            new PropertyMetadata(TextWrapping.NoWrap, OnTextWrappingChanged));
    }

public static DependencyProperty TextWrappingProperty { private set; get; }

public WrapOptionsDialog()
{
    this.InitializeComponent();
}

public TextWrapping TextWrapping
{
    set { SetValue(TextWrappingProperty, value); }
    get { return (TextWrapping)GetValue(TextWrappingProperty); }
}

static void OnTextWrappingChanged(DependencyObject obj,
                         DependencyPropertyChangedEventArgs args)
{
    (obj as WrapOptionsDialog).OnTextWrappingChanged(args);
}

void OnTextWrappingChanged(DependencyPropertyChangedEventArgs args)
{
```

```
    foreach (UIElement child in stackPanel.Children)
    {
      RadioButton radioButton = child as RadioButton;
      radioButton.IsChecked =
        (TextWrapping)radioButton.Tag == (TextWrapping)args.NewValue;
    }
}

void OnRadioButtonChecked(object sender, RoutedEventArgs args)
{
    this.TextWrapping = (TextWrapping)(sender as RadioButton).Tag;
}

}
```

Wrap Options 应用栏按钮的事件处理程序在 MainPage 隐藏代码文件中。事件处理程序实例化 WrapOptionsDialog 对象，并从 TextBox 的 TextWrapping 属性中初始化 TextWrapping 属性。代码再定义两个 TextWrapping 属性之间的绑定。用户可以直接在 TextBox 里看到变化属性的结果。WrapOptionsDialog 对象创建为一个新 Popup 对象的子对象。

```
项目: AppBarPad | 文件: MainPage.xaml.cs(片段)
void OnWrapOptionsAppBarButtonClick(object sender, RoutedEventArgs args)
{
   // Create dialog
   WrapOptionsDialog wrapOptionsDialog = new WrapOptionsDialog
   {
      TextWrapping = txtbox.TextWrapping
   };

   // Bind dialog to TextBox
   Binding binding = new Binding
   {
      Source = wrapOptionsDialog,
      Path = new PropertyPath("TextWrapping"),
      Mode = BindingMode.TwoWay
   };
   txtbox.SetBinding(TextBox.TextWrappingProperty, binding);

   // Create popup
   Popup popup = new Popup
   {
      Child = wrapOptionsDialog,
      IsLightDismissEnabled = true
   };

   // Adjust location based on content size
   wrapOptionsDialog.Loaded += (dialogSender, dialogArgs) =>
   {
      // Get Button location relative to screen
      Button btn = sender as Button;
      Point pt = btn.TransformToVisual(null).TransformPoint(new Point(btn.ActualWidth / 2,
                                                                 btn.ActualHeight / 2));
      popup.HorizontalOffset = pt.X - wrapOptionsDialog.ActualWidth / 2;
      popup.VerticalOffset = this.ActualHeight - wrapOptionsDialog.ActualHeight
                           - this.BottomAppBar.ActualHeight - 48;
   };

   // Open the popup
   popup.IsOpen = true;
}
```

一般而言，这种弹窗就定位在应用栏上面，也就是说，需要知道弹窗高度和页面高度，才能正确显示应用栏高度。我还想水平定位弹窗使其与启用它的按钮保持对齐。这就需要用 TransformToVisual 方法(第 10 章会进行讨论)来获取按钮中心相对于屏幕的坐标。可以在

Popup 子对象的 Loaded 或 SizeChanged 事件中执行此类计算。

Click 处理程序把 Popup 的 IsOpen 设置属性为 true，结果如下图所示。

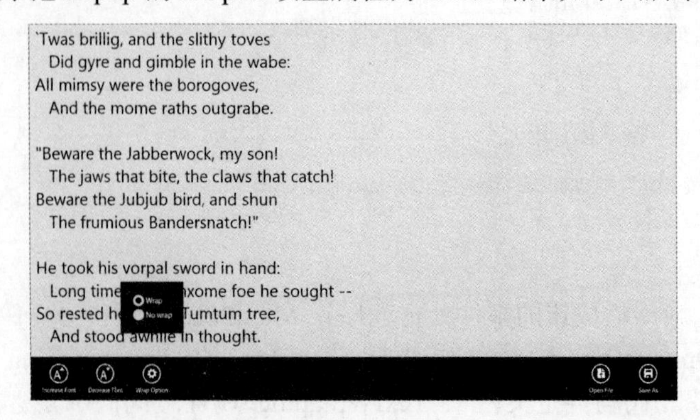

轻击 Popup 之外的任何地方，Popup 会自动消失，用户需要再拍一次才能关闭应用栏。AppBar 和 Popup 执行初始化或清除时，会执行 Opened 和 Closed 事件，因此，可以安装处理程序用于 Popup 的 Closed 事件，用来把 AppBar 的 IsOpen 属性设置为 false(例如)。

文件 I/O 逻辑使用简单的静态 FileIO 方法，但不包含异常处理。

项目：AppBarPad | 文件：MainPage.xaml.cs(片段)

```
public sealed partial class MainPage : Page
{
    ...
    async void OnOpenAppBarButtonClick(object sender, RoutedEventArgs args)
    {
        FileOpenPicker picker = new FileOpenPicker();
        picker.FileTypeFilter.Add(".txt");
        StorageFile storageFile = await picker.PickSingleFileAsync();

        // If user presses Cancel, result is null
        if (storageFile == null)
            return;

        txtbox.Text = await FileIO.ReadTextAsync(storageFile);
    }

    async void OnSaveAsAppBarButtonClick(object sender, RoutedEventArgs args)
    {
        FileSavePicker picker = new FileSavePicker();
        picker.DefaultFileExtension = ".txt";
        picker.FileTypeChoices.Add("Text", new List<string> { ".txt" });
        StorageFile storageFile = await picker.PickSaveFileAsync();

        // If user presses Cancel, result is null
        if (storageFile == null)
            return;

        await FileIO.WriteTextAsync(storageFile, txtbox.Text);
    }
}
```

8.8 XamlCruncher 入门

即使熟悉了 Windows Runtime 的各种特性，综合利用这些特性来创建应用仍然是一项挑战。但如果能创建应用栏和对话框，现在就可以构建真正像应用的东西了。

XamlCruncher 允许在 TextBox 中输入 XAML 并查看结果。XamlCruncher 所用的神奇方法是 XamlReader.Load，第 2 章的 PathMarkupSyntaxCode.project 有过简短介绍。XAML 经过 XamlReader.Load 处理，不能引用事件处理程序和外部组件，但诸如 XamlCruncher 等工具对于 XAML 交互体验和学习非常有用。我不会假装把这个程序说成是商业级，但它真的利用了 Windows 8 的真正特性。

在以下程序视图中，左侧编辑器里有一些 XAML 代码，而右侧显示区域是结果对象。

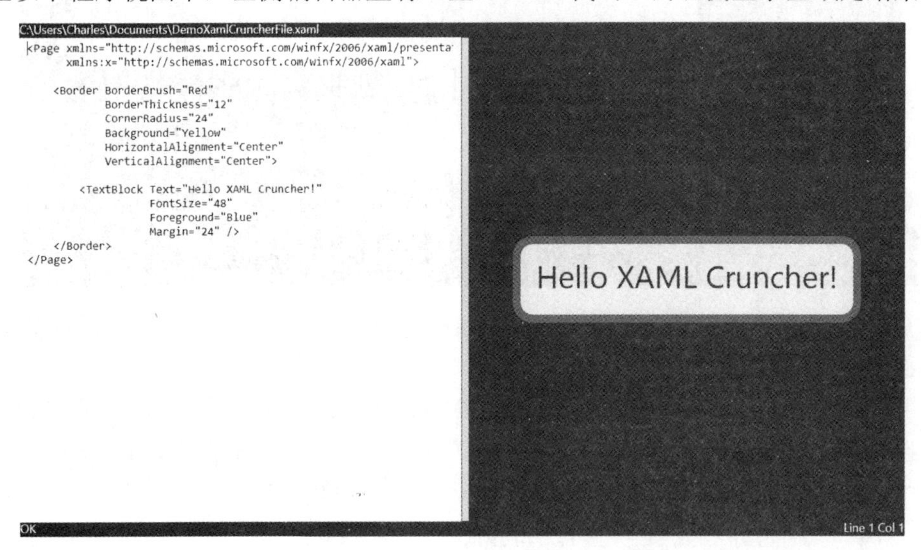

编辑器没有提供便利功能。输入起始标记时，编辑器也不会自动生成结束标记；编辑器不使用不同颜色来区分元素、属性和字符串，它一点儿都不智能。不过，可以改变页面配置，具体做法是把编辑窗口放在顶部、右边或底部。

应用栏里有 Add、Open、Save 和 Save As 按钮，还有 Refresh 按钮以及应用设置按钮。

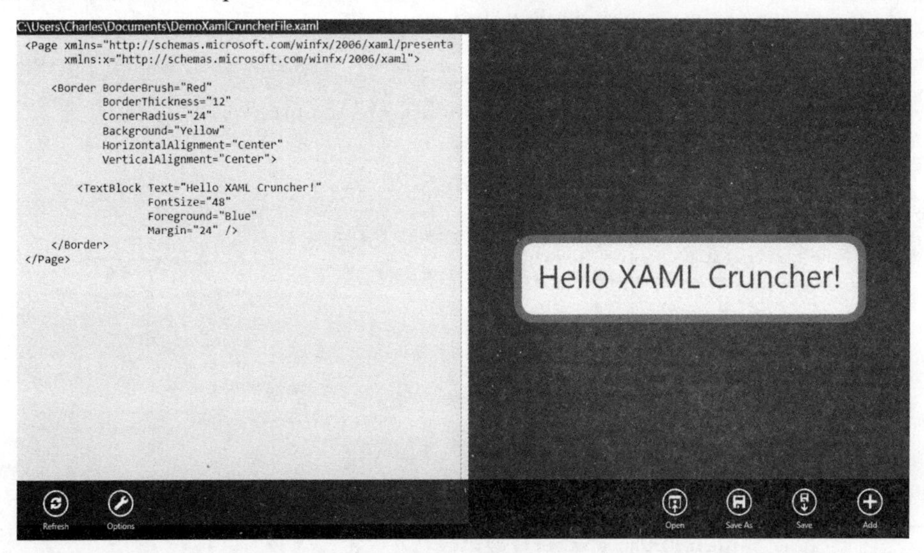

可以选择 XamlCrunche 根据每次按键还是仅根据 Refresh 按钮来重新解析 XAML。点击 Setting 按钮时所启用的对话框会显示可用选项，如下图所示。

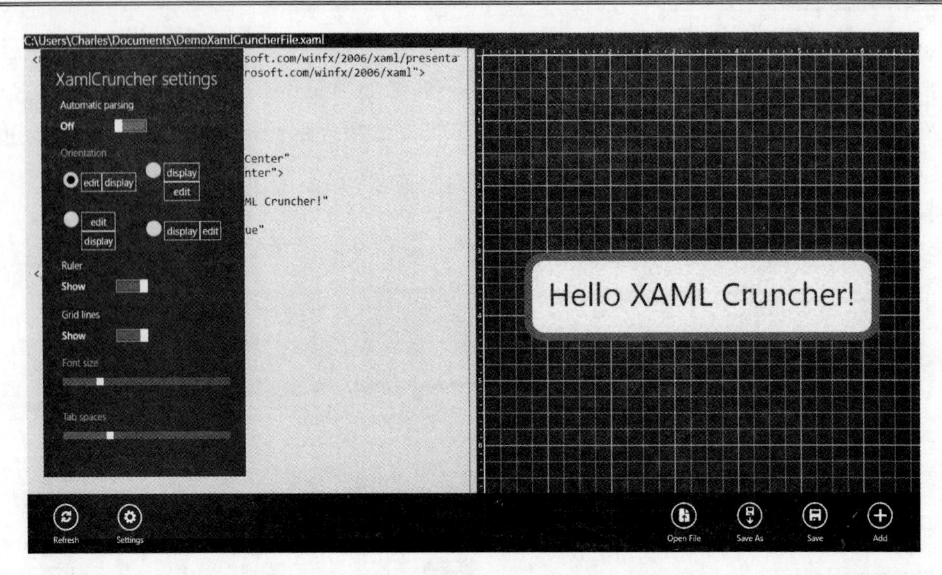

我打开右边显示区域的 Ruler 和 Grid Lines 选项来展示结果。所有这些设置都被保存下来，供程序下次运行时使用。

页面的大部分是一个自定义 UserControl 派生类 SplitContainer。中心是一个 Thumb 控件，可以选择左右两边面板(或者上下面板)的空间比例。在屏幕截图中，Thumb 是一个位于屏幕中心的浅灰色竖栏。SplitContainer 的 XAML 文件包括了定义横向和纵向配置的 Grid。

项目：XamlCruncher | 文件：SplitContainer.xaml

```
<UserControl
    x:Class="XamlCruncher.SplitContainer"
    xmlns="http://schemas.microsoft.com/winfx/2006/xaml/presentation"
    xmlns:x="http://schemas.microsoft.com/winfx/2006/xaml"
    xmlns:local="using:XamlCruncher">

    <Grid>
        <!-- Default Orientation is Horizontal -->
        <Grid.ColumnDefinitions>
            <ColumnDefinition x:Name="coldef1" Width="*" MinWidth="100" />
            <ColumnDefinition Width="Auto" />
            <ColumnDefinition x:Name="coldef2" Width="*" MinWidth="100" />
        </Grid.ColumnDefinitions>

        <!-- Alternative Orientation is Vertical -->
        <Grid.RowDefinitions>
            <RowDefinition x:Name="rowdef1" Height="*" />
            <RowDefinition Height="Auto" />
            <RowDefinition x:Name="rowdef2" Height="0" />
        </Grid.RowDefinitions>

        <Grid Name="grid1"
              Grid.Row="0"
              Grid.Column="0" />

        <Thumb Name="thumb"
               Grid.Row="0"
               Grid.Column="1"
               Width="12"
               DragStarted="OnThumbDragStarted"
               DragDelta="OnThumbDragDelta" />

        <Grid Name="grid2"
              Grid.Row="0"
```

```
        Grid.Column="2" />
    </Grid>
</UserControl>
```

第 5 章的 OrientableColorScroll 程序里有类似标记，横竖屏变化时，界面宽高比改变，Grid 也随之改变。

隐藏代码文件依靠依赖属性(dependency property)定义五个属性。通常情况下，Child1 和 Child2 属性用于设置出现在控件左边和右边的元素，但其实际出现则取决于 Orientation 和 SwapChildren 属性。

```
项目: XamlCruncher | 文件: SplitContainer.xaml.cs(片段)
public sealed partial class SplitContainer : UserControl
{
    // Static constructor and properties
    static SplitContainer()
    {
        Child1Property =
            DependencyProperty.Register("Child1",
                typeof(UIElement), typeof(SplitContainer),
                new PropertyMetadata(null, OnChildChanged));

        Child2Property =
            DependencyProperty.Register("Child2",
                typeof(UIElement), typeof(SplitContainer),
                new PropertyMetadata(null, OnChildChanged));

        OrientationProperty =
            DependencyProperty.Register("Orientation",
                typeof(Orientation), typeof(SplitContainer),
                new PropertyMetadata(Orientation.Horizontal, OnOrientationChanged));

        SwapChildrenProperty =
            DependencyProperty.Register("SwapChildren",
                typeof(bool), typeof(SplitContainer),
                new PropertyMetadata(false, OnSwapChildrenChanged));

        MinimumSizeProperty =
            DependencyProperty.Register("MinimumSize",
                typeof(double), typeof(SplitContainer),
                new PropertyMetadata(100.0, OnMinSizeChanged));
    }

    public static DependencyProperty Child1Property { private set; get; }
    public static DependencyProperty Child2Property { private set; get; }
    public static DependencyProperty OrientationProperty { private set; get; }
    public static DependencyProperty SwapChildrenProperty { private set; get; }
    public static DependencyProperty MinimumSizeProperty { private set; get; }

    // Instance constructor and properties
    public SplitContainer()
    {
        this.InitializeComponent();
    }

    public UIElement Child1
    {
        set { SetValue(Child1Property, value); }
        get { return (UIElement)GetValue(Child1Property); }
    }

    public UIElement Child2
    {
        set { SetValue(Child2Property, value); }
        get { return (UIElement)GetValue(Child2Property); }
    }
```

```
    public Orientation Orientation
    {
        set { SetValue(OrientationProperty, value); }
        get { return (Orientation)GetValue(OrientationProperty); }
    }

    public bool SwapChildren
    {
        set { SetValue(SwapChildrenProperty, value); }
        get { return (bool)GetValue(SwapChildrenProperty); }
    }

    public double MinimumSize
    {
        set { SetValue(MinimumSizeProperty, value); }
        get { return (double)GetValue(MinimumSizeProperty); }
    }
    ...
}
```

Orientation 属性是 Orientation 类型，和用于 StackPanel 和 VariableSizedWrapGrid 的枚举一样。应该使用现有类型的依赖属性，而不是自行发明。注意，MinimumType 是 double 类型，因此会初始化为 100.0 而不是 100 以免运行时类型不匹配。

属性更改处理程序显示了两种不同方法，程序员用来从静态处理程序调用实例属性更改事件处理程序。我已经展示了一种方法，用同样的 ependencyPropertyChangedEventArgs 对象，静态处理程序直接调用实例处理程序。有时候，就像 Orientation、SwapChildren 和 MinimumSize 属性的处理程序一样，静态处理程序为属性类型赋予旧值和新值，因而调用实例处理程序更方便。

```
项目: XamlCruncher | 文件: SplitContainer.xaml.cs(片段)
public sealed partial class SplitContainer : UserControl
{
    ...
    // Property-changed handlers
    static void OnChildChanged(DependencyObject obj,
                    DependencyPropertyChangedEventArgs args)
    {
        (obj as SplitContainer).OnChildChanged(args);
    }

    void OnChildChanged(DependencyPropertyChangedEventArgs args)
    {
        Grid targetGrid = (args.Property == Child1Property ^ this.SwapChildren) ? grid1 : grid2;
        targetGrid.Children.Clear();

        if (args.NewValue != null)
            targetGrid.Children.Add(args.NewValue as UIElement);
    }

    static void OnOrientationChanged(DependencyObject obj,
                    DependencyPropertyChangedEventArgs args)
    {
        (obj as SplitContainer).OnOrientationChanged((Orientation)args.OldValue,
                                    (Orientation)args.NewValue);
    }

    void OnOrientationChanged(Orientation oldOrientation, Orientation newOrientation)
    {
        // Shouldn't be necessary, but...
        if (newOrientation == oldOrientation)
            return;
```

```
    if  (newOrientation == Orientation.Horizontal)
    {
        coldef1.Width = rowdef1.Height;
        coldef2.Width = rowdef2.Height;

        coldef1.MinWidth = this.MinimumSize;
        coldef2.MinWidth = this.MinimumSize;

        rowdef1.Height = new GridLength(1, GridUnitType.Star);
        rowdef2.Height = new GridLength(0);

        rowdef1.MinHeight = 0;
        rowdef2.MinHeight = 0;

        thumb.Width = 12;
        thumb.Height = Double.NaN;

        Grid.SetRow(thumb, 0);
        Grid.SetColumn(thumb, 1);

        Grid.SetRow(grid2, 0);
        Grid.SetColumn(grid2, 2);
    }
    else
    {
        rowdef1.Height = coldef1.Width;
        rowdef2.Height = coldef2.Width;

        rowdef1.MinHeight = this.MinimumSize;
        rowdef2.MinHeight = this.MinimumSize;

        coldef1.Width = new GridLength(1, GridUnitType.Star);
        coldef2.Width = new GridLength(0);

        coldef1.MinWidth = 0;
        coldef2.MinWidth = 0;

        thumb.Height = 12;
        thumb.Width = Double.NaN;

        Grid.SetRow(thumb, 1);
        Grid.SetColumn(thumb, 0);

        Grid.SetRow(grid2, 2);
        Grid.SetColumn(grid2, 0);
    }
}

static void OnSwapChildrenChanged(DependencyObject obj,
                         DependencyPropertyChangedEventArgs args)
{
    (obj as SplitContainer).OnSwapChildrenChanged((bool)args.OldValue,
                                    (bool)args.NewValue);
}

void OnSwapChildrenChanged(bool oldOrientation, bool newOrientation)
{
    grid1.Children.Clear();
    grid2.Children.Clear();

    grid1.Children.Add(newOrientation ? this.Child2 : this.Child1);
    grid2.Children.Add(newOrientation ? this.Child1 : this.Child2);
}

static void OnMinSizeChanged(DependencyObject obj,
                    DependencyPropertyChangedEventArgs args)
{
    (obj as SplitContainer).OnMinSizeChanged((double)args.OldValue,
```

```
                    (double)args.NewValue);
    }

    void OnMinSizeChanged(double oldValue, double newValue)
    {
        if (this.Orientation == Orientation.Horizontal)
        {
            coldef1.MinWidth = newValue;
            coldef2.MinWidth = newValue;
        }
        else
        {
            rowdef1.MinHeight = newValue;
            rowdef2.MinHeight = newValue;
        }
    }
    ...
}
```

Orientation 属性更改处理程序的原始版本假定 Orientation 属性实际会变化，只要属性改变，处理程序被调用，就会发生这种情况。但我发现，有时候把属性值设置为现有值，也会调用属性更改处理程序。

SplitContainer 剩下要做的事情是检查 Thumb 的事件处理程序。这里的想法是基于星星规格来分布两列(或两行)Grid，Grid 的大小和长宽比变化时，列(或行)的相对尺寸会保持一致。然而，为了确保 Thumb 的拖拽逻辑合理简单，如果和星星规格相关的数字比例是实际像素维度，这样做就有效。OnThumbDragStarted 方法初始化并在 OnDragThumbDelta 改变。

```
项目: XamlCruncher | 文件: SplitContainer.xaml.cs(片段)
public sealed partial class SplitContainer : UserControl
{
    ...
    // Thumb event handlers
    void OnThumbDragStarted(object sender, DragStartedEventArgs args)
    {
        if (this.Orientation == Orientation.Horizontal)
        {
            coldef1.Width = new GridLength(coldef1.ActualWidth, GridUnitType.Star);
            coldef2.Width = new GridLength(coldef2.ActualWidth, GridUnitType.Star);
        }
        else
        {
            rowdef1.Height = new GridLength(rowdef1.ActualHeight, GridUnitType.Star);
            rowdef2.Height = new GridLength(rowdef2.ActualHeight, GridUnitType.Star);
        }
    }

    void OnThumbDragDelta(object sender, DragDeltaEventArgs args)
    {
        if (this.Orientation == Orientation.Horizontal)
        {
            double newWidth1 = Math.Max(0, coldef1.Width.Value + args.HorizontalChange);
            double newWidth2 = Math.Max(0, coldef2.Width.Value - args.HorizontalChange);

            coldef1.Width = new GridLength(newWidth1, GridUnitType.Star);
            coldef2.Width = new GridLength(newWidth2, GridUnitType.Star);
        }
        else
        {
            double newHeight1 = Math.Max(0, rowdef1.Height.Value + args.VerticalChange);
            double newHeight2 = Math.Max(0, rowdef2.Height.Value - args.VerticalChange);

            rowdef1.Height = new GridLength(newHeight1, GridUnitType.Star);
            rowdef2.Height = new GridLength(newHeight2, GridUnitType.Star);
```

```
            }
        }
    }
```

前面 XamlCruncher 屏幕截图展示了显示区域的标尺和网格线。标尺以英寸为单位，96 像素为 1 英寸，因此网格线相距 24 像素。如果要设计一些矢量图形或者其他精确布局，标尺和网格线很有用。

标尺和网格线都独立可选。显示两者的 UserControl 派生类称为 RulerContainer。从下面的代码可知，在构建 XamlCruncher 页面时，RulerContainer 的一个实例被设置为 SplitContaine 对象的 Child2 属性。以下是 RulerContainer 的 XAML 文件。

```
项目: XamlCruncher | 文件: RulerContainer.xaml(片段)
<UserControl ... >
    <Grid SizeChanged="OnGridSizeChanged">
        <Canvas Name="rulerCanvas" />
        <Grid Name="innerGrid">
            <Grid Name="gridLinesGrid" />
            <Border Name="border" />
        </Grid>
    </Grid>
</UserControl>
```

RulerContainer 控件有 Child 属性，该控件的子控件设置为 Border 的 Child 属性。Border 视觉效果是由水平线条和垂直线条组成的网格，是 Grid 对象的子对象，名为 gridLinesGrid。如果标尺也呈现，名为 innerGrid 的 Grid 的左边界和上边界将被赋予非 0 值，以满足该标尺。组成标尺的刻度线标记和数字是 Canvas 的子对象，名为 rulerCanvas。

以下是隐藏代码文件中所有依赖属性定义的情况。

```
项目: XamlCruncher | 文件: RulerContainer.xaml.cs(片段)
public sealed partial class RulerContainer : UserControl
{
    ...
    static RulerContainer()
    {
      ChildProperty =
          DependencyProperty.Register("Child",
              typeof(UIElement), typeof(RulerContainer),
              new PropertyMetadata(null, OnChildChanged));

      ShowRulerProperty =
          DependencyProperty.Register("ShowRuler",
              typeof(bool), typeof(RulerContainer),
              new PropertyMetadata(false, OnShowRulerChanged));

       ShowGridLinesProperty =
           DependencyProperty.Register("ShowGridLines",
              typeof(bool), typeof(RulerContainer),
              new PropertyMetadata(false, OnShowGridLinesChanged));
     }

public static DependencyProperty ChildProperty { private set; get; }
public static DependencyProperty ShowRulerProperty { private set; get; }
public static DependencyProperty ShowGridLinesProperty { private set; get; }

public RulerContainer()
{
   this.InitializeComponent();
}

public UIElement Child
{
```

```
        set { SetValue(ChildProperty, value); }
        get { return (UIElement)GetValue(ChildProperty); }
    }

    public bool ShowRuler
    {
        set { SetValue(ShowRulerProperty, value); }
        get { return (bool)GetValue(ShowRulerProperty); }
    }

    public bool ShowGridLines
    {
        set { SetValue(ShowGridLinesProperty, value); }
        get { return (bool)GetValue(ShowGridLinesProperty); }
    }

    // Property changed handlers
    static void OnChildChanged(DependencyObject obj,
                    DependencyPropertyChangedEventArgs args)
    {
        (obj as RulerContainer).border.Child = (UIElement)args.NewValue;
    }

    static void OnShowRulerChanged(DependencyObject obj,
                        DependencyPropertyChangedEventArgs args)
    {
        (obj as RulerContainer).RedrawRuler();
    }

    static void OnShowGridLinesChanged(DependencyObject obj,
                            DependencyPropertyChangedEventArgs args)
    {
        (obj as RulerContainer).RedrawGridLines();
    }

    void OnGridSizeChanged(object sender, SizeChangedEventArgs args)
    {
        RedrawRuler();
        RedrawGridLines();
    }

    ...
}
```

这里也给出了属性更改处理程序(足够简单，可以用于静态版本)和 Grid 的 SizeChanged 处理程序。两个重绘方法处理所有绘画，在两个面板中创建并组织 Line 元素和 TextBlock 元素：

```
项目: XamlCruncher | 文件: RulerContainer.xaml.cs(片段)
public sealed partial class RulerContainer : UserControl
{
    const double RULER_WIDTH = 12;
    ...
    void RedrawGridLines()
    {
        gridLinesGrid.Children.Clear();

        if (!this.ShowGridLines)
            return;

        // Vertical grid lines every 1/4"
        for (double x = 24; x < gridLinesGrid.ActualWidth; x += 24)
        {
            Line line = new Line
            {
                X1 = x,
```

```
                Y1 = 0,
                X2 = x,
                Y2 = gridLinesGrid.ActualHeight,
                Stroke = this.Foreground,
                StrokeThickness = x % 96 == 0 ? 1 : 0.5
            };
            gridLinesGrid.Children.Add(line);
        }

        // Horizontal grid lines every 1/4"
        for (double y = 24; y < gridLinesGrid.ActualHeight; y += 24)
        {
            Line line = new Line
            {
                X1 = 0,
                Y1 = y,
                X2 = gridLinesGrid.ActualWidth,
                Y2 = y,
                Stroke = this.Foreground,
                StrokeThickness = y % 96 == 0 ? 1 : 0.5
            };
            gridLinesGrid.Children.Add(line);
        }
    }

    void RedrawRuler()
    {
        rulerCanvas.Children.Clear();
        if (!this.ShowRuler)
        {
            innerGrid.Margin = new Thickness();
            return;
        }

        innerGrid.Margin = new Thickness(RULER_WIDTH, RULER_WIDTH, 0, 0);

        // Ruler across the top
        for (double x = 0; x < gridLinesGrid.ActualWidth - RULER_WIDTH; x += 12)
        {
            // Numbers every inch
            if (x > 0 && x % 96 == 0)
            {
                TextBlock txtblk = new TextBlock
                {
                    Text = (x / 96).ToString("F0"),
                    FontSize = RULER_WIDTH - 2
                };

                txtblk.Measure(new Size());
                Canvas.SetLeft(txtblk, RULER_WIDTH + x - txtblk.ActualWidth / 2);
                Canvas.SetTop(txtblk, 0);
                rulerCanvas.Children.Add(txtblk);
            }
            // Tick marks every 1/8"
            else
            {
                Line line = new Line
                {
                    X1 = RULER_WIDTH + x,
                    Y1 = x % 48 == 0 ? 2 : 4,
                    X2 = RULER_WIDTH + x,
                    Y2 = x % 48 == 0 ? RULER_WIDTH - 2 : RULER_WIDTH - 4,
                    Stroke = this.Foreground,
                    StrokeThickness = 1
                };
                rulerCanvas.Children.Add(line);
            }
        }
```

```
    // Heavy line underneath the tick marks
    Line topLine = new Line
    {
        X1 = RULER_WIDTH - 1,
        Y1 = RULER_WIDTH - 1,
        X2 = rulerCanvas.ActualWidth,
        Y2 = RULER_WIDTH - 1,
        Stroke = this.Foreground,
        StrokeThickness = 2
    };
    rulerCanvas.Children.Add(topLine);

    // Ruler down the left side
    for (double y = 0; y < gridLinesGrid.ActualHeight - RULER_WIDTH; y += 12)
    {
        // Numbers every inch
        if (y > 0 && y % 96 == 0)
        {
            TextBlock txtblk = new TextBlock
            {
                Text = (y / 96).ToString("F0"),
                FontSize = RULER_WIDTH - 2,
            };
            txtblk.Measure(new Size());
            Canvas.SetLeft(txtblk, 2);
            Canvas.SetTop(txtblk, RULER_WIDTH + y - txtblk.ActualHeight / 2);
            rulerCanvas.Children.Add(txtblk);
        }
        // Tick marks every 1/8"
        else
        {
            Line line = new Line
            {
                X1 = y % 48 == 0 ? 2 : 4,
                Y1 = RULER_WIDTH + y,
                X2 = y % 48 == 0 ? RULER_WIDTH - 2 : RULER_WIDTH - 4,
                Y2 = RULER_WIDTH + y,
                Stroke = this.Foreground,
                StrokeThickness = 1
            };
            rulerCanvas.Children.Add(line);
        }
    }

    Line leftLine = new Line
    {
        X1 = RULER_WIDTH - 1,
        Y1 = RULER_WIDTH - 1,
        X2 = RULER_WIDTH - 1,
        Y2 = rulerCanvas.ActualHeight,
        Stroke = this.Foreground,
        StrokeThickness = 2
    };
    rulerCanvas.Children.Add(leftLine);
    }
}
```

这两种方法广泛使用 Line 元素，在点 (X1,Y1)和(X2,Y2) 之间渲染了一条直线。

RedrawRuler 代码也演示了获取 TextBlock 尺寸的技巧：创建新 TextBlock 时，ActualWidth 和 ActualHeight 属性都为 0。TextBlock 成为可视树的一部分后，才会计算这些属性，并会随布局而变化。然而，可以迫使 TextBlock 调用 Measure 方法来计算其自身大小。该方法由 UIElement 定义，也是布局系统的重要组成部分。

Measure 方法参数是 Size 值，表明供元素使用的尺寸，但是出于此目的可以设置大小为 0：

```
txtblk.Measure(new Size());
```

如果需要找到 TextBlock 大小来进行文本换行，必须为 Size 构造函数的第一个参数提供非 0 值，使 TextBlock 知道换行文本的宽度。

调用 Measure 后，TextBlock 的 ActualWidth 和 ActualHeight 有效，并用于在 Canvas 中定位 TextBlock。只有在 Canvas 中定位 TextBlock 元素的时候，Canvas.SetLeft 和 Canvas.SetTops 属性才有必要调用。在单一 Grid 或者 Canvas 中，Line 元素基于坐标来定位。

从下面可以看出，RulerContainer 实例被设置为控制 XamlCruncher 页面的 SplitContainer 的 Child2 属性。Child1 属性看起来是 TextBox，但实际上它派生自 TextBox 的另一个自定义控件 TabbableTextBox 实例。

标准 TextBox 不响应 Tab 键，但是在编辑器中输入 XAML 时，的确会用到 Tab 键。这就是 TabbableTextBox 的主要特性，全部代码如下所示。

```
项目: XamlCruncher | 文件: TabbableTextBox.cs
using Windows.System;
using Windows.UI.Xaml;
using Windows.UI.Xaml.Controls;
using Windows.UI.Xaml.Input;

namespace XamlCruncher
{
    public class TabbableTextBox : TextBox
    {
        static TabbableTextBox()
        {
            TabSpacesProperty =
                DependencyProperty.Register("TabSpaces",
                    typeof(int), typeof(TabbableTextBox),
                    new PropertyMetadata(4));
        }

        public static DependencyProperty TabSpacesProperty { private set; get; }

        public int TabSpaces
        {
            set { SetValue(TabSpacesProperty, value); }
            get { return (int)GetValue(TabSpacesProperty); }
        }

        public bool IsModified { set; get; }

        protected override void OnKeyDown(KeyRoutedEventArgs args)
        {
            this.IsModified = true;

            if (args.Key == VirtualKey.Tab)
            {
                int line, col;
                GetPositionFromIndex(this.SelectionStart, out line, out col);
                int insertCount = this.TabSpaces - col % this.TabSpaces;
                this.SelectedText = new string(' ', insertCount);
                this.SelectionStart += insertCount;
                this.SelectionLength = 0;
                args.Handled = true;
                return;
            }
            base.OnKeyDown(args);
        }
```

```
            public void GetPositionFromIndex(int index, out int line, out int col)
            {
                if  (index > Text.Length)
                {
                  line = col = -1;
                  return;
                }

                line = col = 0;
                int i = 0;

                while  (i < index)
                {
                  if  (Text[i] == '\n')
                  {
                     line++;
                     col = 0;
                  }
                  else if (Text[i] == '\r')
                  {
                     index++;
                  }
                  else
                  {
                     col++;
                  };
                  i++;
                }
            }
        }
    }
```

TabbableTextBox 类截获 OnKeyDown 方法以确定用户是否按下了 Tab 键。如果按下，就在 Text 对象中插入空格，光标移动到 TabSpaces 属性整数倍的一个文本列。此项计算需要知道当前行光标的字符位置。使用类定义中的 GetPositionFromIndex 方法，可以获取该信息。(TextBox 的 Text 属性的行数受回车和换行限制，但 SelectionStart 指针基于行尾字符进行计算。)此方法是公共方法，XamlCruncher 也用它来显示光标的当前位置和当前所选内容(如果有的话)。

另一个属性(不依赖于依赖属性)也由 TabbableTextBox 定义。该属性是 IsModified，只要发生 KeyDown 事件，IsModified 值就设置为 true。

像许多文本处理程序一样，如果文本文件自上次保存后发生了改变，XamlCruncher 会保持记录。如果用户启动操作来新建文件或者打开现有文件，当前文件就会进入修改状态，程序会询问用户是否要保存文档。

这种逻辑经常发生在 TextBox 控件之外。载入新文件或者保存文件时，程序把 IsModified 的标志设置为 false，在收到 TextChanged 事件后标志为 true。然而，通过程序设置 TextBox 的 Text 属性时，会触发 TextChanged 事件，因此，即使把 TextBox 设置给一个新载入的文件，也会触发 TextChanged 事件，TextChanged 处理程序会给 IsModified 打上标志。你可能会认为在这种情况下设置 IsModified 标记，通过程序设置 Text 属性，设置一个标记可以避免这种情况。然而，设置 Text 属性的方法返回控制给操作系统后，才会调用 TextChanged 处理程序，而这会导致逻辑混乱。在 TextBox 派生类中实施 IsModified 标记有用。

8.9　应用设置和视图模式

不同程序互相启动的时候，许多应用都会维持用户设置和偏好。正如前面提到的，Windows Runtime 提供了应用数据存储的独立区域来存储设置或者所有文件。

在本程序中，我将用户设置统一到名为 AppSettings 类中。该类实现 INotifyPropertyChanged，以用于数据绑定。该类基本上是一个视图模式，或者(在较大应用中)是视图模式的一部分。

有一个应该保存的程序选项，即编辑和显示区域的横竖屏。前面说过，SplitContainer 有两个属性，Orientation 和 SwapChildren。为了保存用户设置，我要给应用增加一些更具体的东西。TextBox(或者更确切地说是 TabbableTextBox)可以在左边、上边、右边或者底部，以下枚举代码封装了这些选项。

```
项目: XamlCruncher | 文件: EditOrientation.cs
namespace XamlCruncher
{
    public enum EditOrientation
    {
        Left, Top, Right, Bottom
    }
}
```

AppSettings 显示了组成程序设置的所有属性。构造函数载入并用 Save 方法保存设置。所有属性值可通过程序默认设置初始化的字段进行恢复。注意，EditOrientation 属性基于 EditOrientation 枚举。

```
项目: XamlCruncher | 文件: AppSettings.cs
public class AppSettings : INotifyPropertyChanged
{
    // Application settings initial values
    EditOrientation editOrientation = EditOrientation.Left;
    Orientation orientation = Orientation.Horizontal;
    bool swapEditAndDisplay = false;
    bool autoParsing = false;
    bool showRuler = false;
    bool showGridLines = false;
    double fontSize = 18;
    int tabSpaces = 4;

    public event PropertyChangedEventHandler PropertyChanged;

    public AppSettings()
    {
        ApplicationDataContainer appData = ApplicationData.Current.LocalSettings;

        if (appData.Values.ContainsKey("EditOrientation"))
            this.EditOrientation = (EditOrientation)(int)appData.Values["EditOrientation"];

        if (appData.Values.ContainsKey("AutoParsing"))
            this.AutoParsing = (bool)appData.Values["AutoParsing"];

        if (appData.Values.ContainsKey("ShowRuler"))
            this.ShowRuler = (bool)appData.Values["ShowRuler"];

        if (appData.Values.ContainsKey("ShowGridLines"))
            this.ShowGridLines = (bool)appData.Values["ShowGridLines"];
```

```
        if (appData.Values.ContainsKey("FontSize"))
            this.FontSize = (double)appData.Values["FontSize"];

        if (appData.Values.ContainsKey("TabSpaces"))
            this.TabSpaces = (int)appData.Values["TabSpaces"];
    }

    public EditOrientation EditOrientation
    {
        set
        {
            if (SetProperty<EditOrientation>(ref editOrientation, value))
            {
                switch (editOrientation)
                {
                    case EditOrientation.Left:
                        this.Orientation = Orientation.Horizontal;
                        this.SwapEditAndDisplay = false;
                        break;

                    case EditOrientation.Top:
                        this.Orientation = Orientation.Vertical;
                        this.SwapEditAndDisplay = false;
                        break;

                    case EditOrientation.Right:
                        this.Orientation = Orientation.Horizontal;
                        this.SwapEditAndDisplay = true;
                        break;

                    case EditOrientation.Bottom:
                        this.Orientation = Orientation.Vertical;
                        this.SwapEditAndDisplay = true;
                        break;
                }
            }
        }
        get { return editOrientation; }
    }

    public Orientation Orientation
    {
        protected set { SetProperty<Orientation>(ref orientation, value); }
        get { return orientation; }
    }

    public bool SwapEditAndDisplay
    {
        protected set { SetProperty<bool>(ref swapEditAndDisplay, value); }
        get { return swapEditAndDisplay; }
    }

    public bool AutoParsing
    {
        set { SetProperty<bool>(ref autoParsing, value); }
        get { return autoParsing; }
}

    public bool ShowRuler
    {
        set { SetProperty<bool>(ref showRuler, value); }
        get { return showRuler; }
    }

    public bool ShowGridLines
    {
        set { SetProperty<bool>(ref showGridLines, value); }
        get { return showGridLines; }
```

```
        }

        public double FontSize
        {
            set { SetProperty<double>(ref fontSize, value); }
            get { return fontSize; }
        }

        public int TabSpaces
        {
            set { SetProperty<int>(ref tabSpaces, value); }
            get { return tabSpaces; }
        }

        public void Save()
        {
            ApplicationDataContainer appData = ApplicationData.Current.LocalSettings;
            appData.Values.Clear();
            appData.Values.Add("EditOrientation", (int)this.EditOrientation);
            appData.Values.Add("AutoParsing", this.AutoParsing);
            appData.Values.Add("ShowRuler", this.ShowRuler);
            appData.Values.Add("ShowGridLines", this.ShowGridLines);
            appData.Values.Add("FontSize", this.FontSize);
            appData.Values.Add("TabSpaces", this.TabSpaces);
        }

        protected bool SetProperty<T>(ref T storage, T value,
                            [CallerMemberName] string propertyName = null)
        {
            if (object.Equals(storage, value))
                return false;

            storage = value;
            OnPropertyChanged(propertyName);
            return true;
        }

        protected void OnPropertyChanged(string propertyName)
        {
            if (PropertyChanged != null)
                PropertyChanged(this, new PropertyChangedEventArgs(propertyName));
        }
    }
```

除了 EditOrientatio 之外，AppSettings 还定义了两个额外的属性，能够进一步直接对应 SplitContainerOrientation 和 SwapEditAndDisplay。set 访问器存为保护状态，并且只有通过 EditOrientation 的 set 访问器来设置属性。其他应用的设置没有保存这两个属性，但很容易从应用设置中派生出两者，绑定比较容易。

8.10　XamlCruncher 页面

我们已经写了足够多的代码，现在可以“组装”成应用了。MainPage.xaml 文件如下所示。

```
项目: XamlCruncher | 文件: MainPage.xaml(片段)
<Page ... >

    <Grid Background="{StaticResource ApplicationPageBackgroundThemeBrush}">
        <Grid.RowDefinitions>
            <RowDefinition Height="Auto" />
            <RowDefinition Height="*" />
```

```
        <RowDefinition Height="Auto" />
    </Grid.RowDefinitions>

    <Grid.ColumnDefinitions>
        <ColumnDefinition Width="*" />
        <ColumnDefinition Width="Auto" />
    </Grid.ColumnDefinitions>

    <TextBlock Name="filenameText"
               Grid.Row="0"
               Grid.Column="0"
               Grid.ColumnSpan="2"
               FontSize="18"
               TextTrimming="WordEllipsis" />

    <local:SplitContainer x:Name="splitContainer"
                          Orientation="{Binding Orientation}"
                          SwapChildren="{Binding SwapEditAndDisplay}"
                          MinimumSize="200"
                          Grid.Row="1"
                          Grid.Column="0"
                          Grid.ColumnSpan="2">
        <local:SplitContainer.Child1>
            <local:TabbableTextBox x:Name="editBox"
                                   AcceptsReturn="True"
                                   FontSize="{Binding FontSize}"
                                   TabSpaces="{Binding TabSpaces}"
                                   TextChanged="OnEditBoxTextChanged"
                              SelectionChanged="OnEditBoxSelectionChanged"/>
            </local:SplitContainer.Child1>

        <local:SplitContainer.Child2>
            <local:RulerContainer x:Name="resultContainer"
                                  ShowRuler="{Binding ShowRuler}"
                                  ShowGridLines="{Binding ShowGridLines}" />
            </local:SplitContainer.Child2>
        </local:SplitContainer>

        <TextBlock Name="statusText"
                   Text="OK"
                   Grid.Row="2"
                   Grid.Column="0"
                   FontSize="18"
                   TextWrapping="Wrap" />

         <TextBlock Name="lineColText"
                    Grid.Row="2"
                    Grid.Column="1"
                    FontSize="18" />
</Grid>

<Page.BottomAppBar>
    <AppBar>
        <Grid>
            <StackPanel Orientation="Horizontal" HorizontalAlignment="Left">
                <Button Style="{StaticResource RefreshAppBarButtonStyle}"
                        Click="OnRefreshAppBarButtonClick" />

                <Button Style="{StaticResource SettingsAppBarButtonStyle}"
                        Click="OnSettingsAppBarButtonClick" />
            </StackPanel>

            <StackPanel Orientation="Horizontal" HorizontalAlignment="Right">
                <Button Style="{StaticResource OpenAppBarButtonStyle}"
                        Click="OnOpenAppBarButtonClick" />

                <Button Style="{StaticResource SaveLocalAppBarButtonStyle}"
                        AutomationProperties.Name="Save As"
```

```
                                Click="OnSaveAsAppBarButtonClick" />

                        <Button Style="{StaticResource SaveAppBarButtonStyle}"
                                Click="OnSaveAppBarButtonClick" />

                        <Button Style="{StaticResource AddAppBarButtonStyle}"
                                Click="OnAddAppBarButtonClick" />
                    </StackPanel>
                </Grid>
            </AppBar>
        </Page.BottomAppBar>
    </Page>
```

主 Grid 有三行。

- 载入文件名(TextBlock 名为“filenameText”)。
- SplitContainer。
- 底部状态栏。

状态栏包括两个 TextBlock 元素，分别为 statusText(用来表示可能的 XAML 解析错误)和 lineColText(TabbableTextBox 的行列)。Grid 进一步拆分成两列，用于状态栏的两个组件。

SplitContainer 占了大部分页面，你会发现它包含了绑定到 AppSettings 的 Orientation 和 SwapEditAndDisplay 属性。SplitContainer 包含一个 TabbableTextBox(绑定到 AppSettings 的 FontSize 和 TabSpaces 属性)和 RulerContainer(绑定到 ShowRuler 和 ShowGridLines)。所有绑定都强烈暗示要把 MainPage 的 DataContext 设置为 AppSettings 实例。

XAML 文件末尾定义的是应用栏中的 Button。

和期望的一样，隐藏代码文件在项目里是最长的文件。但我会分成多个模块进行讨论，让读者不至于感觉太累。以下是构造函数、Loaded 处理程序和一些简单方法。

```
项目: XamlCruncher | 文件: MainPage.xaml.cs(片段)
public sealed partial class MainPage : Page
{
    ...
    AppSettings appSettings;
    StorageFile loadedStorageFile;

    public MainPage()
    {
        this.InitializeComponent();

        ...

        // Why aren't these set in the generated C# files?
        editBox = splitContainer.Child1 as TabbableTextBox;
        resultContainer = splitContainer.Child2 as RulerContainer;

        // Set a fixed-pitch font for the TextBox
        Language language =
            new Language(Windows.Globalization.Language.CurrentInputLanguageTag);
        LanguageFontGroup languageFontGroup = new    LanguageFontGroup(language.LanguageTag);
        LanguageFont languageFont = languageFontGroup.FixedWidthTextFont;
        editBox.FontFamily = new FontFamily(languageFont.FontFamily);

        Loaded += OnLoaded;
        Application.Current.Suspending += OnApplicationSuspending;
    }

    async void OnLoaded(object sender, RoutedEventArgs args)
    {
        // Load AppSettings and set to DataContext
        appSettings = new AppSettings();
```

```
    this.DataContext = appSettings;

    // Load any file that may have been saved
    StorageFolder localFolder = ApplicationData.Current.LocalFolder;
    StorageFile storageFile = await localFolder.CreateFileAsync("XamlCruncher.xaml",
CreationCollisionOption.OpenIfExists);
    editBox.Text = await FileIO.ReadTextAsync(storageFile);

    if (editBox.Text.Length == 0)
      await SetDefaultXamlFile();

    // Other initialization
    ParseText();
    editBox.Focus(FocusState.Programmatic);
    DisplayLineAndColumn();
    ...
  }

  async void OnApplicationSuspending(object sender, SuspendingEventArgs args)
  {
    // Save application settings
    appSettings.Save();

    // Save text content
    SuspendingDeferral deferral = args.SuspendingOperation.GetDeferral();
    await PathIO.WriteTextAsync("ms-appdata:///local/XamlCruncher.xaml", editBox.Text);
    deferral.Complete();
  }

  async Task SetDefaultXamlFile()
  {
    editBox.Text =
      "<Page xmlns=\"http://schemas.microsoft.com/winfx/2006/xaml/presentation\"\r\n" +
      "  xmlns:x=\"http://schemas.microsoft.com/winfx/2006/xaml\">\r\n\r\n" +
      "  <TextBlock Text=\"Hello, Windows 8!\"\r\n" +
      "    FontSize=\"48\" />\r\n\r\n" +
      "</Page>";

    editBox.IsModified = false;
    loadedStorageFile = null;
    filenameText.Text = "";
  }
  ...
  void OnEditBoxSelectionChanged(object sender, RoutedEventArgs args)
  {
    DisplayLineAndColumn();
  }

  void DisplayLineAndColumn()
  {
    int line, col;
    editBox.GetPositionFromIndex(editBox.SelectionStart, out line, out col);
    lineColText.Text = String.Format("Line {0} Col {1}", line + 1, col + 1);

    if (editBox.SelectionLength > 0)
    {
      editBox.GetPositionFromIndex(editBox.SelectionStart + editBox.SelectionLength - 1,
        out line, out col);
      lineColText.Text += String.Format(" - Line {0} Col {1}", line + 1, col + 1);
    }
  }
  ...
}
```

构造函数首先修复一些涉及 editBox 字段和 resultContainer 字段的错误。XAML 解析程序在编译时会创建这些字段，但在运行时，InitializeComponent 调用不进行设定。

构造函数的剩余代码基于 LanguageFontGroup 类所提供的预定义字体，为 TabbableTextBox 设置固定间距字体。这显然是从 Windows Runtime 获得实际字体系列的唯一方法。(第 15 章将演示如何使用 DirectWrite 获得在系统上安装的字体集合)。

剩下的初始化发生在 Loaded 事件处理程序中。页面的 DataContext 设置为 AppSettings 实例，可从 MainPage.xaml 文件中的数据绑定部分预计到这些。

OnLoaded 方法继续载入以前保存的文件或者(如果不存在的话)TabbableTextBox 中的一段默认 XAML，并调用 ParseText 进行解析。(稍后会说明具体过程。)TabbableTextBox 赋予键盘输入焦点，OnLoaded 显示初始行列，只要 TextBox 选择发生变化，这些行列就会随之更新。

你可能想知道为什么把 SetDefaultXamlFile 定义为 async(异步)并且在不包含任何异步代码时返回 Task。稍后你会发现，此方法是作为文件 I/O 逻辑中另一种方法的参数，必须得这么定义。因为不包含任何 await 逻辑，所以编译器会生成一条警告消息。

8.11　解析 XAML

XamlCruncher 的主要功能是将一段 XAML 传递给 XamlReader.Load 并获取对象。AppSettings 类的 AutoParsing 属性允许每个按键都有此功能或按压应用栏中的刷新按钮，程序才会执行。

如果 XamlReader.Load 遇到错误，会报异常，程序会在页面底部状态栏用红色显示错误，同时也把 TabbableTextBox 中的文本标记为红色。

```
项目: XamlCruncher | 文件: MainPage.xaml.cs(片段)
public sealed partial class MainPage : Page
{
  Brush textBlockBrush, textBoxBrush, errorBrush;
  ...
  public MainPage()
  {
    ...
    // Set brushes
    textBlockBrush = Resources["ApplicationForegroundThemeBrush"] as SolidColorBrush;
    textBoxBrush = Resources["TextBoxForegroundThemeBrush"] as SolidColorBrush;
    errorBrush = new SolidColorBrush(Colors.Red);
    ...
  }

  ...

  void OnRefreshAppBarButtonClick(object sender, RoutedEventArgs args)
  {
     ParseText();
     this.BottomAppBar.IsOpen = false;
  }
  ...
  void OnEditBoxTextChanged(object sender, RoutedEventArgs e)
  {
     if (appSettings.AutoParsing)
        ParseText();
  }

  void ParseText()
  {
     object result = null;
```

```
        try
        {
            result = XamlReader.Load(editBox.Text);
        }
        catch (Exception exc)
        {
            SetErrorText(exc.Message);
            return;
        }

        if (result == null)
        {
            SetErrorText("Null result");
        }
        else if (!(result is UIElement))
        {
            SetErrorText("Result is " + result.GetType().Name);
        }
        else
        {
            resultContainer.Child = result as UIElement;
            SetOkText();
            return;
        }
    }

    void SetErrorText(string text)
    {
        SetStatusText(text, errorBrush, errorBrush);
    }

    void SetOkText()
    {
        SetStatusText("OK", textBlockBrush, textBoxBrush);
    }

    void SetStatusText(string text, Brush statusBrush, Brush editBrush)
    {
        statusText.Text = text;
        statusText.Foreground = statusBrush;
        editBox.Foreground = editBrush;
    }
}
```

一段 XAML 有可能成功通过 XamlReaderr.Load 而没有报错，但随后就出现异常。特别是涉及到 XAML 动画的时候可能就会发生这种情况，因为要先加载可视树，才能使动画启动。

唯一真正的解决方案是为 Application 对象所定义的 UnhandledException 事件安装处理程序，而这由 Loaded 处理程序来完成。

项目: XamlCruncher | 文件: MainPage.xaml.cs(片段)

```
async void OnLoaded(object sender, RoutedEventArgs args)
{
    ...
    Application.Current.UnhandledException += (excSender, excArgs) =>
        {
            SetErrorText(excArgs.Message);
            excArgs.Handled = true;
        };
}
```

类似这种问题是要确保程序不会包含其他类型的未处理异常，而这些异常不是由不确定代码所引起的。

另外，Visual Studio 用调试器运行程序时，会找到并报告未处理异常。使用 Debug(调试)菜单里的 Exceptions(异常)对话框，表明要 Visual Studio 拦截哪些异常，哪些留给程序。

8.12　XAML 文件的输入和输出

不管什么时候处理涉及载入和保存文档的代码，我总是想得容易。有一个基本问题：每次新建或者打开文档，都需要检查当前文档是否修改而没有保存。如果是，就显示消息框，询问用户是否希望保存文件。选项包括 Save(保存)、Don't Save(不保存)和 Cancel(取消)。

简单回答是 Cancel，此时程序不需要更多功能。如果用户选择 Don't Save，程序就抛弃当前文档，然后执行新建或者打开命令。

如果用户选择 Save，则需要以其文件名保存现有文档。但如果文档不是从磁盘加载或者之前未保存过，可能会不存在文件名。如果是这种情况，则需要显示 Save As(另存为)对话框。但用户也可以从对话框中选择 Cancel，新建和打开操作就会终止。否则，当前文件会首次保存。

我们先看看保存文档方法。应用有 Save 和 Save As 按钮，但如果文档没有文件名， 则 Save 按钮需要启用 Save As 对话框。

项目：XamlCruncher | 文件：MainPage.xaml.cs(片段)

```
async void OnSaveAsAppBarButtonClick(object sender, RoutedEventArgs args)
{
    StorageFile storageFile = await GetFileFromSavePicker();

    If (storageFile == null)
      return;

      await SaveXamlToFile(storageFile);
}

async void OnSaveAppBarButtonClick(object sender, RoutedEventArgs args)
{
    Button button = sender as Button;
    button.IsEnabled = false;

    if (loadedStorageFile != null)
    {
       await SaveXamlToFile(loadedStorageFile);
    }
    else
    {
       StorageFile storageFile = await GetFileFromSavePicker();

       If (storageFile != null)
      {
          await SaveXamlToFile(storageFile);
      }
    }
    button.IsEnabled = true;
}

async Task<StorageFile> GetFileFromSavePicker()
{
   FileSavePicker picker = new FileSavePicker();
   picker.DefaultFileExtension = ".xaml";
   picker.FileTypeChoices.Add("XAML", new List<string> { ".xaml" });
   picker.SuggestedSaveFile = loadedStorageFile;
   return await picker.PickSaveFileAsync();
```

```
}

async Task SaveXamlToFile(StorageFile storageFile)
{
    loadedStorageFile = storageFile;
    string exception = null;

    try
    {
        await FileIO.WriteTextAsync(storageFile, editBox.Text);
    }
    catch (Exception exc)
    {
        exception = exc.Message;
    }

    if (exception != null)
    {
        string message = String.Format("Could not save file {0}: {1}",
                                       storageFile.Name, exception);
        MessageDialog msgdlg = new MessageDialog(message, "XAML Cruncher");
        await msgdlg.ShowAsync();
    }
    else
    {
        editBox.IsModified = false;
        filenameText.Text = storageFile.Path;
    }
}
```

对于 Save 按钮，处理程序会禁用此按钮，保存完成后再启用。我担心保存文件时可能会重复按 Save 按钮，而且如果第一次保存未完成，甚至会引起重入性问题。

最后一个方法中，FileIO.WriteTextAsync 调用是在一个 try 块中。如果保存文件时发生异常，程序就要使用 MessageDialog 通知用户。但 catch 块不能调用 ShowAsync 之类的异步方法，因此直接保存异常，供以后检查用。

对于 Add 和 Open 按钮，XamlCruncher 都需要检查文件是否已修改。如果已修改，就必须显示消息框来通知用户，并要求进一步指示。这发生在 CheckIfOkToTrashFile 方法中。由于此方法同时适用于 Add 和 Open 按钮，我给这个方法赋予一个 Func<Task>类型参数 commandAction，即返回 Task 的不带参数方法的代理。Click 处理程序通过调用 LoadFileFromOpenPicker 方法来处理 Open 按钮，而 Add 按钮处理程序采用的则是之前提到的 SetDefaultXamlFile。

```
项目: XamlCruncher | 文件: MainPage.xaml.cs(片段)
async void OnAddAppBarButtonClick(object sender, RoutedEventArgs args)
{
    Button button = sender as Button;
    button.IsEnabled = false;
    await CheckIfOkToTrashFile(SetDefaultXamlFile);
    button.IsEnabled = true;
    this.BottomAppBar.IsOpen = false;
}

async void OnOpenAppBarButtonClick(object sender, RoutedEventArgs args)
{
    Button button = sender as Button;
    button.IsEnabled = false;
    await CheckIfOkToTrashFile(LoadFileFromOpenPicker);
    button.IsEnabled = true;
    this.BottomAppBar.IsOpen = false;
}
```

```
async Task CheckIfOkToTrashFile(Func<Task> commandAction)
{
    if (!editBox.IsModified)
    {
        await commandAction();
        return;
    }

    string message =
        String.Format("Do you want to save changes to {0}?",
            loadedStorageFile == null ? "(untitled)" : loadedStorageFile.Name);

    MessageDialog msgdlg = new MessageDialog(message, "XAML Cruncher");
    msgdlg.Commands.Add(new UICommand("Save", null, "save"));
    msgdlg.Commands.Add(new UICommand("Don't Save", null, "dont"));
    msgdlg.Commands.Add(new UICommand("Cancel", null, "cancel"));
    msgdlg.DefaultCommandIndex = 0;
    msgdlg.CancelCommandIndex = 2;
    IUICommand command = await msgdlg.ShowAsync();

    if ((string)command.Id == "cancel")
        return;

    if ((string)command.Id == "dont")
    {
        await commandAction();
        return;
    }

    if (loadedStorageFile == null)
    {
        StorageFile storageFile = await GetFileFromSavePicker();

        if (storageFile == null)
            return;

        loadedStorageFile = storageFile;
    }

    await SaveXamlToFile(loadedStorageFile);
    await commandAction();
}

async Task LoadFileFromOpenPicker()
{
    FileOpenPicker picker = new FileOpenPicker();
    picker.FileTypeFilter.Add(".xaml");
    StorageFile storageFile = await picker.PickSingleFileAsync();

    if (storageFile != null)
    {
        string exception = null;

        try
        {
            editBox.Text = await FileIO.ReadTextAsync(storageFile);
        }
        catch (Exception exc)
        {
            exception = exc.Message;
        }

        if (exception != null)
        {
            string message = String.Format("Could not load file {0}: {1}",
                                            storageFile.Name, exception);
            MessageDialog msgdlg = new MessageDialog(message, "XAML Cruncher");
```

```
        await msgdlg.ShowAsync();
      }
      else
      {
        editBox.IsModified = false;
        loadedStorageFile = storageFile;
        filenameText.Text = loadedStorageFile.Path;
      }
    }
}
```

8.13 设置对话框

用户点击 Settings(设置)按钮，处理程序实例化 UserControl 派生类 SettingsDialog 并使其成为 Popup 子类。这些选项包含横竖屏显示。前面说过，我为四种可能性定义了 EditOrientation 枚举。因此，本项目还包含一个 EditOrientationRadioButton，用于把四个值的其中之一存储为自定义标记。

```
项目: XamlCruncher | 文件: EditOrientationRadioButton.cs
using Windows.UI.Xaml.Controls;

namespace XamlCruncher
{
    public class EditOrientationRadioButton : RadioButton
    {
      public EditOrientation EditOrientationTag { set; get; }
    }
}
```

SettingsDialog.xaml 文件在 StackPanel 里排列所有控件。

```
项目: XamlCruncher | 文件: SettingsDialog.xaml(片段)
<UserControl ... >

   <UserControl.Resources>
      <Style x:Key="DialogCaptionTextStyle"
          TargetType="TextBlock"
          BasedOn="{StaticResource CaptionTextStyle}">
        <Setter Property="FontSize" Value="14.67" />
        <Setter Property="FontWeight" Value="SemiLight" />
        <Setter Property="Margin" Value="7 0 0 0" />
     </Style>
  </UserControl.Resources>

   <Border Background="{StaticResource ApplicationPageBackgroundThemeBrush}"
        BorderBrush="{StaticResource ApplicationForegroundThemeBrush}"
        BorderThickness="1">
      <StackPanel Margin="24">
        <TextBlock Text="XamlCruncher settings"
              Style="{StaticResource SubheaderTextStyle}"
              Margin="0 0 0 12" />

         <!-- Auto parsing -->
         <ToggleSwitch Header="Automatic parsing"
                IsOn="{Binding AutoParsing, Mode=TwoWay}" />

         <!-- Orientation -->
         <TextBlock Text="Orientation"
               Style="{StaticResource DialogCaptionTextStyle}" />

         <Grid Name="orientationRadioButtonGrid"
```

```
    Margin="7 0 0 0">
  <Grid.RowDefinitions>
    <RowDefinition Height="Auto" />
    <RowDefinition Height="Auto" />
</Grid.RowDefinitions>

<Grid.ColumnDefinitions>
    <ColumnDefinition Width="Auto" />
    <ColumnDefinition Width="Auto" />
</Grid.ColumnDefinitions>

<Grid.Resources>
    <Style TargetType="Border">
       <Setter Property="BorderBrush"
           Value="{StaticResource ApplicationForegroundThemeBrush}" />
       <Setter Property="BorderThickness" Value="1" />
       <Setter Property="Padding" Value="3" />
    </Style>

    <Style TargetType="TextBlock">
        <Setter Property="TextAlignment" Value="Center" />
    </Style>

     <Style TargetType="local:EditOrientationRadioButton">
        <Setter Property="Margin" Value="0 6 12 6" />
     </Style>
 </Grid.Resources>

 <local:EditOrientationRadioButton Grid.Row="0" Grid.Column="0"
                           EditOrientationTag="Left"
         Checked="OnOrientationRadioButtonChecked">
       <StackPanel Orientation="Horizontal">
          <Border>
             <TextBlock Text="edit" />
          </Border>
          <Border>
             <TextBlock Text="display" />
          </Border>
       </StackPanel>
    </local:EditOrientationRadioButton>

     <local:EditOrientationRadioButton Grid.Row="0" Grid.Column="1"
                            EditOrientationTag="Bottom"
          Checked="OnOrientationRadioButtonChecked">
        <StackPanel>
           <Border>
             <TextBlock Text="display" />
           </Border>
           <Border>
              <TextBlock Text="edit" />
           </Border>
        </StackPanel>
     </local:EditOrientationRadioButton>

     <local:EditOrientationRadioButton Grid.Row="1" Grid.Column="0"
                            EditOrientationTag="Top"
           Checked="OnOrientationRadioButtonChecked">
        <StackPanel>
            <Border>
               <TextBlock Text="edit" />
            </Border>
            <Border>
               <TextBlock Text="display" />
            </Border>
        </StackPanel>
     </local:EditOrientationRadioButton>

     <local:EditOrientationRadioButton Grid.Row="1" Grid.Column="1"
```

```
                                EditOrientationTag="Right"
                        Checked="OnOrientationRadioButtonChecked">
                    <StackPanel Orientation="Horizontal">
                        <Border>
                            <TextBlock Text="display" />
                        </Border>
                        <Border>
                            <TextBlock Text="edit" />
                        </Border>
                    </StackPanel>
                </local:EditOrientationRadioButton>
            </Grid>

            <!-- Ruler -->
            <ToggleSwitch Header="Ruler"
                          OnContent="Show"
                          OffContent="Hide"
                          IsOn="{Binding ShowRuler, Mode=TwoWay}" />

            <!-- Grid lines -->
            <ToggleSwitch Header="Grid lines"
                          OnContent="Show"
                          OffContent="Hide"
                          IsOn="{Binding ShowGridLines, Mode=TwoWay}" />

            <!-- Font size -->
            <TextBlock Text="Font size"
                       Style="{StaticResource DialogCaptionTextStyle}" />

            <Slider Value="{Binding FontSize, Mode=TwoWay}"
                    Minimum="10"
                    Maximum="48"
                    Margin="7 0 0 0" />

            <!-- Tab spaces -->
            <TextBlock Text="Tab spaces"
                       Style="{StaticResource DialogCaptionTextStyle}" />

            <Slider Value="{Binding TabSpaces, Mode=TwoWay}"
                    Minimum="1"
                    Maximum="12"
                    Margin="7 0 0 0" />
        </StackPanel>
    </Border>
</UserControl>
```

所有双向绑定都强烈暗示，应该像 MainPage 一样，把 DataContext 设置为 AppSettings 实例。它实际上是 AppSettings 的同一个实例，也就是说，在对话框中任何改动都会自动应用到程序。

也就是说，不能在对话框中做很多改变后再点击 Cancel 按钮。这里没有 Cancel 按钮。为了弥补这一点，合理的做法是在对话框中放一个恢复所有信息到到出厂状态的默认按钮。

XAML 文件中有相当一部分用于四个 EditOrientationRadioButton 控件。每一个控件内容包含两个有边界 TextBlock 元素的 StackPanel，用来创建类似于之前屏幕截图的小图形，以便组成四个布局选项。

对话框包含三个 ToggleSwitch 实例。默认情况下，OnContent 和 OffContent 属性设置为文本字符串“On”和“Off”，但我认为 Show 和 Hide 更有利于标尺和网格显示。

ToggleSwitch 也有 Header 属性，文本显示在开关上方。标签“Automatic parsing”、“Ruler”和“Grid lines”都通过 ToggleSwitch 来显示。我觉得标签看起来很不错，所以用 DialogCaptionTextStyle 的 Style 来复制字体和位置。

Slider 用来设置字体大小，似乎合理，但我也用 Slider 来设置 tab 键空格数量，我承认这似乎根本不合理。尽管 AppSettings 类用整数定义 TabSpaces 属性，而绑定 Slider 的 Value 属性无论如何工作，都证明 Slider 是一种改变属性值的简便方法。

隐藏代码文件的剩余部分用来管理 RadioButton 控件。

```
项目: XamlCruncher | 文件: SettingsDialog.xaml.cs
using Windows.UI.Xaml;
using Windows.UI.Xaml.Controls;

namespace XamlCruncher
{
   public sealed partial class SettingsDialog : UserControl
   {
      public SettingsDialog()
      {
         this.InitializeComponent();
         Loaded += OnLoaded;
      }

      // Initialize RadioButton for edit orientation
      void OnLoaded(object sender, RoutedEventArgs args)
      {
         AppSettings appSettings = DataContext as AppSettings;

      if (appSettings != null)
       {
          foreach  (UIElement child in orientationRadioButtonGrid.Children)
          {
             EditOrientationRadioButton radioButton =
               child as EditOrientationRadioButton;
             radioButton.IsChecked =
               appSettings.EditOrientation == radioButton.EditOrientationTag;
          }
       }
      }

     // Set EditOrientation based on checked RadioButton
     void OnOrientationRadioButtonChecked(object sender, RoutedEventArgs args)
     {
        AppSettings appSettings = DataContext as AppSettings;
        EditOrientationRadioButton radioButton = sender as EditOrientationRadioButton;

        if  (appSettings != null)
          appSettings.EditOrientation = radioButton.EditOrientationTag;
       }
     }
}
```

设置对话框的显示和 AppBarPad 程序非常相似。

```
项目: XamlCruncher | 文件: MainPage.xaml.cs(片段)
public sealed partial class MainPage : Page
{
   ...
   void OnSettingsAppBarButtonClick(object sender, RoutedEventArgs args)
   {
      SettingsDialog settingsDialog = new SettingsDialog();
      settingsDialog.DataContext = appSettings;

      Popup popup = new Popup
     {
         Child = settingsDialog,
         IsLightDismissEnabled = true
     };
```

```
        settingsDialog.Loaded += (dialogSender, dialogArgs) =>
            {
                popup.VerticalOffset = this.ActualHeight - settingsDialog.ActualHeight
                                        - this.BottomAppBar.ActualHeight - 24;
                popup.HorizontalOffset = 24;
            };

        popup.Closed += (popupSender, popupArgs) =>
            {
                this.BottomAppBar.IsOpen = false;
            };
        popup.IsOpen = true;
    }
    ...
}
```

Popup 的 Closed 事件处理程序会关闭应用栏。前面提过， Suspending 事件处理程序会保存新设置。

8.14 超越 Windows Runtime

前面提到在 XamlCruncher 中输入的时候，对 XMAL 有一些限制。元素不能有自己的事件集，因为事件需要事件处理程序，而又必须在代码中实施事件处理程序，XAML 也不能包含引用外部类和组件。

然而，经过解析的 XAML 能在 XamlCruncher 过程中运行，也就是说，能访问任何 XamlCruncher 能访问的类，包括我为程序所创建的自定义类。下图所示的这段 XAML 包括为 local 定义的命名空间。这样一来，就可以使用 SplitContainer，并嵌套两个实例。

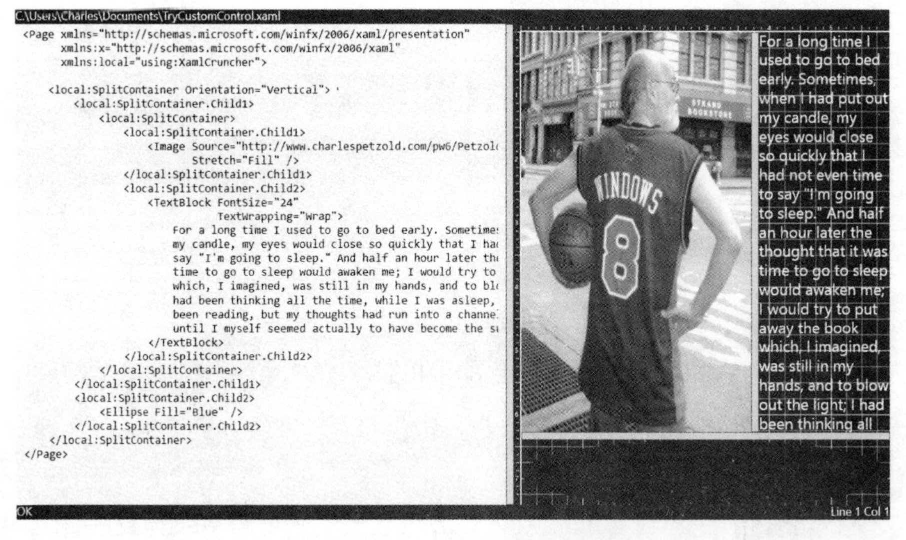

本章这段 XAML 代码提供下载，之前的 XAML 屏幕截屏也可以下载。

XamlCruncher 真的可以超越 Windows Runtime，你也能亲自体验定制类，这很有趣。

第 9 章　动　　画

初看起来，动画主题好像也许更适合于从事游戏或者物理仿真工作的高级程序员。动画似乎并不适用于稳重严肃的商业应用(除了偶然星期五)。

但是动画在 Windows 8 应用中所起的作用，比想象的更重要。第 11 章要展示如何使用 XAML 创建完全重定义控件外观的 ControlTemplate 对象，从中会体现动画的部分作用。尽管可视树是 ControlTemplate 最重要的组成部分，但模板还必须表明在一定条件下控件外观如何变化。例如，Button(按钮)按下的时候会高亮显示，禁用时会变成灰色。ControlTemplate 中所有这些外观变化都定义为动画，哪怕只是瞬时变化，看起来也不真正像动画。

如果要定义不同应用视图之间的过渡或者定义集合变化过程中的条目移动，动画也能发挥作用。试试在开始屏幕上把图块从一个磁贴移动到另一个地方，就会看到邻近磁贴也会相应移动。这些都是动画，也是 Windows 8 美学流畅性质的重要组成部分。

9.1　Windows.UI.Xaml.Media.Animation 命名空间

第 3 章演示了如何使用 CompositionTarget.Rendering 事件使对象变成动画，我把这个方法称为“手动”动画。尽管手动动画很强大，但也有一些局限。其回调方法总是在用户界面线程上运行，也就是说，动画会干扰程序响应用户输入。

同时，用 CompositionTarget.Rendering 演示的动画都是线性的，也就是说，在一段时间内，值的增长或减少都是线性的。如果动画加点变化，往往会更令人愉悦，通常开始时加速，临近结束时减速，或许能再加一点点超现实写实“反弹”。当然，也可以用 CompositionTarget.Rendering 做这类动画，但数学运算会很有挑战性。

而本章将展示内置的 Windows Runtime 动画工具，包含 Windows.UI.Xaml.Media.Animation 命名空间里的 71 个类、4 个枚举项和 2 个结构。这些动画通常在后台线程上运行，并支持若干复杂效果特性。通常可以完全在 XAML 中定义动画，并用代码或者(在特定但常见的情况下)直接从 XAML 触发。

当然，掌握 71 个类的动画工具令人生畏。幸运的是，这些类可以分为几大类型，读到本章结尾，你应该能完全理解该命名空间。

动画涉及到变化，而动画改变的则是对象属性。该属性通常称为动画“目标”。Windows Runtime 动画要求依赖项属性支持目标属性，因此可以用源自 DependencyObject 的类来对它进行定义。

一些图形环境包含基于帧的动画，即动画步调基于视频帧率。不同视频帧速率在不同硬件平台上可能产生不同速度的动画。而 Windows Runtime 运行库动画则基于时间，也就是说，动画是基于时钟的实际持续时间：秒和毫秒。

如果运行动画的线程需要执行一些任务，而动画停止几秒，会发生什么情况？基于帧的动画通常从断点处继续进行。而基于时间的 Windows Runtime 动画则基于时钟时间调整到动画开始的地方。

9.2 动画基础

我们从 TextBlock 的 FontSize 属性开始了解动画，就像第 3 章的 ExpandingText 程序一样。SimpleAnimation 项目有一个两行 Grid，一行是 TextBlock，还有一行是开始运行动画的 Button(按钮)。通常，动画定义在 XAML 文件根元素的 Resources 部分。该简单动画由 Storyboard 和 DoubleAnimation 组成，如下所示：

```
项目: SimpleAnimation | 文件: MainPage.xaml(片段)
<Page ... >

   <Page.Resources>
      <Storyboard x:Key="storyboard">
         <DoubleAnimation Storyboard.TargetName="txtblk"
                          Storyboard.TargetProperty="FontSize"
                          EnableDependentAnimation="True"
                          From="1" To="144" Duration="0:0:3" />
      </Storyboard>
   </Page.Resources>

   <Grid Background="{StaticResource ApplicationPageBackgroundThemeBrush}">
      <Grid.RowDefinitions>
         <RowDefinition Height="*" />
         <RowDefinition Height="*" />
      </Grid.RowDefinitions>

      <TextBlock Name="txtblk"
                 Text="Animated Text"
                 Grid.Row="0"
                 FontSize="48"
                 HorizontalAlignment="Center"
                 VerticalAlignment="Center" />

      <Button Content="Trigger!"
              Grid.Row="1"
              HorizontalAlignment="Center"
              VerticalAlignment="Center"
              Click="OnButtonClick" />
   </Grid>
</Page>
```

DoubleAnimation 类的名称并不意味着执行两个动画！该动画以 Double 类型属性为目标。正如所看到的，Windows Runtime 还支持以 Point、Color 和 Object 类型的属性为目标的动画。(看起来只需要以 Object 类型的属性为目标的动画，但实际上限制为设置离散属性值，而不是变为流畅动画。)

Windows Runtime 要求类似 DoubleAnimation 的动画对象是 Storyboard 的子对象。Storyboard 可以有多个执行并行动画的子对象，Storyboard 的工作则是提供同步其子对象的框架。

Storyboard 还定义了两个附加属性，名为 TargetName 和 TargetProperty。可以在动画对象中设置这些属性，表明目标对象的名称和所希望动画的对象属性：

```
<Storyboard x:Key="storyboard">
   <DoubleAnimation Storyboard.TargetName="txtblk"
```

```
                Storyboard.TargetProperty="FontSize"
                ... />
</Storyboard>
```

默认情况下，辅助线程执行动画，而用户界面线程则仍然可以自由响应用户输入状态。然而，以 TextBlock 的 FontSize 属性为目标的动画，必须在用户界面线程运行，因为字体大小的改变会触发布局改变。Windows Runtime 不喜欢用用户界面线程运行动画，甚至默认不允许这么做！要让 Windows Runtime 知道你的意图(是的，你希望动画运行，即使是采用用户界面线程)，就必须把 EnableDependentAnimation 属性设置为 true：

```
<Storyboard x:Key="storyboard">
   <DoubleAnimation Storyboard.TargetName="txtblk"
                Storyboard.TargetProperty="FontSize"
                EnableDependentAnimation="True"
                ... />
</Storyboard>
```

此时，“依赖”一词意味着“依赖于用户界面线程”。

这个特定动画的剩余部分是用 3 秒钟时间的动画把 FontSize 属性值从 1 变为 144。

```
<Storyboard x:Key="storyboard">
    <DoubleAnimation Storyboard.TargetName="txtblk"
                Storyboard.TargetProperty="FontSize"
                EnableDependentAnimation="True"
                From="1" To="144" Duration="0:0:3" />
</Storyboard>
```

动画持续时间以时、分、秒表示。这三项及两个冒号均为必需。如果只指定一个数字，则视为小时；如果只指定两个数字，冒号则解释为小时和分钟。秒可以是小数秒。如果需要动画运行一天以上，可以在小时数字前写上天数和时期。

第一次运行本程序时，TextBlock 显示为高度 48 像素(见下图)，正如 XAML 文件中 TextBlock 元素所指定的那样。

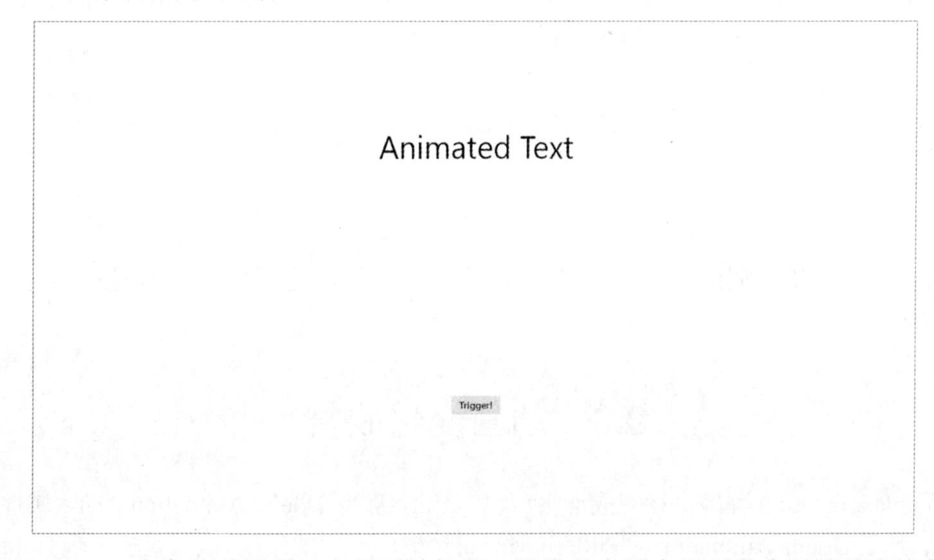

Storyboard 自身并不运行。它需要被触发，通常是在用户界面发生了某些事情之后。在本程序中，Button 的 Click 处理程序访问 Resources 集合，获得对 Storyboard 的引用，然后调用 Begin。

```
项目: SimpleAnimation | 文件: MainPage.xaml.cs
using Windows.UI.Xaml;
using Windows.UI.Xaml.Controls;
using Windows.UI.Xaml.Media.Animation;

namespace SimpleAnimation
{
    public sealed partial class MainPage : Page
    {
        public MainPage()
        {
            this.InitializeComponent();
        }

        void OnButtonClick(object sender, RoutedEventArgs args)
        {
            (this.Resources["storyboard"] as Storyboard).Begin();
        }
    }
}
```

注意其中对 Windows.UI.Xaml.Media 的使用。Visual Studio 模板并未自动提供该指令。

Storyboard 启动后，TextBlock 的 FontSize 立即跳到 1(From 值在 DoubleAnimation 中)，3 秒钟后，FontSize 增加到 144。增加过程为线性：第 1 秒钟时，FontSize 为 48-2/3 像素，2 秒钟时，FontSize 为 96-1/3 像素，3 秒钟之后，动画停止，TextBlock 保持为 144 像素(见下图)。

可以再次点击该按钮，动画会再次开始。事实上，动画运行时可以单击按钮重复动画，每次都从 1 像素大小开始。

9.3 动画变化欣赏

SimpleAnimation 程序执行完成动画后，FontSize 保持 DoubleAnimation 的 To 属性所指定的值。这是 DoubleAnimation 的 FillBehavior 属性值所产生的结果，默认是枚举项 HoldEnd。可以选择将其设置为 Stop。

```
<Storyboard x:Key="storyboard">
    <DoubleAnimation Storyboard.TargetName="txtblk"
                     Storyboard.TargetProperty="FontSize"
```

```
                EnableDependentAnimation="True"
                FillBehavior="Stop"
                From="1" To="144" Duration="0:0:3" />
</Storyboard>
```

动画结束时，目标属性释放动画，FontSize 恢复到动画前的值 48 像素。

另一种变化是省去 From 或 To 的值。

```
<Storyboard x:Key="storyboard">
   <DoubleAnimation Storyboard.TargetName="txtblk"
                Storyboard.TargetProperty="FontSize"
                EnableDependentAnimation="True"
                From="1" Duration="0:0:3" />
</Storyboard>
```

现在，动画从 1 像素开始，但只增加到动画前的 48 像素。而持续时间仍然是 3 秒钟，因此大小增加速度较慢。

以下动画导致 FontSize3 秒钟时间从其当前值增加到 144。

```
<Storyboard x:Key="storyboard">
   <DoubleAnimation Storyboard.TargetName="txtblk"
                Storyboard.TargetProperty="FontSize"
                EnableDependentAnimation="True"
                To="144" Duration="0:0:3" />
</Storyboard>
```

这里所说的 FontSize 从“其当前值”，是因为该值不一定是动画前的值 48。单击按钮，TextBloc 的尺寸仍在增加的时候，再次单击该按钮。每次成功点击，都会有效终止现有动画，并从当前的 FontSize 开始新动画。每一次新的点击都会减慢增加速度，因为动画时长仍然是 3 秒钟。

你可能会假定 DoubleAnimation 类把 To 和 From 属性定义为 double 类型。基本正确。它们实际上是可为空值的 double 类型，而 null 是默认值。这是 DoubleAnimation 如何确定这些属性是否已设置的方法。

另一种方法是 By。

```
<Storyboard x:Key="storyboard">
   <DoubleAnimation Storyboard.TargetName="txtblk"
                Storyboard.TargetProperty="FontSize"
                EnableDependentAnimation="True"
                By="100" Duration="0:0:3" />
</Storyboard>
```

现在，每次点击按钮就会触发动画，该动画在 3 秒钟内，FontSize 会额外增加 100 像素。文本则会越来越大，越来越大。

试试回到初始设置，添加一个属性，把 AutoReverse 设置为 true：

```
<Storyboard x:Key="storyboard">
   <DoubleAnimation Storyboard.TargetName="txtblk"
                Storyboard.TargetProperty="FontSize"
                EnableDependentAnimation="True"
                From="1" To="144" Duration="0:0:3"
                AutoReverse="True" />
</Storyboard>
```

动画触发后，FontSize 降为 1，3 秒钟后上升到 144，再过 3 秒钟，重新跌回到 1，动画结束。整个动画时长为 6 秒钟。设置 FillBehavior 为 Stop，6 秒钟后，FontSize 跳回到动画前的值 48。

也可以把 RepeatBehavior 属性设置为带或者不带 AutoReverse。如果想要执行增加和减少 FontSize 的三个完整周期，则可以像下面这样。

```
<Storyboard x:Key="storyboard">
   <DoubleAnimation Storyboard.TargetName="txtblk"
                Storyboard.TargetProperty="FontSize"
                EnableDependentAnimation="True"
                From="1" To="144" Duration="0:0:3"
                AutoReverse="True"
                RepeatBehavior="3x" />
</Storyboard>
```

整个动画持续 18 秒钟。

也可以把 RepeatBehavior 设置为一段时间。

```
<Storyboard x:Key="storyboard">
   <DoubleAnimation Storyboard.TargetName="txtblk"
                Storyboard.TargetProperty="FontSize"
                EnableDependentAnimation="True"
                From="1" To="144" Duration="0:0:3"
                AutoReverse="True"
                RepeatBehavior="0:0:7.5" />
</Storyboard>
```

整个动画持续 7.5 秒钟。3 秒钟时间，FontSize 从 1 增加到 144，再过 3 秒钟，从 144 降低到 1，然后再次增加并停止。FontSize 的最终值是 73.5。

也可以像下面这样把 RepeatBehavior 设置为 Forever。

```
<Storyboard x:Key="storyboard">
   <DoubleAnimation Storyboard.TargetName="txtblk"
                Storyboard.TargetProperty="FontSize"
                EnableDependentAnimation="True"
                From="1" To="144" Duration="0:0:3"
                AutoReverse="True"
                RepeatBehavior="Forever" />
</Storyboard>
```

代码就是这样精确(除非你看烦了，终止了程序)。

可以像下面这样通过 BeginTime 属性来推迟动画的开始时间。

```
<Storyboard x:Key="storyboard">
   <DoubleAnimation Storyboard.TargetName="txtblk"
                Storyboard.TargetProperty="FontSize"
                EnableDependentAnimation="True"
                BeginTime="0:0:1.5"
                From="1" To="144" Duration="0:0:3" />
</Storyboard>
```

单击按钮，开始的 1.5 秒似乎什么都没有发生，然后 TextBlock 跳至 1 像素大小，并开始增大。点击按钮 4.5 秒钟后，动画结束。

目前所有这些变化，动画都是线性。FontSize 总是每秒钟线性增加或减少某个特定值。有个简单方法能创建非线性动画，即设置 DoubleAnimation 所定义的 EasingFunction 属性。把属性值变成属性元素，并指定其为继承自 EasingFunctionBase 的 11 个类的其中之一。以下是 ElasticEase。

```
<Storyboard x:Key="storyboard">
   <DoubleAnimation Storyboard.TargetName="txtblk"
                Storyboard.TargetProperty="FontSize"
                EnableDependentAnimation="True"
                From="1" To="144" Duration="0:0:3">
```

```
        <DoubleAnimation.EasingFunction>
            <ElasticEase />
        </DoubleAnimation.EasingFunction>
    </DoubleAnimation>
</Storyboard>
```

真的需要试试看看效果。TextBlock越变越大，实际上超过了144像素大小，然后减少到144以下，反复几次，最终定格在To值上。(这种行为真正延伸了“缓动”一词的含义！)

EasingFunctionBase定义了EasingMode属性，所有11个派生类都继承该属性。默认设置是枚举项EasingMode.EaseOut，即动画一开始是线性，结束时应用特效。可以在动画开始时指定EaseIn应用效果，或者在动画开始和结束的时候指定EaseInOut应用效果。

一些EasingFunctionBase衍生类定义了自己的属性用于一些变化。ElasticEase定义了Oscillations属性(以整数来表示值来回波动的次数，默认值是3)和Springiness属性(double类型，默认也为3)。Springiness的值越低，效果越极端。试试以下代码。

```
<Storyboard x:Key="storyboard">
   <DoubleAnimation Storyboard.TargetName="txtblk"
                    Storyboard.TargetProperty="FontSize"
                    EnableDependentAnimation="True"
                    From="1" To="144" Duration="0:0:3">
      <DoubleAnimation.EasingFunction>
         <ElasticEase Oscillations="10"
                      Springiness="0" />
      </DoubleAnimation.EasingFunction>
   </DoubleAnimation>
</Storyboard>
```

探索缓动函数的程序即将形成。

前面提到，类似DoubleAnimation的动画对象必须是Storyboard的子对象。有趣的是，Storyboard和DoubleAnimation在类层级结构中是同层级关系：

```
Object
  DependencyObject
    Timeline
      Storyboard
      DoubleAnimation
      . . .
```

Storyboard定义了TimelineCollection类型的Children属性、附加属性TargetName和TargetProperty，以及暂停和恢复动画的方法。DoubleAnimation定义From、To、By、EnableDependentAnimation和EasingFunction。

目前为止，所有其他属性(AutoReverse、BeginTime、Duration、FillBehavior和RepeatBehavior)都用Timeline定义，即可以在Storyboard设置这些属性，以定义Storyboard上所有子对象的行为。

Timeline还定义了名为SpeedRatio的属性：

```
<Storyboard x:Key="storyboard">
   <DoubleAnimation Storyboard.TargetName="txtblk"
                    Storyboard.TargetProperty="FontSize"
                    EnableDependentAnimation="True"
                    SpeedRatio="10"
                    From="1" To="144" Duration="0:0:3" />
</Storyboard>
```

SpeedRatio设置使动画运行快10倍！当然，也可以在DoubleAnimation设置SpeedRatio，但更常见的做法是在Storyboard进行设置，以便应用于Storyboard中的所有动画子对象。

可以使用 SpeedRatio 来微调动画速度,而不改变所有单个 Duration 次数或调试复杂的动画集合。例如，把 SpeedRatio 设置为 0.1，以减缓动画，这样可以更好看清楚动画。

Timeline 还定义了 Completed 事件，可以在 Storyboard 或者 DoubleAnimation 设置以获取动画完成通知。

也可以完全用代码定义动画。SimpleAnimationCode 项目的 XAML 文件有一个 Grid，包含共享同一个 Click 事件处理程序的 9 个 Button 元素。XAML 文件里没有 Storyboard，也没有 DoubleAnimation：

```
项目: SimpleAnimationCode | 文件: MainPage.xaml(片段)
<Page ... >
   <Page.Resources>
        <Style TargetType="Button">
            <Setter Property="Content" Value="Trigger!" />
            <Setter Property="FontSize" Value="48" />
            <Setter Property="HorizontalAlignment" Value="Center" />
            <Setter Property="VerticalAlignment" Value="Center" />
            <Setter Property="Margin" Value="12" />
        </Style>
</Page.Resources>

    <Grid Background="{StaticResource ApplicationPageBackgroundThemeBrush}">
        <Grid HorizontalAlignment="Center"
            VerticalAlignment="Center">
            <Grid.RowDefinitions>
                <RowDefinition Height="Auto" />
                <RowDefinition Height="Auto" />
                <RowDefinition Height="Auto" />
            </Grid.RowDefinitions>

            <Grid.ColumnDefinitions>
                <ColumnDefinition Width="Auto" />
                <ColumnDefinition Width="Auto" />
                <ColumnDefinition Width="Auto" />
            </Grid.ColumnDefinitions>

            <Button Grid.Row="0" Grid.Column="0" Click="OnButtonClick" />
            <Button Grid.Row="0" Grid.Column="1" Click="OnButtonClick" />
            <Button Grid.Row="0" Grid.Column="2" Click="OnButtonClick" />
            <Button Grid.Row="1" Grid.Column="0" Click="OnButtonClick" />
            <Button Grid.Row="1" Grid.Column="1" Click="OnButtonClick" />
            <Button Grid.Row="1" Grid.Column="2" Click="OnButtonClick" />
            <Button Grid.Row="2" Grid.Column="0" Click="OnButtonClick" />
            <Button Grid.Row="2" Grid.Column="1" Click="OnButtonClick" />
            <Button Grid.Row="2" Grid.Column="2" Click="OnButtonClick" />
        </Grid>
    </Grid>
</Page>
```

在代码隐藏文件中，可以一次创建 Storyboard 和 DoubleAnimation，而无论何时需要触发动画或者根据需要重新创建，都可以进行复用。第一种方法只适用于动画目标始终是同一个对象时。本程序潜在需要 9 个独立动画用于 9 个按钮，因此根据需求来创建会比较容易。一切都在 Click 处理程序中。

```
项目: SimpleAnimationCode | 文件: MainPage.xaml.cs(片段)
void OnButtonClick(object sender, RoutedEventArgs args)
{
   DoubleAnimation anima = new DoubleAnimation
   {
       EnableDependentAnimation = true,
       To = 96,
       Duration = new Duration(new TimeSpan(0, 0, 1)),
```

```
        AutoReverse = true,
        RepeatBehavior = new RepeatBehavior(3)
    };
    Storyboard.SetTarget(anima, sender as Button);
    Storyboard.SetTargetProperty(anima, "FontSize");

    Storyboard storyboard = new Storyboard();
    storyboard.Children.Add(anima);
    storyboard.Begin();
}
```

在前面的 DoubleAnimation 的 XAML 定义中，附加属性 Storyboard.TargetName 和 Storyboard.TargetProperty 表明了动画的对象和属性。代码有一些不同：继续使用静态方法 Storyboard.SetTargetProperty 来设置属性名，但是用 Storyboard.SetTarget(而不是 Storyboard.SetTargetName)来设置目标对象，而不是设置目标对象的 XAML 名称。如果目标对象是 XAML 文件里的 TextBlock，名称为“txtblk”，SetTarget 调用则如下所示。

```
Storyboard.SetTarget(anima, txtblk);
```

它是对象的变量名，不是文本名。在代码示例中，目标对象设置为产生 Click 事件的 Button。

还要注意如何设置 Duration 属性。使用 TimeSpan 是最常见方法，但 Duration 也有两个静态属性：Automatic(这里是 1 秒)和 Forever (不推荐，因为动画会慢得要死)。默认值是 Automatic，如果忘了指定会非常方便。

每个 FontSize 的变化都会影响每个 Button 的大小(见下图)，因此，Grid 需要重新计算单元格的宽度和高度。马上运行所有动画，并观察 Grid 如何改变大小，你会发现很有趣。

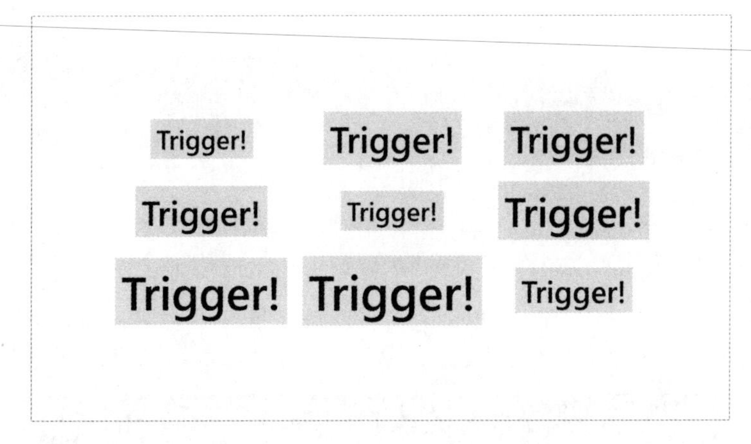

9.4 双 动 画

DoubleAnimation 可以使任何依赖于从属属性的 double 类型属性形成动画，例如 Width 或 Height(或两者)。

项目：EllipseBlobAnimation | 文件：MainPage.xaml(片段)

```
<Page ... >

    <Page.Resources>
        <Storyboard x:Key="storyboard"
                RepeatBehavior="Forever"
```

```
            AutoReverse="True">
        <DoubleAnimation Storyboard.TargetName="ellipse"
                         Storyboard.TargetProperty="Width"
                         EnableDependentAnimation="True"
                         From="100" To="600" Duration="0:0:1" />

        <DoubleAnimation Storyboard.TargetName="ellipse"
                         Storyboard.TargetProperty="Height"
                         EnableDependentAnimation="True"
                         From="600" To="100" Duration="0:0:1" />
      </Storyboard>
    </Page.Resources>

      <Grid Background="{StaticResource ApplicationPageBackgroundThemeBrush}">
        <Ellipse Name="ellipse">
          <Ellipse.Fill>
            <LinearGradientBrush>
              <GradientStop Offset="0" Color="Pink" />
              <GradientStop Offset="1" Color="LightBlue" />
            </LinearGradientBrush>
          </Ellipse.Fill>
        </Ellipse>
      </Grid>
</Page>
```

两个动画并行运行。第一个动画把 Ellipse 的 Width 从 100 到增加 600，而第二个则把 Ellipse 的 Height 从 600 到减少 100。两维只是中间短暂相遇，并构成一个圆。AutoReverse 和 RepeatBehavior 可以在 Storyboard 上(如我所做)，也可以在单个动画里进行设置。

页面加载时，触发动画，以 forever 方式运行：

项目: EllipseBlobAnimation | 文件: MainPage.xaml.cs(片段)

```
public sealed partial class MainPage : Page
{
    public MainPage()
    {
        this.InitializeComponent();

        Loaded += (sender, args) =>
            {
                (this.Resources["storyboard"] as Storyboard).Begin();
            };
    }
}
```

为 Ellipse 着色的 LinearGradientBrush 从矩形边框的左上角到右下角默认渐变，因此，在动画播放期间，渐变实际是会移动的。

Width 和 Height 并不是唯一能做动画效果的 Ellipse 属性。Shape 所定义的 StrokeThickness 属性也是 double 类型，而且有从属属性支持。如下面的 Ellipse，边界有虚线围绕，动画以虚线的 thickness 为目标。

```
项目: AnimateStrokeThickness | 文件: MainPage.xaml(片段)
<Page ... >
   <Page.Resources>
      <Storyboard x:Key="storyboard">
         <DoubleAnimation Storyboard.TargetName="ellipse"
                          Storyboard.TargetProperty="StrokeThickness"
                          EnableDependentAnimation="True"
                          From="1" To="100" Duration="0:0:4"
                          AutoReverse="True"
                          RepeatBehavior="Forever" />
      </Storyboard>
    </Page.Resources>

    <Grid Background="{StaticResource ApplicationPageBackgroundThemeBrush}">
       <Ellipse Name="ellipse"
            Stroke="Red"
            StrokeDashCap="Round"
            StrokeDashArray="0 2" />
    </Grid>
</Page>
```

Loaded 事件中触发动画，代码与之前程序相同。

StrokeDashArray 的“0 2”值表明虚线由 0 单位长的破折号，后面跟着 2 单位长的空白，而单位表明 StrokeThickness 的倍数。由于 StrokeDashCap 属性，破折号的值四舍五入，而四舍五入的值增加到破折号，因此破折号实际上变成了直径和 StrokeThickness 相等的点。这些点的中心被一个长度等于两倍于 StrokeThickness 的空白所分割。

在动画中，点的数量实际随着动画里 StrokeThickness 的增加和减少而减少和增加。点好像是在 Ellipse 的最右边消失和重现(见下图)。

你能找到另一个 double 类型的 Ellipse 属性吗？ StrokeDashOffset 是不是？StrokeDashOffset 表明在一条虚线里破折号和虚线构成的空白的起点。以下 XAML 语句通过一条贝塞尔曲线 Path 用虚线来画一个无穷大符号。动画以 StrokeDashOffset 为目标，而点看上去像是环绕着符号。

```
项目: AnimateDashOffset | 文件: MainPage.xaml(片段)
<Page ... >
    <Page.Resources>
        <Storyboard x:Key="storyboard">
            <DoubleAnimation Storyboard.TargetName="path"
```

```
                    Storyboard.TargetProperty="StrokeDashOffset"
                    EnableDependentAnimation="True"
                    From="0" To="1.5" Duration="0:0:1"
                    RepeatBehavior="Forever" />
        </Storyboard>
    </Page.Resources>

    <Grid Background="{StaticResource ApplicationPageBackgroundThemeBrush}">
        <Viewbox>
            <Path Name="path"
                  Margin="12"
                  Stroke="{StaticResource ApplicationForegroundThemeBrush}"
                  StrokeThickness="24"
                  StrokeDashArray="0 1.5"
                  StrokeDashCap="Round"
                  Data="M 100   0
                        C 45   0,   0  45,  0 100
                        S 45  200, 100 200
                        S 200 150, 250 100
                        S 345   0, 400   0
                        S 500  45, 500 100
                        S 455 200, 400 200
                        S 300 150, 250 100
                        S 155   0, 100   0" />
        </Viewbox>
    </Grid>
</Page>
```

不幸的是，不能在打印纸上展示点环游无穷大符号(见下图)。

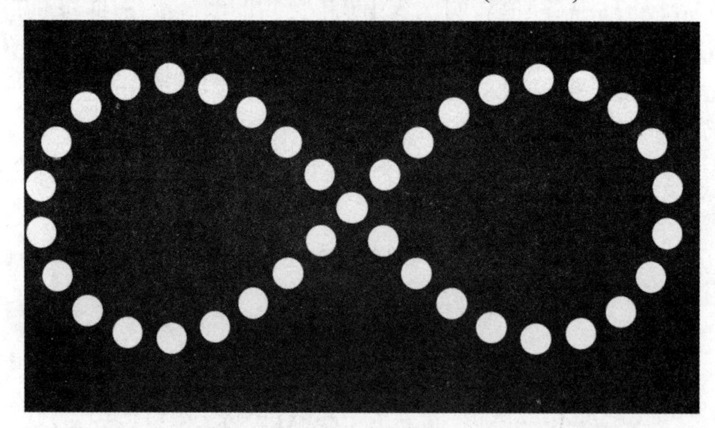

程序中的 Path 定义包含一条著名的贝塞尔曲线，近似四分之一圆。圆心在点(0，0)，右下四分之一圆弧起始于(100，0)，结束于(0，100)。这很近似于贝塞尔曲线，也从(100，0)开始，在(0，100)结束，外加两个控点(100，55)和(55，100)。可以用 4 个这样的“贝塞尔 55”弧线画出一个完整的圆。

因此，从左上角开始画无穷大符号的四分之一圆弧起始于(100，0)，结束于(0，100)，不过圆心在(100,100)，而不是(0,0)，所以，第一个控点离(100，0)左边有 55 单位，第二个控点在(100，0)之上 55 单位，即(45，0)和(0，45)。接下来的贝塞尔曲线应该绕着左下角继续画，从(0，100)开始(也就是上一条贝塞尔曲线的结束位置)在(100，200)结束，控点为(0，155)和(45，200)。但剩下的几何学标记路径不是用 C，即“三次贝塞尔曲线”，而是用 S，即“平滑贝塞尔曲线” 继续画。众所周知，如果有两条相连的贝塞尔曲线，其交点和和两个相邻控点是共线(也就是说，在同一条直线上)，则会有一个平滑连接。路径标记语法中的 S 符号导致自动推倒第一个控点，因此起始点和先前的控点为共线，同前一个交点一样，到起始点的距离一样。因此，基于第一次贝塞尔曲线里的点(0，45)和(0，100)，第一个 S 图像得出第一个控点(0，155)。

画一个首尾相连的虚线时，很有可能会在起点处不连续，只显示部分虚线。用 StrokeThickness 为 24 是实验而得，并不一定是整数。对于这个 Windows Phone 版本的程序，使用的 StrokeThickness 是 23.98。

如果探索 Shapes 库的其他内容，寻找可以成为动画的 double 类型属性，还会发现 Line 的 X1、Y1、X2 和 Y2 属性。本章后面将演示如何使 Point 类型属性形成动画，而这些属性会出现在许多 PathSegment 衍生类中。

Opacity 属性是一种很常见的动画目标，用于元素的淡进和淡出。可以从 0(透明)到 1(不透明)设置 Opacity 值。以下 Opacity 动画来自 John Tenniel 动画插图柴郡猫(Cheshire Cat)，原版来自刘易斯·卡罗尔的《爱丽丝梦游仙境》(1865)。

```
项目: CheshireCat | 文件: MainPage.xaml(片段)
<Page ... >

    <Page.Resources>
        <Storyboard x:Key="storyboard">
            <DoubleAnimation Storyboard.TargetName="image2"
                             Storyboard.TargetProperty="Opacity"
                             From="0" To="1" Duration="0:0:2"
                             AutoReverse="True"
                             RepeatBehavior="Forever" />
        </Storyboard>

    </Page.Resources>

    <!-- Images from Project Gutenberg Book #114
        http://www.gutenberg.org/ebooks/114
        John Tenniel's illustrations for Lewis Carroll's "Alice in Wonderland" -->

    <Grid Background="{StaticResource ApplicationPageBackgroundThemeBrush}">
        <Viewbox>
            <Grid>
                <Image Source="Images/alice23a.gif"
                       Width="640" />

                <TextBlock FontFamily="Century Schoolbook"
                           FontSize="24"
                           Foreground="Black"
                           TextWrapping="Wrap"
                           TextAlignment="Justify"
                           Width="320"
                           Margin="0 0 24 60"
                           HorizontalAlignment="Right"
                           VerticalAlignment="Bottom">
                  "All right," said the Cat; and this
                time it vanished quite slowly, beginning with the end
                of the tail, and ending with the grin, which
                remained some time after the rest of it had gone.
                <LineBreak />
                <LineBreak />
                  "Well! I've often seen a cat without a
                grin," thought Alice; "but a grin without a cat! It's
                the most curious thing I ever saw in all my life!"
              </TextBlock>

                <Image Name="image2"
                       Source="Images/alice24a.gif"
                       Stretch="None"
                       VerticalAlignment="Top">
```

```
                <Image.Clip>
                    <RectangleGeometry Rect="320 70 320 240" />
                </Image.Clip>
            </Image>
        </Grid>
    </Viewbox>
  </Grid>
</Page>
```

正如 XAML 文件中的注释，我从古登堡(Gutenberg)项目获得了图片。在原版《爱丽丝梦游仙境》中，这两张图片均为页面宽度，但为了显示爱丽丝是站在树上的，第一张图片的高度还要扩展到页面高度。然而，古登堡项目中的图并没有同样宽。第一张图(alice23a.gif)是 342×480 像素，第二张(alice24a.gif)是 640×435 像素。我强制用相同宽度渲染图片，所以两张图片看起来是两种绝对不同的画法。我还是决定用一个矩形剪切区域限制第二张图，只显示要消失的猫。此处添加的文字和原版中的不一样。

派生自 Transform 的类开始动画时，DoubleAnimation 的效果会愈加明显。这是下一章的主题(第 10 章)。你可能记得第 3 章中的 RainbowEight 程序，该程序依次使 15 个 GradientStop 对象的 Offset 属性成为动画。可以使用 15 个 DoubleAnimation 对象编写类似程序，但下一章将展示如何使用 DoubleAnimation 来编写类似程序，DoubleAnimation 使设置在 TranslateTransform 上的 LinearGradientBrush 成为动画。

9.5 附加属性动画

下一章要探讨的转换有一种简单的用法涉及在屏幕上移动对象。但不需要为此应用转换。可以把对象放在 Canvas 上并使 Canvas.Lef 和 Canvas.Top 附加属性成为动画。附加属性动画需要 Storyboard.TargetProperty 的特殊语法，如下所示。

项目：AttachedPropertyAnimation | 文件：MainPage.xaml(片段)

```
<Page ... >

    <Page.Resources>
        <Storyboard x:Key="storyboard">
            <DoubleAnimation Storyboard.TargetName="ellipse"
```

```
                        Storyboard.TargetProperty="(Canvas.Left)"
                        From="0" Duration="0:0:2.51"
                        AutoReverse="True"
                        RepeatBehavior="Forever" />

            <DoubleAnimation Storyboard.TargetName="ellipse"
                        Storyboard.TargetProperty="(Canvas.Top)"
                        From="0" Duration="0:0:1.01"
                        AutoReverse="True"
                        RepeatBehavior="Forever" />
        </Storyboard>
    </Page.Resources>

    <Grid Background="{StaticResource ApplicationPageBackgroundThemeBrush}">
        <Canvas SizeChanged="OnCanvasSizeChanged"
                Margin="0 0 48 48">
            <Ellipse Name="ellipse"
                     Width="48"
                     Height="48"
                     Fill="Red" />
        </Canvas>
    </Grid>
</Page>
```

Canvas.Left 和 Canvas.Top 附加属性就在圆括号里。目标是一个红色 Ellipse，这很容易被看成一个球。

请注意，这里没有 EnableDependentAnimation 设置。这表明这些动画不是发生在用户界面线程上的。如果不确定是否要使用 EnableDependentAnimation，就试试不用。如果动画正常，则没有问题。

Storyboard 有两个同步运行的 DoubleAnimation 子对象。注意，每个 DoubleAnimation 定义把 AutoReverse 设置为 True，把 RepeatBehavior 设置为 Forever，Duration 的值分别设置为 1.01 秒和 2.51 秒。此处选择了素数(101 和 251)以免重复模式。两个动画包括 From 值，但没有 To 值。它包含在代码隐藏文件中。

```
项目: AttachedPropertyAnimation | 文件: MainPage.xaml.cs(片段)
public sealed partial class MainPage : Page
{
    public MainPage()
    {
        this.InitializeComponent();

        Loaded += (sender, args) =>
            {
                (this.Resources["storyboard"] as Storyboard).Begin();
            };
    }

    void OnCanvasSizeChanged(object sender, SizeChangedEventArgs args)
    {
        Storyboard storyboard = this.Resources["storyboard"] as Storyboard;

        // Canvas.Left animation
        DoubleAnimation anima = storyboard.Children[0] as DoubleAnimation;
        anima.To = args.NewSize.Width;

        // Canvas.Top animation
        anima = storyboard.Children[1] as DoubleAnimation;
        anima.To = args.NewSize.Height;
    }
}
```

Loaded 事件处理程序启动 Storyboard。每当 Canvas 的大小发生变化(发生在窗口大小改变的时候)，会根据 Canvas 的高度和宽度计算新的 To 值，而 Canvas 在 XAML 文件中有 Margin 设置，以补偿 Ellipse 的大小。你可能会认为不能更改正在进行的动画的值，但这似乎没问题。动画效果是一个球在屏幕四壁反弹(见下图)。

两个 DoubleAnimation 定义包括相同的 AutoReverse 和 RepeatBehavior 设置。如前所述，这些属性用 Timeline 进行定义，Timeline 也是 Storyboard 的父类。这两项设置可能成为 Storyboard 标签吗？试试看。

```
<Storyboard x:Key="storyboard"
        AutoReverse="True"
        RepeatBehavior="Forever">
    <DoubleAnimation Storyboard.TargetName="ellipse"
                Storyboard.TargetProperty="(Canvas.Left)"
                From="0" Duration="0:0:2.51" />

    <DoubleAnimation Storyboard.TargetName="ellipse"
                Storyboard.TargetProperty="(Canvas.Top)"
                From="0" Duration="0:0:1.01" />
</Storyboard>
```

完全合法，但和前面的标记作用不一样。Storyboard 的持续时间是其持续时间最长的动画子对象，本例中是 2.51 秒钟。动画开始时，球在水平和垂直两个方向移动。但在 1.01 秒钟时，球撞到边缘。横屏模式中为底部边缘。Canvas.Top 属性的动画已经完成，但 Canvas.Left 属性的动画继续在另一个 1.5 秒钟里在水平方向移动球，此时球在屏幕右下角。然后两个动画都完成，因此，Storyboard 会回退刚刚看到的动画，直到球又出现在左上角。此后，动画永远重复同样的模式。

只有 Storyboard 中的所有动画时长都相同时，AutoReverse 和 RepeatBehavior 属性才能移到 Storyboard 上。

9.6 缓动函数

假设 DoubleAnimation 的 From 值是 100，To 的值是 500，Duration 是 5 秒钟。默认情况下，DoubleAnimation 是线性的，即目标属性在运行期间基于线性关系值从 100 到 500，如下所示。

Time	Value
0 sec	100
1	180
2	260
3	340
4	420
5	500

或者，以下公式更清楚：

$$\text{Value} = \text{From} + \frac{\text{Time}}{\text{Duration}} \times (\text{To} - \text{From})$$

该缓动函数的作用是让动画变得更有趣。

我原本计划通过展示如何继承 EasingFunctionBase 创建自定义缓动函数来开始讨论，但因为我有很好的理由认为不能继承 EasingFunctionBase。如果能继承，就可以创建你自己的缓动函数，只需重写 Ease 方法，并实施转换函数即可。Ease 方法有 double 类型参数，范围在 0 到 1 之间。该方法返回一个 double 类型值。如果参数是 0，方法返回 0。如果参数为 1，方法返回 1。在 0 和 1 之间的值都可以。通过这种方式，easing 函数有效弯曲了时间，运行时间和动画值之间的关系从而成为非线性。

缓动函数生效时，持续时间除以 Duration(如上述公式)，归一到在 0 和 1 之间。调用 Ease 函数，返回值用来计算值：

$$\text{Value} = \text{From} + \text{Ease}\left(\frac{\text{Time}}{\text{Duration}}\right) \times (\text{To} - \text{From})$$

例如，ExponentialEase 函数，默认 EasingMode 设置为 EaseOut，有以下转换函数：

$$t' = \frac{1 - e^{-Nt}}{1 - e^{-N}}$$

其中，t 是 Ease 函数的参数，t '是结果，N 是 Exponent 属性设置。如果 N 等于 2(默认值)，上表所示动画可以替换为：

Time	t	t'	Value
0 sec	0.0	0.000	100
1	0.2	0.381	252
2	0.4	0.637	355
3	0.6	0.808	423
4	0.8	0.923	469
5	1.0	1.000	500

动画开始时较快，然后逐渐慢了下来。

AnimationEaseGrapher 程序以视觉化方式展现缓动函数，可以用它来做实验。

从下图可以看出图形为转换函数，横坐标代表 t 从 0 到 1，纵坐标代表 t' 从 0 到 1，0 在顶部，1 在底部。从左上角到右下角的虚线是线性转换函数，蓝线是所选择的转换函数。

从代码隐藏文件指定的 Polyline 上的点反复调用被选缓动类的 Ease 方法。按下 Demo 按钮，左上角的小红球水平动画呈现出规则线性，垂直动画呈现的缓动函数，而且足以令人惊讶的是动画遵循图形。

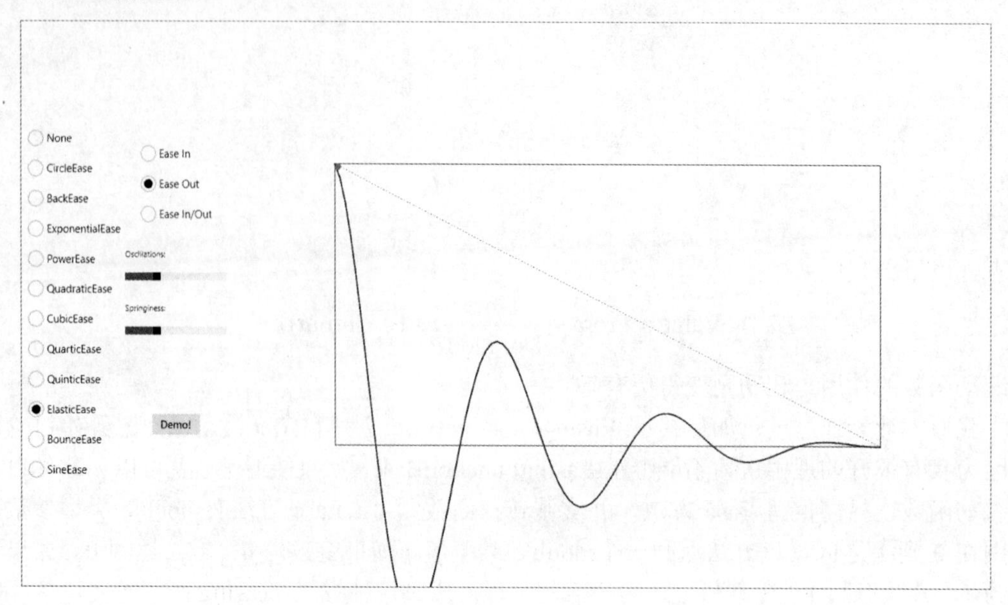

以下是该程序的 XAML 文件，一开始定义红球动画。该动画的缓动函数由代码隐藏文件指定。To 和 From 的值基于球的 6 像素半径(底部出现下降)进行调整。

项目: AnimationEaseGrapher | 文件: MainPage.xaml(片段)

```
<Page ... >
  <Page.Resources>
    <Storyboard x:Key="storyboard"
            FillBehavior="Stop">
      <DoubleAnimation Storyboard.TargetName="redBall"
                       Storyboard.TargetProperty="(Canvas.Left)"
                       From="-6" To="994" Duration="0:0:3" />

       <DoubleAnimation x:Name="anima2"
                        Storyboard.TargetName="redBall"
                        Storyboard.TargetProperty="(Canvas.Top)"
                        From="-6" To="494" Duration="0:0:3" />
    </Storyboard>
  </Page.Resources>

  <Grid Background="{StaticResource ApplicationPageBackgroundThemeBrush}">
     <Grid.ColumnDefinitions>
        <ColumnDefinition Width="Auto" />
        <ColumnDefinition Width="*" />
     </Grid.ColumnDefinitions>

     <!-- Control panel -->
     <Grid Grid.Column="0"
         VerticalAlignment="Center">
       <Grid.RowDefinitions>
          <RowDefinition Height="*" />
          <RowDefinition Height="*" />
          <RowDefinition Height="*" />
       </Grid.RowDefinitions>

       <Grid.ColumnDefinitions>
          <ColumnDefinition Width="Auto" />
```

```
      <ColumnDefinition Width="Auto" />
    </Grid.ColumnDefinitions>

    <!-- Easing function (populated by code) -->
    <StackPanel Name="easingFunctionStackPanel"
                Grid.Row="0"
                Grid.RowSpan="3"
                Grid.Column="0"
                VerticalAlignment="Center">
        <RadioButton Content="None"
                     Margin="6"
                     Checked="OnEasingFunctionRadioButtonChecked" />
    </StackPanel>

    <!-- Easing mode -->
    <StackPanel Name="easingModeStackPanel"
                Grid.Row="0"
                Grid.Column="1"
                HorizontalAlignment="Center"
                VerticalAlignment="Center">
        <RadioButton Content="Ease In"
                     Margin="6"
                     Checked="OnEasingModeRadioButtonChecked">
        <RadioButton.Tag>
            <EasingMode>EaseIn</EasingMode>
        </RadioButton.Tag>
        </RadioButton>

        <RadioButton Content="Ease Out"
                     Margin="6"
                     Checked="OnEasingModeRadioButtonChecked">
            <RadioButton.Tag>
                <EasingMode>EaseOut</EasingMode>
            </RadioButton.Tag>
        </RadioButton>

        <RadioButton Content="Ease In/Out"
                     Margin="6"
                     Checked="OnEasingModeRadioButtonChecked">
            <RadioButton.Tag>
                <EasingMode>EaseInOut</EasingMode>
            </RadioButton.Tag>
        </RadioButton>
    </StackPanel>

    <!-- Easing properties (populated by code) -->
    <StackPanel Name="propertiesStackPanel"
                Grid.Row="1"
                Grid.Column="1"
                HorizontalAlignment="Center"
                VerticalAlignment="Center" />

    <!-- Demo button -->
    <Button Grid.Row="2"
            Grid.Column="1"
            Content="Demo!"
            HorizontalAlignment="Center"
            VerticalAlignment="Center"
            Click="OnDemoButtonClick" />
</Grid>

<!-- Graph using arbitrary coordinates and scaled to window -->
<Viewbox Grid.Column="1">
    <Grid Width="1000"
          Height="500"
          Margin="0 250 0 250">

    <!-- Rectangle outline -->
```

```
            <Polygon Points="0 0, 1000 0, 1000 500, 0 500"
                     Stroke="{StaticResource ApplicationForegroundThemeBrush}"
                     StrokeThickness="3" />

            <Canvas>
                <!-- Linear transfer -->
                <Polyline Points="0 0, 1000 500"
                          Stroke="{StaticResource ApplicationForegroundThemeBrush}"
                          StrokeThickness="1"
                          StrokeDashArray="3 3" />

                <!-- Points set by code based on easing function -->
                <Polyline Name="polyline"
                          Stroke="Blue"
                          StrokeThickness="3" />

                <!-- Animated ball -->
                <Ellipse Name="redBall"
                         Width="12"
                         Height="12"
                         Fill="Red" />
            </Canvas>
        </Grid>
    </Viewbox>
  </Grid>
</Page>
```

代码隐藏文件通过镜像来获取继承自 EasingFunctionBase 的所有类，并为每一个类创建 RadioButton 元素。选中一个时，映像也用于帮助获得类的无参数构造函数。类可以实例化。额外的镜像允许程序获得所有公共属性，而特定 EasingFunctionBase 自身定义了这些公共属性。幸运的是，所有公共属性都限定为 int 或 double 类型，因此，可以为每一个属性创建 Slider 控件。

```
项目: AnimationEaseGrapher | 文件: MainPage.xaml.cs(片段)
public sealed partial class MainPage : Page
{
   EasingFunctionBase easingFunction;

   public MainPage()
   {
      this.InitializeComponent();
      Loaded += OnMainPageLoaded;
   }

   void OnMainPageLoaded(object sender, RoutedEventArgs args)
   {
      Type baseType = typeof(EasingFunctionBase);
      TypeInfo baseTypeInfo = baseType.GetTypeInfo();
      Assembly assembly = baseTypeInfo.Assembly;

      // Enumerate through all Windows Runtime types
      foreach (Type type in assembly.ExportedTypes)
      {
         TypeInfo typeInfo = type.GetTypeInfo();

         // Create RadioButton for each easing function
         if (typeInfo.IsPublic &&
            baseTypeInfo.IsAssignableFrom(typeInfo) &&
            type != baseType)
         {
            RadioButton radioButton = new RadioButton
            {
               Content = type.Name,
               Tag = type,
```

```
                Margin = new Thickness(6),

            };
            radioButton.Checked += OnEasingFunctionRadioButtonChecked;
            easingFunctionStackPanel.Children.Add(radioButton);
        }
    }

    // Check the first RadioButton in the StackPanel (the one labeled "None")
    (easingFunctionStackPanel.Children[0] as RadioButton).IsChecked = true;
}

void OnEasingFunctionRadioButtonChecked(object sender, RoutedEventArgs args)
{
    RadioButton radioButton = sender as RadioButton;
    Type type = radioButton.Tag as Type;
    easingFunction = null;
    propertiesStackPanel.Children.Clear();

    // type is only null for "None" button
    if (type != null)
    {
        TypeInfo typeInfo = type.GetTypeInfo();

        // Find a parameterless constructor and instantiate the easing function
        foreach (ConstructorInfo constructorInfo in typeInfo.DeclaredConstructors)
        {
            if(constructorInfo.IsPublic && constructorInfo.GetParameters().Length == 0)
            {
                easingFunction = constructorInfo.Invoke(null) as EasingFunctionBase;
                break;
            }
        }

        // Enumerate the easing function properties
        foreach (PropertyInfo property in typeInfo.DeclaredProperties)
        {
            // We can only deal with properties of type int and double
            if (property.PropertyType != typeof(int) &&
                property.PropertyType != typeof(double))
            {
                continue;
            }

            // Create a TextBlock for the property name
            TextBlock txtblk = new TextBlock
            {
                Text = property.Name + ":"
            };
            propertiesStackPanel.Children.Add(txtblk);

            // Create a Slider for the property value
            Slider slider = new Slider
            {
                Width = 144,
                Minimum = 0,
                Maximum = 10,
                Tag = property
            };

            if (property.PropertyType == typeof(int))
            {
                slider.StepFrequency = 1;
                slider.Value = (int)property.GetValue(easingFunction);
            }
            else
            {
                slider.StepFrequency = 0.1;
```

```
            slider.Value = (double)property.GetValue(easingFunction);
        }

        // Define the Slider event handler right here
        slider.ValueChanged += (sliderSender, sliderArgs) =>
            {
                Slider sliderChanging = sliderSender as Slider;
                PropertyInfo propertyInfo = sliderChanging.Tag as PropertyInfo;

                if (property.PropertyType == typeof(int))
                    property.SetValue(easingFunction, (int)sliderArgs.NewValue);
                else
                    property.SetValue(easingFunction, (double)sliderArgs.NewValue);

                DrawNewGraph();
            };
        propertiesStackPanel.Children.Add(slider);
    }
}

// Initialize EasingMode radio buttons
foreach (UIElement child in easingModeStackPanel.Children)
{
    RadioButton easingModeRadioButton = child as RadioButton;
    easingModeRadioButton.IsEnabled = easingFunction != null;

    easingModeRadioButton.IsChecked =
    easingFunction != null &&
    easingFunction.EasingMode == (EasingMode)easingModeRadioButton.Tag;
}
    DrawNewGraph();
}

void OnEasingModeRadioButtonChecked(object sender, RoutedEventArgs args)
{
    RadioButton radioButton = sender as RadioButton;
    easingFunction.EasingMode = (EasingMode)radioButton.Tag;
    DrawNewGraph();
}

void OnDemoButtonClick(object sender, RoutedEventArgs args)
{
    // Set the selected easing function and start the animation
    Storyboard storyboard = this.Resources["storyboard"] as Storyboard;
    (storyboard.Children[1] as DoubleAnimation).EasingFunction = easingFunction;
    storyboard.Begin();
}

void DrawNewGraph()
{
    polyline.Points.Clear();

    if (easingFunction == null)
    {
        polyline.Points.Add(new Point(0, 0));
        polyline.Points.Add(new Point(1000, 500));
        return;
    }

    for (decimal t = 0; t <= 1; t += 0.01m)
    {
        double x = (double)(1000 * t);
        double y = 500 * easingFunction.Ease((double)t);
        polyline.Points.Add(new Point(x, y));
    }
}
}
```

这些缓动函数里有一些冗余：QuadraticEase，CubicEase，QuarticEase 和 QuinticEase 都是 PowerEase 类的特殊情况，通过分别设置 Power 属性为 2、3、4 和 5，即可复制 PowerEase。

前面的 ElasticEase 屏幕截图，表明该特定 Ease 函数返回值在 0 和 1 的范围之外。BackEase 也是如此。因为转换函数可能返回值小于 0 或大于 1，当其 From 和 To 属性设置值落在范围以外时，也能呈现动画。

对于许多属性而言，这样做都不会有问题。但对于某些属性而言，可能会报异常。例如，Opacity 不能设置为小于 0 或大于 1 的值。Width 和 Height 不能设置为负值，FontSize 必须大于 0。动画应用这些导致非法值的属性时，会报运行异常。

尽管缓动函数通常用各种方法导致动画放慢和加速，但是可以用某些非常规方式使用缓动函数。例如，如果设置 EasingMode 为默认值 EaseOut 时，SineEase 有如下转换函数

$$t' = \sin\left(\frac{\pi}{2}t\right)$$

这是正弦曲线的第一个四分之一，开始快，然后慢。对于 EaseIn，如下余弦曲线的第一个四分之一，很快从 0 变成 1：

$$t' = 1 - \cos\left(\frac{\pi}{2}t\right)$$

开始慢，然后快。

设置 EasingMode 为 EaseOut，SineEase 是余弦曲线的前半部分，从 0 调整到 1

$$t' = \frac{1 - \cos(\pi t)}{2}$$

开始慢，然后变快，再变慢。如果把 SineEase 的 EaseInOut 变量和 DoubleAnimation 应用到 Ellipse 的 Canvas.Left 属性，并把 AutoReverse 设置为 True，RepeatBehavior 设置为 Forever，就会得到类似钟摆的运动，即反向运动时很慢，在中间时较快。

如果把类似动画应用到 Canvas.Top，但平移半个周期，则可以绕着圆圈移动对象，如以下程序演示的那样。

项目：CircleAnimation | 文件：MainPage.xaml(片段)

```
<Page ... >
   <Page.Resources>
      <Storyboard x:Key="storyboard" SpeedRatio="3">
         <DoubleAnimation Storyboard.TargetName="ball"
                          Storyboard.TargetProperty="(Canvas.Left)"
                          From="-350" To="350" Duration="0:0:2"
                          AutoReverse="True"
                          RepeatBehavior="Forever">
            <DoubleAnimation.EasingFunction>
               <SineEase EasingMode="EaseInOut" />
            </DoubleAnimation.EasingFunction>
         </DoubleAnimation>

         <DoubleAnimation Storyboard.TargetName="ball"
                          Storyboard.TargetProperty="(Canvas.Top)"
                          BeginTime="0:0:1"
                          From="-350" To="350" Duration="0:0:2"
                          AutoReverse="True"
                          RepeatBehavior="Forever">
            <DoubleAnimation.EasingFunction>
               <SineEase EasingMode="EaseInOut" />
            </DoubleAnimation.EasingFunction>
         </DoubleAnimation>
```

```
        </Storyboard>
    </Page.Resources>

    <Grid Background="{StaticResource ApplicationPageBackgroundThemeBrush}">
        <Canvas HorizontalAlignment="Center"
                VerticalAlignment="Center"
                Margin="0 0 48 48">
            <Ellipse Name="ball"
                     Width="48"
                     Height="48"
                     Fill="Red" />
        </Canvas>
    </Grid>
</Page>
```

Canvas 居中对齐，但椭圆会抵消大小，也就是说，相对于 Canvas 而言，点(0，0)距窗口左 24 像素、中心以上 24 像素的位置。Ellipse 的 Canvas.Left 和 Canvas.Top 默认为零，位于中心。动画在上下和左右 350 像素之间移动 Ellipse。

注意，第二个动画有 1 秒种的 BeginTime，因此，加载程序后的最初 1 秒钟，第一个动画水平移动椭圆从-350 像素到 0，然后第二个动画开始生效，垂直移动球从-350 到 0，此时，第一个动画水平移动椭圆从 0 像素到 350。虽然缓动函数旨在减速和加速动画，但 Ellipse 在绕圆周游时有恒定角速度。

下一章将通过 RotateTransform 展示一个更直接的方式来实现绕转。

9.7 完整的 XAML 动画

本章目前演示的几个程序，都是触发页面 Loaded 事件处理程序中的 Storyboard。如果需要程序或页面加载或“演示” 一直运行的动画，这种方法非常方便。

实际上完全可以在 XAML 里执行在 Loaded 事件里触发动画，使用名为 Triggers 的传统属性即可，该属性继承自 Windows Presentation Foundation(WPF)。在从 WPF 到 Windows Runtime 的漫长变迁中，虽然 Triggers 属性几乎已经失去了以前所有的功能，但仍然可以触发 storyboard：

```
<Page.Triggers>
    <EventTrigger>
        <BeginStoryboard>
            <Storyboard ... >
                ...
            </Storyboard>
        </BeginStoryboard>
    </EventTrigger>
</Page.Triggers>
```

Triggers 属性元素通常出现在 XAML 文件的根元素，习惯放在文件底部，但实际上可以在任何动画目标的祖先元素里定义 Triggers 属性元素。

注意 EventTrigger 和 BeginStoryboard。现在是看到这些标签的唯一环境。EventTrigger 有 RoutedEvent 属性，但如果试着将其设置为任意值(包括合理的“Loaded”或者“Page.Loaded”)，就会产生运行时错误。BeginStoryboard 可以有多个 Storyboard 子对象。

以下程序类似于第 3 章的 ManualColorAnimation。Grid 的背景和 TextBlock 的前景以动画形式呈现，从不同方向从黑到白变化。两个 ColorAnimation 对象的目标是两个

SolidColorBrush 对象的 Color 属性。

```
项目: ForeverColorAnimation | 文件: MainPage.xaml(片段)
<Page ... >
   <Grid>
      <Grid.Background>
         <SolidColorBrush x:Name="gridBrush" />
      </Grid.Background>

       <TextBlock Text="Color Animation"
                  FontFamily="Times New Roman"
                  FontSize="96"
                  FontWeight="Bold"
                  HorizontalAlignment="Center"
                  VerticalAlignment="Center">
         <TextBlock.Foreground>
            <SolidColorBrush x:Name="txtblkBrush" />
         </TextBlock.Foreground>
       </TextBlock>
     </Grid>

     <Page.Triggers>
        <EventTrigger>
           <BeginStoryboard>
              <Storyboard RepeatBehavior="Forever"
                          AutoReverse="True">
                <ColorAnimation Storyboard.TargetName="gridBrush"
                                Storyboard.TargetProperty="Color"
                                From="Black" To="White" Duration="0:0:2" />

                 <ColorAnimation Storyboard.TargetName="txtblkBrush"
                                 Storyboard.TargetProperty="Color"
                                 From="White" To="Black" Duration="0:0:2" />
              </Storyboard>
           </BeginStoryboard>
         </EventTrigger>
      </Page.Triggers>
</Page>
```

ColorAnimation 可能是第二个最常见的动画类，仅次于 DoubleAnimation。它被严格限定为以 SolidColorBrush 和 GradientStop 的 Color 属性为目标，但这些画刷经常出现，所以比看起来有更多功能。注意 Storyboard 里 RepeatBehavior 和 AutoReverse 的设置。

代码隐藏文件只包含页面构造函数里的 InitializeComponent 调用。也就是说，可以把该 XAML 文件复制到第 8 章 XamlCruncher 程序的编辑器里，删除 x:Class 属性，并运行动画，而不需要任何代码的帮助。XamlCruncher(或另一个 XAML 编辑器)是实验动画的好方式。

还可以去掉 Point 类型的属性的动画效果。Point 类型的属性并不很常见，但 EllipseGeometry 有 Point 类型的 Center 属性。如果使用 Path 和 EllipseGeometry 而非 Ellipse 类创建圆或椭圆，就可以通过动画 Center 属性绕着屏幕移动圆或椭圆。与动画 Canvas.Left 和 Canvas.Top 不同，Path 不需要在 Canvas 里，图形位置指定为相对于中心，而不是左上角。

然而，不能单独使 Point 值的 X 和 Y 属性成为动画，因为 Point 是结构而不是类，也就是说，不能继承 DependencyObject，X 和 Y 属性不受依赖属性支持。

Point 类型属性也出现在一些 PathSegment 派生类里，ArcSegment、BezierSegment、LineSegment 和 QuadraticBezierSegment 都有 Point 类型属性。这些 Point 属性成为动画，可以动态改变图形。以下程序使用前面讨论过的贝塞尔曲线画圆，然后 13 个点都会成为动画，最后圆变成正方形。这样只是为了演示 Triggers 属性元素不需要定义在 XAML 文件的根元素，定义在 Path 中即可。

项目: SquaringTheCircle | 文件: MainPage.xaml(片段)

```
<Page ... >
    <Grid Background="{StaticResource ApplicationPageBackgroundThemeBrush}">
        <Canvas HorizontalAlignment="Center"
                VerticalAlignment="Center">
            <Path Fill="{StaticResource ApplicationPressedForegroundThemeBrush}"
                  Stroke="{StaticResource ApplicationForegroundThemeBrush}"
                  StrokeThickness="3" >
                <Path.Data>
                    <PathGeometry>
                        <PathFigure x:Name="bezier1" IsClosed="True">
                            <BezierSegment x:Name="bezier2" />
                            <BezierSegment x:Name="bezier3" />
                            <BezierSegment x:Name="bezier4" />
                            <BezierSegment x:Name="bezier5" />
                        </PathFigure>
                    </PathGeometry>
                </Path.Data>

                <Path.Triggers>
                    <EventTrigger>
                        <BeginStoryboard>
                            <Storyboard RepeatBehavior="Forever">
                                <PointAnimation Storyboard.TargetName="bezier1"
                                                Storyboard.TargetProperty="StartPoint"
                                                EnableDependentAnimation="True"
                                                From="0 200" To="0 250"
                                                AutoReverse="True" />

                                <PointAnimation Storyboard.TargetName="bezier2"
                                                Storyboard.TargetProperty="Point1"
                                                EnableDependentAnimation="True"
                                                From="110 200" To="125 125"
                                                AutoReverse="True" />

                                <PointAnimation Storyboard.TargetName="bezier2"
                                                Storyboard.TargetProperty="Point2"
                                                EnableDependentAnimation="True"
                                                From="200 110" To="125 125"
                                                AutoReverse="True" />

                                <PointAnimation Storyboard.TargetName="bezier2"
                                                Storyboard.TargetProperty="Point3"
                                                EnableDependentAnimation="True"
                                                From="200 0" To="250 0"
                                                AutoReverse="True" />

                                <PointAnimation Storyboard.TargetName="bezier3"
                                                Storyboard.TargetProperty="Point1"
                                                EnableDependentAnimation="True"
                                                From="200 -110" To="125 -125"
                                                AutoReverse="True" />

                                <PointAnimation Storyboard.TargetName="bezier3"
                                                Storyboard.TargetProperty="Point2"
                                                EnableDependentAnimation="True"
                                                From="110 -200" To="125 -125"
                                                AutoReverse="True" />

                                <PointAnimation Storyboard.TargetName="bezier3"
                                                Storyboard.TargetProperty="Point3"
                                                EnableDependentAnimation="True"
                                                From="0 -200" To="0 -250"
                                                AutoReverse="True" />
```

```
<PointAnimation Storyboard.TargetName="bezier4"
                Storyboard.TargetProperty="Point1"
                EnableDependentAnimation="True"
                From="-110 -200" To="-125 -125"
   AutoReverse="True" />

          <PointAnimation Storyboard.TargetName="bezier4"
                          Storyboard.TargetProperty="Point2"
                          EnableDependentAnimation="True"
                          From="-200 -110" To="-125 -125"
                          AutoReverse="True" />

                     <PointAnimation Storyboard.TargetName="bezier4"
                                     Storyboard.TargetProperty="Point3"
                                     EnableDependentAnimation="True"
                                     From="-200 0" To="-250 0"
                                     AutoReverse="True" />

                     <PointAnimation Storyboard.TargetName="bezier5"
                                     Storyboard.TargetProperty="Point1"
                                     EnableDependentAnimation="True"
                                     From="-200 110" To="-125 125"
                                     AutoReverse="True" />

                     <PointAnimation Storyboard.TargetName="bezier5"
                                     Storyboard.TargetProperty="Point2"
                                     EnableDependentAnimation="True"
                                     From="-110 200" To="-125 125"
                                     AutoReverse="True" />

                     <PointAnimation Storyboard.TargetName="bezier5"
                                     Storyboard.TargetProperty="Point3"
                                     EnableDependentAnimation="True"
                                     From="0 200" To="0 250"
                                     AutoReverse="True" />

                </Storyboard>
              </BeginStoryboard>
            </EventTrigger>
          </Path.Triggers>
        </Path>
    </Canvas>
  </Grid>
</Page>
```

下图介于正方形和圆之间。

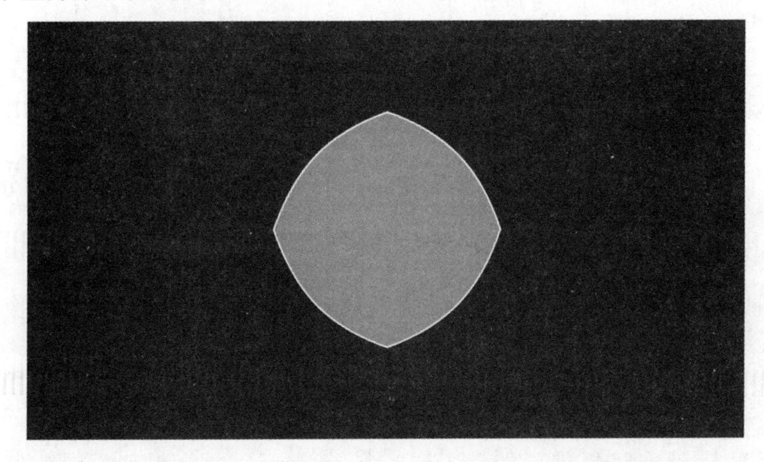

9.8 自定义类动画

是的，可以把自定义类属性也变成动画，但可成为动画的属性必须受依赖属性支持。

以下类名为 PieSlice，继承自 Path，用来渲染饼图里的扇形。自定义属性是 Center、Radius、StartAngle(角度，从 12:00 的顺时针测量)和 SweepAngle(角度，从 StartAngle 的顺时针测量)。

项目: AnimatedPieSlice | 文件: PieSlice.cs

```
using System;
using Windows.Foundation;
using Windows.UI.Xaml;
using Windows.UI.Xaml.Media;
using Windows.UI.Xaml.Shapes;

namespace AnimatedPieSlice
{
    public class PieSlice : Path
    {
        PathFigure pathFigure;
        LineSegment lineSegment;
        ArcSegment arcSegment;

        static PieSlice()
        {
            CenterProperty = DependencyProperty.Register("Center",
                typeof(Point), typeof(PieSlice),
                new PropertyMetadata(new Point(100, 100), OnPropertyChanged));

            RadiusProperty = DependencyProperty.Register("Radius",
                typeof(double), typeof(PieSlice),
                new PropertyMetadata(100.0, OnPropertyChanged));

            StartAngleProperty = DependencyProperty.Register("StartAngle",
                typeof(double), typeof(PieSlice),
                new PropertyMetadata(0.0, OnPropertyChanged));

            SweepAngleProperty = DependencyProperty.Register("SweepAngle",
                typeof(double), typeof(PieSlice),
                new PropertyMetadata(90.0, OnPropertyChanged));
        }

        public PieSlice()
        {
            pathFigure = new PathFigure { IsClosed = true };
            lineSegment = new LineSegment();
            arcSegment = new ArcSegment { SweepDirection = SweepDirection.Clockwise };
            pathFigure.Segments.Add(lineSegment);
            pathFigure.Segments.Add(arcSegment);

            PathGeometry pathGeometry = new PathGeometry();
            pathGeometry.Figures.Add(pathFigure);

            this.Data = pathGeometry;
            UpdateValues();
        }

        public static DependencyProperty CenterProperty { private set; get; }

        public static DependencyProperty RadiusProperty { private set; get; }

        public static DependencyProperty StartAngleProperty { private set; get; }
```

```
        public static DependencyProperty SweepAngleProperty { private set; get; }

        public Point Center
        {
            set { SetValue(CenterProperty, value); }
            get { return (Point)GetValue(CenterProperty); }
        }

        public double Radius
        {
            set { SetValue(RadiusProperty, value); }
            get { return (double)GetValue(RadiusProperty); }
        }

        public double StartAngle
        {
            set { SetValue(StartAngleProperty, value); }
            get { return (double)GetValue(StartAngleProperty); }
        }

        public double SweepAngle
        {
            set { SetValue(SweepAngleProperty, value); }
            get { return (double)GetValue(SweepAngleProperty); }
        }

        static void OnPropertyChanged(DependencyObject obj,
                                      DependencyPropertyChangedEventArgs args)
        {
            (obj as PieSlice).UpdateValues();
        }

        void UpdateValues()
        {
            pathFigure.StartPoint = this.Center;

            double x = this.Center.X + this.Radius * Math.Sin(Math.PI * this.StartAngle / 180);
            double y = this.Center.Y - this.Radius * Math.Cos(Math.PI * this.StartAngle / 180);
            lineSegment.Point = new Point(x, y);

            x = this.Center.X + this.Radius * Math.Sin(Math.PI * (this.StartAngle +this.SweepAngle)/180);

            y = this.Center.Y - this.Radius * Math.Cos(Math.PI * (this.StartAngle +this.SweepAngle)/180);

            arcSegment.Point = new Point(x, y);
            arcSegment.IsLargeArc = this.SweepAngle >= 180;

            arcSegment.Size = new Size(this.Radius, this.Radius);
        }
    }
}
```

该类里的一切都用于依赖属性，除了 UpdateValues 方法(该方法至关重要)。只要四个属性里的任意一个发生变化，就会调用 UpdateValues。四个属性里的任意一个都可以是动画目标，也就是说，对于无限时间，每秒钟调用 60 次 UpdateValues。

对于调用如此频繁的方法，应该小心创建需要在堆上分配内存的对象。可以创建新的 double 类型和 Point 类型的值，因为它们都存储在栈里。但每次调用都要创建新的 PathFigure、LineSegment 和 ArcSegment 对象不是好办法，因为会生成大量分配内存及随后必须释放的活动。可以试试重用或缓存对象，不要重新创建。

PieSlice 类是 AnimatedPieSlice 项目的一部分，其中包括 MainPage.xaml，该文件实例化和初始化 PieSlice 类，并将其变成动画。

项目: AnimatedPieSlice | 文件: MainPage.xaml(片段)

```
<Page ... >
   <Grid Background="{StaticResource ApplicationPageBackgroundThemeBrush}">
      <local:PieSlice x:Name="pieSlice"
                      Center="400 400"
                      Radius="200"
                      Stroke="Red"
                      StrokeThickness="3"
                      Fill="Yellow" />
</Grid>

      <Page.Triggers>
         <EventTrigger>
            <BeginStoryboard>
               <Storyboard>
                  <DoubleAnimation Storyboard.TargetName="pieSlice"
                                   Storyboard.TargetProperty="SweepAngle"
                                   EnableDependentAnimation="True"
                                   From="1" To="359" Duration="0:0:3"
                                   AutoReverse="True"
                                   RepeatBehavior="Forever" />
                     </Storyboard>
                  </BeginStoryboard>
         </EventTrigger>
      </Page.Triggers>
</Page>
```

结果是下图所示的一张饼图，范围从 1 度到 359 度，重复不断。

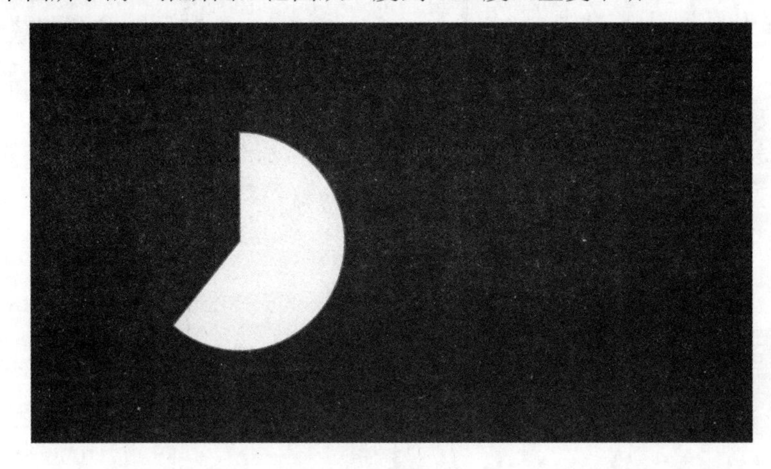

9.9 关键帧动画

到目前为止，所有程序都是用动画把属性从一个值变为另一个值，通常通过指定 DoubleAnimation、ColorAnimation 和 PointAnimation 的 From 和 To 属性来完成，而唯一变化涉及到非线性方法，获得从 From 到 To 的动画，然后再从 To 到 From 倒放动画。

如果需要用动画把属性从一个值变到另一个值，再到第三个值，甚至超越前三个值，该怎么办？你想到的解决方案可能会是在 storyboard 里针对相同属性定义一些动画并使用 BeginTime 延迟其中一些动画，使其不会产生重叠。但这样做是非法的。针对特定属性，storyboard 不能有多个动画。

正确的解决方案是关键帧(key frame)动画，之所以这么叫，是因为要通过一系列关键帧来定义动画进度。每个关键帧表明在特定运行时间属性的值应该是什么，如何从先前的关键帧值变为关键帧的新值。

下面所示为关键帧动画的简单例子，该动画目标是 EllipseGeometry 的 Center 属性，绕着屏幕移动圆。

```
项目: SimpleKeyFrameAnimation | 文件: MainPage.xaml(片段)
<Page ... >
   <Grid Background="{StaticResource ApplicationPageBackgroundThemeBrush}">
      <Path Fill="Blue">
         <Path.Data>
           <EllipseGeometry x:Name="ellipse"
                            RadiusX="24"
                            RadiusY="24" />
</Path.Data>
     </Path>
   </Grid>

   <Page.Triggers>
   <EventTrigger>
      <BeginStoryboard>
         <Storyboard>
            <PointAnimationUsingKeyFrames Storyboard.TargetName="ellipse"
                                          Storyboard.TargetProperty="Center"
                                          EnableDependentAnimation="True"
                                          RepeatBehavior="Forever">
                  <DiscretePointKeyFrame KeyTime="0:0:0" Value="100 100" />
                  <LinearPointKeyFrame KeyTime="0:0:2" Value="700 700" />
                  <LinearPointKeyFrame KeyTime="0:0:2.1" Value="700 100" />
                  <LinearPointKeyFrame KeyTime="0:0:4.1" Value="100 700" />
                  <LinearPointKeyFrame KeyTime="0:0:4.2" Value="100 100" />
               </PointAnimationUsingKeyFrames>
            </Storyboard>
         </BeginStoryboard>
      </EventTrigger>
    </Page.Triggers>
</Page>
```

Storyboard 包含 PointAnimationUsingKeyFrames，没有 PointAnimation。PointAnimationUsingKeyFrames 包含 DiscretePointKeyFrame 和 LinearPointKeyFrame 类型的子对象，而没有在 PointAnimation 指定 From、To 和 Duration 属性。

集合中的每个关键帧指定了从动画开始后的特定时间你想要的目标属性值。关键帧集合通常从 KeyTime 为 0 的 Discrete 条目开始，基本上是把该属性初始化为该值。

```
<DiscretePointKeyFrame KeyTime="0:0:0" Value="100 100" />
```

集合中的下一个关键帧如下：

```
<LinearPointKeyFrame KeyTime="0:0:2" Value="700 700" />
```

也就是说，目标属性在 2 秒钟从先前的(100，100)线性增加到(700，700)。在两秒钟时，值为(700，700)。

下一个关键帧指定运行更快动画：

```
<LinearPointKeyFrame KeyTime="0:0:2.1" Value="700 100" />
```

从 2 秒到 2.1 秒，点从(700，700)变到(700，100)。接下来的 2 秒，动画又变慢：

```
<LinearPointKeyFrame KeyTime="0:0:4.1" Value="100 700" />
```

最后一个关键帧如下：

```
<LinearPointKeyFrame KeyTime="0:0:4.2" Value="100 100" />
```

运行完 4.2 秒，目标属性的值是(100，100)，动画结束。此时，动画可以倒放(如果 AutoReverse 为 true)或从头再开始(如果 RepeatBehavior 值设置适当)。

程序员有可能对关键帧想得过多，而以下两条规则超级简单，可以防止混乱。

- 关键帧总是表明在运行时刻所期望的属性值。
- 动画的持续时间是集合里最大的关键时间。

为了存储关键帧集合，PointAnimationUsingKeyFrames 类定义 PointKeyFrameCollection 类型的 KeyFrames 属性，PointKeyFrameCollection 是 PointKeyFrame 对象的集合。PointKeyFrame 定义 KeyTime 和 Value 属性。四个类都继承自 PointKeyFrame，前面已经讨论过其中两个。

- DiscretePointKeyFrame 跳转到特定值。
- LinearPointKeyFrame 执行线性动画。
- SplinePointKeyFrame 可以加速或减速。
- EasingPointKeyFrame 用缓动函数进行动画。

同样，Windows Runtime 包含 DoubleAnimationUsingKeyFrames 类，该类有 DoubleKeyFrame 子类型，和包含 ColorKeyFrame 类型子对象的 ColorAnimationUsingKeyFrames 有 Discrete、Linear、Spline 和 Easing 派生类似，DoubleKeyFrame 也有 Discrete、Linear、Spline 和 Easing 类派生。

以下项目使用 ColorAnimationUsingKeyFrames 对网格背景进行着色，而颜色是贯穿整个彩虹的动画。

项目: RainbowAnimation | 文件: MainPage.xaml(片段)

```
<Page ... >
<Grid>
    <Grid.Background>
        <SolidColorBrush x:Name="brush" />
    </Grid.Background>
</Grid>

<Page.Triggers>
    <EventTrigger>
        <BeginStoryboard>
            <Storyboard RepeatBehavior="Forever">
                <ColorAnimationUsingKeyFrames Storyboard.TargetName="brush"
                                              Storyboard.TargetProperty="Color">
                    <DiscreteColorKeyFrame KeyTime="0:0:0" Value="#FF0000" />
                    <LinearColorKeyFrame KeyTime="0:0:1" Value="#FFFF00" />
                    <LinearColorKeyFrame KeyTime="0:0:2" Value="#00FF00" />
                    <LinearColorKeyFrame KeyTime="0:0:3" Value="#00FFFF" />
                    <LinearColorKeyFrame KeyTime="0:0:4" Value="#0000FF" />
                    <LinearColorKeyFrame KeyTime="0:0:5" Value="#FF00FF" />
                    <LinearColorKeyFrame KeyTime="0:0:6" Value="#FF0000" />
                </ColorAnimationUsingKeyFrames>
            </Storyboard>
        </BeginStoryboard>
    </EventTrigger>
</Page.Triggers>
</Page>
```

动画时长 6 秒，结束值和开始值相同，也就是说，动画从开始处重新开始时，不会有任何不连续。

以下一对 PointAnimationUsingKeyFrames 对象，把 LinearGradientBrush 对象的 StartPoint 和 EndPoint 属性变成了动画，从而绕圈渐变。

```
项目: GradientBrushPointAnimation | 文件: MainPage.xaml(片段)
<Page ... >
<Grid>
    <Grid.Background>
        <LinearGradientBrush x:Name="gradientBrush">
            <GradientStop Offset="0" Color="Red" />
            <GradientStop Offset="1" Color="Blue" />
        </LinearGradientBrush>
    </Grid.Background>
</Grid>

<Page.Triggers>
    <EventTrigger>
        <BeginStoryboard>
            <Storyboard RepeatBehavior="Forever">
                <PointAnimationUsingKeyFrames Storyboard.TargetName="gradientBrush"
                                              Storyboard.TargetProperty="StartPoint"
                                              EnableDependentAnimation="True">
                    <LinearPointKeyFrame KeyTime="0:0:0" Value="0 0" />
                    <LinearPointKeyFrame KeyTime="0:0:1" Value="1 0" />
                    <LinearPointKeyFrame KeyTime="0:0:2" Value="1 1" />
                    <LinearPointKeyFrame KeyTime="0:0:3" Value="0 1" />
                    <LinearPointKeyFrame KeyTime="0:0:4" Value="0 0" />
                </PointAnimationUsingKeyFrames>

                <PointAnimationUsingKeyFrames Storyboard.TargetName="gradientBrush"
                                              Storyboard.TargetProperty="EndPoint"
                                              EnableDependentAnimation="True">
                    <LinearPointKeyFrame KeyTime="0:0:0" Value="1 1" />
                    <LinearPointKeyFrame KeyTime="0:0:1" Value="0 1" />
                    <LinearPointKeyFrame KeyTime="0:0:2" Value="0 0" />
                    <LinearPointKeyFrame KeyTime="0:0:3" Value="1 0" />
                    <LinearPointKeyFrame KeyTime="0:0:4" Value="1 1" />
                    </PointAnimationUsingKeyFrames>
                </Storyboard>
            </BeginStoryboard>
        </EventTrigger>
</Page.Triggers>
</Page>
```

SplineColorKeyFrame、SplineDoubleKeyFrame 和 SplinePointKeyFrame 对象没有以前用得那么多，因为 EasingDoubleKeyFrame、EasingColorKeyFrame 和 EasingPointKeyFrame 取代了其大部分功能。随着关键帧的 Spline 变化，可以使用 KeySpline 对象来定义贝塞尔曲线的两个控点，该贝塞尔曲线开始于点(0，0)，终止于点(1，1)。本曲线(spline)执行和缓动函数弯曲时间的相同作用，以加速和减速动画。下一章会举例说明。

9.10　Object 动画

Windows Runtime 动画系统还能对 Object 类型的属性进行动画，似乎能对一切进行动画，但有一个问题：没有带 From 和 To 属性的 ObjectAnimation 类。而只有 ObjectAnimationUsingKeyFrames 类，而且唯一继承自 ObjectKeyFrame 类是 DiscreteObjectKeyFrame。

换句话说，确实可以定义目标为任何类型的属性定义动画(只要该属性受依赖属性支持)，但动画只能用来把该属性设置为离散值。

在实践中，对象动画主要用于枚举类型或 Brush 类型的属性为目标，允许把属性设置为预定义的刷资源。大多用在控件模板里，第 11 章会进行说明。

但在下例中，Ellipse 绕着屏幕移动，其 Visibility 属性带枚举项 Visible 和 Collapsed，而其 Fill 属性则采用预定义刷。这些动画会引起 Ellipse 闪烁，由于使用了不同离散颜色，本项目称为 FastNotFluid。

项目: FastNotFluid | 文件: MainPage.xaml(片段)

```
<Page ... >

<Grid Background="Gray">
    <Canvas SizeChanged="OnCanvasSizeChanged"
        Margin="0 0 96 96">
    <Ellipse Name="ellipse"
        Width="96"
        Height="96" />
    </Canvas>
</Grid>

<Page.Triggers>
    <EventTrigger>
        <BeginStoryboard>
            <Storyboard>
                    <DoubleAnimation x:Name="horzAnima"
                                     Storyboard.TargetName="ellipse"
                                     Storyboard.TargetProperty="(Canvas.Left)"
                                     From="0" Duration="0:0:2.51"
                                     AutoReverse="True"
                                     RepeatBehavior="Forever" />

                    <DoubleAnimation x:Name="vertAnima"
                                     Storyboard.TargetName="ellipse"
                                     Storyboard.TargetProperty="(Canvas.Top)"
                                     From="0" Duration="0:0:1.01"
                                     AutoReverse="True"
                                     RepeatBehavior="Forever" />

                <ObjectAnimationUsingKeyFrames
                        Storyboard.TargetName="ellipse"
                        Storyboard.TargetProperty="Visibility"
                        RepeatBehavior="Forever">
                    <DiscreteObjectKeyFrame KeyTime="0:0:0" Value="Visible" />
                    <DiscreteObjectKeyFrame KeyTime="0:0:0.2" Value="Collapsed" />
                    <DiscreteObjectKeyFrame KeyTime="0:0:0.25" Value="Visible" />
                    <DiscreteObjectKeyFrame KeyTime="0:0:0.3" Value="Collapsed" />
                    <DiscreteObjectKeyFrame KeyTime="0:0:0.45" Value="Visible" />
                </ObjectAnimationUsingKeyFrames>

                <ObjectAnimationUsingKeyFrames
                        Storyboard.TargetName="ellipse"
                        Storyboard.TargetProperty="Fill"
                        RepeatBehavior="Forever">
                    <DiscreteObjectKeyFrame KeyTime="0:0:0"
                        Value="{StaticResource ApplicationPageBackgroundThemeBrush}" />
                    <DiscreteObjectKeyFrame KeyTime="0:0:0.2"
                        Value="{StaticResource ApplicationForegroundThemeBrush}" />
                    <DiscreteObjectKeyFrame KeyTime="0:0:0.4"
                        Value="{StaticResource ApplicationPressedForegroundThemeBrush}" />
                    <DiscreteObjectKeyFrame KeyTime="0:0:0.6"
                        Value="{StaticResource ApplicationPageBackgroundThemeBrush}" />
                </ObjectAnimationUsingKeyFrames>
            </Storyboard>
        </BeginStoryboard>
    </EventTrigger>
</Page.Triggers>
</Page>
```

有趣的是，DiscreteObjectKeyFrame 的 Value 属性可以直接设置为枚举项的名称或者设置为 StaticResource，而不会造成类型混淆。

代码隐藏文件可以通过名字来访问某个动画，这是在 Triggers 部分定义 Storyboard 和动画的另一个优点。

项目：FastNotFluid | 文件：MainPage.xaml.cs(片段)

```
void OnCanvasSizeChanged(object sender, SizeChangedEventArgs args)
{
    horzAnima.To = args.NewSize.Width;
    vertAnima.To = args.NewSize.Height;
}
```

9.11　预定义动画和过渡

一开始就说过 Windows.UI.Xaml.Media.Animation 包含 71 个类，如果你一直数着，可能发现现在还没有到这个数字。

除了目前为止提到的类以外，命名空间还包括 14 个预定义动画，均继承自 Timeline，名称以 ThemeAnimation 结尾。这些动画已经设置了属性和目标属性，只需要带 TargetName 属性的目标对象。因此，为了让你试试这些预定义动画，我创建了一个项目，其中有 12 个动画(不包括 SplitOpenThemeAnimation 和 SplitCloseThemeAnimation，不太适合本程序的机制)和其自己的 Storyboard 对象相关联，而 TargetName 设置为名称为“button”的元素：

项目：PreconfiguredAnimations | 文件：MainPage.xaml(片段)

```
<Page ... >
<Page.Resources>
    <Style TargetType="Button">
        <Setter Property="Margin" Value="0 6" />
    </Style>

    <Storyboard x:Key="fadeIn">
        <FadeInThemeAnimation TargetName="button" />
    </Storyboard>

    <Storyboard x:Key="fadeOut">
        <FadeOutThemeAnimation TargetName="button" />
    </Storyboard>

    <Storyboard x:Key="popIn">
        <PopInThemeAnimation TargetName="button" />
    </Storyboard>

    <Storyboard x:Key="popOut">
        <PopOutThemeAnimation TargetName="button" />
    </Storyboard>

    <Storyboard x:Key="reposition">
        <RepositionThemeAnimation TargetName="button" />
    </Storyboard>

    <Storyboard x:Key="pointerUp">
        <PointerUpThemeAnimation TargetName="button" />
    </Storyboard>

    <Storyboard x:Key="pointerDown">
        <PointerDownThemeAnimation TargetName="button" />
    </Storyboard>
```

```
    <Storyboard x:Key="swipeBack">
        <SwipeBackThemeAnimation TargetName="button" />
    </Storyboard>

    <Storyboard x:Key="swipeHint">
        <SwipeHintThemeAnimation TargetName="button" />
    </Storyboard>

    <Storyboard x:Key="dragItem">
        <DragItemThemeAnimation TargetName="button" />
    </Storyboard>

    <Storyboard x:Key="dropTargetItem">
        <DropTargetItemThemeAnimation TargetName="button" />
    </Storyboard>

    <Storyboard x:Key="dragOver">
        <DragOverThemeAnimation TargetName="button" />
    </Storyboard>
</Page.Resources>

<Grid Background="{StaticResource ApplicationPageBackgroundThemeBrush}">
    <Grid.ColumnDefinitions>
        <ColumnDefinition Width="Auto" />
        <ColumnDefinition Width="*" />
    </Grid.ColumnDefinitions>

    <StackPanel Name="animationTriggersStackPanel"
        Grid.Column="0"
        VerticalAlignment="Center">

    <Button Content="Fade In"
        Tag="fadeIn"
        Click="OnButtonClick" />

    <Button Content="Fade Out"
        Tag="fadeOut"
        Click="OnButtonClick" />

    <Button Content="Pop In"
        Tag="popIn"
        Click="OnButtonClick" />

    <Button Content="Pop Out"
        Tag="popOut"
        Click="OnButtonClick" />

    <Button Content="Reposition"
        Tag="reposition"
        Click="OnButtonClick" />

    <Button Content="Pointer Up"
        Tag="pointerUp"
        Click="OnButtonClick" />

    <Button Content="Pointer Down"
        Tag="pointerDown"
        Click="OnButtonClick" />

    <Button Content="Swipe Back"
        Tag="swipeBack"
        Click="OnButtonClick" />

    <Button Content="Swipe Hint"
        Tag="swipeHint"
        Click="OnButtonClick" />
```

```
        <Button Content="Drag Item"
             Tag="dragItem"
             Click="OnButtonClick" />

        <Button Content="Drop Target Item"
             Tag="dropTargetItem"
             Click="OnButtonClick" />

        <Button Content="Drag Over"
             Tag="dragOver"
             Click="OnButtonClick" />
    </StackPanel>

    <!-- Animation target -->
    <Button Name="button"
        Grid.Column="1"
        Content="Big Button"
        FontSize="48"
        HorizontalAlignment="Center"
        VerticalAlignment="Center" />
    </Grid>
    </Page>
```

除了名为“button”的 Button 以外，XAML 文件还为每一个预配置动画定义了 Button。代码隐藏文件使用 Tag 属性触发相应 Storyboard。

```
项目: PreconfiguredAnimations | 文件: MainPage.xaml.cs(片段)
void OnButtonClick(object sender, RoutedEventArgs args)
{
Button btn = sender as Button;
string key = btn.Tag as string;
Storyboard storyboard = this.Resources[key] as Storyboard;
storyboard.Begin();
}
```

小心！其中一些动画会导致目标 Button 消失，而另外一些动画则相当微妙，但你会看到要添加到自己的应用程序的其中某些效果。

预定义动画的另一种设置继承自 Transition 的 8 个类。这些都是更复杂的动画设置，可以设置如下 TransitionCollection 类型的属性：

- UIElement 所定义的 Transitions 属性
- ContentControl 所定义的 ContentTransitions 属性
- Panel 所定义的 ChildrenTransitions 属性
- ItemsControl 所定义的 ItemContainerTransitions 属性

例如，试试用以下语句在 PreconfiguredAnimations 程序更换 StackPanel 标签：

```
<StackPanel Name="animationTriggersStackPanel"
      Grid.Column="0"
           VerticalAlignment="Center">
<StackPanel.ChildrenTransitions>
     <TransitionCollection>
          <EntranceThemeTransition />
     </TransitionCollection>
</StackPanel.ChildrenTransitions>
```

现在，页面加载时，按钮似乎从实际位置平移了一点点并移动到位。

第 11 章和第 12 章将针对这些过渡进行更多讨论。

第 10 章　变　　换

在第 9 章中，你看到了如何使用动画在屏幕上移动对象、改变其大小、颜色或透明度，甚至在虚线上移动点。但是，没有提到某些类型的动画。如果点击按钮，想用动画来旋转按钮，该怎么做？我不是说故意让按钮疯狂旋转，但也许可以轻轻摇晃按钮，好像在说："我根本无法压制热情，就是想要执行你要的命令。"

此项任务(和其他类似任务)所需要的就是变换。过去，变换称为"图形变换"(甚至可能吓跑外行的叫法"矩阵变换")。而近年来，变换已从图形专家的魔掌中获得解放，所有程序员都能用了。

这并不是暗示变换不再和数学相关。(是的，还是和数学有关。)但在 Windows Runtime 中可以使用变换，不涉及变换的数学知识。

10.1　简 短 回 顾

变换基本上就是数学公式，应用于点(x, y)来创建新点(x', y')。如果把同样的公式应用于可视化对象的所有点，则可以有效移动对象，或改变其大小，或者旋转对象，或者甚至以不同的方式扭曲对象。

Windows Runtime 的变换受 UIElement 定义的三项属性支持：RenderTransform、RenderTransformOrigin 和 Projection。这些属性由 UIElement 定义，因此，变换并不像过去一样仅限于矢量图形。可以对任何元素使用转换，包括 Image、TextBlock 和 Button。如果对 Panel 衍生对象应用变换，比如 Grid，则可应用于该 Panel 上的所有子对象。

要把变换应用到元素，可以使用属性元素(property-element)语法把 RenderTransform 属性设置为继承自 Transform 的类实例，例如 RotateTransform。

```
项目: SimpleRotate | 文件: MainPage.xaml(片段)
<Grid Background="{StaticResource ApplicationPageBackgroundThemeBrush}">
<Image Source="http://www.charlespetzold.com/pw6/PetzoldJersey.jpg"
      Stretch="None"
      HorizontalAlignment="Right"
      VerticalAlignment="Bottom">
   <Image.RenderTransform>
      RotateTransform Angle="135" />
   </Image.RenderTransform>
</Image>
</Grid>
```

RotateTransform 的 Angle 属性表示顺时针旋转 135 度，如下图所示。

但结果只是看起来合理，因为我知道 Image 会相对于其左上角旋转，所以有意把 Image 元素定位在页面右下角。二维旋转以总是围绕着某个特定点(就像是把照片钉在软木板上的大头钉)而正确设置该点则是处理变换比较需要技巧的一个方面。

可以把 RenderTransform 属性设置为继承自 Transform 的 7 个类中的任何一个，根据数学复杂程度由简到繁大致如下排列。

Object
　DependencyObject
　　GeneralTransform
　　　Transform
　　　　TranslateTransform
　　　　ScaleTransform
　　　　RotateTransform
　　　　SkewTransform
　　　　CompositeTransform
　　　　MatrixTransform
　　　　TransformGroup

这些类定义了传统二维仿射变换。“仿射”一词暗示已变换对象和未变换对象有密切的联系：一条直线总是变换成另一条直线。位置、大小或方位可能不同，但仍然是一条直线。仿射变换之前平行的直线在变换后仍然保持平行。仿射变换不会导致东西变为无穷大。仿射的数学定义也的确是“保留有限性。”

Windows Runtime 还支持常用于三维透视图的特定类型非仿射。把 UIElement 定义的 Projection 属性设置为派生自 Projection 两个类其中之一的实例，就可以使用 Windows Runtime 实现三维透视图效果。

Object
　DependencyObject
　　Projection

PlaneProjection

Matrix3DProjection

三维旋转总是围绕轴进行。围绕 Y(垂直)轴的旋转如以下 SimpleProjection 项目所示。

```
项目: SimpleProjection | 文件: MainPage.xaml(片段)
<Grid Background="{StaticResource ApplicationPageBackgroundThemeBrush}">
    <Image Source="http://www.charlespetzold.com/pw6/PetzoldJersey.jpg"
           HorizontalAlignment="Center"
           VerticalAlignment="Center">
        <Image.Projection>
            <PlaneProjection RotationY="-60" />
        </Image.Projection>
    </Image>
</Grid>
```

这样就产生了一种截然不同的旋转，看上去像是给屏幕的两个维度增加了第三个维度，如下图所示。

显然，这种变换不会保留平行线，所以看起来好像是在 3D 空间中。

Projection 变换有时也称作伪 3D 变换，旨在为 Windows Runtime 提供 3D 效果。可以定义一个动画，让元素看起来像是一扇门摇晃着进入视线，或者像是扑克牌在翻转，但是元素本身不是 3D 的。Projection 类之一指的是“平面”。基本上是拿一个 2D 元素在 3D 空间内移动。

有数学基础的程序员也许能够让 Matrix3DProjection 在 Windows Runtime 显示实际的 3D 对象。但是 Windows Runtime 会丢失 3D 的一些重要特性，比如基于光源表面渲染以及一个对象部分被另一个对象遮挡时所发生的裁剪。如果要给 Windows 8 应用带来真正的 3D 图形，则需要使用 Direct3D，而 Direct3D 只能从 C++获得，而且(我很难过)超出了本书讨论范围。

10.2 旋　　转

教程里常常通过数学上简单的变换来开始讨论变换：TranslateTransform 移动对象，而 ScaleTransform 变大或变小对象。但这些都不会令人印象非常深刻，因为你已经看过用动画在屏幕上移动对象或改变大小。所以，我要用一些其他方法做不到的事情来进行讨论。

我刚刚演示了可以直接在 XAML 里设置 RotateTransform 的 Angle 属性，但是用数据绑定或动画动态改变 Angle 属性更加有趣，而且其结果更能揭示实际要发生的事情。如下 XAML 文件所示，RotateTransform 的 Angle 属性值绑定了 Slider 的 Value 属性，范围从 0 到 360。

```
项目: RoateTheText | 文件: MainPage.xaml(片段)
<Grid Background="{StaticResource ApplicationPageBackgroundThemeBrush}">
    <Border BorderBrush="{StaticResource ApplicationForegroundThemeBrush}"
            BorderThickness="1"
            HorizontalAlignment="Center"
            VerticalAlignment="Center">
        <Grid>
            <Grid.RowDefinitions>
                <RowDefinition Height="Auto" />
                <RowDefinition Height="Auto" />
            </Grid.RowDefinitions>

            <Slider Name="slider"
                    Grid.Row="0"
                    Minimum="0"
                    Maximum="360" />

            TextBlock Text="Rotate Text with Slider"
                      Grid.Row="1"
                      FontSize="48">
            <TextBlock.RenderTransform>
                <RotateTransform Angle="{Binding ElementName=slider, Path=Value}" />
            </TextBlock.RenderTransform>
        </TextBlock>
    </Grid>
    </Border>
</Grid>
```

Slider 和 TextBlock 占据 Grid 的两行，而 Grid 在 Border 中。屏幕第一次打开时如下图所示。

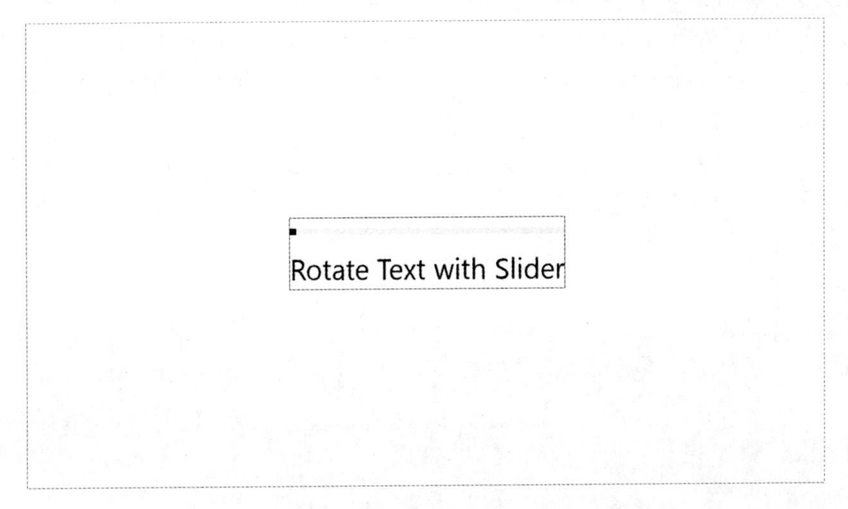

TextBlock 宽度决定了 Grid 宽度，而后者又决定了 Slider 宽度和 Border 宽度。

如果用鼠标或手指改变 Slider 值，TextBlock 会顺时针方向旋转。如果旋转 120 度，结果如下图所示。

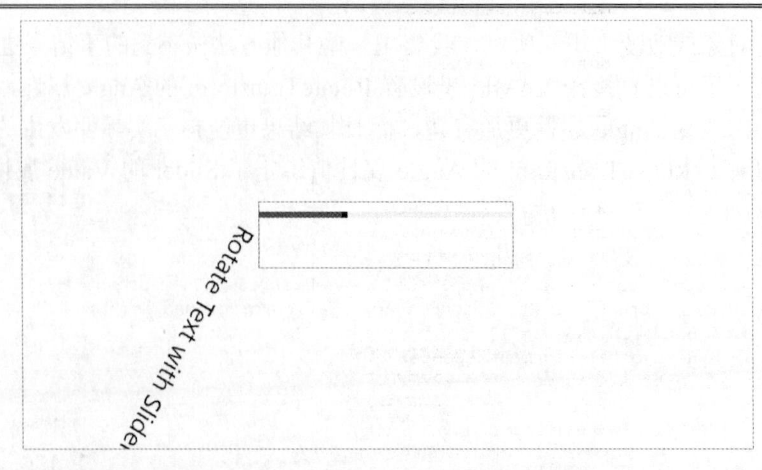

很明显，Grid 和 Border 的大小继续基于未旋转的 TextBlock，而已旋转的 TextBlock 则打破了可视树上其祖先的边界。

设置给 RotateTransform 的 UIElement 属性称为 RenderTransform，需要仔细考虑一下该属性名。“render”(渲染)一词意味着转换影响的只是元素的渲染方式，而不是元素的布局系统。这里既有好事也有坏事。

变换发生在图形组成系统的较深层次，这是好事。旋转 TextBlock 不要求整个视觉树受到更新布局的影响。布局系统不参与变换，因此，变换动画可以发生在辅助线程中，而且性能很好。布局系统完全不关注 TextBlock 正在旋转。

而坏事是，布局系统完全不知道 TextBlock 正在旋转。例如，想要旋转 90 度来显示一个侧向的 TextBlock，也许是对图的旁注。如果布局系统能计算 TextBlock 的旋转维度，则可以直接将其放在入 Gird 的单元格中，并用 Grid 正确定位，这样最方便。但 Windows Runtime 不太容易实现。

相比之下，Windows Presentation Foundation(WPF)所提供的 UIElement 版本既定义了 RenderTransform 属性(像 Windows Runtime 一样)，也定义了 LayoutTransform 属性，后者允许指定布局系统可识别的变换。WPF 在向 Silverlight 和 Windows Runtime 过渡时失去了 LayoutTransform 属性，模仿它需要花一点气力。

我们回到正在运行的 RotateTheText 程序。操纵 Slider，以便 TextBlock 可以有一部分位于 Slider 上方，如下图所示。

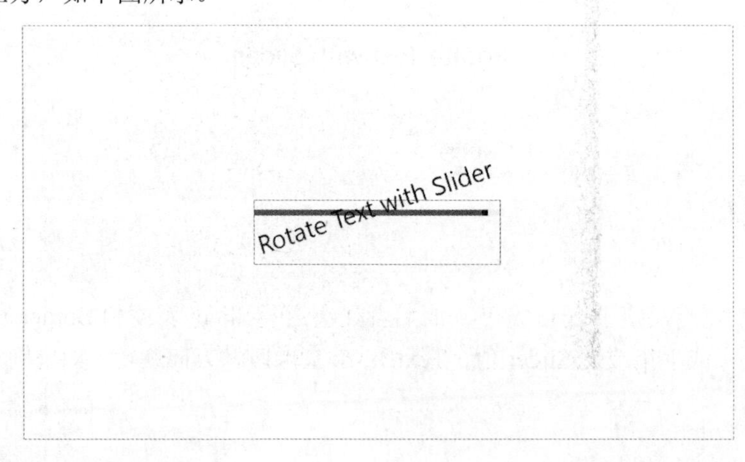

现在，从屏幕上拿开所有手指(或释放鼠标按钮)，试试触摸或点击 Slider 和 TextBlock 重叠的任何点。Slider 没有反应，因为 TextBlock 阻碍了鼠标或触摸输入。这里有个经验：尽管布局系统不知道 TextBlock 已经移动了，但点击测试逻辑会持续关注 TextBlock 的准确位置。(另一方面，实际操作 Slider 的过程中，TextBlock 不会妨碍操作，因为 Slider 捕获了操作输入，第 13 章将讨论捕获概念。)

你还会注意到，TextBlock 的旋转是相对于其左上角，从概念上讲是 TextBlock 的源点(0，0)。在许多图形系统中，图形变换通常是相对于画布的源点，而图形对象则定位在画布上。在 Windows Runtime 中，所有变换都是相对于其应用的元素。

通常都会偏好旋转相对于某个点，而不是相对于左上角。该点有时称为“旋转中心”，可以用三种不同方法来指定。

第一种方法是变换所蕴含数学的最启发性方法，但以后再讲。

第二种方法涉及 RotateTransform 类本身。该类定义了 CenterX 和 CenterY 属性，默认值为 0。如果想要该 TextBlock 相对于其中心旋转，可以把 CenterX 设置为 TextBlock 宽度的一半，而把 CenterY 设置为其高度的一半。Loaded 处理程序可以获得此信息，所以可以添加以下类似代码到后台代码文件的构造函数中。我给 TextBlock 的名称幸好还没有在 XAML 文件中使用过：

```
public MainPage()
{
    this.InitializeComponent();

    Loaded += (sender, args) =>
        {
            RotateTransform rotate = txtblk.RenderTransform as RotateTransform;
            rotate.CenterX = txtblk.ActualWidth / 2;
            rotate.CenterY = txtblk.ActualHeight / 2;
        };
}
```

你可能会觉得这种方法有点麻烦，因此会高兴地发现第三种方法更简单。该方法涉及 UIElement 所定义的 RenderTransformOrigin 属性。该属性是 Point 类型，但将其设置为相对坐标点，通常是范围从 0 到 1 的 X 和 Y 值。默认值为点(0, 0)，即左上角。点(1, 0)为右上角，点(0, 1)为左下角，点(1, 1)为右下角。要指定元素中心的源点，可以使用点(0.5, 0 .5)：

```
<TextBlock Name="txtblk"
           ext="Rotate Text with Slider"
           Grid.Row="1"
           FontSize="48"
           RenderTransformOrigin="0.5 0.5">
    <TextBlock.RenderTransform>
        <RotateTransform Angle="{Binding ElementName=slider, Path=Value}" />
    </TextBlock.RenderTransform>
</TextBlock>
```

注意，CenterX 和 CenterY 是 RotateTransform 的属性，但 RenderTransformOrigin 属性则由 UIElement 所定义，并为全部元素所共有。如果除了 CenterX 和 CenterY,，还设置了 RenderTransformOrigin，则会产生混合效果。在本例中，两个例子的混合效果导致 TextBox 围绕着其右下角旋转。

可以指定位于元素以外的旋转中心。如下 XAML 文件将 TextBlock 定位在页面顶部中心，并启动 Forever 动画来旋转它。

```
项目: RotateAroundCenter | 文件: MainPage.xaml(片段)
<Page ... >
    <Grid Background="{StaticResource ApplicationPageBackgroundThemeBrush}">
        <TextBlock Name="txtblk"
                   Text="Rotated Text"
                   FontSize="48"
                   HorizontalAlignment="Center"
                   VerticalAlignment="Top">
            <TextBlock.RenderTransform>
                <RotateTransform x:Name="rotate" />
            </TextBlock.RenderTransform>
        </TextBlock>
    </Grid>

    <Page.Triggers>
        <EventTrigger>
            <BeginStoryboard>
                <Storyboard RepeatBehavior="Forever">
                    <DoubleAnimation Storyboard.TargetName="rotate"
                                     Storyboard.TargetProperty="Angle"
                                     From="0" To="360" Duration="0:0:2" />
                </Storyboard>
            </BeginStoryboard>
        </EventTrigger>
    </Page.Triggers>
</Page>
```

不需要任何额外代码，该程序就会绕着左上角旋转 TextBlock，并在动画期间的某些时刻清除屏幕。隐藏代码文件里的构造函数定义了两个事件处理程序以设置 RotateTransform 的 CenterX 和 CenterY 属性。

```
项目: RotateAroundCenter | 文件: MainPage.xaml.cs(片段)
public sealed partial class MainPage : Page
{
    public MainPage()
    {
        this.InitializeComponent();

        Loaded += (sender, args) =>
            {
                rotate.CenterX = txtblk.ActualWidth / 2;
            };

        SizeChanged += (sender, args) =>
            {
                rotate.CenterY = args.NewSize.Height / 2;
            };
    }
}
```

旋转中心设置为与 TextBlock 水平居中而又低于 TextBlock、等于页面高度一半的一个点。其结果是 TextBlock 绕着页面中心做圆周运动，如下图所示。

10.3 可视化反馈

动画变换能有效提醒用户屏幕上的情况，而这些情况要求关注或确认操作已启动。在 JiggleButtonDemo 程序中，我添加了一个新的 UserControl，名为 JiggleButton，但后来我把 XAML 和 C#文件中的基类从 UserControl 改成了 Button。以下是完整的 JiggleButton.xaml 文件。

```
项目: JiggleButtonDemo | 文件: JiggleButton.xaml
<Button
    x:Class="JiggleButtonDemo.JiggleButton"
    xmlns="http://schemas.microsoft.com/winfx/2006/xaml/presentation"
    xmlns:x="http://schemas.microsoft.com/winfx/2006/xaml"
    RenderTransformOrigin="0.5 0.5"
    Click="OnJiggleButtonClick">

    <Button.Resources>
        <Storyboard x:Key="jiggleAnimation">
            <DoubleAnimation Storyboard.TargetName="rotate"
                             Storyboard.TargetProperty="Angle"
                             From="0" To="10" Duration="0:0:0.33"
                             AutoReverse="True">
                <DoubleAnimation.EasingFunction>
                    <ElasticEase EasingMode="EaseIn" />
                </DoubleAnimation.EasingFunction>
            </DoubleAnimation>
        </Storyboard>
    </Button.Resources>

    <Button.RenderTransform>
        <RotateTransform x:Name="rotate" />
    </Button.RenderTransform>
</Button>
```

该 XAML 文件没有定义 Button 的内容，而是设置了三个 Button 属性：RenderTransformOrigin(在根标签里)、Resources 和 RenderTransform。一般情况下，如果想用旋转来摇晃元素，则需要用关键帧，因为首先要从 0 度旋转到 10 度，然后从 10 度到-10 度，并来回几次，然后再回到 0 度。但使用带有 EaseIn 的 EasingMode 的 ElasticEase 是另外一个好办法。DoubleAnimation 定义为旋转按钮 10 度然后回 0，但 ElasticEase 函数包含宽幅的反向摆动，所以动画范围实际上是从-10 到 10 度。

JiggleButton 的隐藏代码文件直接在 Click 事件处理程序中触发动画。

```
项目: JiggleButtonDemo | 文件: JiggleButton.xaml.cs
using Windows.UI.Xaml;
using Windows.UI.Xaml.Controls;
using Windows.UI.Xaml.Media.Animation;

namespace JiggleButtonDemo
{
    public sealed partial class JiggleButton : Button
    {
        public JiggleButton()
        {
            this.InitializeComponent();
        }

        void OnJiggleButtonClick(object sender, RoutedEventArgs args)
        {
```

```
            (this.Resources["jiggleAnimation"] as Storyboard).Begin();
        }
    }
}
```

MainPage.xaml 实例化 JiggleButton，以便使用：

```
项目: JiggleButtonDemo | 文件 MainPage.xaml(片段)
<Grid Background="{StaticResource ApplicationPageBackgroundThemeBrush}">
    <local:JiggleButton Content="JiggleButton Demo"
                        FontSize="24"
                        HorizontalAlignment="Center"
                        VerticalAlignment="Center" />
</Grid>
```

记住，JiggleButton 派生自 Button，可以像用其他 Button 一样来使用 JiggleButton，不过不应该对其设置 RenderTransform 或 RenderTransformOrigin 属性，否则会干扰摇晃动画。

10.4 平　　移

TranslateTransform 定义 X 和 Y 两个属性，其结果是元素渲染偏移原位置。TranslateTransform 有一个简单应用，用“浮雕”、“雕刻”或阴影来显示文本，如下图所示。

EMBOSS

ENGRAVE

Drop Shadow

光线一般来自上部，而且也许因为我们习惯于左上方光源照亮电脑屏幕上的 3D 对象的惯例，因此，顶部文本看起来在右边和底部都有阴影，这些字母就像从屏幕向外投射。雕刻效果则相反：阴影在左上边上顶部，字母似乎是雕刻而成。

显示这三个文本字符串的页面实际上由 6 个 TextBlock 元素组成。第一对中有一个 TextBlock 带默认前景刷，被另一个带默认背景刷的 TextBlock 所覆盖，后者在纵横方向上位移 2 个像素。

```
项目: TextEffects | 文件: MainPage.xaml(片段)
<Page ... >
    <Page.Resources>
        <Style TargetType="TextBlock">
            <Setter Property="FontFamily" Value="Times New Roman" />
            <Setter Property="FontSize" Value="192" />
            <Setter Property="HorizontalAlignment" Value="Center" />
            <Setter Property="VerticalAlignment" Value="Center" />
        </Style>
    </Page.Resources>
```

```
    <Grid Background="{StaticResource ApplicationPageBackgroundThemeBrush}">
        <Grid.RowDefinitions>
            <RowDefinition Height="*" />
            <RowDefinition Height="*" />
            <RowDefinition Height="*" />
        </Grid.RowDefinitions>

        <TextBlock Text="EMBOSS"
                   Grid.Row="0" />

        <TextBlock Text="EMBOSS"
                   Grid.Row="0"
                   Foreground="{StaticResource ApplicationPageBackgroundThemeBrush}">
            <TextBlock.RenderTransform>
                <TranslateTransform X="-2" Y="-2" />
            </TextBlock.RenderTransform>
        </TextBlock>

        <TextBlock Text="ENGRAVE"
                   Grid.Row="1" />
        <TextBlock Text="ENGRAVE"
                   Grid.Row="1"
                   Foreground="{StaticResource ApplicationPageBackgroundThemeBrush}">
            <TextBlock.RenderTransform>
                <TranslateTransform X="2" Y="2" />
            </TextBlock.RenderTransform>
        </TextBlock>

        <TextBlock Text="Drop Shadow"
                   Grid.Row="2"
                   Foreground="Gray">
            <TextBlock.RenderTransform>
                <TranslateTransform X="6" Y="6" />
            </TextBlock.RenderTransform>
        </TextBlock>

        <TextBlock Text="Drop Shadow"
                   Grid.Row="2" />
    </Grid>
</Page>
```

注意，浮雕效果需要负偏移(因此顶部的 TextBlock 向左向上移动)，而雕刻效果则要求正偏移。如果在暗色主题中使用这些相同效果，可能不太明显，但必须交换 X 值和 Y 值的符号。

阴影效果与此相类似，只不过一般顶部文本着色，而下方有灰色阴影偏移。

我不推荐长期使用以下方法，但可以把屏幕上的文字稍稍加深一点，即视觉深度而不是智力深度，使用一堆 TextBlock 元素的后果是因为一个像素而产生偏移，如下图所示。

这些元素的生成处理完全由代码隐藏文件来处理。

```
项目: DepthText | 文件: MainPage.xaml.cs(片段)
public sealed partial class MainPage : Page
{
    const int COUNT = 48; // ~1/2 inch

    public MainPage()
    {
        this.InitializeComponent();

        Grid grid = this.Content as Grid;
        Brush foreground = this.Resources["ApplicationForegroundThemeBrush"] as Brush;
        Brush grayBrush = new SolidColorBrush(Colors.Gray);

        for (int i = 0; i < COUNT; i++)
        {
            bool firstOrLast = i == 0 || i == COUNT - 1;

            TextBlock txtblk = new TextBlock
            {
                Text = "DEPTH",
                FontSize = 192,
                FontWeight = FontWeights.Bold,
                HorizontalAlignment = HorizontalAlignment.Center,
                VerticalAlignment = VerticalAlignment.Center,
                RenderTransform = new TranslateTransform
                {
                    X = COUNT - i - 1,
                    Y = i - COUNT + 1,
                },
                Foreground = firstOrLast ? foreground : grayBrush
            };
            grid.Children.Add(txtblk);
        }
    }
}
```

TranslateTransform 不错，能将东西从布局系统所定位的位置移开一点点。StandardStyles.xaml 文件有好几个例子都是通过这种方式来使用 TranslateTransform 的。

第 9 章有一个例子，用 Canvas.Left 和 Canvas.Top 附加属性在屏幕上移动对象。通过定义一个相同类型的动画，可以在希望移动的元素中定义 TranslateTransform 元素，并使动画以 X 和 Y 属性为目标。这样做有一个优点，动画元素不必是 Canvas 的子对象，但在性能上似乎并没有差异。这两种动画都可以在辅助线程中执行。

10.5 变换组

前面提到过，有三种方法可以设置中心旋转，但我把第一种方法留到以后再讨论，现在是讨论的时候了。第一种方法比较复杂一点，因为涉及到由其他变换所构成的变换。

TransformGroup 派生自 Transform，TransformGroup 有一个名为 Children 的 TransformCollection 类型属性，可以用它来从多个 Transform 派生构建复合变换。

可以定义如下类似的 RotateTransform。

```
<RotateTransform Angle="A" CenterX="CX" CenterY="CY" />
```

A、CX 和 CY 可以是实际数字，也可能是数据绑定。其变换等同于以下 TransformGroup。

```
<TransformGroup>
    <TranslateTransform X="-CX" Y="-CY" />
    <RotateTransform Angle="A" />
    <TranslateTransform X="CX" Y="CY" />
</TransformGroup>
```

两个 TranslateTransform 标记似乎相互抵消，但两者围绕着一个 RotateTransform。我用两种方法来演示该变换组本身相当于第一个 RotateTransform。

以下 ImageRotate 程序引用了我网站上的一张位图，320 像素宽，400 像素高。为了使其围绕中心旋转，RotateTransform 通常将 CenterX 和 CenterY 设置成一半值(即 160 和 200)，而我则使用了一对 TranslateTransform 对象。

```
项目: ImageRotate | 文件: MainPage.xaml(片段)
<Page ... >
    <Grid Background="{StaticResource ApplicationPageBackgroundThemeBrush}">
        <Image Source="http://www.charlespetzold.com/pw6/PetzoldJersey.jpg"
               Stretch="None"
               HorizontalAlignment="Center"
               VerticalAlignment="Center">
            <Image.RenderTransform>
                <TransformGroup>
                    <TranslateTransform X="-160" Y="-200" />
                    <RotateTransform x:Name="rotate" />
                    <TranslateTransform X="160" Y="200" />
                </TransformGroup>
            </Image.RenderTransform>
        </Image>
    </Grid>

    <Page.Triggers>
        <EventTrigger>
            <BeginStoryboard>
                <Storyboard RepeatBehavior="Forever">
                    <DoubleAnimation Storyboard.TargetName="rotate"
                                     Storyboard.TargetProperty="Angle"
                                     From="0" To="360" Duration="0:0:3">
                        <DoubleAnimation.EasingFunction>
                            <ElasticEase EasingMode="EaseInOut" />
                        </DoubleAnimation.EasingFunction>
                    </DoubleAnimation>
                </Storyboard>
            </BeginStoryboard>
        </EventTrigger>
    </Page.Triggers>
</Page>
```

带 EaseInOut 模式的 ElasticEase 动画导致图片在实际旋转前后疯狂地来回摇摆，但可以清楚地看到旋转是围绕着图片中心进行的。如下图所示。

下图显示了此过程的各个步骤。最亮的 TextBlock 放置在页面中心位置。下一个最黑的 TextBlock 显示 TranslateTransform 的效果，后者移动 TextBlock，往左一半宽度，往上一半高度。下一个最黑的 TextBlock 相对于其源点旋转，也就是最初的 TextBlock 左上角。最后一个黑色 TextBlock 一半宽度、一半高度移动。最终结果就是最初的 TextBlock 围绕其中心旋转。

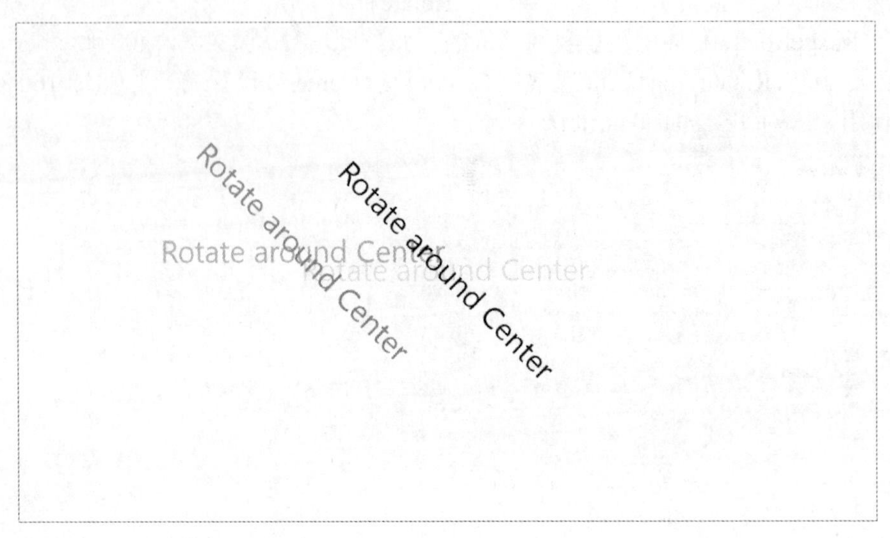

上图的 XAML 文件如下所示。

项目: RotationCenterDemo | 文件: MainPage.xaml(片段)
```
<Page ... >
    <Page.Resources>
        <Style TargetType="TextBlock">
            <Setter Property="Text" Value="Rotate around Center" />
            <Setter Property="FontSize" Value="48" />
            <Setter Property="HorizontalAlignment" Value="Center" />
            <Setter Property="VerticalAlignment" Value="Center" />
        </Style>
    </Page.Resources>

    <Grid Background="{StaticResource ApplicationPageBackgroundThemeBrush}">
        <TextBlock Name="txtblk"
                   Foreground="#D0D0D0" />

        <TextBlock Foreground="#A0A0A0">
            <TextBlock.RenderTransform>
                <TranslateTransform x:Name="translateBack1" />
            </TextBlock.RenderTransform>
        </TextBlock>

        <TextBlock Foreground="#707070">
            <TextBlock.RenderTransform>
                <TransformGroup>
                    <TranslateTransform x:Name="translateBack2" />
                    <RotateTransform Angle="45" />
                </TransformGroup>
            </TextBlock.RenderTransform>
        </TextBlock>

        <TextBlock Foreground="{StaticResource ApplicationForegroundThemeBrush}">
            <TextBlock.RenderTransform>
                <TransformGroup>
                    <TranslateTransform x:Name="translateBack3" />
                    <RotateTransform Angle="45" />
```

```
                    <TranslateTransform x:Name="translate" />
                </TransformGroup>
            </TextBlock.RenderTransform>
        </TextBlock>
    </Grid>
</Page>
```

Loaded 处理程序设置所有 TranslateTransform 标记的 X 和 Y 值：

```
项目: RotationCenterDemo | 文件: MainPage.xaml.cs(片段)
public MainPage()
{
    this.InitializeComponent();

    Loaded += (sender, args) =>
        {
            translateBack1.X =
            translateBack2.X =
            translateBack3.X = -(translate.X = txtblk.ActualWidth / 2);
            translateBack1.Y =
            translateBack2.Y =
            translateBack3.Y = -(translate.Y = txtblk.ActualHeight / 2);
        };
}
```

变换可以结合一些非常有趣的效果，超出非数学、非图形程序员的领域。如下 XAML 文件使用 Polygon 元素定义了一个简易的螺旋桨形状并对其他应用三个变换。RotateTransform、TranslateTransform 和另一个 RotateTransform。

```
项目: Propeller | 文件: MainPage.xaml(片段)
<Page ... >
    <Grid Background="{StaticResource ApplicationPageBackgroundThemeBrush}">
        <Polygon Points=" 40   0,   60   0,  53 47,
                         100  40,  100  60,  53 53,
                          60 100,   40 100,  47 53,
                           0  60,    0  40,  47 47"
                 Stroke="{StaticResource ApplicationForegroundThemeBrush}"
                 Fill="SteelBlue"
                 HorizontalAlignment="Center"
                 VerticalAlignment="Center"
                 RenderTransformOrigin="0.5 0.5">
            <Polygon.RenderTransform>
                <TransformGroup>
                    <RotateTransform x:Name="rotate1" />
                    <TranslateTransform X="300" />
                    <RotateTransform x:Name="rotate2" />
                </TransformGroup>
            </Polygon.RenderTransform>
        </Polygon>
    </Grid>

    <Page.Triggers>
        <EventTrigger>
            <BeginStoryboard>
                <Storyboard>
                    <DoubleAnimation Storyboard.TargetName="rotate1"
                                     Storyboard.TargetProperty="Angle"
                                     From="0" To="360" Duration="0:0:0.5"
                                     RepeatBehavior="Forever" />

                    <DoubleAnimation Storyboard.TargetName="rotate2"
                                     Storyboard.TargetProperty="Angle"
                                     From="0" To="360" Duration="0:0:6"
                                     RepeatBehavior="Forever" />
                </Storyboard>
            </BeginStoryboard>
```

```
        </EventTrigger>
    </Page.Triggers>
</Page>
```

Storyboard 包含两个 DoubleAnimation 对象。第一个 DoubleAnimation 以第一个 RotateTransform 对象为目标，旋转螺旋桨围绕其中心自转，速度为每秒 2 圈。TranslateTransform 移动旋转的螺旋桨到离页面右边中心 300 像素的地方，而第二个 DoubleAnimation 以第二个 RotateTransform 为目标，再次旋转螺旋桨。但该旋转则相对于螺旋桨原来的中心，也就是说，螺旋桨围绕着页面中心做圆周运动，半径为 300 像素，速度为每分钟 10 转。如下图所示。

现在，RenderTransformOrigin 的作用也许就清楚了：RenderTransformOrigin 相当于通过变换前 RenderTransform 属性所指定的负 X 值和 Y 值执行一个 TranslateTransform，并通过 RenderTransform 之后的 X 值和 Y 值执行另一个 TranslateTransform。

10.6 缩放变换

ScaleTransform 类定义 ScaleX 和 ScaleY 属性，两者用于在水平和垂直方向独立增加或减少元素的大小。如果想保留目标的正确长宽比，ScaleX 和 ScaleY 则需要使用相同的值。如果是动画，则需要两个动画对象。

ScaleTransform 并不影响元素的 ActualWidth 和 ActualHeight 属性。

你已经看到了如何使用 Viewbox 以破坏字体正确宽高比的方式来拉伸 TextBlock。而如何用 ScaleTransform 来实现，则如下所示。

```
项目: OppositelyScaledText | 文件: MainPage.xaml(片段)
<Page ... >
    <Grid Background="{StaticResource ApplicationPageBackgroundThemeBrush}">
        <TextBlock Text="Scaled Text"
                   FontSize="144"
                   HorizontalAlignment="Center"
                   VerticalAlignment="Center"
                   RenderTransformOrigin="0.5 0.5">
            <TextBlock.RenderTransform>
```

```
                <ScaleTransform x:Name="scale" />
            </TextBlock.RenderTransform>
        </TextBlock>
    </Grid>

    <Page.Triggers>
        <EventTrigger>
            <BeginStoryboard>
                <Storyboard>
                    <DoubleAnimation Storyboard.TargetName="scale"
                                     Storyboard.TargetProperty="ScaleX"
                                     BeginTime="0:0:2"
                                     From="1" To="0.01" Duration="0:0:2"
                                     AutoReverse="True"
                                     RepeatBehavior="Forever" />

                    <DoubleAnimation Storyboard.TargetName="scale"
                                     Storyboard.TargetProperty="ScaleY"
                                     From="10" To="0.1" Duration="0:0:2"
                                     AutoReverse="True"
                                     RepeatBehavior="Forever" />
                </Storyboard>
            </BeginStoryboard>
        </EventTrigger>
    </Page.Triggers>
</Page>
```

实际上，我并不是很想用这种方式来写这个程序。我原本给 TextBlock 设置了 FontSize 大小为 1，然后以动画形式把 ScaleX 从 1 变到 144、ScaleY 从 144 变到 1，两者再倒过来，一直重复下去。这么做应该有用，但会导致 144 因子增加 1 像素高度，而不是变成 144 像素高的字体。为了能使程序按照我想要的方式工作，我把 TextBlock 设置为 144 像素大小，并用动画效果彼此叠加。TextBlock 在水平和垂直交替拉伸，如下图所示。

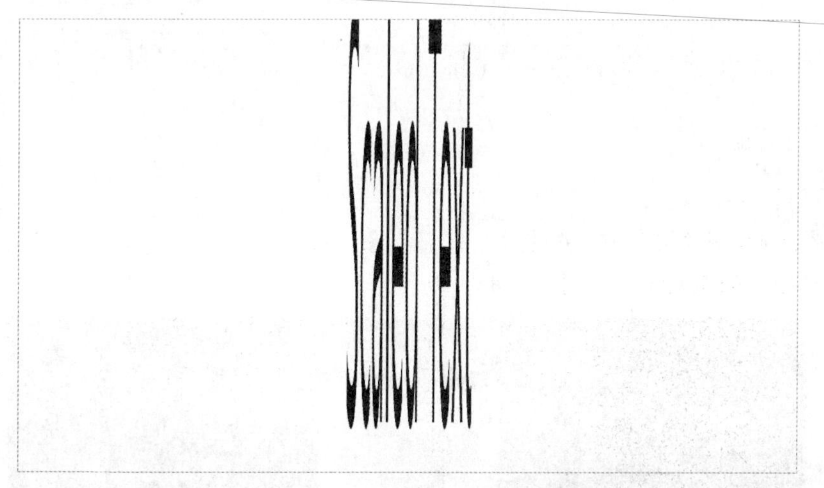

缩放与旋转有一些相似，总是参照中心点。就像 RotateTransform 一样，ScaleTransform 类定义了 CenterX 和 CenterY 属性，或者像我在 OppositelyScaledText 程序里做的一样，也可以设置 RenderTransformOrigin。缩放中心是发生缩放时依然保持在同一位置的点。

缩放和旋转中心在手指操作屏幕对象(如照片)的过程中发挥着重要作用。伸缩、捏压和旋转照片时，缩放和旋转中心会随着手指之间的相互移动而改变。第 13 章会讨论计算旋转中心的技巧。

负缩放因子会围绕水平轴或垂直轴翻转元素。这一技巧对于创建反射效果尤其有用。

不幸的是，Windows Runtime 缺少此效果的一个重要因素：Brush 类型名为 OpacityMask 的 UIElement 属性，OpacityMask 允许定义基于渐变刷颜色的阿尔法通道的渐变透明。在 Windows Runtime 中，必须通过用一个元素遮挡另一种元素来模拟渐变淡出，而后一个元素则有包含透明度和背景颜色的渐变刷。

如以下 ReflectedFadeOutImage 项目所示，Grid 的上半部分由两项共享：一个 Image 和另一个 Grid。第二个 Grid 包含被一个 Recangle 覆盖的相同 Image，该 Recangle 的 LinearGradientBrush 从顶部的背景色渐变到底部的透明色。

项目 Project: ReflectedFadeOutImage | 文件: MainPage.xaml(片段)

```
<Page ... >
    <Grid Background="{StaticResource ApplicationPageBackgroundThemeBrush}">
        <Grid.RowDefinitions>
            <RowDefinition Height="*" />
            <RowDefinition Height="*" />
        </Grid.RowDefinitions>

        <Image Source="http://www.charlespetzold.com/pw6/PetzoldJersey.jpg"
               HorizontalAlignment="Center" />

        <Grid RenderTransformOrigin="0 1"
              HorizontalAlignment="Center">
          <Grid.RenderTransform>
                <ScaleTransform ScaleY="-1" />
          </Grid.RenderTransform>
          <Image Source="http://www.charlespetzold.com/pw6/PetzoldJersey.jpg" />
          <Rectangle>
             <Rectangle.Fill>
                  <LinearGradientBrush StartPoint="0 0" EndPoint="0 1" >
                       <GradientStop Offset="0"
                            Color="{Binding
                                 Source={StaticResource ApplicationPageBackgroundThemeBrush},
                                 Path=Color}" />
                            <GradientStop Offset="1" Color="Transparent" />
                       </LinearGradientBrush>
                  </Rectangle.Fill>
             </Rectangle>
        </Grid>
    </Grid>
</Page>
```

如下图所示，内部的 Grid 也反射在底部边缘。RenderTransformOrigin 在左下方分配了一个变换中心，ScaleTransform 把 ScaleY 设置为-1，将围绕水平轴翻转元素。

在第 14 章中，我将演示另一种方法，通过访问位图像素和适当设置透明度来实现这种效果。

10.7　建立模拟时钟

模拟时钟是圆形的。这一项简单的事实意味着如果使用任意坐标(也就是说，坐标不是以像素为单位，而是你方便选择的单位)以中心为原点，从数学上讲画时钟可能是最简单的。选择以中心原点还意味着可能并不需要弄乱给 RotateTransform 对象的 CenterX 或 CenterY 设置，而 RotateTransform 对象是定位时钟的指针，因为原点也是旋转中心。

在图形环境中，无论赋予的大小，传统模拟时钟都能够自适应。很容易想用 Viewbox 来完成这项工作，但对于模拟时钟而言，这么做可能会有问题。布局系统(和 Viewbox)认为矢量图形对象的大小会取决于坐标点上的最大 X 和 Y 值。负坐标会被忽略，包括以中心原点的四分之三个模拟时钟。

布局系统(和 Viewbox)无法正确确定带负坐标的图形对象大小，它需要一点点“帮助”。变换幸好能从父级传递到子级。可以在一个 Grid 设置一个转换，并将其应用到 Grid 中的所有对象。而 Grid 里的内容可以有自己的变换。

我在 AnalogClock 程序中就是这么做的。Grid 里的所有图形都是固定尺寸，200 像素的宽度和高度，即半径为 100 像素：

```
<Grid Width="200" Height="200">

    ... clock graphics go here

</Grid>
```

Grid 里有五个 Path 元素，用来渲染时钟圆周的刻度线，同时也渲染时针、分针、秒针。这些都是基于坐标系统，X 和 Y 值从-100 到 100。如果能看到 Grid 网格(红色标出)和时钟，效果如下图所示。

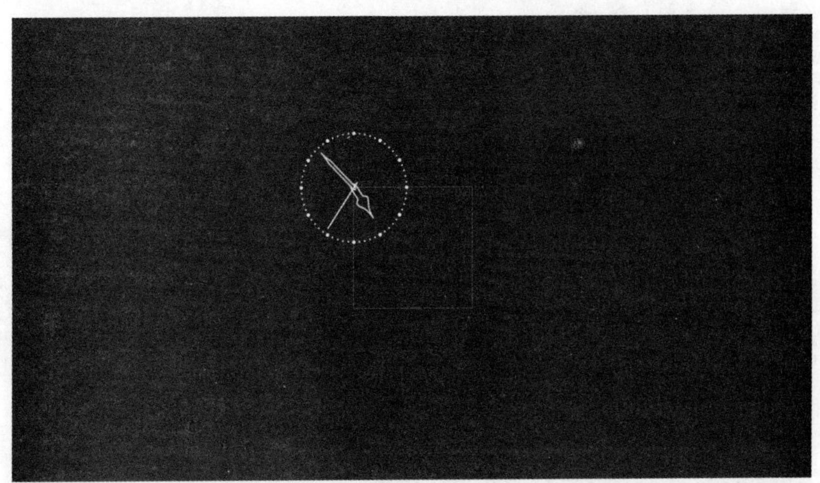

由于默认对齐，所以 Grid 定位在页面中心，但时钟的中心定位在 Grid 的左上角，因为点(0，0)在那里。

现在，把 Grid 放入 Viewbox，如下所示。

```
<Viewbox>
    <Grid Width="200" Height="200">

        ... clock graphics go here

    </Grid>
</Viewbox>
```

Viewbox 可以正确处理原点位于左上角的元素，但不能处理带负坐标的图形(见下图)。

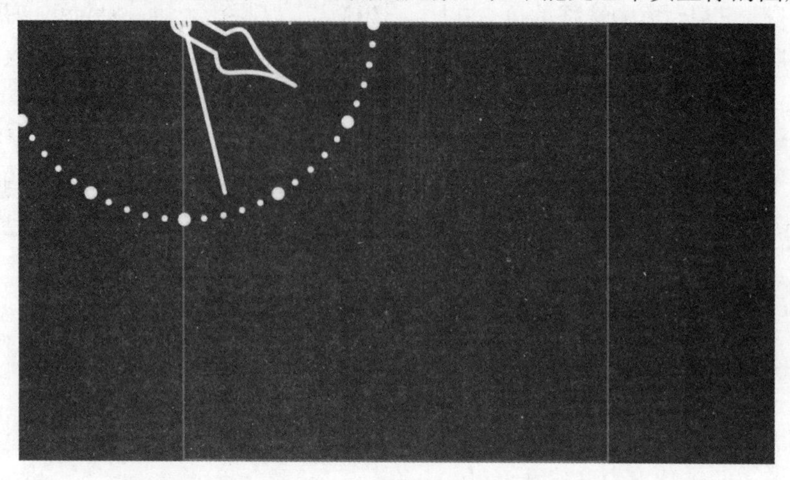

幸运的是，很容易解决这个问题。所要做的就是移动 Grid 和时钟。Viewbox 找到该元素，然后再发生变换，所以仅仅是 100 像素：

```
<Viewbox>
    <Grid Width="200" Height="200">
        <Grid.RenderTransform>
            <TranslateTransform X="100" Y="100" />
        </Grid.RenderTransform>

        ... clock graphics go here

    </Grid>
</Viewbox>
```

如下图所示。

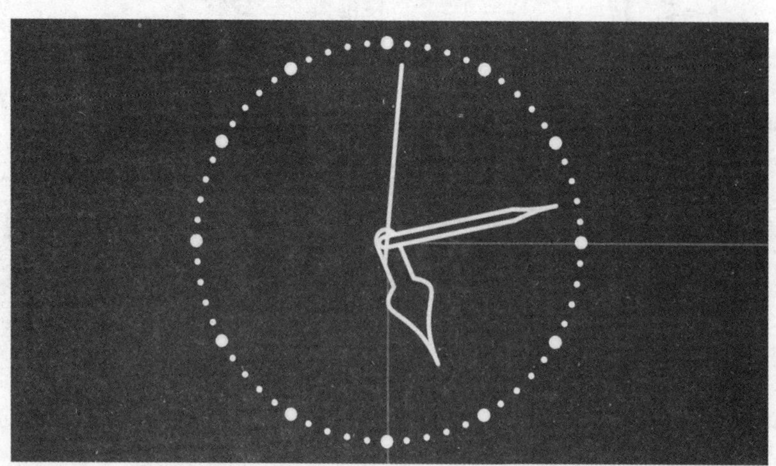

现在，我们要去掉红色边界。

时钟包含五个 Path 元素。三个指针均由包含直线和贝塞尔曲线的路径标记语法定义。时针要指向 12:00 点位置。时针最初主要是在时钟的上半部分，因此当它围绕着中心循环时，其大部分是负 Y 坐标，只有少数正 Y 坐标。

```
<Path Data="M 0 -60 C 0 -30, 20 -30, 5 -20 L 5 0
                    C 5 7.5, - 5 7.5, -5 0 L -5 -20
                    C -20 -30, 0 -30, 0 -60">
    <Path.RenderTransform>
        <RotateTransform x:Name="rotateHour" />
    </Path.RenderTransform>
</Path>
```

刻度线实际上是虚线。以下是小刻度线的 Path 元素。

```
<Path Fill="{x:Null}"
        StrokeThickness="3"
        StrokeDashArray="0 3.14159">
    <Path.Data>
        <EllipseGeometry RadiusX="90" RadiusY="90" />
    </Path.Data>
</Path>
```

这样便创建一个半径为 90 的圆，周长为 2π90，也就是说，60 刻度被 3π分隔，这并非巧合，既是 StrokeThickness 的产品，又是 StrokeDashArray 里的数字，表明 StrokeThickness 中单位点之间的距离。

有了以上的小刻度线 Path，如下是大刻度 Path。

```
<Path Fill="{x:Null}"
        StrokeThickness="6"
        StrokeDashArray="0 7.854">
    <Path.Data>
        <EllipseGeometry RadiusX="90" RadiusY="90" />
    </Path.Data>
</Path>
```

再说一次，周长是 2π90，但只有 12 个刻度，所以相距 15π，足够接近 6 和 7.854 之间的结果。下面把代码放在一起。

```
项目：AnalogClock | 文件：MainPage.xaml(片段)
<Page ... >

    <Page.Resources>
        <Style TargetType="Path">
            <Setter Property="Stroke"
                    Value="{StaticResource ApplicationForegroundThemeBrush}" />
            <Setter Property="StrokeThickness" Value="2" />
            <Setter Property="StrokeStartLineCap" Value="Round" />
            <Setter Property="StrokeEndLineCap" Value="Round" />
            <Setter Property="StrokeLineJoin" Value="Round" />
            <Setter Property="StrokeDashCap" Value="Round" />
            <Setter Property="Fill" Value="Blue" />
        </Style>
    </Page.Resources>

    <Grid Background="{StaticResource ApplicationPageBackgroundThemeBrush}">

        <Viewbox>
            <!-- Grid containing all graphics based on (0, 0) origin, 100-pixel radius -->
            <Grid Width="200" Height="200">
```

```
                <!-- Transform for entire clock -->
                <Grid.RenderTransform>
                    <TranslateTransform X="100" Y="100" />
                </Grid.RenderTransform>

                <!-- Small tick marks -->
                <Path Fill="{x:Null}"
                      StrokeThickness="3"
                      StrokeDashArray="0 3.14159">
                    <Path.Data>
                        <EllipseGeometry RadiusX="90" RadiusY="90" />
                    </Path.Data>
                </Path>

                <!-- Large tick marks -->
                <Path Fill="{x:Null}"
                      StrokeThickness="6"
                      StrokeDashArray="0 7.854">
                <Path.Data>
                    <EllipseGeometry RadiusX="90" RadiusY="90" />
                </Path.Data>
                </Path>

                <!-- Hour hand pointing straight up -->
                <Path Data="M 0 -60 C 0 -30, 20 -30, 5 -20 L 5 0
                                    C 5 7.5, -5 7.5, -5 0 L -5 -20
                                    C -20 -30, 0 -30, 0 -60">
                    <Path.RenderTransform>
                        <RotateTransform x:Name="rotateHour" />
                    </Path.RenderTransform>
                </Path>

                <!-- Minute hand pointing straight up -->
                <Path Data="M 0 -80 C 0 -75, 0 -70, 2.5 -60 L 2.5 0
                                        C 2.5 5, -2.5 5, -2.5 0 L -2.55 -60
                                        C 0 -70, 0 -75, 0 -80">
                    <Path.RenderTransform>
                        <RotateTransform x:Name="rotateMinute" />
                    </Path.RenderTransform>
                </Path>

                <!-- Second hand pointing straight up -->
                <Path Data="M 0 10 L 0 -80">
                    <Path.RenderTransform>
                        <RotateTransform x:Name="rotateSecond" />
                    </Path.RenderTransform>
                </Path>
            </Grid>
        </Viewbox>
    </Grid>
</Page>
```

代码隐藏文件负责计算从12:00点开始顺时针方向的三个RotateTransform的角度测量。

项目: AnalogClock | 文件: MainPage.xaml.cs(片段)

```
public sealed partial class MainPage : Page
{
    public MainPage()
    {
        this.InitializeComponent();
        CompositionTarget.Rendering += OnCompositionTargetRendering;
    }

    void OnCompositionTargetRendering(object sender, object args)
    {
        DateTime dt = DateTime.Now;
        rotateSecond.Angle = 6 * (dt.Second + dt.Millisecond / 1000.0);
        rotateMinute.Angle = 6 * dt.Minute + rotateSecond.Angle / 60;
```

```
            rotateHour.Angle = 30 * (dt.Hour % 12) + rotateMinute.Angle / 12;
        }
    }
```

该时钟有个长秒针，好像在不断移动。如果你喜欢每表一跳的滴答秒针，则可以直接从计算器里删除毫秒。不过，更好的解决方案是使用一个间隔为 1 秒的 DispatcherTimer，而不用一直根据视频刷新率的 CompositionTarget.Rendering。

10.8　倾　　斜

前面讨论过，派生自 Transform 的所有类限于定义二维仿射变换，而仿射变换的特征之一是保留平行线。然而，仿射变换不一定要保留线之间的角度。例如，仿射变换能将正方形变换成平行四边形(见下图)。

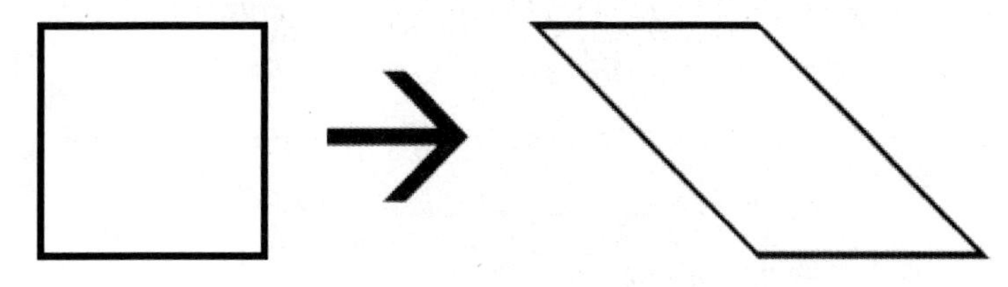

在 Windows Runtime 中，这种类型的变换称为“倾斜”(Skew)，但是在其他图形环境中则称为“切变”(Shear)。图像在水平或垂直方向上正向或者反向逐步移位。在某种意义上，倾斜是最极端的仿射变换，但仍保留了大部分原始几何图形。应用于圆或者椭圆的倾斜变换只能产生椭圆(见下图)。

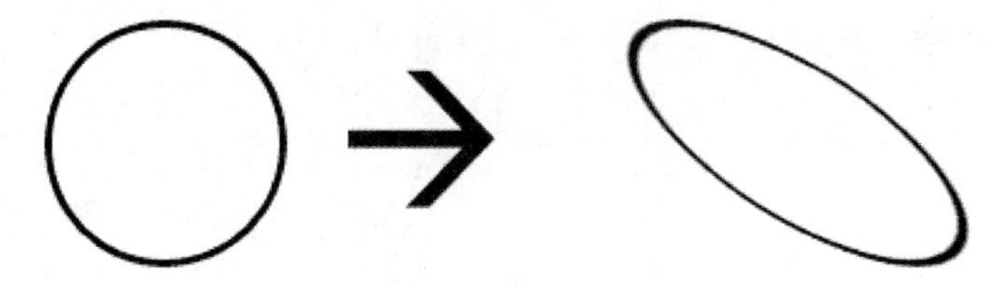

与之类似，贝塞尔曲线倾斜后仍然是贝塞尔曲线。

SkewTransform 有 AngleX 和 AngleY 属性，可以以度将其设置为角。下图所示例子通过 SkewTransform 创建，其 AngleX 设置为 45 度，从底部向右边倾斜。可以设置角度为负，则从底部向左边倾斜。对于文本，负 AngleX 值会造成倾斜效果(类似于斜体，但字符没有任何印刷变化)。如下所示，AngleX 设置为-30 度。

Text → *Text*

AngleY 非零设置会导致垂直方向的倾斜。AngleY 的正值则导致图形右侧向下倾斜(见下图)。

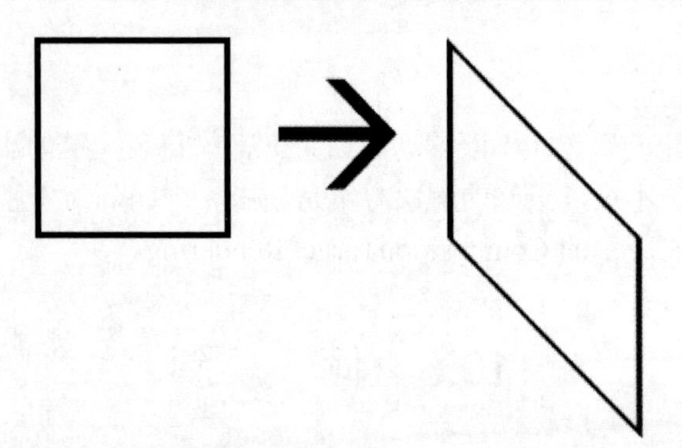

负值导致右侧向上倾斜。在默认情况下，图形左上角和倾斜保持在在同一位置，但可以用 CenterX 和 CenterY 属性或 RenderTransformOrigin 来进行改变。

以下程序演示了把 AngleX 和 AngleY 倾斜结合在一起所发生的情况。

项目：SkewPlusSkew | 文件：MainPage.xaml(片段)

```
<Page ... >
    <Grid Background="{StaticResource ApplicationPageBackgroundThemeBrush}">
        <TextBlock Text="SKEW"
                   FontSize="288"
                   FontWeight="Bold"
                   HorizontalAlignment="Center"
                   VerticalAlignment="Center"
                   RenderTransformOrigin="0.5 0.5">
            <TextBlock.RenderTransform>
                <SkewTransform x:Name="skew" />
            </TextBlock.RenderTransform>
        </TextBlock>
    </Grid>

    <Page.Triggers>
        <EventTrigger>
            <BeginStoryboard>
                <Storyboard SpeedRatio="0.5" RepeatBehavior="Forever">
                    <DoubleAnimationUsingKeyFrames Storyboard.TargetName="skew"
                                                   Storyboard.TargetProperty="AngleX">
                        <!-- Back and forth for 4 seconds -->
                        <DiscreteDoubleKeyFrame KeyTime="0:0:0" Value="0" />
                        <LinearDoubleKeyFrame KeyTime="0:0:1" Value="90" />
                        <LinearDoubleKeyFrame KeyTime="0:0:2" Value="0" />
                        <LinearDoubleKeyFrame KeyTime="0:0:3" Value="-90" />
                        <LinearDoubleKeyFrame KeyTime="0:0:4" Value="0" />

                        <!-- Do nothing for 4 seconds -->
                        <DiscreteDoubleKeyFrame KeyTime="0:0:8" Value="0" />

                        <!-- Back and forth for 4 seconds -->
                        <LinearDoubleKeyFrame KeyTime="0:0:9" Value="90" />
                        <LinearDoubleKeyFrame KeyTime="0:0:10" Value="0" />
                        <LinearDoubleKeyFrame KeyTime="0:0:11" Value="-90" />
                        <LinearDoubleKeyFrame KeyTime="0:0:12" Value="0" />
                    </DoubleAnimationUsingKeyFrames>

                    <DoubleAnimationUsingKeyFrames Storyboard.TargetName="skew"
                                                   Storyboard.TargetProperty="AngleY">

                        <!-- Do nothing for 4 seconds -->
                        <DiscreteDoubleKeyFrame KeyTime="0:0:0" Value="0" />
                        <DiscreteDoubleKeyFrame KeyTime="0:0:4" Value="0" />

                        <!-- Back and forth for 4 seconds -->
```

```
                        <LinearDoubleKeyFrame KeyTime="0:0:5" Value="-90" />
                        <LinearDoubleKeyFrame KeyTime="0:0:6" Value="0" />
                        <LinearDoubleKeyFrame KeyTime="0:0:7" Value="90" />
                        <LinearDoubleKeyFrame KeyTime="0:0:8" Value="0" />

                        <!-- Back and forth for 4 seconds -->
                        <LinearDoubleKeyFrame KeyTime="0:0:9" Value="-90" />
                        <LinearDoubleKeyFrame KeyTime="0:0:10" Value="0" />
                        <LinearDoubleKeyFrame KeyTime="0:0:11" Value="90" />
                        <LinearDoubleKeyFrame KeyTime="0:0:12" Value="0" />
                    </DoubleAnimationUsingKeyFrames>
                </Storyboard>
            </BeginStoryboard>
        </EventTrigger>
    </Page.Triggers>
</Page>
```

我在 Storyboard 里把 SpeedRatio 设置为 0.5，这样效果更好一些，但我要用关键帧时间来进行讨论。在第一个 4 秒，第一个动画把 AngleX 属性变为 90 度、回 0、到-90 度、再回 0。在下一个 4 秒，第二个动画把 AngleY 属性在-90 度和 90 度之间变化。在最后一个 4 秒，两个动画一起运行。

如下图所示，这种方式结合 AngleX 和 AngleY 而导致旋转，你可能会惊讶，也可能不会。

然而，作为数学结果，图形也变得更大。

倾斜常用来给元素增加一点点 3 D 效果，但如果结合非倾斜元素使用，效果会更好，本章后面将演示。

10.9 制作开场

有时候，你想要一个元素在页面第一次加载时发生动画变换。例如，元素会从侧面滑入，然后静止，或者增大，或者向上旋转。

首先把元素定位在其不会变换的最终位置上，这样着手处理一般最容易。然后能定义变换和动画，因此，元素会最终结束在那个地方。经常可以在转换中直接省去

DoubleAnimation 的 To 值，因为 To 值和动画前的默认值一样。

如下 SkewSlideInText 项目所演示，可以看到，TextBlock 定义了一些转换，但根据默认值，元素就在显示屏幕中心。这就是 TextBlock 的最终位置和导向，而动画则结束于该点。

```
项目: SkewSlideInText | 文件: MainPage.xaml(片段)
<Page ... >
    <Grid Background="{StaticResource ApplicationPageBackgroundThemeBrush}">
        <TextBlock Text="Hello!"
                   FontSize="192"
                   HorizontalAlignment="Center"
                   VerticalAlignment="Center"
                   RenderTransformOrigin="0.5 1">
                 <TextBlock.RenderTransform>
                 <TransformGroup>
                     <SkewTransform x:Name="skew" />
                     <TranslateTransform x:Name="translate" />
                 </TransformGroup>
             </TextBlock.RenderTransform>
        </TextBlock>
    </Grid>

    <Page.Triggers>
        <EventTrigger>
            <BeginStoryboard>
                <Storyboard>
                    <DoubleAnimation Storyboard.TargetName="translate"
                                     Storyboard.TargetProperty="X"
                                     From="-1000" Duration="0:0:1" />

                    <DoubleAnimationUsingKeyFrames
                                     Storyboard.TargetName="skew"
                                     Storyboard.TargetProperty="AngleX">
                        <DiscreteDoubleKeyFrame KeyTime="0:0:0" Value="15" />
                        <LinearDoubleKeyFrame KeyTime="0:0:1" Value="30" />
                        <EasingDoubleKeyFrame KeyTime="0:0:1.5" Value="0">
                            <EasingDoubleKeyFrame.EasingFunction>
                                <ElasticEase />
                            </EasingDoubleKeyFrame.EasingFunction>
                        </EasingDoubleKeyFrame>
                    </DoubleAnimationUsingKeyFrames>
                </Storyboard>
            </BeginStoryboard>
        </EventTrigger>
    </Page.Triggers>
</Page>
```

应用于 TranslateTransform 的 DoubleAnimation 有 From 值，From 值让 TextBlock 从位于最终位置左边 1000 像素的地方开始。To 值不存在，意味着动画结束于动画前，即 0。

上述情况正在发生的同时，DoubleAnimationUsingKeyFrames 开始倾斜进程，把 AngleX 值从 15 度变到 30 度，就好像 TextBlock 被拉到屏幕中心。最后关键帧通过动画将 AngleX 回到动画前的 0 值，并在此过程中进行摇晃。

10.10 变换数学

本章一开始就强调变换是公式，把点(x, y)转换到(x', y')，并且将元素上的所有点都进行转换。现在要来看看数学了。

假设 TranslateTransform 把 *X* 和 *Y* 属性设置为 *TX* 和 *TY*。其变换公式把这些平移因子添加到 x 和 y：

$$x' = x + TX$$
$$y' = y + TY$$

如果 ScaleTransform 的 ScaleX 和 ScaleY 属性设置 *SX* 和 *SY*，转换公式也非常明确：

$$x' = SX \bullet x$$
$$y' = SY \bullet y$$

现在，有了这些基础，我们开始组合变换，比如在 TransformGroup 里。如果 ScaleTransform 先发生，接着是 TranslateTransform，则公式如下：

$$x' = SX \bullet x + TX$$
$$y' = SY \bullet y + TY$$

但如果平移转换先应用，接着是比例转换，则公式有点儿不同：

$$x' = SX \bullet (x + TX)$$
$$y' = SX \bullet (y + TX)$$

平移因子现在实际上是乘以缩放因子。

ScaleTransform 不仅定义 ScaleX 和 ScaleY 属性，还定义 CenterX 和 CenterY。前面曾讨论了如何使用中心点构造两个平移。第一个平移是负的，接着是缩放或旋转，再后是正向平移。假设把 CenterX 和 CenterY 设置为值 *CX* 和和 *CY*。复合缩放公式如下：

$$x' = SX \bullet (x - CX) + CX$$
$$y' = SX \bullet (y - CY) + CY$$

可以很容易确认点(*CX*，*CY*)变换成了点(*CX*，*CY*)，点(*CX*，*CY*)具备缩放中心特征：变换后未改变的点。

在目前为止的所有情况中，*x'*完全依赖于乘以 *x* 再加上 *x* 的常数，而 *y'*只依赖于乘以 *y* 再加上 *y* 的常数。旋转的时候有点复杂，因为 *x'*同时依赖于 *x* 和 *y*，*y'*也同时依赖于 *x* 和 *y*。如果 RotateTransform 的 Angle 属性设置为 *A*，则转换公式如下：

$$x' = \cos(A) \bullet x - \sin(A) \bullet y$$
$$y' = \sin(A) \bullet x + \cos(A) \bullet y$$

这些公式很容易确认简单情况。如果 *A* 为 0，则公式如下：

$$x' = x$$
$$y' = y$$

如果 *A* 是 90 度，正弦为 1，而余弦为 0，则

$$x' = -y$$
$$y' = x$$

例如，点(1，0)变换为(0，1)，则(0，1)变转为(1，0)。如果 *A* 为 180 度，正弦为 0，余弦为-1，则有：

$$x' = -y$$
$$y' = -x$$

这是围绕着原点的反射，通过把 ScaleTransform 的 ScaleX 和 ScaleY 都设置为-1，可以得到同样效果。如果 *A* 为 270 度，则有：

$$x' = y$$
$$y' = -x$$

之前的第一张倾斜变换如下图所示。

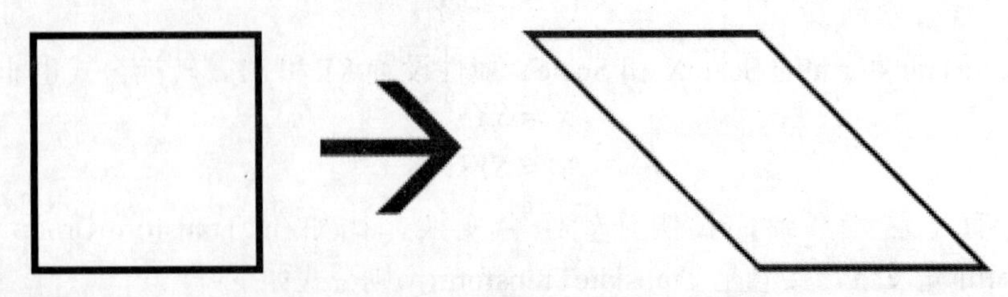

该倾斜的变换公式为(AngleX 设置为 45 度):

$$x' = x + y$$
$$y' = y$$

如果 y 等于 0(在图形顶部)，则 x'就等于 x，y'就等于 y。但如果向下移动图形，y 会变大，因此 x'会比 x 越来越大。如果把 AngleX 设置为 AX、把 AngleY 设置为 AY，SkewTransform 的一般公式如下：

$$x' = x + \sin(AX) \cdot y$$
$$y' = \sin(AX) \cdot x + y$$

如果要探索把旋转与其他变换结合在一起，这种类型符的号会变得相当笨拙。幸运的是，矩阵代数能发挥作用。如果把单个变换表达为矩阵，则可以通过已有的矩阵乘法把变换组合起来。

我们用一个 2×1 矩阵来表示点(x, y):

$$\begin{vmatrix} x & y \end{vmatrix}$$

用一个 2×2 矩阵来表示变换：

$$\begin{vmatrix} M11 & M12 \\ M21 & M22 \end{vmatrix}$$

可以用矩阵乘法来表示应用变换。其结果是变换点：

$$\begin{vmatrix} x & y \end{vmatrix} \times \begin{vmatrix} M11 & M12 \\ M21 & M22 \end{vmatrix} = \begin{vmatrix} x' & y' \end{vmatrix}$$

矩阵乘法的规则意味着有以下公式：

$$x' = M11 \cdot x + M21 \cdot y$$
$$y' = M12 \cdot x + M22 \cdot y$$

如果 M11 是 ScaleX 的值，M22 是 ScaleY 的值，M21 和 M12 为 0，则该过程适用于缩放。也适用于旋转和倾斜，两者都涉及到乘以 x 和 y 的因子。

但该过程不适用于平移。平移公式类似如下所示：

$$x' = x + TX$$
$$y' = y + TY$$

平移因子自身相加，而不是乘以 x 或 y。如果不允许平移，而平移又可能是所有类型中最简单的变化类型，则如何用矩阵表达一般变换？

有趣的解决方案是引入第三维度。除了电脑屏幕上的 X 和 Y 轴外，概念性 Z 轴会从屏幕上延伸出来。我们假设在二维平面上画图，但平面存在于三维空间之中，常量 Z 坐标等于 1。

也就是说，点(x, y)实际上是点$(x,\ y,\ 1)$，我们可以用一个 3×1 矩阵来表示：

$$\begin{vmatrix} x & y & 1 \end{vmatrix}$$

矩阵变换为 3×3 矩阵，乘法如下：

$$\begin{vmatrix} x & y & 1 \end{vmatrix} \times \begin{vmatrix} M11 & M12 & M13 \\ M21 & M22 & M23 \\ M31 & M32 & M33 \end{vmatrix} = \begin{vmatrix} x' & y' & z' \end{vmatrix}$$

矩阵乘法意味着公式为：

$$x' = M11 \cdot x + M21 \cdot y + M31$$
$$y' = M12 \cdot x + M22 \cdot y + M23$$
$$z' = M13 \cdot x + M23 \cdot y + M33$$

现在成功了一部分，因为变换公式包含了 M31 和 M32 的平移因子。这两个数字都没有乘以 x 或者 y。

没有完全成功，因为 z'一般不等于 1，也就是说，我们已经移出了 z 总是等于 1 的平面。回到该平面的一个方法是直接把所有偏离的 z'值都设置为 1。但是变换后远离 z 等于 1 的平面的那些点难道不应该和变换后靠近该平面的点区别出来吗？

有个好办法能让 z 值为 1，而又不会忽略，即采用 3×1 矩阵的值，并把所有三个坐标都除以 z'：

$$\left(\frac{x'}{z'}, \frac{y'}{z'}, \frac{z'}{z'}\right) = \left(\frac{x'}{z'}, \frac{y'}{z'}, 1\right)$$

这种用三维坐标表示二维变换的方法称为“齐次坐标”(Homogenous Coordinates)，由奥古斯特 • 莫比乌斯(August Möbius)在 19 世纪 20 年代发展起来，用来表示当 z'为零时，结果是无穷大的一种方法。但对我们来说，无穷大坐标是一个问题。如果想避免无穷大坐标，则 z'不能为 0。事实上，只要确保 z 总是为 1，则完全可以避免被 z'除。

通过在矩阵里把 M13 和 M23 设置为 0、M33 设置为 1，则可能实现上述要求。现在，变换可以用完全保持在同一平面上的公式表达如下：

$$\begin{vmatrix} x & y & 1 \end{vmatrix} \times \begin{vmatrix} M11 & M12 & 0 \\ M21 & M22 & 0 \\ M31 & M32 & 1 \end{vmatrix} = \begin{vmatrix} x' & y' & 1 \end{vmatrix}$$

这是二维仿射变换的标准矩阵表达式。(允许第三列有其他值会导致非仿射变换。因为这类矩阵能把平行线变换为非平行线，有时也称为“锥形变换”。)

采用我之前使用的符号，把 ScaleX 设置为 SX，把 ScaleY 设置为 SY，则 ScaleTransform 为：

$$\begin{vmatrix} x & y & 1 \end{vmatrix} \times \begin{vmatrix} SX & 0 & 0 \\ 0 & SY & 0 \\ 0 & 0 & 1 \end{vmatrix} = \begin{vmatrix} x' & y' & 1 \end{vmatrix}$$

带 TX 和 TY 因子的 TranslateTransform 如下

$$\begin{vmatrix}x & y & 1\end{vmatrix}\times\begin{vmatrix}1 & 0 & 0\\ 0 & 1 & 0\\ TX & TY & 1\end{vmatrix}=\begin{vmatrix}x' & y' & 1\end{vmatrix}$$

以(CX，CY)为中心的 ScaleTransform 为三个 3×3 变换的乘法：

$$\begin{vmatrix}x & y & 1\end{vmatrix}\times\begin{vmatrix}1 & 0 & 0\\ 0 & 1 & 0\\ -CX & -CY & 1\end{vmatrix}\times\begin{vmatrix}SX & 0 & 0\\ 0 & SY & 0\\ 0 & 0 & 1\end{vmatrix}\times\begin{vmatrix}1 & 0 & 0\\ 0 & 1 & 0\\ CX & CY & 1\end{vmatrix}=\begin{vmatrix}x' & y' & 1\end{vmatrix}$$

类似，以角 A 和(CX，CY)为中心的 RotateTransform 也表达三个变换：

$$\begin{vmatrix}x & y & 1\end{vmatrix}\times\begin{vmatrix}1 & 0 & 0\\ 0 & 1 & 0\\ -CX & -CY & 1\end{vmatrix}\times\begin{vmatrix}\cos(A) & \sin(A) & 0\\ -\sin(A) & \cos(A) & 0\\ 0 & 0 & 1\end{vmatrix}\times\begin{vmatrix}1 & 0 & 0\\ 0 & 1 & 0\\ CX & CY & 1\end{vmatrix}=\begin{vmatrix}x' & y' & 1\end{vmatrix}$$

有角 AX、AY 及一个中心的 SkewTransform，如下所示：

$$\begin{vmatrix}x & y & 1\end{vmatrix}\times\begin{vmatrix}1 & 0 & 0\\ 0 & 1 & 0\\ -CX & -CY & 1\end{vmatrix}\times\begin{vmatrix}1 & \sin(AY) & 0\\ \sin(AX) & 1 & 0\\ 0 & 0 & 1\end{vmatrix}\times\begin{vmatrix}1 & 0 & 0\\ 0 & 1 & 0\\ CX & CY & 1\end{vmatrix}=\begin{vmatrix}x' & y' & 1\end{vmatrix}$$

矩阵乘法有一个众所周知的属性，即不可交换。乘法顺序会造成影响。平移和缩放变换都演示了这点。如果首先是平移，平移因子自身也可以被缩放因子而缩放。

然而，某些类型的变换可以以任意顺序相乘。

- 多个 TranslateTransform 对象。所有平移是每个平移因子之和。
- 具有同样缩放中心的多个 ScaleTransform 对象。所有缩放都是单个缩放因子的结果。
- 具有相同旋转中心的多个 RotateTransforms。所有旋转是每个旋转角度之和。

此外，如果 ScaleTransform 具有相等的 ScaleX 和 ScaleY 属性，则其可以被 RotateTransform 或 SkewTransform 以任意顺序相乘。

Windows Runtime 定义了有 6 个属性的 Matrix 结构，6 个属性对应类似如下的矩阵单元格：

$$\begin{vmatrix}M11 & M12 & 0\\ M21 & M22 & 0\\ OffsetX & OffsetY & 1\end{vmatrix}$$

该矩阵的最后一行是固定的。不能使用该矩阵结构来定义锥形变换或“更疯狂”的东西。OffsetX 和 OffsetY 是平移属性，M11 和 M22 的默认值为 1，其他四项属性的默认值为 0。对角线为 1s 的单位矩阵如下：

$$\begin{vmatrix}1 & 0 & 0\\ 0 & 1 & 0\\ 0 & 0 & 1\end{vmatrix}$$

该矩阵结构有返回值的静态 Identity 属性，以及如果矩阵的值是单位矩阵，返回 true 的 IsIdentity 属性。

通过 ScaleTransform 和 RotateTransform 这样的“简单”Transform 派生，也有可供选择的低级 MatrixTransform，而 MatrixTransform 有 Matrix 类型的属性。如果知道想要的矩阵变换，则可以直接按顺序指定六个数字 M11、M12、M21、M22、OffsetX、OffsetY。以下是设置此变换的一种方法。

```
<TextBlock ... >
    <TextBlock.RenderTransform>
        <MatrixTransform Matrix="10 0 0 5 0 100" />
    </TextBlock.RenderTransform>
</TextBlock>
```

该变换在水平方向的缩放因子是 10(M11)，在垂直方向是 5(M22)，然后移动 TextBlock 下降 100 像素(OffsetY)。但是也可以直接把变换设置为 RenderTransform 属性。

```
<TextBlock ...
        RenderTransform="10 0 0 5 0 100"
        ... />
```

Visual Studio 预览设计视图并不特别关心该语法，但在其编译器或 Windows 8 中则没有问题。

使用这种 MatrixTransform 隐式形式在几种常见旋转变换是很方便的，如以下程序所示。每个 TextBlock 显示了应用于其的变换。

```
项目: CommonMatrixTransforms | 文件: MainPage.xaml(片段)
<Page ... >
    <Page.Resources>
        <Style TargetType="TextBlock">
            <Setter Property="FontSize" Value="24" />
            <Setter Property="RenderTransformOrigin" Value="0 0.5" />
        </Style>
    </Page.Resources>

    <Grid Background="{StaticResource ApplicationPageBackgroundThemeBrush}">

        <!-- Move origin to center -->
        <Canvas HorizontalAlignment="Center"
                VerticalAlignment="Center">

            <TextBlock Text=" RenderTransform='1 0 0 1 0 0'"
                       RenderTransform="1 0 0 1 0 0" />

            <TextBlock Text=" RenderTransform='.7 .7 -.7 .7 0 0'"
                       RenderTransform=".7 .7 -.7 .7 0 0" />

            <TextBlock Text=" RenderTransform='0 1 -1 0 0 0'"
                       RenderTransform="0 1 -1 0 0 0" />

            <TextBlock Text=" RenderTransform='-.7 .7 -.7 -.7 0 0"
                       RenderTransform="-.7 .7 -.7 -.7 0 0" />

            <TextBlock Text=" RenderTransform='-1 0 0 -1 0 0'"
                       RenderTransform="-1 0 0 -1 0 0" />

            <TextBlock Text=" RenderTransform='-.7 -.7 .7 -.7 0 0'"
                       RenderTransform="-.7 -.7 .7 -.7 0 0" />

            <TextBlock Text=" RenderTransform='0 -1 1 0 0 0'"
                       RenderTransform="0 -1 1 0 0 0" />

            <TextBlock Text=" RenderTransform='.7 -.7 .7 .7 0 0"
                       RenderTransform=".7 -.7 .7 .7 0 0" />
        </Canvas>
```

```
    </Grid>
</Page>
```

频繁引用的.7 更精确的应该是.707，45 度的正弦和余弦，以及(并非巧合)2 的平方根的一半。这 8 个变换的结果是每个 TextBlock 从前一个变化又旋转了 45 度，如下图所示。

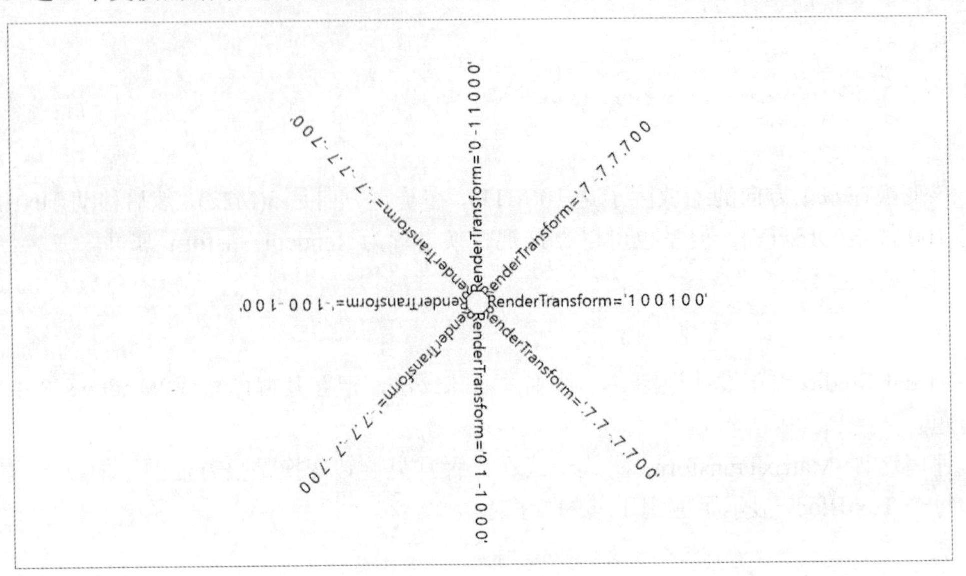

如果看代码，会发现该 Matrix 结构有 Transform 方法，该方法把变换应用为一个点的值，并返回变换后的点。

然而，Matrix 结构很不方便。缺少乘法操作符，无法轻松在代码中执行你自己的矩阵乘法。你可以自己写乘法代码，或者可以使用 TransformGroup，能内部执行矩阵乘法，而结果只提供 Matrix 类型的只读 Value 属性。如果需要执行矩阵乘法，可以在代码中创建 TransformGroup，添加一对初始化的 Transform 派生，并访问 Value 属性。

第 13 章会有一个重要例子。如果要触摸操作屏幕上的对象，矩阵变换计算在计算缩放和旋转中心中的作用会变得至关重要。

10.11 复合变换

如果结合各种类型变换，顺序是会产生影响的。然而在实际应用中，通常就是想要按非常标准的顺序来应用各种变换。

例如，假设你想要旋转、缩放以及平移元素。ScaleTransform 通常首先出现，因为一般情况下要根据未旋转元素来指定缩放比例。而 TranslateTransform 最后出现，因为一般情况下不会希望缩放或旋转会影响平移因子。也就是说，RotateTransform 出现在中间，即顺序是：缩放、旋转、平移。

如果这是你想要的顺序，则可以使用 CompositeTransform。CompositeTransform 有一群已定义属性，能用于按顺序执行变换：

- 缩放
- 倾斜

- 旋转
- 平移

这些属性如下所示：

- CenterX 和 CenterY，用于缩放、倾斜和旋转中心
- ScaleX 和 ScaleY
- SkewX 和 SkewY
- Rotation
- TranslateX 和 TranslateY

如下小程序使用 CompositeTransform 作为结合缩放和倾斜的简便方法。

项目 Project: TiltedShadow | 文件: MainPage.xaml(片段)

```
<Page ... >
    <Page.Resources>
        <Style TargetType="TextBlock">
            <Setter Property="Text" Value="quirky" />
            <Setter Property="FontFamily" Value="Times New Roman" />
            <Setter Property="FontSize" Value="192" />
            <Setter Property="HorizontalAlignment" Value="Center" />
            <Setter Property="VerticalAlignment" Value="Center" />
        </Style>
    </Page.Resources>

    <Grid Background="{StaticResource ApplicationPageBackgroundThemeBrush}">
        <!-- Shadow TextBlock -->
        <TextBlock Foreground="Gray"
                   RenderTransformOrigin="0 1">
            <TextBlock.RenderTransform>
                <CompositeTransform ScaleY="1.5" SkewX="-60" />
            </TextBlock.RenderTransform>
        </TextBlock>

        <!-- TextBlock with all styled properties -->
        <TextBlock />
    </Grid>
</Page>
```

XAML 实例化两个 TextBlock 元素属性，在 Style 里指定大部分相同属性，包括 Text 属性，只要和布局系统有关，两个元素就占据相同空间。底部为灰色，并应用缩放和倾斜变换，如下图所示。

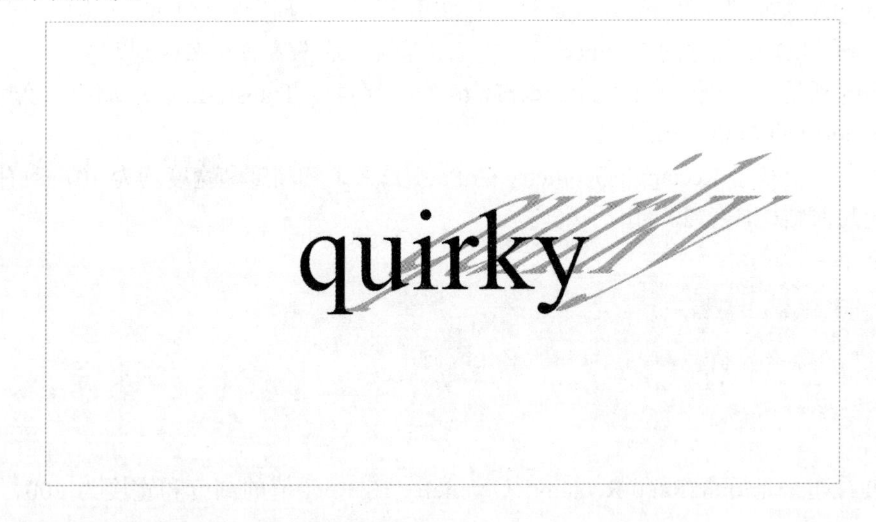

注意，RenderTransformOrigin 设置为点(0, 1)，也就是说，变换是相对于左下角。然而，该点可以指定为(1, 1)或两者之间的任意一点，作用都一样。只需要两个 TextBlock 元素共享同一个底部边缘。ScaleY 为 1.5，用于把阴影高度提高 50%。SkewX 为-60 度，应该会把底部移动到左边，但因为底部是缩放和倾斜的中心，因此顶部向右倾斜。

如果仔细观察，会注意到下降部分的底部不十分符合要求。这是因为 TextBlock 实际上在下降部分的底部延伸了一点。可以把 RenderTransformOrigin 变为(0, 0.96)，会能更符合要求。

如果想要类似文本效果，但不要下降，该怎么办？如下图所示。

需要用 RenderTransformOrigin，其 Y 值等于基线以上文字的相对高度，这是问题。而这依赖于字体。对于该特定截图，我不断尝试，一直到得到了(0，0 .78)，但只适合 Times New Roman 字体。如果想用一般方式去处理这种问题，需要访问字体参数，而在 Windows 8 应用程序只能通过 DirectX 获得字体参数。第 15 章将展示如何做到这一点。

10.12 几 何 变 换

Geometry 类定义了 Transform 属性，自然引发了如下问题：对 Path 元素应用变换、对设置了 Path 的 Data 属性的 Geometry 对象应用变换，这两者有什么区别？

大的区别是，应用于 Path 的 RenderTransform 属性的 Transform 会增加笔画宽度，而应用到 Geometry 的 Transform 则不会。

以下有一个基于 RectangleGeometry 的 Path 元素，其高度和宽度均为 10，但有一个变换应用到几何结构，按 20 的因子来增加：

```
<Path Stroke="Black"
      StrokeThickness="1"
      StrokeDashArray="1 1">
    <Path.Data>
        <RectangleGeometry Rect="0 0 10 10"
                Transform="20 0 0 20 0 0" />
    </Path.Data>
</Path>
```

结果看起来是好像 Rect RectangleGeometry 的 Rect 值的高度和宽度为 200，如下图所示。

该 XAML 具有相同的初始 RectangleGeometry，但将变换应用于 Path：

```
<Path Stroke="Black"
      StrokeThickness="1"
      StrokeDashArray="1 1"
      RenderTransform="20 0 0 20 0 0">
    <Path.Data>
        <RectangleGeometry Rect="0 0 10 10" />
    </Path.Data>
</Path>
```

结果则大不相同，如下图所示。

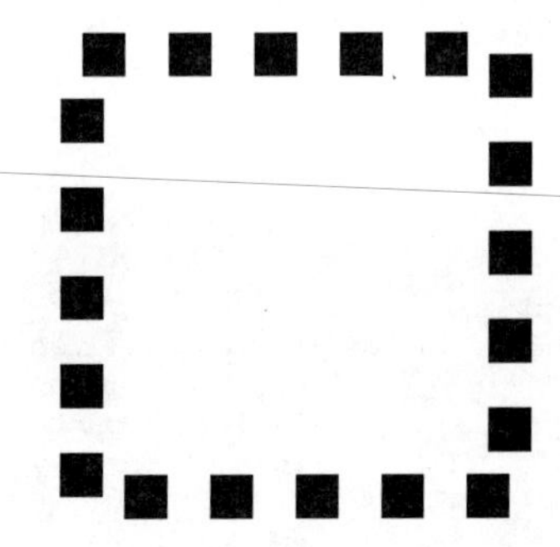

然而对布局系统来说，这些元素则似乎相同。两个 Path 元素都被认为宽度和高度为 10。

10.13 画笔变换

Brush 类定义了两个与变换相关的属性：Transform 和 RelativeTransform，两者之前的区别在于是让你基于画笔的像素大小还是相对尺寸而指定平移因子。RelativeTransform 往往更好用，除非元素被赋予了指定像素大小。

以下复制了第 3 章中的 RainbowEight 程序，但使用的是动画画笔变换。我没有用 TextBlock，而取代了 8 的 Path 演示，因为无法获得画笔通过 Repeat 的 SpreadMethod 属性对 TextBlock 进行重复。

```
项目：RainbowEightTransform | 文件：MainPage.xaml(片段)
<Page ... >
    <Grid Background="{StaticResource ApplicationPageBackgroundThemeBrush}">
        <Viewbox>
            <Path StrokeThickness="50"
```

```
                Margin="0 25 0 0">
            <Path.Data>
                <PathGeometry>
                    <PathFigure StartPoint="110 0">
                        <ArcSegment Size="90 90" Point="110 180"
                                    SweepDirection="Clockwise" />
                        <ArcSegment Size="110 110" Point="110 400"
                                    SweepDirection="Counterclockwise" />
                        <ArcSegment Size="110 110" Point="110 180"
                                    SweepDirection="Counterclockwise" />
                        <ArcSegment Size="90 90" Point="110 0"
                                    SweepDirection="Clockwise" />
                    </PathFigure>
                </PathGeometry>
            </Path.Data>
            <Path.Stroke>
                <LinearGradientBrush StartPoint="0 0" EndPoint="1 1"
                                     SpreadMethod="Repeat">
                    <LinearGradientBrush.RelativeTransform>
                        <TranslateTransform x:Name="translate" />
                    </LinearGradientBrush.RelativeTransform>

                    <GradientStop Offset="0.00" Color="Red" />
                    <GradientStop Offset="0.14" Color="Orange" />
                    <GradientStop Offset="0.28" Color="Yellow" />
                    <GradientStop Offset="0.43" Color="Green" />
                    <GradientStop Offset="0.57" Color="Blue" />
                    <GradientStop Offset="0.71" Color="Indigo" />
                    <GradientStop Offset="0.86" Color="Violet" />
                    <GradientStop Offset="1.00" Color="Red" />
                </LinearGradientBrush>
            </Path.Stroke>
        </Path>
    </Viewbox>
</Grid>

<Page.Triggers>
    <EventTrigger>
        <BeginStoryboard>
            <Storyboard>
                <DoubleAnimation Storyboard.TargetName="translate"
                                 Storyboard.TargetProperty="Y"
                                 EnableDependentAnimation="True"
                                 From="0" To="-1.36" Duration="0:0:10"
                                 RepeatBehavior="Forever" />
            </Storyboard>
        </BeginStoryboard>
    </EventTrigger>
</Page.Triggers>
</Page>
```

图片如下图所示。

标记里有一个神奇数字，即 DoubleAnimation 的 To 值。它应用于 TranslateTransform 的 Y 属性，选择该值的目的，是使带该值的被平移画笔相当于未平移画笔。你可以看到神奇数字是 1.36，我猜你想要知道其从何而来。

如果 LinearGradientBrush 自上而下：StartPoint 为(0，0)，EndPoint 为(0，1)，To 值则为-1。如果渐变从左到右：StartPoint 为(0，0)，EndPoint 为(1，0)，则 TranslateTransform 的 X 属性为动画目标，而将再次使用 1 或者-1 的 To 值。

但如果渐变从一个角落到对面角落，默认 StartPoint 为(0，0)，EndPoint 为(1，1)，则上述就不正确了。用画笔覆盖 Path 元素时，Windows Runtime 会计算边界矩形，该矩形包括元素的几何尺寸加上笔划宽度。画笔会延展到该边界矩形，如下图所示。

渐变线沿着对角线，也就是说，常量色彩的线条和渐变线呈直角。

如果画笔有 Repeat 的 SpreadMethod，画笔在概念上会在指定偏移量之外重复。如果把 TranslateTransform 应用到画笔上，则 SpreadMethod 设置会非常有用，因为似乎无论怎么移动，画笔都会重复。

如果用元素高度来提高画笔(即，TranslateTransform 的 Y 值为-1)，未变换的画笔底部边缘变成了已变换的画笔顶部边缘，但可以看到如下图所示结果，和前面的图不一样。

如果要得到平滑动画，需要再向上移一点。但应该移动多少呢？

我们扩展该图以显示重复画笔的一部分(见下图)，并把元素宽度标记为“w”、高度标记为“h”、对角线标记为“d”，而高度增加标记为“Δh”。

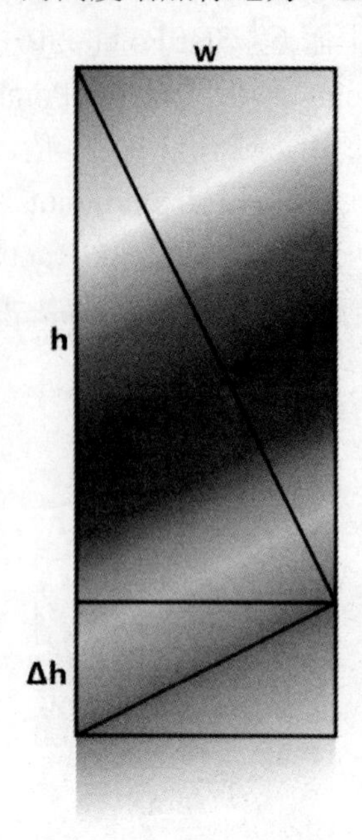

可以有多种方法算出Δh，但最直接的方法也许是用相似三角形：

$$\frac{d}{h}=\frac{h+\Delta h}{d}$$

由此很容易得出

$$\Delta h=\frac{w^2}{h}$$

或者，我们真正想要的数字应当如下所示：

$$\frac{h+\Delta h}{h}=1+\left(\frac{w}{h}\right)^2$$

试试从之前所示的 Path 插入数字。需要把 StrokeThickness 添加到几何图形的宽度和高度。宽度为 270，高度为 450，Δh 是 162。加上 h，并除以 h，就这样得到了神奇数字 1.36。

想知道一个更简单的方法吗？在 Storyboard 里只使用两个 DoubleAnimation 对象，其中一个针对 Y 属性，而另一个针对 X 属性。把两者的 To 值设置为-1，而每次循环时画笔同时向上和向左移动。

10.14 老兄，元素在哪里？

之前提到过，从 TransformGroup 可以计算得到 Matrix 值，但无法从你会期望的其他来源而得到。例如，你会期望 GeneralTransform(Transform 和其他变换类派生于此类)有 Matrix

属性，但并没有。

然而，GeneralTransform 类有 TransformBounds 和 TransformPoint 方法，可以把变换应用到 Rect 值，在某些情况下这会很方便。

假设有一个元素是面板的子对象。该面板负责相对于面板本身来定位元素，不过该元素也可以有应用了平移、缩放、旋转或倾斜的 RenderTransform。出于命中测试目的，该元素的位置和方向对系统内部为已知。但你自己的程序能够找到元素实际定位在哪里吗？

当然能找到！最基本(但隐藏)的方法由 UIElement 定义，称为 TransformToVisual。通常可以带参数对元素调用该方法，而参数是元素的父类或者其他祖先类：

```
GeneralTransform xform = element.TransformToVisual(parent);
```

该方法返回的 GeneralTransform 对象把元素坐标映射到父类坐标。但实际上，你看不出这个变换！方法不会给你 Matrix 值。而你能做的是调用 TransformPoint 或 TransformBounds，或者使用 Inverse 属性。而你常常就需要这些。

如下 XAML 文件把 CompositeTransform 属性变成动画，使 TextBlock 在屏幕上疯狂地运动。

```
项目: WheresMyElement | 文件: MainPage.xaml(片段)
<Page ... >
    <Grid Name="contentGrid"
          Background="{StaticResource ApplicationPageBackgroundThemeBrush}">
        <TextBlock Name="txtblk"
                   Text="Tap to Find"
                   FontSize="96"
                   HorizontalAlignment="Center"
                   VerticalAlignment="Center"
                   RenderTransformOrigin="0.5 0.5">

            <TextBlock.RenderTransform>
                <CompositeTransform x:Name="transform" />
            </TextBlock.RenderTransform>
        </TextBlock>

        <Polygon Name="polygon" Stroke="Blue" />
        <Path Name="path" Stroke="Red" />
    </Grid>

    <Page.Triggers>
        <EventTrigger>
            <BeginStoryboard>
                <Storyboard x:Name="storyboard">
                    <DoubleAnimation Storyboard.TargetName="transform"
                                     Storyboard.TargetProperty="TranslateX"
                                     From="-300" To="300" Duration="0:0:2.11"
                                     AutoReverse="True" RepeatBehavior="Forever" />
                    <DoubleAnimation Storyboard.TargetName="transform"
                                     Storyboard.TargetProperty="TranslateY"
                                     From="-300" To="300" Duration="0:0:2.23"
                                     AutoReverse="True" RepeatBehavior="Forever" />
                    <DoubleAnimation Storyboard.TargetName="transform"
                                     Storyboard.TargetProperty="Rotation"
                                     From="0" To="360" Duration="0:0:2.51"
                                     AutoReverse="True" RepeatBehavior="Forever" />
                    <DoubleAnimation Storyboard.TargetName="transform"
                                     Storyboard.TargetProperty="ScaleX"
                                     From="1" To="2" Duration="0:0:2.77"
                                     AutoReverse="True" RepeatBehavior="Forever" />
                    <DoubleAnimation Storyboard.TargetName="transform"
                                     Storyboard.TargetProperty="ScaleY"
```

```
                                          From="1" To="2" Duration="0:0:3.07"
                                          AutoReverse="True" RepeatBehavior="Forever" />
                        <DoubleAnimation Storyboard.TargetName="transform"
                                          Storyboard.TargetProperty="SkewX"
                                          From="-30" To="30" Duration="0:0:3.31"
                                          AutoReverse="True" RepeatBehavior="Forever" />
                        <DoubleAnimation Storyboard.TargetName="transform"
                                          Storyboard.TargetProperty="SkewY"
                                          From="-30" To="30" Duration="0:0:3.53"
                                          AutoReverse="True" RepeatBehavior="Forever" />

                    </Storyboard>
                </BeginStoryboard>
            </EventTrigger>
        </Page.Triggers>
</Page>
```

注意，Grid 还包含蓝色 Polygon 和红色 Path，但没有实际坐标点。

代码隐藏文件使用 Tapped 事件，调用 TransformToVisual，暂停 Storyboard(再下次点击恢复)，并对 TextBlock 进行“快照”。TransformToVisual 返回 GeneralTransform 对象来描述 TextBlock 和 Grid 的关系。该程序用此来 Polygon 把 TextBlock 的四个角变换为 Grid 坐标，并围绕着 TextBlock 有效地画了一个矩形。

```
项目: WheresMyElement | 文件: MainPage.xaml.cs(片段)
public sealed partial class MainPage : Page
{
    bool storyboardPaused;

    public MainPage()
    {
        this.InitializeComponent();
    }

    protected override void OnTapped(TappedRoutedEventArgs args)
    {
        if (storyboardPaused)
        {
            storyboard.Resume();
            storyboardPaused = false;
            return;
        }

        GeneralTransform xform = txtblk.TransformToVisual(contentGrid);

        // Draw blue polygon around element
        polygon.Points.Clear();
        polygon.Points.Add(xform.TransformPoint(new Point(0, 0)));
        polygon.Points.Add(xform.TransformPoint(new Point(txtblk.ActualWidth, 0)));
        polygon.Points.Add(xform.TransformPoint(new Point(txtblk.ActualWidth,
                                                          txtblk.ActualHeight)));
        polygon.Points.Add(xform.TransformPoint(new Point(0, txtblk.ActualHeight)));

        // Draw red bounding box
        path.Data = new RectangleGeometry
        {
            Rect = xform.TransformBounds(new Rect(new Point(0, 0), txtblk.DesiredSize))
        };

        storyboard.Pause();
        storyboardPaused = true;
        base.OnTapped(args);
    }
}
```

调用 TransformBounds 的结果会有所不同：描述边界框的矩形，长宽与水平和垂直方向平行，并且大到足以包含该元素。如下图所示。

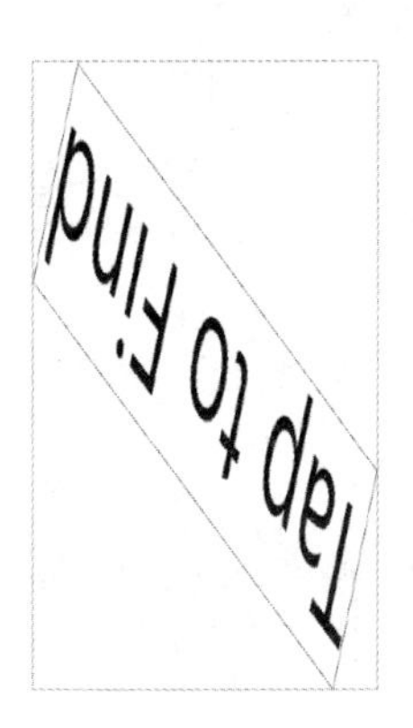

很容易通过被变换四个角的最大和最小的 X 和 Y 坐标计算得到边界矩形，不过方便可用是很好的事情。

10.15　投影变换

本章稍前讨论了为什么在数学上把二维图形变换描述为 3×3 矩阵，而且需要加入第三维度。通过类推，三维图形变换可以用 4×4 矩阵来表达，而 Windows Runtime 就有此矩阵。

Windows.UI.Xaml.Media.Media3D 命名空间只包含两项：Matrix3D 结构以及 Matrix3DHelper 类，前者所有程序员都能用，而后者主要针对 C++程序员，因为 C++程序员不能访问 Matrix3D 定义的任何方法。除了矩阵的每一个单元格都可用之外，Matrix3D 的属性都类似于常规 Matrix 结构。

$$\begin{vmatrix} M11 & M12 & M13 & M14 \\ M21 & M22 & M23 & M24 \\ M31 & M32 & M33 & M34 \\ OffsetX & OffsetY & OffsetZ & M44 \end{vmatrix}$$

然而，很少有程序员真正深入过该矩阵。大多数程序员都满足于使用 PlaneProjection 类，而在本章开始我就简要演示过该类。

PlaneProjection 的主要目的是让你能够在三维空间旋转二维元素。三维空间里的旋转总是围绕轴，而 PlaneProjection 能让元素围绕着水平轴(使用 RotationX 属性)、垂直轴(使用 RotationY 属性)或概念上穿透屏幕的 Z 轴而旋转。围绕 Z 轴旋转仅仅是二维旋转，而远没有其他两个轴那样令人印象深刻。

可以预期右手规则旋转的方向：把右手拇指朝向正轴方向，右为 X 轴、下为 Y 轴，而 Z 轴是朝屏幕外面。手指曲线则表明正向角度的旋转方向。PlaneProjection 按 X、Y、Z 的顺序应用旋转，但通常只使用其中之一。

稍微用一用 PlaneProjection，就能使元素摇摆进入视图，甚至能从概念上翻转元素，以显示“另一面”(稍后将演示)。

接下来是不用 PlaneProjection 的效果。ThreeDeeSpinningText 程序允许独立对 RotationX、RotationY 和 RotationZ 属性进行动画，在三维空间内旋转 TextBlock。如下 XAML 文件中，Begin/Stop 和 Play/Pause 一组按钮在结尾。

项目: ThreeDeeSpinningText | 文件: MainPage.xaml(片段)

```
<Page ... >
    <Page.Resources>
        <Storyboard x:Key="xAxisAnimation" RepeatBehavior="Forever">
            <DoubleAnimation Storyboard.TargetName="projection"
                             Storyboard.TargetProperty="RotationX"
                             From="0" To="360" Duration="0:0:1.9" />
        </Storyboard>

        <Storyboard x:Key="yAxisAnimation" RepeatBehavior="Forever">
            <DoubleAnimation Storyboard.TargetName="projection"
                             Storyboard.TargetProperty="RotationY"
                             From="0" To="360" Duration="0:0:3.1" />
        </Storyboard>

        <Storyboard x:Key="zAxisAnimation" RepeatBehavior="Forever">
            <DoubleAnimation Storyboard.TargetName="projection"
                             Storyboard.TargetProperty="RotationZ"
                             From="0" To="360" Duration="0:0:4.3" />
        </Storyboard>
    </Page.Resources>

    <Grid Background="{StaticResource ApplicationPageBackgroundThemeBrush}">
        <Grid.RowDefinitions>
            <RowDefinition Height="*" />
            <RowDefinition Height="Auto" />
        </Grid.RowDefinitions>

        <TextBlock Text="3D-ish"
                   FontSize="384"
                   HorizontalAlignment="Center"
                   VerticalAlignment="Center">
            <TextBlock.Projection>
                <PlaneProjection x:Name="projection" />
            </TextBlock.Projection>
        </TextBlock>

        <!-- Control Panel -->
        <Grid Grid.Row="1" HorizontalAlignment="Center">
            <Grid.RowDefinitions>
                <RowDefinition Height="Auto" />
                <RowDefinition Height="Auto" />
                <RowDefinition Height="Auto" />
            </Grid.RowDefinitions>

            <Grid.ColumnDefinitions>
                <ColumnDefinition Width="Auto" />
                <ColumnDefinition Width="Auto" />
                <ColumnDefinition Width="Auto" />
            </Grid.ColumnDefinitions>

            <Grid.Resources>
                <Style TargetType="TextBlock">
                <Setter Property="FontSize"
                        Value="{StaticResource ControlContentThemeFontSize}" />
                <Setter Property="VerticalAlignment" Value="Center" />
            </Style>

            <Style TargetType="Button">
                <Setter Property="Width" Value="120" />
                <Setter Property="Margin" Value="12" />
            </Style>
```

```
        </Grid.Resources>

        <TextBlock Text="X Axis: " Grid.Row="0" Grid.Column="0"
                   Tag="xAxisAnimation" />
        <Button Content="Begin" Grid.Row="0" Grid.Column="1"
                Click="OnBeginStopButton" />
        <Button Content="Pause" Grid.Row="0" Grid.Column="2"
                IsEnabled="False"
                Click="OnPauseResumeButton" />

        <TextBlock Text="Y Axis: " Grid.Row="1" Grid.Column="0"
                   Tag="yAxisAnimation" />
        <Button Content="Begin" Grid.Row="1" Grid.Column="1"
                Click="OnBeginStopButton" />
        <Button Content="Pause" Grid.Row="1" Grid.Column="2"
                IsEnabled="False"
                Click="OnPauseResumeButton" />

        <TextBlock Text="Z Axis: " Grid.Row="2" Grid.Column="0"
                   Tag="zAxisAnimation" />
        <Button Content="Begin" Grid.Row="2" Grid.Column="1"
                Click="OnBeginStopButton" />
        <Button Content="Pause" Grid.Row="2" Grid.Column="2"
                IsEnabled="False"
                Click="OnPauseResumeButton" />
        </Grid>
    </Grid>
</Page>
```

单个 DoubleAnimation 对象的持续时间都有一点不同，以避免一起运行时会出现重复模式。代码隐藏文件中的按钮使用 Storyboard 的 Begin、Stop、Pause 和 Resume 方法来控制行动。

```
项目: ThreeDeeSpinningText | 文件: MainPage.xaml.cs(片段)
public sealed partial class MainPage : Page
{
    public MainPage()
    {
        this.InitializeComponent();
    }

    void OnBeginStopButton(object sender, RoutedEventArgs args)
    {
        Button btn = sender as Button;
        string key = GetSibling(btn, -1).Tag as string;
        Storyboard storyboard = this.Resources[key] as Storyboard;
        Button pauseResumeButton = GetSibling(btn, 1) as Button;
        pauseResumeButton.Content = "Pause";

        if  (btn.Content as string == "Begin")
        {
            storyboard.Begin();
            btn.Content = "Stop";
            pauseResumeButton.IsEnabled = true;
        }
        else
        {
            storyboard.Stop();
            btn.Content = "Begin";
            pauseResumeButton.IsEnabled = false;
        }
    }

    void OnPauseResumeButton(object sender, RoutedEventArgs args)
    {
        Button btn = sender as Button;
```

```
        string key = GetSibling(btn, -2).Tag as string;
        Storyboard storyboard = this.Resources[key] as Storyboard;

        if (btn.Content as string == "Pause")
        {
            storyboard.Pause();
            btn.Content = "Resume";
        }
        else
        {
            storyboard.Resume();
            btn.Content = "Pause";
        }
    }

    FrameworkElement GetSibling(FrameworkElement element, int relativeIndex)
    {
        Panel parent = element.Parent as Panel;
        int index = parent.Children.IndexOf(element);
        return parent.Children[index + relativeIndex] as FrameworkElement;
    }
}
```

例如下图中的示例图片。

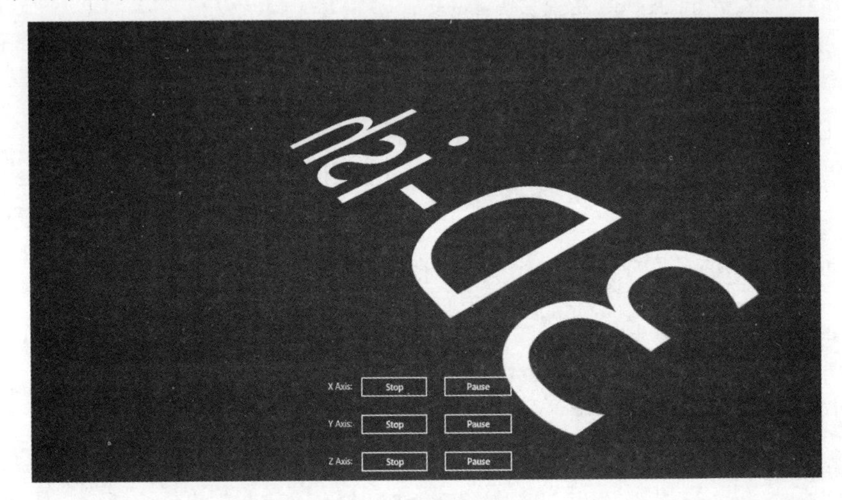

PlaneProjection 类有一堆额外属性。CenterOfRotationX 和 CenterOfRotationY 属性的坐标都是相对于元素的。默认值为 0.5，是元素中心，通常也就是你想要的。CenterOfRotationZ 属性以像素为单位，默认值为 0，和屏幕表面相对应。为了内部计算目的，假设“相机”(或者是你，用户)查看屏幕的距离是 1000 像素，即约 10 英寸。

PlaneProjection 还为 X、Y、Z 维度定义了三个 LocalOffset 属性和三个 GlobalOffset 属性。这些都是以像素为单位的平移因子。LocalOffset 值在旋转之前应用，而 GlobalOffset 值在旋转之后应用。通常，需要设置 GlobalOffset 属性。

以下有一个“翻转面板”的小例子，这项技巧曾经非常难，而且涉及到真正的 3D 编程。其想法如下：面板上有一些控件集合，要通过一组不同(但相关)的控件方法来翻转面板。在本例中，我用两个 Grid 面板表示面板的正面和反面，两个 Grid 面板用不同背景颜色，每一个都包含 TextBlock。

项目：TapToFlip | 文件：MainPage.xaml(片段)

```
<Grid Background="{StaticResource ApplicationPageBackgroundThemeBrush}">
    <Grid HorizontalAlignment="Center"
          VerticalAlignment="Center"
          Tapped="OnGridTapped">

        <Grid Name="grid1"
              Background="Cyan"
              Canvas.ZIndex="1">
            <TextBlock Text="Hello"
                       HorizontalAlignment="Center"
                       FontSize="192" />
        </Grid>

        <Grid Name="grid2"
              Background="Yellow"
              Canvas.ZIndex="0">
            <TextBlock Text="Windows 8"
                       FontSize="192" />
        </Grid>

        <Grid.Projection>
            <PlaneProjection x:Name="projection" />
        </Grid.Projection>
    </Grid>
</Grid>
```

请注意 Canvas.ZIndex 设置。这些设置确保 grid1 在视觉上保持 grid2 上面，即使 grid1 在其共同父类的子集中较早出现。

Resources 部分包含两个 Storyboard 定义，一个是翻转，另一个是翻回。

项目：TapToFlip | 文件：MainPage.xaml(片段)

```
<Page.Resources>
    <Storyboard x:Key="flipStoryboard">
        <DoubleAnimationUsingKeyFrames
                        Storyboard.TargetName="projection"
                        Storyboard.TargetProperty="RotationY">
            <DiscreteDoubleKeyFrame KeyTime="0:0:0" Value="0" />
            <LinearDoubleKeyFrame KeyTime="0:0:0.99" Value="90" />
            <DiscreteDoubleKeyFrame KeyTime="0:0:1.01" Value="-90" />
            <LinearDoubleKeyFrame KeyTime="0:0:2" Value="0" />
        </DoubleAnimationUsingKeyFrames>

        <DoubleAnimation Storyboard.TargetName="projection"
                         Storyboard.TargetProperty="GlobalOffsetZ"
                         From="0" To="-1000" Duration="0:0:1"
                         AutoReverse="True" />

        <ObjectAnimationUsingKeyFrames
                        Storyboard.TargetName="grid1"
                        Storyboard.TargetProperty="(Canvas.ZIndex)">
            <DiscreteObjectKeyFrame KeyTime="0:0:0" Value="1" />
            <DiscreteObjectKeyFrame KeyTime="0:0:1" Value="0" />
        </ObjectAnimationUsingKeyFrames>

        <ObjectAnimationUsingKeyFrames
                        Storyboard.TargetName="grid2"
                        Storyboard.TargetProperty="(Canvas.ZIndex)">
            <DiscreteObjectKeyFrame KeyTime="0:0:0" Value="0" />
            <DiscreteObjectKeyFrame KeyTime="0:0:1" Value="1" />
        </ObjectAnimationUsingKeyFrames>
    </Storyboard>

    <Storyboard x:Key="flipBackStoryboard">
        <DoubleAnimationUsingKeyFrames
                        Storyboard.TargetName="projection"
```

```
                              Storyboard.TargetProperty="RotationY">
                <DiscreteDoubleKeyFrame KeyTime="0:0:0" Value="0" />
                <LinearDoubleKeyFrame KeyTime="0:0:0.99" Value="-90" />
                <DiscreteDoubleKeyFrame KeyTime="0:0:1.01" Value="90" />
                <LinearDoubleKeyFrame KeyTime="0:0:2" Value="0" />
            </DoubleAnimationUsingKeyFrames>

            <DoubleAnimation Storyboard.TargetName="projection"
                             Storyboard.TargetProperty="GlobalOffsetZ"
                             From="0" To="-1000" Duration="0:0:1"
                             AutoReverse="True" />

            <ObjectAnimationUsingKeyFrames
                              Storyboard.TargetName="grid1"
                              Storyboard.TargetProperty="(Canvas.ZIndex)">
                <DiscreteObjectKeyFrame KeyTime="0:0:0" Value="0" />
                <DiscreteObjectKeyFrame KeyTime="0:0:1" Value="1" />
            </ObjectAnimationUsingKeyFrames>

            <ObjectAnimationUsingKeyFrames
                               Storyboard.TargetName="grid2"
                               Storyboard.TargetProperty="(Canvas.ZIndex)">
                <DiscreteObjectKeyFrame KeyTime="0:0:0" Value="1" />
                <DiscreteObjectKeyFrame KeyTime="0:0:1" Value="0" />
            </ObjectAnimationUsingKeyFrames>
        </Storyboard>
    </Page.Resources>
```

两个 storyboards 非常相似。两者都包含 DoubleAnimationUsingKeyFrames 以 PlaneProjection 对象的 RotationY 属性为目标。属性从 0 旋转到 90 度或-90 度(此时对用户成直角)，然后转 180 度，这样动画可以继续在同一方向返回到 0。

同时，GlobalOffsetZ 属性以动画形式从 0 旋转到-1000，然后回到 0。看起来好像是面板掉到了屏幕后面，以准备执行翻转(也许这样，翻转面板就不会碰到用户的鼻子)。

每个 Storyboard 执行到一半的时候，对 Canvas.ZIndex 索引进行交换。Canvas.ZIndex 属性是 ObjectAnimationUsingKeyFrames 的另一个合适目标。

动画由点击触发，而触发由代码隐藏文件处理：

```
项目: TapToFlip | 文件: MainPage.xaml.cs(片段)
public sealed partial class MainPage : Page
{
    Storyboard flipStoryboard, flipBackStoryboard;
    bool flipped = false;

    public MainPage()
    {
        this.InitializeComponent();
        flipStoryboard = this.Resources["flipStoryboard"] as Storyboard;
        flipBackStoryboard = this.Resources["flipBackStoryboard"] as Storyboard;
    }

    void OnGridTapped(object sender, TappedRoutedEventArgs args)
    {
        if (flipStoryboard.GetCurrentState() == ClockState.Active ||
            flipBackStoryboard.GetCurrentState() == ClockState.Active)
        {
            return;
        }

        Storyboard storyboard = flipped ? flipBackStoryboard : flipStoryboard;
        storyboard.Begin();
        flipped ^= true;
    }
}
```

大部分逻辑的目的是防止在前一个 Storyboard 尚未结束时就启动下一个 Storyboard。而根据 storyboard 定义的方式，则会造成导致中断。(试试从 OnGridTapped 删除 return 语句，你会看到令人不满意的结果。)在动画过程中，我宁愿用点击直接逆转操作，只不过需要一些较复杂的逻辑。

10.16 推导 Matrix3D

我们来看一些有点难的数学知识，好吗?

就象前面看到的，二维图形需要 3×3 变换矩阵来进行平移、缩放、旋转和倾斜。从概念上讲，点(x, y)当做好像存在于三维空间，坐标为(x, y, 1)。

一般的二维仿射变换应用如下所示：

$$\begin{vmatrix} x & y & 1 \end{vmatrix} \times \begin{vmatrix} M11 & M12 & 0 \\ M21 & M22 & 0 \\ OffsetX & OffsetY & 1 \end{vmatrix} = \begin{vmatrix} x' & y' & 1 \end{vmatrix}$$

出于此目的，这些是矩阵结构的实际字段。固定第三列限制为仿射变换。矩阵乘法隐含的转换公式为

$$x' = M11 \bullet x + M21 \bullet y + OffsetX$$
$$y' = M12 \bullet x + M22 \bullet y + OffsetY$$

由于是仿射变换，因此正方形总是变换成平行四边形。该平行四边形由三个角定义，而第四个角由其他三个角决定。

是否可以推导出仿射变换，能把一个单位正方形映射为一个任意平行四边形？我们想要的映射如下所示：

$$(0,0) \rightarrow (x_0, y_0)$$
$$(0,0) \rightarrow (x_1, y_1)$$
$$(0,0) \rightarrow (x_2, y_2)$$

如果把这些点代入变换公式，很容易得出所需矩阵的以下单元格：

$$M11 = x_2 - x_0$$
$$M12 = y_2 - y_0$$
$$M21 = x_1 - x_0$$
$$M22 = y_1 - y_0$$
$$OffsetX = x_0$$
$$OffsetY = y_0$$

在 3D 图形编程中，需要 4×4 变换矩阵，而点(*x*, *y*, *z*)则处理为存在于四维空间，其坐标是(*x*, *y*, *z*, 1)。*x*、*y* 和 *z* 之后没有剩余的字母，因此第四维度通常表达为指字母 *w*。转换的应用如下所示：

$$\begin{vmatrix} x & y & z \end{vmatrix} \times \begin{vmatrix} M11 & M12 & M13 & M14 \\ M21 & M22 & M23 & M24 \\ M31 & M32 & M33 & M34 \\ OffsetX & OffsetY & OffsetZ & M44 \end{vmatrix} = \begin{vmatrix} x' & y' & z' & w' \end{vmatrix}$$

以上为 Matrix3D 结构的实际字段。

所得结果的 4×1 矩阵通过所有坐标除以 w'变换回三维空间的一个点：

$$\left| x' \quad y' \quad z' \quad w' \right| \rightarrow \left(\frac{x'}{w'}, \frac{y'}{w'}, \frac{z'}{w'} \right)$$

传统 2D 图形中，通常不希望有被 0 除的情况。但在 3D 图形中，被 0 除的值则是必要的，因为透视图就是这样实现的。你想让平行线在无穷远处相遇，因为那是现实生活中的真实情况。

Windows Runtime 中，Matrix3D 结构的唯一目的是设置为 Matrix3DProjection 对象的 ProjectionMatrix 属性，再可以设置为元素的 Projection 属性，以替代 PlaneProjection。如以下 XAML 所示：

```
<Image ... >
    <Image.Projection>
        <Matrix3DProjection>
            <Matrix3DProjection.ProjectionMatrix>
                1 0 0 0, 0 1 0 0, 0 0 1 0, 0 0 0 1
            </Matrix3DProjection.ProjectionMatrix>
        </Matrix3DProjection>
    </Image.Projection>
</Image>
```

我们实际上不能在 XAML 里实例化 Matrix3D 值，所以需要从第一行开始指定组成矩阵的 16 个数字。本例显示了特征对角线为 1s 的单位矩阵。

在这种环境下，该 4×4 矩阵完全不能使用，因为其应用的元素为平面，Z 坐标为 0，因此矩阵的应用实际如下所示：

$$\left| x \quad y \quad 0 \quad 1 \right| \times \begin{vmatrix} M11 & M12 & M13 & M14 \\ M21 & M22 & M23 & M24 \\ M31 & M32 & M33 & M34 \\ OffsetX & OffsetY & OffsetZ & M44 \end{vmatrix} = \left| x' \quad y' \quad z' \quad w' \right|$$

也就是说，构成整个第三行的单元格(M31、M32、M33 和 M34 值)没有关联。它们被 0 乘，因而不进入计算。

此外，由该方法得到的 3D 点在 Z 轴上折叠，以获得 2D 点映射到视频显示：

$$\left(\frac{x'}{w'}, \frac{y'}{w'}, \frac{z'}{w'} \right) \rightarrow \left(\frac{x'}{w'}, \frac{y'}{w'}, \right)$$

这也是发生在标准 3D 图形上的过程，但通常涉及到较多工作，因为 Z 值也表明对相机来说什么可见、什么被遮挡。

此外，标准 3D 图形只保留了 Z 值范围。“附近的平面”和“远处的平面”依照 Z 来定义，而只有两个平面之间的坐标可见。其余的则直接抛弃，因为这些坐标在概念上要么离相机太近或要么太远。在 Windows Runtime 里，只保留 Z 值在 0 和 1 之间的坐标。为了避免失去一部分变换元素，M13 和 M23 应该设置为 0。OffsetZ 可以设置为 0 和 1 之间的任何值，不过将其设置为 0 也很方便。

如果应用 Matrix3DProjection 到二维元素，转换公式则如下：

$$x' = M11 \cdot x + M21 \cdot y + OffsetX$$
$$y' = M12 \cdot x + M22 \cdot y + OffsetY$$
$$w' = M14 \cdot x + M24 \cdot y + M44$$

如果 M14 和 M24 为 0，M44 为 1，则是一个简单的二维仿射变换。非 0 值的 M14 和 M24 是这些公式的非仿射部分。M44 可以是 1 之外的数，但如果为是 0，则总是可以找到一个 M44 等于 1 的等效变换，只需要把所有字段乘以 1/M44。

在非仿射变换中，正方形不一定要变换为平行四边形。然而，非仿射变换仍有局限。非仿射变换不能把正方形变换成任意四边形。变换后的线不能互相交叉，四个角一定要凸起。

我们尝试推导一个非仿射变换，能把正方形的四个角映射为任意点：

$$(0,0) \to (x_0, y_0)$$
$$(0,1) \to (x_1, y_1)$$
$$(1,0) \to (x_2, y_2)$$
$$(1,1) \to (x_3, y_3)$$

如果分解成两个变换，练习会更容易：

$$(0,0) \to (0,0) \to (x_0, y_0)$$
$$(0,1) \to (0,1) \to (x_1, y_1)$$
$$(1,0) \to (1,0) \to (x_2, y_2)$$
$$(1,1) \to (a,b) \to (x_3, y_3)$$

第一个变换显然是非仿射变换，我把它称为 **B**，第二个变换我们强制成为仿射变换，称为 **A**。强制为仿射变换的方法是把从 a 和 b 推导出值。复合变换则为 **B×A**。

我已经向你展示了仿射变换的推导，如果从 3×3 矩阵变换到 4×4 矩阵，甚至都不需要改变符号。但我们也希望该仿射变换能把点(a,b)映射到任意点(x_3,y_3)。通过应用推导出的仿射变换到(a,b)，同时为了解决 a 和 b，我们得到：

$$a = \frac{M22 \cdot x_3 - M21 \cdot y_3 + M21 \cdot OffsetY - M22 \cdot PffsetX}{M11 \cdot M22 - M12 \cdot M21}$$
$$b = \frac{M11 \cdot y_3 - M12 \cdot x_3 + M12 \cdot OffsetY - M11 \cdot PffsetX}{M11 \cdot M22 - M12 \cdot M21}$$

现在，我们试试非仿射变换，需要能产生以下映射：

$$(0,0) \to (0,0)$$
$$(0,1) \to (0,1)$$
$$(1,0) \to (1,0)$$
$$(1,1) \to (a,b)$$

以下是之前的变换公式：

$$x' = M11 \cdot x + M21 \cdot y + OffsetX$$
$$y' = M12 \cdot x + M22 \cdot y + OffsetY$$
$$w' = M14 \cdot + M24 \cdot y + M44$$

请记住，x'和 y'必须除以 w'，以得到被变换的点。

如果(0，0)映射到(0，0)，则 OffsetX 和 OffsetY 为 0，M44 为非 0。我们来冒下险，把

M44 设定为 1。

如果(0，1)映射到(0，1)，则 M21 必定为 0(以计算 *x'*的零值)，*y'*除以 *w'*必定是 1，也就是说，M24 等于 M22 减 1。

如果(1，0)映射到(1，0)，则 M12 为 0(y'的零值)，*x'*除以 *w'*必定为 1，或者说 M14 等于 M11 减 1。

如果(1,1)映射到(*a*,*b*)，则能得到：

$$M11 = \frac{a}{a+b-1}$$

$$M22 = \frac{b}{a+b-1}$$

a 和 *b* 已经推导过。

现在来写代码。我想显示由该过程推导出的实际矩阵。这就是名为 DisplayMatrix3D 的 UserControl 派生的目的。XAML 文件内容要比 4×4 Grid 的 TextBlock 元素的多一点。

项目：NonAffineStretch | 文件：DisplayMatrix3D.xaml(片段)

```
<UserControl ... >
    <UserControl.Resources>
        <Style TargetType="TextBlock">
            <Setter Property="TextAlignment" Value="Right" />
            <Setter Property="Margin" Value="6 0" />
        </Style>
    </UserControl.Resources>

    <Border BorderBrush="{StaticResource ApplicationForegroundThemeBrush}"
            BorderThickness="1 0">
        <Grid>
            <Grid.RowDefinitions>
                <RowDefinition Height="Auto" />
                <RowDefinition Height="Auto" />
                <RowDefinition Height="Auto" />
                <RowDefinition Height="Auto" />
            </Grid.RowDefinitions>

            <Grid.ColumnDefinitions>
                <ColumnDefinition Width="Auto" />
                <ColumnDefinition Width="Auto" />
                <ColumnDefinition Width="Auto" />
                <ColumnDefinition Width="Auto" />
            </Grid.ColumnDefinitions>

            <TextBlock Name="m11" Grid.Row="0" Grid.Column="0" />
            <TextBlock Name="m12" Grid.Row="0" Grid.Column="1" />
            <TextBlock Name="m13" Grid.Row="0" Grid.Column="2" />
            <TextBlock Name="m14" Grid.Row="0" Grid.Column="3" />

            <TextBlock Name="m21" Grid.Row="1" Grid.Column="0" />
            <TextBlock Name="m22" Grid.Row="1" Grid.Column="1" />
            <TextBlock Name="m23" Grid.Row="1" Grid.Column="2" />
            <TextBlock Name="m24" Grid.Row="1" Grid.Column="3" />

            <TextBlock Name="m31" Grid.Row="2" Grid.Column="0" />
            <TextBlock Name="m32" Grid.Row="2" Grid.Column="1" />
            <TextBlock Name="m33" Grid.Row="2" Grid.Column="2" />
            <TextBlock Name="m34" Grid.Row="2" Grid.Column="3" />

            <TextBlock Name="m41" Grid.Row="3" Grid.Column="0" />
            <TextBlock Name="m42" Grid.Row="3" Grid.Column="1" />
            <TextBlock Name="m43" Grid.Row="3" Grid.Column="2" />
            <TextBlock Name="m44" Grid.Row="3" Grid.Column="3" />
```

```
        </Grid>
    </Border>
</UserControl>
```

代码隐藏文件定义了 Matrix3D 类型的依赖属性，因此每当该属性发生变化，都会收到通知。请注意：如果现有 Matrix3D 结构的属性发生改变，不会产生通知。必须替换整个结构。

```
项目: NonAffineStretch | 文件: DisplayMatrix3D.xaml.cs(片段)
public sealed partial class DisplayMatrix3D : UserControl
{
    static DependencyProperty matrix3DProperty =
        DependencyProperty.Register("Matrix3D",
            typeof(Matrix3D), typeof(DisplayMatrix3D),
            new PropertyMetadata(Matrix3D.Identity, OnPropertyChanged));

    public DisplayMatrix3D()
    {
        this.InitializeComponent();
    }

    public static DependencyProperty Matrix3DProperty
    {
        get { return matrix3DProperty; }
    }

    public Matrix3D Matrix3D
    {
        set { SetValue(Matrix3DProperty, value); }
        get { return (Matrix3D)GetValue(Matrix3DProperty); }
    }

    static void OnPropertyChanged(DependencyObject obj,
                                  DependencyPropertyChangedEventArgs args)
    {
        (obj as DisplayMatrix3D).OnPropertyChanged(args);
    }

    void OnPropertyChanged(DependencyPropertyChangedEventArgs args)
    {
        m11.Text = this.Matrix3D.M11.ToString("F3");
        m12.Text = this.Matrix3D.M12.ToString("F3");
        m13.Text = this.Matrix3D.M13.ToString("F3");
        m14.Text = this.Matrix3D.M14.ToString("F6");

        m21.Text = this.Matrix3D.M21.ToString("F3");
        m22.Text = this.Matrix3D.M22.ToString("F3");
        m23.Text = this.Matrix3D.M23.ToString("F3");
        m24.Text = this.Matrix3D.M24.ToString("F6");

        m31.Text = this.Matrix3D.M31.ToString("F3");
        m32.Text = this.Matrix3D.M32.ToString("F3");
        m33.Text = this.Matrix3D.M33.ToString("F3");
        m34.Text = this.Matrix3D.M34.ToString("F6");

        m41.Text = this.Matrix3D.OffsetX.ToString("F0");
        m42.Text = this.Matrix3D.OffsetY.ToString("F0");
        m43.Text = this.Matrix3D.OffsetZ.ToString("F0");
        m44.Text = this.Matrix3D.M44.ToString("F0");
    }
}
```

格式规范的选择基于这些单元格常见范围内的一点经验。

MainPage 的 XAML 文件包括 DisplayMatrix3D 控件实例，但也引用我网站里的图片，并且用四个 Thumb 控件进行装饰。Thumb 控件允许拖动任意角到任意位置。前缀 ul、ur、ll 和 lr 代表 upper-left(左上)、upper-left(右上)、lower-left(左下)和 lower-right(右下)。

项目: NonAffineStretch | 文件: MainPage.xaml(片段)

```
<Page ... >
    <Page.Resources>
        <Style TargetType="Thumb">
            <Setter Property="Width" Value="48" />
            <Setter Property="Height" Value="48" />
            <Setter Property="HorizontalAlignment" Value="Left" />
            <Setter Property="VerticalAlignment" Value="Top" />
        </Style>
    </Page.Resources>

    <Grid Background="{StaticResource ApplicationPageBackgroundThemeBrush}">
        <Image Source="http://www.charlespetzold.com/pw6/PetzoldJersey.jpg"
               Stretch="None"
               HorizontalAlignment="Left"
               VerticalAlignment="Top">
            <Image.Projection>
                <Matrix3DProjection x:Name="matrixProjection" />
            </Image.Projection>
        </Image>

        <Thumb DragDelta="OnThumbDragDelta">
            <Thumb.RenderTransform>
                <TransformGroup>
                    <TranslateTransform X="-24" Y="-24" />
                    <TranslateTransform x:Name="ulTranslate" X="100" Y="100" />
                </TransformGroup>
            </Thumb.RenderTransform>
        </Thumb>

        <Thumb DragDelta="OnThumbDragDelta">
            <Thumb.RenderTransform>
                <TransformGroup>
                    <TranslateTransform X="-24" Y="-24" />
                    <TranslateTransform x:Name="urTranslate" X="420" Y="100" />
                </TransformGroup>
            </Thumb.RenderTransform>
        </Thumb>

        <Thumb DragDelta="OnThumbDragDelta">
            <Thumb.RenderTransform>
                <TransformGroup>
                    <TranslateTransform X="-24" Y="-24" />
                    <TranslateTransform x:Name="llTranslate" X="100" Y="500" />
                </TransformGroup>
            </Thumb.RenderTransform>
        </Thumb>

        <Thumb DragDelta="OnThumbDragDelta">
            <Thumb.RenderTransform>
                <TransformGroup>
                    <TranslateTransform X="-24" Y="-24" />
                    <TranslateTransform x:Name="lrTranslate" X="420" Y="500" />
                </TransformGroup>
            </Thumb.RenderTransform>
        </Thumb>

        <local:DisplayMatrix3D HorizontalAlignment="Right"
                               VerticalAlignment="Bottom"
                               FontSize="24"
                               Matrix3D="{Binding ElementName=matrixProjection,
                                                  Path=ProjectionMatrix}" />
    </Grid>
</Page>
```

代码隐藏文件实现了刚刚向你展示的数学，只不过需要另一个矩阵把图片的实际大小和位置映射到单位正方形。在 CalculateNewTransform 代码中是称为 S 的矩阵。

项目：NonAffineStretch | 文件：MainPage.xaml.cs(片段)

```
public sealed partial class MainPage : Page
{
    // Location and Size of Image with no transform
    Rect imageRect = new Rect(0, 0, 320, 400);

    public MainPage()
    {
        this.InitializeComponent();

        Loaded += (sender, args) =>
            {
                CalculateNewTransform();
            };
    }

    void OnThumbDragDelta(object sender, DragDeltaEventArgs args)
    {
        Thumb thumb = sender as Thumb;
        TransformGroup xformGroup = thumb.RenderTransform as TransformGroup;
        TranslateTransform translate = xformGroup.Children[1] as TranslateTransform;
        translate.X += args.HorizontalChange;
        translate.Y += args.VerticalChange;
        CalculateNewTransform();
    }

    void CalculateNewTransform()
    {
        Matrix3D matrix = CalculateNewTransform(imageRect,
                                  new Point(ulTranslate.X, ulTranslate.Y),
                                  new Point(urTranslate.X, urTranslate.Y),
                                  new Point(llTranslate.X, llTranslate.Y),
                                  new Point(lrTranslate.X, lrTranslate.Y));
        matrixProjection.ProjectionMatrix = matrix;
    }

    // The returned transform maps the points (0, 0),
    // (0, 1), (1, 0), and (1, 1) to the points
    // ptUL, ptUR, ptLL, and ptLR normalized based on rect.
    static Matrix3D CalculateNewTransform(Rect rect, Point ptUL, Point ptUR,
                                          Point ptLL, Point ptLR)
    {
        // Scale and translate normalization transform
        Matrix3D S = new Matrix3D()
        {
            M11 = 1 / rect.Width,
            M22 = 1 / rect.Height,
            OffsetX = -rect.Left / rect.Width,
            OffsetY = -rect.Top / rect.Height,
            M44 = 1
        };

        // Affine transform: Maps
        // (0, 0) --> ptUL
        // (1, 0) --> ptUR
        // (0, 1) --> ptLL
        // (1, 1) --> (x2 + x1 + x0, y2 + y1 + y0)
        Matrix3D A = new Matrix3D()
        {
            OffsetX = ptUL.X,
            OffsetY = ptUL.Y,
            M11 = (ptUR.X - ptUL.X),
            M12 = (ptUR.Y - ptUL.Y),
```

```
                M21 = (ptLL.X - ptUL.X),
                M22 = (ptLL.Y - ptUL.Y),
                M44 = 1
            };

            // Non-affine transform
            Matrix3D B = new Matrix3D();
            double den = A.M11 * A.M22 - A.M12 * A.M21;
            double a = (A.M22 * ptLR.X - A.M21 * ptLR.Y +
            A.M21 * A.OffsetY - A.M22 * A.OffsetX) / den;
            double b = (A.M11 * ptLR.Y - A.M12 * ptLR.X +
            A.M12 * A.OffsetX - A.M11 * A.OffsetY) / den;

            B.M11 = a / (a + b - 1);
            B.M22 = b / (a + b - 1);
            B.M14 = B.M11 - 1;
            B.M24 = B.M22 - 1;
            B.M44 = 1;

            // Product of three transforms
            return S * B * A;
        }
    }
```

不同于二维 Matrix 结构，Matrix3D 结构做的是乘法操作，因而使数组操作容易得多。

当然，也可以把拇指拖动到图片消失的位置，因为至少有一个角度成凹行或线条相互交叉。但在这些限制条件下，确实可以把图片拉伸为非仿射形状(如下图所示)。

显然，让 Windows Runtime 能应用锥形变换需要花一些精力，但变形照片会让人们觉得很好玩，这种乐趣是对工作的报偿。

第 11 章　三 个 模 板

“模板”一词一般指用于创建相同或类似对象的模式或模具。在 Windows Runtime 中，模板是一段 XAML，Windows 可用来创建可视树元素。看上去可能没那么惊人。本书从第一页开始就在讨论 Windows 把 XAML 变为可视树。但模板几乎总是包含数据绑定，因此，单个模板会产生许多可视树，并根据绑定来源呈现不同外观。出于此原因，模板被经常定义为资源，因此可以共享以及多次使用。

本章标题所指三个模板，分别对应派生自 FrameworkTemplate 的三个类：

```
Object
    DependencyObject
        FrameworkTemplate (非可实例化)
            DataTemplate
            ControlTemplate
            ItemsPanelTemplate
```

不能在代码中定义模板，而必须使用 XAML，也不要期望查询 Windows Runtime 文档能得到有关这些类的更深层次知识。DataTemplate 只定义了一个公共方法，ControlTemplate 只定义了一个公共属性，而 ItemsPanelTemplate 只定义自己。与模板类的机制有联系的一切东西几乎都在 Windows Runtime 内部。

用 DataTemplate 可以给不一定本来就要有视觉效果的数据对象加上可视化外观。我会先通过派生自 ContentControl 的控件来演示 DataTemplate，但初看起来似乎适用性有限。但 DataTemplate 对于展示集合中单个条目是必不可少的，这涉及到派生自 ItemsControl 的控件。

通过 ControlTemplate 可以重定义标准控件外观，如果要定制应用的可视化效果，ControlTemplate 算得上是一件强大的工具。

ItemsPanelTemplate 比其他两个要简单得多，且只在 ItemsControl 派生类中发挥作用。

从对通用工具的期望来看，定义作为 DataTemplate 或者 ControlTemplate 一部分的模板会很复杂。许多程序员都喜爱 Expression Blend 对设计模板所起到的帮助作用。然而，像往常一样，我将演示如何“手工”创建模板。即使还是用 Expression Blend，你也会更好理解 Expression Blend 所生成的 XAML。

本章结束时，Visual Studio 会生成 StandardStyles.xaml 文件作为标准项目的一部分，而你应该完全能理解里面的所有内容。

11.1　按 钮 数 据

Windows Runtime 的几种常见元素和控件，可以有可视化子对象。最明显的例子是 Panel，Panel 可以通过 UIElementCollection 类型的 Children 属性支持多个子对象。Border

可以有一个子对象；其 Child 经常为 UIElement 类型。如果要创建源自 UserControl 的自定义控件，可以把可视树设置为其 Content 属性，Content 属性也为 UIElement 类型。

Button 也有 Content 属性，不过为 Object 类型。为什么会这样？

简而言之，这是因为 Button 派生自 ContentControl，而 ContentControl 定义了 Object 类型的 Content：

Object
  DependencyObject
    UIElement
      FrameworkElement
        Control
          ContentControl
            ButtonBase
              Button

不过，这真的不算是一个好答案。

大多数时候，不需要把 Button 的 Content 属性设置为任何旧对象。大多数时候，可以把 Content 属性设置为文本，你很可能会(而且正确)假定幕后正在创建 TextBlock 用来显示该文本。

对于更漂亮的按钮，可以把 Content 属性设置为派生自 UIElement 的任何东西。例如，如下按钮有一个面板，包含位图和格式化文本：

```
<Button HorizontalAlignment="Center">
    <StackPanel>
        <Image Source="http://www.charlespetzold.com/pw6/PetzoldJersey.jpg"
               Width="100" />
        <TextBlock>
            <Italic>Tap</Italic> to shoot the basket
        </TextBlock>
    </StackPanel>
</Button>
```

效果如下图所示。

但如果 Button 的 Content 属性的确是 Object 类型，则应该可以将其设置为非派生自 UIElement 的东西。这种情况下应该怎么办？试一试把按钮内容设置为 LinearGradientBrush，例如：

```
<Button HorizontalAlignment="Center">
    <LinearGradientBrush>
```

```
            <GradientStop Offset="0" Color="Red" />
            <GradientStop Offset="1" Color="Blue" />
        </LinearGradientBrush>
    </Button>
```

这样做完全合法，即使并不是很清楚你的想法。一般把画笔设置为元素的不同属性(比如 Button 的 Background 或 Foreground 属性)以便用不同方式进行着色，但画笔本身并没有任何可视化表示。出于此原因，在按钮上看到的东西为画笔的 ToString 表示。ToString 会返回对某些类有意义的东西，但默认实施则直接返回全部合格的类名，如下图所示。

效果不是非常令人满意。

这个问题是可以解决的！ContentControl 不仅仅定义(而且 Button 也继承)Content 属性，还有 ContentTemplate 属性。把 ContentTemplate 属性设置为 DataTemplate 类型的对象，并在 DataTemplate 里定义可视树。可视树通常包含引用对象的绑定，而该对象设置为 Content 属性。

我们首先来给 Button 的 ContentTemplate 属性添加属性-元素标签，并设置为 DataTemplate：

```
<Button HorizontalAlignment="Center">
    <LinearGradientBrush>
        <GradientStop Offset="0" Color="Red" />
        <GradientStop Offset="1" Color="Blue" />
    </LinearGradientBrush>
    <Button.ContentTemplate>
        <DataTemplate>

        </DataTemplate>
    </Button.ContentTemplate>
</Button>
```

在 DataTemplate 标签里，可以定义元素的可视树，这些元素通过某些方法来使用按钮内容。我们来试一个 Ellipse：

```
<Button HorizontalAlignment="Center">
    <LinearGradientBrush>
        <GradientStop Offset="0" Color="Red" />
        <GradientStop Offset="1" Color="Blue" />
    </LinearGradientBrush>

    <Button.ContentTemplate>
        <DataTemplate>
            <Ellipse Width="120"
                Height="144"
                Fill="{Binding}" />
        </DataTemplate>
    </Button.ContentTemplate>
</Button>
```

注意 Ellipse 的 Fill 属性的 Binding 标记扩展。这种绑定显然非常简单。不需要 Source，因为已经把模板的 DataContext 来源设置为按钮内容。绑定没有 Path，因为我们想把 Fill 属性直接设置为按钮内容。模板会使按钮内容可见(见下图)。

把 Ellipse 设置为按钮内容和直接在 Fill 属性上定义 LinearGradientBrush，在视觉上是一样的，如下所示：

```
<Button HorizontalAlignment="Center">
     <Ellipse Width="120"
              Height="144">
          <Ellipse.Fill>
               <LinearGradientBrush>
                    <GradientStop Offset="0" Color="Red" />
                    <GradientStop Offset="1" Color="Blue" />
               </LinearGradientBrush>
          </Ellipse.Fill>
     </Ellipse>
</Button>
```

然而，模板可以是样式的一部分，而多个按钮可以共享样式，因此，模板方式绝对更加灵活和通用。

DataTemplate 中的数据绑定不需要像我展示的那样简单。下面有一个更广泛的模板，模板在按钮内容里引用第二个 GradientStop 对象的 Color 属性，并且通过它来设置 SolidColorBrush 的颜色，SolidColorBrush 画出椭圆的圆周：

```
<Button HorizontalAlignment="Center">
     <LinearGradientBrush>
          <GradientStop Offset="0" Color="Red" />
          <GradientStop Offset="1" Color="Blue" />
     </LinearGradientBrush>

     <Button.ContentTemplate>
          <DataTemplate>
               <Ellipse Width="120"
                        Height="144"
                        Fill="{Binding}"
                        StrokeThickness="6">
                    <Ellipse.Stroke>
                         <SolidColorBrush Color="{Binding Path=GradientStops[1].Color}" />
                    </Ellipse.Stroke>
               </Ellipse>
          </DataTemplate>
     </Button.ContentTemplate>
</Button>
```

SolidColorBrush 的 Color 属性的 Binding 使用 Path 来引用 LinearGradientBrush 的 GradientStops 属性，即通过索引来获得特定 GradientStop 对象，然后通过 Color 来获得该对象属性：

```
<SolidColorBrush Color="{Binding Path=GradientStops[1].Color}" />
```

DataTemplate 里的 Binding 通常没有 ElementName 或 Source 设置，因为源作为数据情景而提供。Path 是 Binding 中的第一个(也是唯一)条目，因此，可以删除 Path=部分：

```
<SolidColorBrush Color="{Binding GradientStops[1].Color}" />
```

这就是在数据模板里几乎总会看到的绑定，效果如下图所示。

当然，模板要依赖 LinearGradientBrush 内容。如果没有，绑定则不起作用。

可以在页面(或其他 XAML 文件)的 Resources 节定义 DataTemplate：

```
<Page.Resources>
    <DataTemplate x:Key="ellipseTemplate">
        <Ellipse Width="120"
            Height="144"
            Fill="{Binding}"
            StrokeThickness="6">
            <Ellipse.Stroke>
                <SolidColorBrush Color="{Binding GradientStops[1].Color}" />
            </Ellipse.Stroke>
        </Ellipse>
    </DataTemplate>
</Page.Resources>
```

在按钮中引用该模板，则只需要标准 StaticResource 标记扩展：

```
<Button HorizontalAlignment="Center"
        ContentTemplate="{StaticResource ellipseTemplate}">
    <LinearGradientBrush>
        <GradientStop Offset="0" Color="Red" />
        <GradientStop Offset="1" Color="Blue" />
    </LinearGradientBrush>
</Button>
```

多个按钮(或其他 ContentControl 派生类)可以共享模板。通常不能共享可视树，因为可视化元素不能有多个父类。但该模板的工作方式完全不同。共享模板的时候，可用来为每个引用该模板的控件生成一个唯一的可视树。如果有 100 个按钮把 ContentTemplate 属性设置为该模板，则会创建 100 个 Ellipse 元素。

通常在 Style 中定义模板，以便可以同时把其他属性应用到控件上。SharedStyleWithDataTemplate 项目在页面的 Resources 节定义了一个隐式样式。

```
项目：SharedStyleWithDataTemplate | 文件：MainPage.xaml(片段)
<Page ... >
    <Page.Resources>
        <Style TargetType="Button">
            <Setter Property="HorizontalAlignment" Value="Center" />
            <Setter Property="VerticalAlignment" Value="Center" />
            <Setter Property="ContentTemplate">
```

```
                <Setter.Value>
                    <DataTemplate>
                        <Ellipse Width="144"
                                 Height="192"
                                 Fill="{Binding}" />
                    </DataTemplate>
                </Setter.Value>
            </Setter>
        </Style>
    </Page.Resources>

    <Grid Background="{StaticResource ApplicationPageBackgroundThemeBrush}">
        <Grid.ColumnDefinitions>
            <ColumnDefinition Width="*" />
            <ColumnDefinition Width="*" />
            <ColumnDefinition Width="*" />
        </Grid.ColumnDefinitions>

        <Button Grid.Column="0">
            <SolidColorBrush Color="Green" />
        </Button>

        <Button Grid.Column="1">
            <LinearGradientBrush>
                <GradientStop Offset="0" Color="Blue" />
                <GradientStop Offset="1" Color="Red" />
            </LinearGradientBrush>
        </Button>

        <Button Grid.Column="2">
            <ImageBrush ImageSource="http://www.charlespetzold.com/pw6/ PetzoldJersey.jpg" />
        </Button>
    </Grid>
</Page>
```

该隐式样式自动设置每个 Button 的属性，包括 ContentTemplate 属性。而单个按钮要做的事情是定义 Brush 派生类作为内容：

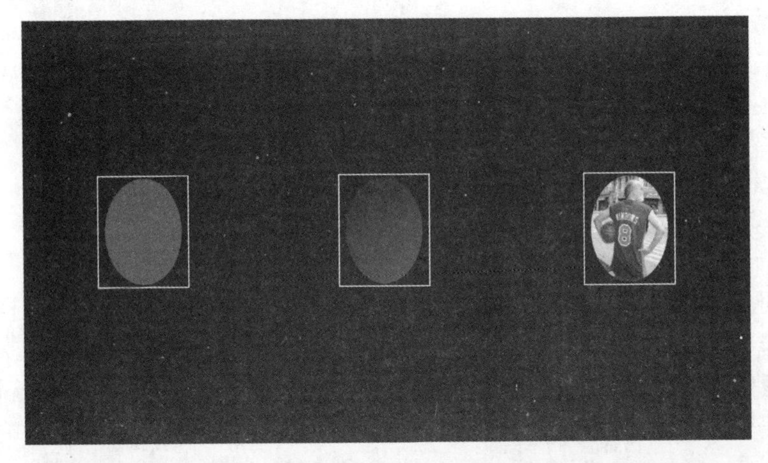

模板通过普通数据绑定以引用对象，因此，如果源对象实施了通知机制(如 INotifyPropertyChanged)，则会自动更新视觉效果。例如，假设创建 Clock 类，该类使用 CompositionTarget.Rendering 事件来获取当前时间，并设置一些属性，每一个属性都会触发 PropertyChanged 事件。

```
项目: ClockButton | 文件: Clock.cs
using System;
using System.ComponentModel;
```

```
using System.Runtime.CompilerServices;
using Windows.UI.Xaml.Media;

namespace ClockButton
{
    public class Clock : INotifyPropertyChanged
    {
        bool isEnabled;
        int hour, minute, second;
        int hourAngle, minuteAngle, secondAngle;

        public event PropertyChangedEventHandler PropertyChanged;

        public bool IsEnabled
        {
            set
            {
                if (SetProperty<bool>(ref isEnabled, value, "IsEnabled"))
                {
                    if  (isEnabled)
                        CompositionTarget.Rendering += OnCompositionTargetRendering;
                    else
                        CompositionTarget.Rendering -= OnCompositionTargetRendering;
                }
            }
            get
            {
                return isEnabled;
            }
        }

        public int Hour
        {
            set { SetProperty<int>(ref hour, value); }
            get { return hour; }
        }

        public int Minute
        {
            set { SetProperty<int>(ref minute, value); }
            get { return minute; }
        }

        public int Second
        {
            set { SetProperty<int>(ref second, value); }
            get { return second; }
        }

        public int HourAngle
        {
            set { SetProperty<int>(ref hourAngle, value); }
            get { return hourAngle; }
        }

        public int MinuteAngle
        {
            set { SetProperty<int>(ref minuteAngle, value); }
            get { return minuteAngle; }
        }

        public int SecondAngle
        {
            set { SetProperty<int>(ref secondAngle, value); }
            get { return secondAngle; }
        }

        void OnCompositionTargetRendering(object sender, object args)
```

```
        {
            DateTime dateTime = DateTime.Now;
            this.Hour = dateTime.Hour;
            this.Minute = dateTime.Minute;
            this.Second = dateTime.Second;

            this.HourAngle = 30 * dateTime.Hour + dateTime.Minute / 2;
            this.MinuteAngle = 6 * dateTime.Minute + dateTime.Second / 10;
            this.SecondAngle = 6 * dateTime.Second + dateTime.Millisecond / 166;
        }

        protected bool SetProperty<T>(ref T storage, T value,
                        [CallerMemberName] string propertyName = null)
        {
            if (object.Equals(storage, value))
                return false;

            storage = value;
            OnPropertyChanged(propertyName);
            return true;
        }

        protected virtual void OnPropertyChanged(string propertyName)
        {
            if (PropertyChanged != null)
                PropertyChanged(this, new PropertyChangedEventArgs(propertyName));
        }
    }
}
```

可以把该类实例设置为 Button 内容并通过 DataTemplate 来定义如何渲染该对象。

```
项目: ClockButton | MainPage.xaml(片段)
<Grid Background="{StaticResource ApplicationPageBackgroundThemeBrush}">
    <Button HorizontalAlignment="Center"
            VerticalAlignment="Center">

        <local:Clock IsEnabled="True" />

        <Button.ContentTemplate>
            <DataTemplate>
                <Grid Width="144" Height="144">
                    <Grid.Resources>
                        <Style TargetType="Polyline">
                            <Setter Property="Stroke"
                                    Value="{StaticResource ApplicationForegroundThemeBrush}" />
                        </Style>
                    </Grid.Resources>

                    <Polyline Points="72 80, 72 24"
                              StrokeThickness="6">
                        <Polyline.RenderTransform>
                            <RotateTransform Angle="{Binding HourAngle}"
                                             CenterX="72"
                                             CenterY="72" />
                        </Polyline.RenderTransform>
                    </Polyline>

                    <Polyline Points="72 88, 72 12"
                              StrokeThickness="3">
                        <Polyline.RenderTransform>
                            <RotateTransform Angle="{Binding MinuteAngle}"
                                             CenterX="72"
                                             CenterY="72" />
                        </Polyline.RenderTransform>
                    </Polyline>

                    <Polyline Points="72 88, 72 6"
```

```
                        StrokeThickness="1">
                <Polyline.RenderTransform>
                    <RotateTransform Angle="{Binding SecondAngle}"
                                     CenterX="72"
                                     CenterY="72" />
                </Polyline.RenderTransform>
            </Polyline>
        </Grid>
    </DataTemplate>
  </Button.ContentTemplate>
 </Button>
</Grid>
```

注意，我在 DataTemplate 的可视树中给 Polyline 定义了隐式 Style。这会应用于该可视树里所有 Polyline 元素。这些 Polyline 元素会将其 RenderTransform 属性设置为 RotateTransform，而 RotateTransform 的 Angle 则受限于 Clock 类的不同属性。三个 Polyline 元素共同构成了一个原始时钟，再加上有完整功能 Button 的一部分，就能指示时间(见下图)。

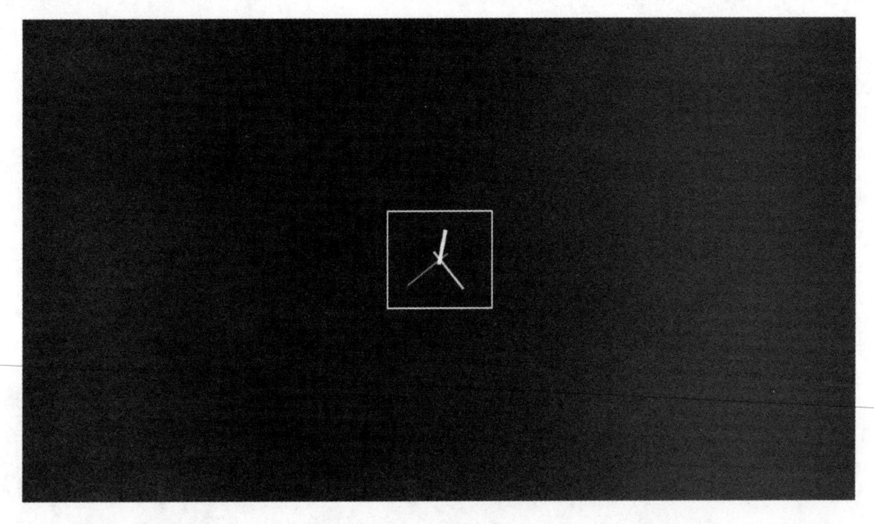

记住，设置为 Button 的 ContentTemplate 属性的 DataTemplate 只定义了按钮内容的外观，而没有定义按钮的镶边。按钮仍然有矩形边界，例如，(在黑色主题中)鼠标经过时，仍然会有灰色外观，而点击按钮，按钮则会呈现白色背景。要改变按钮外观的这些方面需要把 ControlTemplate 对象设置为按钮的 Template 属性，本章稍后会进行讨论。

11.2　决　策

XAML 并不是真正的编程语言，因为它没有循环和 if 语句。XAML 没有决策能力，因此不能包含条件执行的标记块。

但我们试一试总是可以的。

我们要在前面项目中 Clock 类的基础上进行扩展，让时钟能够区分上午和下午。要做到这一点，我们要从 Clock 类中派生新类，同时带有一个新属性，命名为 Hour12，范围从 1 到 12。我们还会赋予新类一对 Boolean 属性，命名为 IsAm 和 IsPm，我们希望通过这些属性可以根据值来显示一点点不同的内容。

ConditionalClockButton 项目包含一个到 ClockButton 项目里 Clock.cs 文件的链接，并

定义了派生自 Clock 的 TwelveHourClock 类：

```
项目: ConditionalClockButton | 文件: TwelveHourClock.cs
namespace ConditionalClockButton
{
    public class TwelveHourClock : ClockButton.Clock
    {
        // Initialize for Hour value of 0
        int hour12 = 1;
        bool isAm = true;
        bool isPm = false;

        public int Hour12
        {
            set { SetProperty<int>(ref hour12, value); }
            get { return hour12; }
        }

        public bool IsAm
        {
            set { SetProperty<bool>(ref isAm, value); }
            get { return isAm; }
        }

        public bool IsPm
        {
            set { SetProperty<bool>(ref isPm, value); }
            get { return isPm; }
        }

        protected override void OnPropertyChanged(string propertyName)
        {
            if (propertyName == "Hour")
            {
                this.Hour12 = (this.Hour - 1) % 12 + 1;
                this.IsAm = this.Hour < 12;
                this.IsPm = !this.IsAm;
            }

            base.OnPropertyChanged(propertyName);
        }
    }
}
```

幸运的是，我在 Clock 中定义了 OnPropertyChanged 方法，因此，新类可以覆写该方法，并检查 propertyName 参数是否等于“Hour”。如果等于，则设置所有三个新属性，这些属性也会调用 SetProperty，因此，OnPropertyChanged 会触发自身的 PropertyChanged 事件。

假设你想要一个按钮，按钮上写着：“It’s after 9 in the morning”或“It’s after 3 in the afternoon”。以下 TwelveHourClock 类就有你需要的所有信息，可以这样开始定义按钮：

```
<Button>

    <local:TwelveHourClock />

    <Button.ContentTemplate>
        <DataTemplate>
            <StackPanel Orientation="Horizontal">
                <TextBlock Text="It's after " />
                <TextBlock Text="{Binding Hour12}" />
                <TextBlock Text=" o'clock" />
                <TextBlock Text=" in the morning!" />
                <TextBlock Text=" in the afternoon!" />
            </StackPanel>
        </DataTemplate>
```

```
        </Button.ContentTemplate>
    </Button>
```

然而，需要禁止最后两个 TextBlock 元素中的一个。只有当 IsAm 属性为 true 时，其中第一个才显示，而只有当 IsPm 属性是 true 时，第二个才显示。你会记得，元素有 Visibility 属性，可以设置为 Visibility 枚举项，Visible 或 Collapsed 都可以。如果有方法可以把 TwelveHourClock 布尔型属性转换为 Visibility 枚举项，那就太好了。

第 4 章已经介绍过绑定转换器，而最受欢迎的绑定转换器之一通常被命名为 BooleanToVisibilityConverter。事实上，如果在 Visual Studio 里创建 Grid App 或者 Split App 类型的项目，则可以在 Common 文件夹中免费获得那些转换器，但写一个也并不困难：

```
项目: ConditionalClockButton | 文件: BooleanToVisibilityConverter.cs
using System;
using Windows.UI.Xaml;
using Windows.UI.Xaml.Data;

namespace ConditionalClockButton
{
    public sealed class BooleanToVisibilityConverter : IValueConverter
    {
        public object Convert(object value, Type targetType, object parameter, string language)
        {
            return (bool)value ? Visibility.Visible : Visibility.Collapsed;
        }

        public object ConvertBack(object value, Type targetType, object parameter, string lang)
        {
            return (Visibility)value == Visibility.Visible;
        }
    }
}
```

Visual Studio 生成的版本要复杂一些：检查 value 参数是被实际抛出的类型。但如果要限制转换器对某些特定标记的使用，则可以放宽类型检查。以下程序当然有限制使用的情况，转换器是在模板的 Resources 节、而不是在 Page 的 Resources 节进行实例化。

```
项目: ConditionalClockButton | 文件: MainPage.xaml(片段)
<Grid Background="{StaticResource ApplicationPageBackgroundThemeBrush}">
    <Button HorizontalAlignment="Center"
            VerticalAlignment="Center"
            FontSize="24">

        <local:TwelveHourClock IsEnabled="True" />

        <Button.ContentTemplate>
            <DataTemplate>
                <StackPanel Orientation="Horizontal">
                    <StackPanel.Resources>
                        <local:BooleanToVisibilityConverter x:Key="booleanToVisibility" />
                    </StackPanel.Resources>

                    <TextBlock Text="It's after " />
                    <TextBlock Text="{Binding Hour12}" />
                    <TextBlock Text=" o'clock" />
                    <TextBlock Text=" in the morning!"
                               Visibility="{Binding IsAm,
                                       Converter={StaticResource booleanToVisibility}}" />
                    <TextBlock Text=" in the afternoon!"
                               Visibility="{Binding IsPm,
                                       Converter={StaticResource booleanToVisibility}}" />
                </StackPanel>
            </DataTemplate>
```

```
        </Button.ContentTemplate>
    </Button>
</Grid>
```

最后两个 TextBlock 的 Visibility 属性现在受限于 TwelveHourClock 的 IsAm 和 IsPm 属性，而 BooleanToVisibilityConverter 则决定哪一项可见，从下图可以看出。

11.3 集合控件和实际使用 DataTemplate

我演示了如何通过派生自 ContentControl 的代表类来使用 DataTemplate，但这种类没有那么多，老实说，通过这些类来使用 DataTemplate 也并不常见。

实际使用 DataTemplate 是通过派生自 ItemsControl 的控件，这些控件保存通常为相同类型的对象集合：

```
Object
    DependencyObject
        UIElement
            FrameworkElement
                Control
                    ItemsControl
                        Selector (非可实例化)
                            ComboBox
                            FlipView
                            ListBox
                            ListViewBase (非可实例化)
                                GridView
                                ListView
```

当然，其中最著名的是 ListBox，ListBox 从 Windows 最开始的时候就存在(以一种形式或者另一种形式)。典型 ListBox 显示为一个竖向的列表，用户可以使用键盘或鼠标进行滚动和选择。(现代 ListBox 则更灵活，能进行触控操作。)Windows 后来出现了 ComboBox，因组合文本编辑字段和下拉列表而得名。FlipView 是 Windows 8 的控件。

GridView 和 ListView 比其他控件更复杂，我们把它们留到第 12 章进行讨论。

处理这些不同控件时，很容易忽视 ItemsControl 本身，然而，其他所有控件都是由其派生而来的。ItemsControl 只为展示的目而显示条目集合，并没有选择的概念。Selector 类添加选择逻辑，其他所有其他类由此派生而来。我一般把整个控件组称为条目控件。它们显示了条目集合。

通过下列四种方法之一可以把对象放入条目控件：单独在 XAML 中、单独在代码中、在散装代码中以及在散装 XMAL 中，而且通常使用数据绑定。能放入条目控件的对象通常不是派生自 UIElement。这些条目通常是业务对象或视图模型。

对于简短的列表，可以在 XAML 文件进行指定：

```
<Grid Background="{StaticResource ApplicationPageBackgroundThemeBrush}">
    <ItemsControl FontSize="24">
        <x:String>One potato</x:String>
        <x:String>Two potato</x:String>
        <x:String>Three potato</x:String>
        <x:String>Four</x:String>
        <x:String>Five potato</x:String>
        <x:String>Six potato</x:String>
        <x:String>Seven potato</x:String>
        <x:String>More</x:String>
    </ItemsControl>
</Grid>
```

ItemsControl 的内容属性是 Items，即 ItemCollection 类型的对象，该类实施 IList、IEnumerable 和 IObservableVector。(稍后将进一步讨论这些接口)。

添加到 ItemsControl 的条目为 String 类型的对象，因此可以显示为文本(见下图)。

ItemsControl 的这种特定使用并不见得就比 StackPanel 好出很多，除了可以用 String 类型的条目而不是用 TextBlock 进行填充。当然，为每一项生成 TextBlock。

不同于派生自 ItemsControl 的控件，ItemsControl 自身并没有内置滚动功能。如果需要滚动一堆条目，你会想把 ItemsControl 放在 ScrollViewer 里，如下所示：

```
<Grid Background="{StaticResource ApplicationPageBackgroundThemeBrush}">
    <ScrollViewer>
        <ItemsControl FontSize="24">
            <Color>AliceBlue</Color>
            <Color>AntiqueWhite</Color>
            <Color>Aqua</Color>
            ...
            <Color>WhiteSmoke</Color>
            <Color>Yellow</Color>
            <Color>YellowGreen</Color>
        </ItemsControl>
    </ScrollViewer>
</Grid>
```

列表可以滚动，但并不是十分有用，因为每个 Color 项都用其 ToString 表示来显示(见下图)。

```
Windows.Foundation.IReference`1<Windows.UI.Color>
Windows.Foundation.IReference`1<Windows.UI.Color>
Windows.Foundation.IReference`1<Windows.UI.Color>
Windows.Foundation.IReference`1<Windows.UI.Color>
Windows.Foundation.IReference`1<Windows.UI.Color>
Windows.Foundation.IReference`1<Windows.UI.Color>
Windows.Foundation.IReference`1<Windows.UI.Color>
Windows.Foundation.IReference`1<Windows.UI.Color>
Windows.Foundation.IReference`1<Windows.UI.Color>
Windows.Foundation.IReference`1<Windows.UI.Color>
Windows.Foundation.IReference`1<Windows.UI.Color>
Windows.Foundation.IReference`1<Windows.UI.Color>
Windows.Foundation.IReference`1<Windows.UI.Color>
Windows.Foundation.IReference`1<Windows.UI.Color>
Windows.Foundation.IReference`1<Windows.UI.Color>
Windows.Foundation.IReference`1<Windows.UI.Color>
Windows.Foundation.IReference`1<Windows.UI.Color>
Windows.Foundation.IReference`1<Windows.UI.Color>
Windows.Foundation.IReference`1<Windows.UI.Color>
Windows.Foundation.IReference`1<Windows.UI.Color>
Windows.Foundation.IReference`1<Windows.UI.Color>
Windows.Foundation.IReference`1<Windows.UI.Color>
Windows.Foundation.IReference`1<Windows.UI.Color>
Windows.Foundation.IReference`1<Windows.UI.Color>
Windows.Foundation.IReference`1<Windows.UI.Color>
Windows.Foundation.IReference`1<Windows.UI.Color>
```

无论何时看见条目控件有类型名称列表，都不要担心！如果看到绑定生效，实际上应该十分高兴才是，因为这意味着可以更好地显示这些条目。要渲染这些项，就要把 ItemsControl 的 ItemTemplate 属性设置为 DataTemplate 绑定：

```
<Grid Background="{StaticResource ApplicationPageBackgroundThemeBrush}">
    <ScrollViewer>
        <ItemsControl>
            <ItemsControl.ItemTemplate>
                <DataTemplate>
                    <Rectangle Width="144"
                               Height="72"
                               Margin="12">
                        <Rectangle.Fill>
                            <SolidColorBrush Color="{Binding}" />
                        </Rectangle.Fill>
                    </Rectangle>
                </DataTemplate>
            </ItemsControl.ItemTemplate>

            <Color>AliceBlue</Color>
            <Color>AntiqueWhite</Color>
            <Color>Aqua</Color>
            ...
            <Color>WhiteSmoke</Color>
```

```
                <Color>Yellow</Color>
                <Color>YellowGreen</Color>
            </ItemsControl>
        </ScrollViewer>
    </Grid>
```

ItemsControl 的 ItemTemplate 属性类似于 ContentControl 的 ContentTemplate 属性。两者都是 DataTemplate 类型。然而，通过 ItemTemplate 属性，模板能为每一条目生成可视树。效果如下图所示。

构造 ItemsControl 的时候，DataTemplate 用于生成 141 个 Rectangle 元素和 141 个 SolidColorBrush 对象，每一个对应着控件里的每一个条目。

当然，你可能不想在 XAML 文件中有一个含有 141 个 Color 条目的整个列表。你可能只想在代码中生成。在 ColorItems 项目里，XAML 文件不包含任何条目，却有一个更精巧的模板，它也显示了颜色组件。

```
项目: ColorItems | 文件: MainPage.xaml(片段)
<Grid Background="{StaticResource ApplicationPageBackgroundThemeBrush}">
    <ScrollViewer>
        <ItemsControl Name="itemsControl"
                      FontSize="24">
            <ItemsControl.ItemTemplate>
                <DataTemplate>
                    <Grid Width="240"
                     Margin="0 12">
                        <Grid.ColumnDefinitions>
                            <ColumnDefinition Width="144" />
                            <ColumnDefinition Width="Auto" />
                        </Grid.ColumnDefinitions>

                        <Grid.RowDefinitions>
                            RowDefinition Height="Auto" />
                            <RowDefinition Height="Auto" />
                            <RowDefinition Height="Auto" />
                            <RowDefinition Height="Auto" />
                        </Grid.RowDefinitions>

                        <Rectangle Grid.Column="0"
                                   Grid.Row="0"
```

```
                            Grid.RowSpan="4"
                            Margin="12 0">
                    <Rectangle.Fill>
                        <SolidColorBrush Color="{Binding}" />
                    </Rectangle.Fill>
                </Rectangle>

                <StackPanel Grid.Column="1"
                            Grid.Row="0"
                            Orientation="Horizontal">
                    <TextBlock Text="A = " />
                    <TextBlock Text="{Binding A}" />
                </StackPanel>

                <StackPanel Grid.Column="1"
                            Grid.Row="1"
                            Orientation="Horizontal">
                    <TextBlock Text="R = " />
                    <TextBlock Text="{Binding R}" />
                </StackPanel>

                <StackPanel Grid.Column="1"
                            Grid.Row="2"
                            Orientation="Horizontal">
                    <TextBlock Text="G = " />
                    <TextBlock Text="{Binding G}" />
                </StackPanel>

                <StackPanel Grid.Column="1"
                            Grid.Row="3"
                            Orientation="Horizontal">
                    <TextBlock Text="B = " />
                    <TextBlock Text="{Binding B}" />
                </StackPanel>
            </Grid>
        </DataTemplate>
    </ItemsControl.ItemTemplate>
  </ItemsControl>
 </ScrollViewer>
</Grid>
```

条目本身在代码中生成。正如现在你会期望的，代码隐藏文件通过反射来获取由静态 Colors 类所定义的所有 Color 属性。通过由 ItemCollection 所定义的 Add 方法，将每个 Color 值添加到 ItemsControl。以下代码展示了把条目放条目控件的第二种方法。

```
项目: ColorItems | 文件: MainPage.xaml(片段)
public MainPage()
{
    this.InitializeComponent();
    IEnumerable<PropertyInfo> properties = typeof(Colors).GetTypeInfo().DeclaredProperties;

    foreach (PropertyInfo property in properties)
    {
        Color clr = (Color)property.GetValue(null);
        itemsControl.Items.Add(clr);
    }
}
```

现在我们就得到了十进制值显示的每种颜色(见下图)。

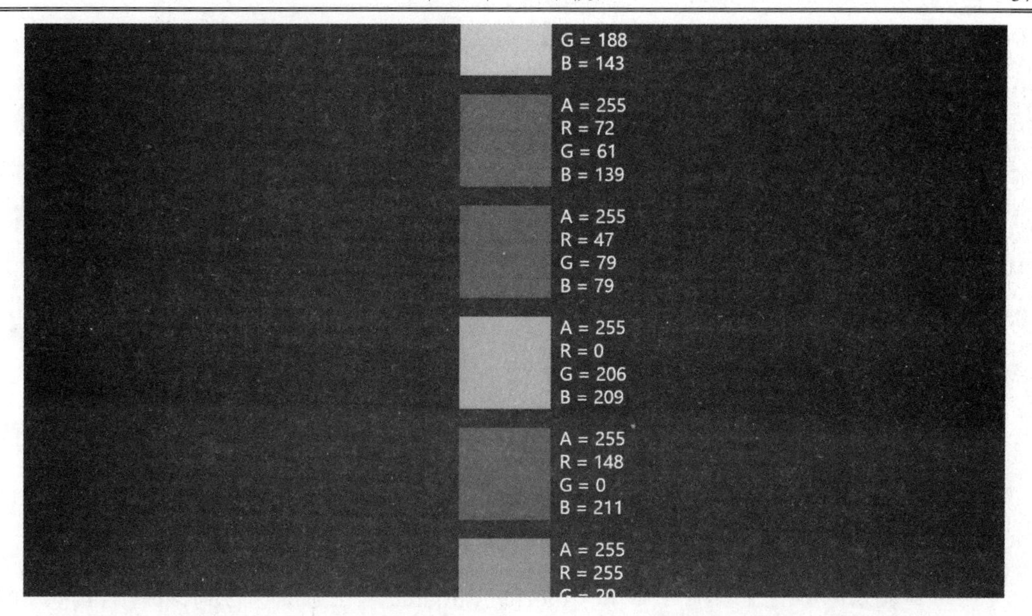

不幸的是，我们不能使用相同方法来显示每个颜色的名字，因为这不是 Color 结构的一部分。如果想要随着颜色而显示名称，需要用提供该名称的类的实例来填充 ItemsControl。

项目：Petzold.ProgrammingWindows6.Chapter11 | 文件：NamedColor.cs(片段)

```
public class NamedColor
{
    static NamedColor()
    {
        List<NamedColor> colorList = new List<NamedColor>();
        IEnumerable<PropertyInfo> properties = typeof(Colors).GetTypeInfo().DeclaredProperties;

        foreach (PropertyInfo property in properties)
        {
            NamedColor namedColor = new NamedColor
            {
                Name = property.Name,
                Color = (Color)property.GetValue(null)
            };

            colorList.Add(namedColor);
        }

        All = colorList;
    }

    public static IEnumerable<NamedColor> All { private set; get; }

    public string Name { private set; get; }

    public Color Color { private set; get; }
}
```

NamedColor 类有两个公共属性：string 类型的 Name 属性、Color 类型的 Color 属性。NamedColor 类还定义了 IEnumerable <NamedColor >类型的静态属性，称为 All。静态构造函数把属性设置为由所有 NamedColor 对象组成的集合，而 NamedColor 对象通过静态 Colors 类反射获取。

我没有把 NamedColor 类定义为实施 INotifyPropertyChanged，因为任何 NamedColor 对象的属性在对象初始化后没有改变。

为了显示十六进制的颜色值，Petzold.ProgrammingWindows6.Chapter11 库还包含了 ByteToHexStringConverter：

```
项目: Petzold.ProgrammingWindows6.Chapter11 | 文件: ByteToHexStringConverter.cs
using System;
using Windows.UI.Xaml.Data;

namespace Petzold.ProgrammingWindows6.Chapter11
{
    public class ByteToHexStringConverter : IValueConverter
    {
        public object Convert(object value, Type targetType, object parameter, string language)
        {
            return ((byte)value).ToString("X2");
        }
        public object ConvertBack(object value, Type targetType, object parameter, string lang)
        {
            return value;
        }
    }
}
```

与本章剩余部分的许多项目一样，ColorItemsSource 解决方案包含到该库项目的一个链接：对于 Visual Studio 里的 ColorItemsSource 解决方案，我右击解决方案资源管理器里的解决方法名称，并选择“添加”和“现有项目”。我导航到了 Petzold.ProgrammingWindows6.Chapter11.csproj 文件。我再定义了一个对该项目的引用：右击 ColorItemsSource 项目下的“引用”，并且在“引用管理器”对话框中选择左边的“项目”和右边的库。MainPage.xaml 包含一个库声明的 XML 命名空间：

```
xmlns:ch11="using:Petzold.ProgrammingWindows6.Chapter11"
```

MainPage.xaml.cs 文件包含该命名空间的 using 指令：

```
using Petzold.ProgrammingWindows6.Chapter11;
```

项目之所以称为 ColorItemsSource，是因为我已经展示了如何用 XAML 或代码访问 ItemsControl 的 Items 属性来填充 ItemCollection 对象。可替代的方法是 ItemsSource 属性。该属性定义为 object 类型，但毫无疑问你会把 ItemsSource 设置为能实现 IEnumerable 接口的东西。设置为 ItemsSource 的对象变成了 ItemsControl 集合，此时的 Items 属性为只读。

可以从代码或 XAML 中设置 ItemsSource 属性。我先展示代码方法。以下 XAML 文件中，大部分内容是 DataTemplate，为集合里的每一个 NamedColor 条目定义可视树。

```
项目: ColorItemsSource | 文件: MainPage.xaml(片段)
<Page ...
    xmlns:ch11="using:Petzold.ProgrammingWindows6.Chapter11"
    ... >
    <Page.Resources>
        <ch11:ByteToHexStringConverter x:Key="byteToHexString" />
    </Page.Resources>

    <Grid Background="{StaticResource ApplicationPageBackgroundThemeBrush}">
        <ScrollViewer>
            <ItemsControl Name="itemsControl">

                <ItemsControl.ItemTemplate>
                    <DataTemplate>
                        <Border BorderBrush="{StaticResource ApplicationForegroundThemeBrush}"
                                BorderThickness="1"
                                Width="336"
```

```
                            Margin="6">
                    <Grid>
                        <Grid.ColumnDefinitions>
                            <ColumnDefinition Width="Auto" />
                            <ColumnDefinition Width="*"/>
                        </Grid.ColumnDefinitions>

                        <Rectangle Grid.Column="0"
                                   Height="72"
                                   Width="72"
                                   Margin="6">
                            <Rectangle.Fill>
                                <SolidColorBrush Color="{Binding Color}" />
                            </Rectangle.Fill>
                        </Rectangle>

                        <StackPanel Grid.Column="1"
                                    VerticalAlignment="Center">
                            <TextBlock FontSize="24"
                                   Text="{Binding Name}" />

                            <ContentControl FontSize="18">
                                <StackPanel Orientation="Horizontal">
                                    <TextBlock Text="{Binding Color.A,
                                        Converter={StaticResource
                                        byteToHexString}}" />
                                    <TextBlock Text="-" />
                                    <TextBlock Text="{Binding Color.R,
                                        Converter={StaticResource
                                        byteToHexString}}" />
                                    <TextBlock Text="-" />
                                    <TextBlock Text="{Binding Color.G,
                                        Converter={StaticResource
                                        byteToHexString}}" />
                                    <TextBlock Text="-" />
                                    <TextBlock Text="{Binding Color.B,
                                        Converter={StaticResource
                                        byteToHexString}}" />
                                </StackPanel>
                            </ContentControl>
                        </StackPanel>
                    </Grid>
                </Border>
            </DataTemplate>
        </ItemsControl.ItemTemplate>
    </ItemsControl>
</ScrollViewer>
</Grid>
</Page>
```

注意显示 Color 组件的 7 个 TextBlock 元素。这些元素都在一个横向 StackPanel 内，而 StackPanel 在 ContentControl 内。ContentControl 存在的唯一原因是提供 7 个 TextBlock 元素所继承的 FontSize。隐式 Style 也会很好地发挥作用。

SolidColorBrush 和 TextBlock 元素的绑定明显暗示正在显示 NamedColor 类型的对象，但在 XAML 文件中，没有实例化 NamedColor 对象。而 ItemsControl 的 ItemsSource 属性在代码隐藏文件的构造函数中进行设置。

项目：ColorItemsSource | 文件：MainPage.xaml.cs（片段）

```
public MainPage()
{
    this.InitializeComponent();

    itemsControl.ItemsSource = NamedColor.All;
}
```

设置好了 ItemsSource，ItemsControl 即对集合里的所有条目生成可视树(见下图)。

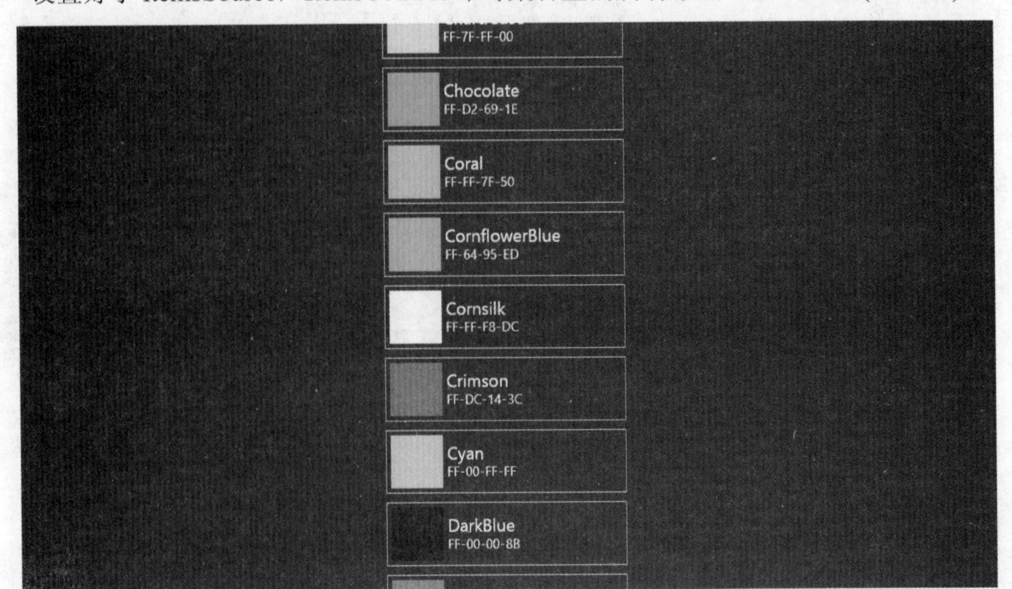

还可以把 ItemsSource 属性绑定到集合来实施一个 XAML 解决方案。ColorItemsSourceWithBinding 项目非常类似于 ColorItemsSource 项目，也使用 Petzold.ProgrammingWindows6.Chapter11 库，同时在 XAML 文件中定义相同 DataTemplate。但 NamedColor 对象被实例化为资源，到 All 属性的绑定是在 ItemsControl 的 ItemsSource 属性里定义的。

```
项目: ColorItemsSourceWithBinding | 文件: MainPage.xaml(片段)
<Page ...
     xmlns:ch11="using:Petzold.ProgrammingWindows6.Chapter11"
     ... >
     <Page.Resources>
          <ch11:NamedColor x:Key="namedColor" />
          <ch11:ByteToHexStringConverter x:Key="byteToHexString" />
     </Page.Resources>

     <Grid Background="{StaticResource ApplicationPageBackgroundThemeBrush}">
          <ScrollViewer>
               <ItemsControl ItemsSource="{Binding Source={StaticResource namedColor},
                                                   Path=All}">
                    <ItemsControl.ItemTemplate>
                         <DataTemplate>
                         ...
                         </DataTemplate>
                    </ItemsControl.ItemTemplate>
               </ItemsControl>
          </ScrollViewer>
     </Grid>
</Page>
```

如果资源本身是集合对象，则可以把 ItemsSource 设置为资源的 StaticResourc 标记扩展，但由于只能从 NamedColor 的 All 属性访问集合，因此需要 Binding 标记扩展以引用 NamedColor 对象和 All 属性。

还记得第 4 章有两个程序吧！它们用不同的方法显示颜色列表，我说过要等到第 11 章才能看到最佳方式。现在是时候了。类定义你想显示的条目类型，而 ItemsControl 上的

DataTemplate 则定义要渲染的条目。

这就是结合使用集合、绑定和模板的情况，代表 Windows Runtime 程序设计的本质。

11.4　集合和接口

在正常情况下，构造 NamedColor 类的时候，我会把实例构造函数定义为 protected 或 private，因为对单个 NamedColor 而言，类的外部实例化并不是非常合理。ColorItemsSource 项目这么做能发挥作用，但 ColorItemsSourceWithBinding 就不能这么做。在第二个程序里，NamedColor 需要一个公共无参数构造函数，因为必须在 XAML 中实例化类为资源。否则无法使用该特定 NamedColor 实例，提供一种方法在后台去访问静态 All 属性。在大部分程序中，可能有一个实例化了一次的视图模型类(称为单例模式)，它提供了特定集合的实例属性。(下一章将定义此类。)

在 NamedColor 里，我在定义 All 属性的类型时是有选择余地的。我本可以把它定义为它本来的样子：List<NamedColor>。我也可以通过另一个极端方式将其定义为对象。这都不是问题。如果设置好了条目控件的 ItemsSource 属性，控件自己就会检查设置为 ItemsSource 的对象是否实现了 IEnumerable。这就可以访问集合中的实际条目。这就是我为什么把属性定义为 IEnumerable <NamedColor>的原因。无论后来我怎么改变 NamedColor 类的内部，我都知道属性总能实施 IEnumerable，因为它必须为条目控件提供合适的集合源。

如果你开始阅读集合和接口文档，很容易产生困惑。.NET 程序员可以识别出 System.Collections.Generic 命名空间里定义的 IEnumerable<T>接口。但在某些情况下，该接口称为 IIterable<T>，由 Windows.Foundations.Collections 命名空间定义。这是同一个接口，但 C#和 Visual Basic 程序员将其称为 IEnumerable，C++程序员使用的则是 IIterable。

C#和 Visual Basic 程序员也习惯了处理两种基本集合类型：List<T>(即 T 类型对象的有序集合)和 Dictionary<TKey,TValue>(即唯一的非空键和相应值的有序集合)。然而，C++程序员知道这两种基本类型集合的名字为 vector 和 map。出于此原因，Windows.Foundations.Collections 命名空间包含 IVector<T>和 IMap<K,V>接口，但 C#和 Visual Basic 程序员把这些接口看成 IList<T>和 IDictionary<TKey,TValue>，两者都在 System.Collections.Generic 里进行定义。

如果就记住“vector 是列表，而 map 是字典”，自然会少一点困惑。

你已经熟悉了 System.ComponentModel 里定义的 INotifyPropertyChanged 接口。(C++程序员使用相同的名称接口，不过由 Windows.UI.Xaml.Data 定义。)如果把一个集合中的条目设置为 ItemsSource，条目实施了 INotifyPropertyChanged 接口，条目属性的任何变化都将反映到绑定到这些属性的可视化元素上。换句话说，DataTemplate 里的绑定能响应属性变化。这一点你在 ClockButton 项目中 Clock 类型的单一条目里看到过。它也能处理集合项，详见下一章的讨论。

处理集合和条目控件时，还有一个重要接口叫 INotifyCollectionChanged，它由 System.Collections.Specialized 定义。该接口定义 CollectionChanged 事件，如果条目自身发生变化时，也就是说，集合中的条目增加、删除或重排序时，即被触发。如果设置为条目控件的 ItemsSource 属性的集合实施了 INotifyCollectionChanged，条目控件就会知道这些变化并从显示中自动增加或删除条目。

对于 C#程序员，ObservableCollection<T>类实施 INotifyCollectionChanged。

11.5 轻击和选择

在 ColorItemsSource 和 ColorItemsSourceWithBinding 这两个项目中，每一项的视觉效果都由 DataTemplate 定义，但并不禁止从单个项中获得输入事件。在 ColorItemsSource 或 ColorItemsSourceWithBinding 中，启用 DataTemplate 的 Border 非空背景，并且定义 Tapped 事件的处理程序：

```
<Border BorderBrush="{StaticResource ApplicationForegroundThemeBrush}"
        BorderThickness="1"
        Width="336"
        Margin="6"
        Background="{StaticResource ApplicationPageBackgroundThemeBrush}"
        Tapped="OnItemTapped">
```

这将为 144 个 Border 元素中的每一个指定同一个 Tapped 处理程序。在处理程序中，sender 参数为 Border，事件参数的 OriginalSource 属性可以是 Border 或模板里的另一个元素。无论如何，元素的 DataContext 都是关联到该项的特定 NamedColor 对象，也就是说，可以提取 Color 值并用于对背景进行着色：

```
void OnItemTapped(object sender, TappedRoutedEventArgs args)
{
    object dataContext = (args.OriginalSource as FrameworkElement).DataContext;
    Color clr = (dataContext as NamedColor).Color;
     (this.Content as Grid).Background = new SolidColorBrush(clr);
}
```

轻击 Brown 项时的结果，如下图所示。

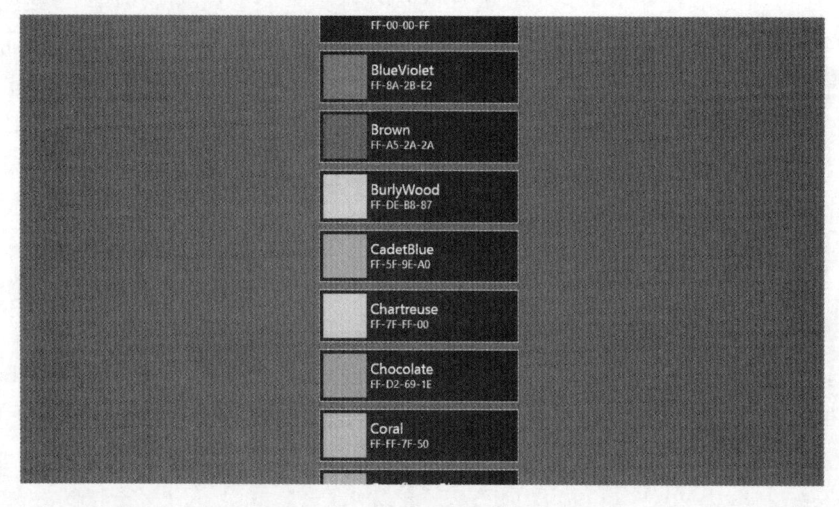

考虑到可以很容易在 ItemsControl 里实现轻击或者点击界面，你可能想知道为什么需要派生自 Selector 的控件，尤其是 ListBox。

有一个简单的回答：轻击不是选择。如果选择 ListBox 中的一项，该项就会有不同的视觉外观。此外，使用键盘上的方向键可以移动选项。如果这些都不是你需要的，

ItemsControl 显然就是令人满意的解决方案。

为了表明当前所选项，Selector 定义了三种不同(但显然相关)的属性。

- SelectedIndex，所选项在集合中的索引，如果当前没有选择项目，则为-1。
- SelectedItem，所选项自身，如果当前没有选择项目，则为 null。
- SelectedValue，通常为所选项的属性值，用 SelectedValuePath 表示。(稍后有更多讨论。)

如果 SelectedIndex 不是-1，则 SelectedItem 和用 SelectedIndex 索引 Items 属性所得到的就是同一个对象。所有这些属性能通过编程或在 XAML 中进行设置。如果 ListBox 首先用条目进行填充，则其 SelectedIndex 为-1，而 SelectedItem 将为 null，直到这些属性明确改变或直到用户用手指或鼠标选择一项。

Selector 定义选择变化时会触发的 SelectionChanged 事件，然后处理程序通过这些属性之一获取所选项。

SelectedItem 由依赖属性支持，也就是说，可以是数据绑定的目标，但在更常用的情况下是被用作绑定源。SimpleListBox 项目使用 NamedColor.All 作为 Items 属性的绑定源，但没有定义模板。相反，项目使用不同的方法来显示条目。

项目: SimpleListBox | 文件: MainPage.xaml(片段)

```
<Page ... >
    <Page.Resources>
        <ch11:NamedColor x:Key="namedColor" />
    </Page.Resources>

    <Grid>
        <ListBox Name="lstbox"
                 ItemsSource="{Binding Source={StaticResource namedColor},
                                       Path=All}"
                 DisplayMemberPath="Name"
                 Width="288"
                 HorizontalAlignment="Center" />

        <Grid.Background>
            <SolidColorBrush Color="{Binding ElementName=lstbox,
                                             Path=SelectedItem.Color}" />
        </Grid.Background>
    </Grid>
</Page>
```

ListBox 包含自身的 ScrollViewer，但往往会试图占据尽可能多的屏幕空间，不管 HorizontalAlignment 和 VerticalAlignment 的设置如何。你可能想给 ListBox 赋予具体的 Width，就像我所做的一样。一会儿你就会明白，为什么 ListBox 不能基于其条目的最大宽度来确定其自身宽度。

我把 DisplayMemberPath 设置为“Name”，即指 ListBox 中条目的 Name 属性，并没有定义 DataTemplate 用于显示 NamedColor 项。这些项为 NamedColor 类型。幸运的是，NamedColor 包含 Name 属性，ListBox 用于显示条目属性。一开始的时候，因为没有选中项，所以设置为 Grid 的 SolidColorBrush 会引用默认 Color 值，但一旦选择了一项，颜色就会形成窗口背景(见下图)。

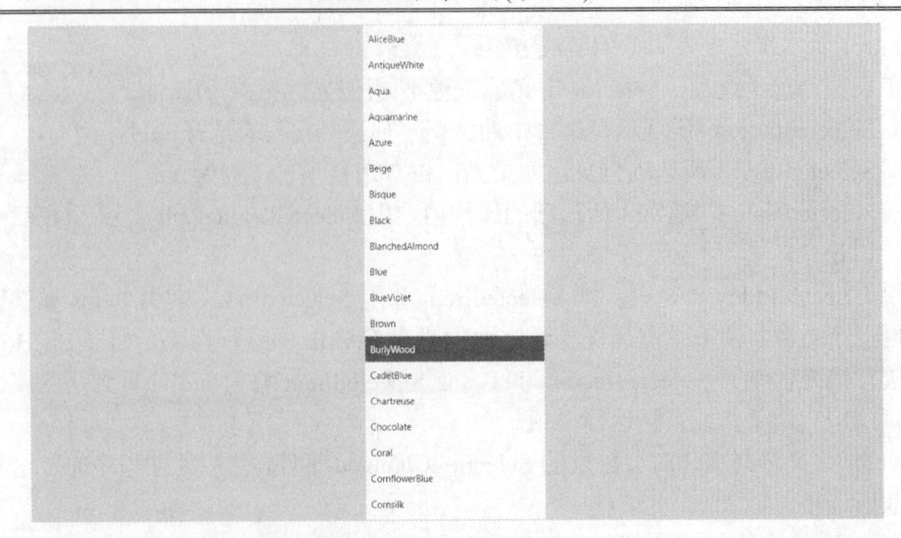

该程序为黑色主题。ListBox 条目的亮背景和所选项高亮表示 ListBox 的默认行为。本章稍后会演示如何改变这种高亮显示。如果尝试运行程序，你会发现可以使用键盘的方向键、PageUp 和 PageDown、Home 和 End 等按键来移动选项。

有另一种替代方法可以用来定义所选项的绑定。SelectedValuePath 类似于用来表明想要显示的条目属性的 DisplayMemberPath，SelectedValuePath 是作为 SelectedValue 显示的属性名称：

```
<Grid>
    <ListBox Name="lstbox"
             ItemsSource="{Binding Source={StaticResource namedColor},
                                   Path=All}"
             DisplayMemberPath="Name"
             SelectedValuePath="Color"
             Width="288"
             HorizontalAlignment="Center" />

    <Grid.Background>
        <SolidColorBrush Color="{Binding ElementName=lstbox,
                                         Path=SelectedValue}" />
    </Grid.Background>
</Grid>
```

ListBox 的 SelectedValuePath 属性表明 ListBox 条目的 Color 属性应作为 SelectedValue 属性而显示，因此简化了 SolidColorBrush 的绑定。

很容易混淆 SelectedItem 和 SelectedValue。如果没有设置 SelectedValuePath 属性，两者就是一回事。如果设置了，SelectedItem 就是集合里的一个对象，而 SelectedValue 则是该对象的属性。

更常见的做法是把 ListBox 的 ItemTemplate 属性设置为 DataTemplate，就像下面一样。我将模板简化为不显示颜色的十六进制表示，但在其他地方和前面的程序一样。

```
项目: ListBoxWithItemTemplate | 文件: MainPage.xaml(片段)
<Page ... >
    <Page.Resources>
        <ch11:NamedColor x:Key="namedColor" />
    </Page.Resources>

    <Grid>
        <ListBox Name="lstbox"
```

```
                    ItemsSource="{Binding Source={StaticResource namedColor},
                                   Path=All}"
                    Width="388">
               <ListBox.ItemTemplate>
                    <DataTemplate>
                         <Border
                              BorderBrush="{Binding RelativeSource={RelativeSource TemplatedParent},
                                            Path=Foreground}"
                              BorderThickness="1"
                              Width="336"
                              Margin="6"
                              Loaded="OnItemLoaded">
                         <Grid>
                              <Grid.ColumnDefinitions>
                                   <ColumnDefinition Width="Auto" />
                                   <ColumnDefinition Width="*"/>
                              </Grid.ColumnDefinitions>

                              <Rectangle Grid.Column="0"
                                         Height="72"
                                         Width="72"
                                         Margin="6">
                              <Rectangle.Fill>
                                   <SolidColorBrush Color="{Binding Color}" />
                              </Rectangle.Fill>
                         </Rectangle>

                         <TextBlock Grid.Column="1"
                                    FontSize="24"
                                    Text="{Binding Name}"
                                    VerticalAlignment="Center" />
                         </Grid>
                    </Border>
               </DataTemplate>
          </ListBox.ItemTemplate>
     </ListBox>

     <Grid.Background>
          <SolidColorBrush Color="{Binding ElementName=lstbox,
                                   Path=SelectedItem.Color}" />
          </Grid.Background>
     </Grid>
</Page>
```

你可能会注意到，在该条目模板的很前面，如何定义 Border 着色很不同。在使用 ItemsControl 的程序里，Border 能够引用主题前台画笔来着色：

```
BorderBrush="{StaticResource ApplicationForegroundThemeBrush}"
```

而在黑色主题中(本章一直都使用黑色主题)，画笔为白色。

然而，我们发现 ListBox 条目默认为白色背景，也就是说在该背景下白色画笔就会消失。我们真的想把它设置为已模板化的祖先元素的 Foreground 属性，就只能通过特定绑定语法：

```
BorderBrush="{Binding RelativeSource={RelativeSource TemplatedParent},
                      Path=Foreground}"
```

RelativeSource 只有两个选项 Self 和 TemplatedParent，此处不能使用 Self，因为 Border 没有 Foreground 属性。

TemplatedParent 到底是什么？在这种情况下，它是 ContentPresenter，ContentPresenter 是一个罕见类，除非你正在写另一种模板类型(控件模板)，本章后面将进行讨论。TextBlock 不需要任何绑定来获得恰当颜色，因为其直接继承 Foreground 属性，当选中项时，这两个元素会正确翻转(见下图)。

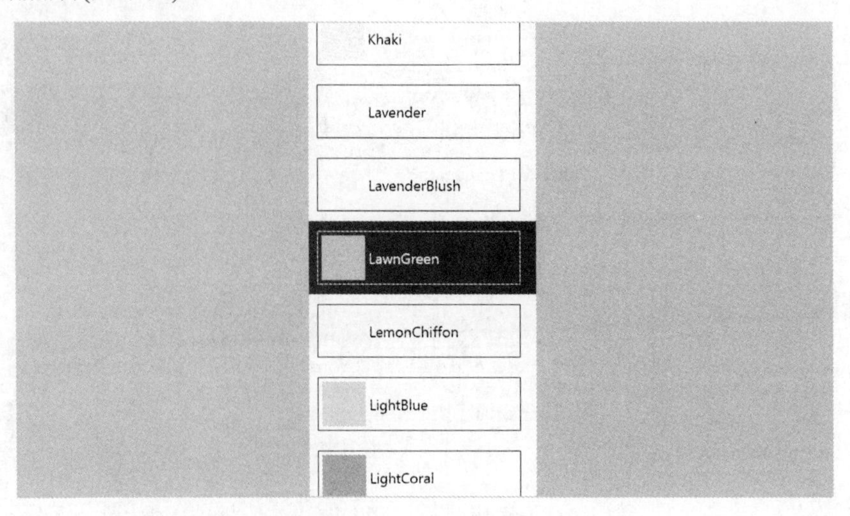

如果需要，ListBox 可以支持多项选中。可以把 SelectionMode 属性设置为 Multiple 或 Extended 并使用 SelectedItems 属性获取所选项。

11.6 面板和虚拟化面板

我用 ListBoxWithItemTemplate 所做的事情在本书中并不经常这么做，我留了一些调试代码。这样做的唯一目的是让你能直观体会 ListBox 内部运行的重要过程。

围绕每项的 Border 元素定义 Loaded 事件处理程序：

```
<Border BorderBrush="{Binding RelativeSource={RelativeSource TemplatedParent},
                              Path=Foreground}"
        BorderThickness="1"
        Width="336"
        Margin="6"
        Loaded="OnItemLoaded">
```

在 Loaded 处理程序中，调用 System.Diagnostics.Debug.WriteLine 显示 NamedColor 对象的 Name 属性，Name 属性设置为加载元素的 DataContext 属性：

```
void OnItemLoaded(object sender, RoutedEventArgs args)
{
    System.Diagnostics.Debug.WriteLine("Item Loaded: " +
        ((sender as FrameworkElement).DataContext as NamedColor).Name);
}
```

在 Visual Studio 调试器中运行该程序并观察 Output 窗口。程序首次加载的时候，你会看到只有几种颜色，绝对没有全部 141 种颜色。我的平板电脑，屏幕高 768 像素，可以全部显示 6 项(从 AliceBlue 到 Beige)，还能显示下一项的一半。Visual Studio Output 窗口中的列表显示已经加载 11 项的可视树，从 AliceBlue 到 BlueViolet。

现在开始滚动列表。你会在 Output 窗口里会看到更多项，我看到了 Brown、BurlyWood 和 CadetBlue，但然后列表停止了。到底是怎么回事？

ListBox 讲究效率。ListBox 只为最初要显示的条目(再加几个)构建可视树，如果一些项滚出视图，而另外一些项就会滚入视图，ListBox 会重用这些可视树。为什么不重用呢？ListBox 要做的就是改变绑定。

如果集合里有成百上千项绑定 ListBox 控件，这种虚拟处理至关重要。但也意味着 ListBox 不能确定需要显示所有项的宽度。

注意：在集合里，条目可能有一些特殊地方，虚拟化处理时会产生问题，第 16 章将讨论这种情况，涉及到包含彼此链接的可视树。在这种情况下，基本上可以关闭虚拟化功能。条目控件总是通过某种形式的 Panel 来显示条目，可以指定其使用哪个 Panel 派生类或自己提供。ItemsControl 定义(并且 ListBox 继承)名为 ItemsPanel 的属性，可以设置为 ItemsPanelTemplate 类型的一个对象。这就是本章标题所指的三个模板中的第二个，当然也是其中最简单的一个。ItemsPanelTemplate 只需要一项，即 Panel 派生类。条目控件用于托管子项的面板。在常规 ItemsControl 中，为 StackPanel。而在 ListBox 中，为 VirtualizingStackPanel。

在 ListBoxWithItemTemplate 中，可以通过以下标记把 ListBox 的 ItemsPanel 属性设置为默认值：

```
<ListBox Name="lstbox"
         ItemsSource="{Binding Source={StaticResource namedColor},
                               Path=All}"
         Width="380">
    <ListBox.ItemTemplate>
    ...
    </ListBox.ItemTemplate>

    <ListBox.ItemsPanel>
        <ItemsPanelTemplate>
            <VirtualizingStackPanel />
        </ItemsPanelTemplate>
    </ListBox.ItemsPanel>
</ListBox>
```

我们来试试将其变为常规 StackPanel：

```
<ListBox.ItemsPanel>
    <ItemsPanelTemplate>
        <StackPanel />
    </ItemsPanelTemplate>
</ListBox.ItemsPanel>
```

现在，ListBox 首次加载时，会创建所有项目，看一看 Visual Studio 的 Output 窗口，就能得到验证。

不过，也可以像下面这样做：

```
<ListBox.ItemsPanel>
    <ItemsPanelTemplate>
        <VirtualizingStackPanel Orientation="Horizontal" />
    </ItemsPanelTemplate>
</ListBox.ItemsPanel>
```

这样可以把垂直 ListBox 变为水平 ListBox。

这样还不行。还需要对 ListBox 的大小和内部的 ScrollViewer 属性做一些调整，就像我在 HorizontalListBox 项目里所做的那样。

```
项目: HorizontalListBox | 文件: MainPage.xaml(片段)
<Page ... >
    <Page.Resources>
```

```
        <ch11:NamedColor x:Key="namedColor" />
    </Page.Resources>

    <Grid>
        <ListBox Name="lstbox"
            ItemsSource="{Binding Source={StaticResource namedColor},
                                  Path=All}"
            Height="120"
            ScrollViewer.HorizontalScrollMode="Enabled"
            ScrollViewer.HorizontalScrollBarVisibility="Auto"
            ScrollViewer.VerticalScrollMode="Disabled"
            ScrollViewer.VerticalScrollBarVisibility="Disabled">
        <ListBox.ItemTemplate>
            <DataTemplate>
                <Border
                    BorderBrush="{Binding RelativeSource={RelativeSource TemplatedParent},
                                          Path=Foreground}"
                        BorderThickness="1"
                        Width="336"
                        Margin="6"
                        Loaded="OnItemLoaded">
                    <Grid>
                        <Grid.ColumnDefinitions>
                            <ColumnDefinition Width="Auto" />
                            <ColumnDefinition Width="*"/>
                        </Grid.ColumnDefinitions>

                        <Rectangle Grid.Column="0"
                                   Height="72"
                                   Width="72"
                                   Margin="6">
                            <Rectangle.Fill>
                                <SolidColorBrush Color="{Binding Color}" />
                            </Rectangle.Fill>
                        </Rectangle>

                        <TextBlock Grid.Column="1"
                                   FontSize="24"
                                   Text="{Binding Name}"
                                   VerticalAlignment="Center" />
                    </Grid>
                </Border>
            </DataTemplate>
        </ListBox.ItemTemplate>

        <ListBox.ItemsPanel>
            <ItemsPanelTemplate>
                <VirtualizingStackPanel Orientation="Horizontal" />
            </ItemsPanelTemplate>
        </ListBox.ItemsPanel>
    </ListBox>

    <Grid.Background>
        <SolidColorBrush Color="{Binding ElementName=lstbox,
                                 Path=SelectedItem.Color}" />
        </Grid.Background>
    </Grid>
</Page>
```

ScrollViewer 定义若干属性用于管理控件的外观和功能，但有时 ScrollViewer 自身会无法访问，当 ScrollViewer 在 ListBox 里的时候就会出现这种情况。在类似情况下，ScrollViewer 简单定义了几个附加属性，可以在 ListBox 标记里方便地设置。

程序和前一个 ListBox 之间的差别仅在于水平方向使用了 VirtualizingStackPanel，另外还改变了 ListBox 标记用于改变控件方向以及提供水平滚动。其结果是一个功能齐全的水平 ListBox(见下图)。

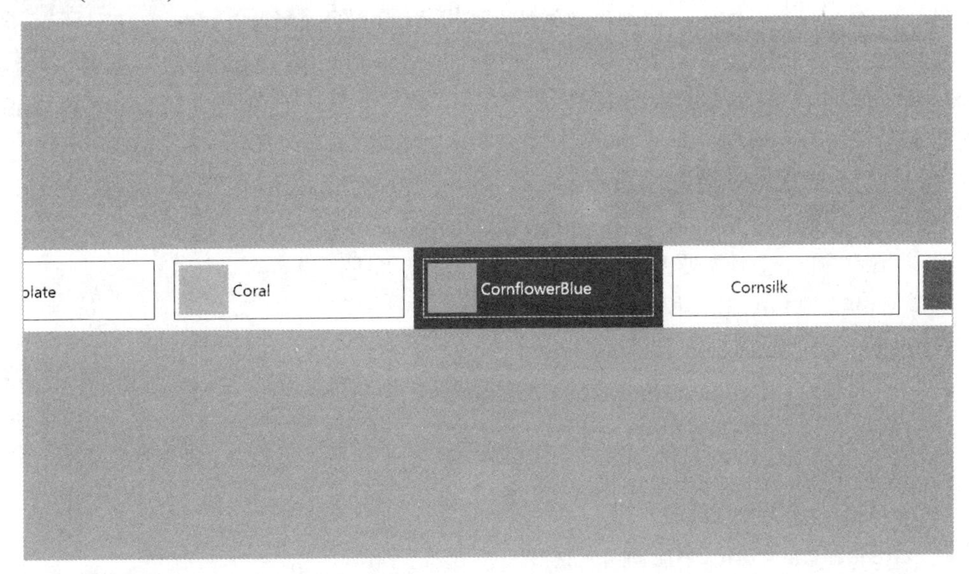

奇怪的是，我没有能够成功使用带 ListBox 的 WrapGrid 或 VariableSizedWrapGrid 面板。试图这么做会引发异常消息“正在使用的 Control Panel 不允许作为 Control 的 ItemsPanel”。但下一节我会提供类似东西来包装面板。

11.7　自定义面板

写自定义 Panel 派生类的主要原因可能是将其当成条目控件的 ItemsPanelTemplate 来使用。每种类型的自定义 Panel 都能以不同方式来布局子对象。

以不寻常方式来布局子对象——例如在圆里——如果所有项和屏幕都非常匹配，也不需要滚动，写这种 Panel 派生类也许最容易了。如果需要滚动，布局需要有利于发挥 ScrollViewer 的能力。或者，ScrollViewer 本身需要换为自定义的滚动机制。

Panel 派生类可以定义依赖项属性和附加属性，例如，Grid 和 Canvas 都定义附加属性。然而，带附加属性的 Panel 派生类一般不能用作 ItemsPanelTemplate，因为从 DataTemplate 设置这些附加属性通常都不合理。

Panel 派生类总是覆写两个虚拟保护方法：MeasureOverride 和 ArrangeOverride。两个方法对应两种布局传递。在 MeasureOverride 方法中，Panel 派生类对所有子对象调用 Measure 方法，计算其自身所需尺寸。而在 ArrangeOverride 方法中，Panel 派生类对所有子对象调用 Arrange，而子对象的大小和位置则相对于其自身。

两个名为 MeasureOverride 和 ArrangeOverride 的方法似乎有点特别。方法名称起源于 Windows Presentation Foundation，涉及 UIElement 和 FrameworkElement 类的 WPF 版本之间的区别。UIElements 实现的是相对简单的布局系统，涉及 Measure 和 Arrange 方法。然而，相对于 WPF UIElement，FrameworkElement 添加了 HorizontalAlignment、VerticalAlignment 和 Margin 属性，使布局变得更复杂。因此，虽然 Measure 和 Arrange 仍然继续在布局中发挥作用，但 FrameworkElement 也定义 MeasureOverride 和 ArrangeOverride

以取代 Measure 和 Arrange 方法。

总之，Panel 派生类覆写 MeasureOverride 和 ArrangeOverride，这些方法对所有子类调用 Measure 和 Arrange。在内部，子类里的 Measure 和 Arrange 方法调用子类的 MeasureOverride 和 ArrangeOverride 方法。然后，子类会对其子类调用 Measure 和 Arrange，而进程会沿着可视树继续。

可以在任何 FrameworkElement 派生类里覆写 MeasureOverride 和 ArrangeOverride，但除了 Panel 派生类，为 Windows Runtimes 所写的程序一般都不这样做。

Panel 派生类自己不需要为其自身或子类进行以下属性设置：

- Width、MinWidth 和 MaxWidth
- Height、MinHeight 和 MaxHeight
- HorizontalAlignment 和 VerticalAlignment
- Margin
- Visibility
- Opacity(不影响布局)
- RenderTransform(不影响布局)

这些属性都会自动处理。

在 Panel 派生类里，MeasureOverride 方法如下所示：

```
protected override Size MeasureOverride(Size availableSize)
{
    ...

    return desiredSize;
}
```

availableSize 参数为 Size 类型，(正如你所知道的)它有两个 double 类型的属性，分别为 Width 和 Height。availableSize 参数有时非常简单。例如，如果面板是 Page 的内容，则 availableSize 表示页面大小，通常为应用窗口的大小。如果面板在 Grid 单元格内，而 Grid 单元格具有特定像素维度，则 availableSize 就是单元格大小。

然而，Width 或 Height 或两者都可能为无限，这也是常见情况。在 MeasureOverride 方法里，可以通过静态 Double.IsPositiveInfinity 方法测试 Width 或 Height 为无限的情况。

如果 availableSize 的 Width 或 Height 属性值为无限，也就是说，panel 的父类要提供给 panel 所需的尽可能多的水平或垂直空间。如果 panel 是垂直 StackPanel 的子对象，则 Height 属性可以为无限；如果 panel 是水平 StackPanel 的子对象，则 Width 属性可以为无限。如果 panel 在 Grid 单元格内，而单元格的宽度和高度都为 Auto，则 availableSize 的 Width 或 Height 属性值可以为无限。

MeasureOverride 方法必须能正确处理这些情况。从 MeasureOverride 方法返回的 desiredSize 不能有无限 Width 或 Height 属性，换句话说，MeasureOverride 不能简单返回 availableSize。这样行不通。

MeasureOverride 方法必须对其每一个子对象调用 Measure，否则，子对象不可见。每一个 Measure 调用返回，子对象的 DesiredSize 属性为有效，panel 可以通过每个子对象所需大小来计算自身所需大小。

如果 MeasureOverride 对子对象调用 Measure，则会为子对象提供可用大小。其一两个属性可以为无限值。例如，垂直 StackPanel 对其所有子对象调用 Measure，可用的 Width 等于其自身可用的 Width，可用的 Height 为无限值：

```
protected override Size MeasureOverride(Size availableSize)
{
    double maxWidth = 0;
    double totalHeight = 0;

    foreach (UIElement child in this.Children)
    {
        child.Measure(new Size(availableSize.Width, Double.PositiveInfinity));
        maxWidth = Math.Max(maxWidth, child.DesiredSize.Width);
        totalHeight += child.DesiredSize.Height;
    }
    return new Size(maxWidth, totalHeight);
}
```

垂直堆叠的 MeasureOverride 方法累积所有子对象的最大宽度和总高度。两者成为其所需大小。

ArrangeOverride 方法有一个参数表示为面板计算大小。对于垂直堆叠的面板，方法会再次遍历所有子对象，进行叠加，为每个子对象指定自己的宽度以及子对象所需的高度：

```
protected override Size ArrangeOverride(Size finalSize)
{
    double y = 0;

    foreach (UIElement child in this.Children)
    {
        child.Arrange(new Rect(0, y, finalSize.Width, child.DesiredSize.Height));
        y += child.DesiredSize.Height;
    }
    return base.ArrangeOverride(finalSize);
}
```

用 WBF 编程时，我发现了一个有用的 Panel 派生类，称为 UniformGrid。它和常规 Grid 类似，不过每个单元格有同样大小。其子对象直接分配到一个单元格，这样就不需要附加属性。虽然子对象的大小可以不同，但 UniformGrid 会把子对象当成大小都相同来进行处理。大小是基于最大子对象的大小或 UniformGrid 的可用空间大小。

我的 UniformGrid 版本定义两个 int 类型的属性 Rows 和 Columns，但默认值是-1，表示没有默认值。如果没有设置这两个属性，UniformGrid 会试图确定一个最优的行数和列数，否则，如果设置了两个属性之一，另一个属性就会根据子对象的数量计算出来。不建议设置两个属性，如果产品的行和列数量少于子对象数，有些子对象就可能不会出现在面板里。

在 UniformGrid 的大多数用途中，Rows 和 Columns 都保留默认值-1 或者把其中之一设置为 1。如果 Rows 或 Columns 设置为 1，UniformGrid 就像一个只有单列或单行的 Grid，或者像一个每个子对象都有相同大小的 StackPanel。

如果 availableSize 的 Width 和 Height 属性都为有限值，UniformGrid 就会尝试让所有子对象适合该空间，否则，会使用最大子对象的大小来布局其子对象。UniformGrid 不能处理的唯一情况是行和列都为-1，而且可用的 Width 和 Height 都为无限。这种情况会造成异常。

像 StackPanel 一样，UniformGrid 还定义了 Orientation 属性。以下是其共享的属性定义和属性变更处理程序。

项目: Petzold.ProgrammingWindows6.Chapter11 | 文件: UniformGrid.cs(片段)

```
public class UniformGrid : Panel
{
    // Set by MeasureOverride, used in ArrangeOverride
    protected int rows, cols;

    static UniformGrid()
    {
        RowsProperty = DependencyProperty.Register("Rows",
            typeof(int),
            typeof(UniformGrid),
            new PropertyMetadata(-1, OnPropertyChanged));

        ColumnsProperty = DependencyProperty.Register("Columns",
            typeof(int),
            typeof(UniformGrid),
            new PropertyMetadata(-1, OnPropertyChanged));

        OrientationProperty = DependencyProperty.Register("Orientation",
            typeof(Orientation),
            typeof(UniformGrid),
            new PropertyMetadata(Orientation.Vertical, OnPropertyChanged));
        }

        public static DependencyProperty RowsProperty { private set; get; }

        public static DependencyProperty ColumnsProperty { private set; get; }

        public static DependencyProperty OrientationProperty { private set; get; }

        public int Rows
        {
            set { SetValue(RowsProperty, value); }
            get { return (int)GetValue(RowsProperty); }
        }

        public int Columns
        {
            set { SetValue(ColumnsProperty, value); }
            get { return (int)GetValue(ColumnsProperty); }
        }

        public Orientation Orientation
        {
            set { SetValue(OrientationProperty, value); }
            get { return (Orientation)GetValue(OrientationProperty); }
        }
        ...

        static void OnPropertyChanged(DependencyObject obj, DependencyPropertyChangedEventArgs args)
        {
        if  (args.Property == UniformGrid.OrientationProperty)
        {
            (obj as UniformGrid).InvalidateArrange();
        }
        else
        {
             (obj as UniformGrid).InvalidateMeasure();
        }
    }
}
```

在属性更改处理程序中，InvalidateMeasure 和 InvalidateArrange 调用告知布局系统需要新布局。调用 InvalidateMeasure 会触发测量以及安排传递，InvalidateArrange 调用只触发安排传递，而忽略测量传递。在这种情况下，一切都保持着相同的大小，但子对象可能会移动到不同位置。

当然，这并不是让布局无效的唯一方式，例如，面板里子对象数量的任何变化都会触发一个新布局。

MeasureOverride 方法首先执行两个有效性检查，然后用 Rows 和 Column 属性以及子元素数量来计算 rows 和 cols 字段：

项目：Petzold.ProgrammingWindows6.Chapter11 | 文件：UniformGrid.cs(片段)

```
protected override Size MeasureOverride(Size availableSize)
{
    // Only bother if children actually exist
    if (this.Children.Count == 0)
        return new Size();

    // Throw exceptions if the properties aren't OK
    if (this.Rows != -1 && this.Rows < 1)
        throw new ArgumentOutOfRangeException("UniformGrid Rows must be greater than zero");

    if (this.Columns != -1 && this.Columns < 1)
        throw new ArgumentOutOfRangeException("UniformGrid Columns must be greater than zero");

    // Determine the actual number of rows and columns
    // ----------------------------------------
    // This option is discouraged
    if (this.Rows != -1 && this.Columns != -1)
    {
        rows = this.Rows;
        cols = this.Columns;
    }
    // These two options often appear with values of 1
    else if (this.Rows != -1)
    {
        rows = this.Rows;
        cols = (int)Math.Ceiling((double)this.Children.Count / rows);
    }
    else if (this.Columns != -1)
    {
        cols = this.Columns;
        rows = (int)Math.Ceiling((double)this.Children.Count / cols);
    }
    // No values yet if both Rows and Columns are both -1, but
    // check for infinite availableSize
    else if (Double.IsInfinity(availableSize.Width) &&
        Double.IsInfinity(availableSize.Height))
    {
        throw new NotSupportedException("Completely unconstrained UniformGrid " +
                                        "requires Rows or Columns property to be set");
    }
    ...
}
```

MeasureOverride 进程继续计算最大子元素的大小。以下代码列举了通过 Children 集合为每一个子元素循环调用 Measure 方法。如果没有调用 Measure，子元素的大小则为 0。Measure 调用后，子元素的 DesiredSize 属性为有效值：

项目：Petzold.ProgrammingWindows6.Chapter11 | 文件：UniformGrid.cs(片段)

```
protected override Size MeasureOverride(Size availableSize)
{
    ...

    // Determine the maximum size of all children
    // ------------------------------------------
    Size maximumSize = new Size();
    Size infiniteSize = new Size(Double.PositiveInfinity,
                                 Double.PositiveInfinity);

    // Find the maximum size of all children
```

```
    foreach (UIElement child in this.Children)
    {
        child.Measure(infiniteSize);
        Size childSize = child.DesiredSize;
        maximumSize.Width = Math.Max(maximumSize.Width, childSize.Width);
        maximumSize.Height = Math.Max(maximumSize.Height, childSize.Height);
    }

    ...
}
```

许多 Panel 派生类都会进行此类计算。然而，Measure 方法并不总是调用无限高度和宽度。在该特定情况下，UniformGrid 要确定每个元素的“自然大小”，方法就这样简单。

前面提到过，Panel 派生类不需要考虑其自身或子类的 Margin 属性设置。可用大小作为参数传递给 MeasureOverride，并不包括对元素设置的任何 Margin 属性。然而，如果面板对子类调用 Measure，可用大小则隐式包含了子类的 Margin。子类的 Measure 方法根据其 Margin 设置减少其可用大小，(当然，如果大小无限，就像这种情况，结果都是一样。)不包含 Margin 的大小传递给子类的 MeasureOverride 方法，子类计算从 MeasureOverride 返回的自身大小。子类的 Measure 方法继续把其 Margin 加上从 MeasureOverride 返回的大小，并把子类的 DesiredSize 属性设置为增加后的大小。

这就是 Margin 如何布局的过程，而 MeasureOverride 则不然。

现在，计算好了最大子类大小，就可以计算 Panel 所需大小。但仍有可能需要相当漫长的计算：如果 Rows 和 Columns 两者都保留默认值，Panel 本身则需要基于可用大小和最大子类大小来计算最优行数和列数。

```
项目: Petzold.ProgrammingWindows6.Chapter11 | 文件: UniformGrid.cs(片段)
protected override Size MeasureOverride(Size availableSize)
{
    ...

    // Find rows and cols if Rows and Colunms are both -1
    if (this.Rows == -1 && this.Columns == -1)
    {
        if (Double.IsInfinity(availableSize.Width))
        {
            rows = (int)Math.Max(1, availableSize.Height / maximumSize.Height);
            cols = (int)Math.Ceiling((double)this.Children.Count / rows);
        }
        else if (Double.IsInfinity(availableSize.Height))
        {
            cols = (int)Math.Max(1, availableSize.Width / maximumSize.Width);
            rows = (int)Math.Ceiling((double)this.Children.Count / cols);
        }
        // Neither dimension is infinite -- the hard one
        else
        {
            double aspectRatio = maximumSize.Width / maximumSize.Height;
            double bestHeight = 0;
            double bestWidth = 0;

            for (int tryRows = 1; tryRows < this.Children.Count; tryRows++)
            {
                int tryCols = (int)Math.Ceiling((double)this.Children.Count / tryRows);
                double childHeight = availableSize.Height / tryRows;
                double childWidth = availableSize.Width / tryCols;

                // Adjust for aspect ratio
                if (childWidth > aspectRatio * childHeight)
                    childWidth = aspectRatio * childHeight;
```

```
                else
                    childHeight = childWidth / aspectRatio;

                // Check if it's larger than other trials
                if (childHeight > bestHeight)
                {
                    bestHeight = childHeight;
                    bestWidth = childWidth;
                    rows = tryRows;
                    cols = tryCols;
                }
            }
        }
    }
    // Return desired size
    Size desiredSize = new Size (Math.Min(cols * maximumSize.Width, availableSize.Width),
                                 Math.Min(rows * maximumSize.Height, availableSize.Height));
    return desiredSize;
}
```

在正常情况下，面板所需大小完全基于子类的大小和其他任何可能需要的开销。大小可能大于 availableSize。ScrollViewer 就是这样知道怎么滚动子元素的。然而，如果 UniformGrid 有非无限值 availableSize，我想把面板刚好限制到该大小。

ArrangeOverride 方法往比 MeasureOverride 简单得多。finalSize 参数是分配给面板的有限大小。对 ArrangeOverride 的唯一要求是对每个子元素调用 Arrange 方法，并传递其 Rect 对象表示子元素相对于面板的位置和子元素的大小。该大小通常是子元素的 DesiredSize 属性，但在这种情况下，我想把面板尺寸分配为行和列。

项目：Petzold.ProgrammingWindows6.Chapter11 | 文件：UniformGrid.cs(片段)

```
protected override Size ArrangeOverride(Size finalSize)
{
    int index = 0;
    double cellWidth = finalSize.Width / cols;
    double cellHeight = finalSize.Height / rows;

    if (this.Orientation == Orientation.Vertical)
    {
        for (int row = 0; row < rows; row++)
        {
            double y = row * cellHeight;

            for (int col = 0; col < cols; col++)
            {
                double x = col * cellWidth;

                if (index < this.Children.Count)
                    this.Children[index].Arrange(new Rect(x, y, cellWidth, cellHeight));

                index++;
            }
        }
    }
    else
    {
        for (int col = 0; col < cols; col++)
        {
            double x = col * cellWidth;

            for (int row = 0; row < rows; row++)
            {
                double y = row * cellHeight;

                if (index < this.Children.Count)
                    this.Children[index].Arrange(new Rect(x, y, cellWidth, cellHeight));
```

```
                index++;
            }
        }
    }

    return base.ArrangeOverride(finalSize);
}
```

这是 Orientation 在 UniformGrid 中唯一发挥作用的地方，并控制着子元素应该首先从左到右定位、还是首先从上到下定位。ArrangeOverrid 方法总是返回 finalSize，这是基本方法返回的。

我们来试试 availableSize 的 Width 和 Height 属性为有限值的情况。ItemsControl 不在 ScrollViewer 时，就像下面这样。

项目: AllColorsItemsControl | 文件: MainPage.xaml(片段)

```
<Page ... >
    <Page.Resources>
        <ch11:NamedColor x:Key="namedColor" />
        <ch11:ColorToContrastColorConverter x:Key="colorConverter" />
    </Page.Resources>

    <Grid Background="{StaticResource ApplicationPageBackgroundThemeBrush}">
        <ItemsControl ItemsSource="{Binding Source={StaticResource namedColor},
                                            Path=All}">
            <ItemsControl.ItemTemplate>
                <DataTemplate>
                    <Border
                        BorderBrush="{Binding RelativeSource={RelativeSource TemplatedParent},
                                              Path=Foreground}"
                            BorderThickness="2"
                            Margin="2">
                        <Border.Background>
                            <SolidColorBrush Color="{Binding Color}" />
                        </Border.Background>

                        <Viewbox>
                            <TextBlock Text="{Binding Name}"
                                       HorizontalAlignment="Center"
                                       VerticalAlignment="Center">
                                <TextBlock.Foreground>
                                    <SolidColorBrush Color="{Binding Color,
                                        Converter={StaticResource colorConverter}}" />
                                </TextBlock.Foreground>
                            </TextBlock>
                        </Viewbox>
                    </Border>
                </DataTemplate>
            </ItemsControl.ItemTemplate>

            <ItemsControl.ItemsPanel>
                <ItemsPanelTemplate>
                    <ch11:UniformGrid />
                </ItemsPanelTemplate>
            </ItemsControl.ItemsPanel>
        </ItemsControl>
    </Grid>
</Page>
```

注意结尾处的 UniformGrid，它被用作控件的 ItemsPanel。

我把条目模板弄得比先前例子简单了一些。现在模板包含一个 Border 和一个 TextBlock 子类，Border 的 Background 属性通过绑定到 NamedColor 的 Color 属性构造，而 TextBlock 显示颜色名称。注意，TextBlock 在 Viewbox 中，这样文本大小应该适应子类的可用大小。还要注意，我把 TextBlock 的 Foreground 绑定到 Color 属性，但通过名为 ColorToContrastColorConverter 的转换器进行传递。转换器计算灰度对应的输入颜色，并选择 Colors.Black 或 Colors.White 进行对比。

```
项目: Petzold.ProgrammingWindows6.Chapter11 | 文件: ColorToContrastColorConverter.cs
public class ColorToContrastColorConverter : IValueConverter
{
    public object Convert(object value, Type targetType, object parameter, string language)
    {
        Color clr = (Color)value;
        double grayShade = 0.30 * clr.R + 0.59 * clr.G + 0.11 * clr.B;
        return grayShade > 128 ? Colors.Black : Colors.White;
    }
    public object ConvertBack(object value, Type targetType, object parameter, string language)
    {
        return value;
    }
}
```

如下图所示，该方法适用于每一种颜色，但 Transparent 除外。

AliceBlue	AntiqueWhite	Aqua	Aquamarine	Azure	Beige	Bisque
Black	BlanchedAlmond	Blue	BlueViolet	Brown	BurlyWood	CadetBlue
Chartreuse	Chocolate	Coral	CornflowerBlue	Cornsilk	Crimson	Cyan
DarkBlue	DarkCyan	DarkGoldenrod	DarkGray	DarkGreen	DarkKhaki	DarkMagenta
DarkOliveGreen	DarkOrange	DarkOrchid	DarkRed	DarkSalmon	DarkSeaGreen	DarkSlateBlue
DarkSlateGray	DarkTurquoise	DarkViolet	DeepPink	DeepSkyBlue	DimGray	DodgerBlue
Firebrick	FloralWhite	ForestGreen	Fuchsia	Gainsboro	GhostWhite	Gold
Goldenrod	Gray	Green	GreenYellow	Honeydew	HotPink	IndianRed
Indigo	Ivory	Khaki	Lavender	LavenderBlush	LawnGreen	LemonChiffon
LightBlue	LightCoral	LightCyan	LightGoldenrodYellow	LightGray	LightGreen	LightPink
LightSalmon	LightSeaGreen	LightSkyBlue	LightSlateGray	LightSteelBlue	LightYellow	Lime
LimeGreen	Linen	Magenta	Maroon	MediumAquamarine	MediumBlue	MediumOrchid
MediumPurple	MediumSeaGreen	MediumSlateBlue	MediumSpringGreen	MediumTurquoise	MediumVioletRed	MidnightBlue
MintCream	MistyRose	Moccasin	NavajoWhite	Navy	OldLace	Olive
OliveDrab	Orange	OrangeRed	Orchid	PaleGoldenrod	PaleGreen	PaleTurquoise
PaleVioletRed	PapayaWhip	PeachPuff	Peru	Pink	Plum	PowderBlue
Purple	Red	RosyBrown	RoyalBlue	SaddleBrown	Salmon	SandyBrown
SeaGreen	SeaShell	Sienna	Silver	SkyBlue	SlateBlue	SlateGray
Snow	SpringGreen	SteelBlue	Tan	Teal	Thistle	Tomato
Transparent	Turquoise	Violet	Wheat	White	WhiteSmoke	Yellow
YellowGreen						

141 种颜色在窗口里都很合适，也就是程序的目标。如果用辅屏模式看程序，单元格和文本会变小(见下图)。

AliceBlue	AntiqueWhite	Aqua	Aquamarine	Azure	Beige
Bisque	Black	BlanchedAlmond	Blue	BlueViolet	Brown
BurlyWood	CadetBlue	Chartreuse	Chocolate	Coral	CornflowerBlue
Cornsilk	Crimson	Cyan	DarkBlue	DarkCyan	DarkGoldenrod
DarkGray	DarkGreen	DarkKhaki	DarkMagenta	DarkOliveGreen	DarkOrange
DarkOrchid	DarkRed	DarkSalmon	DarkSeaGreen	DarkSlateBlue	DarkSlateGray
DarkTurquoise	DarkViolet	DeepPink	DeepSkyBlue	DimGray	DodgerBlue
Firebrick	FloralWhite	ForestGreen	Fuchsia	Gainsboro	GhostWhite
Gold	Goldenrod	Gray	Green	GreenYellow	Honeydew
HotPink	IndianRed	Indigo	Ivory	Khaki	Lavender
LavenderBlush	LawnGreen	LemonChiffon	LightBlue	LightCoral	LightCyan
LightGoldenrodYellow	LightGray	LightGreen	LightPink	LightSalmon	LightSeaGreen
LightSkyBlue	LightSlateGray	LightSteelBlue	LightYellow	Lime	LimeGreen
Linen	Magenta	Maroon	MediumAquamarine	MediumBlue	MediumOrchid
MediumPurple	MediumSeaGreen	MediumSlateBlue	MediumSpringGreen	MediumTurquoise	MediumVioletRed
MidnightBlue	MintCream	MistyRose	Moccasin	NavajoWhite	Navy
OldLace	Olive	OliveDrab	Orange	OrangeRed	Orchid
PaleGoldenrod	PaleGreen	PaleTurquoise	PaleVioletRed	PapayaWhip	PeachPuff
Peru	Pink	Plum	PowderBlue	Purple	Red
RosyBrown	RoyalBlue	SaddleBrown	Salmon	SandyBrown	SeaGreen
SeaShell	Sienna	Silver	SkyBlue	SlateBlue	SlateGray
Snow	SpringGreen	SteelBlue	Tan	Teal	Thistle
Tomato	Transparent	Turquoise	Violet	Wheat	White
WhiteSmoke	Yellow	YellowGreen			

然而，如果单元格变得太小，视觉上会有一点点断裂效果。

现在我们试试把 UniformGrid 放入 ListBox。我保留了简化后的数据模板，但给 Border 和 TextBlock 赋予了特定大小。

```
项目: ListBoxWithUniformGrid | 文件: MainPage.xaml(片段)
<Page ... >
    <Page.Resources>
        <ch11:NamedColor x:Key="namedColor" />
        <ch11:ColorToContrastColorConverter x:Key="colorConverter" />
    </Page.Resources>

    <Grid Background="{StaticResource ApplicationPageBackgroundThemeBrush}">
        <ListBox ItemsSource="{Binding Source={StaticResource namedColor},
                                       Path=All}">
            <ListBox.ItemTemplate>
                <DataTemplate>
                    <Border
                        BorderBrush="{Binding RelativeSource={RelativeSource TemplatedParent},
                                              Path=Foreground}"
                            Width="288"
                            Height="72"
                            BorderThickness="3"
                            Margin="3">
                        <Border.Background>
                            <SolidColorBrush Color="{Binding Color}" />
                        </Border.Background>

                            <TextBlock Text="{Binding Name}"
                                       FontSize="24"
                                       HorizontalAlignment="Center"
                                       VerticalAlignment="Center">
                            <TextBlock.Foreground>
                                <SolidColorBrush Color="{Binding Color,
                                    Converter={StaticResource colorConverter}}" />
                            </TextBlock.Foreground>
                        </TextBlock>
                    </Border>
                </DataTemplate>
            </ListBox.ItemTemplate>

            <ListBox.ItemsPanel>
                <ItemsPanelTemplate>
                    <ch11:UniformGrid />
                </ItemsPanelTemplate>
            </ListBox.ItemsPanel>
        </ListBox>
    </Grid>
</Page>
```

在这种情况下，传给 UniformGrid 里 MeasureOverride 的 availableSize 参数有无限 Height 属性值用于垂直滚动。UniformGrid 根据可用宽度和最大子对象宽度来计算列数，行数也由此进行计算。UniformGrid 所需大小是基于其总高度，面板则可垂直滚动(见下图)。

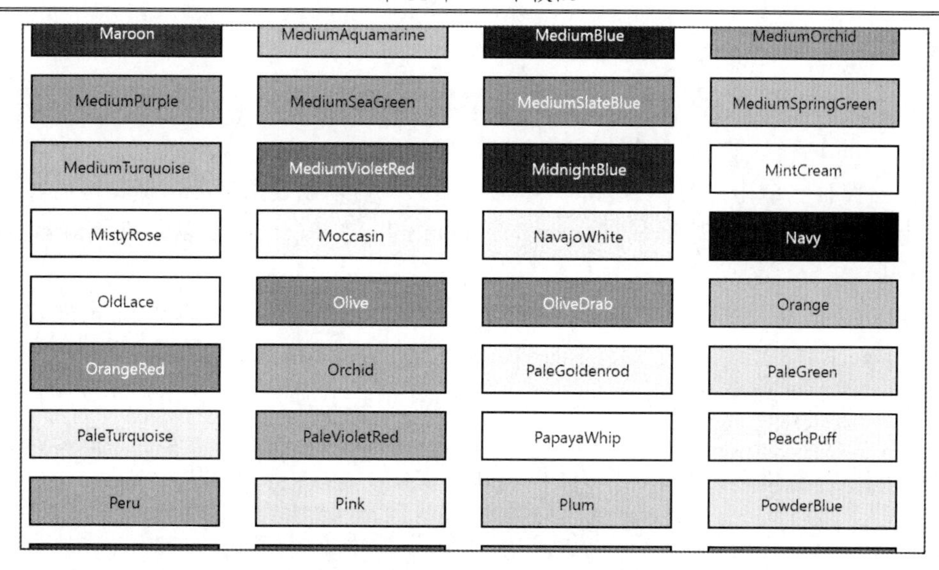

很容易把它转为水平滚动。正如前面在 HorizontalListBox 项目看到的，可以直接设置 ScrollViewer 附加属性，然后把 UniformGrid 的 Orientation 属性设置为 Horizontal：

```
<ListBox ItemsSource="{Binding Source={StaticResource namedColor},
                               Path=All}"
         ScrollViewer.HorizontalScrollMode="Enabled"
         ScrollViewer.HorizontalScrollBarVisibility="Auto"
         ScrollViewer.VerticalScrollMode="Disabled"
         ScrollViewer.VerticalScrollBarVisibility="Disabled">
    <ListBox.ItemTemplate>
        <DataTemplate>
        ...
        </DataTemplate>
    </ListBox.ItemTemplate>

    <ListBox.ItemsPanel>
        <ItemsPanelTemplate>
            <ch11:UniformGrid Orientation="Horizontal" />
        </ItemsPanelTemplate>
    </ListBox.ItemsPanel>
</ListBox>
```

虽然没有严格要求有 Orientation 的 Horizontal 设置，但会引起子元素排序差异，先从上到下，然后从左到右(见下图)。

11.8 条目模板条形图

条目控件有一个伟大的“小把戏”，无需费力就可以创建条形图。你真正需要的是数据项，包含适合绑定的数据属性、ItemTemplate 里的 Rectangle 以及 UniformGrid 或 StackPanel。

RgbBarChart 项目演示了该方法。ItemsControl 的 ItemsSource 当然是 NamedColor 对象的集合。DataTemplate 是一个垂直 StackPanel，包含三个 Rectangle 元素，每个 Rectangle 元素的 Height 属性都绑定到条目的 Color 属性的 R、G、B 属性。一般情况下，这将创建三个 Rectangle 元素的一个堆叠，Rectangle 元素从 StackPanel 的顶部开始，而我想要一个从底部开始的更传统堆叠条形图，所以用 RenderTransform 从上到下翻转条形。

项目：RgbBarChart | 文件：MainPage.xaml(片段)

```
<Page ... >
    <Page.Resources>
        <ch11:NamedColor x:Key="namedColor" />
    </Page.Resources>

    <Grid Background="{StaticResource ApplicationPageBackgroundThemeBrush}">
        <ItemsControl ItemsSource="{Binding Source={StaticResource namedColor},
                                            Path=All}">

            <ItemsControl.ItemTemplate>
                <DataTemplate>
                    <StackPanel Name="stackPanel"
                                Height="765"
                                RenderTransformOrigin="0.5 0.5"
                                Margin="1 0">
                        <StackPanel.RenderTransform>
                            <ScaleTransform ScaleY="-1" />
                        </StackPanel.RenderTransform>

                        <Rectangle Fill="Red"
                                   Height="{Binding Color.R}" />
                        <Rectangle Fill="Green"
                                   Height="{Binding Color.G}" />
                        <Rectangle Fill="Blue"
                                   Height="{Binding Color.B}" />

                        <ToolTipService.ToolTip>
                            <ToolTip x:Name="tooltip"
                                     PlacementTarget="{Binding ElementName=stackPanel}">

                                <!-- Set DataContext to StackPanel containing items-->
                                <Grid DataContext="{Binding ElementName=tooltip,
                                                            Path=PlacementTarget}">
                                    <!-- Set DataContext to NamedColor -->
                                    <StackPanel DataContext="{Binding DataContext}">
                                        <TextBlock Text="{Binding Name}"
                                                   HorizontalAlignment="Center" />
                                        <StackPanel DataContext="{Binding Color}"
                                                    Orientation="Horizontal"
                                                HorizontalAlignment="Center">
                                            <TextBlock Text="R=" />
                                            <TextBlock Text="{Binding R}" />
                                            <TextBlock Text=" G=" />
                                            <TextBlock Text="{Binding G}" />
                                            <TextBlock Text=" B=" />
                                            <TextBlock Text="{Binding B}" />
                                        </StackPanel>
                                    </StackPanel>
                                </Grid>
```

```
                        </ToolTip>
                    </ToolTipService.ToolTip>
                </StackPanel>
            </DataTemplate>
        </ItemsControl.ItemTemplate>

        <ItemsControl.ItemsPanel>
            <ItemsPanelTemplate>
                <ch11:UniformGrid Rows="1" />
            </ItemsPanelTemplate>
        </ItemsControl.ItemsPanel>
    </ItemsControl>
</Grid>
</Page>
```

当然，条形图自身很模糊，而辨认它们有个好方法，查看鼠标悬停时所激活的提示信息即可。但这样会更加混乱。可以把 ToolTipService.ToolTip 附加属性设置作为子元素，并且把 ToolTip 控件定义为其子元素，这样可以在元素上附上提示信息。ToolTip 派生自 ContentControl。然而，ToolTip 元素实际上并不是可视树的一部分，因为它“漂浮”在树外。ToolTip 元素没有通过可视树而继承属性，包括最重要的 DataContext 属性。我不得不通过 ToolTip 的 PlacementTarget 属性来进行处理。

以下条形图显示了所有 141 种颜色的红色、绿色和蓝色相对构成，颜色提示信息显示了颜色名称和 RGB 值。

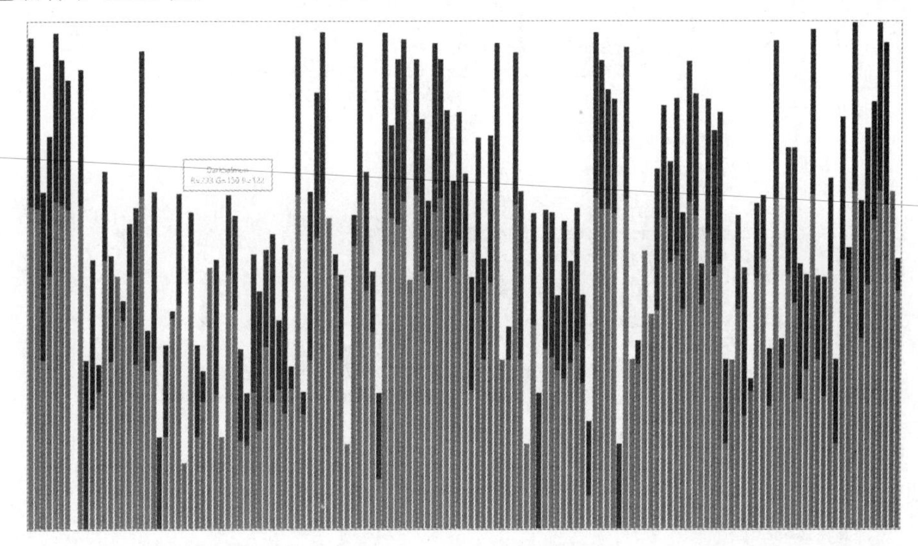

对于 White，颜色成分的总和为 765，只比一个触摸屏 768 像素高度的屏幕小一点点。当然，用 RenderTransform 可以缩短条形图，占据屏幕更小(或更大值)。

11.9　FlipView 控件

我最喜欢 Windows Runtime 引入的控件之一 FlipView，FlipView(和 ListBox 一样)通过 Selector 方式派生自 ItemsControl。FlipView 一次只显示一项，该项即为所选项，因此，在大多数应用中不能代替 ListBox。但 FlipView 有一个好的触控接口，记住 FlipView 很适合用于一些特殊目的。

像本章中的许多其他项目一样，FlipViewColors 项目使用 Petzold.ProgrammingWindows.Chapter11 库。MainPage.xaml 的 Resources 节包含对 NamedColor 类的通常引用，同时定义了 DataTemplate Items 和 PanelTemplate，然后在 Style 中定义中引用了两者，Style 定义还包括 ItemsSource 绑定。Border 和 TextBlock 都有 SolidColorBrush 定义并绑定到两个 FlipView 控件。

项目：FlipViewColors | 文件：MainPage.xaml(片段)

```
<Page ... >
    <Page.Resources>
        <ch11:NamedColor x:Key="namedColor" />
        <ch11:ColorToContrastColorConverter x:Key="colorConverter" />

        <DataTemplate x:Key="colorTemplate">
            <Border BorderBrush="{Binding RelativeSource={RelativeSource TemplatedParent},
                                          Path=Foreground}"
                    Width="288"
                    Height="72"
                    BorderThickness="3"
                    Margin="3">
                <Border.Background>
                    <SolidColorBrush Color="{Binding Color}" />
                </Border.Background>

                <TextBlock Text="{Binding Name}"
                           FontSize="24"
                           HorizontalAlignment="Center"
                           VerticalAlignment="Center">
                    <TextBlock.Foreground>
                        <SolidColorBrush Color="{Binding Color,
                                  Converter={StaticResource colorConverter}}" />
                    </TextBlock.Foreground>
                </TextBlock>
            </Border>
        </DataTemplate>

        <ItemsPanelTemplate x:Key="panelTemplate">
            <VirtualizingStackPanel />
        </ItemsPanelTemplate>

        <Style TargetType="FlipView">
            <Setter Property="Width" Value="300" />
            <Setter Property="Height" Value="100" />
            <Setter Property="ItemsSource" Value="{Binding Source={StaticResource namedColor},
                                                           Path=All}" />
            <Setter Property="ItemTemplate" Value="{StaticResource colorTemplate}" />
            <Setter Property="ItemsPanel" Value="{StaticResource panelTemplate}" />
            <Setter Property="SelectedValuePath" Value="Color" />
        </Style>
    </Page.Resources>

        <Grid Background="{StaticResource ApplicationPageBackgroundThemeBrush}">
            <Grid.RowDefinitions>
                <RowDefinition Height="Auto" />
                <RowDefinition Height="*" />
            </Grid.RowDefinitions>

            <Grid.ColumnDefinitions>
                <ColumnDefinition Width="*" />
                <ColumnDefinition Width="*" />
            </Grid.ColumnDefinitions>

            <Border Grid.Row="0"
                    Grid.Column="0"
                    Grid.ColumnSpan="2"
                    BorderThickness="12"
                    CornerRadius="48"
                    Margin="48"
```

```
                        Padding="48"
                        HorizontalAlignment="Center">
                    <Border.Background>
                        <SolidColorBrush Color="{Binding ElementName=flipView1,
                                                 Path=SelectedValue}" />
                    </Border.Background>

                    <Border.BorderBrush>
                        <SolidColorBrush Color="{Binding ElementName=flipView2,
                                                 Path=SelectedValue}" />
                    </Border.BorderBrush>

                    <TextBlock FontFamily="Times New Roman"
                               FontSize="96">
                        The <Italic>FlipView</Italic> Control
                        <TextBlock.Foreground>
                            <SolidColorBrush Color="{Binding ElementName=flipView2,
                                                     Path=SelectedValue}" />
                    </TextBlock.Foreground>
                </TextBlock>
            </Border>

            <FlipView Name="flipView1"
                      Grid.Row="1"
                      Grid.Column="0" />

            <FlipView Name="flipView2"
                      Grid.Row="1"
                      Grid.Column="1" />
        </Grid>
    </Page>
```

默认情况下，FlipView 的 ItemsPanelTemplate 为 VirtualizingStackPanel，就像 ListBox 一样，不过为水平方向。我用一个垂直 VirtualizingStackPanel 代替了它。像 ListBox 一样，FlipView 控件会扩展到覆盖所有可用空间，因此最好要设置明确的 Height 和 Width 属性。这里的想法是：“拨号”控制两种不同颜色。第一种颜色控制 Border 的背景，而第二种控制边界自身和文本(见下图)。

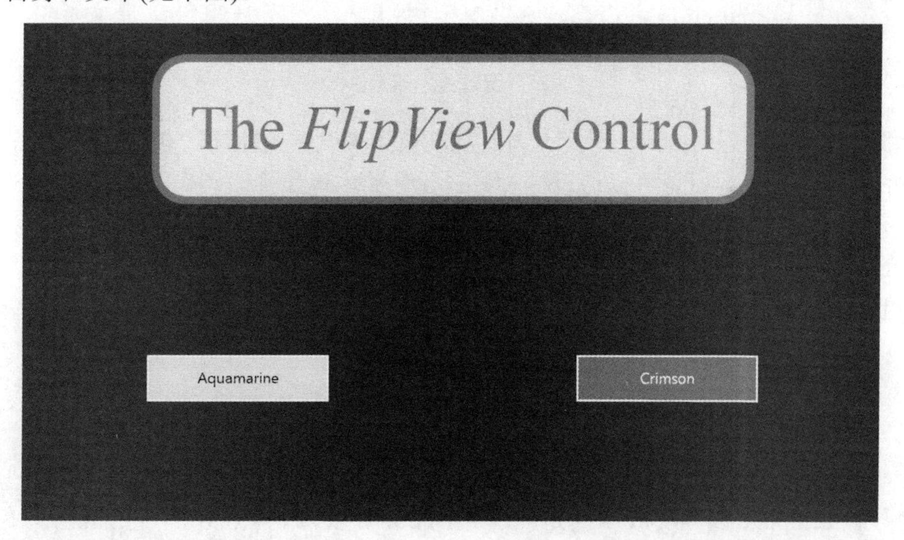

第 16 章将演示如何把 FlipView 控件当成一个简单的电子书阅读器。我之所以有这个想法，是因为标准打印机对话框使用 FlipView 做预览页面，第 17 章会演示此功能，并使用能够进行日期选择的 FlipView 控件。

11.10 基本控件模板

你已经看到了如何把 DataTemplate 设置为 ContentControl 派生类的 ContentTemplate 属性，或者设置为 ItemsControl 派生类的 ItemTemplate，以用于格式化显示数据对象。

你还看到了如何定义 ItemsPanelTemplate，以用于设置 ItemsControl 派生类的 ItemsPanel 供面板托管条目之用。

第三种类型模板为 ControlTemplate 类型。Control 类定义了 ControlTemplate 类型的 Template 属性，允许完全重新定义控件本身的视觉效果，不是指控件内容，而是指通常称为“镶边”的控件部分。

是否存在 Template 属性，可能是 Control 派生类和仅作为 FrameworkElement 派生类之间最重要的区别。控件有镶边，而且镶边外观完全可调。

无论何时你认为需要自定义控件，都应该问问自己真的是新控件，还是只是具有不同外观的现有控件。有时会很幸运，会发现通过 Style 就可以采用现有控件。不过其他时候，则需要 ControlTemplate。

和 Style 一样，常常把 ControlTemplate 定义为资源，以便进行共享。还是跟 Style 一样，ControlTemplate 有控件类型的 TargetType，可用于模板设计。Control 所定义的 Template 属性受依赖属性所支持，也就是说，可以在 Style 里设置 Template 属性。这种做法很常见，如下 Resources 节的代码所示：

```
<Style x:Key="buttonStyle" TargetType="Button">
    <Setter Property="Margin" Value="12" />
    <Setter Property="Template">
        <Setter.Value>
            <ControlTemplate TargetType="Button">
                ...
            </ControlTemplate>
        </Setter.Value>
    </Setter>
</Style>
```

一般来说，你会想用 Setter 对象，通过定义模板来设置控件的一些属性。Setter 标签定义控件的新默认属性，但是可以通过这种样式控件的本地属性设置覆写。

为了明确下面几页内容的目的，我要对控件自身定义 ControlTemplate。而为了演示控件模板的基本知识，我要重定义 Button 外观，只不过不会完全不同于现有 Button。

以下标准 Button 会出现在可视树里。Button 有内容、事件处理程序和一些常见的属性：

```
<Button Content="Click me!"
        Click="OnButtonClick"
        HorizontalAlignment="Center"
        VerticalAlignment="Center" />
```

我们把 Template 属性变成属性元素，对 Button 定义新的 ControlTemplate：

```
<Button Content="Click me!"
        Click="OnButtonClick"
        HorizontalAlignment="Center"
        VerticalAlignment="Center">
    <Button.Template>
        <ControlTemplate TargetType="Button">
```

```
            </ControlTemplate>
        </Button.Template>
    </Button>
```

注意 ControlTemplate 的 TargetType。有时可以不需要，而模板仍然起作用，除非模板引用了 target 控件所定义而非 Control 所定义的属性。

带空 ControlTemplate 的 Button 仍然可以实例化，但不再有任何视觉外观。因为没有视觉效果，因此也没有办法能让用户看到，更不用说点击了。为了确保不会造成太多损害，我们把一个临时 TextBlock 放入 ControlTemplate 标签之间：

```
<Button Content="Click me!"
        Click="OnButtonClick"
        HorizontalAlignment="Center"
        VerticalAlignment="Center">
    <Button.Template>
        <ControlTemplate TargetType="Button">
            <TextBlock Text="temporary" />
        </ControlTemplate>
    </Button.Template>
</Button>
```

现在，Button 有了只包含文本的视觉外观，也有了功能。如果轻击或点击 TextBlock，肯定会触发 Click 事件。然而，视觉效果是静态的。鼠标指针悬停在 Button 之上或者点击 Button 的过程中，都不会有任何特殊的外观指示。有了标准视觉效果，可以在模板中定义这些特殊外观。

可以围绕 TextBlock 放置 Border：

```
<Button Content="Click me!"
        Click="OnButtonClick"
        HorizontalAlignment="Center"
        VerticalAlignment="Center">
    <Button.Template>
        <ControlTemplate TargetType="Button">
            <Border BorderBrush="Red"
                    BorderThickness="3">
                <TextBlock Text="temporary" />
            </Border>
        </ControlTemplate>
    </Button.Template>
</Button>
```

效果如下图所示。

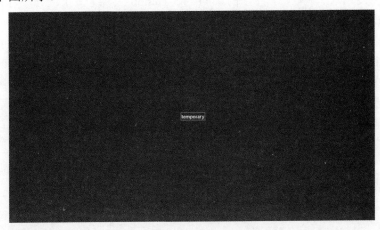

但你真的想在模板中通过硬编码写出红色画笔吗？如果为单个 Button 定义模板，就像我现在所做的这样，可以这么做。但在一般情况下，可以把模板定义为共享资源，有时你希望 Border 为红色，其他时候你又希望是别的颜色。

Control 本身定义了 BorderBrush 和 BorderThickness 属性，而 Button 继承这些属性，因此，对 Button 本身定义这些属性更合理一些：

```
<Button Content="Click me!"
        Click="OnButtonClick"
        HorizontalAlignment="Center"
        VerticalAlignment="Center"
        BorderBrush="Red"
        BorderThickness="3">
    <Button.Template>
        <ControlTemplate TargetType="Button">
            <Border>
                <TextBlock Text="temporary" />
            </Border>
        </ControlTemplate>
    </Button.Template>
</Button>
```

但 Border 现在从 Button 的视觉中完全消失了！模板里的 Border 并没有神奇般地获得 Button 上设置的属性。模板里的 Border 需要几种绑定来引用 Button 所定义的属性。

以下是一种非常特殊类型的绑定，称为 TemplateBinding，它有自己的标记扩展：

```
<Button Content="Click me!"
        Click="OnButtonClick"
        HorizontalAlignment="Center"
        VerticalAlignment="Center"
        BorderBrush="Red"
        BorderThickness="3">
    <Button.Template>
        <ControlTemplate TargetType="Button">
            <Border BorderBrush="{TemplateBinding BorderBrush}"
                    BorderThickness="{TemplateBinding BorderThickness}">
                <TextBlock Text="temporary" />
            </Border>
        </ControlTemplate>
    </Button.Template>
</Button>
```

TemplateBinding 所做的是把 ControlTemplate 可视树里的元素属性绑定到应用 ControlTemplate 控件的属性上。与以前不同，该 Button 的视觉效果包含了红色 Border。

TemplateBinding 语法非常简单，它总是以 ControlTemplate 的可视树里的元素依赖属性为目标，总是引用应用模板的控件的属性。TemplateBinding 标记不能运行其他东西，TemplateBinding 只出现在 ControlTemplate 里的可视树上。

TemplateBinding 实际上是 RelativeSource 绑定的捷径。以下绑定也能发挥作用，但语法明显地较混乱：

```
<Button Content="Click me!"
        Click="OnButtonClick"
        HorizontalAlignment="Center"
        VerticalAlignment="Center"
        BorderBrush="Red"
        BorderThickness="3">
    <Button.Template>
        <ControlTemplate TargetType="Button">
```

```
            <Border BorderBrush="{Binding RelativeSource={RelativeSource TemplatedParent},
                                          Path=BorderBrush}"
                    BorderThickness="{Binding RelativeSource={RelativeSource TemplatedParent},
                                              Path=BorderThickness}">
                <TextBlock Text="temporary" />
            </Border>
        </ControlTemplate>
    </Button.Template>
</Button>
```

如果在 ControlTemplate 里需要建立双向绑定，就可以使用这种冗长的语法。TemplateBinding 为单向，且不接受 Mode 设置。

现在，假设你想要把这种红色边界作为新按钮默认值，但又想让单个按钮覆写默认值。在这种情况下，可以把 ControlTemplate 定义为 Style 的一部分。请记住，一般情况下 Style 被定义为资源，供多个按钮共享。但在本练习中，我的做法是直接把 Style 附加到按钮上：

```
<Button Content="Click me!"
        Click="OnButtonClick"
        HorizontalAlignment="Center"
        VerticalAlignment="Center">
    <Button.Style>
        <Style TargetType="Button">
            <Setter Property="BorderBrush" Value="Red" />
            <Setter Property="BorderThickness" Value="3" />
            <Setter Property="Template">
                <Setter.Value>
                    <ControlTemplate TargetType="Button">
                        <Border BorderBrush="{TemplateBinding BorderBrush}"
                                BorderThickness="{TemplateBinding BorderThickness}">
                            <TextBlock Text="temporary" />
                        </Border>
                    </ControlTemplate>
                </Setter.Value>
            </Setter>
        </Style>
    </Button.Style>
</Button>
```

现在，可以对 Button 自身设置 BorderBrush 和 BorderThickness 属性，覆写 Style 里的那些设置。把默认的 Background 和 Foreground 属性添加到 Style，同时添加 FontSize，让文本大一点：

```
<Button Content="Click me!"
        Click="OnButtonClick"
        HorizontalAlignment="Center"
        VerticalAlignment="Center">
    <Button.Style>
        <Style TargetType="Button">
            <Setter Property="Background" Value="White" />
            <Setter Property="Foreground" Value="Blue" />
            <Setter Property="BorderBrush" Value="Red" />
            <Setter Property="BorderThickness" Value="3" />
            <Setter Property="FontSize" Value="24" />
            <Setter Property="Template">
                <Setter.Value>
                    <ControlTemplate TargetType="Button">
                        <Border Background="{TemplateBinding Background}"
                                BorderBrush="{TemplateBinding BorderBrush}"
                                BorderThickness="{TemplateBinding BorderThickness}">
                            <TextBlock Text="temporary" />
                        </Border>
                    </ControlTemplate>
                </Setter.Value>
```

```
            </Setter>
        </Style>
    </Button.Style>
</Button>
```

注意 Border 的 Background 属性的 TemplateBinding。不过，TextBlock 不需要 Foreground 或 FontSize 属性的 TemplateBinding，因为这些属性通过可视树继承。TextBlock 现在显示蓝色文本，比以前大了一点(见下图)。

目前为止，每个 TemplateBinding 都通过控件里相同名称的属性绑定了可视树的元素属性。并不要求一对一对等。在模板里，可以轻松交换 Background 和 BorderBrush 绑定，因为两者都是 Brush 类型。

```
<ControlTemplate TargetType="Button">
    <Border Background="{TemplateBinding BorderBrush}"
            BorderBrush="{TemplateBinding Background}"
            BorderThickness="{TemplateBinding BorderThickness}">
        <TextBlock Text="temporary" />
    </Border>
</ControlTemplate>
```

除了可能让人困惑之外，这么做其实并没有错。

你也许想让新 Button 的 Border 有圆角，然而 Control 或 Button 里没有任何对应属性，因此，除非我们想定义 Button 派生类，而且包含 CornerRadius 属性，否则就不得不写代码。以下仅为 ControlTemplate 部分标记：

```
<ControlTemplate TargetType="Button">
    <Border Background="{TemplateBinding Background}"
            BorderBrush="{TemplateBinding BorderBrush}"
            BorderThickness="{TemplateBinding BorderThickness}"
            CornerRadius="12">
        <TextBlock Text="temporary" />
    </Border>
</ControlTemplate>
```

效果目前如下图所示。

现在我们来解决 TextBlock 使用临时文本这个小问题。根据目前所看到的，你可能会用 TemplateBinding 绑定到 Button 的 Content 属性来取代临时文本：

```
<TextBlock Text="{TemplateBinding Content}" />
```

在该例中这样做有效，不过错得离谱。本章开始时讨论过诸如 Button 的 ContentControl 派生类的 Content 属性如何成为 object 类型以及 TextBlock 怎么只对文本起作用。如果内容设置为位图，这么做甚至不起作用。

幸运的是，有一个特殊类专门用于在 ContentControl 派生类里显示内容。该类就是 ContentPresenter。和 ContentControl 一样，ContentPresenter 有 object 类型的 Content 属性：

```
<ControlTemplate TargetType="Button">
    <Border Background="{TemplateBinding Background}"
            BorderBrush="{TemplateBinding BorderBrush}"
            BorderThickness="{TemplateBinding BorderThickness}"
            CornerRadius="12">
        <ContentPresenter Content="{TemplateBinding Content}" />
    </Border>
</ControlTemplate>
```

大多数时候，在每个 ContentControl 派生类模板里，我们都会发现 ContentPresenter。ContentPresenter 派生自 FrameworkElement，但也会产生自身的可视树来呈现内容。在本特例中，ContentPresenter 创建 TextBlock 来显示其 Content 属性。

ContentPresenter 也被委托建立可视树，用于显示基于控件的 ContentTemplate 属性的任何类型内容。事实上，ContentPresenter 有自己的 ContentTemplate 属性，可以绑定到控件的 ContentTemplate 属性：

```
<ControlTemplate TargetType="Button">
    <Border Background="{TemplateBinding Background}"
            BorderBrush="{TemplateBinding BorderBrush}"
            BorderThickness="{TemplateBinding BorderThickness}"
            CornerRadius="12">
        <ContentPresenter Content="{TemplateBinding Content}"
                          ContentTemplate="{TemplateBinding ContentTemplate}" />
    </Border>
</ControlTemplate>
```

ContentPresenter 上绑定的这两个模板非常标准和重要，以至于实际根本不做要求！ContentPresenter 会从使用它的控件里自动获得这些属性的值。如果想省略，就可以省略。我个人觉得看到它们会更放心。

你可能还记得，Control 定义了名为 Padding 的属性，旨在为控件的镶边和内容之间提供一些空间。我们试试在 Button 里设置 Padding 属性：

```
<Button Content="Click me!"
        Click="OnButtonClick"
        HorizontalAlignment="Center"
        VerticalAlignment="Center"
        Padding="24">
    ...
</Button>
```

什么效果都没有。你需要添加一些内容到 ControlTemplate，以明确为 Border 和 ContentPresenter 之间留出一些空间。可以是 Border 的 Padding 属性上的 TemplateBinding，但更通用的方法是在 ContentPresenter 的 Margin 属性上设置 TemplateBinding：

```
<ControlTemplate TargetType="Button">
    <Border Background="{TemplateBinding Background}"
            BorderBrush="{TemplateBinding BorderBrush}"
            BorderThickness="{TemplateBinding BorderThickness}"
            CornerRadius="12">
        <ContentPresenter Content="{TemplateBinding Content}"
                          ContentTemplate="{TemplateBinding ContentTemplate}"
                          Margin="{TemplateBinding Padding}" />
    </Border>
</ControlTemplate>
```

现在试试把 Button 的 HorizontalAlignment 和 VerticalAlignment 属性设置为 Stretch。Button 会适当扩大，以填充页面(见下图)。

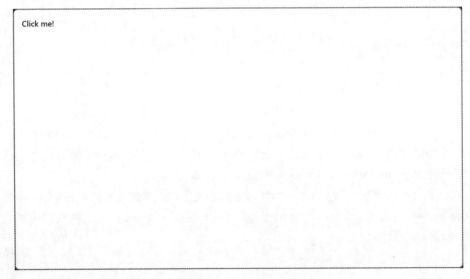

很好，因为这意味着自动处理这些属性。然而，内容在按钮左上角。Control 定义两个属性，分别名为 HorizontalContentAlignment 和 VerticalContentAlignment，两者管理着应该如何定位按钮中内容，但如果试试设置这些属性，就会发现不起作用。

也就是说，必须要添加一些内容到模板才能生效。标准方法是把这些属性绑定到

ContentPresenter 的 HorizontalAlignment 和 VerticalAlignment 属性：

```
<ControlTemplate TargetType="Button">
     <Border Background="{TemplateBinding Background}"
             BorderBrush="{TemplateBinding BorderBrush}"
             BorderThickness="{TemplateBinding BorderThickness}"
             CornerRadius="12">
          <ContentPresenter Content="{TemplateBinding Content}"
                            ContentTemplate="{TemplateBinding ContentTemplate}"
                            Margin="{TemplateBinding Padding}"
                            HorizontalAlignment="{TemplateBinding HorizontalContentAlignment}"
                            VerticalAlignment="{TemplateBinding VerticalContentAlignment}" />
     </Border>
</ControlTemplate>
```

这些属性在其父类里定位 ContentPresenter，在本例中是 Border。我要在 ContentPresenter 再多加一个 TemplateBinding，随后声明就绪即可：

```
<Button Content="Click me!"
        Click="OnButtonClick"
        HorizontalAlignment="Center"
        VerticalAlignment="Center"
        Padding="24">
     <Button.ContentTransitions>
          <TransitionCollection>
               <EntranceThemeTransition />
          </TransitionCollection>
     </Button.ContentTransitions>
     <Button.Style>
          <Style TargetType="Button">
               <Setter Property="Background" Value="White" />
               <Setter Property="Foreground" Value="Blue" />
               <Setter Property="BorderBrush" Value="Red" />
               <Setter Property="BorderThickness" Value="3" />
               <Setter Property="FontSize" Value="24" />
               <Setter Property="Template">
                    <Setter.Value>
                         <ControlTemplate TargetType="Button">
                              <Border Background="{TemplateBinding Background}"
                                      BorderBrush="{TemplateBinding BorderBrush}"
                                      BorderThickness="{TemplateBinding BorderThickness}"
                                      CornerRadius="12">
                                   <ContentPresenter Content="{TemplateBinding Content}"
                                        ContentTemplate="{TemplateBinding ContentTemplate}"
                                        Margin="{TemplateBinding Padding}"
                                        HorizontalAlignment=
                                             "{TemplateBinding HorizontalContentAlignment}"
                                        VerticalAlignment=
                                             "{TemplateBinding VerticalContentAlignment}"
                                        ContentTransitions=
                                             "{TemplateBinding ContentTransitions}" />
                              </Border>
                         </ControlTemplate>
                    </Setter.Value>
               </Setter>
          </Style>
     </Button.Style>
</Button>
```

ContentPresenter 的 ContentTransitions 属性现在绑定到 Button 的 ContentTransitions 属性，我添加了 EntranceThemeTransition 到 Button 来进行测试。现在，如果加载 Button，文本会从右边滑入。

11.11 视觉状态管理器

如果你一直在定义新 Button 的视觉，可能会注意到点击或者轻击 Button，总是会触发 Click 事件。然而，Button 无法有效向用户提供视觉反馈。禁用、获得键盘输入焦点、点击过程或者鼠标滑过时，普通按钮看起来都应该有些不同。

这些不同外观称为视觉状态，在模板里可以通过作为视觉状态管理器一部分的类来进行构建。

Button 有两组 7 种视觉状态：

- CommonStates：Normal、PointerOver、Pressed 和 Disabled
- FocusStates：Focused、Unfocused、PointerFocused

每一组中的状态均相互排斥。例如，被按下的禁用按钮，没有视觉状态。

通过调用 VisualStateManager.GoToState，控件的控制代码负责把这些控件放入状态。这些状态总是用文本名称进行引用。

通常通过模板可视树里的附加元素来实现视觉状态，这些元素一般不可见。通过和背景颜色、Collapsed 的 Visibility 属性或者为 0 的 Opacity 进行颜色匹配，即可实现不可见。动画以该属性为目标使元素可见。这些动画的持续时间通常为 0，也就是说瞬间发生，但如果你愿意，也可以延长动画。

预先警告，负责这些视觉状态无疑是定义模板中最复杂的部分。如果只在特定应用中使用控件，你可能希望除去一些东西。例如，如果知道控件永远不会不可用，则不需要为此提供视觉状态。

在我构建的 ControlTemplate 中，要继续处理 Pressed、Disabled 和 Focused 状态，然后声明完成。

在标准 Button 中，用虚线围绕按钮的边界来表示键盘输入焦点。我要继续把它变成虚线围绕按钮的内容来表示，也就是说它随着 ContentPresenter 进入 Border，也就是说，虚线和 ContentPresenter 两者需要进入只有一个单元格的 Grid。用名为 focusRectangle 的 Rectangle 来实施虚线，如下所示：

```
<ControlTemplate TargetType="Button">
    <Border Background="{TemplateBinding Background}"
            BorderBrush="{TemplateBinding BorderBrush}"
            BorderThickness="{TemplateBinding BorderThickness}"
            CornerRadius="12">
        <Grid>

            <ContentPresenter Content="{TemplateBinding Content}"
                              ContentTemplate="{TemplateBinding ContentTemplate}"
                              Margin="{TemplateBinding Padding}"
                              HorizontalAlignment="{TemplateBinding HorizontalContentAlignment}"
                              VerticalAlignment="{TemplateBinding VerticalContentAlignment}"
                              ContentTransitions="{TemplateBinding ContentTransitions}" />

            <Rectangle Name="focusRectangle"
                       Stroke="{TemplateBinding Foreground}"
                       StrokeThickness="1"
                       StrokeDashArray="2 2"
                       Margin="4"
                       RadiusX="12"
```

```
                    adiusY="12" />
        </Grid>
    </Border>
</ControlTemplate>
```

效果如下图所示。

当然，你不想所有时候都会出现 Rectangle。有一个方法能使 Rectangle 不可见，即赋予其为 0 的 Opacity：

```
<Rectangle Name="focusRectangle"
           Stroke="{TemplateBinding Foreground}"
           Opacity="0"
           StrokeThickness="1"
           StrokeDashArray="2 2"
           Margin="4"
           RadiusX="12"
           RadiusY="12" />
```

然后，一般都是处理可视树中的根元素(组成 ControlTemplate，在本例子中紧跟在 Border 的开始标签之后)，VisualStateGroups 节。里面有每一组的 VisualStateGroups 标签，而在 VisualStateGroups 中有每种状态的 VisualState 标签。所有都用 x: Name 属性来识别。

```
<VisualStateManager.VisualStateGroups>
    <VisualStateGroup x:Name="CommonStates">
        <VisualState x:Name="Normal">
        ...
        </VisualState>

        <VisualState x:Name="PointerOver">
        ...
        </VisualState>

        <VisualState x:Name="Pressed">
        ...
        </VisualState>

        <VisualState x:Name="Disabled">
        ...
        </VisualState>
    </VisualStateGroup>

    <VisualStateGroup x:Name="FocusedStates">
```

```
            <VisualState x:Name="Unfocused">
            ...
            </VisualState>

            <VisualState x:Name="Focused">
            ...
            </VisualState>

            <VisualState x:Name="PointerFocused">
            ...
            </VisualState>
        </VisualStateGroup>
    </VisualStateManager.VisualStateGroups>
```

如果基本模板的视觉部分是为 Normal 和 Unfocused 状态而设计的，则可以构造这些空标签。如果不希望处理各种状态，也可以使这些标签为空：

```
<VisualStateManager.VisualStateGroups>
    <VisualStateGroup x:Name="CommonStates">
        <VisualState x:Name="Normal" />
        <VisualState x:Name="PointerOver" />

        <VisualState x:Name="Pressed">
        ...
        </VisualState>

        <VisualState x:Name="Disabled">
        ...
        </VisualState>
    </VisualStateGroup>

    <VisualStateGroup x:Name="FocusedStates">
        <VisualState x:Name="Unfocused" />

        <VisualState x:Name="Focused">
        ...
        </VisualState>

        <VisualState x:Name="PointerFocused" />
    </VisualStateGroup>
</VisualStateManager.VisualStateGroups>
```

但不要删除这些标签。在特定组里，应该有标签对应于所有状态。如果少一个，就不会过渡回到该状态。

对于要处理的状态，可以在包含动画的两个 VisualState 标签之间放入 Storyboard。例如：

```
<VisualStateGroup x:Name="FocusedStates">
        <VisualState x:Name="Unfocused" />

        <VisualState x:Name="Focused">
            <Storyboard>
                <DoubleAnimation Storyboard.TargetName="focusRectangle"
                                 Storyboard.TargetProperty="Opacity"
                                 To="1" Duration="0" />
            </Storyboard>
    </VisualState>

    <VisualState x:Name="PointerFocused" />
</VisualStateGroup>
```

注意，这里没有 From 属性。你只想表明值在何处终止，而不是从何处开始。

有了这个，如果底层控件接收到输入焦点，就会调用 OnGotFocus 方法，通过调用为

Focused 的 VisualStateManager.GoToState，控件会做出响应。这样会触发 Storyboard，而 Storyboard 把目标 Opacity 属性设置为 1。如果底层控件失去输入焦点，则调用状态为 Unfocused 的 VisualStateManager.GoToState，并撤销动画。

对于禁用状态，我想使整个控件都变灰，这有一个好办法，即用 Visibility 为 Collapsed 的半透明黑色矩形来覆盖整个控件。因此，我们把 Border 放到另一个 Grid，并给位于 Border 上面的 Grid 添加一个已命名的 Rectangle。与此同时，我已经把 Visual State Manager 标记移到最外层 Grid：

```
<ControlTemplate TargetType="Button">
     <Grid>
          <VisualStateManager.VisualStateGroups>
               ...
          </VisualStateManager.VisualStateGroups>

          <Border Name="border" ... >
               <Grid>
                    <ContentPresenter Name="contentPresenter" ... />
                    <Rectangle Name="focusRectangle" ... />
               </Grid>
          </Border>

          <Rectangle Name="disabledRect"
                     Visibility="Collapsed"
                     Fill="Black"
                     Opacity="0.5" />
     </Grid>
</ControlTemplate>
```

我还命名了 Border 和 ContentPresenter，以便在动画里进行引用。对于 Disabled 状态，我定义了动画使 disabledRect 可见；而对于 Pressed 状态，我定义了两个动画来设置控件的背景色和前景色。

可以在 CustomButtonTemplate 项目中看到这些，项目有最终样式和模板。主要为了避免篇幅过长，我在 Resources 字典里已经把 ControlTemplate 定义为单独对象，并从 Style 引用。

项目：CustomButtonTemplate | 文件：MainPage.xaml(片段)

```
<Page ... >
     <Page.Resources>
          <ControlTemplate x:Key="buttonTemplate" TargetType="Button">
               <Grid>
                    <VisualStateManager.VisualStateGroups>
                         <VisualStateGroup x:Name="CommonStates">
                              <VisualState x:Name="Normal" />
                              <VisualState x:Name="PointerOver" />

                              <VisualState x:Name="Pressed">
                                   <Storyboard>
                                        <ObjectAnimationUsingKeyFrames
                                                  Storyboard.TargetName="border"
                                                  Storyboard.TargetProperty="Background">
                                             <DiscreteObjectKeyFrame KeyTime="0"
                                                                     Value="LightGray" />
                                        </ObjectAnimationUsingKeyFrames>

                                        <ObjectAnimationUsingKeyFrames
                                           Storyboard.TargetName="contentPresenter"
                                           Storyboard.TargetProperty="Foreground">
                                             <DiscreteObjectKeyFrame KeyTime="0"
                                                                         Value="Black" />
```

```
                        </ObjectAnimationUsingKeyFrames>
                    </Storyboard>
                </VisualState>

                <VisualState x:Name="Disabled">
                    <Storyboard>
                        <ObjectAnimationUsingKeyFrames
                            Storyboard.TargetName="disabledRect"
                            Storyboard.TargetProperty="Visibility">
                            <DiscreteObjectKeyFrame KeyTime="0"
                                                    Value="Visible" />
                        </ObjectAnimationUsingKeyFrames>
                    </Storyboard>
                </VisualState>
            </VisualStateGroup>

            <VisualStateGroup x:Name="FocusedStates">
                <VisualState x:Name="Unfocused" />

                <VisualState x:Name="Focused">
                    <Storyboard>
                        <DoubleAnimation Storyboard.TargetName="focusRectangle"
                                         Storyboard.TargetProperty="Opacity"
                                         To="1" Duration="0" />
                    </Storyboard>
                </VisualState>

                <VisualState x:Name="PointerFocused" />
            </VisualStateGroup>
        </VisualStateManager.VisualStateGroups>

        <Border Name="border"
                Background="{TemplateBinding Background}"
                BorderBrush="{TemplateBinding BorderBrush}"
                BorderThickness="{TemplateBinding BorderThickness}"
                CornerRadius="12">

            <Grid>
                <ContentPresenter Name="contentPresenter"
                    Content="{TemplateBinding Content}"
                    ContentTemplate="{TemplateBinding ContentTemplate}"
                    Margin="{TemplateBinding Padding}"
                    HorizontalAlignment="{TemplateBinding
                    HorizontalContentAlignment}"
                    VerticalAlignment="{TemplateBinding
                    VerticalContentAlignment}"
                    ContentTransitions="{TemplateBinding
                    ContentTransitions}" />

                <Rectangle Name="focusRectangle"
                           Stroke="{TemplateBinding Foreground}"
                           Opacity="0"
                           StrokeThickness="1"
                           StrokeDashArray="2 2"
                           Margin="4"
                           RadiusX="12"
                           RadiusY="12" />
            </Grid>
        </Border>

        <Rectangle Name="disabledRect"
                   Visibility="Collapsed"
                   Fill="Black"
                   Opacity="0.5" />
    </Grid>
</ControlTemplate>

<Style x:Key="buttonStyle" TargetType="Button">
```

```
                <Setter Property="Background" Value="White" />
                <Setter Property="Foreground" Value="Blue" />
                <Setter Property="BorderBrush" Value="Red" />
                <Setter Property="BorderThickness" Value="3" />
                <Setter Property="FontSize" Value="24" />
                <Setter Property="Padding" Value="12" />
                <Setter Property="Template" Value="{StaticResource buttonTemplate}" />
            </Style>
        </Page.Resources>

        <Grid Background="{StaticResource ApplicationPageBackgroundThemeBrush}">
            <Grid.ColumnDefinitions>
                <ColumnDefinition Width="*" />
                <ColumnDefinition Width="*" />
                <ColumnDefinition Width="*" />
            </Grid.ColumnDefinitions>

            <Button Content="Disable center button"
                    Grid.Column="0"
                    Style="{StaticResource buttonStyle}"
                    Click="OnButton1Click"
                    HorizontalAlignment="Center"
                    VerticalAlignment="Center" />

            <Button Name="centerButton"
                    Content="Center button"
                    Grid.Column="1"
                    Style="{StaticResource buttonStyle}"
                    FontSize="48"
                    Background="DarkGray"
                    Foreground="Red"
                    HorizontalAlignment="Center"
                    VerticalAlignment="Center" />

            <Button Content="Enable center button"
                    Grid.Column="2"
                    Style="{StaticResource buttonStyle}"
                    Click="OnButton3Click"
                    HorizontalAlignment="Center"
                    VerticalAlignment="Center" />
        </Grid>
</Page>
```

XAML 文件结尾有三个按钮，中间的按钮获取一些局部属性值用于覆写 Style。外面的两个按钮分别禁用和启用中间按钮。

项目: CustomButtonTemplate | 文件: MainPage.xaml.cs (片段)

```
void OnButton1Click(object sender, RoutedEventArgs args)
{
    centerButton.IsEnabled = false;
}

void OnButton3Click(object sender, RoutedEventArgs args)
{
    centerButton.IsEnabled = true;
}
```

如下屏幕截图中，中间按钮被禁用，而第三个按钮则获取键盘输入焦点。

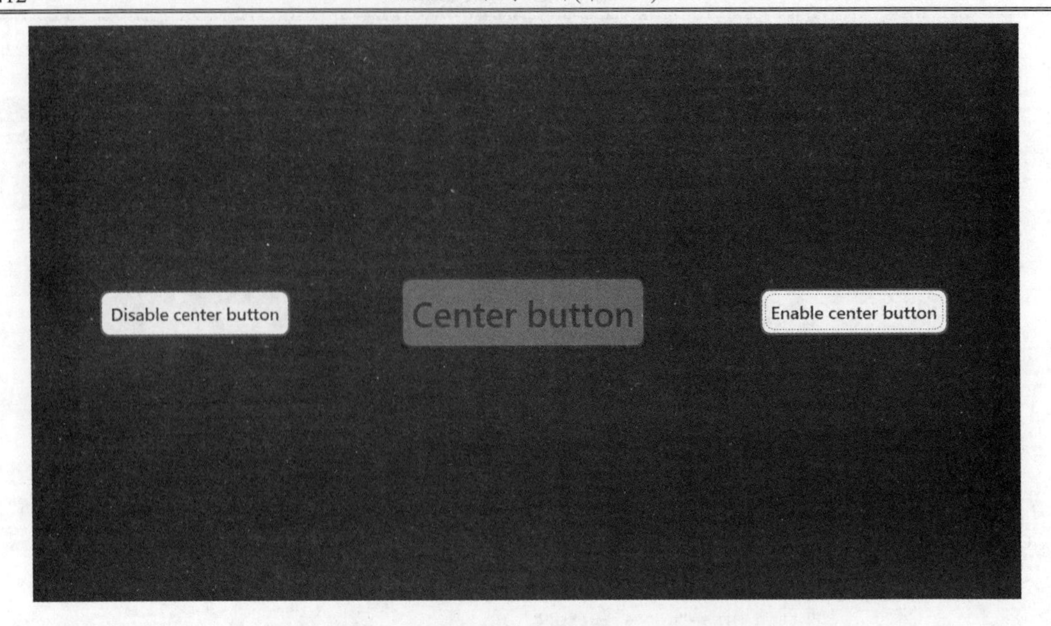

11.12 使用 generic.xaml

在安装了 Visual Studio 的机器上查看以下目录：

C:\Program Files (x86)\Windows Kits\8 .0\Include\winrt\xaml\design

应该会看到两个文件。较小文件为 themeresources.xaml，主要包含 SolidColorBrush 定义，用于供 Windows Runtime 应用使用的标准颜色，包括著名的 ApplicationPageBackgroundThemeBrush 和 ApplicationForegroundThemeBrush 颜色。整套颜色分为三部分：Default(即黑色主题)、Light 和 HighContrast。用户可以从 Ease of Access 小节选择高对比度显示，而从电脑 Settings 程序里可以访问 Ease of Access。

较大文件为 generic.xaml，包含和 themeresources.xaml 一样的定义，还有为所有标准控件定义的默认 Style 和 ControlTemplate 定义。

如果你想擅长为控件设计自定义模板，则有必要研究 generic.xaml 的默认模板。在这些模板里，也是(显然)和每个控件相关的视觉状态的唯一文档以及命名部分，下一节会进行讨论。

要找到特定控件的默认 Style，可以通过 TargetType="后面加上控件名称的方式进行搜索。

模板通常引用在 generic.xaml 开头定义的画笔，而不同视觉状态各自都有特殊的画笔。例如，在默认 Button 模板里，视觉状态动画用名称引用画笔，比如 ButtonPressedBackgroundThemeBrush 和 ButtonPressedForegroundThemeBrush。根据应用所选 Light 或 Dark 主题、以及用户可以选择的 HighContrast 主题，画笔的实际颜色不同。

所有标准控件的 Style 定义没有键名。控件实例化时，基本上是应用于控件的隐式样式。应用提供的其他任何内容都是对该隐式样式的补充。

为控件开发新模板有一个好办法：直接把所有现有 Style 定义从 generic.xaml 复制到你自己的 XAML 文件，然后再开始修改。

11.13 模板部分

在我引导你为 Button 构造模板的整个过程中，你可能想知道它是如何处理更复杂的控件的。以 Slider 为例。Slider 有可移动部分，底层控件如何引用模板的这些部分？

诸如 Slider 之类控件的底层代码会假定组成模板的某个元素具有特定名称。在初始化过程中，通过这些名称调用 GetTemplateChild 方法覆写 OnApplyTemplate 方法，控件代码从而引用这些元素。控件代码可以把这些元素对象保存为字段，给元素安装事件处理程序，并改变其属性，而用户可以操纵控件。

不幸的是，这些已命名部分还没有显示在任何 Windows Runtime 文档中。你得研究 generic.xaml 的默认模板，才能从中找到它们。许多情况下，并不需要了解其中的每一个。如果模板中的某些部分丢失了，但不会引发异常，则可以认为控件正确。

为了达到最基本的必要功能，Slider 模板必须包含水平和垂直方向的模板。这些单独模板一般为 Grid 类型。可以把它们命名为“HorizontalTemplate”和“VerticalTemplate”。

每个 Grid 必须有一个 Rectangle(Rectangle 包含名为“HorizontalTemplate”或“VerticalTemplate”的完整 Slider，每个 Grid 还必须有名为“HorizontalThumb”或“VerticalThumb”的 Thumb)以及第二个 Rectangle(出现在 Thumb 左边，并命名为“HorizontalDecreaseRect”或“VerticalDecreaseRect”)。用户操作 Thumb、点击或轻击 Slider 内的任何地方时，底层控件会改变第二个矩形的大小，以反映 Slider 的值。

我们来看一个几乎是最小功能的 Slider 模板，模板包含几个显式属性设置并忽略了提供可选刻度线的 TickBar 元素。我把该项目称为 BareBonesSlider。

项目：BareBonesSlider | 文件：MainPage.xaml(片段)

```
<Page ... >
    <Page.Resources>
        <ControlTemplate x:Key="sliderTemplate"
                         TargetType="Slider">
            <Grid>
                <Grid Name="HorizontalTemplate"
                      Background="Transparent"
                      Height="48">
                    <Grid.ColumnDefinitions>
                        <ColumnDefinition Width="Auto" />
                        <ColumnDefinition Width="Auto" />
                        <ColumnDefinition Width="*" />
                    </Grid.ColumnDefinitions>

                    <Rectangle Name="HorizontalTrackRect"
                               Grid.Column="0"
                               Grid.ColumnSpan="3"
                               Fill="Blue"
                               Margin="0 12" />

                    <Thumb Name="HorizontalThumb"
                           Grid.Column="1"
                           DataContext="{TemplateBinding Value}"
                           Width="24" />

                    <Rectangle Name="HorizontalDecreaseRect"
                               Grid.Column="0"
                               Fill="Red"
                               Margin="0 12" />
```

```
                </Grid>

                <Grid Name="VerticalTemplate"
                      Visibility="Collapsed"
                      Background="Transparent"
                      Width="48">
                    <Grid.RowDefinitions>
                        <RowDefinition Height="*" />
                        <RowDefinition Height="Auto" />
                        <RowDefinition Height="Auto" />
                    </Grid.RowDefinitions>

                    <Rectangle Name="VerticalTrackRect"
                               Grid.Row="0"
                               Grid.RowSpan="3"
                               Fill="Blue"
                               Margin="12 0" />

                    <Thumb Name="VerticalThumb"
                           Grid.Row="1"
                           DataContext="{TemplateBinding Value}"
                           Height="24" />

                    <Rectangle Name="VerticalDecreaseRect"
                               Grid.Row="2"
                               Fill="Red"
                               Margin="12 0" />
                </Grid>
            </Grid>
        </ControlTemplate>
    </Page.Resources>

    <Grid Background="{StaticResource ApplicationPageBackgroundThemeBrush}">
        <Grid.RowDefinitions>
            <RowDefinition Height="Auto" />
            <RowDefinition Height="*" />
        </Grid.RowDefinitions>

        <Slider Grid.Row="0"
                Template="{StaticResource sliderTemplate}"
                Margin="48" />

        <Slider Grid.Row="1"
                Template="{StaticResource sliderTemplate}"
                Orientation="Vertical"
                Margin="48" />
    </Grid>
</Page>
```

XAML 文件结尾有两个 Slider 控件，一个横向，一个纵向，两者都引用这些模板。

我要描述横向 Slider 模板，纵向的结构与之类似。

名为 HorizontalTemplate 的 Grid 的总宽度为布局中 Slider 控件的宽度。Grid 有三列。名为 HorizontalTrackRect 的 Rectangle 跨越三列，因此，Rectangle 的宽度总是等于 Slider 自身的宽度。名为 HorizontalTrackRect 的 Rectangle 占据 Grid 的第一列，宽度为 Auto，能把 Rectangle 的减少到 0 宽度。Thumb 占据 Grid 的中间一列，宽度同样为 Auto，也就是说该中间列为 Thumb 的大小。

底层代码只允许 Thumb 水平移动，而且不能超过 Slide 的限制。如果用户操纵 Thumb、按压或轻击 Slider 的任何位置，底层代码会相应设置 HorizontalDecreaseRect 元素的 Width 属性。滑块的最小值，宽度属性设置为 0；滑块的最大值，设置为将 HorizontalTrackRect 元素的宽度减去 Thumb 的宽度。我给出了这些组件的大小和边界，因此 Thumb 比矩形大一点点(见下图)。

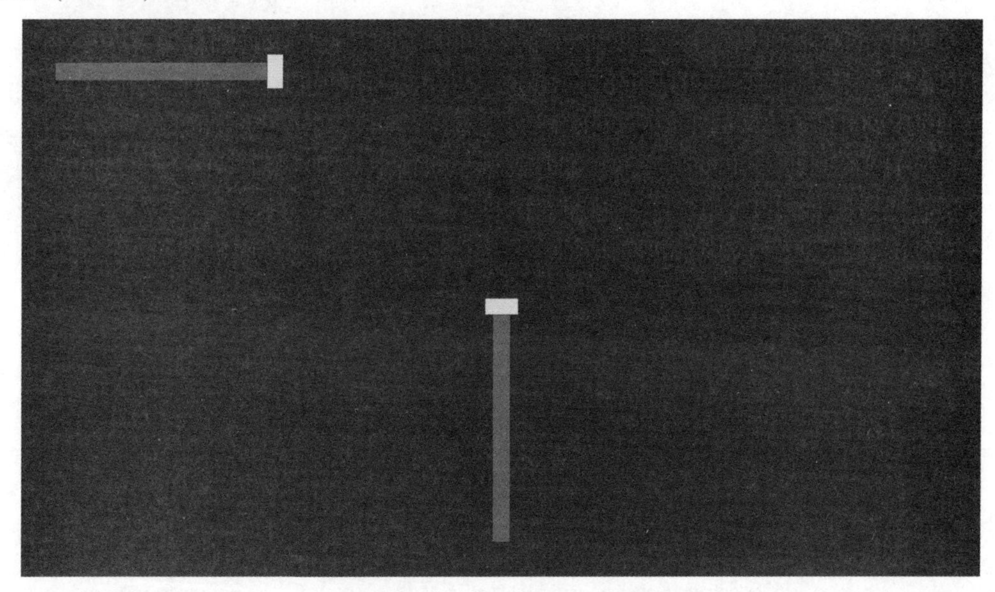

注意，模板包含一个 TemplateBinding，它把 Thumb 的 DataContext 绑定到 Slider 的 Value 属性。这是让 Slider 弹出式工具提示用于显示正确值所需要的。

在 BareBonesSlider 里操作 Thumb 的时候，你会发现如果按压，Thumb 几乎会变成透明黑色。Thumb 派生自 Control，因此可以被赋予自己的模板。这是在 Resources 节的默认 Slider 模块中完成的，而 Resources 附属于模块的最外层 Grid。

只要对 BareBonesSlider 程序做一点小小修改，就可以做出一些花哨的东西，我把以下代码称为 SpringLoadedSlider。

```
项目: SpringLoadedSlider | 文件: MainPage.xaml(片段)
<Page ... >
    <Page.Resources>
        <ControlTemplate x:Key="sliderTemplate"
                         TargetType="Slider">
            <Grid>
                <Grid.Resources>
                    <Style TargetType="Path">
                        <Setter Property="StrokeThickness" Value="6" />
                        <Setter Property="StrokeLineJoin" Value="Round" />
                        <Setter Property="Stretch" Value="Fill" />
                    </Style>
                </Grid.Resources>

                <Grid Name="HorizontalTemplate"
                      Background="Transparent"
                      Height="48">

                    <Grid.ColumnDefinitions>
                        <ColumnDefinition Width="Auto" />
                        <ColumnDefinition Width="Auto" />
                        <ColumnDefinition Width="*" />
                    </Grid.ColumnDefinitions>

                    <Rectangle Name="HorizontalTrackRect"
                               Grid.Column="0"
                               Grid.ColumnSpan="3"
                               Fill="Transparent" />
```

```
                <Thumb Name="HorizontalThumb"
                       Grid.Column="1"
                       DataContext="{TemplateBinding Value}"
                       Width="12" />

                <Rectangle Name="HorizontalDecreaseRect"
                           Grid.Column="0"
                           Fill="Transparent" />

                <Path Stroke="Red"
                      Grid.Column="0"
                      Width="{Binding ElementName=HorizontalDecreaseRect,
                                      Path=Width}"
                    Data="M 0 0 L 100 100, 200 0, 300 100, 400 0,
                          400 100, 300 0, 200 100, 100 0, 0 100 Z" />

                <Path Stroke="Blue"
                      Grid.Column="2"
                      Data="M 0 0 L 100 100, 200 0, 300 100, 400 0,
                            400 100, 300 0, 200 100, 100 0, 0 100 Z" />
            </Grid>

            <Grid Name="VerticalTemplate"
                  Visibility="Collapsed"
                  Background="Transparent"
                  Width="48">
                <Grid.RowDefinitions>
                    <RowDefinition Height="*" />
                    <RowDefinition Height="Auto" />
                    <RowDefinition Height="Auto" />
                </Grid.RowDefinitions>

                <Rectangle Name="VerticalTrackRect"
                           Grid.Row="0"
                           Grid.RowSpan="3"
                           Fill="Transparent" />

                <Thumb Name="VerticalThumb"
                       Grid.Row="1"
                       DataContext="{TemplateBinding Value}"
                       Height="12" />

                <Rectangle Name="VerticalDecreaseRect"
                           Grid.Row="2"
                           Fill="Transparent" />

                <Path Stroke="Red"
                    Grid.Row="2"
                    Height="{Binding ElementName=VerticalDecreaseRect,
                                     Path=Height}"
                    Data="M 0 0 L 100 100, 0 200, 100 300, 0 400,
                          100 400, 0 300, 100 200, 0 100, 100 0 Z" />

                <Path Stroke="Blue"
                    Grid.Row="0"
                    Data="M 0 0 L 100 100, 0 200, 100 300, 0 400,
                          100 400, 0 300, 100 200, 0 100, 100 0 Z" />
            </Grid>
        </Grid>
    </ControlTemplate>
</Page.Resources>

<Grid Background="{StaticResource ApplicationPageBackgroundThemeBrush}">
    <Grid.RowDefinitions>
        <RowDefinition Height="Auto" />
        <RowDefinition Height="*" />
    </Grid.RowDefinitions>
```

```
            <Slider Grid.Row="0"
                    Template="{StaticResource sliderTemplate}"
                    Margin="48" />

            <Slider Grid.Row="1"
                    Template="{StaticResource sliderTemplate}"
                    Orientation="Vertical"
                    Margin="48" />
        </Grid>
    </Page>
```

除了赋予 Rectangle 元素 Transparent 的 Fill 颜色外，两个模板的结构相同。此外，两个 Path 元素添加到每个模板。第一个 Path 位于第一列(水平滑块)，颜色为红色。该 Path 的 Width 绑定到名为“HorizontalDecreaseRect”的元素的 Width。第二个 Path 为蓝色，占据第三列。每个 Path 都有相同几何结构(纵横交错的晶格)，Stretch 模式为 Fill，也就是说，可以填充的空间。

Thumb 两边的弹簧外观，如下图所示。

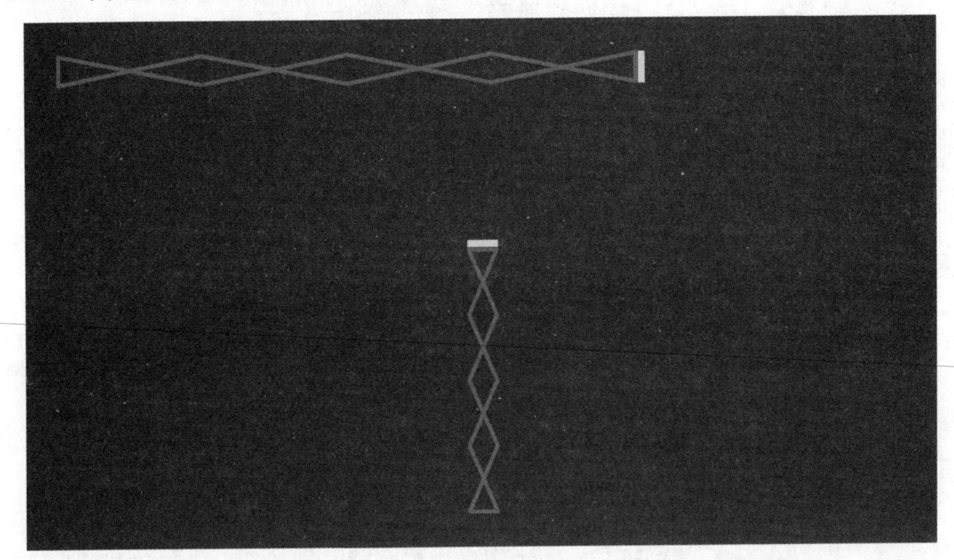

ProgressBar 的默认模板相当精巧，因为需要包含确定的和不确定的外观。然而，如果你限制为只包含确定的 ProgressBar，事情就会变得非常简单：底层代码改变名为 ProgressBarIndicator 元素的宽度，范围从 0 到名为 DeterminateRoot 元素的宽度。在默认模板里，DeterminateRoot 为 Border，包含左对齐的名为 ProgressBarIndicator 的 Rectangle。

在 SpeedometerProgressBar 里，DeterminateRoot 和 ProgressBarIndicator 均为不可见，但 DeterminateRoot 的 Width 硬编码为 180。也就是说，ProgressBarIndicator 的 Width 范围从 0 到 180。ProgressBarIndicator 的 Width 属性的绑定以 RotateTransform 的 Angle 属性为目标，从 0 到 180 度旋转一个箭头指示器。

项目：SpeedometerProgressBar | 文件：MainPage.xaml(片段)

```
<Page ... >
    <Page.Resources>
        <ControlTemplate x:Key="progressTemplate"
                         TargetType="ProgressBar">
            <Grid>
                <Grid.Resources>
                    <Style TargetType="Line">
                        <Setter Property="Stroke" Value="Black" />
```

```
                <Setter Property="StrokeThickness" Value="1" />
                <Setter Property="X1" Value="-85" />
                <Setter Property="X2" Value="-95" />
            </Style>

            <Style TargetType="TextBlock">
                <Setter Property="FontSize" Value="11" />
                <Setter Property="Foreground" Value="Black" />
            </Style>
        </Grid.Resources>

        <Border Width="270" Height="120"
            BorderBrush="{TemplateBinding BorderBrush}"
            BorderThickness="{TemplateBinding BorderThickness}"
            Background="White">

            <!-- Canvas for positioning graphics-->
            <Canvas Width="0" Height="0"
                    RenderTransform="1 0 0 1 0 50" >

            <!-- The required parts of the ProgressBar template -->
            <Border Name="DeterminateRoot"
                    Width="180">
                <Rectangle Name="ProgressBarIndicator"
                           HorizontalAlignment="Left" />
            </Border>

            <Line RenderTransform=" 1.00 0.00 -0.00 1.00 0 0" />
            <Line RenderTransform=" 0.95 0.31 -0.31 0.95 0 0" />
            <Line RenderTransform=" 0.81 0.59 -0.59 0.81 0 0" />
            <Line RenderTransform=" 0.59 0.81 -0.81 0.59 0 0" />
            <Line RenderTransform=" 0.31 0.95 -0.95 0.31 0 0" />
            <Line RenderTransform=" 0.00 1.00 -1.00 0.00 0 0" />
            <Line RenderTransform="-0.31 0.95 0.95 0.31 0 0" />
            <Line RenderTransform="-0.59 0.81 0.81 0.59 0 0" />
            <Line RenderTransform="-0.81 0.59 0.59 0.81 0 0" />
            <Line RenderTransform="-0.95 0.31 0.31 0.95 0 0" />
            <Line RenderTransform="-1.00 0.00 0.00 1.00 0 0" />

            <TextBlock Text="0%" Canvas.Left="-115" Canvas.Top="-6" />
            <TextBlock Text="20%" Canvas.Left="-104" Canvas.Top="-65" />
            <TextBlock Text="40%" Canvas.Left="-42" Canvas.Top="-105" />
            <TextBlock Text="60%" Canvas.Left="25" Canvas.Top="-105" />
            <TextBlock Text="80%" Canvas.Left="82" Canvas.Top="-65" />
            <TextBlock Text="100%" Canvas.Left="100" Canvas.Top="-6" />

            <!-- Arrow to point to percentage -->
            <Polygon Points="5 5 5 -5 -75 0"
                     Stroke="Black"
                          Fill="Red">
                <Polygon.RenderTransform>
                    <RotateTransform
                        Angle="{Binding ElementName=ProgressBarIndicator,
                               Path=Width}" />
                    </Polygon.RenderTransform>
                </Polygon>
            </Canvas>
        </Border>
    </Grid>
  </ControlTemplate>
</Page.Resources>

<Grid Background="{StaticResource ApplicationPageBackgroundThemeBrush}">
    <Grid.RowDefinitions>
        <RowDefinition Height="Auto" />
        <RowDefinition Height="*" />
    </Grid.RowDefinitions>
```

```
        <ProgressBar Grid.Row="0"
                     Template="{StaticResource progressTemplate}"
                     Margin="48"
                     Value="{Binding ElementName=slider, Path=Value}" />

        <Slider Name="slider"
                Grid.Row="1"
                Margin="48"
                VerticalAlignment="Center" />
    </Grid>
</Page>
```

XAML 文件的结尾用该模板实例化 ProgressBar，并为了测试目的绑定到 Slider(见下图)。

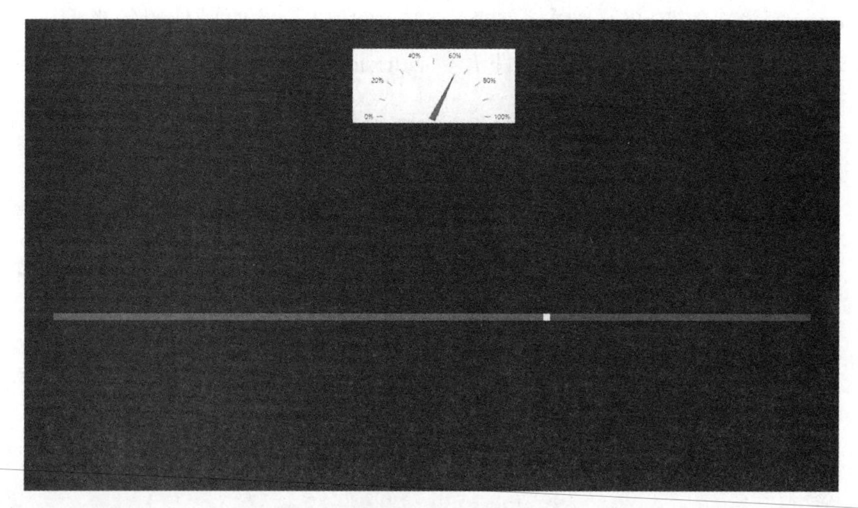

SpringLoadedSlider 和 SpeedometerProgressBar 基于我最初为一篇 WPF 文章所创建的 XAML 文件，该文章发表于 2007 年 1 月的 MSDN 杂志。虽然我需要改变模板的一些东西来说明 WPF 和 Windows Runtime 之间的差异，但大多数情况下两者都非常相似。虽然我们还没有完成所有基于 XAML 环境的可移植性，但六年前完成的这项工作可以很容易适应新平台。

11.14 自定义控件

如果在 Windows Runtime 库中创建自定义控件，你会想让该控件能够用于各种应用程序，甚至能出售给其他程序员。在这种情况下，应该为控件提供默认 Style，包括默认 ControlTemplate。

包含自定义控件类的库也应该在 Themes 文件夹中包含名为 generic.xaml 的文件。就像你已经看过的 generic.xaml 文件，该 generic.xaml 文件有一个 ResourceDictionary 根元素，同时包含 TargetType 的 Style 定义，表明自定义控件名称，并且没有字典键。该 Style 应该包含默认 ControlTemplate。

Visual Studio 会生成 generic.xaml 文件框架。本章已经使用过的 Petzold.ProgrammingWindows6.Chapter11 库里，我激活了“添加新项”对话框并选择“模板控件”，将其命名为 NewToggle。Visual Studio 生成了 NewToggle.cs 文件，包含很多 using

指令及以下类定义：

```
namespace Petzold.ProgrammingWindows6.Chapter11
{
    public sealed class NewToggle : Control
    {
        public NewToggle()
        {
            this.DefaultStyleKey = typeof(NewToggle);
        }
    }
}
```

这不是一个分部类定义！没有相应的 NewToggle.xaml 文件，构造函数也不包含调用 InitializeComponent。DefaultStyleKey 属性表明搜索隐式样式时所使用的类型。

Visual Studio 还生成了 Themes 文件夹和 generic.xaml 文件，包含该隐式样式：

```
<ResourceDictionary
    xmlns="http://schemas.microsoft.com/winfx/2006/xaml/presentation"
    xmlns:x="http://schemas.microsoft.com/winfx/2006/xaml"
    xmlns:local="using:Petzold.ProgrammingWindows6.Chapter11">

    <Style TargetType="local:NewToggle">
        <Setter Property="Template">
            <Setter.Value>
                <ControlTemplate TargetType="local:NewToggle">
                    <Border
                        Background="{TemplateBinding Background}"
                        BorderBrush="{TemplateBinding BorderBrush}"
                        BorderThickness="{TemplateBinding BorderThickness}">
                    </Border>
                </ControlTemplate>
            </Setter.Value>
        </Setter>
    </Style>
</ResourceDictionary>
```

如果库里有多个自定义控件，同一个文件中就会还包含它们所有的默认 Style 定义。该文件有特定名称和位置，是因为文件会永远关联到库里所定义的自定义控件，而且不需要通过任何其他方式进行引用。

NewToggle 控件旨在通过在同一时刻展示两段不同内容来实现触发按钮的功能，一段内容关联未选中状态，另一段关联到已选中状态。可以轻击其中一段来改变选中状态。视觉效果如何变化以反映变化由模板负责。

我从 ContentControl 派生 NewToggle，这样 NewToggle 可以继承 Content 和 ContentTemplate 属性。NewToggle 定义了两个新的依赖属性，分别为 CheckedContent 和 IsChecked。

项目：Petzol.ProgrammingWindows6.Chapter11 | 文件：NewToggle.cs

```
public class NewToggle : ContentControl
{
    public event EventHandler CheckedChanged;
    Button uncheckButton, checkButton;

    static NewToggle()
    {
        CheckedContentProperty = DependencyProperty.Register("CheckedContent",
            typeof(object),
            typeof(NewToggle),
            new PropertyMetadata(null));
```

```
            IsCheckedProperty = DependencyProperty.Register("IsChecked",
                typeof(bool),
                typeof(NewToggle),
                new PropertyMetadata(false, OnCheckedChanged));
        }

        public NewToggle()
        {
            this.DefaultStyleKey = typeof(NewToggle);
        }

        public static DependencyProperty CheckedContentProperty { private set; get; }

        public static DependencyProperty IsCheckedProperty { private set; get; }

        public object CheckedContent
        {
            set { SetValue(CheckedContentProperty, value); }
            get { return GetValue(CheckedContentProperty); }
        }

        public bool IsChecked
        {
            set { SetValue(IsCheckedProperty, value); }
            get { return (bool)GetValue(IsCheckedProperty); }
        }

        protected override void OnApplyTemplate()
        {
            if (uncheckButton != null)
                uncheckButton.Click -= OnButtonClick;

            if (checkButton != null)
                checkButton.Click -= OnButtonClick;

            uncheckButton = GetTemplateChild("UncheckButton") as Button;
            checkButton = GetTemplateChild("CheckButton") as Button;

            if (uncheckButton != null)
                uncheckButton.Click += OnButtonClick;

            if (checkButton != null)
                checkButton.Click += OnButtonClick;

            base.OnApplyTemplate();
        }

        void OnButtonClick(object sender, RoutedEventArgs args)
        {
            this.IsChecked = sender == checkButton;
        }

        static void OnCheckedChanged (DependencyObject obj,
                                DependencyPropertyChangedEventArgs args)
        {
            (obj as NewToggle).OnCheckedChanged(EventArgs.Empty);
        }

        protected virtual void OnCheckedChanged(EventArgs args)
        {
            VisualStateManager.GoToState (this,
                            this.IsChecked ? "Checked" : "Unchecked",
                            true);

            if (CheckedChanged != null)
                CheckedChanged(this, args);
        }
    }
```

OnApplyTemplate 覆写会假设模板有两个 Button 控件，名为“UncheckButton”和“CheckButton”。如果是这样，它们会被保存为字段，并附加上 Click 处理程序。如果点击其中一个，就会改变 IsChecked 属性，触发 CheckedChanged 事件，并且调用静态 VisualStateManager.GoToState，其状态为“已选中”或“未选中”。

generic.xaml 中的模板包含两个按钮，也带有这些名称以及为两个状态定义的 Storyboard 对象。

项目: Petzold.ProgrammingWindows11.Chapter11 | 文件: generic.xaml(片段)

```
<Style TargetType="local:NewToggle">
    <Setter Property="BorderBrush" Value="{StaticResource ApplicationForegroundThemeBrush}" />
    <Setter Property="BorderThickness" Value="1" />
    <Setter Property="Template">
        <Setter.Value>
            <ControlTemplate TargetType="local:NewToggle">
                <Border Background="{TemplateBinding Background}"
                        BorderBrush="{TemplateBinding BorderBrush}"
                        BorderThickness="{TemplateBinding BorderThickness}">

                    <VisualStateManager.VisualStateGroups>
                        <VisualStateGroup x:Name="CheckStates">
                            <VisualState x:Name="Unchecked" />

                            <VisualState x:Name="Checked">
                                <Storyboard>
                                <ObjectAnimationUsingKeyFrames
                                        Storyboard.TargetName="UncheckButton"
                                        Storyboard.TargetProperty="BorderThickness">
                                    <DiscreteObjectKeyFrame KeyTime="0"
                                                            Value="0" />
                            </ObjectAnimationUsingKeyFrames>

                            <ObjectAnimationUsingKeyFrames
                                    Storyboard.TargetName="CheckButton"
                                    Storyboard.TargetProperty="BorderThickness">
                                <DiscreteObjectKeyFrame KeyTime="0"
                                                        Value="8" />
                            </ObjectAnimationUsingKeyFrames>
                        </Storyboard>
                    </VisualState>
                </VisualStateGroup>
            </VisualStateManager.VisualStateGroups>

            <local:UniformGrid Rows="1">
                <Button Name="UncheckButton"
                        Content="{TemplateBinding Content}"
                        ContentTemplate="{TemplateBinding ContentTemplate}"
                        FontSize="{TemplateBinding FontSize}"
                        BorderBrush="Red"
                        BorderThickness="8"
                        HorizontalAlignment="Stretch" />

                <Button Name="CheckButton"
                        Content="{TemplateBinding CheckedContent}"
                        ContentTemplate="{TemplateBinding ContentTemplate}"
                        FontSize="{TemplateBinding FontSize}"
                        BorderBrush="Green"
                        BorderThickness="0"
                        HorizontalAlignment="Stretch" />
                    </local:UniformGrid>
                </Border>
            </ControlTemplate>
        </Setter.Value>
    </Setter>
</Style>
```

记住，在更广泛的模板里，两个按钮本身可以模板化。两者包含绑定到 Content 和 CheckedContent 属性的模板，并共享控件中的相同 ContentTemplate。已选中项用粗边框高亮显示，左边按钮为红色，而右边为绿色。

NewToggleDemo 项目演示了 NewToggle 控件。

项目：NewToggleDemo | 文件：MainPage.xaml(片段)

```
<Page ... >
    <Page.Resources>
        <Style TargetType="ch11:NewToggle">
            <Setter Property="HorizontalAlignment" Value="Center" />
            <Setter Property="VerticalAlignment" Value="Center" />
        </Style>
    </Page.Resources>

    <Grid Background="{StaticResource ApplicationPageBackgroundThemeBrush}">
        <Grid.ColumnDefinitions>
            <ColumnDefinition Width="*" />
            <ColumnDefinition Width="*" />
        </Grid.ColumnDefinitions>

        <ch11:NewToggle Content="Don't do it!"
                        CheckedContent="Let's go for it!"
                        Grid.Column="0"
                        FontSize="24" />

        <ch11:NewToggle Grid.Column="1">
            <ch11:NewToggle.Content>
                <Image Source="Images/MunchScream.jpg" />
            </ch11:NewToggle.Content>

            <ch11:NewToggle.CheckedContent>
                <Image Source="Images/BotticelliVenus.jpg" />
            </ch11:NewToggle.CheckedContent>
        </ch11:NewToggle>
    </Grid>
</Page>
```

第一个 NewToggle 包含两个文本字符串，处于未选中状态。第二个 NewToggle 用了两张名画来表示两种状态，目前状态为已选中。

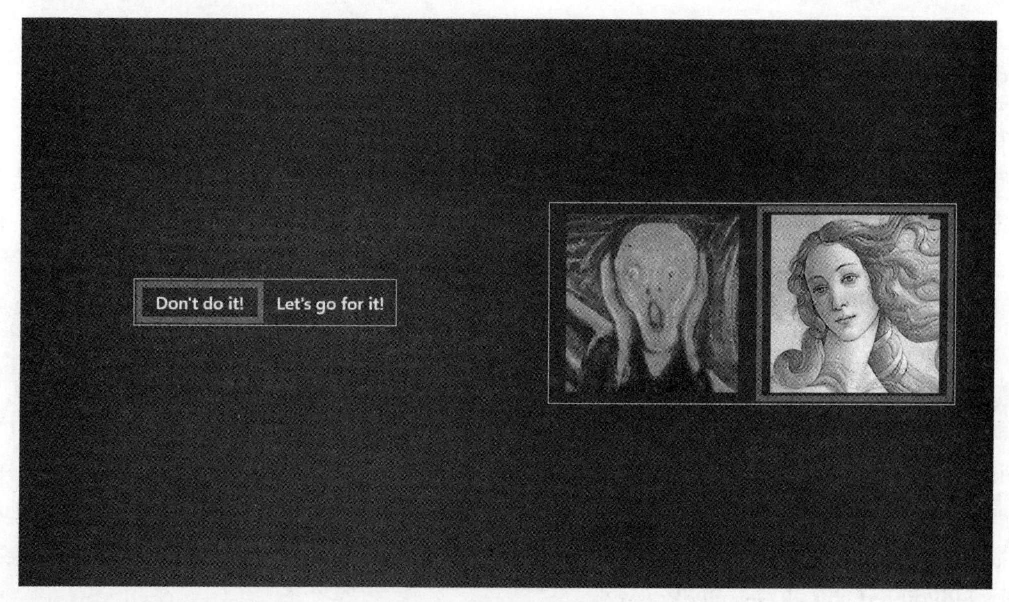

第 13 章将提供另外一个自定义控件 XYSlider 的例子。

如果在单个应用中使用自定义控件，就可以在应用项目里定义控件，默认模板可以放入用于控件分部类定义的 XAML 文件。

11.15 模板和条目容器

模板化 ItemsControl 派生类(比如 ListBox)，和模板化其他类型的控件非常类似，只不过模板会包含 ItemsPresenter 元素。基本上，我们用占位符来代表项列表，而不需要模板绑定。正如查看 ListBox 默认模板可以看到的，模板的大部分是 ScrollViewer。如果你发现编码能更好或者更适合应用，则可以在 ListBox 里替换 ScrollViewer。

轻击或点击 ListBox 条目，或者使用键盘方向键在列表里导航，所选项会被高亮显示。高亮显示从何而来？谁负责？

实际执行高亮显示的类属于 ContentControl 派生类的分类，我还没有讨论到。这些控件均派生自 SelectorItem：

```
Object
    DependencyObject
        UIElement
            FrameworkElement
                Control
                    ContentControl
                        SelectorItem (非可实例化)
                            ComboBoxItem
                            FlipViewItem
                            GridViewItem
                            ListBoxItem
                            ListViewItem
```

映射到派生自 Selector 的可实例化的这五个类，正如本章前面所示，用于在条目控件中控制单个条目。ItemsControl 没有条目类，因为不能选中条目。

你还没看到这些类，因为通常你不会自己进行实例化。相反，Selector 控件自身负责生成条目。这些类派生自 ContentControl，因此有自己的默认模板(由 generic.xaml 定义)，而这些模板都涉及 ContentPresenter。

假设你想提供不同类型的选择高亮。怎么做？怎么对甚至都没有见过的 ListBoxItem 类应用样式？

ItemsControl 定义了一个 ItemContainerStyle 属性，可以设置为 Style 对象。例如，如果要处理 ListBox，你可以用带 ListBoxItem 的 TargetType 的 Style。而该 Style 可以包括 Template 属性的设置。

如果看看 generic.xaml 中的默认 ListBoxItem 样式，就就会看到一个 SelectionStates 视觉状态组，它包含六个互斥状态：Unselected、Selected、SelectedUnfocused、SelectedDisabled、SelectedPointerOver 和 SelectedPressed。

如果想让所有选中状态都一样，可以定义模板以反映选中状态，然后可以为 Unselected 状态定义 Storyboard，CustomListBoxItemStyle 项目使用了这种方法。除了额外包含设置为 ItemContainerStyle 属性的 Style，该项目和 ListBoxWithItemTemplate 项目是相类似的。

项目: CustomListBoxItemStyle | 文件: MainPage.xaml(片段)

```
<Page ... >
    <Page.Resources>
        <ch11:NamedColor x:Key="namedColor" />
    </Page.Resources>

    <Grid>
        <ListBox Name="lstbox"
                 ItemsSource="{Binding Source={StaticResource namedColor},
                                       Path=All}"
                 Width="380">
            <ListBox.ItemTemplate>
                <DataTemplate>
                    ...
                </DataTemplate>
            </ListBox.ItemTemplate>

            <ListBox.ItemContainerStyle>
                <Style TargetType="ListBoxItem">
                    <Setter Property="Background" Value="Transparent" />
                    <Setter Property="TabNavigation" Value="Local" />
                    <Setter Property="Padding" Value="8,10" />
                    <Setter Property="HorizontalContentAlignment" Value="Left" />
                    <Setter Property="Template">
                        <Setter.Value>
                            <ControlTemplate TargetType="ListBoxItem">
                                <Border Background="{TemplateBinding Background}"
                                        BorderBrush="{TemplateBinding BorderBrush}"
                                        BorderThickness="{TemplateBinding BorderThickness}">

                        <VisualStateManager.VisualStateGroups>
                            <VisualStateGroup x:Name="SelectionStates">
                                <VisualState x:Name="Unselected">
                                    <Storyboard>
                                        <ObjectAnimationUsingKeyFrames
                                            Storyboard.TargetName="ContentPresenter"
                                            Storyboard.TargetProperty="FontStyle">
                                            <DiscreteObjectKeyFrame KeyTime="0"
                                                                    Value="Normal" />
                                            </ObjectAnimationUsingKeyFrames>

                                        <ObjectAnimationUsingKeyFrames
                                            Storyboard.TargetName="ContentPresenter"
                                            Storyboard.TargetProperty="FontWeight">
                                            <DiscreteObjectKeyFrame KeyTime="0"
                                                                    Value="Normal" />
                                        </ObjectAnimationUsingKeyFrames>
                                    </Storyboard>
                                </VisualState>

                                <VisualState x:Name="Selected" />
                                <VisualState x:Name="SelectedUnfocused" />
                                <VisualState x:Name="SelectedDisabled" />
                                <VisualState x:Name="SelectedPointerOver" />
                                <VisualState x:Name="SelectedPressed" />
                            </VisualStateGroup>
                        </VisualStateManager.VisualStateGroups>

                        <Grid Background="Transparent">
                            <ContentPresenter x:Name="ContentPresenter"
                                      FontStyle="Italic"
                                      FontWeight="Bold"
```

```
                                        Content="{TemplateBinding Content}"
                                    ContentTransitions=
                                        "{TemplateBinding ContentTransitions}"
                                    ContentTemplate=
                                        "{TemplateBinding ContentTemplate}"
                                    HorizontalAlignment=
                                        "{TemplateBinding HorizontalContentAlignment}"
                                    VerticalAlignment=
                                        "{TemplateBinding VerticalContentAlignment}"
                                    Margin="{TemplateBinding Padding}" />
                            </Grid>
                        </Border>
                    </ControlTemplate>
                </Setter.Value>
            </Setter>
        </Style>
    </ListBox.ItemContainerStyle>
</ListBox>

<Grid.Background>
    <SolidColorBrush Color="{Binding ElementName=lstbox,
                                      Path=SelectedItem.Color}" />
</Grid.Background>
</Grid>
</Page>
```

Style 设置为 ListBox 的 ItemContainerStyle，当然可以定义为资源。我决定把一个选中项显示为粗斜体文本，所以定义了 ContentPresenter 的 FontStyle 和 FontWeight 的属性。如果没有选择项(其实是正常情况)，FontStyle 和 FontWeight 通过动画得到正常状态。效果如下图所示。

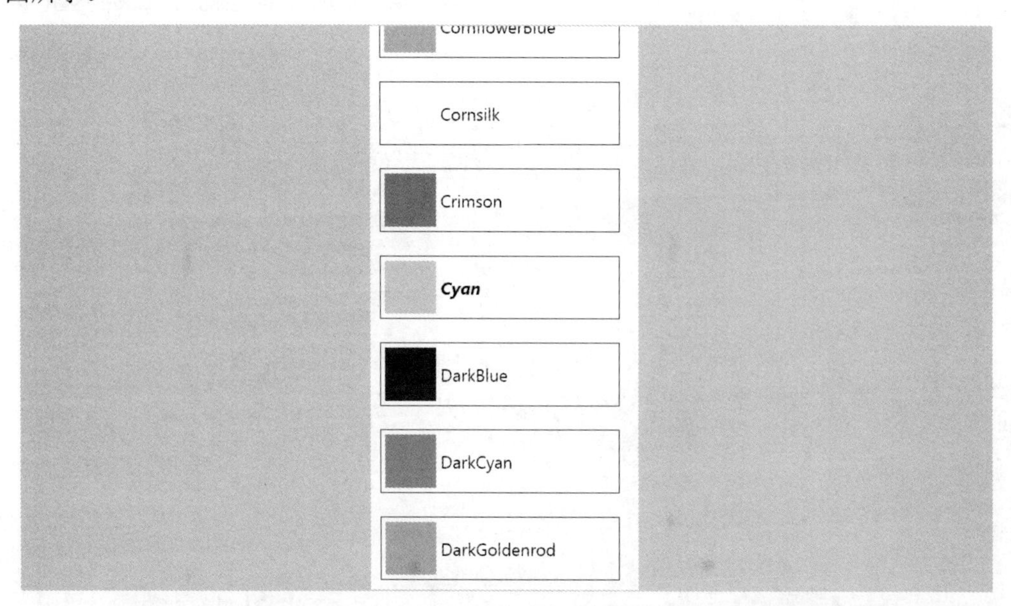

这种高亮条目的方式相当奇怪，但是对于某些应用而言，可能真的需要不寻常的高亮方式。

模板的真正目的并不是为了让控件更不寻常(尽管的确有趣)，而是让控件更有用，即让控件的视觉效果和与功能相适应。

下一章继续讨论如何使用 ListViewBase 派生类(ListView 和 GridView)控件条目并通过视图模型探索控件来使用。

第 12 章　页面及导航

大多数 Windows 8 应用程序都是围绕着 Page 类实例而构建的。这当然不是要求的，但能提供一些便利，例如轻松集成应用栏。在本章之前的内容中，我关注的程序只有一个 Page 派生类 MainPage 的实例，但现在是时候探索可以在多个 Page 派生类中类似 Web 导航的程序了。

Visual Studio 有两个用于多页面应用的项目模板，称为 Grid App 和 Split App。两个模板围绕强大的 ListView 和 GridView 控件而构建，并通过视图模型使用控件。两个模板也可以感知布局，也就是说，如果屏幕方向和辅屏模式改变了，模板就会响应，因此，这里顺理成章地将探索窗口大小调整问题作为本章内容的开头。

响应窗口大小变化对 Windows 程序员来说不是新鲜事。大多数传统 Windows 桌面程序都有大小可变的边界，用户能对大小和应用窗口的长宽比进行大量控制。Windows 程序员已经被教育了 25 年，深知所写程序要能适应用户所选的任何大小。当然，这样并非总是可行：在电子表格程序里，如果用户缩小窗口，一直到看不见单元格，会怎么样？一些程序(比如 Windows 计算器)会直接设定一个固定窗口大小，足以显示程序所有内容。对于传统的桌面应用程序而言，如果要确保窗口小于屏幕，只有这么做才合适。

Windows 8 应用多数以全屏模式运行，实际上在获取最小屏幕尺寸方面有更大的保证。然而，Windows 8 应用也容易受到方向和辅屏模式变化的影响，许多应用都会注意到这些变化。

12.1　屏幕分辨率问题

电脑屏幕有特定的水平和垂直像素大小，而屏幕的物理大小通常用测量对角线所得的英寸数来表示。通过勾股定理，可以结合这些大小计算出每英寸的像素分辨率，也称为每英寸点数(DPI)。

例如，1024×768 像素的屏幕，其对角线为 1280 像素。如果屏幕对角线为 12 英寸，则分辨率为 106 DPI。23 英寸的桌面监视器，其标准高度定义为 1920×1080 像素，对角线约为 2203 像素，96 DPI 分辨率。27 英寸的显示器，分辨率 2560×1440 像素，约 109 DPI。

本书前面说过，可以把屏幕分辨率假设为每英寸 96 像素。正如你所看到的，在上面三个例子的显示器中，该假设就很适用，尽管你可能会遇到显示器会按规则有所延伸：对于本书的大部分内容，我一直在使用三星平板电脑，其像素为 1366×768，对角线约为 11.6 英寸，1567 像素，分辨率为 135 DPI。如果在该屏幕上画一个 96 像素的正方形，我想让它有一平方英寸，接近 7/10 英寸的正方形。

96 DPI 的假设通常不适用于有大量像素的小屏幕。例如，考虑有一个 10.6 英寸的屏幕，1920×1080 像素。该屏幕分辨率为 208 DPI，而程序员所认为的一英寸实际上显示不到半英寸。文本变得更小，尽管因为像素密度高，文本仍然可读，但可能无法提供足够大的触控目标。

出于该原因，Windows 8 试图通过对应用相当透明的方式以弥补高分辨率屏幕，如果一个屏幕的像素大小为 2560×1440 或更高，而物理大小(例如 12 英寸)就会为 240 DPI 分辨率或更高，Windows 会把应用使用或遇到的坐标和尺寸调整到 180%。2560×1440 的屏幕会用 1422×800 像素来显示应用程序。

如果屏幕没那么高的像素密度，而像素为 1920×1080 或更大，但其物理尺寸又小到会导致 174 DPI 分辨率或更高，Windows 8 会调整所有像素尺寸到 140%，因此，1920×1080 显示看起来像是 1371×771 像素。

记住，这些自动调整只出现在屏幕物理尺寸很小但像素很多的情况下。如果屏幕物理尺寸很大，分辨率低于 174 DPI，则不会调整，因此应用会是全尺寸。

Windows Runtime 将假定视频显示分辨率称为“逻辑 DPI”。一般情况下，逻辑 DPI 为 96，但对于高像素密度的显示器，逻辑 DPI 可以是 134.4(即 96 DPI 放大到 140%)或 172.8(即 96 DPI 放大到 180%)。

我们来看看逻辑 DPI 是如何工作的。WhatRes 程序类似于首次在第 3 章介绍的 WhatSize 程序，但除了显示窗口大小(同页面一样大小)，WhatRes 还要获取屏幕分辨率信息。

WhatRes 里的 XAML 文件直接初始化 TextBlock：

```
项目: WhatRes | 文件 MainPage.xaml(片段)
<Grid Background="{StaticResource ApplicationPageBackgroundThemeBrush}">
    <TextBlock Name="textBlock"
               HorizontalAlignment="Center"
               VerticalAlignment="Center"
               FontSize="24" />
</Grid>
```

代码隐藏文件为页面的 SizeChanged 事件设置处理程序，也为 Windows.Graphics.Display 命名空间里所定义的 DisplayProperties 类的静态 LogicalDpiChanged 事件设置处理程序。

```
项目: WhatRes | 文件: MainPage.xaml.cs(片段)
public sealed partial class MainPage : Page
{
    public MainPage()
    {
        this.InitializeComponent();

        this.SizeChanged += OnMainPageSizeChanged;
        DisplayProperties.LogicalDpiChanged += OnLogicalDpiChanged;

        Loaded += (sender, args) =>
        {
            UpdateDisplay();
        };
    }

    void OnMainPageSizeChanged(object sender, SizeChangedEventArgs args)
    {
        UpdateDisplay();
    }

    void OnLogicalDpiChanged(object sender)
    {
        UpdateDisplay();
    }
```

```
        void UpdateDisplay()
        {
            double logicalDpi = DisplayProperties.LogicalDpi;
            int pixelWidth = (int)Math.Round(logicalDpi * this.ActualWidth / 96);
            int pixelHeight = (int)Math.Round(logicalDpi * this.ActualHeight / 96);

            textBlock.Text =
                String.Format("Window size = {0} x {1}\r\n" +
                              "ResolutionScale = {2}\r\n" +
                              "Logical DPI = {3}\r\n" +
                              "Pixel size = {4} x {5}",
                              this.ActualWidth, this.ActualHeight,
                              DisplayProperties.ResolutionScale,
                              DisplayProperties.LogicalDpi,
                              pixelWidth, pixelHeight);
        }
    }
```

现实中不会经常激活 DisplayProperties.LogicalDpiChanged 事件，因为如果程序正在运行，视频显示器就不会改变像素大小或物理尺寸。然而，如果 Windows 8 电脑附加了第二个显示器，就可能激活该事件，因为两个显示器有不同逻辑 DPI 设置，所以程序会从一个显示器跑到另一个显示器。

WhatRes 程序通过页面的 ActualWidth 和 ActualHeight 属性来获取窗口大小，然后根据 DisplayProperties. LogicalDpi 设置来计算实际像素大小。

在 1366×768 的平板电脑上运行该程序，如下图所示，对于本书大部分内容，我一直都使用这个平板电脑。

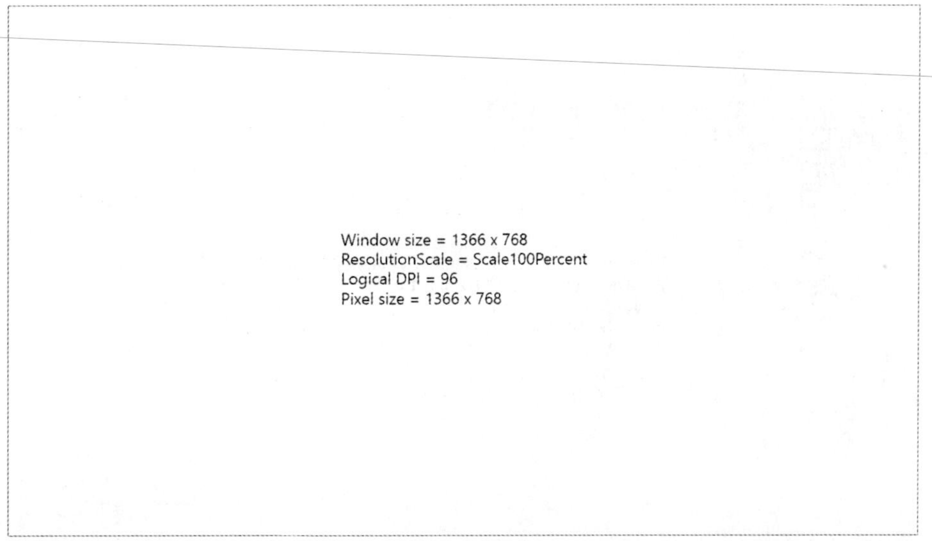

Window size = 1366 x 768
ResolutionScale = Scale100Percent
Logical DPI = 96
Pixel size = 1366 x 768

像本书的大多数屏幕截图一样，该屏幕截图已经缩放到 35%的像素大小，以适合本书篇幅。

为了写这本书，我还一直用一台 1920×1080 的显示器，21.5 英寸，实际分辨率为 102 DPI。下图是程序在屏幕上的运行效果。

Window size = 1920 x 1080
ResolutionScale = Scale100Percent
Logical DPI = 96
Pixel size = 1920 x 1080

该屏幕截图的像素大于前面的截图，因此，我不得不把它变成到 25%，以便能放入本页的相同位置。在现实中，无论是在平板电脑或大屏幕上运行该程序，文本大小都一样，但相对于大屏幕，文本则较小，表明应用有一个更大的区域供其运行。

WhatRes 能够很好在 Windows 8 模拟器里运行，你可以在 Visual Studio 的标准工具栏里选择 Windows 8 模拟器。该模拟器允许在某些普通显示器尺寸下运行 WhatRes。例如，在一个模拟的 1920×1080，10.6 英寸显示器上运行 WhatRes，如下图所示。

Window size = 1371.42858886719 x 771.428588867188
ResolutionScale = Scale140Percent
Logical DPI = 134.4
Pixel size = 1920 x 1080

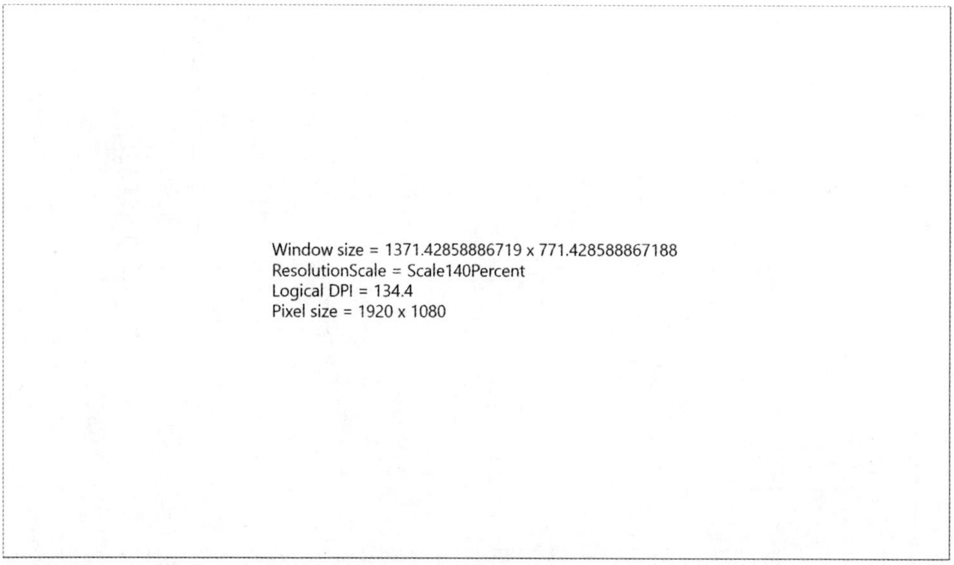

和前面的屏幕截图一样，该屏幕截图已经缩放到 25%以适应本页篇幅。对于 Windows 8 应用，窗口尺寸为 1371×771 像素，所有文本和图形将基于此大小而进行显示。计算所得的像素大小和显示器的像素尺寸相匹配。正如你所见，18 点文本相对于 1366×768 显示器似乎占据相同区域。

在模拟的 2560 × 1440 像素、10.6 英寸的屏幕上运行 WhatRes，如下图所示。

Window size = 1422.22229003906 x 800
ResolutionScale = Scale180Percent
Logical DPI = 172.8
Pixel size = 2560 x 1440

该屏幕截图已经缩放到 19%，以便能复制到本页，但请再注意，应用认为屏幕大小非常接近 1366×768 像素，而文本占据了屏幕的同样相对区域。

现在，在一个物理尺寸很大的显示器上运行模拟器。像素尺寸也是 2560×1440，但模拟屏幕大小是 27 英寸，因此没有进行调整(见下图)。

Window size = 2560 x 1440
ResolutionScale = Scale100Percent
Logical DPI = 96
Pixel size = 2560 x 1440

像前面的屏幕截图一样，我不得不把大小减至 19%，文本会显得非常小。然而，文本在 27 英寸显示器看起来大小合理，而小文本真正表明的是应用在更大的平台上如何有更大的空间。

12.2　缩 放 问 题

Windows 程序员习惯于以像素为单位处理坐标和尺寸。正如你所见，如果在高像素密度的小物理屏幕上运行程序，Windows 会根据显示大小和分辨率，按 140%或 180%对坐标和大小进行缩放。

因此，与其说用像素画画或进行大小控制，不如更正确说是用与设备无关的单位(DIU)

或简称为 units 单位进行处理。一些人把这些单位称为与设备无关的像素，但我觉得很矛盾。

在下表中，第一列显示的是在程序里画画或者改变大小所使用的单位，其他列显示如何转换为视频显示器的实际像素。

	Resolution Scale		
DIUs	100%	140%	180%
5	5	7	9
10	10	14	18
15	15	21	27
20	20	28	36

你可以自己继续填下去。

该图表显示的是如果大小和坐标乘以五单位倍数，这些单位转换为整数像素。这种积分转换有时可以帮助保持图形的保真度。

Windows 进行这些调整时，会缩放文本和矢量图形而不会损失分辨率。例如，如果指定 20 的 FontSize，程序在 180%比例分辨率上运行，则所得字体不会是放大 180%的 20 像素高，有锯齿或模糊。得到的是真正平滑的 36 像素 FontSize 字体。

但位图不同。位图有特定像素大小，如果要显示实际像素大小为 200 像素平方的位图，除了把图像放到为 140%或 180%，Windows 没有其他选择，位图会变得大，但会变得模糊。

为了避免该问题，可以用三个不同尺寸来创建位图(例如，200 像素平方、280 像素平方、360 像素平方)以供应用使用。甚至可以把这些图像存储为程序资产，让 Windows 自动选择正确的一个！

AutoImageSelection 项目演示了如何实现上述内容。我用了一张分辨率相当高的位图，剪裁为 2304 像素平方大小。然后，我把图片调整了三次：640 像素平方、896 像素平方和 1152 像素平方。三张图对应三个分辨率比例：640 的 140%是 896 像素，640 的 180%是 1152 像素。我还用 Windows 画板在每张图片中嵌入一些文本，用来表明实际像素大小。我必须使用三种不同文本大小，使文字在三个图片中大小大约相同。

我通过两种不同命名约定，分两次把三张图添加到 AutoImageSelection 项目两个不同的文件夹下，如下图所示(在 Visual Studio Solution Explorer 中)。

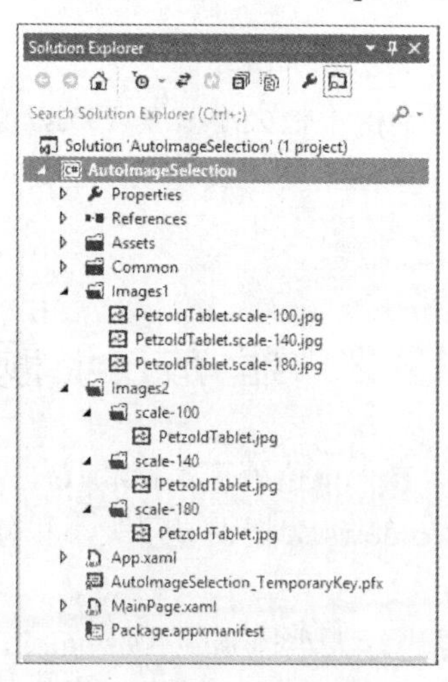

在 Images1 文件夹里，三个位图赋予不同名称。注意，句点把“scale-100”、“scale-140”、“scale-180”和“PetzoldTablet”名称及“jpg”扩展名分开了。

在 Images2 目录里，三个位图有相同的名称，但它们是分别在表明缩放的三个不同子文件夹里。

在两种情况下，scale-100 位图为 640 像素平方，scale-140 位图为 896 像素平方，scale-180 位图为 1152 像素平方。

MainPage.xaml 文件包含两个 Image 元素，引用 Images1 和 Images2 目录中的一张位图。在两种情况下，文件名或文件路径部分表明这些路径没有缩放：

项目：AutoImageSelection | 文件：MainPage.xaml(片段)

```
<Grid Background="{StaticResource ApplicationPageBackgroundThemeBrush}">
    <Grid.ColumnDefinitions>
        <ColumnDefinition Width="*" />
        <ColumnDefinition Width="*" />
    </Grid.ColumnDefinitions>

        <Image Source="Images1/PetzoldTablet.jpg"
               Grid.Column="0"
               Width="640"
               Height="640"
               HorizontalAlignment="Center"
               VerticalAlignment="Center" />

        <Image Source="Images2/PetzoldTablet.jpg"
               Grid.Column="1"
               Width="640"
               Height="640"
               HorizontalAlignment="Center"
               VerticalAlignment="Center" />
</Grid>
```

注意，两个 Image 元素被赋予明确的 Width 和 Height 设置，对应 100%位图的像素大小。这至关重要！不要指望为 None 的 Stretch 模式会强制 Image 元素正确执行缩放。

我在三个不同的 10.6 英寸(模拟)显示器上运行程序。(如果在 Windows 8 模拟器上运行，不要在程序运行时切换分辨率。而要先终止程序，切换分辨率，然后再运行程序。)1366×768 显示器上的运行效果，如下图所示。

和通常一样，1366×768 屏幕截图在本页上缩放至 35%。

在 920 × 1080、10.6 英寸的显示器上运行程序(见下图)。

该屏幕截图大小为 25%以适合本页大小。尽管 Windows 8 应用程序认为这个显示有 1371×771 像素大小，896 像素平方的位图已选中，以其原大小显示：位图的每个像素对应一个显示像素。

在 2560×1440，10.6 英寸对角线的显示器上运行该程序(见下图)。

屏幕截图大小为 19%，但在实际显示器中，位图像素和屏幕像素之间不会像这样一一对应。

如果在相同物理大小的屏幕上运行程序，位图应该也有相同物理大小，就像例子所示那样。但位图渲染在更高精度的屏幕上效果会更好，也和例子一样。在更大物理尺寸的显示器上，图像相对于屏幕小得多，但物理尺寸大致相同。

12.3 辅屏视图

一台 Windows 8 的机器至少需要 1024×768 像素的显示器来运行 Windows Store 应用。显示器尺寸宽高比为 4:3，和 20 世纪 50 年代早期的宽屏电影相一致，也和宽屏之前的传统电视和电脑显示器相一致。

在平板电脑上，屏幕可以进行横屏和竖屏模式切换，因此，机器上运行的应用也会遇到 768×1024 的显示尺寸。但在这么大小的显示器上，Windows Store 应用只需要处理两个尺寸。

下一步是设置 1366×768 显示尺寸，宽高比约为 16∶9，与高清电视相符。该显示为 768×1366 的竖屏模式。

此外，1366×768 是支持辅屏模式的最小显示尺寸。辅屏模式允许两个程序共享屏幕，但只能横屏用。

Windows.UI.ViewManagement 命名空间包含 ApplicationView 类，带名为 Value 的静态属性，该属性为 ApplicationViewState 类型，即表明应用当前辅屏模式的枚举项。没有事件与此信息相对应。如果更改视图时需要通知程序，则需要在 SizeChanged 处理程序检查该值。

除了包含显示 ApplicationView.Value 属性，WhatSnap 程序其他的部分都很像 WhatRes。

```
项目: WhatSnap | 文件: MainPage.xaml.cs(片段)
void UpdateDisplay()
{
    double logicalDpi = DisplayProperties.LogicalDpi;
    int pixelWidth = (int)Math.Round(logicalDpi * this.ActualWidth / 96);
    int pixelHeight = (int)Math.Round(logicalDpi * this.ActualHeight / 96);

    textBlock.Text =
        String.Format("ApplicationViewState = {0}\r\n" +
                      "Window size = {1} x {2}\r\n" +
                      "ResolutionScale = {3}\r\n" +
                      "Logical DPI = {4}\r\n" +
                      "Pixel size = {5} x {6}",
                      ApplicationView.Value,
                      this.ActualWidth, this.ActualHeight,
                      DisplayProperties.ResolutionScale,
                      DisplayProperties.LogicalDpi,
                      pixelWidth, pixelHeight);
}
```

此外，TextBlock 在 Viewbox 里，这样一来，即便屏幕太窄，TextBlock 也仍然是可见的。

```
项目: WhatSnap | 文件: MainPage.xaml(片段)
<Grid Background="{StaticResource ApplicationPageBackgroundThemeBrush}">
    <Viewbox HorizontalAlignment="Center"
             VerticalAlignment="Center"
             StretchDirection="DownOnly"
             Margin="24">

        <TextBlock Name="textBlock"
                   FontSize="24" />
    </Viewbox>
</Grid>
```

ApplicationViewState 枚举有四项。在竖屏模式下，唯一适用的是 FullScreenPortrait，如下图所示。

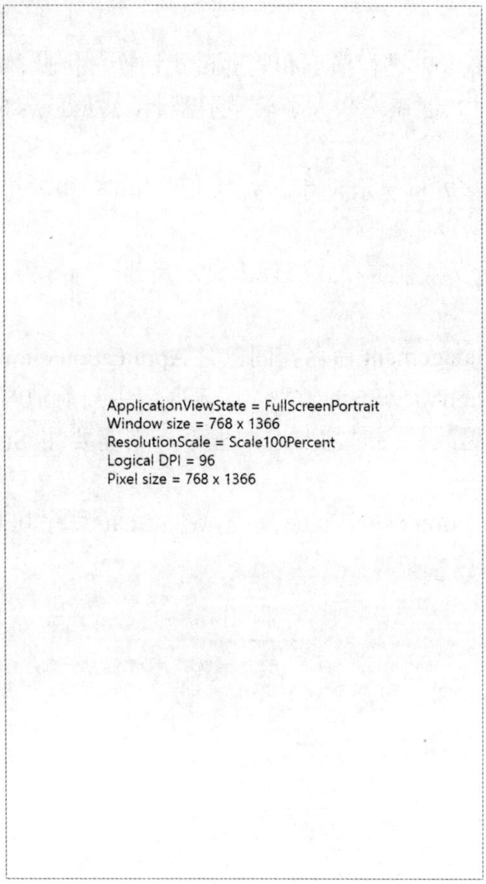

ApplicationViewState = FullScreenPortrait
Window size = 768 x 1366
ResolutionScale = Scale100Percent
Logical DPI = 96
Pixel size = 768 x 1366

辅屏模式只在横屏中发挥作用。如果应用占据整个屏幕，则 ApplicationViewState 的值为 FullScreenLandscape，如下图所示。

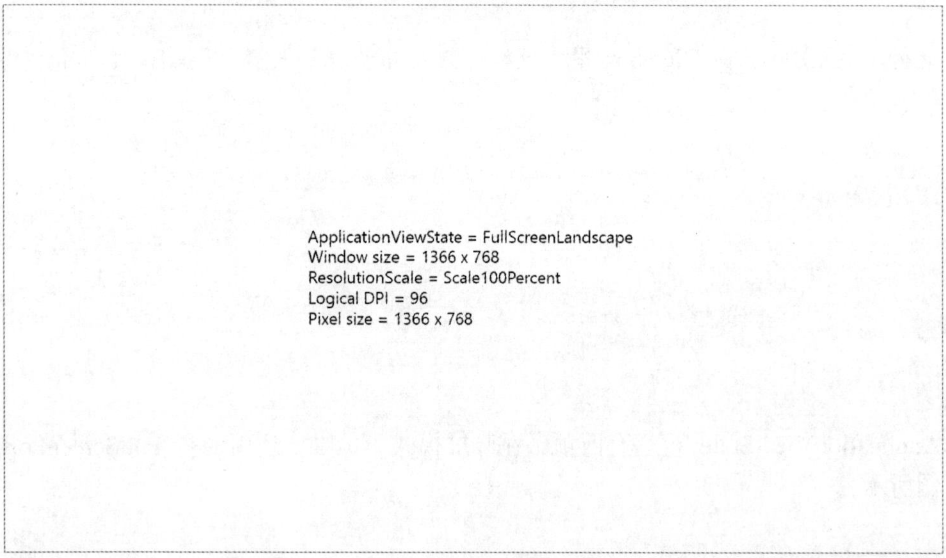

ApplicationViewState = FullScreenLandscape
Window size = 1366 x 768
ResolutionScale = Scale100Percent
Logical DPI = 96
Pixel size = 1366 x 768

如果手指扫过屏幕左边缘，可以得到显示其他应用的柱状图。如果拖动手指，就可以把另一个程序带入局部视图。此时，ApplicationViewState 会变成 Filled，如下图所示。

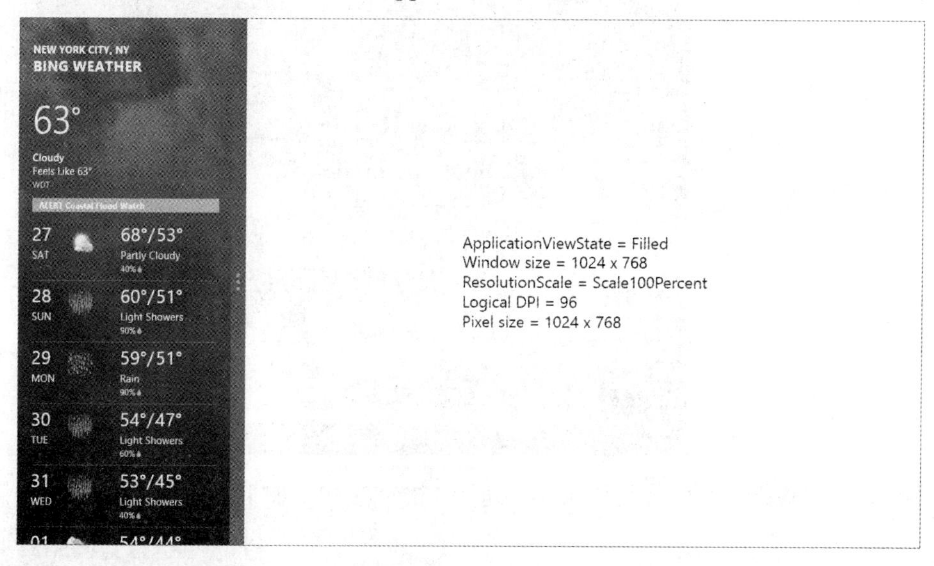

注意，支持辅屏模式的最小屏幕大小 1366×768，而 Filled 大小为 1024×768，这就是运行 Windows Store 应用的最小尺寸屏幕。

进一步向右拖动程序栏，ApplicationViewState 会变成 Snapped，如下图所示。

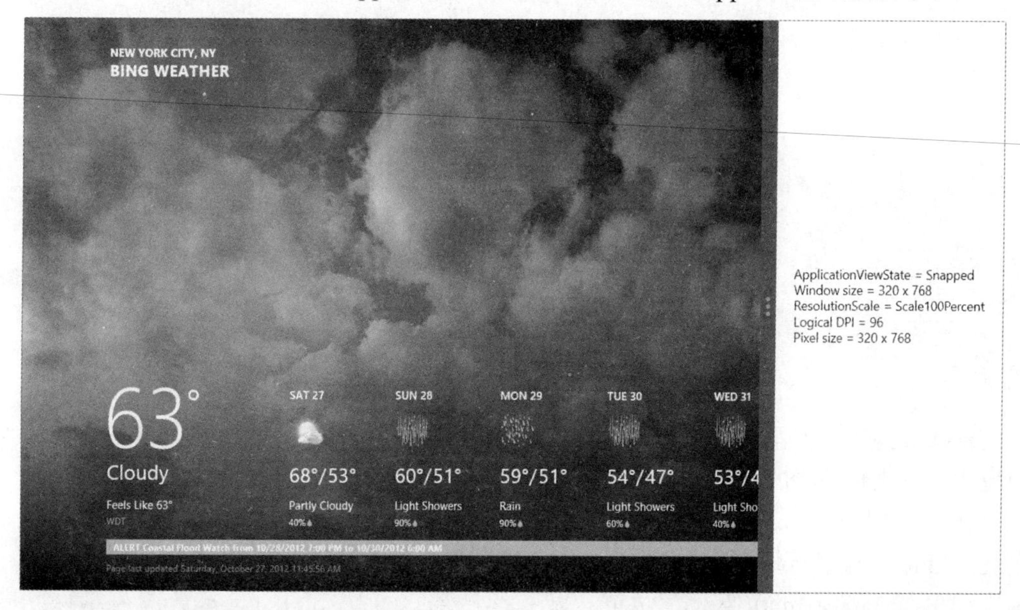

只有四种可能。如果应用在左边而不是在右边，会得到相同的 Snapped 值，如下图所示。

继续把程序栏拖到右边，程序会再次进入 Filled 模式，如下图所示。

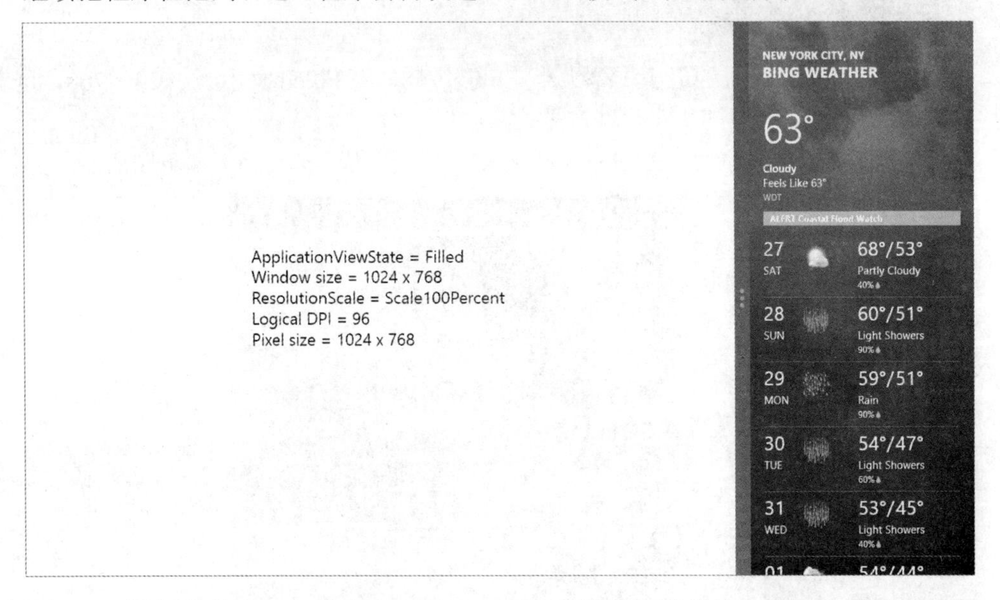

Snapped 视图总是为 320 单位宽，而 Filled 视图总是为屏幕总宽度减去其他应用的 320 单位，再减去拖拽栏的 22 单位。

例如，如果在 2560×1440 像素、10.6 英寸显示器上运行程序，那么屏幕总宽度为 1422 单位，Filled 模式分到 1080 单位，Snapped 模式分到 320 单位，而分隔符占 22 单位。

如果在 2560×1440 像素，27 英寸显示器上运行该程序，DIU 和像素是一样的。屏幕的总宽度为 2560 单位，分配给 Filled 模式 2218 单位、Snapped 模式 320 单位、分隔符 22 单位。

在 Filled 模式下，程序可以通过加上 320 和 22 的宽度确定 DIU 的全屏幕大小。通过进一步整合逻辑 DPI 设置，程序可以用像素来确定全屏幕尺寸。

因为显示模式的数量非常有限(特别是因为 Snapped 模式总是 320 单位宽)我们期望定制应用，能适应每个模式。正如你所见，Bing 的天气应用为 Snapped 模式调整了每日天气预报显示。然而，一般都不太可能需要为 Filled 模式调整应用。

改变 StackPanel 的方向是处理 Snapped 模式的一种简单方法，而另一种方法是对 Grid 的行和列采取一些小技巧，正如第 5 章中 OrientableColorScroll 程序所演示的。在本章稍后的章节中，你会看到可以在显示项集合的 GridView 和 ListView 之间进行切换。

但很显然，没有适合于每个应用的解决方案。这确实是一个问题，需要单独解决。

ApplicationView 类有静态的 TryUnsnap 方法，TryUnsnap 试图不对前台应用采取辅屏，但并不鼓励使用这种方法，而且很难想到需要这样做的理由。

12.4　横屏和竖屏的变化

在让应用适应 Filled 模式和 Snapped 模式的同时，也要让应用适应横屏模式和竖屏模式。即使你相信应用只运行在桌面电脑上，而从不在平板电脑上运行，也应该要意识到一些桌面显示器能翻转成为竖屏模式，从事大量写作的人喜欢这种显示器。

通过前面几章的描述，你已经知道 ApplicationView.Value 属性表明 ApplicationViewState.FullScreenPortrait 枚举项的竖屏模式，但如果你需要更多信息(比如你喜欢让事件告知方向改变了)，你会想用 Windows.Graphics.DisplayProperties 命名空间的 DisplayProperties 类。这和提供逻辑 DPI 缩放信息的类相同。

Windows.Graphics.DisplayProperties 命名空间定义 DisplayOrientations 枚举，有 5 项，如下所示，括号里为其值：

- None (0)，仅用于 DisplayProperties.AutoRotationPreferences
- Landscape (1)，从 PortraitFlipped 顺时针旋转 90 度
- Portrait (2)，从 Landscape 顺时针旋转 90 度
- LandscapeFlipped (4)，从 Portrait 顺时针旋转 90 度
- PortraitFlipped (8)，从 LandscapeFlipped 顺时针旋转 90 度

这里提到的“顺时针旋转 90 度”是指用户将平板电脑(或电脑屏幕)顺时针旋转 90 度。正如你看到的，Windows 8 会自动响应，将反向旋转屏幕内容以保持方向相同。

静态 DisplayProperties.NativeOrientation 属性表明屏幕的“原生” 或“最自然”方向。可以是 Landscape 或 Portrait，一般通过设备的按钮或标识位置进行控制。静态 DisplayProperties.CurrentOrientation 可以为任何非零值。

如果 CurrentOrientation(用户旋转屏幕的结果)或 NativeOrientation 发生变化，就会触发 Displayproperties.OrientationChanged 事件，而如果应用是从一个显示器转移到另一个显示器，则很少发生变化。应用启动时，无论屏幕最初是什么方向，都不会触发 OrientationChanged 事件，因此，在程序初始化时就复制 OrientationChanged 事件处理是一个好主意。

NativeUp 程序中的 XAML 文件显示了一个向上箭头。

```
项目: NativeUp | 文件: MainPage.xaml(片段)
<Grid Background="{StaticResource ApplicationPageBackgroundThemeBrush}">
    <StackPanel HorizontalAlignment="Center"
                VerticalAlignment="Center"
                RenderTransformOrigin="0.5 0.5">
        <Path Data="M 100 0 L 200 100, 150 100, 150 500, 50 500, 50 100, 0 100 Z"
              Stroke="Yellow"
              StrokeThickness="12"
```

```
            Fill="Red"
            RenderTransformOrigin="0.5 0.5"
            HorizontalAlignment="Center" />

        <TextBlock Text="Native Up"
                   FontSize="96" />

        <StackPanel.RenderTransform>
            <RotateTransform x:Name="rotate" />
        </StackPanel.RenderTransform>
    </StackPanel>
</Grid>
```

在正常情况下，如果在平板电脑上运行该程序，并用手旋转平板电脑，Windows 8 会改变显示方向，这样箭头方向总是向上，或如果有 90 度旋转增量，箭头方向几乎也总是向上。

然而，该特定程序的代码隐藏文件通过 OrientationChanged 事件来抵消旋转。结果是，箭头总是朝着电脑顶部，就好像程序并不会受到横竖屏变化的影响。

```
项目: NativeUp | 文件: MainPage.xaml.cs(片段)
public sealed partial class MainPage : Page
{
    public MainPage()
    {
        this.InitializeComponent();

        SetRotation();
        DisplayProperties.OrientationChanged += OnOrientationChanged;
    }

    void OnOrientationChanged(object sender)
    {
        SetRotation();
    }

    void SetRotation()
    {
        rotate.Angle = 90 * (Log2(DisplayProperties.CurrentOrientation) -
                             Log2(DisplayProperties.NativeOrientation));
    }

    int Log2(DisplayOrientations orientation)
    {
        int value = (int)orientation;
        int log = 0;

        while (value > 0 && (value & 1) == 0)
        {
            value >>= 1;
            log += 1;
        }
        return log;
    }
}
```

例如，假设以原生方向启动程序。箭头会向上。顺时针 90 度旋转平板。Windows 会把程序逆时针调整 90 度进行重新定位，但 OrientationChanged 处理程序会将文本和箭头按顺时针旋转 90 度。因为屏幕会略有缩小，但仍然可以看到横竖屏发生了变化，而箭头方向相对于屏幕却并没有改变。

程序依赖 DisplayOrientations 枚举项顺时针旋转值 1、2、4 和 8。以 2 为底的对数值是 0、1、2 和 3，每增加 1，就相当于顺时针方向 90 度的变化。

应用可以请求特定所需方向。有两种方法可以做到这一点。可以在 Visual Studio 打开 Package.appmanifest 文件，选择 Application UI 标签，并选择四个方向中的一个或多个，如下图所示。

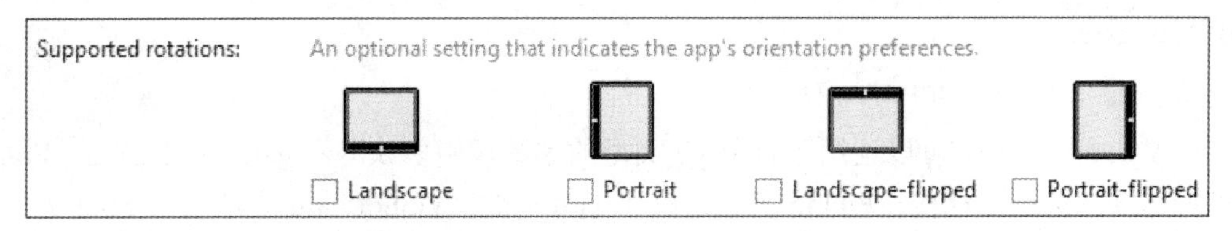

不管选择什么，它都会成为静态 DisplayProperties.AutoRotationPreferences 属性的初始值。但在程序初始化时，可以通过 C#按位 OR 运算符(|)把属性设置为一个或多个 DisplayOrientations 枚举项。

这里的关键词是“偏好”。Windows 8 会忽视请求。例如，如果要求应用只在竖屏模式下运行，但程序恰巧运行在横屏的台式电脑上，程序就会在横屏模式下运行。即使应用运行在平板电脑上，而平板电脑正为横屏模式的扩展中，应用只能继续以横屏运行。

换句话说，如果偏好不能在当前环境中发挥作用，Windows 8 就会覆写程序偏好。这么做是合理的：不管程序想要什么，都不应该让用户侧着头看屏幕。

我建议避免指定横竖屏偏好，而应该通过代码让程序适应所有方向。唯一可能的例外是涉及基于位图图形的游戏，必须以特定方式为方向，或还会涉及到使用方向传感器的程序，比如在第 18 章所提到的程序。

但请记住，把程序限制到特定方向可能会让用户感到困惑。例如，假设要求程序运行在横屏模式，但是程序正在已锁定为竖屏模式的平板电脑上运行。在正常情况下，用户手指滑过屏幕的左边或者右边，会激活应用切换器或符号栏。如果程序运行在横屏模式下，而平板电脑为竖屏模式，则用户必须滑过屏幕的顶部和底部来调用应用切换器或符号栏，而这两项特性会显示在侧面，因为两者和当前应用的方向一样。

12.5　简单页面导航

到目前为止，本书中几乎所有的应用都是围绕着类的单一实例而构建，该类为 MainPage，派生自 Page。MainPage 的实例设置为 Frame 类型对象的 Content 属性，而 Frame 对象设置为 Window 类实例的 Content 属性，我们还没有注意这一点。

可以在标准 App 类的 OnLaunched 方法中看到该层级。实际代码(在本章稍后章节)会检查错误，并确保只进行一次初始化，但基本上都只在简单情况中执行：

```
Frame rootFrame = new Frame();
Window.Current.Content = rootFrame;
rootFrame.Navigate(typeof(MainPage), args.Arguments);
Window.Current.Activate();
```

Frame 派生自 ContentControl，但不能直接设置 Content 属性。而 Navigate 方法会接受引用 Page 派生类的 Type 参数。Navigate 方法实例化该类型 (本例中为 MainPage)，而该实例成为 Frame 对象的 Content 属性以及用户交互的主要焦点。

在程序中，也可以用 Navigate 方法从一个页面跳转到另一个页面。Navigate 有两种版本： OnLaunched 方法中的版本传递一些数据给 Page 对象，但其他版本并不这样。(在本章稍后你会看到这是如何工作的。)

Page 类可以非常方便地定义 Frame 属性，因此，在 Page 派生类中，可以像下面这样调用 Navigate：

```
this.Frame.Navigate(pageType);
```

在多页面应用中，可以运用各种 Page 类型的参数调用多次 Navigate。在其内部，Frame 类维护访问过的页面堆栈。Frame 类还定义了 GoBack 和 GoForward 方法以及 bool 类型的 CanGoBack 和 CanGoForward 属性。

SimplePageNavigation 项目包含两个而不是一个 Page 派生类。我对该项目继续用 Blank App 模板，因此，如往常一样由 Visual Studio 创建 MainPage 类。为了往项目中添加另一个 Page 派生类，我从 Project(项目)菜单中选择 Add New Item(添加新项)，然后从相应的对话框中选择 Blank Page(空白页)而不是 Basic Page(基本页)。我把新的页面类命名为 SecondPage。

SimplePageNavigation 项目演示如何通过各种方法在之间页面进行互相导航。MainPage.xaml 实例化 TextBlock 用来识别页面，实例化 TextBox 用来输入一些文本，并实例化三个按钮，按钮上面的文本分别为“Go to Second Page”、“Go Forward”和“Go Back”。

项目: SimplePageNavigation | 文件: MainPage.xaml(片段)

```
<Page ... >
    <Grid Background="{StaticResource ApplicationPageBackgroundThemeBrush}">
        <StackPanel>
            <TextBlock Text="Main Page"
                       FontSize="48"
                       HorizontalAlignment="Center"
                       Margin="48" />

            <TextBox Name="txtbox"
                     Width="320"
                     HorizontalAlignment="Center"
                     Margin="48" />

            <Button Content="Go to Second Page"
                    HorizontalAlignment="Center"
                    Margin="48"
                    Click="OnGotoButtonClick" />

            <Button Name="forwardButton"
                    Content="Go Forward"
                    HorizontalAlignment="Center"
                    Margin="48"
                    Click="OnForwardButtonClick" />

            <Button Name="backButton"
                    Content="Go Back"
                    HorizontalAlignment="Center"
                    Margin="48"
                    Click="OnBackButtonClick" />
        </StackPanel>
    </Grid>
</Page>
```

代码隐藏文件通过 OnNavigatedTo 覆写使得 forward 和 back 按钮可用，但它依赖的是 Frame 定义的 CanGoForward 和 CanGoBack 属性。三个 Click 处理程序调用 Navigate(引用 SecondPage 对象)、GoForward 和 GoBack。

项目：SimplePageNavigation | 文件：MainPage.xaml.cs(片段)
```
public sealed partial class MainPage : Page
{
    public MainPage()
    {
        this.InitializeComponent();
    }

    protected override void OnNavigatedTo(NavigationEventArgs args)
    {
        forwardButton.IsEnabled = this.Frame.CanGoForward;
        backButton.IsEnabled = this.Frame.CanGoBack;
    }

    void OnGotoButtonClick(object sender, RoutedEventArgs args)
    {
        this.Frame.Navigate(typeof(SecondPage));
    }

    void OnForwardButtonClick(object sender, RoutedEventArgs args)
    {
        this.Frame.GoForward();
    }

    void OnBackButtonClick(object sender, RoutedEventArgs args)
    {
        this.Frame.GoBack();
    }
}
```

除了通过 OnGotoButtonClick 方法导航到 MainPage，SecondPage 的其他部分完全相同。

项目：SimplePageNavigation | 文件：SecondPage.xaml.cs(片段)
```
void OnGotoButtonClick(object sender, RoutedEventArgs args)
{
    this.Frame.Navigate(typeof(MainPage));
}
```

首次运行程序，效果如下图所示。

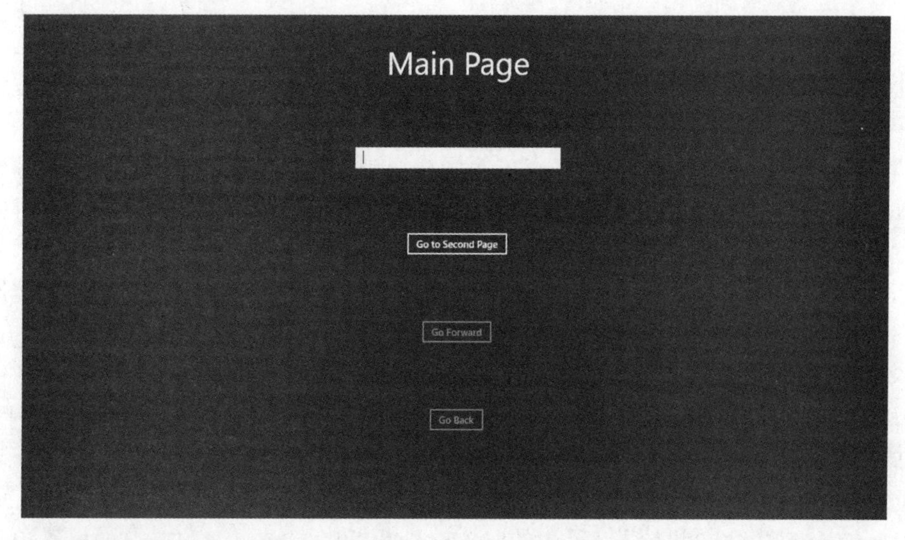

前后翻页的按钮都不可用。如果点击 Go to Second Page 按钮，程序会导航到如下图所示页面。

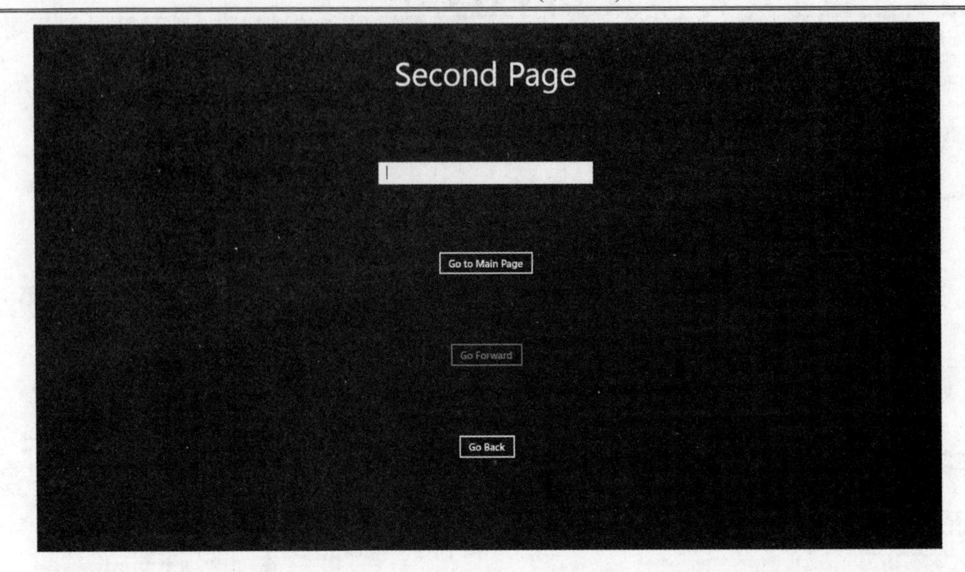

Go Back 按钮现在可用，会把你带回到 MainPage。Go to Second Page 的作用也一样，但有区别：如果按 Go Back 按钮回到 MainPage ，Go Forward 按钮则可用，Go Back 按钮不可用。如果按 Go to Second Page 按钮，Go Back 按钮可用而 Go Forward 不可用。

在探讨细节之前，我想先向你展示另一种方法，能让 Go Forward 按钮和 Go Back 按钮可用。Frame 的 CanGoBack 和 CanGoForward 属性可以绑定源：

```
<Page ... Name="page">
    ...
            <Button Name="forwardButton"
                    IsEnabled="{Binding ElementName=page, Path=Frame.CanGoForward}"
                         ... />

            <Button Name="backButton"
                    IsEnabled="{Binding ElementName=page, Path=Frame.CanGoBack}"
                         ... />
    ...
</Page>
```

这样便可以去除程序对 OnNavigatedTo 方法的需要，但实施页面导航的任何程序都可能利用该方法的其他用途以及其同伴 OnNavigatedFrom 方法。

如果试试 SimplePageNavigation，点击不同按钮进行导航，前进、后退和在遇到的每个 TextBox 中输入几个字符，你会发现一个重要的导航特征。你会发现，无论什么时候从一个页面到另一个页面(不管是调用 Navigate、GoForward 还是 GoBack 方法)，文本框中初始都均为空白。也就是说，每次点击按钮，都会创建新的 MainPage 或 SecondPage 实例。无论在文本框中输入什么都会丢失，因为已经抛弃了包含 TextBox 的 Page 实例。

这多少有点让人惊讶。你可能期望按下 Go to Main Page 或 Go to Second Page 按钮会创建新实例，但也可能期望按下 Go Forward 或 Go Back 按钮导航回到页面之前的实例。但事实并非如此，无论如何，都是在创建了新实例。

Page 类定义三个虚拟方法来协助页面处理导航。三个虚拟方法分别为 OnNavigatingFrom、OnNavigatedFrom(注意这个方法名称中的时态区别！)和 OnNavigatedTo。如果能记录调用这三个方法、Page 类的构造函数以及 Loaded 和 Unloaded 事件的激活，就能从中发现一个页面跳转到另一个页面过程的顺序，如下所示。

From Page	To Page
OnNavigatingFrom	
	Constructor
OnNavigatedFrom	
	OnNavigatedTo
	Loaded
Unloaded	

无论页面跳转是 Navigate、GoForward 还是 GoBack 的结果，都会产生以上顺序。

直到这一章，我们一直让 MainPage 存于应用运行中，就好像把 MainPage 作为唯一的 Page 派生类。然而，一旦开始处理多页面应用程序，就需要考虑创建和丢弃 Page 派生类。构造多个 Page 派生类是一个好主意，这样在 OnNavigatedTo 或 Loaded 期间，Page 派生类可以附加事件处理程序并获取资源，而在 OnNavigatedFrom 和 Unloaded 期间则可以分离事件处理程序并释放这些资源。

无论页面导航到哪里，想创建 Page 派生类的新实例显然都很容易，因为这是默认发生的事情。如果你喜欢有些不同，可以有两种选择，一个非常容易，而另一个不那么容易。

简单方法是把 Page 的 NavigationCacheMode 属性设置为其他值，不是设置为枚举项默认的 Disabled。例如：

```
public MainPage()
{
    this.InitializeComponent();
    this.NavigationCacheMode = NavigationCacheMode.Enabled;
}
```

另一个选项是 Required，但对于该程序，Enabled 和 Required 的作用一样。如果对 Page 对象设置 Enabled 或 Required，每个 Page 派生类只有一个实例被创建和缓存，不管 Navigate、GoForward 或 GoBack 发生什么效果，每次访问该页面，都会重用该实例。方法调用和事件发生的顺序与前面所示第一次导航到某个特定页面类型的表一样；随后顺序是一样的，但没有构造函数。可以给不同 Page 类设置不同 NavigationCacheMode 属性。

这种选项可能适合“中心”架构，MainPage 导航到若干不同二级页面，二级页面可以回到 MainPage。Enabled 和 Required 之间的区别是：如果缓存页面的数量超过 Frame 的 CacheSize 属性，Enabled 可能会导致丢弃实例化的页面，其默认值为 10，不过可以变。

然而，在一般情况下，你可能想创建新实例以用于 Navigate 调用，而把现有实例用于 GoForward 和 GoBack。这种方法也有简单的属性设置方法，稍后我会向你展示怎么做。

12.6　返回堆栈

Web 浏览器有 Back 按钮，但没有 Forward 按钮。浏览器实现 Back 按钮功能的方式非常简单，即通过熟知的称为“堆栈”(stack)的数据结构来保存访问过的页面。在浏览器环境中，称为“回退栈”(Back Stack)：每当浏览器导航到新页面，就把前一页压到栈中。每当用户按 Back 按钮，浏览器就会从栈中弹出页面并导航到该页面。如果堆栈为空，则禁用

Back 按钮。

实现 Forward 按钮会使上述过程复杂化。浏览器并不通过栈来保存访问过的页面，而需要有序列表。(然而，该列表通常仍然称为回退栈。)列表包括当前页。无论何时浏览器导航到新页面，新页面就会添加到列表末尾。然而，如果用户按下 Back 按钮，并没有从列表中移除所导航页面。该页面必须保留在列表中，因为用户可能会按下 Forward 按钮。

例如，用户可能从 Page Zero 页面开始，导航到 Page One，然后是 Page Two、Page Three、Page Four 和 Page Five。回退如下所示，最新最近的页面在顶部，箭头所指为当前页面：

```
Page Five ←
Page Four
Page Three
Page Two
Page One
Page Zero
```

现在，假设用户按四次 Back 按钮，则当前页面为 Page One：

```
Page Five
Page Four
Page Three
Page Two
Page One ←
Page Zero
```

用户然后按 Forward 按钮，则当前页面为 Page Two：

```
Page Five
Page Four
Page Three
Page Two ←
Page One
Page Zero
```

显然，通过 Back 和 Forward 按钮，用户可以在六个页面中任意导航。如果当前页面到达底部，则禁用 Back 按钮。如果当前页面到达顶部，则禁用 Forward 按钮。

但现在假设用户从 Page Two 导航到 Page Six。则必须丢弃一部分完整列表。被丢弃部分之前保存用来点击 Forward 按钮，但这些页面不能再导航到新页面：

```
Page Six ←
Page Two
Page One
Page Zero
```

现在 Forward 按钮被禁用。只有当用户按下 Back 按钮，才会重新启用 Forward 按钮。

Frame 类在内部维护着已访问页面的回退栈。然而，应用不能访问回退栈，甚至不能获得其大小。

但只要在回退栈中获取 BackStackDepth 属性，就可以获得当前页面位置。如果应用开始运行，并导航到初始页面，BackStackDepth 会报告零值。在前面所示的四个例子中，BackStackDepth 分别等于 5、1、2 和 3。

BackStackDepth 是非常重要的信息，因为有了它，特定页面类可以唯一性标识其自身的特定实例。我们来看看是怎么回事。

12.7　导航事件和页面恢复

在正常情况下，如果程序调用 GoBack 或 GoForward 返回特定页面，你想让用户看到和之前访问过页面一样的内容。你已经知道不能自动实现：如果把 NavigationCacheMode 设置为默认值 Disabled，调用 GoBack 和 GoForward 总会导致创建特定 Page 类的新实例。如果设置为 Enabled 或 Required，则会重用现有 Page 类实例，不过同时也重用于 Navigate。

如前所述，Page 类定义了涉及导航的三个虚拟方法。调用 OnNavigatingFrom 开始一个导航序列，这种方法并不常用。事件参数为允许取消导航的 NavigatingCancelEventArgs 类型。

在从一个页面导航到另一个页面的过程中(无论是调用 Navigate、GoBack 还是 GoForward 的结果)，第一个页面的 OnNavigatedFrom 后面紧跟着第二个页面的 OnNavigatedTo 调用，两个方法都有 NavigationEventArgs 类型的事件参数。这些事件参数用于其他情况(比如 WebView 类)，因此，使用这两种覆写时有一部分属性无关。导航事件有以下三大重要属性。

- object 类型的 Parameter 属性。这是 Navigate 方法可选第二个参数的设置，用于把数据从一个页面传到另一个页面。稍后将进一步讨论这个过程。
- Content 和 SourcePageType 属性总是指要被导航到的页面。Content 对象为 Page 派生类的实际实例，而 SourcePageType 为该实例的类型——也就是说，Navigate 调用的第一个参数创建该页面。该信息只对 OnNavigatedFrom 覆写有真正价值。在 OnNavigatedTo 覆写中，Content 属性等同于该值，而 SourcePageType 等同于调用 GetType。
- NavigationMode 属性为 NavigationMode 枚举项，枚举项还有 New、Refresh、Back 和 Forward。由 Navigate 方法发起的该导航值为 New 或 Refresh。如果页面导航到本身，该值为 Refresh，如果导航由 GoBack 或 GoForward 方法发起，该值则分别为 Back 或 Forward。

为 Navigate 调用创建新的页面内容(即 NavigationMode 为 New 时)且该页面之前未被访问过及用户未通过 Back 或 Forward 导航，NavigationMode 属性是实现这一架构的关键。

Page 派生类首先定义一个字段来保存和恢复状态：

```
Dictionary<string, object> pageState;
```

使用该字典的方式与第 7 章所演示的 PrimitivePad 程序中 ApplicationData.LocalSettings 字典的使用方式是一样的。然而，应用再次运行时，没有保存 Application.Suspending 事件和恢复期间的应用设置，而是在 OnNavigatedFrom 覆写时保存页面状态到字典中，并在 OnNavigatedTo 时进行恢复。

什么是页面状态？一般是由用户输入并从输入中得到的任何结果：复选框、单选按钮、滑块，尤其是文本输入状态。在我使用的范例应用中，真正重要的页面状态只有 TextBox 的内容。可以在 OnNavigatedFrom 时用键名保存：

```
pageState.Add("TextBoxText", txtbox.Text);
```

可以在 OnNavigatedTo 时进行恢复：

```
txtbox.Text = pageState["TextBoxText"] as string;
```

TextBox 肯定还有其他可能的属性是你想要保存和恢复的(比如 SelectionStart 和 SelectionLength)，但现在让我们保持简单。

除非为每个导航事件实例化 Page 类，否则在字典里保存和恢复页面状态的过程完全没用，因为 pageState 的新实例会被创建为新页面的一部分！你需要的是把该字典实例保存到另一个字典中，而后者定义为静态，以便页面的所有实例都可以共享：

```
static Dictionary<int, Dictionary<string, object>> pages;
```

字典里的值是我称为 pageState 的 Dictionary 实例。该字典键是 BackStackDepth 的值，因而不同 pageState 字典都可以关联到回退栈里页面实例的固定位置。

如果要让多个页面派生类使用同样的技巧，可以在 Page 派生类中定义两个字典，而 Page 派生类则作为其他页面的基类。应用中的所有页面都可以共享静态 pages 字典。

我们来看看上述内容在一个简单应用中是如何工作的。VisitedPageSave 程序定义了一个名为 SaveStatePage 的类。我用简单的 Class 模板来创建类；没有 XAML 文件与之关联。SaveStatePage 类派生自 Page，两个字典都定义为 protected，这样派生类都可以访问：

项目：VisitedPageSave | 文件：SaveStatePage.cs(片段)

```
public class SaveStatePage : Page
{
    protected Dictionary<string, object> pageState;

    static protected Dictionary<int, Dictionary<string, object>> pages =
                new Dictionary<int, Dictionary<string, object>>();
    ...
}
```

静态字典在其定义中进行实例化(或者也可以在静态构造函数中进行实例化)，而实例字典则不是这样。你会看到，如果 NavigationMode 为 New，则会在 OnNavigatedTo 覆写时实例化实例字典。

我直接在 SimplePageNavigation 里创建 SecondPage 类，但在 MainPage 和 SecondPage 的 XAML 文件和代码隐藏文件中，我把基类从 Page 改成 SaveStatePage。否则，MainPage 和 SecondPage 在 SimplePageNavigation 里是一样的。两个代码隐藏文件基本上彼此相同。MainPage.xaml.cs 显示了之前按钮 Click 处理程序的同一个实现。

项目：VisitedPageSave | 文件：MainPage.xaml.cs(片段)

```
public sealed partial class MainPage : SaveStatePage
{
    public MainPage()
    {
        this.InitializeComponent();
    }

    ...

    void OnGotoButtonClick(object sender, RoutedEventArgs args)
    {
        this.Frame.Navigate(typeof(SecondPage));
    }

    void OnForwardButtonClick(object sender, RoutedEventArgs args)
    {
```

```
            this.Frame.GoForward();
        }

        void OnBackButtonClick(object sender, RoutedEventArgs args)
        {
            this.Frame.GoBack();
        }
    }
```

NavigationCacheMode 保留了默认设置 Disabled，这样一来，在所有导航事件期间都会实例化新页面。

在 OnNavigatedTo 覆写里，静态字典的整数键是 BackStackDepth 属性。如果 NavigationMode 不为 New，方法会直接使用该键获得 pageState 字典，以对应回退栈中的位置，然后使用 pageState 字典来初始化页面，在本例中为 TextBox。

```
项目: VisitedPageSave | 文件: MainPage.xaml.cs(片段)
protected override void OnNavigatedTo(NavigationEventArgs args)
{
    // Enable buttons
    forwardButton.IsEnabled = this.Frame.CanGoForward;
    backButton.IsEnabled = this.Frame.CanGoBack;

    // Construct a dictionary key
    int pageKey = this.Frame.BackStackDepth;

    if (args.NavigationMode != NavigationMode.New)
    {
        // Get the page state dictionary for this page
        pageState = pages[pageKey];

        // Get the page state from the dictionary
        txtbox.Text = pageState["TextBoxText"] as string;
    }

    base.OnNavigatedTo(args);
}
```

然而，如果 NavigationMode 为 New，我们就知道调用 Navigate 可以到达该页面，而且应该把该页面视为新的未初始化页面。这个额外的逻辑发生在 SaveStatePage 里 OnNavigatedTo 的实现中，你会注意到调用在 MainPage 和 SecondPage 里 OnNavigatedTo 重写结束时进行的。以下代码创建新的 pageState 字典并将它添加到静态 pages 字典。

```
项目: VisitedPageSave | 文件: SaveStatePage.cs(片段)
public class SaveStatePage : Page
{
    ...
    protected override void OnNavigatedTo(NavigationEventArgs args)
    {
        if (args.NavigationMode == NavigationMode.New)
        {
            // Construct a dictionary key
            int pageKey = this.Frame.BackStackDepth;

            // Remove page key and higher page keys
            for (int key = pageKey; pages.Remove(key); key++) ;

            // Create a new page state dictionary and save it
            pageState = new Dictionary<string, object>();
            pages.Add(pageKey, pageState);
        }

        base.OnNavigatedTo(args);
```

```
        }
    }
```

然而，静态 pages 字典还必须清除任何等于或高于 BackStackDepth 键的记录。这些记录是由于未经 GoForward 调用平衡的 GoBack 调用而造成的。如果你明白键不存在时，Dictionary 的 Remove 方法会返回 false，就更容易理解为什么要移除这些记录的 for 语句：

```
for (int key = pageKey; pages.Remove(key); key++) ;
```

在 MainPage 和 SecondPage 里，OnNavigatedFrom 覆写要简单得多，只是在现有 pageState 字典中保存页面状态：

项目：VisitedPageSave | 文件：MainPage.xaml.cs(片段)

```
protected override void OnNavigatedFrom(NavigationEventArgs args)
{
    pageState.Clear();

    // Save the page state in the dictionary
    pageState.Add("TextBoxText", txtbox.Text);

    base.OnNavigatedFrom(args);
}
```

也要记住，pageState 字典可以保存更多项，多到需要重建整个页面状态。

在连续页面上在 TextBox 中输入 1、2、3 等等，也许是检查程序是否正常工作的最简单方法。按下 Go Forward 和 Go Back 按钮，会看到恢复的记录。

如果在 Visual Studio 暂停再继续应用，你会发现一切都会正确恢复。然而，应用暂停的时候并不保存任何东西，因此，如果应用暂停后终止，下次启动后则会出现新的原始状态。你可能不希望出现这种情况，但即使出现了，也不难解决。

12.8 保存和恢复应用状态

如果应用终止后重新启动，比如 VisitedPageSave，你可能想让应用看起来得好像从未终止过。你可能想让之前创建的所有页面都恢复到以前，你还可能想让应用展示用户上次访问的相同页面。

换句话说，不仅需要保存(并恢复)每个页面状态，还需要保存(并恢复)回退栈状态。如果不恢复回退栈，恢复每个页面实际上就没用，因为没有回退栈，就没有必要恢复页面的记录！

前面提到过，回退栈完全在 Frame 对象的内部。幸运的是，Frame 提供两个方法来保存和恢复回退栈，而不需要知道其内部结构：GetNavigationState 返回一个字符串，可以保存到应用设置里；程序下一次运行时，可以获取字符串并将其作为参数传递给 SetNavigationState。

这个字符串是什么？如果愿意的话，你可以看看。你会发现该字符串用数字包含在回退栈里的页面类名称。没有文档说明这些数字是什么，将来可能也会改变，因此，真的应该是只能用字符串从 GetNavigationState 传递给 SetNavigationState。

GetNavigationState 实际上不仅仅只返回对回退栈态进行编码的字符串。调用 GetNavigationState 会导致当前页面获得 OnNavigatedFrom 调用，同时 NavigationMode 为

Forward。这样，当前页面能保存其页面状态，但这也意味着无法在任何想要的时候调用 GetNavigationState。只能在中断应用的时候调用 GetNavigationState。可以在 App.xaml.cs 的 OnSuspending 事件处理程序里这么做。

以下是我在 ApplicationStateSave 程序里所做的事情。

```
项目: ApplicationStateSave | 文件: App.xaml.cs(片段)
private void OnSuspending(object sender, SuspendingEventArgs e)
{
    var deferral = e.SuspendingOperation.GetDeferral();
    //TODO: Save application state and stop any background activity

    // Code added for ApplicationStateSave project
    ApplicationDataContainer appData = ApplicationData.Current.LocalSettings;
    appData.Values["NavigationState"] = (Window.Current.Content as Frame).GetNavigationState();
    // End of code added for ApplicationStateSave project

    deferral.Complete();
}
```

我保留了 Visual Studio 生成的完整 OnSuspending 版本，只添加了前后有注释的两行代码。代码从 Frame 获取 GetNavigationState 字符串，并用“NavigationState”的名称将其保存到应用设置。

本书前面的一些程序从 MainPage 保存应用设置。这里为什么不也这么做？回想一下，多页面环境的 Page 派生类应该在 OnNavigatedTo 或 Loaded 时附加所需的任何事件处理程序，并在 OnNavigatedFrom 或 Unloaded 期间再分离，也就是说，应用中每一个 Page 派生类都需要设置 Suspending 处理程序来执行该工作。但并不真的是 Page 派生类的工作。这项工作涉及保存定义多个页面之间导航关系的导航状态，因此，应该是应用本身的责任。

这就是 App 类中代码这么写的一个原因。而另一个原因是，恢复导航状态也需要在 App 类中运行，实际上是在 App 类的特定位置，因为在那里可以有效覆写默认逻辑编码。

为了恢复回退栈，可以通过从 GetNavigationState 获得的已保存字符串来调用 SetNavigationState。而调用 SetNavigationState 会导致导航至之前的当前页。通过设置为 Back 的 NavigationMode，调用该页面的 OnNavigatedTo 方法，页面重新加载自身页面设置，不认为它是新页面。

至关重要的是在 App.xaml.cs 里 OnLaunched 方法的特定位置调用 SetNavigationState。ApplicationStateSave 项目完整保留了 OnLaunched 生成的所有代码和注释，如下所示。

```
项目: ApplicationStateSave | 文件: App.xaml.cs(片段)
protected override void OnLaunched(LaunchActivatedEventArgs args)
{
    Frame rootFrame = Window.Current.Content as Frame;

    // Do not repeat app initialization when the Window already has content,
    // just ensure that the window is active
    if (rootFrame == null)
    {
        // Create a Frame to act as the navigation context and navigate to the first page
        rootFrame = new Frame();

        if (args.PreviousExecutionState == ApplicationExecutionState.Terminated)
        {
            //TODO: Load state from previously suspended application

            // Code added for ApplicationStateSave project
            ApplicationDataContainer appData = ApplicationData.Current.LocalSettings;
```

```
            if (appData.Values.ContainsKey("NavigationState"))
                rootFrame.SetNavigationState(appData.Values["NavigationState"] as string);
            // End of code added for ApplicationStateSave project
        }

        // Place the frame in the current Window
        Window.Current.Content = rootFrame;
    }

    if (rootFrame.Content == null)
    {
        // When the navigation stack isn't restored navigate to the first page,
        // configuring the new page by passing required information as a navigation
        // parameter
        if (!rootFrame.Navigate(typeof(MainPage), args.Arguments))
        {
            throw new Exception("Failed to create initial page");
        }
    }
    // Ensure the current window is active
    Window.Current.Activate();

    ...
}
```

我又用注释来识别添加到项目的代码。(注意结尾的省略号。下一节会讨论添加到 App.xaml.cs 里的额外代码。)

在 OnLaunched 方法的底部，通过 MainPage 类来调用 Navigate 方法。如果正在恢复回退状态，你肯定不希望发生 Navigate 调用，因为这会离开之前的当前页，并可能导致回退栈的一部分在 MainPage 的 OnNavigatedTo 方法中被删除。出于该原因，必须该调用之前恢复回退栈，确保 Frame 对象的 Content 属性设置为之前的当前页，并确保跳过到 MainPage 的导航。

目前为止的所有代码都涉及保存和恢复回退栈。接下来牵涉保存和恢复所有页面状态。在之前的项目中，我定义了一个类来维护两个字典，称为 SaveStatePage(一个字典实例字典，一个是静态字典)，均用于保存页面状态。MainPage 和 SecondPage 均派生于该类。

我为程序保留了上述架构。MainPage 和 SecondPage 和之前项目的类也相同。但 SaveStatePage 得到增强，用来在应用本地存储中保存并检索所有页面的所有设置。

如果一个特定回退栈引用 4 个 MainPage 实例和 3 个 SecondPage 实例，则总共有 7 个键名“TextBoxText”的设置。7 个设置必须有区分。幸运的是，用于存储应用设置的 ApplicationDataContainer 有“容器”功能，有点类似于文件夹或子目录。该功能似乎适合隔离每个页面的设置。容器由名称来识别，我为每个 Page 实例所选的名称都表明实例在回退栈中的位置，和转换为字符串的 pages 字典整数键一样。

以下是增强版 SaveStatePage 的静态构造函数和 Suspending 处理程序。Suspending 事件的处理程序由静态构造函数附加，因此只执行一次保存所有页面设置，不用知道设置内容。

```
项目: ApplicationStateSave | 文件: SaveStatePage.cs(片段)
public class SaveStatePage : Page
{
    protected Dictionary<string, object> pageState;

    static protected Dictionary<int, Dictionary<string, object>> pages =
                new Dictionary<int, Dictionary<string, object>>();
```

```
    static SaveStatePage()
    {
        // Set handler for Suspending event
        Application.Current.Suspending += OnApplicationSuspending;

        ApplicationDataContainer appData = ApplicationData.Current.LocalSettings;

        // Loop through containers, one for each page in the back stack
        foreach (ApplicationDataContainer container in appData.Containers.Values)
        {
            // Create a page state dictionary for that page
            Dictionary<string, object> pageState = new Dictionary<string, object>();

            // Fill it up with saved values
            foreach (string key in container.Values.Keys)
            {
                pageState.Add(key, container.Values[key]);
            }

            // Save in static dictionary
            int pageKey = Int32.Parse(container.Name);
            pages[pageKey] = pageState;
        }
    }

    static void OnApplicationSuspending(object sender, SuspendingEventArgs args)
    {
        ApplicationDataContainer appData = ApplicationData.Current.LocalSettings;

        foreach (int pageKey in pages.Keys)
        {
            // Create container based on location within back state
            string containerName = pageKey.ToString();

            // Get container with that name and clear it
            ApplicationDataContainer container =
                appData.CreateContainer(containerName,
                        ApplicationDataCreateDisposition.Always);
            container.Values.Clear();

            // Save settings for each page in that container
            foreach (string key in pages[pageKey].Keys)
                container.Values.Add(key, pages[pageKey][key]);
        }
    }
    ...
}
```

静态构造函数结束的时候，pages 字典为回退栈上的每个页面包含一条记录。这些单个页面都不会进行实例化。然而，每个 SaveStatePage 派生类会实例化，在 OnNavigatedTo 覆写时期间获取自己的 pageState 字典，要么从 pages 字典中检索而得，要么创建新的。

12.9　导航加速器和鼠标按钮

你用 5 个按钮(而不是通常 3 个按钮)的鼠标吗？我不用，不过有人用，而且有些人习惯用额外两个按钮在 Internet Explorer 里进行前进和后退导航，其他 Internet Explorer 用户已经习惯了使用左右箭头键及 Alt 键进行后退和前进导航。一些键盘有特殊按键来执行这些操作。

你可能想实现同样的快捷键，允许用户在应用页面之间导航。要做到这一点，需要你还没有见过的两个事件：PointerPressed 和 AcceleratorKeyActivated。

AcceleratorKeyActivated 事件在 Page 类、Frame 类甚至支撑 Frame 的 Window 类里都不可用。但它在 CoreWindow 中可用，CoreWindow 是支持 Window 输入事件的对象，可从当前 Window 对象获得 CoreWindow 对象。

AcceleratorKeyActivated 处理程序获取第一次按键，如果该处理程序把一个特定键识别为命令加速器，则可以在应用里通过把事件参数的 Handled 属性设置为 true 来进一步禁止该键可见。

正如第 13 章所述，鼠标按钮按压、手指或手写笔触摸屏幕时会触发 PointerPressed 鼠标事件。该事件由 UIElement 定义，而由 Frame 和 Page 进行继承，但为了获取页面导航的按钮点击，也可以为该事件给 CoreWindow 定义一个处理程序。

键盘和鼠标加速器的功能高于页面级别，因此，可以很方便地把它们放入 App 类。

前面我在 ApplicationStateSave 项目的 App 类里展示过 OnLaunched 方法。该方法结尾用省略号表明该方法包含更多代码，如下所示。

项目：ApplicationStateSave | 文件：App.xaml.cs(片段)

```
protected override void OnLaunched(LaunchActivatedEventArgs args)
{
    ...
    // Code added for ApplicationStateSave project
    Window.Current.CoreWindow.Dispatcher.AcceleratorKeyActivated += OnAcceleratorKeyActivated;
    Window.Current.CoreWindow.PointerPressed += OnPointerPressed;
    // End of code added for ApplicationStateSave project
}
```

PointerPressed 事件处理程序是两者中较简单的一个，我们先来看看。从事件参数的 CurrentPoint 属性的 Properties 属性可以得到 5 个鼠标按钮的状态，把常用于导航的两个额外的按钮识别为 XButton1 和 XButton2。我们只对这种情况感兴趣：不按压所有常规按钮，只按压一个额外按钮(也就是说，它们的状态彼此不相等)。

项目：ApplicationStateSave | 文件：App.xaml.cs(片段)

```
void OnPointerPressed(CoreWindow sender, PointerEventArgs args)
{
    PointerPointProperties props = args.CurrentPoint.Properties;

    if (!props.IsLeftButtonPressed &&
        !props.IsMiddleButtonPressed &&
        !props.IsRightButtonPressed &&
         props.IsXButton1Pressed != props.IsXButton2Pressed)
    {
        if (props.IsXButton1Pressed)
            GoBack();
        else
            GoForward();

        args.Handled = true;
    }
}

void GoBack()
{
    Frame frame = Window.Current.Content as Frame;

    if (frame != null && frame.CanGoBack)
        frame.GoBack();
}
```

```
void GoForward()
{
    Frame frame = Window.Current.Content as Frame;

    if (frame != null && frame.CanGoForward)
        frame.GoForward();
}
```

如果事件导致调用 GoBack 或 GoForward 方法，事件处理程序就会把事件参数的 Handled 属性设置为 true。

对于键盘加速器，事件处理程序可以对左右箭头键使用 VirtualKey，此项没有表示特殊浏览器键的枚举项。在 Win32 API 中，特殊浏览器键被识别为 VK_BROWSER_BACK 和 VK_BROWSER_FORWARD，值分别为 166 和 167。

项目: ApplicationStateSave | 文件: App.xaml.cs(片段)

```
void OnAcceleratorKeyActivated(CoreDispatcher sender, AcceleratorKeyEventArgs args)
{
    if ((args.EventType == CoreAcceleratorKeyEventType.SystemKeyDown ||
         args.EventType == CoreAcceleratorKeyEventType.KeyDown) &&
        (args.VirtualKey == VirtualKey.Left ||
         args.VirtualKey == VirtualKey.Right ||
        (int)args.VirtualKey == 166 ||
        (int)args.VirtualKey == 167))
    {
        CoreWindow window = Window.Current.CoreWindow;
        CoreVirtualKeyStates down = CoreVirtualKeyStates.Down;

        // Ignore key combinations where Shift or Ctrl is down
        if ((window.GetKeyState(VirtualKey.Shift) & down) == down ||
            (window.GetKeyState(VirtualKey.Control) & down) == down)
        {
            return;
        }

        // Get alt key state
        bool alt = (window.GetKeyState(VirtualKey.Menu) & down) == down;

        // Go back for Alt-Left key or browser left key
        if (args.VirtualKey == VirtualKey.Left && alt ||
            (int)args.VirtualKey == 166 && !alt)
        {
            GoBack();
            args.Handled = true;
        }

        // Go forward for Alt-Right key or browser right key
        if (args.VirtualKey == VirtualKey.Right && alt ||
            (int)args.VirtualKey == 167 && !alt)
        {
            GoForward();
            args.Handled = true;
        }
    }
}
```

只有当 Alt 键(也称为“菜单键”)也被同时按下(而不是 Shift 或者 Ctrl 键被按下时)，左右箭头键才有加速器功能，而且只有在没有按下辅助键的时候，才接受特殊浏览器键。

GetKeyState 方法用起来有点笨拙，因为 GetKeyState 可以返回 CoreVirtualKeyStates 的三个枚举项：None(等于 0)、Down(等于 1)、Locked(等于 2)。在内部，所有键都视为开关，枚举项为标识。键弹起时状态为 0，按下时状态为 3，释放时状态为 2。再次按下时状态为 1，释放状态时再次回到 0。

12.10 传递和返回数据

页面经常需要共享数据，例如，页面共享视图模式就很常见。App 类是维护页面共享数据的好地方。往该类添加方法和属性时不要犹豫。例如，可以添加名为 ViewModel 的公共属性，该属性有公共 get 访问器和私有 set 访问器，这样可以在 App 构造函数中进行初始化。

另一方面，从数据应该只对需要知道的类可见这一哲学而言，不要把所有东西都放进 App 类。导航是把数据从一个页面传递到另一个页面以及把数据从第二个页面返回到第一页面，一种非常结构化的方法。

DataPassingAndReturning 项目通过两个简单页面来演示这些方法。第一个页面与通常一样称为 MainPage，而第二个页面称为 DialogPage，因为其功能很像一个对话框。MainPage 只能导航到 DialogPage，而 DialogPage 只能返回到 MainPage。

两个页面之间的导航受限，因此页面不需要保存页面状态。为了使程序更加简单，在暂停期间不保存导航状态或页面状态，也不实现键盘或鼠标快捷方式。尽管非常简单，但仍然是以解释基本的数据传递方法。

DialogPage 的 XAML 文件有三个 RadioButton 控件，分别是 Red，Green 和 Blue，还有一个标签为“Finished”的常规 Button 控件。

项目: DataPassingAndReturning | 文件: DialogPage.xaml(片段)

```
<Grid Background="{StaticResource ApplicationPageBackgroundThemeBrush}">
    <StackPanel>
        <TextBlock Text="Color Dialog"
                   FontSize="48"
                   HorizontalAlignment="Center"
                   Margin="48" />

        <StackPanel Name="radioStack"
                    HorizontalAlignment="Center"
                    Margin="48">
            <RadioButton Content="Red" Margin="12">
                <RadioButton.Tag>
                    <Color>Red</Color>
                </RadioButton.Tag>
            </RadioButton>

            <RadioButton Content="Green" Margin="12">
                <RadioButton.Tag>
                    <Color>Green</Color>
                </RadioButton.Tag>
            </RadioButton>

            <RadioButton Content="Blue" Margin="12">
                <RadioButton.Tag>
                    <Color>Blue</Color>
                </RadioButton.Tag>
            </RadioButton>
        </StackPanel>

        <Button Content="Finished"
                HorizontalAlignment="Center"
                Margin="48"
                Click="OnReturnButtonClick" />
    </StackPanel>
</Grid>
```

注意，每个 RadioButton 都把其 Tag 属性设置为与该按钮相对应的 Color 值。DialogPage 的代码隐藏文件负责从这些按钮获取选定的 Color 并返回给 MainPage。

有趣的是，MainPage.xaml 和 DialogPage.xaml 非常相似，只不过 Grid 有名称，中间的 RadioButton 为选中，Button 的标签为“Get Color”。

项目：DataPassingAndReturning | 文件：MainPage.xaml（片段）

```
<Grid Name="contentGrid"
      Background="{StaticResource ApplicationPageBackgroundThemeBrush}">
    <StackPanel>
        <TextBlock Text="Main Page"
                   FontSize="48"
                   HorizontalAlignment="Center"
                   Margin="48" />

        <StackPanel Name="radioStack"
                    HorizontalAlignment="Center"
                    Margin="48">
            <RadioButton Content="Red" Margin="12">
                <RadioButton.Tag>
                    <Color>Red</Color>
                </RadioButton.Tag>
            </RadioButton>

            <RadioButton Content="Green" Margin="12"
                         IsChecked="True">
                <RadioButton.Tag>
                    <Color>Green</Color>
                </RadioButton.Tag>
            </RadioButton>

            <RadioButton Content="Blue" Margin="12">
            <RadioButton.Tag>
                <Color>Blue</Color>
                    </RadioButton.Tag>
                </RadioButton>
        </StackPanel>

        <Button Content="Get Color"
                HorizontalAlignment="Center"
                Margin="48"
                Click="OnGotoButtonClick" />
    </StackPanel>
</Grid>
```

这里的想法是，通过 MainPage 中的 RadioButton 控件，为 DialogPage 中的 RadioButton 控件选择初始值，也就是说，MainPage 需要传递数据给 DialogPage。

MainPage 和 DialogPage 相互之间传递的数据只有 Color 值，但对实际应用而言可以是多得多的值。我们通过定义专门在页面之间传递数据的类来反映这种可能性，MainPage 传递给 DialogPage 数据的类如下所示。

项目：DataPassingAndReturning | 文件：PassData.cs

```
using Windows.UI;

namespace DataPassingAndReturning
{
    public class PassData
    {
        public Color InitializeColor { set; get; }
    }
}
```

对本例而言，从 DialogPage 返回到 MainPage 的数据非常相似。

```
项目: DataPassingAndReturning | 文件: ReturnData.cs
using Windows.UI;

namespace DataPassingAndReturning
{
    public class ReturnData
    {
        public Color ReturnColor { set; get; }
    }
}
```

我在本例中可以使用相同的类，当然在一般情况下，两项任务可以使用不同的类。

小心！为了和逻辑及数据流保持一致，我会在 MainPage 和 DialogPage 代码隐藏文件之间反复讨论。

简单的数据传输是从 MainPage 到 DialogPage。点击 MainPage 里的“Get Color”按钮，代码隐藏文件就会创建 PassData 类型的对象，然后扫描整个 RadioButton 控件集合，看看哪个被选中，是分配给 PassData 的 InitializeColor 属性的 Color 值。PassData 对象则成为 Navigate 的第二个参数：

```
项目: DataPassingAndReturning | 文件: MainPage.xaml.cs(片段)
void OnGotoButtonClick(object sender, RoutedEventArgs args)
{
    // Create PassData object
    PassData passData = new PassData();

    // Set the InitializeColor property from the RadioButton controls
    foreach (UIElement child in radioStack.Children)
        if ((child as RadioButton).IsChecked.Value)
            passData.InitializeColor = (Color)(child as RadioButton).Tag;

    // Pass that object to Navigate
    this.Frame.Navigate(typeof(DialogPage), passData);
}
```

如果调用 DialogPage 中的 OnNavigatedTo 覆写，事件参数的 Parameter 属性则为传递给 Navigate 第二个参数的对象。DialogPage 通过该属性初始化自己 RadioButton 控件的设置：

```
项目: DataPassingAndReturning | 文件: DialogPage.xaml.cs(片段)
protected override void OnNavigatedTo(NavigationEventArgs args)
{
    // Get the object passed as the second argument to Navigate
    PassData passData = args.Parameter as PassData;

    // Use that to initialize the RadioButton controls
    foreach (UIElement child in radioStack.Children)
        if ((Color)(child as RadioButton).Tag == passData.InitializeColor)
            (child as RadioButton).IsChecked = true;
    base.OnNavigatedTo(args);
}
```

现在,可以单击三个 RadioButton 控件来选择 Color 值。如果对选择满意,则按下 Finished 按钮。处理程序直接调用 GoBack，并返回 MainPage。

```
项目: DataPassingAndReturning | 文件: DialogPage.xaml.cs(片段)
void OnReturnButtonClick(object sender, RoutedEventArgs args)
{
    this.Frame.GoBack();
}
```

如果 GoBack 有一个可选参数能够被设置为返回目标页面的数据，那就太好了。但事实并非如此。没有机制做这件事，必须考虑其他方法。

有这样一种可能性：DialogPage 调用 GoBack，然后调用 DialogPage 中的 OnNavigatedFrom 覆写。事件参数的 Content 属性为(将要导航到的)MainPage 的实例。也就是说，MainPage 可以定义公用属性和方法来专门从 DialogPage 获取信息，而 DialogPage 可以在 OnNavigatedFrom 覆写时设置该属性或者调用该方法。

只不过在架构上有点儿粗糙，因为 DialogPage 必须熟悉导航到自身的页面类型。一般情况下，这不是一个好的解决方案。

DialogPage 有一个更好的解决方案，即通过它需要返回的数据类型来定义 Completed 事件：

```
项目: DataPassingAndReturning | 文件: DialogPage.xaml.cs(片段)
public sealed partial class DialogPage : Page
{
    public event EventHandler<ReturnData> Completed;
    ...
}
```

MainPage 需要为该事件设置事件处理程序。MainPage 中只能在 OnNavigatedFrom 方法里设置，因为事件参数包含 Content 属性，而 Content 属性是将被导航到的 DialogPage 的实例：

```
项目: DataPassingAndReturning | 文件: MainPage.xaml.cs(片段)
protected override void OnNavigatedFrom(NavigationEventArgs args)
{
    if (args.SourcePageType.Equals(typeof(DialogPage)))
        (args.Content as DialogPage).Completed += OnDialogPageCompleted;

    base.OnNavigatedFrom(args);
}
```

MainPage 知道 DialogPage，因为它要被导航到 DialogPage。但 MainPage 也可能导航到其他页面，因此要检查事件参数的 SourcePageType 属性，以确保 MainPage 知道特定 OnNavigatedFrom 事件所表明的页面。

有了该结构，DialogPage 就不需要知道 MainPage，也应该这样。在面向对象的编程环境中，事件的主要目的之一就是不让信息提供者知道信息使用者。

DialogPage 在 Button 的 Click 处理程序中触发 Completed 事件，但我选择在 OnNavigatedFrom 中实现该逻辑。

```
项目: DataPassingAndReturning | 文件: DialogPage.xaml.cs(片段)
protected override void OnNavigatedFrom(NavigationEventArgs args)
{
    if (Completed != null)
    {
        // Create ReturnData object
        ReturnData returnData = new ReturnData();

        // Set the ReturnColor property from the RadioButton controls
        foreach (UIElement child in radioStack.Children)
            if ((child as RadioButton).IsChecked.Value)
                returnData.ReturnColor = (Color)(child as RadioButton).Tag;

        // Fire the Completed event
        Completed(this, returnData);
    }
```

```
        base.OnNavigatedFrom(args);
    }
```

如果确实有 Completed 事件的处理程序，DialogPage 就会实例化 ReturnData 对象，然后从 RadioButton 控件的集合里设置 ReturnColor 属性。

在 Completed 处理程序中，MainPage 通过来自 DialogPage 的数据来设置其 Grid 的 Background 属性，并选中一个 RadioButton：

```
项目: DataPassingAndReturning | 文件: MainPage.xaml.cs(片段)
void OnDialogPageCompleted(object sender, ReturnData args)
{
    // Set background from returned color
    contentGrid.Background = new SolidColorBrush(args.ReturnColor);

    // Set RadioButton for returned color
    foreach (UIElement child in radioStack.Children)
        if ((Color)(child as RadioButton).Tag == args.ReturnColor)
            (child as RadioButton).IsChecked = true;

     (sender as DialogPage).Completed -= OnDialogPageCompleted;
}
```

处理程序把自己从发送方分离出去，然后结束。

但我所给出的代码有一个缺陷，即在默认情况下，MainPage 的实例在 DialogPage 里设置 Completed 事件的处理程序，但并不是 DialogPage 返回到的那个 MainPage 的实例！解决这个小问题需要把 NavigationCacheMode 设置为 Disabled 以外的值。

```
项目: DataPassingAndReturning | 文件: MainPage.xaml.cs(片段)
public sealed partial class MainPage : Page
{
    public MainPage()
    {
        this.InitializeComponent();
        this.NavigationCacheMode = NavigationCacheMode.Enabled;
    }
    ...
}
```

在 MainPage 这样做可以保证单个实例对以应用为架构中心的页面是非常合理的。引发 DialogPage 的 MainPage 实例和从其获取数据的实例应该是同一个。

12.11 Visual Studio 标准模板

我坦白，这几年来我一直在各种 Windows 环境中写页面导航逻辑代码，从来没碰到过要实现加速键或鼠标按钮快捷键的情况。之前展示的代码改写自 Visual Studio 生成的一个类，名为 LayoutAwarePage，派生自 Page，它实现了若干有用功能。

如果调用 Add New Item 对话框并添加一项 Basic Page 而不是 Blank Page，LayoutAwarePage 和其他组合类会自动添加到 Visual Studio 项目。这些文件也是 Grid App 和 Split App 模板的一部分。页面类通过选择派生自 LayoutAwarePage 的 Basic Page 而不是通过 Page 创建。LayoutAwarePage 定义 SaveState 和 LoadState 虚拟方法，页面实例能保存和加载状态，再结合另一个生成的 SuspensionManager 类，能完成很多工作。

如果程序暂停并在重启后重新加载其状态，那么 LayoutAwarePage 结合

SuspensionManager 还能保存应用状态(包括回退栈)。

LayoutAwarePage 如此命名，是因为它通过 SizeChanged 方法检查 ApplicationView.Value 属性，并通过与 ApplicationViewState 枚举项“FullScreenLandscape”、“FullScreenPortrait”、“Filled”和“Snapped”相对应的字符串调用 VisualStateManager.GoToState。这些状态允许 XAML 文件通过视觉状态管理器标记来查看自身视图变化。

是否想使用这些类或实施这些功能(或类似功能)，由你自己决定。不管是否使用，研究这些类并看看能学到什么，不会有坏处。

在 Visual Studio 里创建新项目时，我一直都用 Blank App，但有两个替代品 Grid App 和 Split App。这些模板利用 LayoutAwarePage 和 SuspensionManager 以及 DataModel 文件夹中的示例视图模型。它们演示了在屏幕上布局数据的推荐方法。也许最重要的是，Grid App 和 Split App 模板演示使用剩下的两个 ItemsControl 派生类基本方法：GridView 和 ListView。

GridView 和 ListView 通过 Selector 和 ListViewBase 派生自 ItemsControl。GridView 和 ListView 都不为自身定义任何公用属性和方法，但从 ListViewBase 共享很多属性和方法。如果检查 generic.xaml，你还会发现 GridView、ListView、GridViewItem 和 ListViewItem 的模板其实并不一样。尤其在默认情况下，GridView 使用 WrapGrid 来显示项目，而 ListView 则使用 VirtualizingStackPanel。

GridView 和 ListView 也适合用于分组项目。定义条目如何分组以及划分组间隔的头部外观。Grid App 和 Split App 中有相应的例子。

Windows 8 自身的启动屏幕就是 GridView 或者非常类似于 GridView。你可能知道，可以在启动屏幕上点击项目来进行选择。这种选择方式由 ListViewBase 所支持(因此 GridView 和 ListView 也支持)，但在 Visual Studio 模板里不可用。

Windows 8 启动屏幕允许移动项目。该功能受 ListViewBase 支持(但有趣的是，不支持项目分组。)Windows 8 启动屏幕支持语义缩放：如果用手指捏住启动屏幕，屏幕会被折叠起来使我们看到更大的分组，然后选择所有组。可以通过 SemanticZoom 类在你自己的应用中这样做。

现在，我们来仔细看一看 Grid App 模板。(可以自学 Split App。)该项目包含 3 个 LayoutAwarePage 派生类。

如下图所示，Grid App 通过显示 GroupedItemsPage 来初始化。

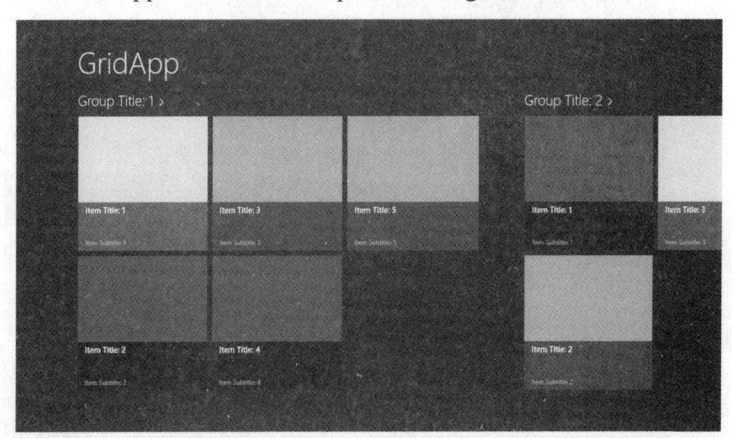

在实际应用中，这些灰色盒子可能是图片或其他图形。

页面有一个标题和一个能横向滚动的 GridView 控件。单个项目由 StandardStyles.xaml 里名为“Standard250×250ItemTemplate”的 DataTemplate 资源定义。头部外观(“Group Title: 1 >”等等)在 GroupedItemsPage.xaml 里通过 GridView 的 GroupStyle 属性的 HeaderTemplate 属性定义。

页面在 Filled 视图中具有同样外观，但在 Snapped 视图中，则切换为一个垂直滚动的 ListView，如下图所示。

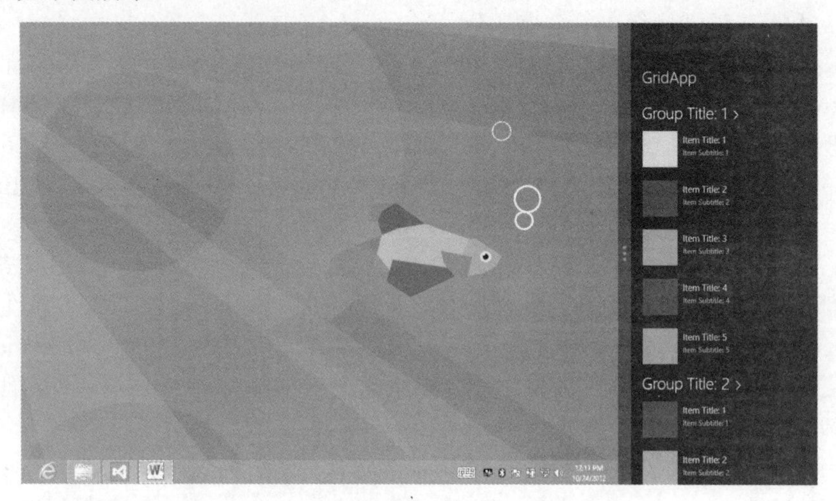

ItemTemplate 属性现在为 DataTemplate 资源“Standard80ItemTemplate”。注意，网页标题格式也不同。其格式为“SnappedPageHeaderTextStyle”，而不是通常的“PageHeaderTextStyle”，两者都在 StandardStyles.xaml 中定义。

如果程序处于下图所示的 Snapped 模式，GridView 和 ListView 之间的切换就发生在 GroupedItemsPage.xaml 文件中并基于 LayoutAwarePage 里的 VisualStateManager 调用。GroupedItemsPage.xaml 文件包含一个视觉状态管理器小节，能响应 Snapped 状态以及 FullScreenPortrait 状态。

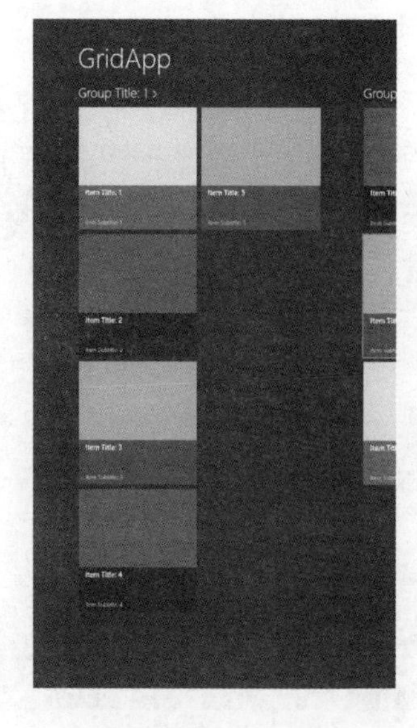

以下是 GridView 在更宽的横向视图，但你会注意到两边少了一点边距。像这样在 XAML 中定义变化，是使用视觉状态管理器来告知不同视图的优点之一。

如果点击其中一个头部标题，会导航到 GroupDetailPage，见下图。

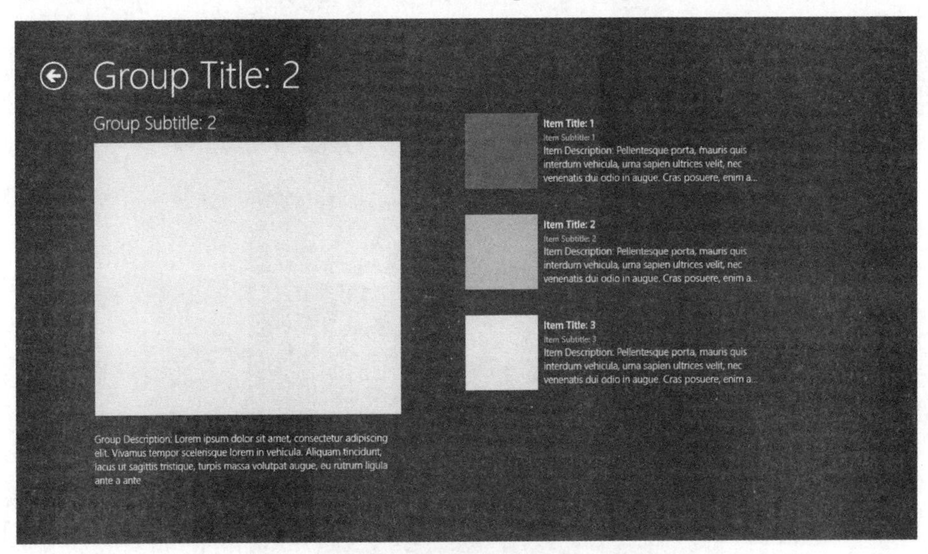

注意，Back 按钮由左上角的一个圆圈箭头来实现。Button 将其 Style 设置为 StandardStyles.xaml 里所定义的 BackButtonStyle 资源。这仍然是 GridView，除了头部非常大并且出现在左边。单个项目现在通过 ItemTemplate 来显示，并基于在 StandardStyles.xaml 的 Standard500x130ItemTemplate 资源。

页面再次切换到 Snapped 状态中的 ListView，见下图。

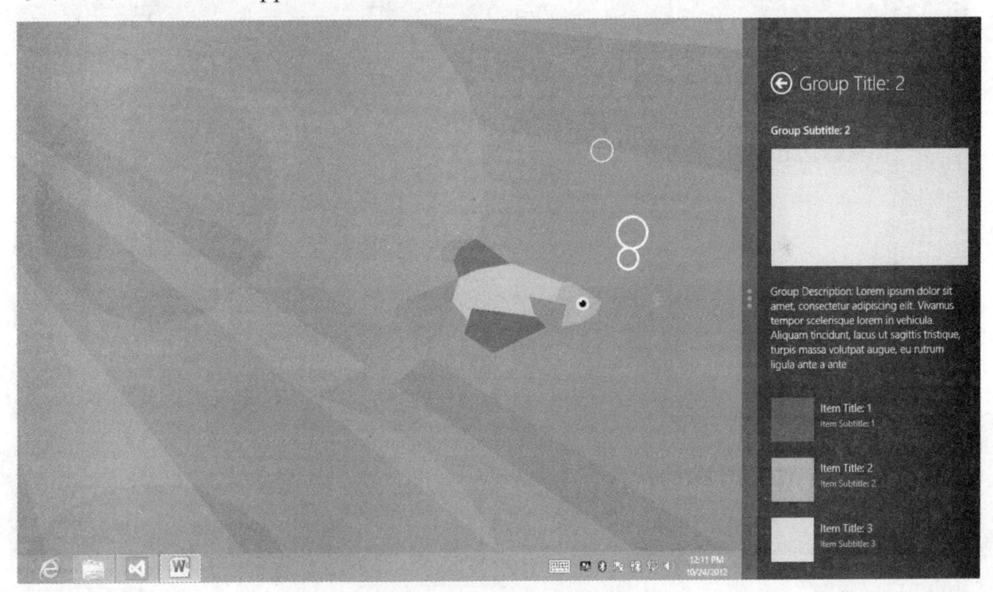

注意，Button 外观也改变了。StandardStyles.xaml 有 SnappedBackButtonStyle 和 PortraitBackButtonStyle。下图为竖屏视图。

无论从 GroupedItemsPage 还是从 GroupDetailPage，都可以导航到单个项目页面，见下图。

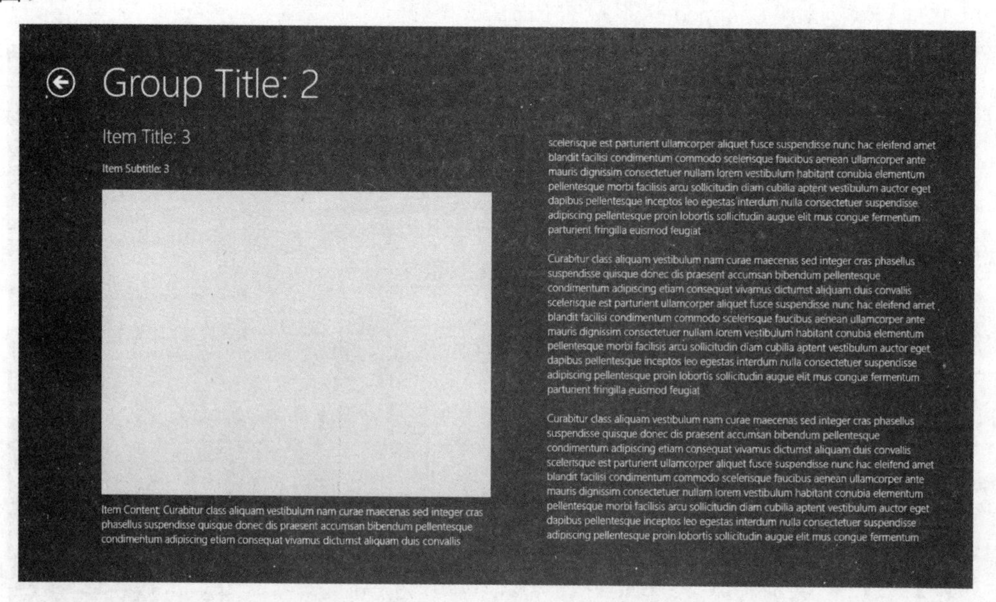

起初看起来是一个条目，然而，水平滚动可以查看同一组的其他条目。页面主体实际上一个 FlipView。而其中每个条目都是 ScrollViewer，包含 RichTextBlock 元素集合。我会在第 16 章进行讨论。在 Grid App 模板中，RichTextBlock 元素由 RichTextColumns 类生成，可以在 Common 文件夹里找到 RichTextColumns 类。

条目视图在 Snapped 状态里也不一样，如下图所示。

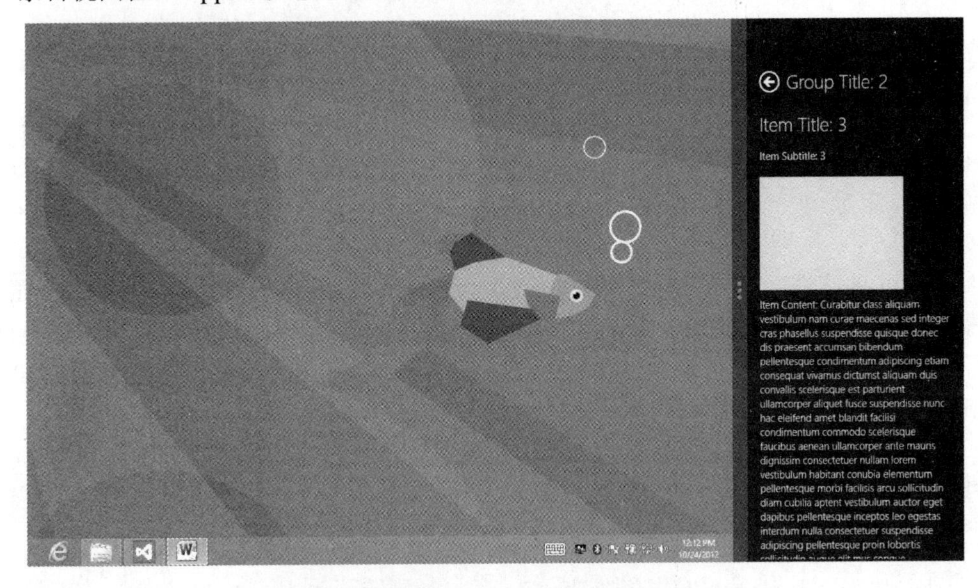

Portrait 视图也不一样，如下图所示。

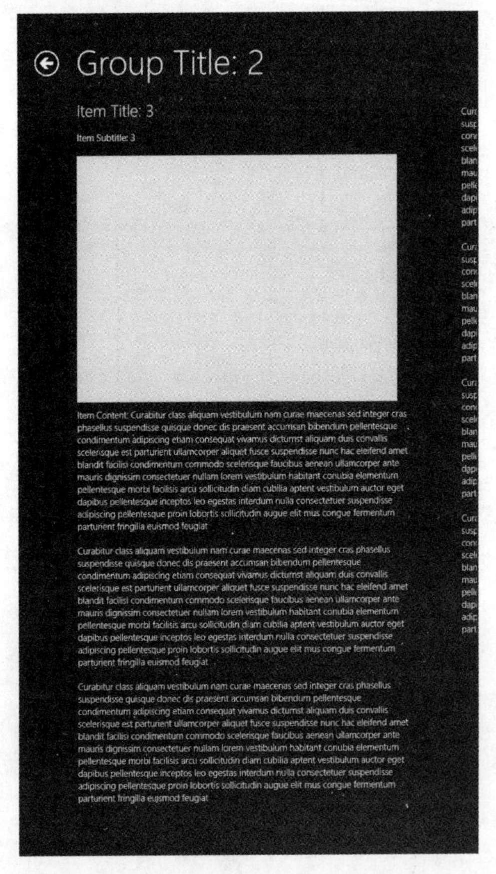

虽然我会在本书的项目中继续使用 Blank App 和 Blank Page 模板，但要以更易于理解的(我希望)简化方式在更复杂的模板中实现一些功能。

12.12 视图模式和集合

正如第 11 章所示，Colors 类提供了便捷对象源，用于在 ItemsControl 和 ListBox 中显示。然而，如果谈到 GridView 和 ListView 控件，则需要找一些更高级、更真实的示例数据。

为此，我的网站 http://www.charlespetzold.com/Students 目录包含有一个名为 students.xml 的文件，包含一所高中 69 名学生的信息。目录里还包含这些学生的靓照，取自得克萨斯州埃尔帕索 1912 年到 1914 年的高中年鉴。年鉴属于公共领域，由埃尔帕索公众图书馆进行数字化，并在 http://www.elpasotexas.gov/library/ourlibraries/main_library/yearbooks/yearbooks.asp 上提供给公众。

ElPasoHighSchool 项目是一个库，能访问该 XML 文件并构造视图模型，向应向提供该信息。以下 Student 类代表一个学生。注意，该类实现 INotifyPropertyChanged 以适合数据绑定：

项目：ElPasoHighSchool | 文件：Student.cs

```
using System.ComponentModel;
using System.Runtime.CompilerServices;

namespace ElPasoHighSchool
{
    public class Student : INotifyPropertyChanged
    {
        string fullName, firstName, middleName, lastName, sex, photoFilename;
        double gradePointAverage;

        public event PropertyChangedEventHandler PropertyChanged;

        public string FullName
        {
            set { SetProperty<string>(ref fullName, value); }
            get { return fullName; }
        }

        public string FirstName
        {
            set { SetProperty<string>(ref firstName, value); }
            get { return firstName; }
        }

        public string MiddleName
        {
            set { SetProperty<string>(ref middleName, value); }
            get { return middleName; }
        }

        public string LastName
        {
            set { SetProperty<string>(ref lastName, value); }
            get { return lastName; }
        }

        public string Sex
        {
            set { SetProperty<string>(ref sex, value); }
            get { return sex; }
        }
```

```
        public string PhotoFilename
        {
            set { SetProperty<string>(ref photoFilename, value); }
            get { return photoFilename; }
        }

        public double GradePointAverage
        {
            set { SetProperty<double>(ref gradePointAverage, value); }
            get { return gradePointAverage; }
        }

        protected bool SetProperty<T>(ref T storage, T value,
                [CallerMemberName] string propertyName = null)
        {
            if  (object.Equals(storage, value))
                return false;

            storage = value;
            OnPropertyChanged(propertyName);
            return true;
        }

        protected void OnPropertyChanged(string propertyName)
        {
            if  (PropertyChanged != null)
                PropertyChanged(this, new PropertyChangedEventArgs(propertyName));
        }
    }
}
```

以下 StudentBody 类也实现 INotifyPropertyChanged。该类包含学校名称和 Student 类型的 ObservableCollection，用于存储所有 Student 对象。

项目: ElPasoHighSchool | 文件: StudentBody.cs

```
using System.Collections.ObjectModel;
using System.ComponentModel;
using System.Runtime.CompilerServices;

namespace ElPasoHighSchool
{
    public class StudentBody : INotifyPropertyChanged
    {
        string school;
        ObservableCollection<Student> students = new ObservableCollection<Student>();

        public event PropertyChangedEventHandler PropertyChanged;

        public string School
        {
            set { SetProperty<string>(ref school, value); }
            get { return school; }
        }

        public ObservableCollection<Student> Students
        {
            set { SetProperty<ObservableCollection<Student>>(ref students, value); }
            get { return students; }
        }

        protected bool SetProperty<T>(ref T storage, T value,
                [CallerMemberName] string propertyName = null)
        {
            if  (object.Equals(storage, value))
                return false;
```

```
            storage = value;
            OnPropertyChanged(propertyName);
            return true;
        }

        protected void OnPropertyChanged(string propertyName)
        {
            if (PropertyChanged != null)
                PropertyChanged(this, new PropertyChangedEventArgs(propertyName));
        }
    }
}
```

ObservableCollection 实现 INotifyCollectionChanged 接口，该接口定义 CollectionChanged 事件。无论什么时候添加条目到集合、从集合删除条目，或者现有条目重新排序，ObservableCollection 都会触发该事件。如果把对象设置为条目控件的 ItemsSource 属性，控件就会检查对象是否实现了 INotifyCollectionChanged。如果已实施，则为 CollectionChanged 事件附加处理程序，并且如果发生条目增加、删除或重新排序，则修改其显示。

我网站上的 student.xml 文件，如下所示。

文件: http://www.charlespetzold.com/Students/students.xml(excerpt)

```
<?xml version="1.0" encoding="utf-8" ?>
<StudentBody xmlns:xsd="http://www.w3.org/2001/XMLSchema"
        xmlns:xsi="http://www.w3.org/2001/XMLSchema-instance">
    <School>El Paso High School</School>-<Students>
        <Student>
            <FullName>Adkins Bowden</FullName>
            <FirstName>Adkins</FirstName>
            <MiddleName/>
            <LastName>Bowden</LastName>
            <Sex>Male</Sex>
            <PhotoFilename>
                http://www.charlespetzold.com/Students/AdkinsBowden.png
            </PhotoFilename>
            <GradePointAverage>2.71</GradePointAverage>
        </Student>
    <Student>
        <FullName>Alfred Black</FullName>
        <FirstName>Alfred</FirstName>
        <MiddleName/>
        <LastName>Black</LastName>
        <Sex>Male</Sex>
        <PhotoFilename>
            http://www.charlespetzold.com/Students/AlfredBlack.png
        </PhotoFilename>
        <GradePointAverage>2.87</GradePointAverage>
    </Student>
    <Student>
        <FullName>Alice Bishop</FullName>
        <FirstName>Alice</FirstName>
        <MiddleName/>
        <LastName>Bishop</LastName>
        <Sex>Female</Sex>
        <PhotoFilename>
            http://www.charlespetzold.com/Students/AliceBishop.png
        </PhotoFilename>
        <GradePointAverage>3.68</GradePointAverage>
    </Student>
    ...
    <Student>
        <FullName>William Sheley Warnock</FullName>
        <FirstName>William</FirstName>
        <MiddleName>Sheley</MiddleName>
        <LastName>Warnock</LastName>
```

```
            <Sex>Male</Sex>
            <PhotoFilename>
                http://www.charlespetzold.com/Students/WilliamSheleyWarnock.png
            </PhotoFilename>
            <GradePointAverage>1.82</GradePointAverage>
            </Student>
        </Students>
    </StudentBody>
```

Student 和 StudentBody 元素标签对应于前面的 Student 和 StudentBody 类。我通过 XmlSerializer 类使用.NET 序列化创建了该 XML 文件，可以用同样方式进行反序列化。这是 StudentBodyPresenter 类的目的，StudentBodyPresenter 类又实施了 INotifyPropertyChanged，但只有一个 StudentBody 类型的属性：

项目：ElPasoHighSchool | 文件：StudentBodyPresenter.cs（片段）

```
public class StudentBodyPresenter : INotifyPropertyChanged
{
    StudentBody studentBody;
    Random rand = new Random();
    Window currentWindow = Window.Current;

    public event PropertyChangedEventHandler PropertyChanged;

    public StudentBodyPresenter()
    {
        // Download XML file
        HttpClient httpClient = new HttpClient();
        Task<string> task =
            httpClient.GetStringAsync("http://www.charlespetzold.com/ Students/students.xml");
        task.ContinueWith(GetStringCompleted);
    }

    async void GetStringCompleted(Task<string> task)
    {
        if  (task.Exception == null && !task.IsCanceled)
        {
            string xml = task.Result;

            // Deserialize XML
            StringReader reader = new StringReader(xml);
            XmlSerializer serializer = new XmlSerializer(typeof(StudentBody));

            await currentWindow.Dispatcher.RunAsync(CoreDispatcherPriority.Normal, () =>
                {
                    this.StudentBody = serializer.Deserialize(reader) as StudentBody;

                    // Set a timer for random changes
                    DispatcherTimer timer = new DispatcherTimer
                    {
                        Interval = TimeSpan.FromMilliseconds(100)
                    };
                    timer.Tick += OnTimerTick;
                    timer.Start();
                });
        }
    }

    public StudentBody StudentBody
    {
        set { SetProperty<StudentBody>(ref studentBody, value); }
        get { return studentBody; }
    }

    // Mimic changing grade point averages
    void OnTimerTick(object sender, object args)
```

```
        {
            int index = rand.Next(studentBody.Students.Count);
            Student student = this.StudentBody.Students[index];
            double factor = 1 + (rand.NextDouble() - 0.5) / 5;
            student.GradePointAverage =
                Math.Max(0.0,
                    Math.Min(5.0, (int)(100 * factor * student.GradePointAverage) / 100.0));
        }

        protected bool SetProperty<T>(ref T storage, T value,
                [CallerMemberName] string propertyName = null)
        {
            if (object.Equals(storage, value))
                return false;

            storage = value;
            OnPropertyChanged(propertyName);
            return true;
        }

        protected void OnPropertyChanged(string propertyName)
        {
            if (PropertyChanged != null)
                PropertyChanged(this, new PropertyChangedEventArgs(propertyName));
        }
    }
```

这个类带来了一些问题。我想从类的构造函数来启动加载和反序列化 XML 文件，但无法将构造函数声明为 async，因此，我需要一个显式附加处理程序。然而，附加处理程序不在用户界面线程运行。这是一个问题，因为这种方法需要设置 StudentBody 属性，但这会导致触发 PropertyChanged 事件，并可能导致用户界面对象绑定更新。这个类需要在用户界面线程通过 CoreDispatcher 设置 StudentBody 属性，但 CoreDispatcher 对象从何而来？该类无法访问在用户界面线程创建对象，而这些对象通常又是 CoreDispatcher 的来源。

幸运的是，Window 对象有 Dispatcher 属性，很容易通过 Window.Current 静态属性来获取当前窗口。StudentBodyPresenter 类还设置了 DispatcherTimer，用于模拟学生 GPA 的实时变化，给 PropertyChanged 事件做做试验。

我们来创建一个新的解决方案和项目，名为 DisplayHighSchoolStudents。把现有 ElPasoHighSchool 项目添加到该解决方案。在 DisplayHighSchoolStudents 项目的 References 节里，设置对 ElPasoHighSchool 项目的引用，并在 MainPage.xaml 文件里创建新的 XML 命名空间前缀：

```
xmlns:elpaso="using:ElPasoHighSchool"
```

然后可以在 MainPage.xaml 的 Resources 节实例化 StudentBodyPresenter 类：

```
<Page.Resources>
    <elpaso:StudentBodyPresenter x:Key="presenter" />
</Page.Resources>
```

现在试试从该视图模式访问条目。

例如，以下标记会让 TextBlock 显示全限定类名 ElPasoHighSchool.StudentBodyPresenter：

```
<Grid Background="{StaticResource ApplicationPageBackgroundThemeBrush}">
    <TextBlock Text="{Binding Source={StaticResource presenter}}"
               FontSize="24" />
</Grid>
```

以下标记显示全限类名 ElPasoHighSchool.StudentBody：

```
<TextBlock Text="{Binding Source={StaticResource presenter},
                          Path=StudentBody}"
           FontSize="24" />
```

试试更深入地另一个属性，如下所示：

```
<TextBlock Text="{Binding Source={StaticResource presenter},
                          Path=StudentBody.School}"
           FontSize="24" />
```

现在你会得到一些真实数据：School 属性值或“El Paso High School”。

StudentBody 的另一个属性是 Students，试试如下标记：

```
<TextBlock Text="{Binding Source={StaticResource presenter},
                          Path=StudentBody.Students}"
           FontSize="24" />
```

显示的文本非常长，是另一个全限定类名 “System.Collections.ObjectModel.ObservableCollection`1[ElPasoHighSchool.Student]”。

然而，可以在标记里索引 Students 属性：

```
<TextBlock Text="{Binding Source={StaticResource presenter},
                          Path=StudentBody.Students[23]}"
           FontSize="24" />
```

结果是另一个全限定类名“ElPasoHighSchool.Student”，但现在我们可以看到该类的实际属性。

Student 类的一个属性是 FullName，试试如下代码：

```
<TextBlock Text="{Binding Source={StaticResource presenter},
                          Path=StudentBody.Students[23].FullName}"
           FontSize="24" />
```

结果是学生姓名：“Elizabeth Barnes”。

试试用 Image 元素替换 TextBlock，并引用 Student 的 PhotoFilename 属性：

```
<Grid Background="{StaticResource ApplicationPageBackgroundThemeBrush}">
    <Image Source="{Binding Source={StaticResource presenter},
                          Path=StudentBody.Students[23].PhotoFilename}" />
</Grid>
```

结果如下图所示。

现在我们来试试把 ItemsSource 属性设置为 StudentBody 的 Students 属性，并用 GridView 替换 Image 元素：

```
<Grid Background="{StaticResource ApplicationPageBackgroundThemeBrush}">
    <GridView ItemsSource="{Binding Source={StaticResource presenter},
                                    Path=StudentBody.Students}" />
</Grid>
```

结果是显示 Student 对象，如下图所示。

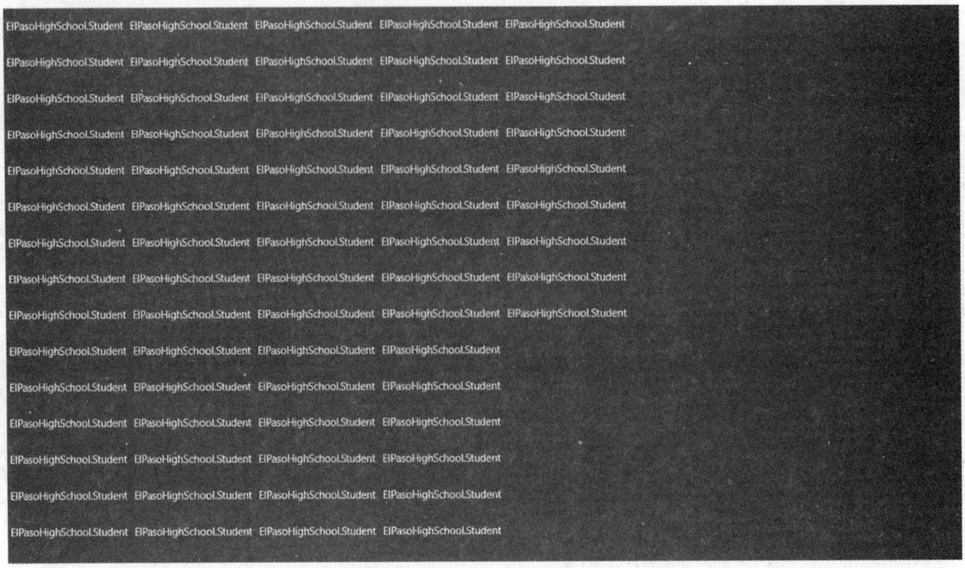

尽管 Student 对象只显示为全限定类名，但我们还是可以发现 GridView 的一些作用。如果想滚动，显示条目会弹回一点，而且可以选择单个条目，如下图所示。

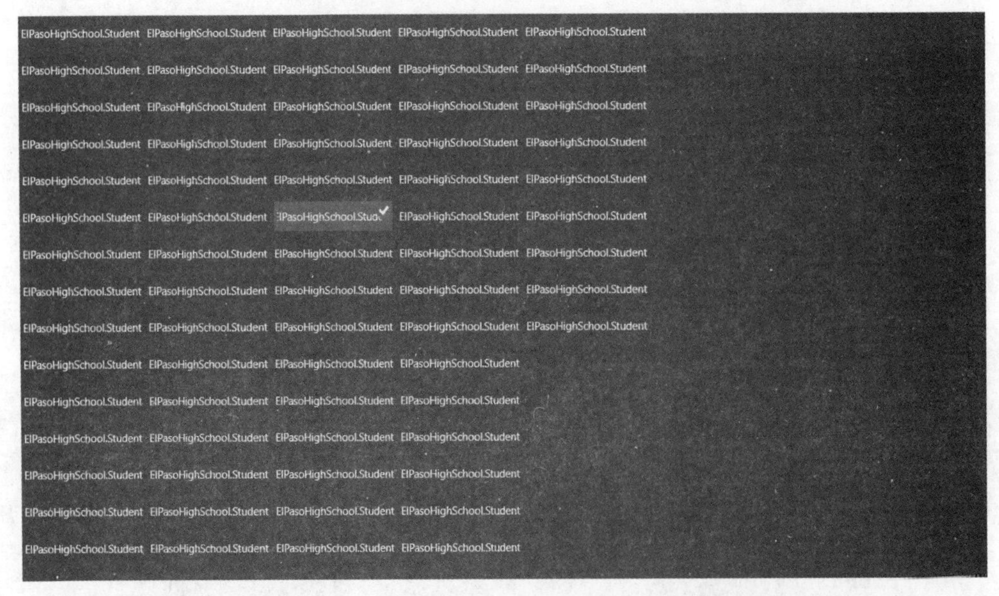

把 Binding 的一部分移作 DataContext 属性的 Grid，如此进行简化：

```
<Grid Background="{StaticResource ApplicationPageBackgroundThemeBrush}"
      DataContext="{Binding Source={StaticResource presenter},
```

```
                              Path=StudentBody}">

    <Grid.RowDefinitions>
        <RowDefinition Height="Auto" />
        <RowDefinition Height="*" />
    </Grid.RowDefinitions>

    <TextBlock Text="{Binding School}"
               Grid.Row="0"
               Style="{StaticResource PageHeaderTextStyle}" />

    <GridView ItemsSource="{Binding Students}"
              Grid.Row="1" />
</Grid>
```

Grid 中的任何元素现在都可以通过很简单的绑定来访问 StudentBody 类的属性。Grid 中的 TextBlock 会引用 School 属性。

现在需要为 Student 条目把 DataTemplate 添加到 GridView：

```
<Grid Background="{StaticResource ApplicationPageBackgroundThemeBrush}"
      DataContext="{Binding Source={StaticResource presenter},
                              Path=StudentBody}">
    <Grid.RowDefinitions>
        <RowDefinition Height="Auto" />
        <RowDefinition Height="*" />
    </Grid.RowDefinitions>

    <TextBlock Text="{Binding School}"
               Grid.Row="0"
               Style="{StaticResource PageHeaderTextStyle}" />

    <GridView ItemsSource="{Binding Students}"
               Grid.Row="1">
        <GridView.ItemTemplate>
            <DataTemplate>
                <Border BorderBrush="{StaticResource ApplicationForegroundThemeBrush}"
                        BorderThickness="1">
                    <Grid Height="120">
                        <Grid.ColumnDefinitions>
                            <ColumnDefinition Width="80" />
                            <ColumnDefinition Width="200" />
                        </Grid.ColumnDefinitions>

                        <Image Source="{Binding PhotoFilename}"
                               Grid.Column="0" />

                        <TextBlock Text="{Binding FullName}"
                                   Grid.Column="1"
                                   VerticalAlignment="Center"
                                   Margin="5 0" />
                    </Grid>
                </Border>
            </DataTemplate>
        </GridView.ItemTemplate>
    </GridView>
</Grid>
```

效果如下图所示，当然是横屏滚动。

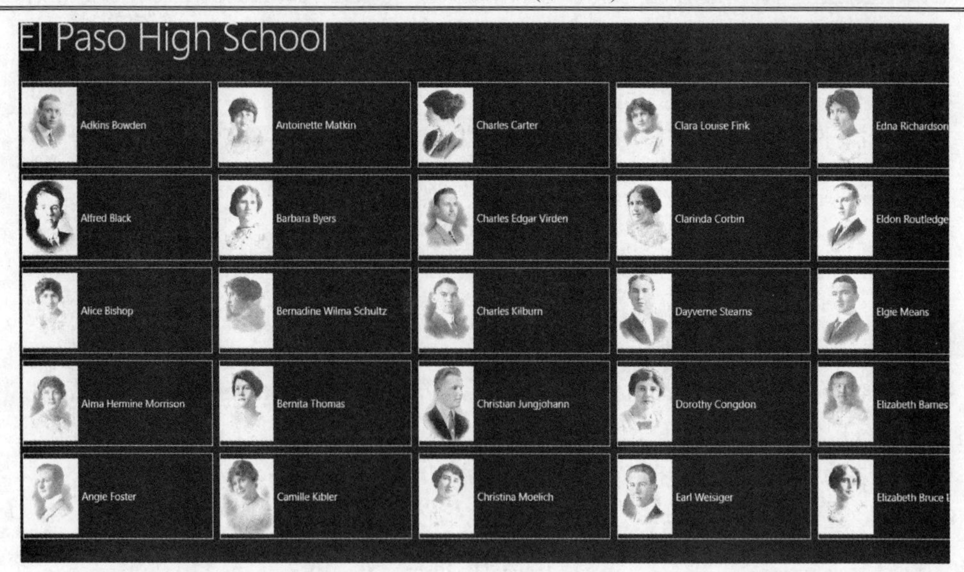

ListViewBase 会对点击条目进行区分，就好像是一个按钮，并且选择一个条目。大部分选中支持都继承自 Selector，并且类似于 ListBox。默认情况下，如果轻击一个条目，则选中该条目。用彩色背景和钩号来显示该条目，控件同时触发 SelectionChanged 事件。

默认情况下，禁用条目点击，但是可以通过设置属性和事件处理程序来启用：

```
<GridView ItemsSource="{Binding Students}"
          Grid.Row="1"
          IsItemClickEnabled="True"
          ItemClick="OnGridViewItemClick">
```

现在如果轻击一个条目，则不会选中条目，而是触发 ItemClick 事件。ItemClick 处理程序的事件参数包括条目，在本例中为 Student 类型的对象。

然而，用户仍然可以通过轻扫或右击条目来进行选择和取消选择。把 SelectionMode 设置为 None，可以完全关闭选择功能：

```
<GridView ItemsSource="{Binding Students}"
          Grid.Row="1"
          SelectionMode="None"
          IsItemClickEnabled="True"
          ItemClick="OnGridViewItemClick">
```

也可以把 SelectionMode 设置为 Multiple，但很显然，如果程序不能对选中条目进行任何操作，则完全不需要实施选中功能。

即使把 SelectionMode 设置为 None，仍然可以轻扫条目，使其可以移动但不会选中任何东西。如果要用 AllowDrop 和 CanRecorderItems 属性实现拖动和重新排序功能，你可能要保持轻扫操作可用：

```
<GridView ItemsSource="{Binding Students}"
          Grid.Row="1"
          SelectionMode="None"
          AllowDrop="True"
          CanReorderItems="True"
          IsItemClickEnabled="True"
          ItemClick="OnGridViewItemClick">
```

然而，如果不想允许选中或重新排序，最好完全禁用轻扫：

```
<GridView ItemsSource="{Binding Students}"
          Grid.Row="1"
          SelectionMode="None"
          IsSwipeEnabled="False"
          IsItemClickEnabled="True"
          ItemClick="OnGridViewItemClick" />
```

在完整的 DisplayHighSchoolStudents 项目中，我使用的仍然是 Blank App，同时试图模仿 Visual Studio 标准 Grid App 的一般布局。MainPage 的代码隐藏文件使用 SizeChanged 处理程序，基于当前视图来设置可视状态。

```
项目: DisplayHighSchoolStudents | 文件: MainPage.xaml.cs(片段)
public sealed partial class MainPage : Page
{
     public MainPage()
     {
          this.InitializeComponent();
          SizeChanged += OnPageSizeChanged;
     }

     void OnPageSizeChanged(object sender, SizeChangedEventArgs args)
     {
          VisualStateManager.GoToState(this, ApplicationView.Value.ToString(), true);
     }
     ...
}
```

XAML 文件有 GridView 显示除 Snapped 之外的所有视图状态，还有 ListView 用于 Snapped。两个控件共享 DataTemplate 来显示条目。视图模型以及 XAML 文件的 Resources 节对此进行定义。

```
项目: DisplayHighSchoolStudents | 文件: MainPage.xaml(片段)
<Page ... >
     <Page.Resources>
          <elpaso:StudentBodyPresenter x:Key="presenter" />

          <DataTemplate x:Key="studentTemplate">
               <Border Height="120"
                       Width="280">
                    <Grid>
                         <Grid.RowDefinitions>
                              <RowDefinition Height="*" />
                              <RowDefinition Height="*" />
                         </Grid.RowDefinitions>

                         <Grid.ColumnDefinitions>
                              <ColumnDefinition Width="Auto" />
                              <ColumnDefinition Width="*" />
                         </Grid.ColumnDefinitions>

                         <Image Grid.Row="0" Grid.Column="0" Grid.RowSpan="2"
                                Source="{Binding PhotoFilename}"
                                Height="120"
                                Margin="5" />

                         <TextBlock Text="{Binding FullName}"
                                    Grid.Row="0" Grid.Column="1"
                                    VerticalAlignment="Center"
                                    Margin="5 0" />

                         <StackPanel Grid.Row="1" Grid.Column="1"
                                     Orientation="Horizontal"
                                     VerticalAlignment="Center"
                                     Margin="5 0">
```

```
                        <TextBlock Text="GPA = " />
                        <TextBlock Text="{Binding GradePointAverage}" />
                    </StackPanel>
                </Grid>
            </Border>
        </DataTemplate>
    </Page.Resources>
    ...
</Page>
```

DataTemplate 在 GridView 和 ListView 之间共享，ListView 用于 Snapped 模式，而且 Snapped 模式总是意味着 320 单位宽度，因此，为 Snapped 所定义的模板需要小于 320 单位。当然，总是可以给两个控件使用不同的条目模板，就像 Grid App 一样。

页面分为两行，上面一行专门用于一个不可见的 Back 按钮和页面标题。

```
项目: DisplayHighSchoolStudents | 文件: MainPage.xaml(片段)
<Page ... >
    ...
    <Grid Background="{StaticResource ApplicationPageBackgroundThemeBrush}"
        DataContext="{Binding Source={StaticResource presenter},
                Path=StudentBody}">
        <Grid.RowDefinitions>
            <RowDefinition Height="140" />
            <RowDefinition Height="*" />
        </Grid.RowDefinitions>

        <Grid Grid.Row="0">
            <Grid.ColumnDefinitions>
                <ColumnDefinition Width="Auto" />
                <ColumnDefinition Width="*" />
            </Grid.ColumnDefinitions>

            <Button Name="backButton"
                    Grid.Column="0"
                    Style="{StaticResource BackButtonStyle}"
                    IsEnabled="False" />

            <TextBlock Name="pageTitle"
                       Text="{Binding School}"
                       Grid.Column="1"
                       Style="{StaticResource PageHeaderTextStyle}" />
        </Grid>
        ...
    </Grid>
</Page>
```

注意，是主页，因此 Button 被禁用。如果按钮禁用，其标准样式就会完全隐藏该按钮。TextBlock 也基于标准样式，并绑定到 School 属性。

Grid 的第二行包含 GridView 和 ListView，但 ListView 把 Visibility 属性设置为 Collapsed。

```
项目: DisplayHighSchoolStudents | 文件: MainPage.xaml(片段)
<Page ... >
    ...
    <Grid Background="{StaticResource ApplicationPageBackgroundThemeBrush}"
        DataContext="{Binding Source={StaticResource presenter},
                            Path=StudentBody}">
        <Grid.RowDefinitions>
            <RowDefinition Height="140" />
            <RowDefinition Height="*" />
        </Grid.RowDefinitions>
        ...
        <GridView Name="gridView"
                Grid.Row="1"
                ItemsSource="{Binding Students}"
```

```
                    Padding="116 0 40 46"
                    SelectionMode="None"
                    IsSwipeEnabled="False"
                    IsItemClickEnabled="True"
                    ItemClick="OnGridViewItemClick"
                    ItemTemplate="{StaticResource studentTemplate}" />

          <ListView Name="listView"
                    Grid.Row="1"
                    ItemsSource="{Binding Students}"
                    Visibility="Collapsed"
                    SelectionMode="None"
                    IsSwipeEnabled="False"
                    IsItemClickEnabled="True"
                    ItemClick="OnGridViewItemClick"
                    ItemTemplate="{StaticResource studentTemplate}" />
          ...
     </Grid>
</Page>
```

很明显，GridView 和 ListView 共享很多属性。可以用 ListViewBase 的 TargetType 在 Style 里对此进行定义。已经禁用选中，但两个控件都把 ItemClick 事件设置为代码隐藏文件中的一个处理程序。

最后，MainPage 有一部分用于 Visual State Manager 标记，主要目的是如果应用处于 Snapped 状态，则隐藏 GridView，并显示 ListView。

```
项目: DisplayHighSchoolStudents | 文件: MainPage.xaml(片段)
<Page ... >
     ...
     <Grid Background="{StaticResource ApplicationPageBackgroundThemeBrush}"
          DataContext="{Binding Source={StaticResource presenter},
                                Path=StudentBody}">
          ...
          <VisualStateManager.VisualStateGroups>
               <VisualStateGroup x:Name="ApplicationViewStates">

                    <VisualState x:Name="FullScreenLandscape" />

                    <VisualState x:Name="Filled" />

                    <VisualState x:Name="FullScreenPortrait">
                         <Storyboard>
                              <ObjectAnimationUsingKeyFrames Storyboard.TargetName="backButton"
                                             Storyboard.TargetProperty="Style">
                                   <DiscreteObjectKeyFrame KeyTime="0"
                                             Value="{StaticResource PortraitBackButtonStyle}" />
                              </ObjectAnimationUsingKeyFrames>

                              <ObjectAnimationUsingKeyFrames Storyboard.TargetName="gridView"
                                             Storyboard.TargetProperty="Padding">
                                   <DiscreteObjectKeyFrame KeyTime="0" Value="96 0 10 56" />
                              </ObjectAnimationUsingKeyFrames>
                         </Storyboard>
                    </VisualState>

                    <VisualState x:Name="Snapped">
                         <Storyboard>
                              <ObjectAnimationUsingKeyFrames Storyboard.TargetName="gridView"
                                             Storyboard.TargetProperty="Visibility">
                                   <DiscreteObjectKeyFrame KeyTime="0" Value="Collapsed" />
                              </ObjectAnimationUsingKeyFrames>

                              <ObjectAnimationUsingKeyFrames Storyboard.TargetName="listView"
                                             Storyboard.TargetProperty="Visibility">
                                   <DiscreteObjectKeyFrame KeyTime="0" Value="Visible" />
                              </ObjectAnimationUsingKeyFrames>
```

```
                        <ObjectAnimationUsingKeyFrames Storyboard.TargetName="backButton"
                                                       Storyboard.TargetProperty="Style">
                            <DiscreteObjectKeyFrame KeyTime="0"
                                Value="{StaticResource SnappedBackButtonStyle}" />
                        </ObjectAnimationUsingKeyFrames>

                        <ObjectAnimationUsingKeyFrames Storyboard.TargetName="pageTitle"
                                                       Storyboard.TargetProperty="Style">
                            <DiscreteObjectKeyFrame KeyTime="0"
                                Value="{StaticResource SnappedPageHeaderTextStyle}" />
                        </ObjectAnimationUsingKeyFrames>
                    </Storyboard>
                </VisualState>
            </VisualStateGroup>
        </VisualStateManager.VisualStateGroups>
    </Grid>
</Page>
```

除了交换 GridView 和 ListViews 是否可见外，视觉状态管理器节部分还改变按钮和标题样式以及 GridView 填补。

程序正常运行的效果，如下图所示。

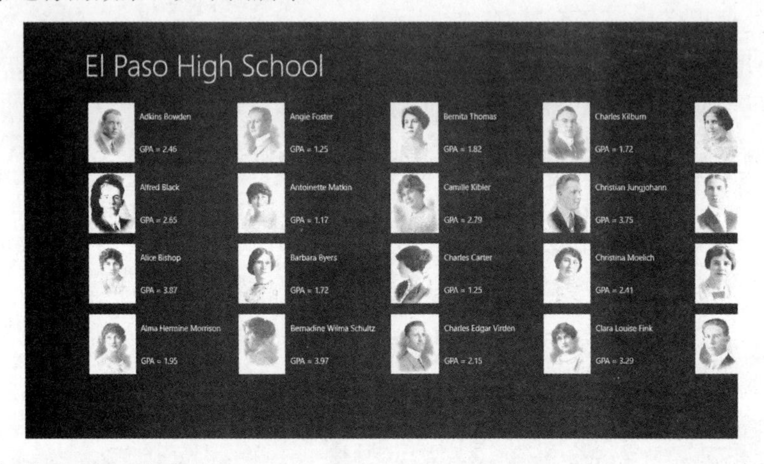

很快就会看出 GPA 变化。

如下图所示，在 Snapped 模式下，程序切换到 ListView，并有一个较小标题。

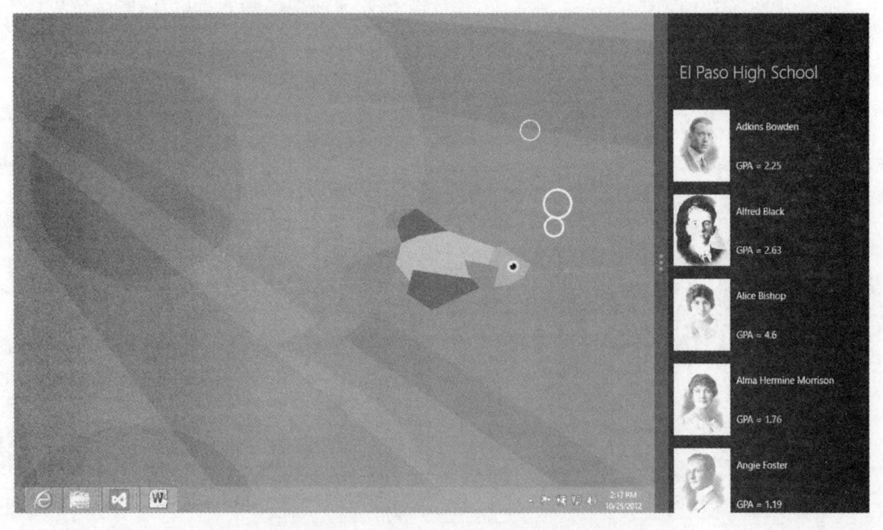

如下图所示，在竖屏模式下，两边的额外空间会稍微缩小一点。

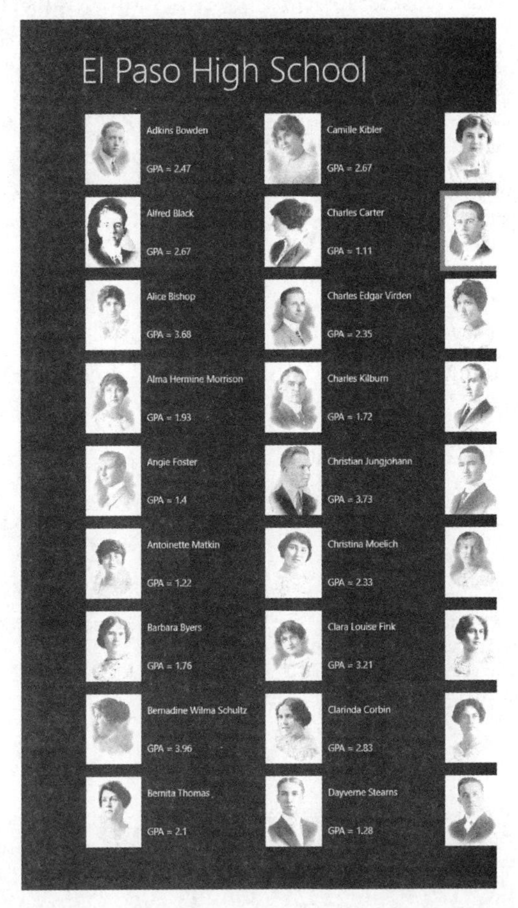

无论任何时候点击某一条目，GridView 或 ListView 会触发 ItemClick 事件。而这会启动到 StudentPage 类型的 Page 派生类的导航，并把事件参数的 ClickedItem 属性传递给 Page 派生类，事件参数为 Student 类型的对象。

项目：DisplayHighSchoolStudents | 文件：MainPage.xaml.cs（片段）

```
public sealed partial class MainPage : Page
{
    ...
    void OnGridViewItemClick(object sender, ItemClickEventArgs args)
    {
        this.Frame.Navigate(typeof(StudentPage), args.ClickedItem);
    }
}
```

在 StudentPage 的 OnNavigatedTo 覆写中，该 Student 对象设置为页面的 DataContext。

项目：DisplayHighSchoolStudents | 文件：StudentPage.xaml.cs（片段）

```
public sealed partial class StudentPage : Page
{
    public StudentPage()
    {
        this.InitializeComponent();
        SizeChanged += OnPageSizeChanged;
    }

    void OnPageSizeChanged(object sender, SizeChangedEventArgs args)
```

```
    {
        VisualStateManager.GoToState(this, ApplicationView.Value.ToString(), true);
    }

    protected override void OnNavigatedTo(NavigationEventArgs args)
    {
        this.DataContext = args.Parameter;
        base.OnNavigatedTo(args);
    }

    void OnBackButtonClick(object sender, RoutedEventArgs args)
    {
        this.Frame.GoBack();
    }
}
```

还要注意对 VisualStateManager.GoToState 的调用以及返回 MainPage 的 Click 处理程序。

StudentPage.xaml 文件显示了 Student 类的一些属性。

项目: DisplayHighSchoolStudents | 文件: StudentPage.xaml(片段)

```
<Page ...
    Name="page"
    FontSize="24">

    <Grid Background="{StaticResource ApplicationPageBackgroundThemeBrush}">
        <Grid.RowDefinitions>
            <RowDefinition Height="140" />
            <RowDefinition Height="*" />
        </Grid.RowDefinitions>

        <Grid Grid.Row="0">
            <Grid.ColumnDefinitions>
                <ColumnDefinition Width="Auto" />
                <ColumnDefinition Width="*" />
            </Grid.ColumnDefinitions>

            <Button Name="backButton"
                    Grid.Column="0"
                    Style="{StaticResource BackButtonStyle}"
                    IsEnabled="{Binding ElementName=page, Path=Frame.CanGoBack}"
                    Click="OnBackButtonClick" />

                <TextBlock Name="pageTitle"
                           Text="{Binding FullName}"
                           Grid.Column="1"
                           Style="{StaticResource PageHeaderTextStyle}" />
            </Grid>

            <StackPanel Grid.Row="1"
                        HorizontalAlignment="Center">
                <Image Source="{Binding PhotoFilename}"
                       Width="240" />

                <TextBlock Text="{Binding Sex}"
                           HorizontalAlignment="Center"
                           Margin="10" />

                <StackPanel Orientation="Horizontal"
                            HorizontalAlignment="Center"
                            Margin="10">
                    <TextBlock Text="GPA = " />
                    <TextBlock Text="{Binding GradePointAverage}" />
                </StackPanel>
            </StackPanel>
```

```
            <VisualStateManager.VisualStateGroups>
                <VisualStateGroup x:Name="ApplicationViewStates">

                    <VisualState x:Name="FullScreenLandscape" />

                    <VisualState x:Name="Filled" />

                    <VisualState x:Name="FullScreenPortrait">
                        <Storyboard>
                          <ObjectAnimationUsingKeyFrames Storyboard.TargetName="backButton"
                                        Storyboard.TargetProperty="Style">
                                <DiscreteObjectKeyFrame KeyTime="0"
                                        Value="{StaticResource PortraitBackButtonStyle}" />
                            </ObjectAnimationUsingKeyFrames>
                        </Storyboard>
                    </VisualState>

                    <VisualState x:Name="Snapped">
                        <Storyboard>
                            <ObjectAnimationUsingKeyFrames Storyboard.TargetName="backButton"
                                        Storyboard.TargetProperty="Style">
                                <DiscreteObjectKeyFrame KeyTime="0"
                                Value="{StaticResource SnappedBackButtonStyle}" />
                            </ObjectAnimationUsingKeyFrames>

                            <ObjectAnimationUsingKeyFrames Storyboard.TargetName="pageTitle"
                                        Storyboard.TargetProperty="Style">
                                <DiscreteObjectKeyFrame KeyTime="0"
                                Value="{StaticResource SnappedPageHeaderTextStyle}" />
                        </ObjectAnimationUsingKeyFrames>
                    </Storyboard>
                </VisualState>
            </VisualStateGroup>
        </VisualStateManager.VisualStateGroups>
    </Grid>
</Page>
```

视觉状态管理器并不像以前那么复杂，因为不再需要切换 GridView 和 ListView。而唯一的真正问题涉及到样式。如下图所示的竖屏模式。

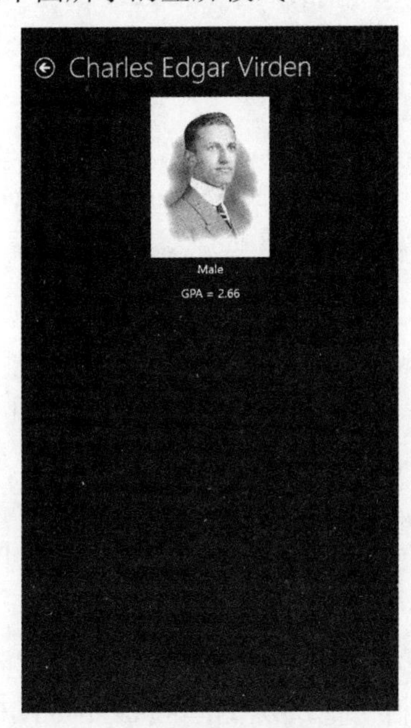

12.13 分组条目

要对 GridView 和 ListView 中的条目进行分组，视图模型需要与组别对应的 ObservableCollection 类型的属性。集合中的条目为类实例，类包括用于识别组的标题以及用于条目自身的 ObservableCollection。结合代理集合类 CollectionViewSource 来使用此视图模型。

StudentBodyPresenter 视图模型无此属性，但是很容易创建一个新类达到此目的。

GroupBySex 项目演示了如何通过男性和女性对学生进行分组。该项目利用几个额外属性对在 ElPasoHighSchool 项目中实施的视图模型进行补充。第一个属性称为 StudentGroup，有两个属性。Title 属性作为组标题，而 Students 属性是 Student 对象的集合。

```
项目: GroupBySex | 文件: StudentGroup.cs
public class StudentGroup : INotifyPropertyChanged
{
    string title;
    ObservableCollection<Student> students = new ObservableCollection<Student>();

    public event PropertyChangedEventHandler PropertyChanged;

    public StudentGroup()
    {
        this.Students = new ObservableCollection<Student>();
    }

    public string Title
    {
        set { SetProperty<string>(ref title, value); }
        get { return title; }
    }

    public ObservableCollection<Student> Students
    {
        set { SetProperty<ObservableCollection<Student>>(ref students, value); }
        get { return students; }
    }

    protected bool SetProperty<T>(ref T storage, T value,
                        [CallerMemberName] string propertyName = null)
    {
        if (object.Equals(storage, value))
            return false;

        storage = value;
        OnPropertyChanged(propertyName);
        return true;
    }

    protected void OnPropertyChanged(string propertyName)
    {
        if (PropertyChanged != null)
            PropertyChanged(this, new PropertyChangedEventArgs(propertyName));
    }
}
```

StudentGroups 类(注意为复数)只有一个可获得的属性称为 Groups，为 StudentGroup 对象的集合。StudentGroups 类还有一个仅设置的 StudentBodyPresenter 类型的 Source 属性，Source 属性构造 StudentGroups 和 StudentGroup 类。我把 Source 作为属性，以便可以在 XAML

进行设置。

项目: GroupBySex | 文件: StudentGroups.cs(片段)

```
public class StudentGroups : INotifyPropertyChanged
{
    StudentBodyPresenter presenter;
    ObservableCollection<StudentGroup> groups = new ObservableCollection<StudentGroup>();

    public event PropertyChangedEventHandler PropertyChanged;
    public StudentBodyPresenter Source
    {
        set
        {
            if (value != null)
            {
                presenter = value;
                presenter.PropertyChanged += OnHighSchoolPropertyChanged;
            }
        }
    }

    void OnHighSchoolPropertyChanged(object sender, PropertyChangedEventArgs args)
    {
        if (args.PropertyName == "StudentBody" && presenter.StudentBody != null)
        {
            this.Groups = new ObservableCollection<StudentGroup>();
            this.Groups.Add(new StudentGroup { Title = "Male" });
            this.Groups.Add(new StudentGroup { Title = "Female" });

            foreach (Student student in presenter.StudentBody.Students)
                if (student.Sex == "Male")
                    this.Groups[0].Students.Add(student);
                else
                    this.Groups[1].Students.Add(student);
        }
    }

    public ObservableCollection<StudentGroup> Groups
    {
        set { SetProperty<ObservableCollection<StudentGroup>>(ref groups, value); }
        get { return groups; }
    }

    protected bool SetProperty<T>(ref T storage, T value,
                         [CallerMemberName] string propertyName = null)
    {
        if (object.Equals(storage, value))
            return false;

        storage = value;
        OnPropertyChanged(propertyName);
        return true;
    }

    protected void OnPropertyChanged(string propertyName)
    {
        if (PropertyChanged != null)
        PropertyChanged(this, new PropertyChangedEventArgs(propertyName));
    }
}
```

如果把 Source 属性设置为 StudentBodyPresenter 的实例，set 访问器则对 PropertyChanged 事件附加处理程序，并等待 StudentBody 属性设置可用。此时，set 访问器会创建两个 StudentGroup 类的实例，并用男性和女性学生进行填充。

为了清楚说明问题，MainPage.xaml 文件我几乎只保留了最基本的必需功能，它只有一个 GridView，而且不会对不同视图改变布局。DataTemplate 几乎没有任何格式化旁白用于显示每个条目，我在代码中未包含该模板，因为和前面的程序相同。

Resources 节包含有助于 GridView 使用集合的三个类。StudentBodyPresenter 类是第一个，和前面的项目一样。接下来，StudentGroups 实例化，其 Source 属性设置为 StudentBodyPresenter 实例。最后，CollectionViewSource(代理集合)将其 Source 属性绑定到 StudentGroups 的 Groups 属性。StudentGroups 对象为 StudentGroup 对象的集合，CollectionViewSource 需要知道来源代表组集合，同时还需要 StudentGroups 类的属性，StudentGroups 类包含实际条目，在本例中为 Students。

项目：GroupBySex | 文件：MainPage.xaml(片段)

```
<Page ... >
    <Page.Resources>
        <elpaso:StudentBodyPresenter x:Key="presenter" />

        <local:StudentGroups x:Key="studentGroups"
                             Source="{StaticResource presenter}" />

        <CollectionViewSource x:Key="collectionView"
                      Source="{Binding Source={StaticResource studentGroups},
                                       Path=Groups}"
                      IsSourceGrouped="True"
                      ItemsPath="Students" />

        <DataTemplate x:Key="studentTemplate">
        ...
        </DataTemplate>
    </Page.Resources>

    <Grid Background="{StaticResource ApplicationPageBackgroundThemeBrush}">
        <GridView ItemsSource="{Binding Source={StaticResource collectionView}}"
                  ItemTemplate="{StaticResource studentTemplate}">

            <!-- The Panel for the groups themselves -->
            <GridView.ItemsPanel>
                <ItemsPanelTemplate>
                    <WrapGrid />
                </ItemsPanelTemplate>
            </GridView.ItemsPanel>

            <GridView.GroupStyle>
                <GroupStyle>
                    <!-- The content of the header -->
                    <GroupStyle.HeaderTemplate>
                        <DataTemplate>
                            <TextBlock Text="{Binding Title}"
                                 Style="{StaticResource GroupHeaderTextStyle}" />
                        </DataTemplate>
                    </GroupStyle.HeaderTemplate>

                    <!-- The panel for the items within each group -->
                    <GroupStyle.Panel>
                        <ItemsPanelTemplate>
                            <VariableSizedWrapGrid Orientation="Vertical" />
                        </ItemsPanelTemplate>
                    </GroupStyle.Panel>
                </GroupStyle>
            </GridView.GroupStyle>
        </GridView>
    </Grid>
</Page>
```

GridView 的 ItemsSource 受限于该 CollectionViewSource，但一些其他属性也在此进行

设置：头部的两个面板及 DataTemplate。两个面板中的第一个(即设置为 ItemsPanel 属性的 WrapGrid)和 GridView 的默认模板一样，这样就不需要标记了。然而，标记有助于明确显示此处的两类面板，一个用于组，而另一个用于每个组里的条目。

滚动后的结果如下图所示，这样可以看到最后面的男孩和最前面的女孩。

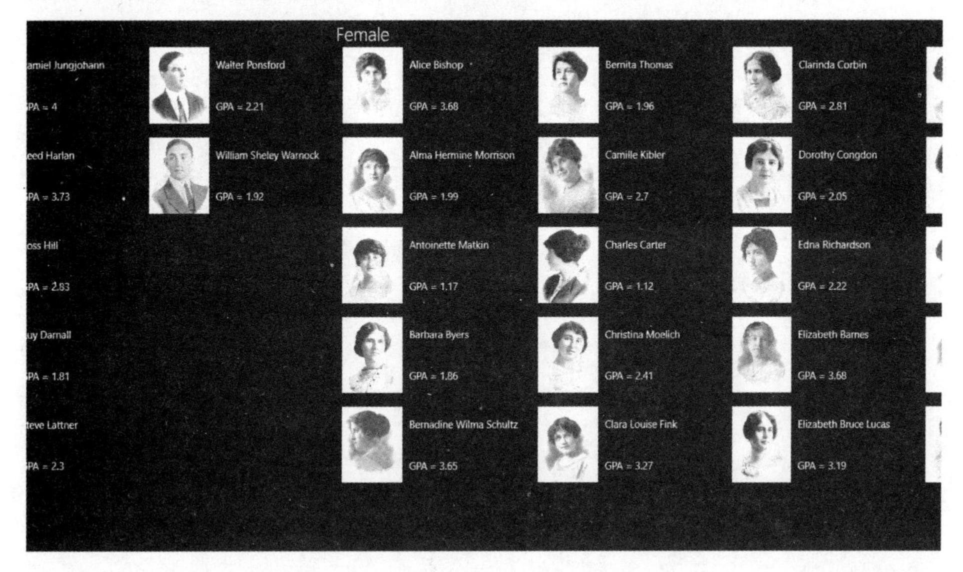

虽然我一直在 XAML 文件中实例化视图模式，但在一般情况下你可能想在多个页面之间分享视图模式。最好在 App 类中实例化视图模式，可以将其作为公共属性供应用的其余部分使用。

第 II 部分
Windows 8 新特性

第13章 触　　控

把触屏、鼠标和手写笔输入三者整合起来，这充分体现了 Windows Runtime 的前瞻性。不再需要为现有的鼠标应用添加触屏支持，也不需要为触屏应用添加支持鼠标。从最初开始，程序员就可以通过相互转换方式来处理所有这些形式的输入。按照 Windows Runtime 编程接口，如果不需要区分实际输入设备，我会使用“指针”一词来指包括触屏、鼠标和手写笔(也称触控笔)在内的所有输入设备。

处理指针输入的最佳方式是通过现有 Windows Runtime 控件。正如你所看到的，标准控件(比如 Button、Slider、ScrollViewer 和 Thumb 等)都会响应指针输入并为应用提供较高级输入。

然而，在某些情况下，程序员需要获取实际指针输入，为此，UIElement 定义了三组不同的事件：

- 以 Pointer 开头的 8 个低级事件
- 以 Manipulation 开头的 5 个高级事件
- Tapped、RightTapped、DoubleTapped 和 Holding 事件

Control 类通过虚拟保护方法来补充这些事件，这些虚拟保护方法都以 On 开头，后面跟着事件名称。

为接受指针输入，FrameworkElement 派生类必须把 isHitTestVisible 属性设置为 true，把 Visibility 属性设置为 Visible。Control 派生类必须把 IsEnabled 属性设置为 true。该元素必须在屏幕上用某种图形来表示；Panel 派生类可以有 Transparent 背景，但不是 null 背景。

所有这些事件都与事件发生时手指、鼠标或手写笔背后的元素有关。唯一例外的是指针被元素“获取”的时候，本章稍后会进行讨论。

如果需要追踪单个手指，则需要使用 Pointer 事件。每个事件都有一个 ID 数字，该编号是触碰屏幕的手指或手写笔的唯一标识，也可以是鼠标或手写笔。本章将演示如何在手工画图程序和钢琴键盘程序(可惜无声音)中使用 Pointer 事件。这两个程序显然都需要处理来自多个手指的同时输入。

从某种意义上说，你只需要 Pointer 事件。例如，如果想实现用户用两个手指来放大/缩小照片的功能，则可以跟踪两个手指的 Pointer 事件，并测量两者间距。而 Manipulation 事件已经提供了这类计算。Manipulation 事件将多个手指整合成单一行为，因此非常适合移动、拉伸、缩小和旋转可视对象。

对有些应用而言，你可能会疑惑到底是用 Pointer 还是用 Manipulation 事件。也许应该首选 Manipulation 事件。特别是如果你以为“我希望用户不要用第二个手指，因为我必须忽略”，肯定就想用 Manipulation 事件。如果用户真的用了两个或更多手指而其实一个手指就足够时，多个手指的使用效果会被平均。

然而，Manipulation 事件存在内在延迟。触碰屏幕时手指需要先动一动，才能确定手指滑动是否有助于操作。如果手指只是敲击或按住屏幕不动，就不会触发 Manipulation 事件。有时，这种延迟会促使你用 Pointer 事件来代替 Manipulation 事件。本章所提到的 XYSlider

自定义控件就是一个好例子。本章中的控件版本是用 Manipulation 事件来写的，因为控件不知道该如何处理多个手指。但延迟时间是一个明显的问题，因此，第 14 章中，我还用 Pointer 事件写了另一个版本。

CoreWindow 对象在窗口级生成 Pointer 事件，可以自己通过 GestureRecognizer 来获取 Manipulation 事件，但本章会忽略这些控件的使用，而是坚持使用由 UIElement 定义的事件以及由 Control 定义的虚拟方法。我不讨论硬件输入设备的有关信息，这些信息可从 Windows.Devices.Input 命名空间的类中获得。

针对涉及选择、擦除和存储笔划以及手写识别，手写笔输入有一些特殊的考虑。这些内容将在第 19 章中进行介绍。微软 Surface 平板电脑不支持手写笔输入。

13.1 Pointer 路线图

在 8 大 Pointer 事件中，其中 5 个很常见。如果用手指触碰一个已激活的可视的 UIElement 派生类、移动，提起，则按以下顺序生成这 5 个 Pointer 事件：

- PointerEntered
- PointerPressed
- PointerMoved (一般会多次出现)
- PointerReleased
- PointerExited

只有当手指触碰屏幕时或刚刚被移开时，Pointer 事件才会产生。并没有所谓的“悬停”触碰。

鼠标则有一些不同。即使没有按鼠标按钮，鼠标也会生成 PointerMoved 事件。假设把鼠标指针移向一个特定元素，按下按钮，移动鼠标，再松开按钮，然后移动鼠标离开该元素。该元素将生成以下一系列事件：

- PointerEntered
- PointerMoved (多个)
- PointerPressed
- PointerMoved (多个)
- PointerReleased
- PointerMoved (多个)
- PointerExited

如果用户按下并释放不同鼠标按钮，也可以生成多个 PointerPressed 和 PointerReleased 事件。

我们试试手写笔。在笔触碰到屏幕之前，元素已经会响应手写笔，因此，首先看到 PointerMoved 事件，然后看到 PointerEntered 事件。当笔触碰屏幕时，会产生 PointerPressed 事件。移动笔，再拿起来。PointerReleased 事件之后，该元素会继续触发 PointerMoved 事件，当笔移动远离屏幕时，通过 PointerExited 元素达到极点。和鼠标触发的事件序列相同。

如果用户旋转鼠标滚轮，则生成下列事件：

- PointerWheelChanged

剩下两个事件较为罕见：

- PointerCaptureLost
- PointerCanceled

本章后面的章节会讨论捕获，那时 PointerCaptureLost 事件更重要。

我还未见过 PointerCanceled 事件，即使在我从电脑上拔下鼠标的时候也没见过，但该事件的存在目的就是报告这类错误。

所有这些事件都伴随着 PointerRoutedEventArgs 实例，由 Windows.UI.Xaml.Input 命名空间进行定义。(注意：Windows.UI.Core 命名空间中还有 PointerEventArgs 类，但用于处理窗口级的指针输入)。正像该类的名称所示，这些 Pointer 事件都是顺着可视树的路由事件。

PointerRoutedEventArgs 定义路由事件的两个共同属性：

- OriginalSource 表明引发事件的元素。
- Handled 可以停止可视树中的下一路由事件。

通过 PointerRoutedEventArgs 对象可以获得大量其他信息。以下说明仅涵盖重点。该类还定义如下成员：

- Pointer 类型的 Pointer 属性
- KeyModifiers 属性，表示 Shift 键、Control 键、Menu 键(或 Alt 键)和 Windows 键的状态
- GetCurrentPoint 方法，返回 PointerPoint 对象

注意：目前我们处理的是 Pointer 类(在 Windows.UI.Xaml.Input 命名空间中进行定义)以及 PointerPoint 类(在 Windows.UI.Input 中进行定义)。

该 Pointer 类有 4 个属性：

- PointerId 属性，识别鼠标，手指或手写笔的无符号整数
- PointerDeviceType，枚举值，为 Touch、Mouse 或 Pen
- IsInRange，布尔值，表示设备是否在屏幕范围内
- IsInContact，布尔值，表示手指或手写笔是否触碰屏幕，或鼠标按钮是否按下

PointerId 属性非常重要。可以用该属性来跟踪各个手指的运动。处理 Pointer 事件的程序几乎可以定义一个字典，而 PointerId 属性在该字典中作为关键字出现。

PointerRoutedEventArgs 的 GetCurrentPoint 方法听起来好像该方法返回指针的当前位置坐标，而且的确也就是这样，除此之外，该方法还提供了更多东西。得到特定元素的位置更方便，因此，GetCurrentPoint 可以接受了 UIElement 类型的参数。该方法返回的 PointerPoint 对象从 Pointer(PointerId 和 IsInContact 属性)复制一些信息，并提供一些其他信息：

- Point 类型的 Position，事件发生时指针的(X，Y)坐标
- ulong 类型的 Timestamp
- Pointer 类型的 PointProperties(在 Windows.UI.Input 中进行定义)

Position 属性总是与传递给 GetCurrentPoint 方法的元素的左上角相关。

PointerRoutedEventArgs 还定义了一个名为 GetIntermediatePoints 的方法，该方法类似于 GetCurrentPoint，只不过返回的是 PointerPoint 对象集合。很多时候，这种集合只有一项(和从 GetCurrentPoint 返回一样返回 PointerPoint 组成)，但对于 PointerMovedevent，返回的

会不止一项，特别是事件处理程序不是很快的时候。我特别注意到一点，在微软 Surface 平板电脑上，GetIntermediatePoints 会返回多个 PointerPoint 对象。

PointerPointProperties 类定义了 22 个属性，主要提供事件相关的详细信息，包括按了鼠标哪个按钮、是否按下笔杆、笔如何倾斜、手指与屏幕的接触矩形(如果有)、手指或手写笔对屏幕的压力(如果有)以及 MouseWheelDelta 等等。

可以根据需要取舍这些信息。当然，其中一些信息并不适用于每个设备，因此会有默认值。

13.2 初 试 手 绘

可以用手指在屏幕绘图上的程序，也许就是典型的多点触屏应用。可以编写这类程序，只处理三个 Pointer 事件，并检查两个来自事件参数的属性，但恐怕结果会留有缺陷，无法补偿其简洁性。

FingerPaint1 的 MainPage.xaml 文件只为标准 Grid 提供命名。

项目：FingerPaint1 | 文件：MainPage.xaml(片段)

```
<Page ... >
    <Grid Name="contentGrid"
          Background="{StaticResource ApplicationPageBackgroundThemeBrush}" />
</Page>
```

该代码隐藏文件做的第一件事情是用 uint 类型的关键字定义 Dictionary。前面提到过，几乎每个处理 Pointer 事件的程序都有这种 Dictionary。Dictionary 的存储项类型依赖于应用，有时应用会专门为此定义类或结构。在简单的手绘应用中，每个手指触碰屏幕都将绘制一条独特 Polyline，使 Dictionary 可以存储 Polyline 实例。

项目：FingerPaint1 | 文件：MainPage.xaml.cs(片段)

```
public sealed partial class MainPage : Page
{
    Dictionary<uint, Polyline> pointerDictionary = new Dictionary<uint, Polyline>();
    Random rand = new Random();
    byte[] rgb = new byte[3];

    public MainPage()
    {
        this.InitializeComponent();
    }

    protected override void OnPointerPressed(PointerRoutedEventArgs args)
    {
        // Get information from event arguments
        uint id = args.Pointer.PointerId;
        Point point = args.GetCurrentPoint(this).Position;

        // Create random color
        rand.NextBytes(rgb);
        Color color = Color.FromArgb(255, rgb[0], rgb[1], rgb[2]);

        // Create Polyline
        Polyline polyline = new Polyline
        {
            Stroke = new SolidColorBrush(color),
            StrokeThickness = 24,
        };
        polyline.Points.Add(point);
```

```
        // Add to Grid
        contentGrid.Children.Add(polyline);

        // Add to dictionary
        pointerDictionary.Add(id, polyline);
        base.OnPointerPressed(args);
    }

    protected override void OnPointerMoved(PointerRoutedEventArgs args)
    {
        // Get information from event arguments
        uint id = args.Pointer.PointerId;
        Point point = args.GetCurrentPoint(this).Position;

        // If ID is in dictionary, add the point to the Polyline
        if (pointerDictionary.ContainsKey(id))
            pointerDictionary[id].Points.Add(point);

        base.OnPointerMoved(args);
    }

    protected override void OnPointerReleased(PointerRoutedEventArgs args)
    {
        // Get information from event arguments
        uint id = args.Pointer.PointerId;

    // If ID is in dictionary, remove it
        if (pointerDictionary.ContainsKey(id))
        pointerDictionary.Remove(id);

    base.OnPointerReleased(args);
    }
}
```

在 OnPointerPressed 覆写中，该程序创建一个 Polyline，并赋予其随机颜色。第一点就是指针的位置。该 Polyline 添加到 Grid 和 Dictionary。

调用后续 OnPointerMoved，PointerId 属性会识别手指，因此，可以通过字典访问与手指相关联的特定 Polyline，同时新的 Point 值也会添加到该 Polyline。由于和 Grid 中的 Polyline 是相同实例，因此在手指移动时，屏幕上的对象会随之变长。

OnPointerReleased 处理程序只移除字典中的条目。同时完成该特定 Polyline。

如果运行该程序，第一件事当然就是整只手扫过屏幕，就像在上纽约州创造手指湖(Finger Lakes)的冰川一样(见下图)。

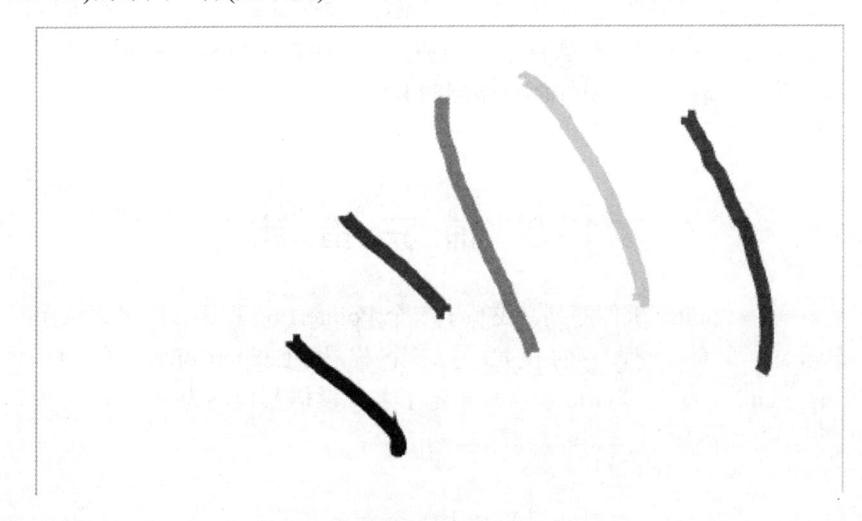

每个手指通过连接点绘出各自的折线，采用单一系列的特定颜色，你会发现，用鼠标和手写笔也可以画。

我之前提到该段代码有缺陷。OnPointerMoved 和 OnPointerReleased 覆盖要仔细检查特定 ID 是否作为关键字存在于字典中，然后再用其访问字典。这对鼠标和手写笔的处理非常重要，因为这些设备会在 OnPointerPressed 之前产生 PointerMoved 事件。

但试试这样做。如下图所示把程序放入快照模式，用手指画一条线，这条线会超出页面，再回到页面。

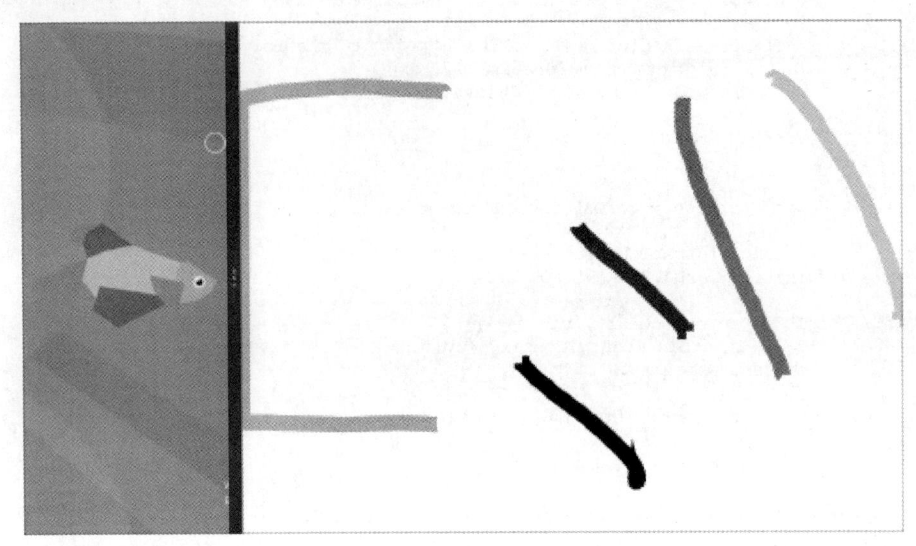

看一看图中页面左侧竖直向下的直线。手指重新进入页面时就画出了这条线，这说明手停留在页面之外时程序没有获得 PointerMoved 事件。用鼠标看看。效果一样。

现在试试这样做：用手指画一条线，从页面内一直延伸到页面外，然后抬起手指。现在，用手指在页面内再画一次。这样做应该可以。

接下来用鼠标试试。在 FingerPaint1 页面上按下鼠标按键，并将鼠标移动到页面之外，然后松开鼠标按钮。现在，移动鼠标再回到 FingerPaint1 页面。虽然已经松开了鼠标按钮，但程序还在继续画线！这显然错了(但我肯定你见过这种“混乱”程序)。如果 OnPointerPressed 方法通过字典中已有的关键字把条目添加到字典中，再按下鼠标按钮就会产生异常。不同于触屏或手写笔，所有的鼠标事件都有相同的 ID。

我们来解决这些问题。

13.3 捕获指针

为了更好理解 Pointer 事件序列，我写了一个 PointerLog 程序，用来记录在屏幕上输入所有的 Pointer 事件。程序的核心是一个称为 LoggerControl 的 UserControl。LoggerControl.xaml 文件中的 Grid 已经被指定了名称，但其他均为空：

```
项目：PointerLog | 文件：LoggerControl.xaml(片段)
<UserControl ... >

    <Grid Name="contentGrid" Background="Transparent" />
```

```
</UserControl>
```

该代码隐藏文件覆写了全部 8 个 Pointer 方法，所有方法都调用一个名为 Log 的方法，并含事件名称和事件参数。和所有 Pointer 程序一样，定义 Dictionary，但其中的值并不是简单的对象。为了用于存储每个手指信息，我在 LoggerControl 开头定义了一个名为 PointerInfo 的类。

项目：PointerLog | 文件：LoggerControl.xaml.cs(片段)

```
public sealed partial class LoggerControl : UserControl
{
    class PointerInfo
    {
        public StackPanel stackPanel;
        public string repeatEvent;
        public TextBlock repeatTextBlock;
    };

    Dictionary<uint, PointerInfo> pointerDictionary = new Dictionary<uint, PointerInfo>();

    public LoggerControl()
    {
        this.InitializeComponent();
    }

    public bool CaptureOnPress { set; get; }

    protected override void OnPointerEntered(PointerRoutedEventArgs args)
    {
        Log("Entered", args);
        base.OnPointerEntered(args);
    }

    protected override void OnPointerPressed(PointerRoutedEventArgs args)
    {
        if (this.CaptureOnPress)
            CapturePointer(args.Pointer);

        Log("Pressed", args);
        base.OnPointerPressed(args);
    }
    protected override void OnPointerMoved(PointerRoutedEventArgs args)
    {
        Log("Moved", args);
        base.OnPointerMoved(args);
    }

    protected override void OnPointerReleased(PointerRoutedEventArgs args)
    {
        Log("Released", args);
        base.OnPointerReleased(args);
    }

    protected override void OnPointerExited(PointerRoutedEventArgs args)
    {
        Log("Exited", args);
        base.OnPointerExited(args);
    }

    protected override void OnPointerCaptureLost(PointerRoutedEventArgs args)
    {
        Log("CaptureLost", args);
        base.OnPointerCaptureLost(args);
    }
```

```
protected override void OnPointerCanceled(PointerRoutedEventArgs args)
{
    Log("Canceled", args);
    base.OnPointerCanceled(args);
}

protected override void OnPointerWheelChanged(PointerRoutedEventArgs args)
{
    Log("WheelChanged", args);
    base.OnPointerWheelChanged(args);
}

void Log(string eventName, PointerRoutedEventArgs args)
{
    uint id = args.Pointer.PointerId;
    PointerInfo pointerInfo;

    if (pointerDictionary.ContainsKey(id))
    {
        pointerInfo = pointerDictionary[id];
    }
    else
    {
        // New ID, so new StackPanel and header
        TextBlock header = new TextBlock
        {
            Text = args.Pointer.PointerId + " - " + args.Pointer.PointerDeviceType,
            FontWeight = FontWeights.Bold
        };
        StackPanel stackPanel = new StackPanel();
        stackPanel.Children.Add(header);

        // New PointerInfo for dictionary
        pointerInfo = new PointerInfo
        {
            stackPanel = stackPanel
        };
        pointerDictionary.Add(id, pointerInfo);

        // New column in the Grid for the StackPanel
        ColumnDefinition coldef = new ColumnDefinition
        {
            Width = new GridLength(1, GridUnitType.Star)
        };
        contentGrid.ColumnDefinitions.Add(coldef);
        Grid.SetColumn(stackPanel, contentGrid.ColumnDefinitions.Count - 1);
        contentGrid.Children.Add(stackPanel);
    }

    // Don't repeat PointerMoved and PointerWheelChanged events
    TextBlock txtblk = null;

    If (eventName == pointerInfo.repeatEvent)
    {
        txtblk = pointerInfo.repeatTextBlock;
    }
    else
    {
        txtblk = new TextBlock();
        pointerInfo.stackPanel.Children.Add(txtblk);
    }

    txtblk.Text = eventName + " ";

    if (eventName == "WheelChanged")
    {
        txtblk.Text += args.GetCurrentPoint(this).Properties.MouseWheelDelta;
    }
```

```
            else
            {
                txtblk.Text += args.GetCurrentPoint(this).Position;
            }

            txtblk.Text += args.Pointer.IsInContact ? " C" : "";
            txtblk.Text += args.Pointer.IsInRange ? " R" : "";

            if  (eventName == "Moved" || eventName == "WheelChanged")
            {
                pointerInfo.repeatEvent = eventName;
                pointerInfo.repeatTextBlock = txtblk;
            }
            else
            {
                pointerInfo.repeatEvent = null;
                pointerInfo.repeatTextBlock = null;
            }
        }

        public void Clear()
        {
            contentGrid.ColumnDefinitions.Clear();
            contentGrid.Children.Clear();
            pointerDictionary.Clear();
        }
    }
```

Log 方法似乎很复杂，但每次碰到事件参数中新的 PointerId 值时，就会给 Grid 新增一列，在顶部放置 TextBlock，用来表明 ID 和设备类型，并把条目添加到字典。而该 ID 的所有后续事件都会存放进该列，但连续的 PointerMoved 和 PointerWheelChanged 事件不会获得额外条目。不能滚动，而且最终有很多列，但是公共 Clear 方法会把一切都恢复到原始状态。

对于该控件， LoggerControl 只得到 Pointer 事件。为了缓解手指在控件之间移动时产生的检查，我做了一个 LoggerControl，页面顶部为程序名称，而底部是三个按钮。

```
项目: PointerLog | 文件: MainPage.xaml(片段)
<Grid Background="{StaticResource ApplicationPageBackgroundThemeBrush}">
    <Grid.RowDefinitions>
        <RowDefinition Height="Auto" />
        <RowDefinition Height="*" />
        <RowDefinition Height="Auto" />
    </Grid.RowDefinitions>

    <TextBlock Text="Pointer Event Log"
               Grid.Row="0"
               Style="{StaticResource HeaderTextStyle}"
               HorizontalAlignment="Center"
               Margin="12" />

    <local:LoggerControl x:Name="logger"
                         Grid.Row="1"
                         FontSize="{StaticResource ControlContentThemeFontSize}" />

    <Grid Grid.Row="2">
        <Grid.ColumnDefinitions>
            <ColumnDefinition Width="*" />
            <ColumnDefinition Width="*" />
            <ColumnDefinition Width="*" />
        </Grid.ColumnDefinitions>

        <Button Content="Clear"
            Grid.Column="0"
```

```
                HorizontalAlignment="Center"
                Click="OnClearButtonClick" />

        <ToggleButton Name="captureButton"
                      Content="Capture on Press"
                      Grid.Column="1"
                      HorizontalAlignment="Center"
                      Checked="OnCaptureToggleButtonChecked"
                      Unchecked="OnCaptureToggleButtonChecked" />

        <Button Content="Release Captures in 5 seconds"
                Grid.Column="2"
                IsEnabled="{Binding ElementName=captureButton, Path=IsChecked}"
                HorizontalAlignment="Center"
                Click="OnReleaseCapturesButtonClick" />
    </Grid>
</Grid>
```

请注意，只有触发 ToggleButton 时，最后一个 Button 才可用。

代码隐藏文件只处理按钮(稍后讨论)。

项目：PointerLog | 文件：MainPage.xaml.cs(片段)

```
public sealed partial class MainPage : Page
{
    DispatcherTimer timer;

    public MainPage()
    {
        this.InitializeComponent();
        timer = new DispatcherTimer { Interval = TimeSpan.FromSeconds(5) };
        timer.Tick += OnTimerTick;
    }

    void OnClearButtonClick(object sender, RoutedEventArgs args)
    {
        logger.Clear();
    }

    void OnCaptureToggleButtonChecked(object sender, RoutedEventArgs args)
    {
        ToggleButton toggle = sender as ToggleButton;
        logger.CaptureOnPress = toggle.IsChecked.Value;
    }

    void OnReleaseCapturesButtonClick(object sender, RoutedEventArgs args)
    {
        timer.Start();
    }

    void OnTimerTick(object sender, object args)
    {
        logger.ReleasePointerCaptures();
        timer.Stop();
    }
}
```

从下图所示的屏幕上可以看到，每次手指按压屏幕都会得到唯一的 ID，并且仅生成 5 个事件。每个新的系列手写笔事件(使用相同编号顺序触屏)也都会得到各自的 ID 和其他几个事件。鼠标的 ID 总是 1。

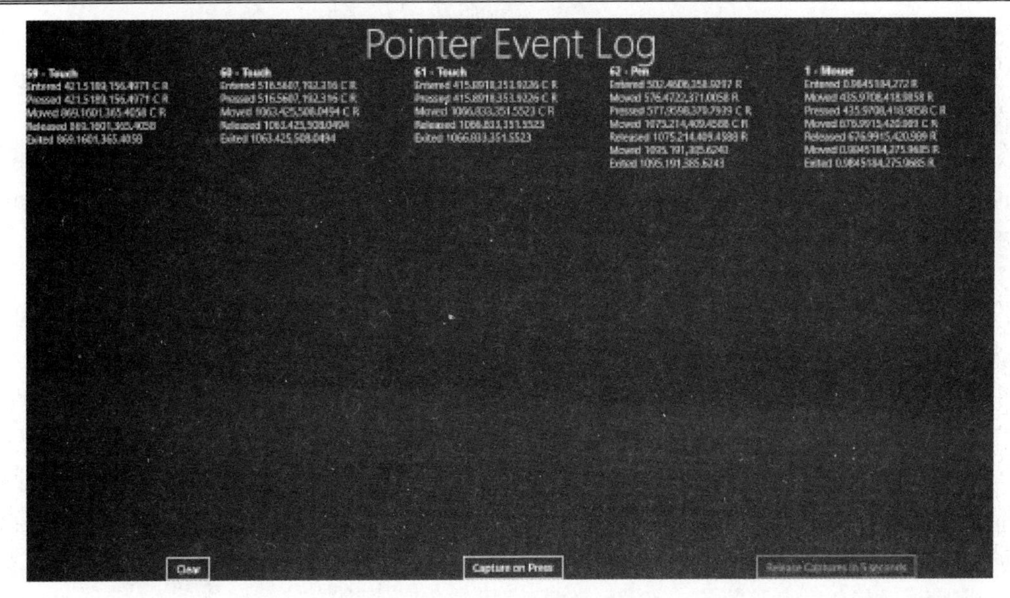

字母 C 和 R 表示 Pointer 对象的 IsInContact 和 IsInRange 属性的 true 值。正如你看到的，手写笔触碰屏幕或按下鼠标按钮时，可以通过 IsInRange 属性来区别将要发生的 PointerMoved 事件。

在默认情况下，只有指针在元素的边界之内时，元素才能获得 Pointer 输入。而这有时可能会导致信息丢失。为了方便演示，我有意设计一个程序，LoggerControl 不会延伸到整个屏幕高度。上方区域是程序标题，而下方是按钮区域。这些地方是 MainPage 领域。这种配置可以让你实验从一个元素到另一个元素的输入。

例如，触碰 PointerLog 屏幕的中间某个地方，手指四处移动，然后将手指移动到顶部标题区域或底部按钮区域，之后手指离开屏幕。程序不会收到 PointerReleased 事件，并且也不知道已释放指针。程序永远不会得到另一个特定 ID 号码的事件，而仍处在无知状态。字典中的条目永远也不会删除。

与之类似，触碰屏幕的顶部或底部区域，然后手指移动到中心区。程序会注册 PointerEntered 和 PointerMoved 事件，而不是 PointerPressed 事件。

跟踪特定指针往往需要继续获取输入，即使指针漂移到元素之外。得不到指针输入就说明 FingerPaint1 程序有缺陷。

通过调用由 UIElement 定义的 CapturePointer 方法，就可以得到你想要的称为“捕捉指针”过程。该方法带 Pointer 类型参数，并返回布尔值，以表明指针捕获是否成功。什么时候不成功？如果在 PointerPressed 之前、之间或者之后调用 CapturePointer，就不成功。

正是出于该原因(同时出于程序的优雅性)只有在 PointerPressed 事件执行过程中调用 CapturePointer 才真正有意义。通过在特定元素上按下手指(手写笔或鼠标按钮)，就表示用户希望与该元素进行交互，即使手指有时移到元素以外的其他地方。

如果手指按压 PointerLog 屏幕底部的 Capture on Press 按钮，程序就会在 OnPointerPressed 覆写期间调用如下指令：

```
CapturePointer(args.Pointer);。
```

现在，如果按压 PointerLog 程序的中心区域，把手指移到顶部或底部，然后放开，程序会记录 PointerReleased 事件、PointerExited 以及之后的最终 PointerCaptureLost 事件。

程序可以得到所有调用 PointerCaptures 捕获指针，并通过调用 ReleasePointerCapture 释放特定捕获，或者通过多个 ReleasePointerCaptures 释放所有指针捕获。

在实际应用中，很容易直接忽略 PointerCaptureLost 事件，但这可不是好事。如果 Windows 有一些紧急情况需要和用户沟通，可以从程序直接拿走指针捕获。我在 Windows 8 中还未见过这种情况，但过去是通过系统模态对话框来显示——该对话框非常重要，要先得到用户的所有输入，再获得释放。

为了演示这种情况，我定义第三个按钮把 DispatcherTimer 设置为五秒钟，然后通过调用 ReleasePointerCaptures 给 LoggerControl 得到结果。发生这种情况时，已被捕获的指针就会触发 PointerCaptureLost 事件。如果指针仍然在元素之上，就会继续接收其他指针事件，但如果移到元素之外，则无法接收。

如果应用收到意外的 PointerCaptureLost，此时怎么办则取决于应用。例如，对于手绘程序，你可能想把 PointerReleased 逻辑移入 PointerCaptureLost，并同等对待意料及非意料的捕获损失。

或者说，完全丢弃该特定的绘图事件，这可能也是合理的。

事实上，可能需要在程序中添加这项功能。比如你决定用户应该能按 Esc 键来停止正在运行的绘图事件。为此，你可以通过直接调用 ReleasePointerCaptures。

以下 FingerPaint2 程序完成的就是这项功能。XAML 文件和 FingerPaint1 相同，因此代码隐藏文件也相同，但以下除外。

```
项目: FingerPaint2 | 文件: MainPage.xaml.cs(片段)
public sealed partial class MainPage : Page
{
    ...
    public MainPage()
    {
        this.InitializeComponent();
this.IsTabStop = true;
    }

    protected override void OnPointerPressed(PointerRoutedEventArgs args)
{
        ...
        // Capture the Pointer
        CapturePointer(args.Pointer);

        // Set input focus
        Focus(FocusState.Programmatic);

        base.OnPointerPressed(args);
    }
    ...
    protected override void OnPointerCaptureLost(PointerRoutedEventArgs args)
    {
        // Get information from event arguments
        uint id = args.Pointer.PointerId;

        // If ID is in dictionary, abandon the drawing operation
        if (pointerDictionary.ContainsKey(id))
        {
            contentGrid.Children.Remove(pointerDictionary[id]);
            pointerDictionary.Remove(id);
        }

        base.OnPointerCaptureLost(args);
```

```
        }

        protected override void OnKeyDown(KeyRoutedEventArgs args)
        {
            if (args.Key == VirtualKey.Escape)
                ReleasePointerCaptures();

            base.OnKeyDown(args);
        }
    }
```

在构造函数中，为了能让元素接收键盘输入，必须把 IsTabStop 属性设置为 true。任何时间，只有一个元素可以接收键盘输入。这就是所谓键盘“焦点”元素，而且有一些控件通过特殊外观来表明包含了键盘焦点，例如虚线。元素被触碰时或者(在这种情况下)在 OnPointerPressed 事件期间，往往可以通过调用 Focus 方法来给元素自己赋予键盘焦点。该覆写方法通过调用 Focus 方法和 CapturePointer 来结束处理。

OnPointerCaptureLost 方法从 Grid 中删除 Polyline 并从字典中删除 ID。然而，手指离开屏幕之后，PointerCaptureLost 事件会正常发生，因此，只要页面没有调用 OnPointerReleased，ID 就还会在字典中。

OnKeyDown 方法得到按键，因而为 Esc 键调用 ReleasePointerCaptures 了。如果没有捕获到指针，调用就没有任何效果。

试试 FingerPaint1 所定义的问题行为，你会发现该版本不存在这个问题了。另外，可以在屏幕上绘图并按下 Esc 键，正在画的东西都会消失，而手指不会起任何作用，直到松开并再次按下。(希望你就是想要这个效果)。

13.4 编辑弹出菜单

我们接下来给程序增加编辑功能。如果鼠标右键点击现有 Polyline(或做一些相当于用手指或手写笔做的事情)就会弹出小菜单，并出现 Change Color 和 Delete 这两个选项。

前两个 FingerPaint 程序中，如下创建、初始化并添加 Polyline 到 Grid，触碰字典：

```
// Create Polyline
Polyline polyline = new Polyline
{
    Stroke = new SolidColorBrush(color),
    StrokeThickness = 24,
};
polyline.Points.Add(point);

// Add to Grid
contentGrid.Children.Add(polyline);

// Add to dictionary
pointerDictionary.Add(id, polyline);
```

对于 FingerPaint3，我们来添加一些额外的代码，设置 Polyline 两个事件的处理程序。目标是通过处理程序为 Polyline 的 RightTapped 事件显示弹出菜单。

```
项目: FingerPaint3 | 文件: MainPage.xaml.cs(片段)
protected override void OnPointerPressed(PointerRoutedEventArgs args)
{
    ...
    // Create Polyline
```

```
        Polyline polyline = new Polyline
        {
            Stroke = new SolidColorBrush(color),
            StrokeThickness = 24,
        };
        polyline.PointerPressed += OnPolylinePointerPressed;
        polyline.RightTapped += OnPolylineRightTapped;
        polyline.Points.Add(point);
        ...
    }
```

虽然我们只关心 Polyline 的 RightTapped 事件，但我也为 PointerPressed 事件设置了处理程序。该处理程序虽然没有什么意思，但非常重要。

项目：FingerPaint3 | 文件：MainPage.xaml.cs(片段)

```
void OnPolylinePointerPressed(object sender, PointerRoutedEventArgs args)
{
    args.Handled = true;
}
```

你一定想试试这个不带特定处理程序的程序，因为如果触发 PointerPressed 事件，该事件就会与最上方已启用用户输入功能的元素相关联。如果点击或右键点击 Polyline 而不是 Mainpage 的表面，就会触发该 Polyline 的 PointerPressed 事件。

然而，PointerPressed 是路由事件，第 3 章曾经讲过，路由事件会顺着可视树“走”，也就是说如果 Polyline 对该事件没有兴趣，则前往 Mainpage，即假定你想画新图形。为了防止程序中发生这种情况，Polyline 通过把事件参数中的 Handled 属性设置为 true 来处理 PointerPressed 事件。这样可以防止事件回到 Mainpage。

这个弹出菜单逻辑发生在 RightTapped 事件中。

项目：FingerPaint3 | 文件：MainPage.xaml.cs(片段)

```
async void OnPolylineRightTapped(object sender, RightTappedRoutedEventArgs args)
{
    Polyline polyline = sender as Polyline;
    PopupMenu popupMenu = new PopupMenu();
    popupMenu.Commands.Add(new UICommand("Change color", OnMenuChangeColor, polyline));
    popupMenu.Commands.Add(new UICommand("Delete", OnMenuDelete, polyline));
    await popupMenu.ShowAsync(args.GetPosition(this));
}
```

正如第 8 章所演示，PopupMenu 很容易使用。创建对象后，最多可以为菜单添加 6 个项目。每个项目包含一个文本标签、回调以及一个帮助回调识别事件的可选对象。ShowAsync 方法把菜单显示在特定位置。

通过加载回调方法的 IUICommand 参数的 Id 属性值，处理程序可以获取传递给 UICommand 构造函数的最终参数。

项目：FingerPaint3 | 文件：MainPage.xaml.cs(片段)

```
void OnMenuChangeColor(IUICommand command)
{
    Polyline polyline = command.Id as Polyline;
    rand.NextBytes(rgb);
    Color color = Color.FromArgb(255, rgb[0], rgb[1], rgb[2]);
    (polyline.Stroke as SolidColorBrush).Color = color;
}

void OnMenuDelete(IUICommand command)
{
    Polyline polyline = command.Id as Polyline;
    contentGrid.Children.Remove(polyline);
}
```

我相信你明白如何用鼠标右键单击 Polyline。触碰时，需要将手指稳定放在 Polyline 上一会儿，然后再松开。如果停留时间足够，就会出现一个正方形。同样，用手写笔的时候要停留一会，直到看到一个圆形之后再松开。出现如下图所示的菜单。

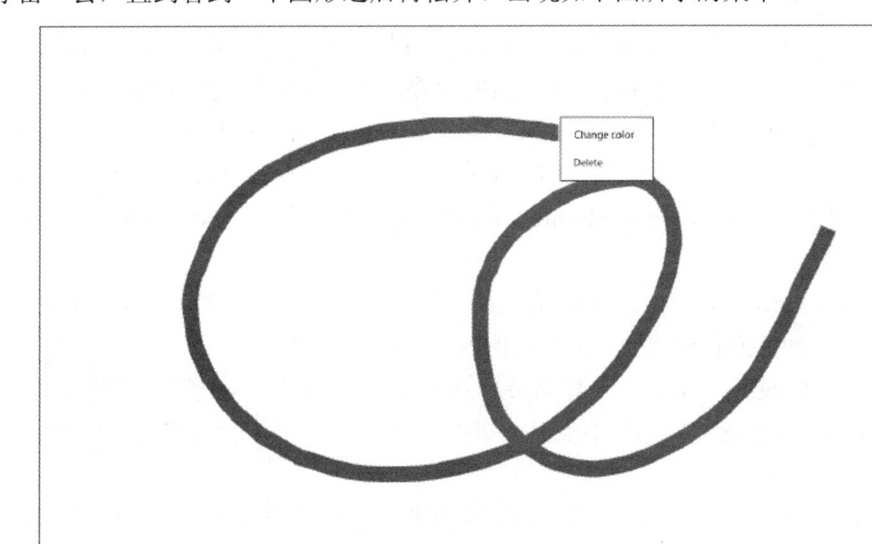

当你的手指或手写笔停留在屏幕上时，你所看见的方形和圆形实际上都和 Holding 事件相关联。如果把 Polyline 的 IsHoldingEnabled 属性设置为 false，就不会出现矩形和圆，用户也可能不确定需要按多长时间。在用户从屏幕上拿起手指或手写笔之前，都不会触发 RightTapped 事件。

FingerPaint3 中的 OnMenuDelete 方法实际上有一个小错误。如果手指画一条线，而另一个手指激活那条线的菜单，会使 OnMenuDelete 从屏幕上删除 Polyline，而不会移除带 Polyline 的字典条目。虽然这样不会有坏结果，但会导致字典中积累一些废弃条目。解决该问题的逻辑必须能够搜索字典中被删除的 Polyline，然后删除。

正如第 3 章所演示的路由事件，无论什么时候处理不同元素产生的事件，都可以用各种方法来进行事件处理。例如，在 Mainpage 上 OnPointerPressed 覆写可以整合我放在 OnPolylinePointerPressed 中的逻辑，你可以在 OnRightTapped 覆写中执行所有 RightTapped 处理。你要做的是检查事件参数的 OriginalSource 属性，以确定输入是来自 Polyline 还是 MainPage。

该程序有一个小缺点就是无法从现有线上的某一点开始画另一条线。Polyline 收到的任何 PointerPressed 事件都会被标记为 Handled，并丢弃。

如果想给用户两种选择，怎么办？如果用户按住一个现有 Polyline 并开始移动，则出现新图形。如果用户按下并保持，则出现菜单。

最简单的方法可能是放弃使用 RightTapped 事件，而通过 Pointer 逻辑来处理。如果 OnPointerPressed 发生于现有 Polyline，则可以把 DispatcherTimer 设置为 1 秒钟，但如果 OnPointerMoved 发生了，即表明手指移动的距离大于预先设定的标准，则取消该计时器(并开始绘画操作)。如果定时器触发，则显示菜单。

13.5 压力灵敏度

通过各个 FingerPaint 程序来绘制的线条都是一种粗细笔画(准确说是 24 像素)，但一些触控设备能区分触碰的轻重，真正好的 FingerPaint 程序要能响应笔划的粗细轻重。

手绘程序中有两个属性会影响线条粗细，两者都由 PointerPointProperties 对象所定义，而 PointerPointProperties 对象则是由 PointerPoint 类的 Properties 属性返回所得(而 Properties 属性又是通过调用 PointerRoutedEventArgs 事件参数的 GetCurrentPoint 方法所得)。

第一个属性是 ContactRect，为 Rect 值，用来显示屏幕上的手指(或手写笔点)接触区域的矩形边框。该属性可能只适用于十分复杂的触控设备。而对本书中我一直使用的平板电脑，无论是什么指针设备，Rect 总是有 Width 和 Height 的 0 值。对于第一代微软 Surface 平板电脑，Width 和 Height 值都是小整数，如 1、2 和 3，似乎不会有更大值。(但我可能错了。)

第二个属性是 Pressure，为 float 值，范围在 0 和 1 之间。我在本书中一直使用的平板电脑，手指和鼠标的 Pressure 默认值是 0.5，但对于手写笔，该值可变，因此，我能有机会试一试。(在第一代微软 Surface 平板电脑中，Pressure 值始终为 0.5。)

为简单起见，FingerPaint4 程序不包括 Esc 键处理和编辑功能，但实现了指针捕获。最大的区别在于必须放弃通过 Polyline 方法画图，因为 Polyline 只有一个 StrokeThickness 属性。在新程序中，每个笔划都必须由颜色相同的单一的短线条构成，而且每个独特 StrokeThickness 都通过 Pressure 值计算而得到，但颜色都相同。也就是说，字典需要包含 Color 类型的值(或更好有 Brush 类型)以及前一个 Point 的值。现在有了它们，来定义我所称的 PointerInfo 的自定义结构。

项目：FingerPaint4 | 文件：MainPage.xaml.cs(片段)

```
public sealed partial class MainPage : Page
{
    struct PointerInfo
    {
        public Brush Brush;
        public Point PreviousPoint;
    }

    Dictionary<uint, PointerInfo> pointerDictionary = new Dictionary<uint, PointerInfo>();
    Random rand = new Random();
    byte[] rgb = new byte[3];

    public MainPage()
    {
        this.InitializeComponent();
    }

    protected override void OnPointerPressed(PointerRoutedEventArgs args)
    {
        // Get information from event arguments
        uint id = args.Pointer.PointerId;
        Point point = args.GetCurrentPoint(this).Position;

        // Create random color
```

```
            rand.NextBytes(rgb);
            Color color = Color.FromArgb(255, rgb[0], rgb[1], rgb[2]);

            // Create PointerInfo
            PointerInfo pointerInfo = new PointerInfo
            {
                PreviousPoint = point,
                Brush = new SolidColorBrush(color)
            };

            // Add to dictionary
            pointerDictionary.Add(id, pointerInfo);

            // Capture the Pointer
            CapturePointer(args.Pointer);

            base.OnPointerPressed(args);
        }

        ...

        protected override void OnPointerReleased(PointerRoutedEventArgs args)
        {
            // Get information from event arguments
            uint id = args.Pointer.PointerId;

            // If ID is in dictionary, remove it
            if (pointerDictionary.ContainsKey(id))
                pointerDictionary.Remove(id);

            base.OnPointerReleased(args);
        }

        protected override void OnPointerCaptureLost(PointerRoutedEventArgs args)
        {
            // Get information from event arguments
            uint id = args.Pointer.PointerId;

            // If ID is still in dictionary, remove it
            if (pointerDictionary.ContainsKey(id))
                pointerDictionary.Remove(id);

            base.OnPointerCaptureLost(args);
        }
    }
```

以前的 PointerPressed 处理程序创建了 Polyline，赋予其初始点，并将其添加到 Grid 和 Dictionary 中。而在该程序中，它只创建 PointerInfo 值，并添加到字典中。更多功能实现在 PointerMoved 处理程序中，因为我决定用 GetIntermediatePoints，而不使用 GetCurrentPoint，结果(至少在理论上)在微软 Surface 平板上的笔触更流畅。但我觉得有点奇怪，这些点都以相反顺序集合在一起。

以下代码通过这些点来进行循环。对于每个新点和前一个点，会构造 Line 元素并添加到 Grid 中。在 PointerInfo 值中，最后一个点取代前一个点。

项目：FingerPaint4 | 文件：MainPage.xaml.cs(片段)

```
protected override void OnPointerMoved(PointerRoutedEventArgs args)
{
    // Get ID from event arguments
    uint id = args.Pointer.PointerId;

    // If ID is in dictionary, start a loop
    if (pointerDictionary.ContainsKey(id))
    {
```

```
            PointerInfo pointerInfo = pointerDictionary[id];
            IList<PointerPoint> pointerpoints = args.GetIntermediatePoints(this);

            for (int i = pointerpoints.Count - 1; i >= 0; i--)
            {
                PointerPoint pointerPoint = pointerpoints[i];

                // For each point, create a new Line element and add to Grid
                Point point = pointerPoint.Position;
                float pressure = pointerPoint.Properties.Pressure;

                Line line = new Line
                {
                    X1 = pointerInfo.PreviousPoint.X,
                    Y1 = pointerInfo.PreviousPoint.Y,
                    X2 = point.X,
                    Y2 = point.Y,
                    Stroke = pointerInfo.Brush,
                    StrokeThickness = pressure * 24,
                    StrokeStartLineCap = PenLineCap.Round,
                    StrokeEndLineCap = PenLineCap.Round
                };
                contentGrid.Children.Add(line);

                // Update PointerInfo
                pointerInfo.PreviousPoint = point;
            }
            // Store PointerInfo back in dictionary
            pointerDictionary[id] = pointerInfo;
        }
        base.OnPointerMoved(args);
    }
```

注意，StrokeThickness 设置为 Pressure 值的 24 倍。其结果是线条最大为 24，而对于非压力敏感设备线条为 12。还要注意，StrokeStartLineCap 和 StrokeEndLineCap 属性设置为 Round。试试注释掉这些属性设置，看看线条突然改变时会发生什么情况。两条短线之间有角度，因此会出现小空白。而线帽会盖住这些空白。

下图这件小艺术品是我完全用手写笔完成的。

请注意，如果使用压感输入设备进行渲染，笔划会优美精妙得多。

就我的经验而言，PointerMoved 事件每秒可触发 100 次，这比视频显示器的帧速率快，但还是没有极端有力的手指快。

13.6　平滑锥度

你有没有发现，解决了一个问题之后常常又会引出另一个问题？压力灵敏度是手绘程序的一个重要特点。然而，如果在 FingerPaint4 中如果用压力感应笔快速画东西，你可能会注意到线条好像没有产生正确的锥度。相反，锥度会随着不连续跳动而增加或减少(见下图)。

当然会这样。曲线右下部分的每个片段都是一个自带 StrokeThickness 的 Line 元素。我用这种速度画曲线时压力变化很大，因此造成了线条不连续跳跃。

如果你认为特定 Line 元素只能有一个恒定的 StrokeThickness，那么解决这个问题就会非常困难。 但解决方案其实很简单(至少在概念上)：不要为每个事件画一个 Line，而是画一个已填充的 Path，该 Path 由两条线所连接的两个不同半径的圆弧组成。

为了更容易一点，可以考虑 Vector 结构，除了 Windows Runtime，所有现代操作系统都有它。以下结构，我称之为 Vector2(“2”指两个维度)，它只是第 14 章中一个更大的库的一部分。因此，采用长命名空间名称。

```
项目: FingerPaint5 | 文件: Vector2.cs
using System;
using Windows.Foundation;
using Windows.UI.Xaml.Media;

namespace Petzold.Windows8.VectorDrawing
{
    public struct Vector2
    {
        // Constructors
        public Vector2(double x, double y)
            : this()
        {
            X = x;
            Y = y;
        }

        public Vector2(Point p)
            : this()
        {
            X = p.X;
```

```
        Y = p.Y;
    }

    public Vector2(double angle)
        : this()
    {
        X = Math.Cos(Math.PI * angle / 180);
        Y = Math.Sin(Math.PI * angle / 180);
    }

    // Properties
    public double X { private set; get; }
    public double Y { private set; get; }

    public double LengthSquared
    {
        get { return X * X + Y * Y; }
    }

    public double Length
    {
        get { return Math.Sqrt(LengthSquared); }
    }

    public Vector2 Normalized
    {
        get
        {
            double length = this.Length;

            if (length != 0)
            {
                return new Vector2(this.X / length,
                                   this.Y / length);
            }
            return new Vector2();
        }
    }

    // Methods
    public Vector2 Rotate(double angle)
    {
        RotateTransform xform = new RotateTransform { Angle = angle };
        Point pt = xform.TransformPoint(new Point(X, Y));
        return new Vector2(pt.X, pt.Y);
    }

    // Static methods
    public static double AngleBetween(Vector2 v1, Vector2 v2)
    {
        return 180 * (Math.Atan2(v2.Y, v2.X) - Math.Atan2(v1.Y, v1.X)) / Math.PI;
    }

    // Operators
    public static Vector2 operator +(Vector2 v1, Vector2 v2)
    {
        return new Vector2(v1.X + v2.X, v1.Y + v2.Y);
    }

    public static Point operator +(Vector2 v, Point p)
    {
        return new Point(v.X + p.X, v.Y + p.Y);
    }

    public static Point operator +(Point p, Vector2 v)
    {
        return new Point(v.X + p.X, v.Y + p.Y);
    }
```

```
        public static Vector2 operator -(Vector2 v1, Vector2 v2)
        {
            return new Vector2(v1.X - v2.X, v1.Y - v2.Y);
        }

        public static Point operator -(Point p, Vector2 v)
        {
            return new Point(p.X - v.X, p.Y - v.Y);
        }

        public static Vector2 operator *(Vector2 v, double d)
        {
            return new Vector2(d * v.X, d * v.Y);
        }

        public static Vector2 operator *(double d, Vector2 v)
        {
            return new Vector2(d * v.X, d * v.Y);
        }

        public static Vector2 operator /(Vector2 v, double d)
        {
            return new Vector2(v.X / d, v.Y / d);
        }

        public static Vector2 operator -(Vector2 v)
        {
            return new Vector2(-v.X, -v.Y);
        }

        public static explicit operator Point(Vector2 v)
        {
            return new Point(v.X, v.Y);
        }

        // Overrides
        public override string ToString()
        {
            return String.Format("({0} {1})", X, Y);
        }
    }
}
```

FingerPaint5 通过前一个点保存之前的半径(根据压力设定)。在下图中，我用两个有独立半径的圆来代表两个连续手指的位置。如下图所示，较小圆的圆心为 *c*0、半径为 *r*0，而较大圆的圆心为 *c*1、半径为 *r*1。

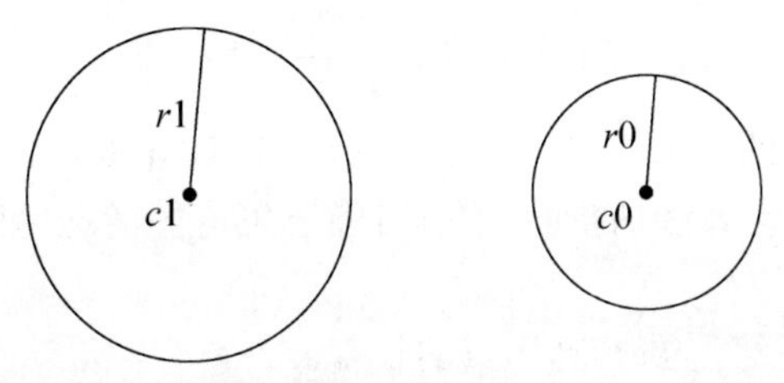

目标是获得 Path，包括两个圆圈和两者之间的区域。要做到这一点，必须用一条线连接两个圆，而同时两个圆相切，这需要一点儿技巧(从数学上讲)。因此，我们先用线条连接两个圆的圆心，标记为 *d*(见下图)。

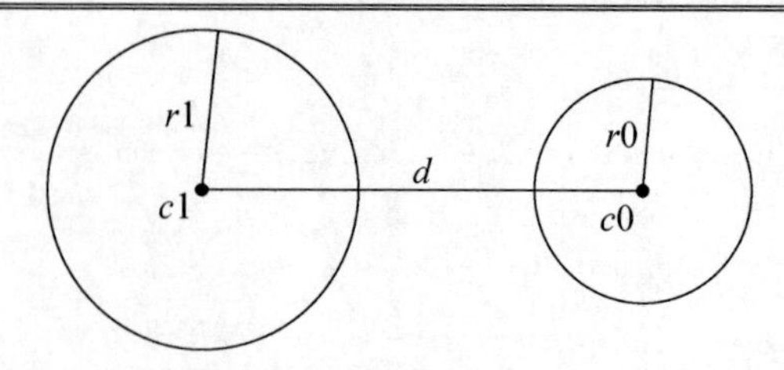

通过 Vector2 值可以获得线条长度以及表示其方向的正规化向量：

```
Vector2 vCenters = new Vector2(c0) - new Vector2(c1);
double d = vCenters.Length;
vCenters = vCenters.Normalized;
```

现在，我们根据 d 和两个圆的半径来定义另一个长度 e。点 F 为 e 到 $c0$ 的距离，其方向和两个圆心之间的矢量相同：

```
double e = d * r0 / (r1 - r0);
Point F = c0 + e * vCenters;
```

如下图所示。

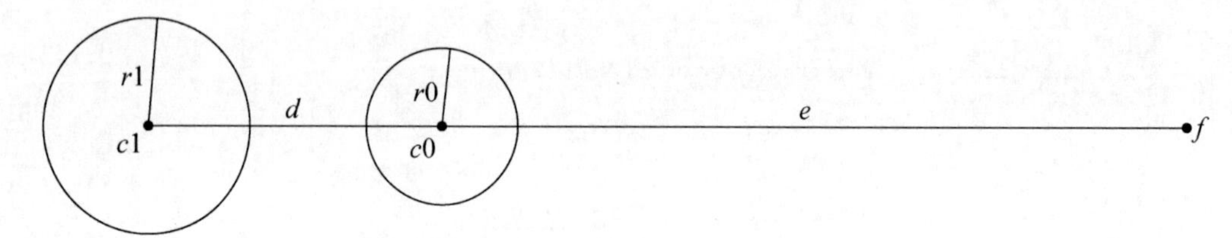

之所以称之为 F，是因为它是一个“焦点”。我认为以 F 为起点的线会与两圆都相切，也就是说，切线与半径线构成的是直角(见下图)。

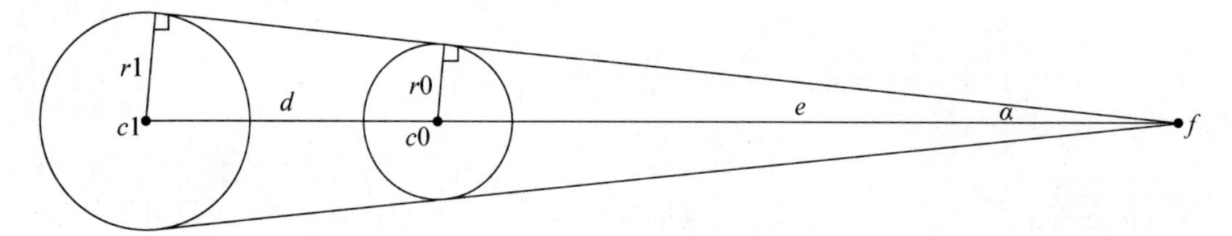

我这么想是因为直线 e 的定义方式。e 与 $r0$ 的比与 d 加 e 与 $r1$ 的比是相同的。可以直接如下计算角α(在上图右侧)：

```
double alpha = 180 * Math.Asin(r0 / e) / Math.PI;
```

如果 Math.Asin 方法的参数大于 1，则方法返回 NaN(非数字)。只有当 $r0$ 加 d 的和小于 r1(也就是小圆被完全包大圆内)时，才有可能发生这种情况。因此，这个问题更容易被预测。

通过勾股定理可以计算得到 F 到切点的长度：

```
double leg0 = Math.Sqrt(e * e - r0 * r0);
double leg1 = Math.Sqrt((e + d) * (e + d) - r1 * r1);
```

Vector2 结构有一个简便的 Rotate 方法，可以通过α和$-\alpha$度来旋转 vCenter 矢量：

```
Vector2 vRight = -vCenters.Rotate(alpha);
Vector2 vLeft = -vCenters.Rotate(-alpha);
```

变量名的“Right”和“Left”两个部分可从 F 视角得到。在图中，vRight 矢量和圆顶部切线相对应，vLeft 和底部切线相对应。有了矢量和长度，就可以计算实际切点：

```
Point t0R = F + leg0 * vRight;
Point t0L = F + leg0 * vLeft;
Point t1R = F + leg1 * vRight;
Point t1L = F + leg1 * vLeft;
```

这些点可以用来构造 PathGeometry，包含 2 个 ArcSegment 对象和 2 个 LineSegment 对象，如下图中的粗线轮廓。

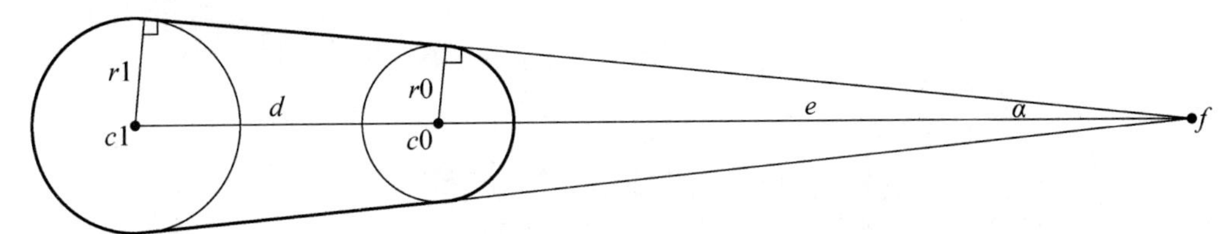

注意，小圆的 ArcSegment 始终小于 180 度，而大圆的 ArcSegment 则始终大于 180 度。这些特点会影响 ArcSegment 的 IsLargeArc 特性。同时请记住，通过规定关闭图，可以暗示创建两个 LineSegment 对象的其中之一。

FingerPaint5 中定义的实际算法如下所示。请注意，算法还必须实现相对简单的情况，即两个半径相同或一个圆完全包含另外一个圆的时候：

```
项目: FingerPaint5 | 文件: MainPage.xaml.cs(片段)
Geometry CreateTaperedLineGeometry(Point c0, double r0, Point c1, double r1)
{
    // Swap the centers and radii so that c0 is
    //        the center of the smaller circle.
    if (r1 < r0)
    {
        Point point = c0;
        c0 = c1;
        c1 = point;

        double radius = r0;
        r0 = r1;
        r1 = radius;
    }

    // Get vector from c1 to c0
    Vector2 vCenters = new Vector2(c0) - new Vector2(c1);

    // Get length and normalized version
    double d = vCenters.Length;
    vCenters = vCenters.Normalized;

    // Determine if one circle is enclosed in the other
    bool enclosed = r0 + d < r1;

    // Define tangent points derived in both algorithms
    Point t0R = new Point();
    Point t0L = new Point();
    Point t1R = new Point();
    Point t1L = new Point();

    // Case for two circles of same size
    if (r0 == r1 || enclosed)
    {
        // Rotate centers vector 90 degrees
        Vector2 vLeft = new Vector2(-vCenters.Y, vCenters.X);
```

```
        // Rotate -90 degrees
        Vector2 vRight = -vLeft;

        // Find tangent points
        t0R = c0 + r0 * vRight;
        t0L = c0 + r0 * vLeft;
        t1R = c1 + r1 * vRight;
        t1L = c1 + r1 * vLeft;
    }
    // A bit more difficult for two circles of unequal size
    else
    {
        // Create focal point F extending from c0
        double e = d * r0 / (r1 - r0);
        Point F = c0 + e * vCenters;

        // Find angle and length of right-triangle legs
        double alpha = 180 * Math.Asin(r0 / e) / Math.PI;
        double leg0 = Math.Sqrt(e * e - r0 * r0);
        double leg1 = Math.Sqrt((e + d) * (e + d) - r1 * r1);

        // Vectors of tangent lines
        Vector2 vRight = -vCenters.Rotate(alpha);
        Vector2 vLeft = -vCenters.Rotate(-alpha);

        // Find tangent points
        t0R = F + leg0 * vRight;
        t0L = F + leg0 * vLeft;
        t1R = F + leg1 * vRight;
        t1L = F + leg1 * vLeft;
    }

    // Create PathGeometry with implied closing line
    PathGeometry pathGeometry = new PathGeometry();
    PathFigure pathFigure = new PathFigure
    {
        StartPoint = t0R,
        IsClosed = true,
        IsFilled = true
    };
    pathGeometry.Figures.Add(pathFigure);

    // Arc around smaller circle
    ArcSegment arc0Segment = new ArcSegment
    {
        Point = t0L,
        Size = new Size(r0, r0),
        SweepDirection = SweepDirection.Clockwise,
        IsLargeArc = false
    };
    pathFigure.Segments.Add(arc0Segment);

    // Line connecting smaller circle to larger circle
    LineSegment lineSegment = new LineSegment
    {
        Point = t1L
    };
    pathFigure.Segments.Add(lineSegment);

    // Arc around larger circle
    ArcSegment arc1Segment = new ArcSegment
    {
        Point = t1R,
        Size = new Size(r1, r1),
        SweepDirection = SweepDirection.Clockwise,
        IsLargeArc = true
    };
    pathFigure.Segments.Add(arc1Segment);
```

```
        return pathGeometry;
    }
```

此时应该能完全理解 FingerPaint5 的其余部分。OnPointerReleased 和 OnPointerCaptureLost 覆写与 FingerPaint4 相同。内部 PointerInfo 类现在包括 PreviousRadius 字段。

项目: FingerPaint5 | 文件: MainPage.xaml.cs (片段)

```
public sealed partial class MainPage : Page
{
    struct PointerInfo
    {
        public Brush Brush;
        public Point PreviousPoint;
        public double PreviousRadius;
    }

    Dictionary<uint, PointerInfo> pointerDictionary = new Dictionary<uint, PointerInfo>();
    Random rand = new Random();
    byte[] rgb = new byte[3];

    public MainPage()
    {
        this.InitializeComponent();
    }

    protected override void OnPointerPressed(PointerRoutedEventArgs args)
    {
        // Get information from event arguments
        uint id = args.Pointer.PointerId;
        PointerPoint pointerPoint = args.GetCurrentPoint(this);

        // Create random color
        rand.NextBytes(rgb);
        Color color = Color.FromArgb(255, rgb[0], rgb[1], rgb[2]);

        // Create PointerInfo
        PointerInfo pointerInfo = new PointerInfo
        {
            PreviousPoint = pointerPoint.Position,
            PreviousRadius = 24 * pointerPoint.Properties.Pressure,
            Brush = new SolidColorBrush(color)
        };

        // Add to dictionary
        pointerDictionary.Add(id, pointerInfo);

        // Capture the Pointer
        CapturePointer(args.Pointer);
        base.OnPointerPressed(args);
    }

    protected override void OnPointerMoved(PointerRoutedEventArgs args)
    {
        // Get ID from event arguments
        uint id = args.Pointer.PointerId;

        // If ID is in dictionary, start a loop
        if (pointerDictionary.ContainsKey(id))
        {
            PointerInfo pointerInfo = pointerDictionary[id];
            IList<PointerPoint> pointerpoints = args.GetIntermediatePoints(this);

            for (int i = pointerpoints.Count - 1; i >= 0; i--)
            {
```

```
                PointerPoint pointerPoint = pointerpoints[i];

                // For each point, create a Path element and add to Grid
                Point point = pointerPoint.Position;
                float pressure = pointerPoint.Properties.Pressure;
                double radius = 24 * pressure;

                Geometry geometry =
                    CreateTaperedLineGeometry(pointerInfo.PreviousPoint,
                                              pointerInfo.PreviousRadius,
                                              point,
                                              radius);

                Path path = new Path
                {
                    Data = geometry,
                    Fill = pointerInfo.Brush
                };
                contentGrid.Children.Add(path);

                // Update PointerInfo
                pointerInfo.PreviousPoint = point;
                pointerInfo.PreviousRadius = radius;
            }

            // Store PointerInfo back in dictionary
            pointerDictionary[id] = pointerInfo;
        }
        base.OnPointerMoved(args);
    }

    protected override void OnPointerReleased(PointerRoutedEventArgs args)
    {
        ...
    }

    protected override void OnPointerCaptureLost(PointerRoutedEventArgs args)
    {
        ...
    }

    Geometry CreateTaperedLineGeometry(Point c0, double r0, Point c1, double r1)
    {
        ...
    }
}
```

现在，在压力感应装置快速画图，线条锥度也会很平滑，不会断裂(见下图)。

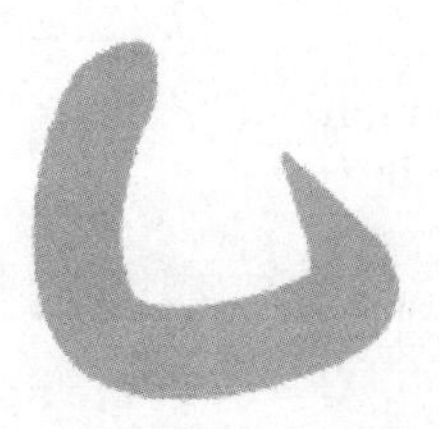

13.7 如何保存图画

以上手绘程序都不能保存图画，如何实现该功能呢？

每个程序都通过把 Polyline、Line 或 Path 元素添加到 Grid 来进行绘画。保存图画的方法可以是访问这些对象并把所有点和其他信息保存到一个文件中，也可以是 XML 格式文件。此后添加功能进行回载并根据这些信息来创建新的 Polyline、Line 或 Path 元素。

但你可能更倾向于保存图画的位图。(就传统而言，画图程序处理矢量，而绘图程序处理位图。)事实上，FingerPaint 程序对位图上执行其绘画是有道理的。

这是能够实现的，但居然那么简单。最简单的方法是用 WriteableBitmap，但你必须实现自己的绘画逻辑，以便能在位图上渲染线条。第 14 章会介绍具体办法。也可以通过 DirectX，再加一些 C++编码来实现。第 15 章将对此展开讨论。

13.8　现实和超现实手绘

最近几年，绘画程序已经尝试模仿现实生活的绘图材料，如铅笔、粉笔和水彩。当然，想这么做需要综合视觉灵敏和编程技能，同时还要有一定的随机性。

当然，你也可以反方向，在屏幕上渲染一些在现实世界永远不会遇到的东西。Whirligig 程序和 FingerPaint 系列程序在结构上非常相似，但前者渲染的螺旋如下图所示。

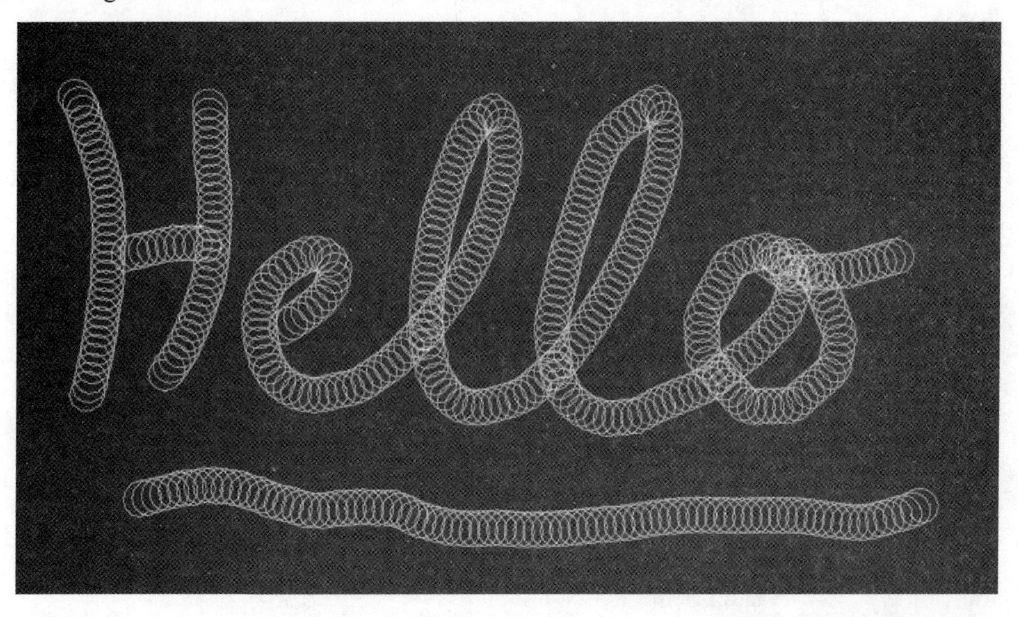

Whirligig 程序实现了指针捕获，但未实现 Esc 键终止，因此，OnPointerReleased 和 OnPointerCaptureLost 覆写和之前几个项目一样。对于每个手指笔画，程序都会像早期版本一样渲染单一 Polyline，但 Polyline 只有一个像素粗且以圆环绕。

项目：Whirligig | 文件：MainPage.xaml.cs(片段)

```
public sealed partial class MainPage : Page
{
    const double Radius = 24;                  // 1/4 inch
    const double AngleIncrement = 0.5;         // radians per pixel

    class TouchInfo
    {
        public Point LastPoint;
        public Polyline Polyline;
        public double Angle;
    }
```

```
Dictionary<uint, TouchInfo> pointerDictionary = new Dictionary<uint, TouchInfo>();

public MainPage()
{
    this.InitializeComponent();
}

protected override void OnPointerPressed(PointerRoutedEventArgs args)
{
    // Get information from event arguments
    uint id = args.Pointer.PointerId;
    Point point = args.GetCurrentPoint(this).Position;

    // Create Polyline
    Polyline polyline = new Polyline
    {
        Stroke = this.Resources["ApplicationForegroundThemeBrush"] as Brush,
        StrokeThickness = 1,
    };

    // Add to Grid
    contentGrid.Children.Add(polyline);

    // Create TouchInfo
    TouchInfo touchInfo = new TouchInfo
    {
        LastPoint = point,
        Polyline = polyline
    };

    // Add to dictionary
    pointerDictionary.Add(id, touchInfo);

    // Capture the Pointer
    CapturePointer(args.Pointer);
    base.OnPointerPressed(args);
}

protected override void OnPointerMoved(PointerRoutedEventArgs args)
{
    // Get information from event arguments
    uint id = args.Pointer.PointerId;
    Point point = args.GetCurrentPoint(this).Position;

    // If ID is not in dictionary, don't do anything
    if (!pointerDictionary.ContainsKey(id))
        return;

    // Get TouchInfo objects
    Polyline polyline = pointerDictionary[id].Polyline;
    Point lastPoint = pointerDictionary[id].LastPoint;
    double angle = pointerDictionary[id].Angle;

    // Distance from last point to this point
    double distance = Math.Sqrt(Math.Pow(point.X - lastPoint.X, 2) +
                                Math.Pow(point.Y - lastPoint.Y, 2));

    int divisions = (int)distance;

    for (int i = 0; i < divisions; i++)
    {
        // Sub-divide the distance between the last point and the new
        double x = (i * point.X + (divisions - i) * lastPoint.X) / divisions;
        double y = (i * point.Y + (divisions - i) * lastPoint.Y) / divisions;
        Point pt = new Point(x, y);

        // Increase the angle
```

```
                angle += distance * AngleIncrement / divisions;

                // Rotate the point
                pt.X += Radius * Math.Cos(angle);
                pt.Y += Radius * Math.Sin(angle);

                // Add to Polyline
                polyline.Points.Add(pt);
            }

            // Save new information
            pointerDictionary[id].LastPoint = point;
            pointerDictionary[id].Angle = angle;

            base.OnPointerMoved(args);
        }
        ...
    }
```

有了 Polyline，TouchInfo 类会存储 LastPoint 值和 Angle 值。对于每个 PointerMoved 事件，程序会把当前点到前一个点的距离划分为像素大小长度。 对于这些像素大小长度，程序会附加约 30 度的圆形图案。(30 度是 AngleIncrement 常量。)程序不会渲染实际点，而是积累的角度来旋转点，并添加到 Polyline。

13.9 触控钢琴

并非所有触屏应用都采用相同模式。例如，想想触屏钢琴键盘。很显然，你想要能用手指演奏和弦，因此，这是 Pointer 事件的工作，而不是 Manipulation 事件的工作。

但你真正想做的是手指上下敲击让触屏钢琴键盘，从而产生滑音。如果做不到这一点，你肯定会认为钢琴坏了。也就是说，你可能并不只关心 PointerPressed 和 PointerReleased。当然，你可以按下一个键而松开另一个键，但除此之外，你会用手指扫在许多其他按键而进行演奏。

基本上，有两种方法来构造钢琴键盘。对于整个键盘，可以使用一个控件，也可以使用许多控件(我说“许多”的意思其实是一个控件对应一个按键)。

单个控件必须能控制所有按键并用按键来比较指针位置，从而评估 PointerMoved 事件。需要能跟踪每根手指，以确定 PointerMoved 事件表明按键何时进入按键边界以及何时离开按键边界。这是经典“命中测试”，即检查指针位置，以确定指针是否位于边界内。

然而，如果每个按键都有独立的控件，则不需要执行命中测试。如果按键会获得 Pointer 事件，则 Pointer 在控件的边界之内(除非控件已捕获指针，否则指针捕获在此应用中没有任何意义)。

实现钢琴键需要什么 Pointer 事件？不要一开始就想着按下再松开。可以先想想滑音。如果我们讨论的只是触屏键盘，则只需要两个 Pointer 事件：PointerEntered 和 PointerExited。

然而，你可能想让键盘能够合理响应鼠标和手写笔。如果还没有按下鼠标按钮，钢琴按键会从鼠标获得 PointerEntered 和 PointerExited 事件，这是一个问题。PointerEntered 处理程序需要检查 IsInContact 属性，才能正确处理鼠标和手写笔。IsInContact 属性对于触碰事件始终为 true，但只有按下鼠标按钮(对于鼠标)才为 true，如果手写笔在与屏幕接触，也为 true。

此外，考虑单一元素时，鼠标和手写笔会在 PointerPressed 之前生成 PointerReleased 事件，而在之后 PointerEntered 生成 PointerExited，因此也必须处理好 PointerPressed 和 PointerReleased。

现在我们从按键自下而上来构造两个八度的钢琴键盘。以下 Key 类为 Control 派生类，不带默认模板，因此没有默认视觉外观。但 Key 类确实定义了 IsPressed 依赖属性以及 IsPressed 的属性变更处理程序，该程序可以切换 Normal 和 Pressed 两个视觉状态。

项目：SilentPiano | 文件：Key.cs(片段)

```
namespace SilentPiano
{
    public class Key : Control
    {
        static readonly DependencyProperty isPressedProperty =
            DependencyProperty.Register("IsPressed",
                typeof(bool), typeof(Key),
                new PropertyMetadata(false, OnIsPressedChanged));

        List<uint> pointerList = new List<uint>();

        public static DependencyProperty IsPressedProperty
        {
            get { return isPressedProperty; }
        }

        public bool IsPressed
        {
            set { SetValue(IsPressedProperty, value); }
            get { return (bool)GetValue(IsPressedProperty);
        }

        protected override void OnPointerEntered(PointerRoutedEventArgs args)
        {
            if (args.Pointer.IsInContact)
                AddToList(args.Pointer.PointerId);
            base.OnPointerEntered(args);
        }

        protected override void OnPointerPressed(PointerRoutedEventArgs args)
        {
            AddToList(args.Pointer.PointerId);
            base.OnPointerPressed(args);
        }

        protected override void OnPointerReleased(PointerRoutedEventArgs args)
        {
            RemoveFromList(args.Pointer.PointerId);
            base.OnPointerReleased(args);
        }

        protected override void OnPointerExited(PointerRoutedEventArgs args)
        {
            RemoveFromList(args.Pointer.PointerId);
            base.OnPointerExited(args);
        }
        void AddToList(uint id)
        {
            if (!pointerList.Contains(id))
                pointerList.Add(id);

            CheckList();
        }

        void RemoveFromList(uint id)
```

```
        {
            if (pointerList.Contains(id))
                pointerList.Remove(id);

            CheckList();
        }

        void CheckList()
        {
            this.IsPressed = pointerList.Count > 0;
        }

        static void OnIsPressedChanged(DependencyObject obj,
                                       DependencyPropertyChangedEventArgs args)
        {
            VisualStateManager.GoToState(obj as Key,
                            (bool)args.NewValue ? "Pressed" : "Normal", false);
        }
    }
}
```

因为可以用两个手指触碰同一个按键，所以控件需要能够跟踪单个手指。但控件并不需要 Dictionary 来保留每个 ID 信息。可以直接采用 List。在 OnPointerEntered 覆写(但只有当 IsInContact 为 true 的时候)和 OnPointerPressed 中，ID 会放进 List 中，在 OnPointerReleased 和 OnPointerExited 中删除，并会触发在视觉状态变化。如果 List 中至少包含一个条目，IsPressed 属性则为 true。PointerPressed 和 PointerReleased 事件处理程序只适用于鼠标和手写笔。

两个模板(一个用于白键，一个用于黑键)都在 Octave.xaml 文件中定义。两个模板的区别在于 Polygon 大小，而 Polygon 定义了按键的形状和默认颜色。(两个键的形状都是矩形。我本来想让白键有不同形状，就像一台真正的钢琴那样，但颜色相同更简单一些，而且所需要的模板少得多。)在 Pressed 状态时，两个模板颜色切换为红色。

```
项目: SilentPiano | 文件: Octave.xaml(片段)
<UserControl ... >
    <UserControl.Resources>
        <ControlTemplate x:Key="whiteKey" TargetType="local:Key">
            <Grid Width="80">
                <Polygon Points="2 0, 78 0, 78 320, 02 320">
                    <Polygon.Fill>
                        <SolidColorBrush x:Name="brush" Color="White" />
                    </Polygon.Fill>
                </Polygon>

                <VisualStateManager.VisualStateGroups>
                    <VisualStateGroup x:Name="CommonStates">
                        <VisualState x:Name="Normal"/>
                        <VisualState x:Name="Pressed">
                            <Storyboard>
                                <ColorAnimationUsingKeyFrames Storyboard.TargetName="brush"
                                            Storyboard.TargetProperty="Color">
                                    <DiscreteColorKeyFrame KeyTime="0" Value="Red" />
                                </ColorAnimationUsingKeyFrames>
                            </Storyboard>
                        </VisualState>
                    </VisualStateGroup>
                </VisualStateManager.VisualStateGroups>
            </Grid>
        </ControlTemplate>

        <ControlTemplate x:Key="blackKey" TargetType="local:Key">
            <Grid>
```

```
                    <Polygon Points="0 0, 40 0, 40 220, 0 220">
                        <Polygon.Fill>
                            <SolidColorBrush x:Name="brush" Color="Black" />
                        </Polygon.Fill>
                    </Polygon>

                    <VisualStateManager.VisualStateGroups>
                        <VisualStateGroup x:Name="CommonStates">
                            <VisualState x:Name="Normal"/>
                            <VisualState x:Name="Pressed">
                                <Storyboard>
                                    <ColorAnimationUsingKeyFrames Storyboard.TargetName="brush"
                                                    Storyboard.TargetProperty="Color">
                                        <DiscreteColorKeyFrame KeyTime="0" Value="Red" />
                                    </ColorAnimationUsingKeyFrames>
                                </Storyboard>
                            </VisualState>
                        </VisualStateGroup>
                    </VisualStateManager.VisualStateGroups>
                </Grid>
            </ControlTemplate>
        </UserControl.Resources>

        <Grid>
            <StackPanel Orientation="Horizontal">
                <local:Key Template="{StaticResource whiteKey}" />
                <local:Key Template="{StaticResource whiteKey}" />
                <local:Key Template="{StaticResource whiteKey}" />
                <local:Key Template="{StaticResource whiteKey}" />
                <local:Key Template="{StaticResource whiteKey}" />
                <local:Key Template="{StaticResource whiteKey}" />
                <local:Key Template="{StaticResource whiteKey}" />
                <local:Key x:Name="lastKey"
                        Template="{StaticResource whiteKey}"
                        Visibility="Collapsed" />
            </StackPanel>
            <Canvas>
                <local:Key Template="{StaticResource blackKey}"
                        Canvas.Left="60" Canvas.Top="0" />
                <local:Key Template="{StaticResource blackKey}"
                        Canvas.Left="140" Canvas.Top="0" />
                <local:Key Template="{StaticResource blackKey}"
                        Canvas.Left="300" Canvas.Top="0" />
                <local:Key Template="{StaticResource blackKey}"
                        Canvas.Left="380" Canvas.Top="0" />
                <local:Key Template="{StaticResource blackKey}"
                        Canvas.Left="460" Canvas.Top="0" />
            </Canvas>
        </Grid>
    </UserControl>
```

8 个白键水平排列在 StackPanel 中，5 个黑键在 Canvas 中。这种配置允许白键可以定义控件大小，而让黑键在白键上方并覆盖白键的一部分。

8 个白键从 C 到 C。小键盘在很多情况下都从 C 开始、以 C 结束，但在两个八度相连的地方，你不想要一对相邻 C 键。这就是为什么最后一个按键有 Visibility 的 Collapsed 的原因。通过代码隐藏文件把 Visibility 属性设置为 Visible 或 Collapsed，代码隐藏文件则基于 LastKeyVisible 依赖项属性的设置。

```
项目: SilentPiano | 文件: Octave.xaml.cs(片段)
public sealed partial class Octave : UserControl
{
    static readonly DependencyProperty lastKeyVisibleProperty =
            DependencyProperty.Register("LastKeyVisible",
                    typeof(bool), typeof(Octave),
```

```
                    new PropertyMetadata(false, OnLastKeyVisibleChanged));

        public Octave()
        {
            this.InitializeComponent();
        }

        public static DependencyProperty LastKeyVisibleProperty
        {
            get { return lastKeyVisibleProperty; }
        }

        public bool LastKeyVisible
        {
            set { SetValue(LastKeyVisibleProperty, value); }
            get { return (bool)GetValue(LastKeyVisibleProperty); }
        }

        static void OnLastKeyVisibleChanged(DependencyObject obj,
                                        DependencyPropertyChangedEventArgs args)
        {
             (obj as Octave).lastKey.Visibility =
                 (bool)args.NewValue ? Visibility.Visible : Visibility.Collapsed;
        }
    }
```

剩下的事情是把 MainPage.xaml 文件中的两个 Octave 对象实例化，第二个对象 LastKeyVisible 设置为 true。

```
项目: SilentPiano | 文件: MainPage.xaml(片段)
<Page ... >
    <Grid Background="Gray">
        <StackPanel Orientation="Horizontal"
                    HorizontalAlignment="Center"
                    VerticalAlignment="Center">
            <local:Octave />
            <local:Octave LastKeyVisible="True" />
        </StackPanel>
    </Grid>
</Page>
```

如下图，我现在可以弹奏我喜欢的和弦了(为主要编程语言和音)。

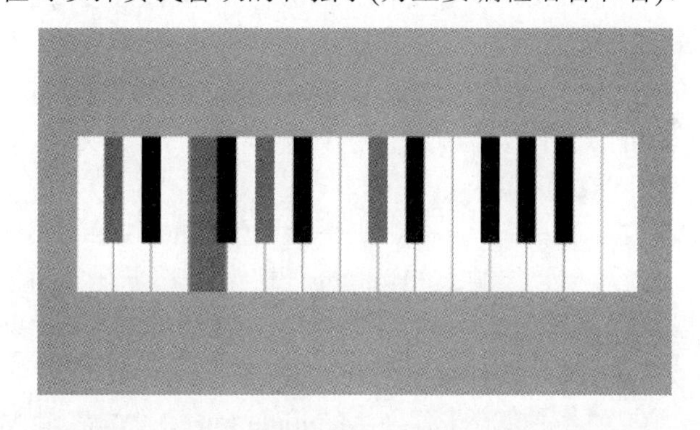

13.10　操控、手指和元素

Pointer 事件的优点是可以追踪单个手指。而 Manipulation 事件的优点则是不能追踪单个手指。

Manipulation 事件把多个手指(“多个”的真正意思经常是指“两个”)就组合成更高层次的手势，例如捏住、旋转。这些手势常见图形变化相对应：平移、缩放(虽然仅限在水平及垂直方向上的相等比例)以及旋转。捕获是操控的固有特性。惯性也可以用。

请记住，把多个手指组合成单一系列 Manipulation 事件，并不是针对整个窗口，而是针对处理这些事件的每个元素。也就是说，可以使用一个或一对手指来操纵一个元素，而使用另一对手指来操作第二个元素。

UIElement 定义了 5 个 Manipulation 事件，每一个元素一般都按照以下顺序进行接收(注意，前两个事件的名称十分相似)：

- ManipulationStarting
- ManipulationStarted
- ManipulationDelta (很多)
- ManipulationInertiaStarting
- ManipulationDelta(更多)
- ManipulationCompleted

Control 类定义了对应五个事件的虚拟方法，命名为 ManipulationStarting 等等。

鼠标或手写笔能生成 Manipulation 事件。当手指第一次触碰一个元素时，只有在按下鼠标按钮或手写笔触碰屏幕时才行。手指首先触碰元素按钮、鼠标点击到元素或手写笔触碰到元素，才会发生 ManipulationStarting 事件。

ManipulationStarted 事件一般发生在 ManipulationStarting 之后(但稍后讨论，这里的关键词是“一般”)。随后，手指在屏幕上移动，触发大量 ManipulationDelta 事件。如果所有手指都离开元素，就会停止 ManipulationInertiaStarting。元素继续生成表示惯性的 ManipulationDelta 事件，但 ManipulationCompleted 表示序列结束。

虽然手指首先触碰元素(单击鼠标或按下手写笔)便会发生 ManipulationStarting 事件，但该事件之后并不一定是 ManipulationStarted 事件，而且 ManipulationStarted 可能也会延迟。问题是，系统必须区分点击或保持实际操作。如果手指(鼠标或手写笔)移动一点点，就会触发 ManipulationStarted。

例如，如果用扫屏动作触碰元素，ManipulationStarting 之后很快就是 ManipulationStarted 和多个 ManipulationDelta 事件。但是，把手指按在一个地方并保持住，就会延迟 ManipulationStarted 事件一段时间。

如果用户点击、右键点击或双击屏幕，则不会发生 ManipulationStarted 事件。然而，ManipulationStarting 之后可能会触发 Holding 事件，用户也可能移动手指并生成 ManipulationStarted 以及事件其余部分。有了表明 Canceled 的 HoldingState 属性，也会触发另一个 Holding 事件。

然而在默认情况下，元素不会产生任何 Manipulation 事件！启用 Manipulation 事件首先必须要基于每个元素。要让程序准确指定需要哪些类型操作，UIElement 会定义枚举类型 ManipulationModes 的 ManipulationMode 属性。(属性名称为单数，枚举名称是复数。) ManipulationMode 的默认设置是 ManipulationModes.System，对于应用而言就相当于 ManipulationModes.None。要启用操作元素，至少需要将其设置为另一个 ManipulationModes 成员。枚举项被定义为位标志，因此，可以用 OR 运算符 (|)其进行位或运算。

虽然一些应用需要处理所有 5 个 Manipulation 事件，但可以写为只检查 ManipulationDelta 的代码。

ManipulationTracker 程序就是这种情况。程序显示了 ManipulationModes 枚举项的很多 CheckBox 控件以及可以操纵的三个 Rectangle 元素。为了简化某些代码和标记，自定义 CheckBox 派生类来存储并显示 ManipulationModes 成员。

项目: ManipulationTracker | 文件: ManipulationModeCheckBox.cs

```
using Windows.UI.Xaml.Controls;
using Windows.UI.Xaml.Input;

namespace ManipulationTracker
{
    public class ManipulationModeCheckBox : CheckBox
    {
        public ManipulationModes ManipulationModes { set; get; }
    }
}
```

自定义 CheckBox 的 10 个实例安排在 Mainpage.XAML 的 StackPanel 中，每个实例都用枚举项名称(名称中插入空格更易读)和整数值进行识别。

项目: ManipulationTracker | 文件: MainPage.xaml(片段)

```
<Page ... >
    <Page.Resources>
        <Style TargetType="local:ManipulationModeCheckBox">
            <Setter Property="Margin" Value="12 6 24 6" />
        </Style>

        <Style TargetType="Rectangle">
            <Setter Property="Width" Value="144" />
            <Setter Property="Height" Value="144" />
            <Setter Property="HorizontalAlignment" Value="Left" />
            <Setter Property="VerticalAlignment" Value="Top" />
            <Setter Property="RenderTransformOrigin" Value="0.5 0.5" />
        </Style>
    </Page.Resources>

    <Grid Background="{StaticResource ApplicationPageBackgroundThemeBrush}">
        <Grid.ColumnDefinitions>
            <ColumnDefinition Width="Auto" />
            <ColumnDefinition Width="*" />
        </Grid.ColumnDefinitions>

        <StackPanel Name="checkBoxPanel"
                    Grid.Column="0">
            <local:ManipulationModeCheckBox Checked="OnManipulationModeCheckBoxChecked"
                                            Unchecked="OnManipulationModeCheckBoxChecked"
                                            Content="Translate X (1)"
                                            ManipulationModes="TranslateX" />

            <local:ManipulationModeCheckBox Checked="OnManipulationModeCheckBoxChecked"
                                            Unchecked="OnManipulationModeCheckBoxChecked"
                                            Content="Translate Y (2)"
                                            ManipulationModes="TranslateY" />

            <local:ManipulationModeCheckBox Checked="OnManipulationModeCheckBoxChecked"
                                            Unchecked="OnManipulationModeCheckBoxChecked"
                                            Content="Translate Rails X (4)"
                                            ManipulationModes="TranslateRailsX" />

            <local:ManipulationModeCheckBox Checked="OnManipulationModeCheckBoxChecked"
                                            Unchecked="OnManipulationModeCheckBoxChecked"
                                            Content="Translate Rails Y (8)"
                                            ManipulationModes="TranslateRailsY" />
```

```
            <local:ManipulationModeCheckBox Checked="OnManipulationModeCheckBoxChecked"
                                            Unchecked="OnManipulationModeCheckBoxChecked"
                                            Content="Rotate (16)"
                                            ManipulationModes="Rotate" />

            <local:ManipulationModeCheckBox Checked="OnManipulationModeCheckBoxChecked"
                                            Unchecked="OnManipulationModeCheckBoxChecked"
                                            Content="Scale (32)"
                                            ManipulationModes="Scale" />

            <local:ManipulationModeCheckBox Checked="OnManipulationModeCheckBoxChecked"
                                            Unchecked="OnManipulationModeCheckBoxChecked"
                                            Content="Translate Inertia (64)"
                                            ManipulationModes="TranslateInertia" />

            <local:ManipulationModeCheckBox Checked="OnManipulationModeCheckBoxChecked"
                                            Unchecked="OnManipulationModeCheckBoxChecked"
                                            Content="Rotate Inertia (128)"
                                            ManipulationModes="RotateInertia" />

            <local:ManipulationModeCheckBox Checked="OnManipulationModeCheckBoxChecked"
                                            Unchecked="OnManipulationModeCheckBoxChecked"
                                            Content="Scale Inertia (256)"
                                            ManipulationModes="ScaleInertia" />

            <local:ManipulationModeCheckBox Checked="OnManipulationModeCheckBoxChecked"
                                            Unchecked="OnManipulationModeCheckBoxChecked"
                                            Content="All (0xFFFF)"
                                            ManipulationModes="All" />
        </StackPanel>

        <Grid Name="rectanglePanel"
              Grid.Column="1">
            <Rectangle Fill="Red">
                <Rectangle.RenderTransform>
                    <CompositeTransform />
                </Rectangle.RenderTransform>
            </Rectangle>

            <Rectangle Fill="Green">
                <Rectangle.RenderTransform>
                    <CompositeTransform />
                </Rectangle.RenderTransform>
            </Rectangle>

            <Rectangle Fill="Blue">
                <Rectangle.RenderTransform>
                    <CompositeTransform />
                </Rectangle.RenderTransform>
            </Rectangle>
        </Grid>
    </Grid>
</Page>
```

Grid 的较大区域有三个 Rectangle 元素，带 Computerstan 的状态标志三种颜色：红色，绿色，和蓝色。

在代码隐藏文件中，通过把与用 OR 运算符选中复选框相关联的枚举项组合在一起，任意选中或未选中自定义 CheckBox 控件会导致计算新的 ManipulationModes 值。这种组合 ManipulationModes 值可以设置为三个 Rectangle 元素的 ManipulationMode 属性。

```
项目: ManipulationTracker | 文件: MainPage.xaml.cs(片段)
public sealed partial class MainPage : Page
{
    public MainPage()
```

```
        {
            this.InitializeComponent();
        }

        void OnManipulationModeCheckBoxChecked(object sender, RoutedEventArgs args)
        {
            // Get composite ManipulationModes value of checked CheckBoxes
            ManipulationModes manipulationModes = ManipulationModes.None;

            foreach (UIElement child in checkBoxPanel.Children)
            {
                ManipulationModeCheckBox checkBox = child as ManipulationModeCheckBox;

                If ((bool)checkBox.IsChecked)
                    manipulationModes |= checkBox.ManipulationModes;
            }

            // Set ManipulationMode property of each Rectangle
            foreach (UIElement child in rectanglePanel.Children)
                child.ManipulationMode = manipulationModes;
            }

            protected override void OnManipulationDelta(ManipulationDeltaRoutedEventArgs args)
            {
                // OriginalSource is always Rectangle because nothing else has its
                // ManipulationMode set to anything other than ManipulationModes.None
                Rectangle rectangle = args.OriginalSource as Rectangle;
                CompositeTransform transform = rectangle.RenderTransform as CompositeTransform;
                    transform.TranslateX += args.Delta.Translation.X;

            transform.TranslateY += args.Delta.Translation.Y;
            transform.ScaleX *= args.Delta.Scale;
            transform.ScaleY *= args.Delta.Scale;
            transform.Rotation += args.Delta.Rotation;
            base.OnManipulationDelta(args);
        }
    }
```

程序的最后一部分是 OnManipulationDelta 覆写，即由 Control 类定义的虚拟方法，通过 Control 类，可以更容易地访问 UIElement 所定义的 ManipulationDelta 事件。ManipulationDelta 是主要 Manipulation 事件，并表明用户手指参与的操作。

注意，OnManipulationDelta 覆写会把事件参数的 OriginalSource 属性投射到 Rectangle，但不会检查投射是否成功。从理论上讲，OriginalSource 属性可以是 MainPage 或 MainPage 子类。然后，只有 Rectangle 元素可操作，因此，只有 Rectangle 元素可以生成 ManipulationDelta 事件。

CompositeTransform 设置为特定 Rectangle 的 RenderTransform 属性，覆写获得 CompositeTransform，并基于事件参数 Delta 属性调整转换的五个属性。Delta 属性的类型为 ManipulationDelta，包含有四个属性的结构。(注意！该结构与提供它的事件具有相同名称！)该值表示上次 ManipulationDelta 事件后的变化。

通过这段代码可以访问四个 ManipulationDelta 属性中的三个。第四个属性是 Expansion，与 Scale 类似，但用像素来表达，而不用乘法缩放因子。ManipulationDelta 结构的 Translation 属性表示从上一次 ManipulationDelta 事件以来手指移动的平均距离，因此这些值会添加到 CompositeTransform 的 translateX 和 translateY 属性。如果没有运动，这些值就是零。

与之类似(但处理方式却有所不同)，ManipulationDelta 结构的 Scale 属性表示从上次事件以来手指之间增加的距离。CompositeTransform 的 ScaleX 和 ScaleY 属性与该系数相乘。(由于 Manipulation 事件不会为水平和垂直缩放提供单独缩放因子，因此，所有操控缩放必须各向同性，即在两个方向都是一样。)如果没有缩放(或尚未启用缩放)，则缩放值为 1。ManipulationDelta 的 Rotate 属性表示转动手指所引起的旋转角变化，并且添加到 CompositeTransform 的 Rotation 属性。

如下图所示，选中几个复选框，就真的可以用鼠标、手写笔或多个手指的移动来移动、缩放和旋转矩形，甚至一次可以操控两三个。

对于使用 Manipulation 事件的程序，规则很简单：始终把 ManipulationMode 属性设置为 Manipulation 元素的或者一个或多个 Manipulation 事件元素的非默认值。这样一来，每个元素都会生成自己独立的 Manipulation 事件数据流。也可以为元素本身的 ManipulationDelta 事件设置处理程序，也可以通过可视树的源头来处理事件。

我说过这种操控会如预期发挥作用，但并不完全正确。你会发现，除了把 RenderTransformOrigin 设置为参考点(0.5，0.5)，无论是代码还是 XAML 都没有引用缩放或旋转中心。因此，所有缩放和旋转都与特定矩形的中心相对应。

这不是正确行为。例如，假设把一个手指放在接近矩形顶角的地方并保持稳定。可以用第二个手指按住对角并进行拖拽或旋转。缩放和旋转结果应该与第一个手指相关。换而言之，矩形的其余部分围绕第一个手指进行缩放或绕旋转，而第一个手指处的矩形部分应保持在原位。

解决该问题需要较为复杂的逻辑，因此我暂时忽略，到本章后面再讨论。

同时，还可以使用一些其他类型的操控。有三类惯性操控(可用于平移、缩放和旋转)，而且真的可以轻弹或旋转矩形离开屏幕。我稍后会讨论一些可以控制惯性程度的方法。

可以用如下代码设置之前屏幕截图的等效 ManipulationMode 属性：

```
rectangle.ManipulationMode = ManipulationModes.TranslateX |
                             ManipulationModes.TranslateY |
                             ManipulationModes.Scale |
                             ManipulationModes.Rotate;
```

但不是在 XAML 中。在 XAML 中设置 ManipulationMode 属性仅限于单一枚举项，而在实际应用中，很可能为 All。

如果想使操控只限定于水平移动，可以指定 ManipulationModes 的成员 translateX 而不是 translateY：

```
rectangle.ManipulationMode = ManipulationModes.TranslateX;
```

同样，要使操控只限定于垂直移动，则需要指定 translateY 而不是 translateX。

ManipulationModes 的两个枚举项称为 TranslateRailsX 和 TranslateRailsY。如果需要同时指定 translateX 和 translateY，两者则只能按预定方式运行。例如，

```
rectangle.ManipulationMode = ManipulationModes.TranslateX |
                             ManipulationModes.TranslateY |
                             ManipulationModes.TranslateRailsX;
```

这种配置可以随意移动水平和垂直方向的元素。然后，如果操控从水平方向上的移动开始，元素就会卡在导轨上(可以这么说)，而且所有接下来的移动都限制在水平方向，直到抬起手指再从头开始。

与之类似，如果操控是从垂直移动开始，这种配置就会限制于垂直移动：

```
rectangle.ManipulationMode = ManipulationModes.TranslateX |
                             ManipulationModes.TranslateY |
                             ManipulationModes.TranslateRailsY;
```

也可以同时指定两者：

```
rectangle.ManipulationMode = ManipulationModes.TranslateX |
                             ManipulationModes.TranslateY |
                             ManipulationModes.TranslateRailsX |
                             ManipulationModes.TranslateRailsY;
```

如果以对角拖动元素，则可以以任意方式来移动。但如果开始时就是水平或垂直移动，元素就会卡在导轨上。

正如在之前代码中所看到的，ManipulationTracker 程序使用 ManipulationDeltaRoutedEventArgs 参数的 Delta 属性来更改 CompositeTransform：

```
transform.TranslateX += args.Delta.Translation.X;
transform.TranslateY += args.Delta.Translation.Y;

transform.ScaleX *= args.Delta.Scale;
transform.ScaleY *= args.Delta.Scale;

transform.Rotation += args.Delta.Rotation;
```

如果你检查 ManipulationDeltaRoutedEventArgs 的属性，就会发现除了 Delta 属性之外还有 Cumulative 属性，也为 ManipulationDelta 类型。Delta 属性表示从上一次 ManipulationDelta 事件的变化，而 Cumulative 则表示从上一次 ManipulationStarted 的变化。

你可能怀疑 Cumulative 属性比 Delta 属性更容易处理。因为可以把值传送到相应的 CompositeTransform 属性，如下所示：

```
transform.TranslateX = args.Cumulative.Translation.X;
transform.TranslateY = args.Cumulative.Translation.Y;

transform.ScaleX = args.Cumulative.Scale;
transform.ScaleY = args.Cumulative.Scale;

transform.Rotation = args.Cumulative.Rotation;
```

有了这些代码在第一次操控元素的时候似乎可以很好发挥作用。但抬起手指并在同一个元素上尝试另一项操控，元素会跳回到屏幕左上角处它原来的位置！

Cumulative 属性并不是从程序开始就进行累计，而是只从特定 ManipulationStarted 事件才进行累积。

13.11 处理惯性

Manipulation 事件支持惯性平移、缩放和旋转，但如果不想要惯性，可以直接不设定 ManipulationModes。

如果想随时停止操控或惯性，ManipulationStarted 和 ManipulationDelta 事件的事件参数有 Complete 方法，它可以触发 ManipulationCompleted 事件。

如果想自己处理惯性，也可以。ManipulationDelta 和 ManipulationInertiaStarting 事件的事件参数有 Velocities 属性，该属性表示线性、缩放和旋转速度。对于线性运动，Velocities 属性为每毫秒像素，我怀疑这不是直觉单位。我试过用手指快速点击屏幕上的对象，可以达到每毫秒 10 个像素，但不会更高。也就是每秒 10 000 个像素，相当于每秒约 100 英寸，或每秒大约 8 英尺，或每小时大约 6 英里。

也提供有默认减速，但如果想自己设定，则需要处理 ManipulationInertiaStarting 事件。ManipulationInertiaStartingRoutedEventArgs 类定义了以下三个属性：

- InertiaTranslationBehavior 类型的 TranslationBehavior
- InertiaExpansionBehavior 类型的 ExpansionBehavior
- InertiaRotationBehavior 类型的 RotationBehavior

InertiaTranslationBehavior 类(例如)能让你按照两种方式来设置线性减速：以像素为单位的 DesiredDisplacement 属性(即你想使物体运行多远)或以每毫秒平方像素为单位的 DesiredDeceleration 属性。两个属性都有 NaN(非数字)默认值。

DesiredDeceleration 值一般都非常小，但可能要来谈谈物理机制了。

根据基本物理知识，我们知道给静止物体施加恒定加速度为 t 时间之后，物体运动的距离为：

$$x = \frac{1}{2}at^2$$

例如，在没有空气阻力的地球表面，物体进行自由落体运动，加速度恒定为每秒 32 英或 32 英尺每平方秒。把 a 设置为 32，则可以计算出该物体 1 秒钟内运行了 16 英尺、2 秒钟运行 64 英尺、3 秒钟运行 144 英尺。

速度 v 为时间相对于距离的一阶导数：

$$v = \frac{\mathrm{d}x}{\mathrm{d}t} = at$$

同样，对于自由落体的物体，第 1 秒钟结束时速度为每秒 32 英尺，第 2 秒钟结束时每秒为 64 英尺，而第 3 秒钟结束时为每秒 96 英尺。每过一秒钟，速度增加每秒 32 英尺。

减速则是反向过程。根据第二个公式，我们知道：

$$a=\frac{v}{t}$$

如果物体的运动速度为 v，经过 t 秒恒定减速，最终速度变为 0。如果屏幕上的物体运动速度为每毫秒 5 个像素，则可以用该公式来计算在固定时间内速度减为零所需的加速度，例如，5 秒钟或 5000 毫秒：

$$a=\frac{5}{5000}=0.001\text{pixels}/\text{m}\sec^2$$

FlickAndBounce 项目也有类似的计算，但可以通过 Slider 来设置减速时间，范围从 1 秒到 60 秒。XAML 文件包含 Slider，已经带 ManipulationMode 设置的 Ellipse 和三个 Manipulation 事件。虽然 ManipulationMode 设置为 All(因为在 XAML 中没有太多替代方法)，但程序只使用平移并通过设置 CanvasLeft 和 CanvasTop 附加属性而不是通过变换来移动 Ellipse。

```
项目: FlickAndBounce | 文件: MainPage.xaml(片段)
<Page ... >
    <Grid Name="contentGrid"
          Background="{StaticResource ApplicationPageBackgroundThemeBrush}">
        <Canvas>
            <Ellipse Name="ellipse"
                     Fill="Red"
                     Width="144"
                     Height="144"
                     ManipulationMode="All"
                     ManipulationStarted="OnEllipseManipulationStarted"
                     ManipulationDelta="OnEllipseManipulationDelta"
                     ManipulationInertiaStarting="OnEllipseManipulationInertiaStarting" />
        </Canvas>

        <Slider x:Name="slider"
                Value="5" Minimum="1" Maximum="60"
                VerticalAlignment="Bottom"
                Margin="24 0" />
    </Grid>
</Page>
```

当然，如果物体正好飞过屏幕边缘，任何减速都没用。由于这个原因，ManipulationDelta 处理程序会检测 Ellipse 何时移出屏幕边缘。ManipulationDelta 还会移回到视图中，就好像从边缘反弹回来并通过 xDirection 和 YDirection 字段对其反转进一步移动。

注意，该逻辑使用了 IsInertial 属性。不会通过拖动 Ellipse 经过屏幕边缘来阻止你操控。

```
项目: FlickAndBounce | 文件: MainPage.xaml.cs(片段)
public sealed partial class MainPage : Page
{
    int xDirection;
    int yDirection;

    public MainPage()
    {
        this.InitializeComponent();
    }

    void OnEllipseManipulationStarted(object sender, ManipulationStartedRoutedEventArgs args)
    {
        // Initialize directions
        xDirection = 1;
        yDirection = 1;
    }
```

```
        void OnEllipseManipulationDelta(object sender, ManipulationDeltaRoutedEventArgs args)
        {
            // Find new position of ellipse regardless of edges
            double x = Canvas.GetLeft(ellipse) + xDirection * args.Delta.Translation.X;
            double y = Canvas.GetTop(ellipse) + yDirection * args.Delta.Translation.Y;

            if (args.IsInertial)
            {
                // Bounce it off the edges
                Size playground = new Size(contentGrid.ActualWidth - ellipse.Width,
                                           contentGrid.ActualHeight - ellipse.Height);

                while (x < 0 || y < 0 || x > playground.Width || y > playground.Height)
                {
                    if (x < 0)
                    {
                        x = -x;
                        xDirection *= -1;
                    }
                    if (x > playground.Width)
                    {
                        x = 2 * playground.Width - x;
                        xDirection *= -1;
                    }
                    if (y < 0)
                    {
                        y = -y;
                        yDirection *= -1;
                    }
                    if (y > playground.Height)
                    {
                        y = 2 * playground.Height - y;
                        yDirection *= -1;
                    }
                }
            }

            Canvas.SetLeft(ellipse, x);
            Canvas.SetTop(ellipse, y);
        }

        voidOnEllipseManipulationInertiaStarting(object sender,
                                ManipulationInertiaStartingRoutedEventArgs args)
        {
            double maxVelocity = Math.Max(Math.Abs(args.Velocities.Linear.X),
                                          Math.Abs(args.Velocities.Linear.Y));

            args.TranslationBehavior.DesiredDeceleration = maxVelocity / (1000 * slider.Value);
        }
    }
```

在 ManipulationInertiaStarting 处理程序的结尾，通过水平和垂直速度绝对值的最大值来，并基于 Slider 值来计算以秒为单位的减速。

13.12 XYSlider 控件

XYSlider 控件类似于一个滑杠，但可以通过改变十字位置(或类似东西)在二维表面上选择点。初一看，Pointer 事件可以用于该控件，直到你意识到该控件并想处理多个手指。如果使用 Manipulation 事件，就能避免这些情况。

我最初就是这么想的，下面来试试看。

我从 ContentControl 派生出 XYSlider，以便只需设置 Content 属性就可以显示想作为背景的任何东西。而在此之上是用手指、鼠标或手写笔左右移动的十字。控件有一个属性、Point 类型的 Value 以及 ValueChanged 事件。Point 属性的 X 和 Y 坐标进行归一化处理，范围为 0 至 1 的内容相对值，这样会减轻定义类似于 RangeBase 或 Slider 的 IsDirectionReversed 属性 Minimum 和 Maximum 值的控件负担。(实际上，需要一对 IsDirectionReversed 属性提供 X 轴和 Y 轴。)

控件定义本身为无模板，但模板中需要两个部分： ContentControl 模板中正常可见的定制 ContentPresenter 以及视觉上类似于十字的东西。通过 Canvas.Left 和 Canvass.Top 附加属性的代码来移动十字，强烈表明模板需要在 Canvas 中定义十字。

```
项目: XYSliderDemo | 文件: XYSlider.cs
namespace XYSliderDemo
{
    public class XYSlider : ContentControl
    {
        ContentPresenter contentPresenter;
        FrameworkElement crossHairPart;

        static readonly DependencyProperty valueProperty =
                DependencyProperty.Register("Value",
                        typeof(Point), typeof(XYSlider),
                        new PropertyMetadata(new Point(0.5, 0.5), OnValueChanged));

        public event EventHandler<Point> ValueChanged;

        public XYSlider()
        {
            this.DefaultStyleKey = typeof(XYSlider);
        }

        public static DependencyProperty ValueProperty
        {
            get { return valueProperty; }
        }

        public Point Value
        {
            set { SetValue(ValueProperty, value); }
            get { return (Point)GetValue(ValueProperty); }
        }

        protected override void OnApplyTemplate()
        {
            // Detach event handlers
            if (contentPresenter != null)
            {
                contentPresenter.ManipulationStarted -= OnContentPresenterManipulationStarted;
                contentPresenter.ManipulationDelta -= OnContentPresenterManipulationDelta;
                contentPresenter.SizeChanged -= OnContentPresenterSizeChanged;
            }

            // Get new parts
            crossHairPart = GetTemplateChild("CrossHairPart") as FrameworkElement;
            contentPresenter = GetTemplateChild("ContentPresenterPart") as ContentPresenter;

            // Attach event handlers
            if (contentPresenter != null)
            {
                contentPresenter.ManipulationMode = ManipulationModes.TranslateX |
                ManipulationModes.TranslateY;
                contentPresenter.ManipulationStarted += OnContentPresenterManipulationStarted;
```

```
            contentPresenter.ManipulationDelta += OnContentPresenterManipulationDelta;
            contentPresenter.SizeChanged += OnContentPresenterSizeChanged;
        }

        // Make cross-hair transparent to touch
        if (crossHairPart != null)
        {
            crossHairPart.IsHitTestVisible = false;
        }

        base.OnApplyTemplate();
    }

    void OnContentPresenterManipulationStarted(object sender,
                                ManipulationStartedRoutedEventArgs args)
    {
        RecalculateValue(args.Position);
    }

    void OnContentPresenterManipulationDelta(object sender,
                                ManipulationDeltaRoutedEventArgs args)
    {
        RecalculateValue(args.Position);
    }

    void OnContentPresenterSizeChanged(object sender, SizeChangedEventArgs args)
    {
        SetCrossHair();
    }

    void RecalculateValue(Point absolutePoint)
    {
        double x = Math.Max(0,Math.Min(1, absolutePoint.X / contentPresenter.ActualWidth));
        double y = Math.Max(0,Math.Min(1, absolutePoint.Y / contentPresenter.ActualHeight));
        this.Value = new Point(x, y);
    }

    void SetCrossHair()
    {
        if (contentPresenter != null && crossHairPart != null)
        {
            Canvas.SetLeft(crossHairPart, this.Value.X * contentPresenter.ActualWidth);
            Canvas.SetTop(crossHairPart, this.Value.Y * contentPresenter.ActualHeight);
        }
    }

    static void OnValueChanged(DependencyObject obj,
                            DependencyPropertyChangedEventArgs args)
    {
        (obj as XYSlider).SetCrossHair();
        (obj as XYSlider).OnValueChanged((Point)args.NewValue);
    }

    protected void OnValueChanged(Point value)
    {
        if (ValueChanged != null)
            ValueChanged(this, value);
    }
  }
}
```

如果用编程方式设置 Value 属性，则类必须通过相对坐标将 ContentPresenter 的宽度和高度相乘并把十字设置到正确位置。这种情况发生在 SetCrossHair 方法中。ManipulationStarted 和 ManipulationDelta 事件处理程序是在 ContentPresenter 对象中进行设置。二者调用 RecalculateValue 方法，将 Value 属性的指针绝对坐标转换为相对坐标。

ManipulationStarted 和 ManipulationDelta 处理程序都引用 Position 事件参数属性，之前没有提到过这个该属性。对于鼠标或手写笔，Position 属性就是与生成 Manipulation 事件(此时为 ContentPresenter)相关的鼠标指针位置或笔尖位置。对于触碰，Position 属性是所有参与操控的手指的平均位置。如果只需要一个手指的位置，Position 属性就可以提供方便的方式来处理多个手指。

MainPage.xaml 文件实例化 XYSlider，并且引用从美国 NASA 网站获得扁平地球地图。但 XAML 文件的大部分专门用于为 XYSlider(特别是十字)定义模板。请注意，我把 ContentPresenter 和 Canvas 放入 Grid，并为 Grid 分配了通常分配给 ContentPresenter 的一些属性。也就是说，ContentPresenter 和 Canvas 的左上角对齐，ContentPresenter 坐标和相对坐标系之间的转换就会容易一些。

```
项目: XYSliderDemo | 文件 2: MainPage.xaml(片段)
<Page ... >
    <Page.Resources>
        <ControlTemplate x:Key="xySliderTemplate" TargetType="local:XYSlider">
            <Border BorderBrush="{TemplateBinding BorderBrush}"
                    BorderThickness="{TemplateBinding BorderThickness}"
                    Background="{TemplateBinding Background}">

                <Grid Margin="{TemplateBinding Padding}"
                      HorizontalAlignment="{TemplateBinding HorizontalContentAlignment}"
                      VerticalAlignment="{TemplateBinding VerticalContentAlignment}">

                    <ContentPresenter Name="ContentPresenterPart"
                                      Content="{TemplateBinding Content}"
                                      ContentTemplate="{TemplateBinding ContentTemplate}" />
                    <Canvas>
                        <Path Name="CrossHairPart"
                              Stroke="{TemplateBinding Foreground}"
                              StrokeThickness="3"
                              Fill="Transparent">
                            <Path.Data>
                                <GeometryGroup FillRule="Nonzero">
                                    <EllipseGeometry RadiusX="48" RadiusY="48" />
                                    <EllipseGeometry RadiusX="6" RadiusY="6" />
                                    <LineGeometry StartPoint="-48 0" EndPoint="-6 0" />
                                    <LineGeometry StartPoint="48 0" EndPoint="6 0" />
                                    <LineGeometry StartPoint="0 -48" EndPoint="0 -6" />
                                    <LineGeometry StartPoint="0 48" EndPoint="0 6" />
                                </GeometryGroup>
                            </Path.Data>
                        </Path>
                    </Canvas>
                </Grid>
            </Border>
        </ControlTemplate>

        <Style TargetType="local:XYSlider">
            <Setter Property="Template" Value="{StaticResource xySliderTemplate}" />
        </Style>
    </Page.Resources>

    <Grid Background="{StaticResource ApplicationPageBackgroundThemeBrush}">
        <Grid.RowDefinitions>
            <RowDefinition Height="*" />
            <RowDefinition Height="Auto" />
        </Grid.RowDefinitions>

        <local:XYSlider x:Name="xySlider"
                        Grid.Row="0"
                        Margin="48"
```

```
                    ValueChanged="OnXYSliderValueChanged">
            <!-- Image courtesy of NASN/JPL-Caltech (http://maps.jpl.nasa.gov) -->
            <Image Source="Images/ear0xuu2.jpg" />
        </local:XYSlider>

        <TextBlock Name="label"
                   Grid.Row="1"
                   Style="{StaticResource SubheaderTextStyle}"
                   HorizontalAlignment="Center" />
    </Grid>
</Page>
```

代码隐藏文件中有一个 XYSlider 之 ValueChanged 事件的处理程序，它可以用来显示对应的经纬度。看看其他工作方式，代码还采用 Geolocator 类来获取正在运行该程序的计算机的当前地理位置。

```
项目: XYSliderDemo | 文件: MainPage.xaml.cs(片段)
public sealed partial class MainPage : Page
{
    bool manualChange = false;

    public MainPage()
    {
        this.InitializeComponent();

        // Initialize position of cross-hair in XYSlider
        Loaded += async (sender, args) =>
            {
                Geolocator geolocator = new Geolocator();

                // Might not have permission!
                try
                {
                    Geoposition position = await geolocator.GetGeopositionAsync();

                    If (!manualChange)
                    {
                        double x = (position.Coordinate.Longitude + 180) / 360;
                        double y = (90 - position.Coordinate.Latitude) / 180;
                        xySlider.Value = new Point(x, y);
                    }
                }
            catch
            {
            }
        };
    }

    void OnXYSliderValueChanged(object sender, Point point)
    {
        double longitude = 360 * point.X - 180;
        double latitude = 90 - 180 * point.Y;
        label.Text = String.Format("Longitude: {0:F0} Latitude: {1:F0}",
                                   longitude, latitude);
        manualChange = true;
    }
}
```

使用 Geolocator 类需要编辑 Package.appxmanifest 类来获得定位能力。在 Visual Studio 中，选择 Package.appxmanifest 文件，再选择 Capabilities 选项卡，然后点击 Location。运行时，Windows 8 会询问用户是否确定程序可以获取电脑位置。如果用户拒绝，GetGeopositionAsync 调用会报异常。

效果如下图所示。

在我为 Windows Phone 7 写的该控件早期版本时，十字使用了模板化 Thumb。该版本并不理想，因为需要用户把 Thumb 从当前位置到拖动新位置。而对新版本，我想通过简单触碰就足以让十字跳到新的位置。

但我也不肯定是否能成功。正如前面提到的(你也会体验到)，仅仅触碰位置并不可以把十字拖拽到该点，因为需要一些动作才能触发 ManipulationStarted 事件。

起初，我认为用 PointerPressed 事件代替 ManipulationStarted 响应速度能够更快。然而，对 PointerRoutedEventArgs 对象直接调用 GetCurrentPoint 显然会禁止 Manipulation 事件。

Pointer 事件最理想的就是出现这种情况，如果多个手指要移动十字，就要平均处理这些手指。如果下一章有更好的 XYSlide 能用于在基于位图的手绘程序中选择颜色，我觉得也很正常。

13.13　中心缩放和旋转

在第一次介绍 Manipulation 事件的缩放和旋转特性时，我提到过通过引用中心点来应用变换。但在许多情况下，这一点很重要。触屏界面的满意程度在很大程度上取决于用户手指与屏幕对象之间的连接有多紧密。

我在前面使用过 Position 属性，而确定该属性的缩放和旋转需要讲一点儿技巧。Position 属性是所有手指相对于被操控元素的位置平均值。Position 属性不是缩放和旋转的中心，但可以用来推导出中心。

CenteredTransforms 项目有一个 XAML 文件，文件引用的是我网站上的位图。

项目：CenteredTransforms | 文件：MainPage.xaml(片段)
```
<Page ... >
    <Grid Background="{StaticResource ApplicationPageBackgroundThemeBrush}">
        <Image Name="image"
               Source="http://www.charlespetzold.com/pw6/PetzoldJersey.jpg"
               Stretch="None"
               HorizontalAlignment="Left"
               VerticalAlignment="Top">
```

```
                <Image.RenderTransform>
                    <TransformGroup x:Name="xformGroup">
                        <MatrixTransform x:Name="matrixXform" />
                        <CompositeTransform x:Name="compositeXform" />
                    </TransformGroup>
                </Image.RenderTransform>
            </Image>
        </Grid>
</Page>
```

请注意，RenderTransform 属性现在被设置为 TransformGroup，它同时包含 MatrixTransform 和 CompositeTransform。

该代码隐藏文件使轨道之外所有形式的 Manipulation 都可行。

项目: CenteredTransforms | 文件: MainPage.xaml.cs(片段)

```
public sealed partial class MainPage : Page
{
    public MainPage()
    {
        this.InitializeComponent();
        image.ManipulationMode = ManipulationModes.All &
                                 ~ManipulationModes.TranslateRailsX &
                                 ~ManipulationModes.TranslateRailsY;
    }

    protected override void OnManipulationDelta(ManipulationDeltaRoutedEventArgs args)
    {
        // Make this the entire transform to date
        matrixXform.Matrix = xformGroup.Value;

        // Use that to transform the Position property
        Point center = matrixXform.TransformPoint(args.Position);

        // That becomes the center of the new incremental transform
        compositeXform.CenterX = center.X;
        compositeXform.CenterY = center.Y;

        // Set the other properties
        compositeXform.TranslateX = args.Delta.Translation.X;
        compositeXform.TranslateY = args.Delta.Translation.Y;
        compositeXform.ScaleX = args.Delta.Scale;
        compositeXform.ScaleY = args.Delta.Scale;
        compositeXform.Rotation = args.Delta.Rotation;

        base.OnManipulationDelta(args);
    }
}
```

OnManipulationDelta 覆写 XAML 文件中定义的三种变换对象。任何时候，TransformGroup 的 Value 属性(为 Matrix 值)代表整个变换，是由 MatrixTransform 和 CompositeTransform 对象所代表的变换的结果。ManipulationDelta 处理程序首先把 Matrix 值从 TransformGroup 设置为 MatrixTransform，也就是说，MatrixTransform 现在是到该点的整个变换。把转换也应用到 Position 属性，然后成为 CompositeTransform 的 CenterX 和 CenterY 属性。可以把从 ManipulationDelta 结构所得新值直接设置为 CompositeTransform 的其他属性。

这样能起作用吗？你肯定想试一下，因为仅靠下面的屏幕截图是看不出来的。

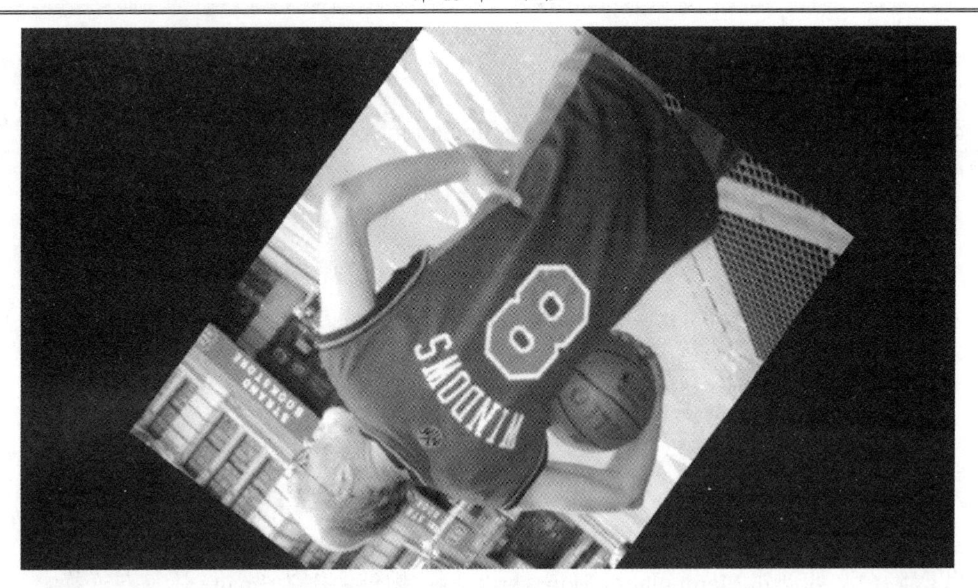

试着把一个手指固定在画面的一个角，同时拉对角或旋转，你会看到图片会随着手指运动，当然要指定同性缩放限制。

为了让你更方便，我写了一个很小的类，称为 ManipulationManager，该类会在自己的私有变换集合中执行计算，变换在构造函数中进行创建并保存在字段中。

项目：ManipulationManagerDemo | 文件：ManipulationManager.cs

```
using Windows.Foundation;
using Windows.UI.Input;
using Windows.UI.Xaml.Media;

namespace ManipulationManagerDemo
{
    public class ManipulationManager
    {
        TransformGroup xformGroup;
        MatrixTransform matrixXform;
        CompositeTransform compositeXform;

        public ManipulationManager()
        {
            xformGroup = new TransformGroup();
            matrixXform = new MatrixTransform();
            xformGroup.Children.Add(matrixXform);
            compositeXform = new CompositeTransform();
            xformGroup.Children.Add(compositeXform);
            this.Matrix = Matrix.Identity;
        }

        public Matrix Matrix { private set; get; }

        public void AccumulateDelta(Point position, ManipulationDelta delta)
        {
            matrixXform.Matrix = xformGroup.Value;
            Point center = matrixXform.TransformPoint(position);
            compositeXform.CenterX = center.X;
            compositeXform.CenterY = center.Y;
            compositeXform.TranslateX = delta.Translation.X;
            compositeXform.TranslateY = delta.Translation.Y;
            compositeXform.ScaleX = delta.Scale;
            compositeX     form.ScaleY = delta.Scale;
            compositeXform.Rotation = delta.Rotation;
            this.Matrix = xformGroup.Value;
```

```
            }
        }
    }
```

公共 AccumulateDelta 方法直接接受 ManipulationDelta 值并计算出新的 Matrix 属性。这样允许必须以这种方式操控的元素只有单一变换。

项目 Project: ManipulationManagerDemo | 文件: MainPage.xaml(片段)

```
<Page ... >
    <Grid Background="{StaticResource ApplicationPageBackgroundThemeBrush}">
        <Image Name="image"
               Source=http://www.charlespetzold.com/pw6/PetzoldJersey.jpg
               Stretch="None"
               HorizontalAlignment="Left"
               VerticalAlignment="Top">
            <Image.RenderTransform>
                <MatrixTransform x:Name="matrixXform" />
            </Image.RenderTransform>
        </Image>
    </Grid>
</Page>
```

代码隐藏文件创建 ManipulationManager 实例并用该实例为 Image 计算新变换。

项目: ManipulationManagerDemo | 文件: MainPage.xaml.cs(片段)

```
public sealed partial class MainPage : Page
{
    ManipulationManager manipulationManager = new ManipulationManager();

    public MainPage()
    {
        this.InitializeComponent();

        image.ManipulationMode = ManipulationModes.All &
                                 ~ManipulationModes.TranslateRailsX &
                                 ~ManipulationModes.TranslateRailsY;
    }

    protected override void OnManipulationDelta(ManipulationDeltaRoutedEventArgs args)
    {
        manipulationManager.AccumulateDelta(args.Position, args.Delta);
        matrixXform.Matrix = manipulationManager.Matrix;
        base.OnManipulationDelta(args);
    }
}
```

如果屏幕上有多个可操控的对象，则需要为每一个对象都创建 ManipulationManager 实例。下一章 PhotoScatter 项目中，我将使用 ManipulationManager 变化，该项目在图片目录中显示图片并且可以通过手指仔细查看。

13.14 单手指旋转

虽然 ManipulationStarting 事件并不一定会通知实际要发生的操控，但它为程序提供了一些方法来初始化操控，它们都涉及 ManipulationStartingRoutedEventArgs 的属性。

- Mode 属性是人们熟悉的枚举类型 ManipulationModes，可以设定成你需要处理的操控类型。但要记住，只有 Mode 将其 ManipulationMode 属性设置为除 ManipulationModes.None 或 ManipulationModes.System 之外的属性时，才会得到 ManipulationStarting 事件。

- Container 属性在其他所有 Manipulation 事件都为只读，但在 ManipulationStarting 事件中为可写。默认情况下，Container 属性与 OriginalSource 属性相同，但在之后的事件中，Container 属性是和 Position 属性相对的元素。如果想让 Position 属性是对于 OriginalSource 以外的元素，则可以把 Container 属性设置为该元素。
- Pivot 属性允许单指旋转，详见后文讨论。

假设桌子上有一张照片。(我指的是真的桌子上有一张真的照片。)用手指碰碰照片一角，将照片拉向自己一方。照片还留在同一个方向吗？不一定会。如果轻轻碰照片，桌面和照片之间的摩擦力会让照片略有旋转，其他部分会留在你所拉那一角的后面。

用单手旋转或得到类似效果，但你需要使用我刚刚描述的围绕中心旋转对象的技巧。事实上，XAML 文件基本上和 CenteredTransforms 项目一样。

项目：SingleFingerRotate | 文件：MainPage.xaml(片段)

```
<Page ... >
    <Grid Background="{StaticResource ApplicationPageBackgroundThemeBrush}">
        <Image Name="image"
               Source="http://www.charlespetzold.com/pw6/PetzoldJersey.jpg"
               Stretch="None"
               HorizontalAlignment="Left"
               VerticalAlignment="Top"
               RenderTransformOrigin="0 0">
            <Image.RenderTransform>
                <TransformGroup x:Name="xformGroup">
                    <MatrixTransform x:Name="matrixXform" />
                    <CompositeTransform x:Name="compositeXform" />
                </TransformGroup>
            </Image.RenderTransform>
        </Image>
    </Grid>
</Page>
```

代码隐藏文件几乎相同，不同的是 OnManipulationStarting 覆写。

项目：SingleFingerRotate | 文件：MainPage.xaml.cs(片段)

```
public sealed partial class MainPage : Page
{
    public MainPage()
    {
        this.InitializeComponent();

        image.ManipulationMode = ManipulationModes.All &
                                 ~ManipulationModes.TranslateRailsX &
                                 ~ManipulationModes.TranslateRailsY;
    }

    protected override void OnManipulationStarting(ManipulationStartingRoutedEventArgs args)
    {
        args.Pivot = new ManipulationPivot(new Point(image.ActualWidth / 2,
                                                     image.ActualHeight / 2),
                                           50);
        base.OnManipulationStarting(args);
    }

    protected override void OnManipulationDelta(ManipulationDeltaRoutedEventArgs args)
    {
        // Make this the entire transform to date
        matrixXform.Matrix = xformGroup.Value;

        // Use that to transform the Position property
        Point center = matrixXform.TransformPoint(args.Position);

        // That becomes the center of the new incremental transform
```

```
            compositeXform.CenterX = center.X;
            compositeXform.CenterY = center.Y;

            // Set the other properties
            compositeXform.TranslateX = args.Delta.Translation.X;
            compositeXform.TranslateY = args.Delta.Translation.Y;
            compositeXform.ScaleX = args.Delta.Scale;
            compositeXform.ScaleY = args.Delta.Scale;
            compositeXform.Rotation = args.Delta.Rotation;

            base.OnManipulationDelta(args);
        }
    }
```

这里的关键是把 ManipulationStartingRoutedEventArgs 对象的 Pivot 属性设置为 ManipulationPivot 对象。该对象提供了以下两项内容：

- 旋转中心，总是所操控对象的中心
- 围绕中心的保护半径，此处设置为 50 像素

如果没有第二项，手指会非常接近元素中心，而小小的移动就可以使元素转一大圈。

这是你真正必须自己尝试的程序之一，感受一下单手旋转如何给拖动操作增加真实感。

还记得第 5 章的 SliderSketch 程序吗？记得你的疑问吗：“这些不应该是刻盘，而应该是滑块？”本章最后的 DialSketch 程序就使用结合了单指旋转的 Dial 控件。

为了比较容易定义 Dial 类，我决定 Dial 类应该派生自与 Slider 相似的 RangeBase。这样能赋予控件所有 double 类型的 Minimum、Maximum 和 Value 属性以及 ValueChanged 事件。然而，控件中的 double 值是旋转角度，唯一启用的操控为旋转。

```
项目: DialSketch | 文件: Dial.cs
using System;
using Windows.Foundation;
using Windows.UI.Xaml.Controls.Primitives;
using Windows.UI.Xaml.Input;

namespace DialSketch
{
    public class Dial : RangeBase
    {
        public Dial()
        {
            ManipulationMode = ManipulationModes.Rotate;
        }

        protected override void OnManipulationStarting ( ManipulationStartingRoutedEventArgs args )
        {
            args.Pivot = new ManipulationPivot(new Point(this.ActualWidth / 2,
                                                         this.ActualHeight / 2),
                                               48);
        base.OnManipulationStarting(args);
        }

        protected override void OnManipulationDelta(ManipulationDeltaRoutedEventArgs args)
        {
            this.Value = Math.Max(this.Minimum,
                         Math.Min(this.Maximum, this.Value + args.Delta.Rotation));

            base.OnManipulationDelta(args);
        }
    }
}
```

就是这样！当然，还没有模板，也无法访问任何变换。只设置一个新的 Value 属性(导

致 RangeBase 触发 ValueChanged 事件)，并预计在其他地方实现。

其中两个 Dial 控件在 DialSketch 的 XAML 文件中进行实例化。Resources 部分专门为这两个控件提供 Style，包括 ControlTemplate。Dial 控件从视觉上让用户知道是在旋转，因此，模板使用非常短的破折号来模拟刻度虚线。

注意对 Dial 设置的 Minimum 和 Maximum 值。这意味着可以在最小位置和最大位置之间旋转 Dial 整整 10 次。为了从 DialSketch 画布一边到另一边画一条线，需要转动转 10 次。

项目: DialSketch | 文件: MainPage.xaml(片段)

```
<Page ... >
    <Page.Resources>
        <Style TargetType="local:Dial">
            <Setter Property="Minimum" Value="-1800" />
            <Setter Property="Maximum" Value="1800" />
            <Setter Property="RenderTransformOrigin" Value="0.5 0.5" />
            <Setter Property="Width" Value="144" />
            <Setter Property="Height" Value="144" />
            <Setter Property="Margin" Value="24" />
            <Setter Property="Template">
                <Setter.Value>
                    <ControlTemplate>
                        <Grid>
                            <Ellipse Fill="DarkRed" />
                            <Ellipse Stroke="Black"
                                     StrokeThickness="12"
                                     StrokeDashArray="0.1 1"
                                     Margin="3" />
                            <Ellipse Fill="Black"
                                     Width="6"
                                     Height="6" />
                        </Grid>
                    </ControlTemplate>
                </Setter.Value>
            </Setter>
        </Style>
    </Page.Resources>

    <Grid Background="{StaticResource ApplicationPageBackgroundThemeBrush}">
        <Grid.RowDefinitions>
            <RowDefinition Height="*" />
            <RowDefinition Height="Auto" />
        </Grid.RowDefinitions>

        <Grid.ColumnDefinitions>
            <ColumnDefinition Width="Auto" />
            <ColumnDefinition Width="*" />
            <ColumnDefinition Width="Auto" />
        </Grid.ColumnDefinitions>

        <Border Grid.Row="0"
                Grid.Column="0"
                Grid.ColumnSpan='3'
                BorderBrush="{StaticResource ApplicationForegroundThemeBrush}"
                BorderThickness="3 0 0 3"
                Background="#C0C0C0"
                Padding="24">
            <Grid Name="drawingGrid">
                <Polyline Name="polyline"
                          Stroke="#404040"
                          StrokeThickness="3" />
            </Grid>
        </Border>

        <local:Dial x:Name="horzDial"
                    Grid.Row="1"
```

```
                    Grid.Column="0"
                    Maximum="1800"
                       ValueChanged="OnDialValueChanged">
            <local:Dial.RenderTransform>
                 <RotateTransform />
            </local:Dial.RenderTransform>
        </local:Dial>

        <Button Content="Clear"
                Grid.Row="1"
                Grid.Column="1"
                HorizontalAlignment="Center"
                VerticalAlignment="Center"
                Click="OnClearButtonClick" />

        <local:Dial x:Name="vertDial"
                    Grid.Row="1"
                    Grid.Column="2"
                    Maximum="1800"
                    ValueChanged="OnDialValueChanged">
        <local:Dial.RenderTransform>
             <RotateTransform />
        </local:Dial.RenderTransform>
    </local:Dial>
  </Grid>
</Page>
```

注意，单个 Dial 控件重复出现了 Maximum 设置。在我使用的 Windows 8 版本中，Style 中的设定似乎并没有“拿走”。还要注意，每个 Dial 控件都附加了 RotateTransform。

代码隐藏文件把 Polyline 初始化为一个中心点。对于 Dial 产生的每个 ValueChanged 事件，都要设置控件中的 RotateTranform 并把新的 Point 添加到 Polyline。

项目: DialSketch | 文件: MainPage.xaml.cs(片段)

```
public sealed partial class MainPage : Page
{
    public MainPage()
    {
        this.InitializeComponent();

        loaded += (sender, args) =>
            {
                polyline.Points.Add(new Point(drawingGrid.ActualWidth / 2,
                                              drawingGrid.ActualHeight / 2));
            };

    }

    void OnDialValueChanged(object sender, RangeBaseValueChangedEventArgs args)
    {
        Dial dial = sender as Dial;
        RotateTransform rotate = dial.RenderTransform as RotateTransform;
        rotate.Angle = args.NewValue;

        double xFraction = (horzDial.Value - horzDial.Minimum) /
                                (horzDial.Maximum - horzDial.Minimum);

        double yFraction = (vertDial.Value - vertDial.Minimum) /
                                (vertDial.Maximum - vertDial.Minimum);

        double x = xFraction * drawingGrid.ActualWidth;
        double y = yFraction * drawingGrid.ActualHeight;
        polyline.Points.Add(new Point(x, y));
    }

    void OnClearButtonClick(object sender, RoutedEventArgs args)
```

```
        {
            polyline.Points.Clear();
        }
    }
```

当然，程序到现在还无法使用，但至少能打招呼了(见下图)。

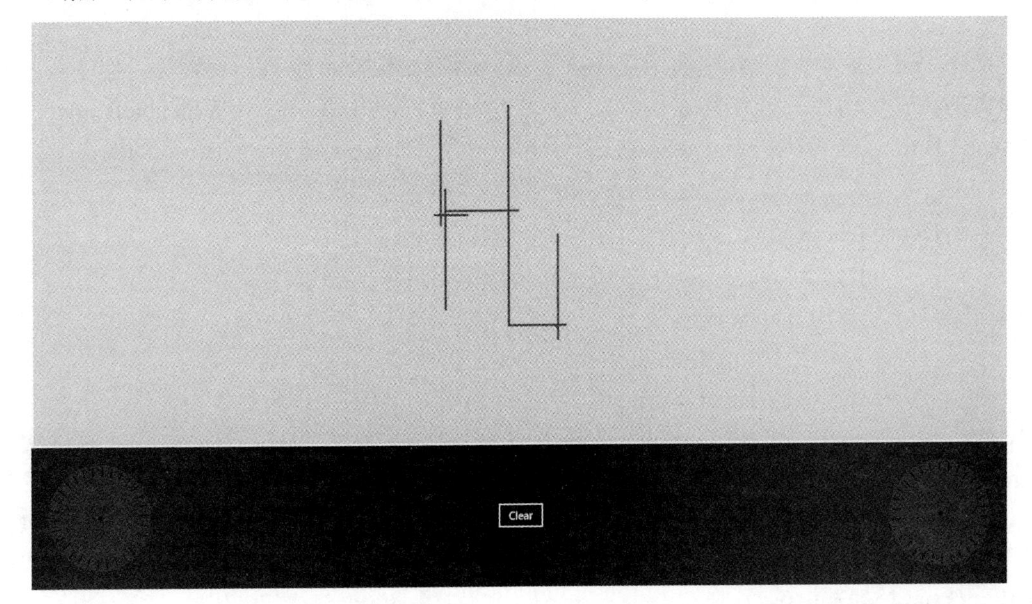

第 14 章　位　　图

从本书开始，我们一直都在用位图来显示、画笔、拉伸、倾斜和旋转等。但本章将介绍位图的内在灵魂以及操作其像素位。本章中的每个程序几乎都使用 WriteableBitmap 类，WriteableBitmap 类派生自 ImageSource，因此可以用作 Image 和 ImageBrush 的源：

Object
　　DependencyObject
　　　　ImageSource
　　　　　　BitmapSource
　　　　　　　　BitmapImage
　　　　　　　　WriteableBitmap

Writeablebitmap 从 Bitmapsource 继承 SetSource 方法，通过该方法可以通过实施 IRrandomAccessStream 的类来加载位图文件。

Writeablebitmap 的不同之处在于其定义 PixelBuffer 属性，允许你访问像素位。可以操控现有图片的像素或从头开始创建完整图片。本章还会讨论基于像素数组读/写多种格式的图片文件(如 PNG 和 JPEG)。

如果你熟悉 WriteableBitmap 的 Silverlight 版本，则可能很失望地发现 Windows Runtime 版本没有实现 Render 方法，而有了它，你可以在图片表面渲染任何 UIElement。这会大大限制 WriteableBitmap 的若干常见目的。

例如在第 13 章，你看到很多手绘程序通过 Line、Polyline 和 Path 元素来渲染指针输入，你可能注意到我没有给你提供如何保存绘画的方式。有一种非常合理的方式能保存绘画，即在位图中渲染 Line、Polyline 和 Path 元素，然后将位图保存为文件。但 WriteableBitmap 中缺少 Render 方法，大大阻碍了这一过程。

在本章中，我会向你展示如何用算法在位图上画线条。这样一来，我可以给出 FingerPaint 程序(项目名中无任何数字)，可以将作品存储为位图。在第 15 章中，我会向你展示如何使用 SurfaceImageSource，SurfaceImageSource 也派生自 ImageSource，并且可以通过 C ++代码的 DirectX 绘图操作进行绘制。

我不会在书中讨论第三方库的 API，但如果你需要在位图上绘制复杂的图形，WriteableBitmapEx 非常有用，可访问 http://writeablebitmapex.codeplex.com 了解详情。

14.1　像　素　位

位图图片的行和列均为整数。对于任何一个派生自 BitmapSource 类的实例，PixelHeight 和 PixelWidth 属性都提供这些尺寸。

从概念上讲，像素位存储在一个大小等于 PixelHeight 和 PixelWidth 的二维数组中。在实际中，数组只有一个维度，但最大的问题是单个像素本身的性质。这有时称为位图的“颜色格式”，范围可以从每像素 1 位(在只有黑和白的位图中)到每像素 1 字节(在灰度位图或一个 256 色的位图中)再到每像素 3 或 4 字节(有或没有透明度的全彩色)，而对于更多颜色分辨率甚至更高。

然而在处理 WriteableBitmap 的时候，已经建立统一颜色格式。在每个 WriteableBitmap 中，每个像素都包含 4 个字节。因此，位图中的像素阵列的总字节数为：

```
PixelHeight * PixelWidth * 4
```

图片从最上面一行开始，每行从左到右。没有行填充。对于每个像素，字节顺序如下：

```
Blue、Green、Red、Alpha
```

Color 值的字节范围从 0 到 255。WriteableBitmap 颜色值根据 sRGB(“标准 RGB”)进行假设，因此会兼容 Windows Runtime 的 Color 值(除了稍后讨论的 Color.Transparent)。

WriteableBitmap 中的像素为预乘 α 通道格式。我稍后简要讨论其含义。

Blue、Green、Red、Alpha 的顺序似乎比我们通常如何指称这些颜色字节(以及它们在 Color.FromArgb 方法中的顺序)落后，但如果你认为 WriteableBitmap 像素实际是一个 32 位无符号整数，α值存储在高字节，而蓝色值存储在低字节，则更合理。操作系统安装在 Intel 微处理器时，整数通常存储在位图的小字节序(最低字节优先)。

我们来创建 writeablebitmap 并用像素进行填充，以构建一个自定义图片。为了便于计算，假设 writeablebitmap 有 256 行和 256 列。左上角设置为黑色，右上角设置为蓝色，左下角设置为红色，而右下角设置为品红色，即蓝色和红色的组合。这是一种渐变形式，但不像 Windows Runtime 中可用的任何渐变。

以下 XAML 文件的 Image 元素准备接受 ImageSource 派生类。

```
项目: CustomGradient | 文件: MainPage.xaml(片段)
<Grid Background="{StaticResource ApplicationPageBackgroundThemeBrush}">
    <Image Name="image" />
</Grid>
```

无法实例化 XAML 中的 WriteableBitmap，因为没有无参数构造函数。代码隐藏文件在 Loaded 事件的处理程序中创建并建立 WriteableBitmap。如下完整文件中，还可以看到 Using 指令。WriteableBitmap 本身是在 Windows.UI.Xaml.Media.Imaging 命名空间中进行定义的。

```
项目: CustomGradient | 文件: MainPage.xaml.cs
using System.IO;
using System.Runtime.InteropServices.WindowsRuntime;
using Windows.UI.Xaml;
using Windows.UI.Xaml.Controls;
using Windows.UI.Xaml.Media.Imaging;

namespace CustomGradient
{
    public sealed partial class MainPage : Page
    {
        public MainPage()
        {
            this.InitializeComponent();
            Loaded += OnMainPageLoaded;
        }

        async void OnMainPageLoaded(object sender, RoutedEventArgs args)
        {
            WriteableBitmap bitmap = new WriteableBitmap(256, 256);
            byte[] pixels = new byte[4 * bitmap.PixelWidth * bitmap.PixelHeight];
```

```
                for  (int y = 0; y < bitmap.PixelHeight; y++)
                    for  (int x = 0; x < bitmap.PixelWidth; x++)
                    {
                        int index = 4 * (y * bitmap.PixelWidth + x);
                        pixels[index + 0] = (byte)x; // Blue
                        pixels[index + 1] = 0;              // Green
                        pixels[index + 2] = (byte)y; // Red
                        pixels[index + 3] = 255;            // Alpha
                    }

                using (Stream pixelStream = bitmap.PixelBuffer.AsStream())
                {
                    await pixelStream.WriteAsync(pixels, 0, pixels.Length);
                }
                bitmap.Invalidate();
                image.Source = bitmap;
            }
        }
    }
```

WriteableBitmap 构造函数需要像素的宽度和高度。程序根据宽度和高度为像素分配字节数组：

```
byte[] pixels = new byte[4 * bitmap.PixelWidth * bitmap.PixelHeight];
```

WriteableBitmap 的数组大小总是照此计算的。

行和列循环到位图的每个像素。可以如下计算引用特定像素的像素数组索引：

```
int index = 4 * (y * bitmap.PixelWidth + x);
```

每个像素可以按照为蓝、绿、红、α 的顺序进行设置。

在这个特定的例子中，两个循环解决像素存储在数组中的顺序，因此，并不需要为每个像素重新计算 index。index 可以初始化为零并像下面这样递增：

```
int index = 0;
for (int y = 0; y < bitmap.PixelHeight; y++)
    for (int x = 0; x < bitmap.PixelWidth; x++)
    {
        pixels[index++] = (byte)x; // Blue
        pixels[index++] = 0; // Green
        pixels[index++] = (byte)y; // Red
        pixels[index++] = 255; // Alpha
    }
```

这肯定比我以前用的方法快，但通用性一般差一点。也可以为 index 定义循环，然后计算 x 和 y。(大多数情况下)重要的是能访问每个像素。

在 byte 数组已满后，像素必须传到 WriteableBitmap。第一次检查时该过程看似有点让人迷惑。WriteableBitmap 所定义的 PixelBuffer 属性为 IBuffer 类型，只定义了两个属性：Capacity 和 Length。我在第 7 章讨论过，IBuffer 对象通常为在操作系统内维护的一个存储区域，用于引用计数，以便可以在不需要的时候删除掉。你需要把字节传到该缓冲区。

幸运的是，有一个称为 AsStream 的扩展方法，它可以把 IBuffer 作为.NET Stream 对象来处理：

```
Stream pixelStream = bitmap.PixelBuffer.AsStream();
```

要使用该扩展方法，必须给 Runtime.interopservices.windowsruntime 命名空间包括 using 指令。如果没有 using 指令，IntelliSense 不会显示存在该方法。

然后可以使用 Stream 所定义的正常 Write 方法来为 Stream 对象写入字节数组，或者可以像我一样使用 WriteaSync。由于该位图不是很大，而且调用只通过 API 来传输字节数组，Write 应该快到足以证明用户界面线程的工作。可以“手动”处理 Stream 或让其自动处理，也可以像我一样把 Stream 逻辑放在 using 语句中：

```
using (Stream pixelStream = bitmap.PixelBuffer.AsStream())
{
    await pixelStream.WriteAsync(pixels, 0, pixels.Length);
}
```

无论什么时候改变 WriteableBitmap 像素，都要习惯对位图调用 Invalidate：

```
bitmap.Invalidate();
```

该调用会请求重绘位图。在这里的特定情况下，并没有做严格要求，但在其他情况下该调用很重要。

最后，不要忘记显示最终位图！该程序直接将其设置为 XAML 文件中 Image 元素的 Source 属性：

```
image.Source = bitmap;
```

效果如下图所示。

如果把 Stream 对象和像素数组保持为位图进一步操作的字段(位图图片可能随时间而变化)，则需要用 Seek 调用开始 WriteAsync，以当前位置设置回到起点：

```
pixelStream.Seek(0, SeekOrigin.Begin);
```

但也要注意，还可以只把部分 byte 数组写入位图。例如，假设只修改对应像素坐标(x1，y1)的像素，但不包括(x2，y2)。首先找到对应两个坐标的字节索引：

```
int index1 = 4 * (y1 * bitmap.PixelWidth + x1);
int index2 = 4 * (y2 * bitmap.PixelWidth + x2);
```

然后表明你想把像素从 index1 更新到 index2：

```
pixelStream.Seek(index1, SeekOrigin.Begin);
pixelStream.Write(pixels, index1, index2 - index1);
bitmap.Invalidate();
```

我们来试试另一个自定义渐变。以下程序我称为 CircularGradient，其渐变基于由特定像素与位图中心所形成的角度。(数学比你想象得容易。)

XAML 文件用粗线定义 Ellipse，并给 Stroke 属性定义 imageBrush。动画围绕其中心旋转 Ellipse。

```
项目: CircularGradient | 文件: MainPage.xaml(片段)
<Page ... >
    <Grid Background="{StaticResource ApplicationPageBackgroundThemeBrush}">
        <Ellipse Width="576"
                 Height="576"
                 StrokeThickness="48"
                 RenderTransformOrigin="0.5 0.5">
            <Ellipse.Stroke>
                <ImageBrush x:Name="imageBrush" />
            </Ellipse.Stroke>

            <Ellipse.RenderTransform>
                <RotateTransform x:Name="rotate" />
            </Ellipse.RenderTransform>
        </Ellipse>
    </Grid>

    <Page.Triggers>
        <EventTrigger>
            <BeginStoryboard>
                <Storyboard>
                    <DoubleAnimation Storyboard.TargetName="rotate"
                                     Storyboard.TargetProperty="Angle"
                                     From="0" To="360" Duration="0:0:3"
                                     RepeatBehavior="Forever" />
                </Storyboard>
            </BeginStoryboard>
        </EventTrigger>
    </Page.Triggers>
</Page>
```

代码隐藏文件中的 Loaded 处理程序类似于以前的程序。两个循环通过位图的行和列进行，每个像素具有一个相对于左上角的位置(x，y)。位于中心的像素坐标为(bitmap.PixelWidth/2，bitmap.PixelHeight/2)。从单个像素减去中心，并除以位图的宽和高，可以把像素坐标转换为－1/2 和 1/2 之间的值，然后可以传递给 math.atan2 方法以准确得到我们所需要的角度。

```
项目: CircularGradient | 文件: MainPage.xaml.cs(片段)
public sealed partial class MainPage : Page
{
    public MainPage()
    {
        this.InitializeComponent();
        Loaded += OnMainPageLoaded;
    }

    async void OnMainPageLoaded(object sender, RoutedEventArgs args)
    {
        WriteableBitmap bitmap = new WriteableBitmap(256, 256);
        byte[] pixels = new byte[4 * bitmap.PixelWidth * bitmap.PixelHeight];
        int index = 0;
        int centerX = bitmap.PixelWidth / 2;
        int centerY = bitmap.PixelHeight / 2;
        for (int y = 0; y < bitmap.PixelHeight; y++)
```

```
            for (int x = 0; x < bitmap.PixelWidth; x++)
            {
                double angle =
                    Math.Atan2(((double)y - centerY) / bitmap.PixelHeight,
                               ((double)x - centerX) / bitmap.PixelWidth);
                double fraction = angle / (2 * Math.PI);
                pixels[index++] = (byte)(fraction * 255); // Blue
                pixels[index++] = 0; // Green
                pixels[index++] = (byte)(255 * (1 - fraction)); // Red
                pixels[index++] = 255; // Alpha
            }

        using (Stream pixelStream = bitmap.PixelBuffer.AsStream())
        {
            await pixelStream.WriteAsync(pixels, 0, pixels.Length);
        }
        bitmap.Invalidate();
        imageBrush.ImageSource = bitmap;
    }
}
```

角度转换为 0 和 1 之间的分数来计算渐变。位图看起来像一个整体，用于把 ImageBrush 设置为 Rectangle 的 Fill 属性，如下图所示。

然而，如果把位图限制在圆圈内，并开始旋转，更有趣，就像是渐变本身在旋转(见下图)。

正如你所看到的，Windows Runtime 中的画笔一般拉伸到正在着色的元素。ImageBrush 也会这么做，从某种意义而言，底层位图的大小并不重要。但当然也的确重要。位图过小可能没有必要的细节，过大则徒然浪费像素。

14.2 透明度和预乘 Alpha

如果要在视频显示器表面渲染位图，位图的像素一般不会直接转移到视频显示器表面。如果位图支持透明像素，则像素必须基于其 α 设置来结合现有表面的颜色。如果 α 值为 255(不透明)，则位图像素可直接复制到表面。如果 α 值为 0(透明)，则根本不需要复制。如果 α 值为 128，其结果是使用位图像素的平均值和复制表面颜色进行渲染。

以下公式是对单个像素的计算。实际上 A、R、G 和 B 的取值范围从 0 到 255，但该简化公式假设已经将值标准化为 0 到 1。下标表示对现有“表明”渲染部分透明的“位图”像素的“结果”：

$$R_{result} = (1 - A_{bitmap}) \bullet R_{surface} + A_{bitmap} \bullet R_{bitmap}$$
$$G_{result} = (1 - A_{bitmap}) \bullet R_{surface} + A_{bitmap} \bullet G_{bitmap}$$
$$B_{result} = (1 - A_{bitmap}) \bullet R_{surface} + A_{bitmap} \bullet B_{bitmap}$$

注意每一行的第二个乘法。第二个乘法只涉及位图像素本身的乘法，而不是表面。也就是说，如果像素的 R、G、B 值已经乘以 A 值，则可以加快整个在表面渲染位图的过程：

$$R_{result} = (1 - A_{bitmap}) \bullet R_{surface} + R_{bitmap}$$
$$G_{result} = (1 - A_{bitmap}) \bullet R_{surface} + G_{bitmap}$$
$$B_{result} = (1 - A_{bitmap}) \bullet R_{surface} + B_{bitmap}$$

这就是所谓的“预乘 alpha”。

例如，假设一个非预乘 alpha 位图包含带有 ARGB 值的像素(192，40，60，255)。alpha 值 192 表示 75%的不透明度(192 除以 255)。带预乘 alpha 的等效像素为(192，30，45，192)。红色、绿色和蓝色的值都乘了 75%。

渲染 WriteableBitmap 的时候，操作系统会假设像素数据已经预乘 alpha。对于任何像素，R、G 和 B 的值都不能大于 A 值。如果大于，则没有东西会“放大”，你也得不到你想要的颜色和透明度级别。

我们来看一些例子。在第 10 章中，我向你展示了如何翻转图像并带有淡出效果，使图片看起来像反射。然而，由于 Windows Runtime 不支持不透明蒙板，我不得不用部分透明矩形进行覆盖，使反射图像有淡出效果。

ReflectedAlphaImage 项目采取的是不同方法。XAML 文件有两个 Image 元素，都占据两行 Grid 的相同顶部单元。第二个 Image 元素有 RenderTransformOrigin 和 ScaleTransform 绕着其底部边缘进行翻转，但没有指定位图：

```
项目: ReflectedAlphaImage | 文件: MainPage.xaml(片段)
<Grid Background="{StaticResource ApplicationPageBackgroundThemeBrush}">
    <Grid.RowDefinitions>
        <RowDefinition Height="*" />
        <RowDefinition Height="*" />
    </Grid.RowDefinitions>
```

```
    <Image Source="http://www.charlespetzold.com/pw6/PetzoldJersey.jpg"
           HorizontalAlignment="Center" />

    <Image Name="reflectedImage"
           RenderTransformOrigin="0 1"
           HorizontalAlignment="Center">
        <Image.RenderTransform>
            <ScaleTransform ScaleY="-1" />
        </Image.RenderTransform>
    </Image>
</Grid>
```

第一个 Image 元素引用的相同位图必须在代码隐藏文件中独立加载。(你可能想知道如果把对象设置第一个 Image 对象的 Source 属性，是否可能基于该对象而获得 WriteableBitmap。但该对象为 BitmapSource 类型，而且不能从 BitmapSource 创建 WriteableBitmap。)如果不必修改已下载位图，构造函数中的代码就如下所示：

```
Loaded += async (sender, args) =>
    {
        Uri uri = new Uri("http://www.charlespetzold.com/pw6/PetzoldJersey.jpg");
        RandomAccessStreamReference streamRef = RandomAccessStreamReference.CreateFromUri(uri);
        IRandomAccessStreamWithContentType fileStream = await streamRef.OpenReadAsync();
        WriteableBitmap bitmap = new WriteableBitmap(1, 1);
        bitmap.SetSource(fileStream);
        reflectedImage.Source = bitmap;
    };
```

由于会牵涉到一些异步处理，因此有必要把该代码放到 Loaded 处理程序。请注意，如果数据来自 SetSource 方法，则可以用“未知”大小来创建 WriteableBitmap。读取 JPEG 流的时候，WriteableBitmap 可以计算出实际像素尺寸。

然而，如果 FileStream 对象传递给 WriteableBitmap 的 SetSource 方法并且把 WriteableBitmap 设置为 Image 元素的 Source 属性，则位图尚未下载。下载发生在 WriteableBitmap 中。也就是说，你还不能开始修改像素，因为像素还没有到达！如果 WriteableBitmap 能像 BitmapImage 一样定义事件来表明 SetSource 何时完成加载位图文件，就好了，但事实并非如此。Image 元素的 ImageOpened 事件也不为 WriteableBitmap 提供此信息。

因此，我们接下来做的事情是在位图文件中加载并进行修改。下面的代码可以通过本章稍后提到的其他类进行简化，但我们来看看如果没有那些类又该如何完成。过程如下所示。

```
项目: ReflectedAlphaImage | 文件: MainPage.xaml.cs(片段)
public sealed partial class MainPage : Page
{
    public MainPage()
    {
        this.InitializeComponent();
        Loaded += OnMainPageLoaded;
    }

    async void OnMainPageLoaded(object sender, RoutedEventArgs args)
    {
        Uri uri = new Uri("http://www.charlespetzold.com/pw6/PetzoldJersey.jpg");
        RandomAccessStreamReference streamRef = RandomAccessStreamReference.CreateFromUri(uri);

        // Create a buffer for reading the stream
        Windows.Storage.Streams.Buffer buffer = null;
```

```
            // Read the entire file
            using (IRandomAccessStreamWithContentType fileStream = await streamRef.OpenReadAsync())
            {
                buffer = new Windows.Storage.Streams.Buffer((uint)fileStream.Size);
                await fileStream.ReadAsync(buffer, (uint)fileStream.Size, InputStreamOptions.None);
            }

            // Create WriteableBitmap with unknown size
            WriteableBitmap bitmap = new WriteableBitmap(1, 1);

            // Create a memory stream for transferring the data
            using (InMemoryRandomAccessStream memoryStream = new InMemoryRandomAccessStream())
            {
                await memoryStream.WriteAsync(buffer);
                memoryStream.Seek(0);
                // Use the memory stream as the Bitmap source
                bitmap.SetSource(memoryStream);
            }
            // Now get the pixels from the bitmap
            byte[] pixels = new byte[4 * bitmap.PixelWidth * bitmap.PixelHeight];
            int index = 0;

            using (Stream pixelStream = bitmap.PixelBuffer.AsStream())
            {
                await pixelStream.ReadAsync(pixels, 0, pixels.Length);

                // Apply opacity to the pixels
                for (int y = 0; y < bitmap.PixelHeight; y++)
                {
                    double opacity = (double)y / bitmap.PixelHeight;
                    for (int x = 0; x < bitmap.PixelWidth; x++)
                        for (int i = 0; i < 4; i++)
                        {
                            pixels[index] = (byte)(opacity * pixels[index]);
                            index++;
                        }
                }

                // Put the pixels back in the bitmap
                pixelStream.Seek(0, SeekOrigin.Begin);
                await pixelStream.WriteAsync(pixels, 0, pixels.Length);
            }
            bitmap.Invalidate();
            reflectedImage.Source = bitmap;
        }
    }
```

Buffer 类需要完全限定名，包括 Windows.Storage.Streams 命名空间，因为 System 命名空间还包括一个名为 Buffer 的类。

这里有个目标，即把 IRandomAccessStream 类型的对象传递到 WriteableBitmap 的 SetSource 方法。然而，这样做之后，我们需要立即处理结果位图的像素。除非已经充分读取文件，否则无法进行。

这就是创建 Buffer 对象用于读取 FileStream 对象的理由，然后使用相同 Buffer 对象将内容写入 InMemoryRandomAccessStream。顾名思义，InMemoryRandomAccessStream 类实现 IRandomAccessStream 接口，以便能传递到 WriteableBitmap 的 SetSource 方法。(但请注意，流位置必须先设置为 0。)

注意，我们在处理两个非常不同的数据块。FileStream 引用 PNG 文件，本例中为 82 824 个字节的压缩图片数据。InMemoryRandomAccessStream 为相同数据块。一旦流传递到

WriteableBitmap 的 SetSource 方法，即解码为行和列像素。pixels 数组的大小为 512 000 个字节，pixelStream 对象会引用这些压缩像素。pixelStream 对象首先把像素读取到 pixels 数组，然后再写回到位图。

两个调用之间是渐变不透明度的实际应用。如果 Windows Runtime 不支持 WriteableBitmap 的像素来有预乘 alpha 格式，修改 Alpha 字节即可。预乘格式会要求颜色字节也要相乘。效果如下图所示。

如果你想看看只调整 Alpha 字节会怎么样，可以在内循环替换为以下代码：

```
for (int i = 0; i < 4; i++)
{
    if (i == 3)
        pixels[index] = (byte)(opacity * pixels[index]);
    index++;
}
```

只有背景为白色时，才能得到想要的透明度。如果背景为黑色，就没有透明度可言！看看公式就知道。

假设你想变一下 CircularGradient 项目，这样渐变从纯色到完全透明。可以用变化后的代码来设置四个字节：

```
pixels[index++] = (byte)(fraction * 255);        // Blue
pixels[index++] = 0;                             // Green
pixels[index++] = 0;                             // Red
pixels[index++] = (byte)(fraction * 255);        // Alpha
```

蓝色成分和 alpha 成分会得到相同设置。蓝色成分永远为 255，没有预乘 Alpha 格式。效果如下图所示。

14.3 径向渐变画笔

Windows Runtime 中有很多神秘失踪的东西，RadialGradientBrush 就是其中之一，RadialGradientBrush 一般用于给圆着色，从圆上某点到圆周会有渐变。RadialGradientBrush 常用来把一个圆分成三维“球”，看起来好像靠近左上角的区域在反射光。

我开始写 RadialGradientBrushSimulator 类的时候，想在 XAML 文件中让该类的 GradientOrigin 属性成为动画。因此，我创建了由 FrameworkElement 派生的 RadialGradientBrushSimulator，虽然其本身并不显示任何东西。通过从 FrameworkElement 派生而创建，我更容易在 XAML 中实例化该类。我还要考虑动画和绑定，因此把所有属性定义为依赖属性。该类代码不只含依赖属性的系统开销。

项目：RadialGradientBrushDemo | 文件：RadialGradientBrushSimulator.cs(片段)

```
public class RadialGradientBrushSimulator : FrameworkElement
{
    ...
    static readonly DependencyProperty gradientOriginProperty =
            DependencyProperty.Register("GradientOrigin",
                    typeof(Point),
                    typeof(RadialGradientBrushSimulator),
                    new PropertyMetadata(new Point(0.5, 0.5), OnPropertyChanged));
    static readonly DependencyProperty innerColorProperty =
            DependencyProperty.Register("InnerColor",
                    typeof(Color),
                    typeof(RadialGradientBrushSimulator),
                    new PropertyMetadata(Colors.White, OnPropertyChanged));

    static readonly DependencyProperty outerColorProperty =
            DependencyProperty.Register("OuterColor",
                    typeof(Color),
                    typeof(RadialGradientBrushSimulator),
                    new PropertyMetadata(Colors.Black, OnPropertyChanged));

    static readonly DependencyProperty clipToEllipseProperty =
            DependencyProperty.Register("ClipToEllipse",
```

```
                typeof(bool),
                typeof(RadialGradientBrushSimulator),
                new PropertyMetadata(false, OnPropertyChanged));

public static DependencyProperty imageSourceProperty =
        DependencyProperty.Register("ImageSource",
                typeof(ImageSource),
                typeof(RadialGradientBrushSimulator),
                new PropertyMetadata(null));

public RadialGradientBrushSimulator()
{
    SizeChanged += OnSizeChanged;
}

public static DependencyProperty GradientOriginProperty
{
    get { return gradientOriginProperty; }
}

public static DependencyProperty InnerColorProperty
{
    get { return innerColorProperty; }
}

public static DependencyProperty OuterColorProperty
{
    get { return outerColorProperty; }
}

public static DependencyProperty ClipToEllipseProperty
{
    get { return clipToEllipseProperty; }
}

public static DependencyProperty ImageSourceProperty
{
    get { return imageSourceProperty; }
}

public Point GradientOrigin
{
    set { SetValue(GradientOriginProperty, value); }
    get { return (Point)GetValue(GradientOriginProperty); }
}

public Color InnerColor
{
    set { SetValue(InnerColorProperty, value); }
    get { return (Color)GetValue(InnerColorProperty); }
}

public Color OuterColor
{
    set { SetValue(OuterColorProperty, value); }
    get { return (Color)GetValue(OuterColorProperty); }
}

public bool ClipToEllipse
{
    set { SetValue(ClipToEllipseProperty, value); }
    get { return (bool)GetValue(ClipToEllipseProperty); }
}

public ImageSource ImageSource
{
    private set { SetValue(ImageSourceProperty, value); }
    get { return (ImageSource)GetValue(ImageSourceProperty); }
```

```
        }

        void OnSizeChanged(object sender, SizeChangedEventArgs args)
        {
            this.RefreshBitmap();
        }

        static void OnPropertyChanged(DependencyObject obj, DependencyPropertyChangedEventArgs args)
        {
            (obj as RadialGradientBrushSimulator).RefreshBitmap();
        }
        ...
    }
```

在稍后展示的 RefreshBitmap 方法中，该类使用 GradientOrigin、InnerColor、OuterColor 和 ClipToEllipse 属性(以及元素的 actualwidth 和 actualheight)来创建 WriteableBitmap，而该类通过 ImageSource 属性显示 WriteableBitmap，从而允许 XAML 文件中的另一个元素通过绑定 ImageBrush 的 ImageSource 属性进行引用。

然后，我发现生成径向渐变画笔图片的算法并不简单。从概念上讲，你是在处理椭圆，尽管你能用位图来为矩形或其他形状着色。椭圆边界的颜色为 OuterColor 属性。Point 类型的 GradientOrigin 属性为相对坐标。例如，值(0.5，0.5)把 GradientOrigin 设置为椭圆中心。GradientOrigin 中的颜色为 InnerColor 属性。

对于位图中的任意点(x, y)，算法需要找到一个内插因子来计算 InnerColor 和 OuterColor 之间的颜色。该内插因子是从 GradientOrigin 通过点(x，y)到椭圆周长的一条直线。点(x，y)划分直线的位置确定了插值因子的值。

为获得最佳性能，我想避开三角学方面的知识。我的策略是找到椭圆圆周和 GradientOrigin 到(x，y)线条的交点。这涉及对位图中的每一点求解一个二次方程。

以下为 RefreshBitmap 方法。

```
项目: RadialGradientBrushDemo | 文件: RadialGradientBrushSimulator.cs(片段)
public class RadialGradientBrushSimulator : FrameworkElement
{
    WriteableBitmap bitmap;
    byte[] pixels;
    Stream pixelStream;
    ...
    void RefreshBitmap()
    {
        if (this.ActualWidth == 0 || this.ActualHeight == 0)
        {
            this.ImageSource = null;
            bitmap = null;
            pixels = null;
            pixelStream = null;
            return;
        }

        if (bitmap == null || (int)this.ActualWidth != bitmap.PixelWidth ||
                              (int)this.ActualHeight != bitmap.PixelHeight)
        {
            bitmap = new WriteableBitmap((int)this.ActualWidth, (int)this.ActualHeight);
            this.ImageSource = bitmap;
            pixels = new byte[4 * bitmap.PixelWidth * bitmap.PixelHeight];
            pixelStream = bitmap.PixelBuffer.AsStream();
        }
        else
        {
            for (int i = 0; i < pixels.Length; i++)
```

```
            pixels[i] = 0;
    }

    double xOrigin = 2 * this.GradientOrigin.X - 1;
    double yOrigin = 2 * this.GradientOrigin.Y - 1;

    byte aOutsideCircle = 0;
    byte rOutsideCircle = 0;
    byte gOutsideCircle = 0;
    byte bOutsideCircle = 0;

    if (!this.ClipToEllipse)
    {
        double opacity = this.OuterColor.A / 255.0;
        aOutsideCircle = this.OuterColor.A;
        rOutsideCircle = (byte)(opacity * this.OuterColor.R);
        gOutsideCircle = (byte)(opacity * this.OuterColor.G);
        gOutsideCircle = (byte)(opacity * this.OuterColor.B);
    }

    int index = 0;
    for (int yPixel = 0; yPixel < bitmap.PixelHeight; yPixel++)
    {
        // Calculate y relative to unit circle
        double y = 2.0 * yPixel / bitmap.PixelHeight - 1;

        for (int xPixel = 0; xPixel < bitmap.PixelWidth; xPixel++)
        {
            // Calculate x relative to unit circle
            double x = 2.0 * xPixel / bitmap.PixelWidth - 1;

            // Check if point is within circle
            if (x * x + y * y <= 1)
            {
                // relative length from gradient origin to point
                double length1 = 0;

                // relative length from point to unit circle
                // (length1 + length2 = 1)
                double length2 = 0;

                if (x == xOrigin && y == yOrigin)
                {
                    length2 = 1;
                }
                else
                {
                    // Remember: xCircle^2 + yCircle^2 = 1
                    double xCircle = 0, yCircle = 0;

                    if (x == xOrigin)
                    {
                        xCircle = x;
                        yCircle = (y < yOrigin ? -1 : 1) * Math.Sqrt(1 - x * x);
                    }
                    else if (y == yOrigin)
                    {
                        xCircle = (x < xOrigin ? -1 : 1) * Math.Sqrt(1 - y * y);
                        yCircle = y;
                    }
                    else
                    {
                        // Express line from origin to point as y = mx + k
                        double m = (yOrigin - y) / (xOrigin - x);
                        double k = y - m * x;

                        // Now substitute (mx + k) for y into x^2 + y^2 = 1
                        // x^2 + (mx + k)^2 = 1
```

```
                        // x^2 + (mx)^2 + 2mxk + k^2 - 1 = 0
                        // (1 + m^2)x^2 + (2mk)x + (k^2 - 1) = 0 is quadratic equation
                        double a = 1 + m * m;
                        double b = 2 * m * k;
                        double c = k * k - 1;

                        // Now solve for x
                        double sqrtTerm = Math.Sqrt(b * b - 4 * a * c);
                        double x1 = (-b + sqrtTerm) / (2 * a);
                        double x2 = (-b - sqrtTerm) / (2 * a);

                        if  (x < xOrigin)
                            xCircle = Math.Min(x1, x2);
                        else
                            xCircle = Math.Max(x1, x2);

                        yCircle = m * xCircle + k;
                    }

                    // Length from origin to point
                    length1 = Math.Sqrt(Math.Pow(x - xOrigin, 2) +
                        Math.Pow(y - yOrigin, 2));

                    // Length from point to circle
                    length2 = Math.Sqrt(Math.Pow(x - xCircle, 2) +
                        Math.Pow(y - yCircle, 2));

                    // Normalize those lengths
                    double total = length1 + length2;
                    length1 /= total;
                    length2 /= total;
                }

                // Interpolate color
                double alpha = length2 * this.InnerColor.A + length1 * this.OuterColor.A;
                double red= alpha * (length2 * this.InnerColor.R +
                                        length1 * this.OuterColor.R) / 255;
                double green = alpha * (length2 * this.InnerColor.G +
                                           length1 * this.OuterColor.G) / 255;
                double blue = alpha * (length2 * this.InnerColor.B +
                                          length1 * this.OuterColor.B) / 255;

                // Store in array
                pixels[index++] = (byte)blue;
                pixels[index++] = (byte)green;
                pixels[index++] = (byte)red;
                pixels[index++] = (byte)alpha;
            }
            else
            {
                pixels[index++] = bOutsideCircle;
                pixels[index++] = gOutsideCircle;
                pixels[index++] = rOutsideCircle;
                pixels[index++] = aOutsideCircle;
            }
        }
    }
    pixelStream.Seek(0, SeekOrigin.Begin);
    pixelStream.Write(pixels, 0, pixels.Length);
    bitmap.Invalidate();
  }
}
```

为了能成为动画，像素数列以及用于将像素转移到位图的 Stream 对象都保存为字段。RefreshBitmap 方法不要求从堆中分配，除非 WriteableBitmap 需要重建，因为该元素的大小已更改。

然而事实证明，即使采用相当小的尺寸，动画的表现也非常差。但如果要避免渐变本身的动画，一定能用该位图把着色对象变成动画。MainPage.xaml 文件实例化一个 RadialGradientBrushSimulator、一个绑定到模拟器的 Ellipse 以及几个动画。

项目：RadialGradientBrushDemo | 文件：MainPage.xaml(片段)

```
<Page ... >
    <Grid Background="{StaticResource ApplicationPageBackgroundThemeBrush}">
        <Canvas SizeChanged="OnCanvasSizeChanged"
                Margin="0 0 96 96">

            <Grid Name="ballContainer"
                  Width="96"
                  Height="96">

                <Ellipse Name="ellipse">
                    <Ellipse.Fill>
                        <ImageBrush ImageSource="{Binding ElementName=brushSimulator,
                                                          Path=ImageSource}" />
                    </Ellipse.Fill>
                </Ellipse>

                <local:RadialGradientBrushSimulator x:Name="brushSimulator"
                                                    InnerColor="White"
                                                    OuterColor="Red"
                                                    GradientOrigin="0.3 0.3" />
            </Grid>
        </Canvas>
    </Grid>

    <Page.Triggers>
        <EventTrigger>
            <BeginStoryboard>
                <Storyboard>
                    <DoubleAnimation x:Name="leftAnima"
                                     Storyboard.TargetName="ballContainer"
                                     Storyboard.TargetProperty="(Canvas.Left)"
                                     From="0" Duration="0:0:2.51"
                                     AutoReverse="True"
                                     RepeatBehavior="Forever" />

                    <DoubleAnimation x:Name="rightAnima"
                                     Storyboard.TargetName="ballContainer"
                                     Storyboard.TargetProperty="(Canvas.Top)"
                                     From="0" Duration="0:0:1.01"
                                     AutoReverse="True"
                                     RepeatBehavior="Forever" />
                </Storyboard>
            </BeginStoryboard>
        </EventTrigger>
    </Page.Triggers>
</Page>
```

注意我如何把 Ellipse 和 RadialGradientBrushSimulator 放到相同 96 像素正方形 Grid 使两个元素都有相同尺寸，而且模拟器生成位图和用于着色的 Ellipse 具有相同大小。代码隐藏文件只对基于 Canvas 大小的动画调整 To 值。

项目：RadialGradientBrushDemo | 文件：MainPage.xaml.cs(片段)

```
void OnCanvasSizeChanged(object sender, SizeChangedEventArgs args)
{
    // Canvas.Left animation
    leftAnima.To = args.NewSize.Width;

    // Canvas.Top animation
    rightAnima.To = args.NewSize.Height;
}
```

如下图所示，模拟光反射使 Ellipse 看起来更像现实世界中的东西。

14.4 加载及保存图片文件

正如你所看到的，可以给 WriteableBitmap 的 SetSource 方法赋予引用 PNG 文件的流，方法可以优雅地解码压缩文件，并转换成行和列的数组。通过 Windows.Graphics.Imaging 命名空间中的类可以更了解该过程。可以加载位图文件作为像素数组，还可以用另外一种方法，即把程序创建的 WriteableBitmap 像素位数组保存为常用图片格式文件。

位图文件格式一般通过其常用的压缩类型(包括没有类型)进行识别，当然还有独特的数据结构、头文件和用于存储数据的压缩方法。读取特定文件格式并转换成像素数组的代码称为“解码器”。解码器允许把图片文件加载到应用。Windows.Graphics.Imaging 命名空间中的 BitmapDecoder 类支持下表中的格式。

文件格式	MIME 类型	文件扩展名
Windows Bitmap	image/bmp	.bmp .dib .rle
Windows Icon	image/ico image/x-icon	.ico .icon
GIF files	image/gif	.gif
JPEG	image/jpeg image/jpe image/jpg	.jpeg .jpe .jpg .jfif .exif

续表

文件格式	MIME 类型	文件扩展名
PNG	image/png	.png
TIFF	image/tiff image/tif	.tiff .tif
WMPhoto	image/vnd .ms-photo	.wdp .jxr

BitmapDecoder 类决定加载哪些类型的文件，如果无法识别则引发异常。

从像素比特数组创建特定格式文件的代码称为“编码器”，在 Windows Runtime 中编码器为 BitmapEncoder 类。使用编码器和使用解码器有一点不同。解码器可以决定要求加载什么文件类型，但编码器不会明白你的思想而确定要保存的文件格式。必须要告知编码器。BitmapEncoder 类和 BitmapDecoder 类支持相同格式，Windows Icon 文件除外。

编码器和解码器有时合称为“编码解码器”，能方便地表示“编码器/解码器”或“压缩器/解压缩器。”

上表所示的 7 个文件格式通过全球唯一 ID(Guid 类型的对象)进行识别，这些 ID 在 BitmapEncoder 和 BitmapDecoder 类中定义为静态属性，但不需要在程序中硬编码这些 ID 或者说也不需要包括很多特定信息。

ImageFileIO 程序要演示如何通过 FileOpenPicker 和 BitmapDecoder 把位图文件加载到应用，如何通过 FileSavePicker 和 BitmapEncoder 从应用中选择文件格式并保存位图文件。在两者之间，还有几个应用栏按钮用于 90 度旋转图片。本程序使用文件选择器来获取 StorageFile 对象，因此在访问用户文件时不需要任何特别的许可。

XAML 文件定义一个 Image 元素、一个用于显示加载位图信息的 TextBlock 以及 AppBar。

```
项目: ImageFileIO | 文件: MainPage.xaml(片段)
<Page ... >
    <Grid Background="Gray">
        <Grid.RowDefinitions>
            <RowDefinition Height="Auto" />
            <RowDefinition Height="*" />
        </Grid.RowDefinitions>

        <TextBlock Name="txtblk"
                   Grid.Row="0"
                   HorizontalAlignment="Center"
                   FontSize="18" />

        <Image Name="image"
               Grid.Row="1" />
    </Grid>

    <Page.BottomAppBar>
        <AppBar IsOpen="True">
            <Grid>
                <StackPanel Orientation="Horizontal"
                            HorizontalAlignment="Left">

                    <Button Name="rotateLeftButton"
```

```
                    IsEnabled="False"
                    Style="{StaticResource AppBarButtonStyle}"
                    Content="&#x21B6;"
                    AutomationProperties.Name="Rotate Left"
                    Click="OnRotateLeftAppBarButtonClick" />

            <Button Name="rotateRightButton"
                    IsEnabled="False"
                    Style="{StaticResource AppBarButtonStyle}"
                    Content="&#x21B7;"
                    AutomationProperties.Name="Rotate Right"
                    Click="OnRotateRightAppBarButtonClick" />
        </StackPanel>

        <StackPanel Orientation="Horizontal"
                    HorizontalAlignment="Right">

            <Button Style="{StaticResource OpenFileAppBarButtonStyle}"
                    Click="OnOpenAppBarButtonClick" />

            <Button Name="saveAsButton"
                    IsEnabled="False"
                    Style="{StaticResource SaveLocalAppBarButtonStyle}"
                    AutomationProperties.Name="Save As"
                    Click="OnSaveAsAppBarButtonClick" />
        </StackPanel>
      </Grid>
    </AppBar>
  </Page.BottomAppBar>
</Page>
```

请注意，AppBar 的 IsOpen 属性初始化为 true。加载文件之前程序不做任何事情。AppBar 中的所有其他按钮都已经被禁用。

为了使程序比较简单，其中没有留存很多信息。程序从硬盘加载的任何位图只作为 Image 元素的 Source 属性而留存。代码隐藏文件中定义的唯一字段只存储位图分辨率信息，然而，这并不重要。

```
项目: ImageFileIO | 文件: MainPage.xaml.cs(片段)
public sealed partial class MainPage : Page
{
    double dpiX, dpiY;

    public MainPage()
    {
        this.InitializeComponent();
    }
    ...
}
```

如果用户点击 AppBar 中的 Open 按钮，程序会创建 FileOpenPicker，并初始化显示 Pictures 文件夹中的文件。

```
项目: ImageFileIO | 文件: MainPage.xaml.cs(片段)
public sealed partial class MainPage : Page
{
    ...
    async void OnOpenAppBarButtonClick(object sender, RoutedEventArgs args)
    {
        // Create FileOpenPicker
        FileOpenPicker picker = new FileOpenPicker();
        picker.SuggestedStartLocation = PickerLocationId.PicturesLibrary;

        // Initialize with filename extensions
        IReadOnlyList<BitmapCodecInformation> codecInfos =
```

```
                    BitmapDecoder.GetDecoderInformationEnumerator();

        foreach (BitmapCodecInformation codecInfo in codecInfos)
            foreach (string extension in codecInfo.FileExtensions)
                picker.FileTypeFilter.Add(extension);

        // Get the selected file
        StorageFile storageFile = await picker.PickSingleFileAsync();

        if (storageFile == null)
            return;
        ...
    }
    ...
}
```

静态 BitmapDecoder.GetDecoderInformationEnumerator 非常有用。它会返回 7 个 BitmapCodecInformation 对象的集合，与前几页表中的 7 种文件格式相对应。每个对象都包含 MIME 类型的集合和文件扩展名的集合。(我用来获取表中所显示的信息。)文件扩展名可以直接进入 FileOpenPicker 对象，FileOpenPicker 显示带扩展名的所有文件。

如果 PickSingleFileAsync 调用返回非空 StorageFile 对象，下一步则是从该文件创建一个 BitmapDecoder。

项目：ImageFileIO | 文件：MainPage.xaml.cs(片段)

```
public sealed partial class MainPage : Page
{
    ...
    async void OnOpenAppBarButtonClick(object sender, RoutedEventArgs args)
    {
        ...
        // Open the stream and create a decoder
        BitmapDecoder decoder = null;

        using (IRandomAccessStreamWithContentType stream = await storageFile.OpenReadAsync())
        {
            string exception = null;

            try
            {
                decoder = await BitmapDecoder.CreateAsync(stream);
            }
            catch (Exception exc)
            {
                exception = exc.Message;
            }

            if (exception != null)
            {
                MessageDialog msgdlg =
                    new MessageDialog("That particular image file could not be loaded. " +
                                      "The system reports an error of: " + exception);
                await msgdlg.ShowAsync();
                return;
            }
            ...
        }
        ...
}
```

如果指定的是非图片文件或其他不能处理的东西，BitmapDecoder.CreateAsync 方法会报异常。

你可能知道，GIF 文件可以包含多张图片，这些图片按顺序播放动画。这些单独的图片称为“帧”，并由 Windows Runtime 支持。创建 BitmapDecoder 对象后，下一步一般是开始提取帧。然而，如果不想使用多帧 GIF 文件(如果你不想用，我也不会怪你)，则可以直接提取第一帧并且调用。我在以下代码中就是这么做的。

```
项目: ImageFileIO | 文件: MainPage.xaml.cs(片段)
public sealed partial class MainPage : Page
{
    ...
    async void OnOpenAppBarButtonClick(object sender, RoutedEventArgs args)
    {
        ...
                // Get the first frame
                BitmapFrame bitmapFrame = await decoder.GetFrameAsync(0);

                // Set information title
                txtblk.Text = String.Format("{0}: {1} x {2} {3} {4} x {5} DPI",
                                    storageFile.Name,
                                    bitmapFrame.PixelWidth, bitmapFrame.PixelHeight,
                                    bitmapFrame.BitmapPixelFormat,
                                    bitmapFrame.DpiX, bitmapFrame.DpiY);
                // Save the resolution
                dpiX = bitmapFrame.DpiX;
                dpiY = bitmapFrame.DpiY;

                // Get the pixels
                PixelDataProvider dataProvider =
                    await bitmapFrame.GetPixelDataAsync(BitmapPixelFormat.Bgra8,
                                        BitmapAlphaMode.Premultiplied,
                                        new BitmapTransform(),
                                        ExifOrientationMode.RespectExifOrientation,
                                        ColorManagementMode.ColorManageToSRgb);

                byte[] pixels = dataProvider.DetachPixelData();
                ...
        }
        ...
    }
    ...
}
```

该方法在页面顶部 TextBlock 显示第一帧信息并把分辨率设置保存为字段。

BitmapFrame 的 BitmapPixelFormat 和 BitmapAlphaMode 属性包含有关像素格式的重要信息。BitmapPixelFormat 是 Rgba16(红、绿、蓝和 α 16 位值)、Rgba8(红、绿、蓝和 alpha8 位值)或 Bgra8(蓝、绿、红和 α 8 位值)中的枚举项，最后一个兼容与 WriteableBitmap 相关联的格式。来自该文件的像素数据转换成这些格式之一。该 BitmapAlphaMode 属性可以表明 Ignore、Straight 或 Premultiplied。

可以直接调用不带任何参数的 GetPixelDataAsync 方法来获取帧像素格式中的像素字节数组。然而，如果想使用位图数据来创建 WriteableBitmap，则应该调用 GetPixelDataAsync 的较长版本，以指定兼容 WriteableBitmap 的格式。

GetPixelDataAsync 获取 WriteableBitmap 支持的相同格式字节数组，创建和显示位图的代码类似于你以前见过的代码。

```
项目: ImageFileIO | 文件: MainPage.xaml.cs(片段)
public sealed partial class MainPage : Page
{
    ...
    async void OnOpenAppBarButtonClick(object sender, RoutedEventArgs args)
    {
```

```
        ...
            // Create WriteableBitmap and set the pixels
            WriteableBitmap bitmap = new WriteableBitmap((int)bitmapFrame.PixelWidth,
                                                         (int)bitmapFrame.PixelHeight);

            using (Stream pixelStream = bitmap.PixelBuffer.AsStream())
            {
                await pixelStream.WriteAsync(pixels, 0, pixels.Length);
            }

            // Invalidate the WriteableBitmap and set as Image source
            bitmap.Invalidate();
            image.Source = bitmap;
        }

        // Enable the other buttons
        saveAsButton.IsEnabled = true;
        rotateLeftButton.IsEnabled = true;
        rotateRightButton.IsEnabled = true;
    }
    ...
}
```

以上就是对 Open 按钮的处理。总之，FileOpenPicker 返回 StorageFile 对象，打开并且传递到 BitmapDecoder.CreateAsync。BitmapDecoder 对象把图片显示为 BitmapFrame 对象，而 GetPixelDataAsync 方法获取用于创建 WriteableBitmap 的字节数组。

以下是显示位图的程序，第 13 章使用过下图中的位图。

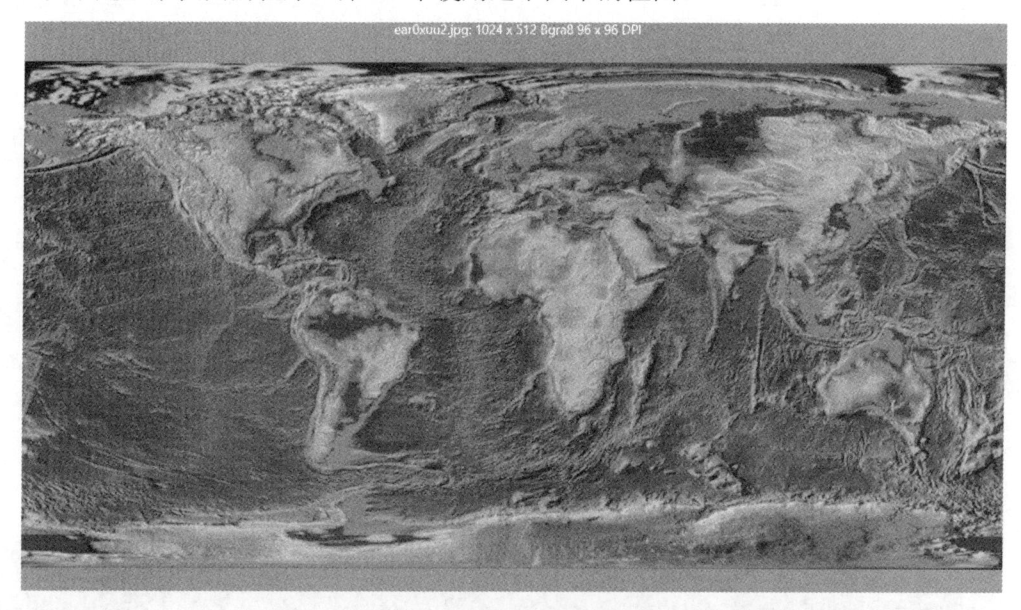

应用栏上的 Save As 按钮执行 OnSaveAsAppBarButtonClick 方法，后者要创建 FileSavePicker 对象。BitmapEncoder.GetEncoderInformationEnumerator 提供 BitmapEncoder 类支持的文件格式的信息，但该信息和 FileOpenPicker 的使用方式有些不同。

FileSavePicker 需要文件类型列表，同时带有每种类型的一个或多个文件扩展名。不幸的是，BitmapCodecInformation 对象的 FriendlyName 属性是一个字符串，类似“JPEG 编码器”，因此，我用 String 的 Split 方法来提取第一个单词(例如“JPEG”)，并且与多个文件扩展名结合在一起。代码还构造了支持 MIME 类型及其关联 GUID 对象的字典。

项目：ImageFileIO | 文件：MainPage.xaml.cs(片段)

```
public sealed partial class MainPage : Page
{
    ...
    async void OnSaveAsAppBarButtonClick(object sender, RoutedEventArgs args)
    {
        FileSavePicker picker = new FileSavePicker();
        picker.SuggestedStartLocation = PickerLocationId.PicturesLibrary;

        // Get the encoder information
        Dictionary<string, Guid> imageTypes = new Dictionary<string, Guid>();
        IReadOnlyList<BitmapCodecInformation> codecInfos =
                        BitmapEncoder.GetEncoderInformationEnumerator();

        foreach (BitmapCodecInformation codecInfo in codecInfos)
        {
            List<string> extensions = new List<string>();
            foreach (string extension in codecInfo.FileExtensions)
                extensions.Add(extension);

            string filetype = codecInfo.FriendlyName.Split(' ')[0];
            picker.FileTypeChoices.Add(filetype, extensions);

            foreach (string mimeType in codecInfo.MimeTypes)
                imageTypes.Add(mimeType, codecInfo.CodecId);
        }

        // Get a selected StorageFile
        StorageFile storageFile = await picker.PickSaveFileAsync();

        if (storageFile == null)
            return;
        ...
    }
    ...
}
```

如下图所示，如果 FileSavePicker 显示自身，用户就可以从弹出框中选择文件类型。

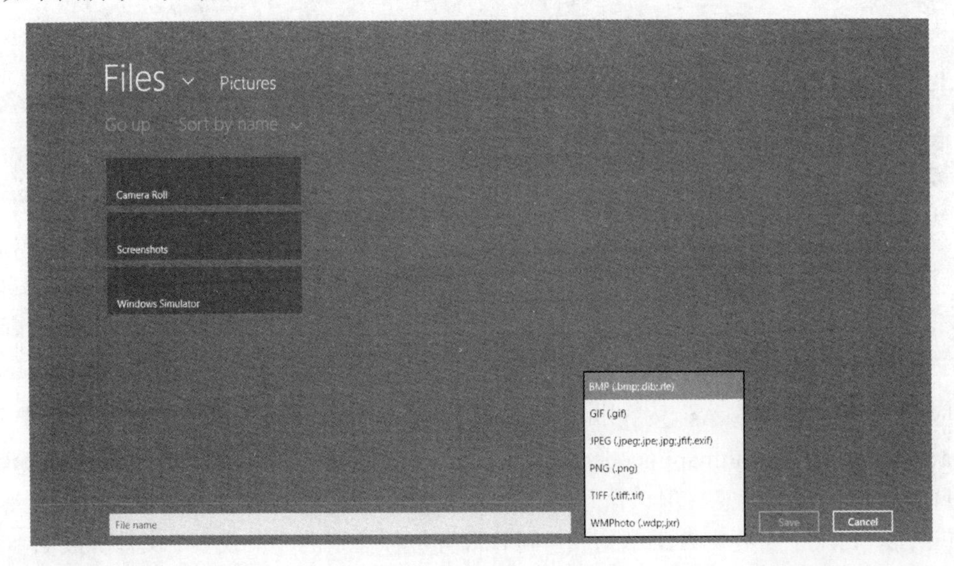

从 FileSavePicker 返回的 StorageFile 对象有 ContentType 字段，为 MIME 类型字符串，并能识别用户在弹出框中选择的文件类型。程序可以使用自己的字典来获得与该类型关联的 GUID 对象。

项目：ImageFileIO | 文件：MainPage.xaml.cs（片段）
```
public sealed partial class MainPage : Page
{
    ...
    async void OnSaveAsAppBarButtonClick(object sender, RoutedEventArgs args)
    {
        ...
        // Open the StorageFile
        using (IRandomAccessStream fileStream =
                        await storageFile.OpenAsync(FileAccessMode.ReadWrite))

        {
            // Create an encoder
            Guid codecId = imageTypes[storageFile.ContentType];
            BitmapEncoder encoder = await BitmapEncoder.CreateAsync(codecId, fileStream);

            // Get the pixels from the existing WriteableBitmap
            WriteableBitmap bitmap = image.Source as WriteableBitmap;
            byte[] pixels = new byte[4 * bitmap.PixelWidth * bitmap.PixelHeight];

            using (Stream pixelStream = bitmap.PixelBuffer.AsStream())
            {
                await pixelStream.ReadAsync(pixels, 0, pixels.Length);
            }

        // Write those pixels to the first frame
        encoder.SetPixelData(BitmapPixelFormat.Bgra8, BitmapAlphaMode.Premultiplied,
                             (uint)bitmap.PixelWidth, (uint)bitmap.PixelHeight,
                             dpiX, dpiY, pixels);

            await encoder.FlushAsync();
        }
    }
    ...
}
```

有了 GUID，静态 BitmapEncoder.CreateAsync 方法会返回一个 BitmapEncoder 对象。该对象包含 SetPixelData 方法，可用于将字节数组传输到新图片文件的第一帧。Save As 操作完成。

程序的其余部分支持 90 度旋转图片。此功能在 BitmapEncoder 类中实际就有。BitmapEncoder 类定义了一个 Transform 属性，可以在保存图片的同时以 90 度的增量来缩放、翻转、裁剪或旋转图片。但如果你想看到转换后的图像，必须自行实现逻辑。

90 度旋转图片涉及三种方法。

项目：ImageFileIO | 文件：MainPage.xaml.cs（片段）
```
public sealed partial class MainPage : Page
{
    ...
    void OnRotateLeftAppBarButtonClick(object sender, RoutedEventArgs args)
    {
        Rotate((BitmapSource bitmap, int x, int y) =>
            {
                return 4 * (bitmap.PixelWidth * x + (bitmap.PixelWidth - y - 1));
        });
    }

    void OnRotateRightAppBarButtonClick(object sender, RoutedEventArgs args)
    {
        Rotate((BitmapSource bitmap, int x, int y) =>
            {
                return 4 * (bitmap.PixelWidth * (bitmap.PixelHeight - x - 1) + y);
            });
    }
```

```
    async void Rotate(Func<BitmapSource, int, int, int> calculateSourceIndex)
    {
        // Get the source bitmap pixels
        WriteableBitmap srcBitmap = image.Source as WriteableBitmap;
        byte[] srcPixels = new byte[4 * srcBitmap.PixelWidth * srcBitmap.PixelHeight];

        using (Stream pixelStream = srcBitmap.PixelBuffer.AsStream())
        {
            await pixelStream.ReadAsync(srcPixels, 0, srcPixels.Length);
        }

        // Create a destination bitmap and pixels array
        WriteableBitmap dstBitmap =
                new WriteableBitmap(srcBitmap.PixelHeight, srcBitmap.PixelWidth);
        byte[] dstPixels = new byte[4 * dstBitmap.PixelWidth * dstBitmap.PixelHeight];

        // Transfer the pixels
        int dstIndex = 0;
        for (int y = 0; y < dstBitmap.PixelHeight; y++)
            for (int x = 0; x < dstBitmap.PixelWidth; x++)
            {
                int srcIndex = calculateSourceIndex(srcBitmap, x, y);

                for (int i = 0; i < 4; i++)
                    dstPixels[dstIndex++] = srcPixels[srcIndex++];
            }

        // Move the pixels into the destination bitmap
        using (Stream pixelStream = dstBitmap.PixelBuffer.AsStream())
        {
            await pixelStream.WriteAsync(dstPixels, 0, dstPixels.Length);
        }
        dstBitmap.Invalidate();

        //  Swap the DPIs
        double dpi = dpiX;
        dpiX = dpiY;
        dpiY = dpi;

        // Display the new bitmap
        image.Source = dstBitmap;
    }
}
```

两项工作大部分是由同一个 Rotate 方法进行处理，只不过该方法有一个参数，参数为一函数并基于目标位图的 x 和 y 像素位置来计算来源索引。如果对大文件尝试该方法，你会发现旋转需要几秒钟，强烈暗示不应该在用户界面线程执行此类程序。

通过把嵌套 for 循环传递到 Task.Run 并等待返回，可以异步执行旋转。然而，异步代码不能访问 WriteableBitmap。需要先获取位图的宽度和高度，再执行异步代码并重新定义 CalculateSourceIndex 来接受位图的宽度和高度，而不是位图。在此期间，也应该禁用应用栏按钮，以免完成之前受到任何干扰。

14.5 色调分离和单色化

大多数图片处理程序都能够对位图进行色调分离。颜色分辨率可以降低到有限调色板，图片看起来更像海报，而不是照片。另一个常见做法是图像转换为单色。这两项功能代表最简单的图片处理操作。

Posterizer 程序像 ImageFileIO 一样有 Open File 和 Save As 按钮，但页面还包含一个“控制面板”(一连串 RadioButton 控件)，可以选择大量颜色分辨率位(独立于三个颜色通道)，还有把图片转换为单色的 CheckBox。

假设用户加载位图并点击 CheckBox 将其转换为单色，程序会综合每个像素的 Red、Green 和 Blue 值变成灰色阴影。用户再取消选中 CheckBox。我们希望你的程序保存有原始图片！这就是为什么 Posterizer 程序会维护两个完整像素数组的原因，一个为原始像素(源像素，命名为 srcPixels)，另一个为修改后的像素(目标像素，命名为 dstPixels)。

XAML 文件包含控制面板、Image 元素和应用栏。

项目：Posterizer | 文件：MainPage.xaml(片段)

```
<Page ... >
    <Page.Resources>
        <Style TargetType="TextBlock">
            <Setter Property="FontSize" Value="18" />
            <Setter Property="TextAlignment" Value="Center" />
        </Style>
    </Page.Resources>

    <Grid Background="{StaticResource ApplicationPageBackgroundThemeBrush}">
        <Grid.ColumnDefinitions>
            <ColumnDefinition Width="Auto" />
            <ColumnDefinition Width="*" />
        </Grid.ColumnDefinitions>

        <Grid Name="controlPanelGrid"
              Grid.Column="0"
              Margin="12 0"
              HorizontalAlignment="Center"
              VerticalAlignment="Center">
            <Grid.ColumnDefinitions>
                <ColumnDefinition Width="*" />
                <ColumnDefinition Width="*" />
                <ColumnDefinition Width="*" />
                <ColumnDefinition Width="*" />
            </Grid.ColumnDefinitions>

            <Grid.RowDefinitions>
                <RowDefinition Height="Auto" />
                <RowDefinition Height="Auto" />
                <RowDefinition Height="Auto" />
                <RowDefinition Height="Auto" />
                <RowDefinition Height="Auto" />
                <RowDefinition Height="Auto" />
                <RowDefinition Height="Auto" />
                <RowDefinition Height="Auto" />
                <RowDefinition Height="Auto" />
                <RowDefinition Height="Auto" />
            </Grid.RowDefinitions>

            <TextBlock Text="Red" Grid.Row="0" Grid.Column="0" />
            <TextBlock Text="Green" Grid.Row="0" Grid.Column="1" />
            <TextBlock Text="Blue" Grid.Row="0" Grid.Column="2" />
            <TextBlock Text="All" Grid.Row="0" Grid.Column="3" />

            <CheckBox Name="monochromeCheckBox"
                      Content="Monochrome"
                      Grid.Row="9"
                      Grid.Column="0"
                      Grid.ColumnSpan="4"
                      Margin="0 12"
                      HorizontalAlignment="Center"
                      Checked="OnCheckBoxChecked"
```

```
                        Unchecked="OnCheckBoxChecked" />
        </Grid>

        <Image Name="image"
               Grid.Column="1" />
    </Grid>

    <Page.BottomAppBar>
        <AppBar>
            <Grid>
                <StackPanel Orientation="Horizontal"
                            HorizontalAlignment="Right">
                    <Button Style="{StaticResource OpenFileAppBarButtonStyle}"
                            Click="OnOpenAppBarButtonClick" />

                    <Button Name="saveAsButton"
                            IsEnabled="False"
                            Style="{StaticResource SaveLocalAppBarButtonStyle}"
                            AutomationProperties.Name="Save As"
                            Click="OnSaveAsAppBarButtonClick" />
                </StackPanel>
            </Grid>
        </AppBar>
    </Page.BottomAppBar>
</Page>
```

然而，XAML 文件缺少实际的 RadioButton 控件。我决定单独控制三个颜色通道，但要有第四列能一键更改所有三个颜色通道。Loaded 处理程序会创建按钮并用简便的 Tag 属性来进行识别。

```
项目: Posterizer | 文件: MainPage.xaml.cs(片段)
public sealed partial class MainPage : Page
{
    ...
    public MainPage()
    {
        this.InitializeComponent();
        Loaded += OnLoaded;
    }

    void OnLoaded(object sender, RoutedEventArgs args)
    {
        // Create the RadioButton controls
        // NOTE: 'a' here means "All" not "Alpha"!
        string[] prefix = { "r", "g", "b", "a" };

        for (int col = 0; col < 4; col++)
            for (int row = 1; row < 9; row++)
            {
                RadioButton radio = new RadioButton
                {
                    Content = row.ToString(),
                    Margin = new Thickness(12, 6, 12, 6),
                    GroupName = prefix[col],
                    Tag = prefix[col] + row,
                    IsChecked = row == 8
                };
                radio.Checked += OnRadioButtonChecked;

                Grid.SetColumn(radio, col);
                Grid.SetRow(radio, row);
                controlPanelGrid.Children.Add(radio);
            }
    }
    ...
}
```

文件 I/O 和 ImageFileIO 项目非常相似，只不过在加载图片时会创建像素的第二个数组，另外还有一个名为 UpdateBitmap 的方法(稍后介绍)负责通过该数组更新 WriteableBitmap。在保存文件时，使用 dstPixels 数组。

项目：Posterizer | 文件：MainPage.xaml.cs(片段)

```
public sealed partial class MainPage : Page
{
    WriteableBitmap bitmap;
    Stream pixelStream;
    byte [] srcPixels;
    byte[] dstPixels;
    ...
    async void OnOpenAppBarButtonClick(object sender, RoutedEventArgs args)
    {
        // Create FileOpenPicker
        FileOpenPicker picker = new FileOpenPicker();
        picker.SuggestedStartLocation = PickerLocationId.PicturesLibrary;

        // Initialize with filename extensions
        IReadOnlyList<BitmapCodecInformation> codecInfos =
                        BitmapDecoder.GetDecoderInformationEnumerator();

        foreach (BitmapCodecInformation codecInfo in codecInfos)
            foreach (string extension in codecInfo.FileExtensions)
                picker.FileTypeFilter.Add(extension);

        // Get the selected file
        StorageFile storageFile = await picker.PickSingleFileAsync();

        if (storageFile == null)
            return;

        // Open the stream and create a decoder
        BitmapDecoder decoder = null;

        using (IRandomAccessStreamWithContentType stream = await storageFile.OpenReadAsync())
        {
            string exception = null;

            try
            {
                decoder = await BitmapDecoder.CreateAsync(stream);
            }
            catch (Exception exc)
            {
                exception = exc.Message;
            }

            if (exception != null)
            {
                MessageDialog msgdlg =
                    new MessageDialog("That particular image file could not be loaded. " +
                        "The system reports on error of: " + exception);

                await msgdlg.ShowAsync();
                return;
            }

            // Get the first frame
            BitmapFrame bitmapFrame = await decoder.GetFrameAsync(0);

            // Get the source pixels
            PixelDataProvider dataProvider =
                await bitmapFrame.GetPixelDataAsync(BitmapPixelFormat.Bgra8,
                                                    BitmapAlphaMode.Premultiplied,
                                                    new BitmapTransform(),
```

```
                                                    ExifOrientationMode.RespectExifOrientation,
                                                    ColorManagementMode.ColorManageToSRgb);

            srcPixels = dataProvider.DetachPixelData();
            dstPixels = new byte[srcPixels.Length];

            // Create WriteableBitmap and set as Image source
            bitmap = new WriteableBitmap((int)bitmapFrame.PixelWidth,
                                         (int)bitmapFrame.PixelHeight);
            pixelStream = bitmap.PixelBuffer.AsStream();
            image.Source = bitmap;

            // Update bitmap from masked pixels
            UpdateBitmap();
        }

        // Enable the Save As button
        saveAsButton.IsEnabled = true;
    }

    async void OnSaveAsAppBarButtonClick(object sender, RoutedEventArgs args)
    {
        FileSavePicker picker = new FileSavePicker();
        picker.SuggestedStartLocation = PickerLocationId.PicturesLibrary;

        // Get the encoder information
        Dictionary<string, Guid> imageTypes = new Dictionary<string, Guid>();
        IReadOnlyList<BitmapCodecInformation> codecInfos =
                        BitmapEncoder.GetEncoderInformationEnumerator();

        foreach (BitmapCodecInformation codecInfo in codecInfos)
        {
            List<string> extensions = new List<string>();

            foreach (string extension in codecInfo.FileExtensions)
                extensions.Add(extension);

            string filetype = codecInfo.FriendlyName.Split(' ')[0];
            picker.FileTypeChoices.Add(filetype, extensions);

            foreach (string mimeType in codecInfo.MimeTypes)
                imageTypes.Add(mimeType, codecInfo.CodecId);
        }

        // Get a selected StorageFile
        StorageFile storageFile = await picker.PickSaveFileAsync();

        If (storageFile == null)
            return;

        // Open the StorageFile
        using (IRandomAccessStream fileStream =
                    await storageFile.OpenAsync(FileAccessMode.ReadWrite))
        {
            // Create an encoder
            Guid codecId = imageTypes[storageFile.ContentType];
            BitmapEncoder encoder = await BitmapEncoder.CreateAsync(codecId, fileStream);

            // Write the destination pixels to the first frame
            encoder.SetPixelData(BitmapPixelFormat.Bgra8, BitmapAlphaMode.Premultiplied,
                                 (uint)bitmap.PixelWidth, (uint)bitmap.PixelHeight,
                                 96, 96, dstPixels);

            await encoder.FlushAsync();
        }
    }
    ...
}
```

由于有第四列按钮，所以 RadioButton 事件处理程序变得比较复杂。我想点击第四列的 RadioButton 的时候也检查其他三个按钮，但我肯定不想多次调用 UpdateBitmap。因此，我把三个字节掩码数列保持为字段并在 RadioButton 事件处理程序中进行设置。只有当至少有一个掩码值被改变，才调用 UpdateBitmap。

```
项目: Posterizer | 文件: MainPage.xaml.cs(片段)
public sealed partial class MainPage : Page
{
    ...
    // Byte masks for blue, green, red
    byte[] masks = { 0xFF, 0xFF, 0xFF };
    ...
    void OnRadioButtonChecked(object sender, RoutedEventArgs args)
    {
        // Decode the RadioButton Tag property
        RadioButton radio = sender as RadioButton;
        string tag = radio.Tag as string;
        int maskIndex = -1;
        int bits = Int32.Parse(tag[1].ToString()); // 1 to 8
        byte mask = (byte)(0xFF << 8 - bits);
        bool needsUpdate;

        // Find the index of the masks array
        switch (tag[0])
        {
            case 'r': maskIndex = 2; break;
            case 'g': maskIndex = 1; break;
            case 'b': maskIndex = 0; break;
        }

        // For "All", check all the other buttons in the row
        if (tag[0] == 'a')
        {
            needsUpdate = masks[0] != mask && masks[1] != mask && masks[2] != mask;

            if (needsUpdate)
                masks[0] = masks[1] = masks[2] = mask;

            foreach (UIElement child in (radio.Parent as Panel).Children)
            {
                if (child != radio &&
                    Grid.GetRow(child as FrameworkElement) == Grid.GetRow(radio))
                {
                    (child as RadioButton).IsChecked = true;
                }
            }
        }
        else
        {
            needsUpdate = masks[maskIndex] != mask;
            if (needsUpdate)
                masks[maskIndex] = mask;
        }

        if (needsUpdate)
            UpdateBitmap();
    }

    void OnCheckBoxChecked(object sender, RoutedEventArgs args)
    {
        UpdateBitmap();
    }
    ...
}
```

剩下的就是 UpdateBitmap 本身。三个掩码值被应用到蓝、绿和红分量，如果选择 Monochrome 按钮，则组合分量以创建灰度。灰度权重为用于 NTSC 和 PAL 彩色电视标准的标准换算系数。

项目：Posterizer | 文件：MainPage.xaml.cs(片段)

```
void UpdateBitmap()
{
    if (bitmap == null)
        return;

    for (int index = 0; index < srcPixels.Length; index += 4)
    {
        // Mask source pixels
        byte B = (byte)(masks[0] & srcPixels[index + 0]);
        byte G = (byte)(masks[1] & srcPixels[index + 1]);
        byte R = (byte)(masks[2] & srcPixels[index + 2]);
        byte A = srcPixels[index + 3];

        // Possibly convert to gray shade
        if (monochromeCheckBox.IsChecked.Value)
            B = G = R = (byte)(0.30 * R + 0.59 * G + 0.11 * B);

        // Save destination pixels
        dstPixels[index + 0] = B;
        dstPixels[index + 1] = G;
        dstPixels[index + 2] = R;
        dstPixels[index + 3] = A;
    }

    // Update bitmap
    pixelStream.Seek(0, SeekOrigin.Begin);
    pixelStream.Write(dstPixels, 0, dstPixels.Length);
    bitmap.Invalidate();
}
```

如果要给该示例程序加点其他东西，我会考虑放到第二个线程中进行。但应该只放改变像素位的循环到线程中。任何对 WriteableBitmap 自身的修改都必须在用户界面线程进行。

把图片的素分辨率减少到两位，效果如下图所示，也就是说，整张图只用了 64 种颜色来显示。

14.6　保存手绘作品

前面讨论的一系列 FingerPaint 程序都有一个很大缺点：无法保存作品。我曾建议过一种保存图片的方法，即在 Grid 中枚举 Line 、Polyline 或 Path 元素，并创建某种文本文件(可能为 XML 格式)，重新加载该文件时，可以重建所有元素。

或者画一张位图进行保存。然而，此时的大问题是 WriteableBitmap 的 Windows Runtime 版本不支持渲染 Line 和 Path 等元素。基本上必须实现你自己的直线和圆弧绘制算法。对于一组绘制线条的坐标和其他参数，这些算法必须计算应该如何设置位图中的像素才能渲染这些图形对象。

假设需要在两点之间的画一条线。

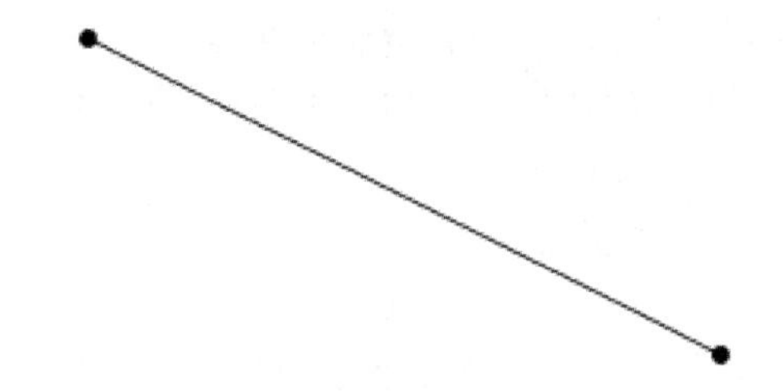

该几何线的宽度为零，但经过渲染的线必须有非零宽度，而且在手绘程序中，宽度可能相当大(例如 24 像素)。我们真的想画一个矩形来渲染该线，矩形在该线两边延伸，为其一半宽度，本例中为每边 12 像素。

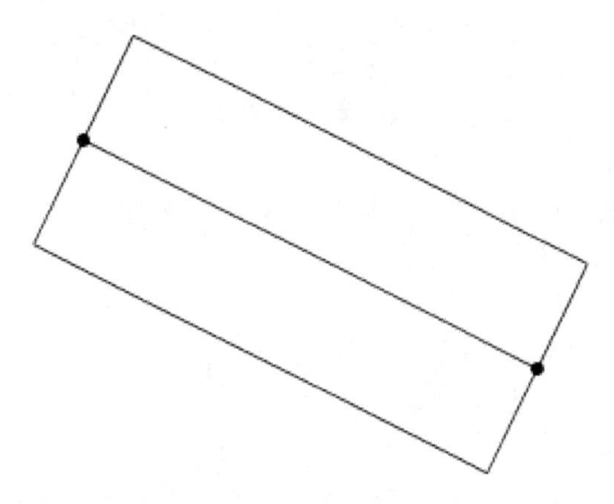

旋转两个几何点之间的标准化矢量 90 和−90 度，并且乘以整个直线长度的一半，而得到矩形的四个角。

但如果要为每个 PointerMoved 事件都画一个矩形，就不会正确通过曲线相连。会出现一些小小的缝隙。为了避免这些缝隙，要在矩形中画出圆角帽。

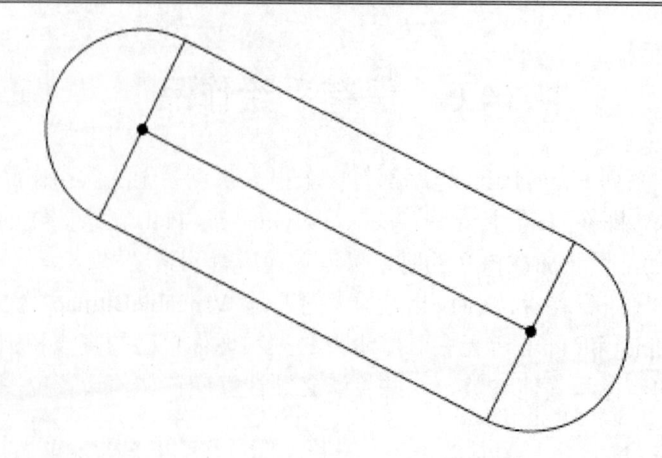

每个圆弧的半径为线长度的一半。

整体形状包含用两条线连接的两段圆弧。(如果两点之间的宽度增加或减少，也会涉及到类似形状，上一章已经讲过)。但我们并不想画出我所画的轮廓。我们需要填充内部，也就是说，要对整个形状内的每个像素进行着色。

高中时候可能你就熟悉斜率直线方程：

$$y = mx + b$$

其中 m 为斜率(“上升”)，b 为 y 在直线截取 Y 轴的值。

然而，在传统计算机图形领域，可以基于水平扫描线填充区域，也称为“光栅线”(该术语来自电视技术)。直线方程把 x 作为 y 的函数：

$$x = ay + b$$

对于从 pt1 到 pt2 的直线，可以通过下列公式计算出 a 和 b：

$$a = \frac{pt2.X - pt1.X}{pt2.Y - pt1.Y}$$

$$b = pt1.X - a \bullet pt1.Y$$

对于任何 y(即任何扫描线)，如果 y 是在 pT1.Y 和 Pt2.Y 之间，y 值就会对应于线上的一个点。可以从直线方程计算得到该点的 x 坐标。

看一看最新的图并设想一下截取图的水平扫描线。对于任意 y，我们可以确定扫描线是否与外部两条线的一条或两条相交。如果是，则可以计算出两个点的 x 值。所有 x 值之间的像素都必须着色。每个 y 都可以如此重复。

如果扫描线通过圆帽，着色过程就会有点乱。半径为 r 的圆以原点为中心，包括满足方程的所有点(x，y)：

$$x^2 + y^2 = r^2$$

如果圆以点(x_c, y_c)为圆心，则方程为：

$$(x - x_c)^2 + (y - y_c)^2 = r^2$$

或者表示为 y 的函数

$$x = x_c \pm \sqrt{r^2 - (y - y_c)^2}$$

对于任意 y，如果表达式的平方根为负，则 y 完全在圆之外，即在圆的上方或下方某处。否则，(一般来说)每个 y 会有两个 x 值。唯一例外是平方根为零时，即 y 为来自 y_c 的 r 单位，在圆顶部和底部的点。

处理部分围绕圆的圆弧，更复杂一些。弧上任一点均为形成来自圆心的角度。该角度可以用 Math.Atan2 方法来计算。如果已知圆弧的起点和终点，Math.Atan2 可以计算出对应两点的角度。也可以使用 Math.Atan2 来计算圆上任意点的角度。如果圆上的某点位于起点和终点之间，则该点位于圆弧上。

一般情况下，对于任意 *y*，我们都可以检查两条线和两个圆弧，并确定与四个图一致的所有点(x, y)。大多数情况下只有两个这样的点：一个位于扫描线进入图形的位置，另一个位于退出的位置。对于该扫描线，两点之间的所有像素都有进行填充。

FingerPaint 解决方案包含库项目，名为 Petzold.Windows8.Vector，该项目包含若干结构来实施在位图上绘画线条。(我采用了结构，而没有用类，因为它们要实例化并被频繁丢弃。)

你已经看过包含在库中的 Vector2 结构。所有其他结构都会实现它。

```
项目: Petzold.Windows8.VectorDrawing | 文件: IGeometrySegment.cs
using System.Collections.Generic;

namespace Petzold.Windows8.VectorDrawing
{
    public interface IGeometrySegment
    {
        void GetAllX(double y, IList<double> xCollection);
    }
}
```

对于任意 *y*，GetAllX 方法把条目添加到 *x* 值集合。在实际应用中，通过库中的结构来实现接口，该集合往往返回空值。有时包含一个条目，有时两个。

以下为 LineSegment 结构。

```
项目: Petzold.Windows8.VectorDrawing | 文件: LineSegment.cs
using System.Collections.Generic;
using Windows.Foundation;

namespace Petzold.Windows8.VectorDrawing
{
    public struct LineSegment : IGeometrySegment
    {
        readonly Point point1, point2;
        readonly double a, b; // as in x = ay + b

        public LineSegment(Point point1, Point point2) : this()
        {
            this.point1 = point1;
            this.point2 = point2;

            a = (point2.X - point1.X) / (point2.Y - point1.Y);
            b = point1.X - a * point1.Y;
        }

        public void GetAllX(double y, IList<double> xCollection)
        {
            if ((point2.Y > point1.Y && y >= point1.Y && y < point2.Y) ||
                (point2.Y < point1.Y && y <= point1.Y && y > point2.Y))
            {
                xCollection.Add(a * y + b);
            }
        }
    }
}
```

请注意，GetAllX 中的 if 语句会检查 *y* 是否位于 point.1Y 和 point2.Y 之间。允许等于 point1.的 *Yy* 值，但不允许等于 point2.Y 的 *Yy* 值。换句话说，定义了直线来自 point1(包含)的点，但不包括来自 point2 的点。这有助于练习这方面的一些严格规则和注意事项。否则，如果处理连接线和圆弧，我们会在集合得到重复 *x* 值，工作更难做。

这里没有特别考虑水平线，水平线是 point1.Y 等于 point2.Y 且 *a* 等于无穷大的直线。在这种情况下，if 语句永远不会满足，且会忽略水平线。扫描线不会穿过水平边界线。

ArcSegment 结构是圆周上的一段弧。

项目：Petzold.Windows8.VectorDrawing | 文件：ArcSegment.cs

```
using System;
using System.Collections.Generic;
using Windows.Foundation;

namespace Petzold.Windows8.VectorDrawing
{
    public struct ArcSegment : IGeometrySegment
    {
        readonly Point center, point1, point2;
        readonly double radius;
        readonly double angle1, angle2;

        public ArcSegment(Point center, double radius,
                Point point1, Point point2) :
            this()
        {
            this.center = center;
            this.radius = radius;
            this.point1 = point1;
            this.point2 = point2;
            this.angle1 = Math.Atan2(point1.Y - center.Y, point1.X - center.X);
            this.angle2 = Math.Atan2(point2.Y - center.Y, point2.X - center.X);
        }

        public void GetAllX(double y, IList<double> xCollection)
        {
            double sqrtArg = radius * radius - Math.Pow(y - center.Y, 2);

            if (sqrtArg >= 0)
            {
                double sqrt = Math.Sqrt(sqrtArg);
                TryY(y, center.X + sqrt, xCollection);
                TryY(y, center.X - sqrt, xCollection);
            }
        }

        void TryY(double y, double x, IList<double> xCollection)
        {
            double angle = Math.Atan2(y - center.Y, x - center.X);

            if ((angle1 < angle2 && (angle1 <= angle && angle < angle2)) ||
                 (angle1 > angle2 && (angle1 <= angle || angle < angle2)))
            {
                xCollection.Add((float)x);
            }
        }
    }
}
```

TryY 中相当复杂(但对称)的 if 语句解释角度值的从π到-π再返回。angle1 及 angle2 的角度比较表明角等于 angle1(而不是等于 angle2)的时候，扫描线和弧线相交。

LineSegment 的 GetAllX 方法可以在集合放入零个或一个 x 值。ArcSegment 的 GetAllX 方法可以在集合放入零个、一个或两个 x 值。对于具有均匀线宽的线条，RoundCappedLine 结构结合了两个 LineSegment 实例和两个 ArcSegment 实例。

```
项目: Petzold.Windows8.VectorDrawing | 文件: RoundCappedLine.cs
using System.Collections.Generic;
using Windows.Foundation;

namespace Petzold.Windows8.VectorDrawing
{
    public struct RoundCappedLine : IGeometrySegment
    {
        LineSegment lineSegment1;
        ArcSegment arcSegment1;
        LineSegment lineSegment2;
        ArcSegment arcSegment2;

        public RoundCappedLine(Point point1, Point point2, double radius) : this()
        {
            Vector2 vector = new Vector2(point2 - new Vector2(point1));
            Vector2 normVect = vector;
            normVect = normVect.Normalized;

            Point pt1a = point1 + radius * new Vector2(normVect.Y, -normVect.X);
            Point pt2a = pt1a + vector;
            Point pt1b = point1 + radius * new Vector2(-normVect.Y, normVect.X);
            Point pt2b = pt1b + vector;

            lineSegment1 = new LineSegment(pt1a, pt2a);
            arcSegment1 = new ArcSegment(point2, radius, pt2a, pt2b);
            lineSegment2 = new LineSegment(pt2b, pt1b);
            arcSegment2 = new ArcSegment(point1, radius, pt1b, pt1a);
        }

        public void GetAllX(double y, IList<double> xCollection)
        {
            arcSegment1.GetAllX(y, xCollection);
            lineSegment1.GetAllX(y, xCollection);
            arcSegment2.GetAllX(y, xCollection);
            lineSegment2.GetAllX(y, xCollection);
        }
    }
}
```

该结构调用两个 LineSegment 实例和两个 ArcSegment 实例的 GetAllX 方法来实施 GetAllX。确保集合之前也被清除，这是在该结构中调用 GetAllX 代码的责任。该方法返回有零个、一个或两个 x 值的集合。对于填充目的，可以忽略零个或一个 x 值的情况。而对于两个 x 值，可以填充两个值之间的像素。

RoundCappedPath 结构也比较相似，只不过允许线条在压力感应触摸屏上开始和结束时可以有不同的宽度。

```
项目: Petzold.Windows8.VectorDrawing | 文件: RoundCappedPath.cs
using System;
using System.Collections.Generic;
using Windows.Foundation;

namespace Petzold.Windows8.VectorDrawing
{
    public struct RoundCappedPath : IGeometrySegment
    {
        LineSegment lineSegment1;
        ArcSegment arcSegment1;
```

```
        LineSegment lineSegment2;
        ArcSegment arcSegment2;

        public RoundCappedPath(Point point1, Point point2, double radius1, double radius2)
            : this()
        {
            Point c0 = point1;
            Point c1 = point2;
            double r0 = radius1;
            double r1 = radius2;

            // Get vector from c1 to c0
            Vector2 vCenters = new Vector2(c0) - new Vector2(c1);

            // Get length and normalized version
            double d = vCenters.Length;
            vCenters = vCenters.Normalized;

            // Create focal point F extending from c0
            double e = d * r0 / (r1 - r0);
            Point F = c0 + e * vCenters;

            // Find angle and length of right-triangle legs
            double alpha = 180 * Math.Asin(r0 / e) / Math.PI;
            double leg0 = Math.Sqrt(e * e - r0 * r0);
            double leg1 = Math.Sqrt((e + d) * (e + d) - r1 * r1);

            // Vectors of tangent lines
            Vector2 vRight = -vCenters.Rotate(alpha);
            Vector2 vLeft = -vCenters.Rotate(-alpha);

            // Find tangent points
            Point t0R = F + leg0 * vRight;
            Point t0L = F + leg0 * vLeft;
            Point t1R = F + leg1 * vRight;
            Point t1L = F + leg1 * vLeft;

            lineSegment1 = new LineSegment(t1R, t0R);
            arcSegment1 = new ArcSegment(c0, r0, t0R, t0L);
            lineSegment2 = new LineSegment(t0L, t1L);
            arcSegment2 = new ArcSegment(c1, r1, t1L, t1R);
        }

        public void GetAllX(double y, IList<double> xCollection)
        {
            arcSegment1.GetAllX(y, xCollection);
            lineSegment1.GetAllX(y, xCollection);
            arcSegment2.GetAllX(y, xCollection);
            lineSegment2.GetAllX(y, xCollection);
        }
    }
}
```

我采用了前一章中 FingerPaint5 程序的逻辑。

在实际程序中使用这些结构并不像实例化 Line、Polyline 或 Path 那么简单！以下是 FingerPaint 中的 RenderOnBitmap 方法。该方法利用 WriteableBitmap 位图、pixels 像素数组以及 pixelStream 的字符串对象。方法首先确定是该使用 RoundCappedLine 还是 RoundCappedPath。

```
项目: FingerPaint | 文件: MainPage.Pointer.cs(片段)
public sealed partial class MainPage : Page
{
    ...
    bool RenderOnBitmap(Point point1, double radius1, Point point2, double radius2, Color color)
    {
```

```
bool bitmapNeedsUpdate = false;

// Define a line between the two points with rounded caps
IGeometrySegment geoseg = null;

// Adjust the points for any bitmap scaling
Point center1 = ScaleToBitmap(point1);
Point center2 = ScaleToBitmap(point2);

// Find the distance between them
double distance = Math.Sqrt(Math.Pow(center2.X - center1.X, 2) +
                    Math.Pow(center2.Y - center1.Y, 2));

// Choose the proper way to render the segment
if (radius1 == radius2)
    geoseg = new RoundCappedLine(center1, center2, radius1);

else if (radius1 < radius2 && radius1 + distance < radius2)
    geoseg = new RoundCappedLine(center1, center2, radius2);

else if (radius2 < radius1 && radius2 + distance < radius1)
    geoseg = new RoundCappedLine(center1, center2, radius1);

else if (radius1 < radius2)
    geoseg = new RoundCappedPath(center1, center2, radius1, radius2);

else
    geoseg = new RoundCappedPath(center2, center1, radius2, radius1);

// Find the minimum and maximum vertical coordinates
int yMin = (int)Math.Min(center1.Y - radius1, center2.Y - radius2);
int yMax = (int)Math.Max(center1.Y + radius1, center2.Y + radius1);

yMin = Math.Max(0, Math.Min(bitmap.PixelHeight, yMin));
yMax = Math.Max(0, Math.Min(bitmap.PixelHeight, yMax));

// Loop through all the y coordinates that contain part of the segment
for (int y = yMin; y < yMax; y++)
{
    // Get the range of x coordinates in the segment
    xCollection.Clear();
    geoseg.GetAllX(y, xCollection);

    if (xCollection.Count == 2)
    {
        // Find the minimum and maximum horizontal coordinates
        int xMin = (int)(Math.Min(xCollection[0], xCollection[1]) + 0.5f);
        int xMax = (int)(Math.Max(xCollection[0], xCollection[1]) + 0.5f);

        xMin = Math.Max(0, Math.Min(bitmap.PixelWidth, xMin));
        xMax = Math.Max(0, Math.Min(bitmap.PixelWidth, xMax));

        // Loop through the X values
        for (int x = xMin; x < xMax; x++)
        {
            {
                // Set the pixel
                int index = 4 * (y * bitmap.PixelWidth + x);
                pixels[index + 0] = color.B;
                pixels[index + 1] = color.G;
                pixels[index + 2] = color.R;
                pixels[index + 3] = 255;

                bitmapNeedsUpdate = true;
            }
        }
    }
}
```

```
        // Update bitmap
        if (bitmapNeedsUpdate)
        {
            // Find the starting index and number of pixels
            int start = 4 * yMin * bitmap.PixelWidth;
            int count = 4 * (yMax - yMin) * bitmap.PixelWidth;

            pixelStream.Seek(start, SeekOrigin.Begin);
            pixelStream.Write(pixels, start, count);
            bitmap.Invalidate();
        }

        return bitmapNeedsUpdate;
    }

    Point ScaleToBitmap(Point pt)
    {
        return new Point((pt.X - imageOffset.X) / imageScale,
                         (pt.Y - imageOffset.Y) / imageScale);
    }
}
```

请注意，renderonbitmap 在最后把更新限制为仅扫描受到特殊绘图操作影响的线条。ScaleToBitmap 方法会调整大于或小于当前页面尺寸的位图的点。

为了在功能模块中组织 FingerPaint 项目的源代码文件，我把 Mainpage 的隐藏代码逻辑分为三个文件：正常的 MainPage.xaml.cs; MainPage.Pointer.cs，包含所有 Pointer 事件处理(包括我刚刚展过的 RenderOnBitmap 方法)；MainPage.File.cs，其中包含文件 I/O 操作。MainPage.Pointer.cs 的剩余部分应该比较熟悉。除了调用 RenderOnBitmap 之外，其他几乎和 FingerPaint5 一样。

```
项目: FingerPaint | 文件: MainPage.Pointer.cs(片段)
public sealed partial class MainPage : Page
{
    struct PointerInfo
    {
        public Brush Brush;
        public Point PreviousPoint;
        public double PreviousRadius;
    }

    Dictionary<uint, PointerInfo> pointerDictionary = new Dictionary<uint, PointerInfo>();
    List<double> xCollection = new List<double>();

    protected override void OnPointerPressed(PointerRoutedEventArgs args)
    {
        // Get information from event arguments
        uint id = args.Pointer.PointerId;
        PointerPoint pointerPoint = args.GetCurrentPoint(this);

        // Create PointerInfo
        PointerInfo pointerInfo = new PointerInfo
        {
            PreviousPoint = pointerPoint.Position,
            PreviousRadius = appSettings.Thickness * pointerPoint.Properties.Pressure,
            Brush = new SolidColorBrush(appSettings.Color)
        };

        // Add to dictionary
        pointerDictionary.Add(id, pointerInfo);

        // Capture the Pointer
        CapturePointer(args.Pointer);
```

```
            base.OnPointerPressed(args);
        }

        protected override void OnPointerMoved(PointerRoutedEventArgs args)
        {
            // Get ID from event arguments
            uint id = args.Pointer.PointerId;

            // If ID is in dictionary, start a loop
            if (pointerDictionary.ContainsKey(id))
            {
                PointerInfo pointerInfo = pointerDictionary[id];

                foreach (PointerPoint pointerPoint in args.GetIntermediatePoints(this).Reverse())
                {
                    // For each point, get new position and pressure
                    Point point = pointerPoint.Position;
                    double radius = appSettings.Thickness * pointerPoint.Properties.Pressure;

                    // Render and flag that it's modified
                    appSettings.IsImageModified =
                        RenderOnBitmap(pointerInfo.PreviousPoint, pointerInfo.PreviousRadius,
                                       point, radius,
                                       appSettings.Color);

                    // Update PointerInfo
                    pointerInfo.PreviousPoint = point;
                    pointerInfo.PreviousRadius = radius;
                }

                // Store PointerInfo back in dictionary
                pointerDictionary[id] = pointerInfo;
            }
            base.OnPointerMoved(args);
        }

        protected override void OnPointerReleased(PointerRoutedEventArgs args)
        {
            // Get information from event arguments
            uint id = args.Pointer.PointerId;

            // If ID is in dictionary, remove it
            if (pointerDictionary.ContainsKey(id))
                pointerDictionary.Remove(id);

            base.OnPointerReleased(args);
        }

        protected override void OnPointerCaptureLost(PointerRoutedEventArgs args)
        {
            // Get information from event arguments
            uint id = args.Pointer.PointerId;

            // If ID is still in dictionary, remove it
            if (pointerDictionary.ContainsKey(id))
                pointerDictionary.Remove(id);

            base.OnPointerCaptureLost(args);
        }
        ...
    }
```

OnPointerPressed 和 OnPointerMoved 都引用 AppSettings 类型称为 appSettings 的字段。程序暂停时，该对象会把设置保存到本地存储并在程序启动时重新加载。该类的整体结构在这一点上和前面描述的差不多。

项目: FingerPaint | 文件: AppSettings.cs

```
using System.ComponentModel;
using System.Runtime.CompilerServices;
using Windows.Storage;
using Windows.UI;

namespace FingerPaint
{
    public class AppSettings : INotifyPropertyChanged
    {
        // Application settings initial values
        string loadedFilePath = null;
        string loadedFilename = null;
        bool isImageModified = false;
        Color color = Colors.Blue;
        double thickness = 16;

        public event PropertyChangedEventHandler PropertyChanged;

        public AppSettings()
        {
            ApplicationDataContainer appData = ApplicationData.Current.LocalSettings;

            if (appData.Values.ContainsKey("LoadedFilePath"))
                this.LoadedFilePath = (string)appData.Values["LoadedFilePath"];

            if (appData.Values.ContainsKey("LoadedFilename"))
                this.LoadedFilename = (string)appData.Values["LoadedFilename"];

            if (appData.Values.ContainsKey("IsImageModified"))
                this.IsImageModified = (bool)appData.Values["IsImageModified"];

            if (appData.Values.ContainsKey("Color.Red") &&
                appData.Values.ContainsKey("Color.Green") &&
                appData.Values.ContainsKey("Color.Blue"))
            {
                this.Color = Color.FromArgb(255,
                                            (byte)appData.Values["Color.Red"],
                                            (byte)appData.Values["Color.Green"],
                                            (byte)appData.Values["Color.Blue"]);
            }

            if (appData.Values.ContainsKey("Thickness"))
                this.Thickness = (double)appData.Values["Thickness"];
        }

        public string LoadedFilePath
        {
            set { SetProperty<string>(ref loadedFilePath, value); }
            get { return loadedFilePath; }
        }

        public string LoadedFilename
        {
            set { SetProperty<string>(ref loadedFilename, value); }
            get { return loadedFilename; }
        }

        public bool IsImageModified
        {
            set { SetProperty<bool>(ref isImageModified, value); }
            get { return isImageModified; }
        }

        public Color Color
        {
            set { SetProperty<Color>(ref color, value); }
            get { return color; }
```

```
        }

        public double Thickness
        {
            set { SetProperty<double>(ref thickness, value); }
            get { return thickness; }
        }

        public void Save()
        {
            ApplicationDataContainer appData = ApplicationData.Current.LocalSettings;
            appData.Values.Clear();
            appData.Values.Add("LoadedFilePath", this.LoadedFilePath);
            appData.Values.Add("LoadedFilename", this.LoadedFilename);
            appData.Values.Add("IsImageModified", this.IsImageModified);
            appData.Values.Add("Color.Red", this.Color.R);
            appData.Values.Add("Color.Green", this.Color.G);
            appData.Values.Add("Color.Blue", this.Color.B);
            appData.Values.Add("Thickness", this.Thickness);
        }

        protected bool SetProperty<T>(ref T storage, T value,
                                      [CallerMemberName] string propertyName = null)
        {
            if (object.Equals(storage, value))
                return false;

            storage = value;
            OnPropertyChanged(propertyName);
            return true;
        }

        protected void OnPropertyChanged(string propertyName)
        {
            if (PropertyChanged != null)
                PropertyChanged(this, new PropertyChangedEventArgs(propertyName));
        }
    }
}
```

可以用 FingerPaint 加载现有文件、绘画并保存。如果你正在做这些事情，那么文件名和整个文件路径都是用户设置的一部分。如果是从空白画布开始，则 LoadedFileNme 和 LoadedFilePath 属性都均为 null。不管怎样，如果图片已经修改，但没有保存到已命名文件，IsImageModified 属性就为 true。

遵循简洁应用概念，MainPage.xaml 文件直接实例化一个 Image 元素并实现应用栏。

项目：FingerPaint | 文件：MainPage.xaml(片段)

```
<Page ... >
    <Grid Background="{StaticResource ApplicationPageBackgroundThemeBrush}">

        <Image Name="image" />

        <!-- Disable file I/O buttons in the Snapped state -->
        <VisualStateManager.VisualStateGroups>
            <VisualStateGroup x:Name="ApplicationViewStates">
                <VisualState x:Name="FullScreenLandscape" />
                <VisualState x:Name="Filled" />
                <VisualState x:Name="FullScreenPortrait" />

                <VisualState x:Name="Snapped">
                    <Storyboard>
                        <ObjectAnimationUsingKeyFrames Storyboard.TargetName="fileButtons"
                                                       Storyboard.TargetProperty="Visibility">
                            <DiscreteObjectKeyFrame KeyTime="0" Value="Collapsed" />
                        </ObjectAnimationUsingKeyFrames>
```

```
                </Storyboard>
            </VisualState>
        </VisualStateGroup>
    </VisualStateManager.VisualStateGroups>
</Grid>

<Page.BottomAppBar>
    <AppBar>
        <Grid>
            <StackPanel Orientation="Horizontal"
                        HorizontalAlignment="Left">

                <Button Style="{StaticResource AppBarButtonStyle}"
                        AutomationProperties.Name="Color"
                        Content="&#x1F308;"
                        Click="OnColorAppBarButtonClick" />

                <Button Style="{StaticResource EditAppBarButtonStyle}"
                        AutomationProperties.Name="Thickness"
                        Click="OnThicknessAppBarButtonClick" />
        </StackPanel>

        <StackPanel Name="fileButtons"
                    Orientation="Horizontal"
                    HorizontalAlignment="Right">

                <Button Style="{StaticResource OpenFileAppBarButtonStyle}"
                        Click="OnOpenAppBarButtonClick" />

                <Button Style="{StaticResource SaveLocalAppBarButtonStyle}"
                        AutomationProperties.Name="Save As"
                        Click="OnSaveAsAppBarButtonClick" />

                <Button Style="{StaticResource SaveAppBarButtonStyle}"
                        Click="OnSaveAppBarButtonClick" />

                <Button Style="{StaticResource AddAppBarButtonStyle}"
                        Click="OnAddAppBarButtonClick" />
            </StackPanel>
        </Grid>
    </AppBar>
</Page.BottomAppBar>
</Page>
```

注意 Visual State Manager 标记，如果应用在辅屏状态，该标记可以使应用栏中的所有文件 I/O 按钮消失。这实际上比避免重复按键更重要：在辅屏状态下，文件选取器是被禁用的。

把该程序组合在一起的时候，我碰到一个小问题，涉及文件 I/O 以及终止时重新启动。假设用户启用 FileOpenPicker，从照片库选择一个现有文件。程序会从 FileOpenPicker 获取 StorageFile 对象，用于打开文件，并把 StorageFile 对象作为字段保留。如果用户按下应用栏中的 Save 按钮，程序会直接使用现有 StorageFile 对象来保存文件。

然而，假设程序暂停、终止，后来又重新启动。StorageFile 对象不能保存在本地应用存储中！程序必须抛弃，通过保存文件的完整文件路径来弥补(正如我在 appSettings 类中所做过的)。程序再次启动，用户按下 Save 按钮，应用没有该文件的 StorageFile 对象。相反，必须使用静态 StorageFile.GetFileFromPathaSync 方法进行创建。但使用该方法则意味着程序会访问文件系统，而不使用从文件选取器所获取的 StorageFile 对象。

为此，FingerPaint 程序需要有访问照片库的权限。在 Visual Studio 中，我显示 Package.appxmanifest 的属性，选择 Capabilities 标签，并选中 Picture Library。我不想程序

有这个特殊权限，但唯一的替代办法是强迫用户使用 FileSavePicker 来保存之前加载过的文件。

以下为 MainPage.File.cs 文件。从本章先前的程序以及第 8 章中 XamlCruncher 应用的逻辑，可以识别出询问用户保存已修改图片的位图加载和保存逻辑。

项目：FingerPaint | 文件：MainPage.File.cs(片段)

```
public sealed partial class MainPage : Page
{
    WriteableBitmap bitmap;
    Stream pixelStream;
    byte [] pixels;

    async Task CreateNewBitmapAndPixelArray()
    {
        bitmap = new WriteableBitmap((int)this.ActualWidth, (int)this.ActualHeight);
        pixels = new byte[4 * bitmap.PixelWidth * bitmap.PixelHeight];

        // Set whole bitmap to white
        for (int index = 0; index < pixels.Length; index++)
            pixels[index] = 0xFF;

        await InitializeBitmap();

        appSettings.LoadedFilePath = null;
        appSettings.LoadedFilename = null;
        appSettings.IsImageModified = false;
    }

    async Task LoadBitmapFromFile(StorageFile storageFile)
    {
        using (IRandomAccessStreamWithContentType stream = await storageFile.OpenReadAsync())
        {
            BitmapDecoder decoder = await BitmapDecoder.CreateAsync(stream);
            BitmapFrame bitmapframe = await decoder.GetFrameAsync(0);
            PixelDataProvider dataProvider =
                await bitmapframe.GetPixelDataAsync(BitmapPixelFormat.Bgra8,
                                                    BitmapAlphaMode.Premultiplied,
                                                    new BitmapTransform(),
                                                    ExifOrientationMode.RespectExifOrientation,
                                                    ColorManagementMode.ColorManageToSRgb);
            pixels = dataProvider.DetachPixelData();
            bitmap = new WriteableBitmap((int)bitmapframe.PixelWidth,
                                         (int)bitmapframe.PixelHeight);
            await InitializeBitmap();
        }
    }

    async Task InitializeBitmap()
    {
        pixelStream = bitmap.PixelBuffer.AsStream();
        await pixelStream.WriteAsync(pixels, 0, pixels.Length);
        bitmap.Invalidate();
        image.Source = bitmap;
    CalculateImageScaleAndOffset();
    }

    async void OnAddAppBarButtonClick(object sender, RoutedEventArgs args)
    {
        Button button = sender as Button;
        button.IsEnabled = false;

        await CheckIfOkToTrashFile(CreateNewBitmapAndPixelArray);

        button.IsEnabled = true;
        this.BottomAppBar.IsOpen = false;
```

```
}

async void OnOpenAppBarButtonClick(object sender, RoutedEventArgs args)
{
    Button button = sender as Button;
    button.IsEnabled = false;

    await CheckIfOkToTrashFile(LoadFileFromOpenPicker);

    button.IsEnabled = true;
    this.BottomAppBar.IsOpen = false;
}

async Task CheckIfOkToTrashFile(Func<Task> commandAction)
{
    if (!appSettings.IsImageModified)
    {
        await commandAction();
        return;
    }

    string message =
        String.Format("Do you want to save changes to {0}?",
            appSettings.LoadedFilePath ?? "(untitled)");

    MessageDialog msgdlg = new MessageDialog(message, "Finger Paint");
    msgdlg.Commands.Add(new UICommand("Save", null, "save"));
    msgdlg.Commands.Add(new UICommand("Don't Save", null, "dont"));
    msgdlg.Commands.Add(new UICommand("Cancel", null, "cancel"));
    msgdlg.DefaultCommandIndex = 0;
    msgdlg.CancelCommandIndex = 2;
    IUICommand command = await msgdlg.ShowAsync();

    If ((string)command.Id == "cancel")
        return;

    if ((string)command.Id == "dont")
    {
        await commandAction();
        return;
    }

    if (appSettings.LoadedFilePath == null)
    {
        StorageFile storageFile = await GetFileFromSavePicker();

        if (storageFile == null)
            return;

        appSettings.LoadedFilePath = storageFile.Path;
        appSettings.LoadedFilename = storageFile.Name;
    }

    string exception = null;

    try
    {
        await SaveBitmapToFile(appSettings.LoadedFilePath);
    }

    catch (Exception exc)
    {
        exception = exc.Message;
    }

    if (exception != null)
    {
```

```
            msgdlg = new MessageDialog("The image file could not be saved. " +
                    "The system reports an error of: " + exception,
                    "Finger Paint");
            await msgdlg.ShowAsync();
            return;
        }

        await commandAction();
    }

    async Task LoadFileFromOpenPicker()
    {
        // Create FileOpenPicker
        FileOpenPicker picker = new FileOpenPicker();
        picker.SuggestedStartLocation = PickerLocationId.PicturesLibrary;

        // Initialize with filename extensions
        IReadOnlyList<BitmapCodecInformation> codecInfos =
                        BitmapDecoder.GetDecoderInformationEnumerator();

        foreach (BitmapCodecInformation codecInfo in codecInfos)
            foreach (string extension in codecInfo.FileExtensions)
                picker.FileTypeFilter.Add(extension);

        // Get the selected file
        StorageFile storageFile = await picker.PickSingleFileAsync();

        If (storageFile == null)
            return;

        string exception = null;

        try
        {
            await LoadBitmapFromFile(storageFile);
        }
        catch (Exception exc)
        {
            exception = exc.Message;
        }

        if (exception != null)
        {
            MessageDialog msgdlg =
                new MessageDialog("The image file could not be loaded. " +
                    "The system reports an error of: " + exception,
                    "Finger Paint");
            await msgdlg.ShowAsync();
            return;
        }

        appSettings.LoadedFilePath = storageFile.Path;
        appSettings.LoadedFilename = storageFile.Name;
        appSettings.IsImageModified = false;
    }

    async void OnSaveAppBarButtonClick(object sender, RoutedEventArgs args)
    {
        Button button = sender as Button;
        button.IsEnabled = false;

        if (appSettings.LoadedFilePath != null)
        {
            await SaveWithErrorNotification(appSettings.LoadedFilePath);
        }
        else
        {
            StorageFile storageFile = await GetFileFromSavePicker();
```

```
            if (storageFile == null)
                return;

            await SaveWithErrorNotification(storageFile);
        }

        button.IsEnabled = true;
    }

    async void OnSaveAsAppBarButtonClick(object sender, RoutedEventArgs args)
    {
        StorageFile storageFile = await GetFileFromSavePicker();

        if (storageFile == null)
            return;

        await SaveWithErrorNotification(storageFile);
    }

    async Task<StorageFile> GetFileFromSavePicker()
    {
        FileSavePicker picker = new FileSavePicker();
        picker.SuggestedStartLocation = PickerLocationId.PicturesLibrary;
        picker.SuggestedFileName = appSettings.LoadedFilename ?? "MyFingerPainting";

        // Get the encoder information
        Dictionary<string, Guid> imageTypes = new Dictionary<string, Guid>();
        IReadOnlyList<BitmapCodecInformation> codecInfos =
                    BitmapEncoder.GetEncoderInformationEnumerator();

        foreach (BitmapCodecInformation codecInfo in codecInfos)
        {
            List<string> extensions = new List<string>();

            foreach (string extension in codecInfo.FileExtensions)
                extensions.Add(extension);

        string filetype = codecInfo.FriendlyName.Split(' ')[0];
        picker.FileTypeChoices.Add(filetype, extensions);

        foreach (string mimeType in codecInfo.MimeTypes)
            imageTypes.Add(mimeType, codecInfo.CodecId);
        }

        // Get a selected StorageFile
        return await picker.PickSaveFileAsync();
    }

    async Task<bool> SaveWithErrorNotification(string filename)
    {
        StorageFile storageFile = await StorageFile.GetFileFromPathAsync(filename);
        return await SaveWithErrorNotification(storageFile);
    }

    async Task<bool> SaveWithErrorNotification(StorageFile storageFile)
    {
        string exception = null;

        try
        {
            await SaveBitmapToFile(storageFile);
        }
        catch (Exception exc)
        {
            exception = exc.Message;
        }
```

```
            if (exception != null)
            {
                MessageDialog msgdlg =
                    new MessageDialog("The image file could not be saved. " +
                                      "The system reports an error of: " + exception,
                                      "Finger Paint");
                await msgdlg.ShowAsync();
                return false;
            }

            appSettings.LoadedFilePath = storageFile.Path;
            appSettings.IsImageModified = false;
            return true;
        }

        async Task SaveBitmapToFile(string filename)
        {
            StorageFile storageFile = await StorageFile.GetFileFromPathAsync(filename);
            await SaveBitmapToFile(storageFile);
        }

        async Task SaveBitmapToFile(StorageFile storageFile)
        {
            using (IRandomAccessStream fileStream =
                        await storageFile.OpenAsync(FileAccessMode.ReadWrite))
            {
                BitmapEncoder encoder =
                    await BitmapEncoder.CreateAsync(BitmapEncoder.PngEncoderId, fileStream);

                encoder.SetPixelData(BitmapPixelFormat.Bgra8, BitmapAlphaMode.Premultiplied,
                                     (uint)bitmap.PixelWidth, (uint)bitmap.PixelHeight,
                                     96, 96, pixels);
                await encoder.FlushAsync();
            }
        }
    }
```

接下来的 MainPage.xaml.cs 文件有几项责任。程序暂停时，文件代码会保存应用设置并在程序启动时恢复应用设置。还会保存当前图片并重新加载。这种逻辑使用了 MainPage.File.cs 文件中的方法，当然会忽略任何可能发生的异常。

该文件还负责处理 SizeChanged 事件，可以在 XAML 文件中设置视觉状态，还可以设置 ImageScale 和 ImageOffset 字段。根据位图的原始大小、屏幕方向和辅屏状态，当前作为手绘画布的位图会大于或小于页面。画布总是最大化显示，且比例不失真，但触摸坐标必须转换为绘画时的位图坐标。

```
项目: FingerPaint | 文件: MainPage.xaml.cs(片段)
public sealed partial class MainPage : Page
{
    AppSettings appSettings = new AppSettings();
    double imageScale = 1;
    Point imageOffset = new Point();

    public MainPage()
    {
        this.InitializeComponent();

        SizeChanged += OnMainPageSizeChanged;
        Loaded += OnMainPageLoaded;
        Application.Current.Suspending += OnApplicationSuspending;
    }

    void OnMainPageSizeChanged(object sender, SizeChangedEventArgs args)
    {
```

```
        VisualStateManager.GoToState(this, ApplicationView.Value.ToString(), true);

        If  (bitmap != null)
        {
            CalculateImageScaleAndOffset();
        }
    }

    void CalculateImageScaleAndOffset()
    {
        imageScale = Math.Min(this.ActualWidth / bitmap.PixelWidth,
                              this.ActualHeight / bitmap.PixelHeight);

        imageOffset = new Point((this.ActualWidth - imageScale * bitmap.PixelWidth) / 2,
                    (this.ActualHeight - imageScale * bitmap.PixelHeight) / 2);
    }

    async void OnMainPageLoaded(object sender, RoutedEventArgs args)
    {
        try
        {
            StorageFolder localFolder = ApplicationData.Current.LocalFolder;
            StorageFile storageFile = await localFolder.GetFileAsync("FingerPaint.png");
            await LoadBitmapFromFile(storageFile);
        }
        catch
        {
            // Ignore any errors
        }

        if  (bitmap == null)
            await CreateNewBitmapAndPixelArray();
    }

    async void OnApplicationSuspending(object sender, SuspendingEventArgs args)
    {
        // Save application settings
        appSettings.Save();

        // Save current bitmap
        SuspendingDeferral deferral = args.SuspendingOperation.GetDeferral();

        try
        {
            StorageFolder localFolder = ApplicationData.Current.LocalFolder;
            StorageFile storageFile = await localFolder.CreateFileAsync("FingerPaint.png",
                                        CreationCollisionOption.ReplaceExisting);

            await SaveBitmapToFile(storageFile);
        }
        catch
        {
            // Ignore any errors
        }

        deferral.Complete();
    }
    ...
}
```

MainPage.aml.cs 文件还负责用名为 ColorSettingDialog 和 ThicknessSettingDialog 的 UserControl 派生类来显示 Popup 对象并处理 Color 和 Thickness 应用栏按钮。

项目: FingerPaint | 文件: MainPage.xaml.cs(片段)

```
public sealed partial class MainPage : Page
{
    ...
```

```
    void OnColorAppBarButtonClick(object sender, RoutedEventArgs args)
    {
        DisplayDialog(sender, new ColorSettingDialog());
    }

    void OnThicknessAppBarButtonClick(object sender, RoutedEventArgs args)
    {
        DisplayDialog(sender, new ThicknessSettingDialog());
    }

    void DisplayDialog(object sender, FrameworkElement dialog)
    {
        dialog.DataContext = appSettings;

        Popup popup = new Popup
        {
            Child = dialog,
            IsLightDismissEnabled = true
        };

        dialog.SizeChanged += (dialogSender, dialogArgs) =>
        {
            // Get Button location relative to screen
            Button btn = sender as Button;
            Point pt = btn.TransformToVisual(null).TransformPoint(
                                            new Point(btn.ActualWidth / 2,
                                                      btn.ActualHeight / 2));

            popup.HorizontalOffset = Math.Max(24, pt.X - dialog.ActualWidth / 2);

            if (popup.HorizontalOffset + dialog.ActualWidth > this.ActualWidth)
                popup.HorizontalOffset = this.ActualWidth - dialog.ActualWidth;

            popup.HorizontalOffset = Math.Max(0, popup.HorizontalOffset);

            popup.VerticalOffset = this.ActualHeight - dialog.ActualHeight
                                 - this.BottomAppBar.ActualHeight - 24;
        };

        popup.Closed += (popupSender, popupArgs) =>
        {
            this.BottomAppBar.IsOpen = false;
        };

        popup.IsOpen = true;
    }
}
```

ThicknessSettingDialog 是两者中较为简单的。它只包含带有一些直线粗细值的 ListBox。我想发挥 2 的力量(比如 2、4、8、16、32)，但我也想要这些值之间的值，因此，值基本上以 2 的立方根增加，同时对冗余进行舍入和消除。

项目: FingerPaint | 文件: ThicknessSettingDialog.xaml(片段)

```
<Grid>
    <Border Background="White"
            BorderBrush="Black"
            BorderThickness="3"
            Padding="32">

        <ListBox SelectedItem="{Binding Thickness, Mode=TwoWay}"
                 Width="150">
            <x:Double>2</x:Double>
            <x:Double>3</x:Double>
            <x:Double>4</x:Double>
            <x:Double>5</x:Double>
            <x:Double>6</x:Double>
```

```
<x:Double>8</x:Double>
<x:Double>10</x:Double>
<x:Double>13</x:Double>
<x:Double>16</x:Double>
<x:Double>20</x:Double>
<x:Double>25</x:Double>
<x:Double>32</x:Double>
<x:Double>40</x:Double>

<ListBox.Foreground>
    <SolidColorBrush Color="{Binding Color}" />
</ListBox.Foreground>

<ListBox.ItemTemplate>
    <DataTemplate>
        <Grid Height="{Binding}"
              Width="120">
            <Canvas VerticalAlignment="Center"
                    HorizontalAlignment="Center">
            <Polyline Points="-36 0 36 0"
                Stroke="{Binding RelativeSource={RelativeSource TemplatedParent},
                                 Path=Foreground}"
                StrokeThickness="{Binding}"
                StrokeStartLineCap="Round"
                StrokeEndLineCap="Round" />
            </Canvas>
        </Grid>
    </DataTemplate>
</ListBox.ItemTemplate>

<ListBox.ItemContainerStyle>
    <Style TargetType="ListBoxItem">
      <Setter Property="Background" Value="Transparent" />
        <Setter Property="Template">
          <Setter.Value>
            <ControlTemplate TargetType="ListBoxItem">
              <Grid>
                <VisualStateManager.VisualStateGroups>
                  <VisualStateGroup x:Name="SelectionStates">
                    <VisualState x:Name="Unselected">
                        <Storyboard>
                            <ObjectAnimationUsingKeyFrames
                                Storyboard.TargetName="border"
                                Storyboard.TargetProperty="BorderBrush">
                                <DiscreteObjectKeyFrame
                                                KeyTime="0"
                                                Value="Transparent" />
                            </ObjectAnimationUsingKeyFrames>
                        </Storyboard>
                    </VisualState>

                        <VisualState x:Name="Selected" />
                        <VisualState x:Name="SelectedUnfocused" />
                        <VisualState x:Name="SelectedDisabled" />
                        <VisualState x:Name="SelectedPointerOver" />
                        <VisualState x:Name="SelectedPressed" />
                    </VisualStateGroup>
                  </VisualStateManager.VisualStateGroups>

                  <Border Name="border"
                          BorderBrush="Black"
                          BorderThickness="1"
                          Background="Transparent"
                          Padding="12">

                    <ContentPresenter Content="{TemplateBinding Content}" />

                  </Border>
```

```
                        </Grid>
                    </ControlTemplate>
                </Setter.Value>
            </Setter>
        </Style>
    </ListBox.ItemContainerStyle>
</ListBox>
</Border>
</Grid>
```

当然，所有“神奇”都发生在模板中。(代码隐藏文件中除了调用 InitializeComponent 以外其他什么都没有)。ItemTemplate 使用 ListBox 中的值作为实际线条宽度，而 ItemContainerStyle 的结果是围绕着所选值画出矩形(见下图)。

MainPage 显示该对话框时，DataContext 设置为 appSettings 实例，因而会通过数据绑定来更新该类的 Thickness 属性值。Thickness 属性值用于所有后续绘画操作，直到出现下一个变化。

14.7　HSL 颜色选择

在本书中，你可能已经见过太多用滑块构建的颜色选择器，你可以选择 Red、Green 和 Blue 值。这是选择颜色的简单方法，因为这就是视频显示器和 Windows Runtime 定义颜色的方式。

然而，这并不是选择颜色的直观方式。人们似乎更喜欢用色调(Hue)、饱和度(Saturation)和亮度(Lightness)值构建的系统。色调基本上是彩虹的一种颜色，艾萨克·牛顿将其命名为红、橙、黄、绿、蓝、靛、紫的颜色。使用更“计算机化”的颜色，色调范围从红色，通过黄色，绿色，青色，蓝色，品红再返回到红色。注意三原色(红、绿、蓝)和三补色(黄、青和品红)，三补色为围绕三原色的配对组合。

色调可以与饱和度值相结合。如果饱和度为最大值，则色彩最鲜艳。如果饱和度为最小值，则颜色为灰色。此时亮度就会发挥作用。增加亮度会冲刷颜色，并最终用最大值变为白色。减少来自于媒介值的亮度会让颜色变黑。

Hue-Saturation-Lightness(或 HSL)颜色选择用于 Windows Paint 和 Microsoft Word 中，有一个二维网格(类似于第 13 章中的 XYSlider)用色调与饱和度，而有一个常规滑块用于色调。

为了模拟这种颜色选择，我创建了一个 HSL 结构来表示 HSL 颜色值。该结构采用了 RGB 和 HSL 之间的转换例程。

```
项目: FingerPaint | 文件: HSL.cs
public struct HSL
{
    public HSL(byte hue, byte saturation, byte lightness) :
        this(360 * hue / 255.0, saturation / 255.0, lightness / 255.0)
    {
    }

    // Hue from 0 to 360, saturation and lightness from 0 to 1
    public HSL(double hue, double saturation, double lightness) : this()
    {
        this.Hue = hue;
        this.Saturation = saturation;
        this.Lightness = lightness;

        double chroma = saturation * (1 - Math.Abs(2 * lightness - 1));
        double h = hue / 60;

        double x = chroma * (1 - Math.Abs(h % 2 - 1));
        double r = 0, g = 0, b = 0;

        if (h < 1)
        {
            r = chroma;
            g = x;
        }
        else if (h < 2)
        {
            r = x;
            g = chroma;
        }
        else if (h < 3)
        {
            g = chroma;
            b = x;
        }
        else if (h < 4)
        {
            g = x;
            b = chroma;
        }
        else if (h < 5)
        {
            r = x;
            b = chroma;
        }
        else
        {
            r = chroma;
            b = x;
        }

        double m = lightness - chroma / 2;
        this.Color = Color.FromArgb(255, (byte)(255 * (r + m)),
                                         (byte)(255 * (g + m)),
                                         (byte)(255 * (b + m)));
    }
```

```
    public HSL(Color color)
        : this()
    {
        this.Color = color;

        double r = color.R / 255.0;
        double g = color.G / 255.0;
        double b = color.B / 255.0;
        double max = Math.Max(r, Math.Max(g, b));
        double min = Math.Min(r, Math.Min(g, b));

        double chroma = max - min;
        this.Lightness = (max + min) / 2;

        if (chroma != 0)
        {
            if (r == max)
                this.Hue = 60 * (g - b) / chroma;

            else if (g == max)
                this.Hue = 120 + 60 * (b - r) / chroma;

            else
                this.Hue = 240 + 60 * (r - g) / chroma;

            this.Hue = (this.Hue + 360) % 360;

            if (this.Lightness < 0.5)
                this.Saturation = chroma / (2 * this.Lightness);
            else
                this.Saturation = chroma / (2 - 2 * this.Lightness);
        }
    }

    public double Hue { private set; get; }

    public double Saturation { private set; get; }

    public double Lightness { private set; get; }

    public Color Color { private set; get; }
}
```

注意两个不同的构造函数，一个使用 byte 参数，另一个采用 double 参数。对于对 HSL 构造函数的特定调用，C# 编译器需要选择使用哪一个构造函数，只有所有参数都为 byte 值时，选择第一个构造函数。第三个构造函数中没有这种歧义，第三个构造函数把 Color 值转换为 HSL。

我在之前的章节展示了 XYSlider 控件，但我也表明如果使用 Pointer 事件而不是 Manipulation 事件，控件会更加实用。以下为修改后的版本。因为要处理 Pointer 事件，所以需要跟踪多个手指，但控件直接使用手指位置的平均值来创建一个复合位置。除此之外，控制基本上相同。

项目：FingerPaint | 文件：XYSlider.cs(片段)

```
public class XYSlider : ContentControl
{
    ContentPresenter contentPresenter;
    FrameworkElement crossHairPart;
    Dictionary<uint, Point> pointerDictionary = new Dictionary<uint, Point>();

    static readonly DependencyProperty valueProperty =
            DependencyProperty.Register("Value",
                    typeof(Point), typeof(XYSlider),
```

```
                                new PropertyMetadata(new Point(), OnValueChanged));

        public event EventHandler<Point> ValueChanged;

        public XYSlider()
        {
            this.DefaultStyleKey = typeof(XYSlider);
        }

        public static DependencyProperty ValueProperty
        {
            get { return valueProperty; }
        }

        public Point Value
        {
            set { SetValue(ValueProperty, value); }
            get { return (Point)GetValue(ValueProperty); }
        }

        protected override void OnApplyTemplate()
        {
            // Detach event handlers
            if (contentPresenter != null)
            {
                contentPresenter.PointerPressed -= OnContentPresenterPointerPressed;
                contentPresenter.PointerMoved -= OnContentPresenterPointerMoved;
                contentPresenter.PointerReleased -= OnContentPresenterPointerReleased;
                contentPresenter.PointerCaptureLost -= OnContentPresenterPointerReleased;
                contentPresenter.SizeChanged -= OnContentPresenterSizeChanged;
            }

            // Get new parts
            crossHairPart = GetTemplateChild("CrossHairPart") as FrameworkElement;
            contentPresenter = GetTemplateChild("ContentPresenterPart") as ContentPresenter;

            // Attach event handlers
            if (contentPresenter != null)
            {
                contentPresenter.PointerPressed += OnContentPresenterPointerPressed;
                contentPresenter.PointerMoved += OnContentPresenterPointerMoved;
                contentPresenter.PointerReleased += OnContentPresenterPointerReleased;
                contentPresenter.PointerCaptureLost += OnContentPresenterPointerReleased;
                contentPresenter.SizeChanged += OnContentPresenterSizeChanged;
            }

            // Make cross-hair transparent to touch
            if (crossHairPart != null)
            {
                crossHairPart.IsHitTestVisible = false;
            }

            base.OnApplyTemplate();
        }
        void OnContentPresenterPointerPressed(object sender, PointerRoutedEventArgs args)
        {
            uint id = args.Pointer.PointerId;
            Point point = args.GetCurrentPoint(contentPresenter).Position;
            pointerDictionary.Add(id, point);
            contentPresenter.CapturePointer(args.Pointer);

            RecalculateValue();
            args.Handled = true;
        }

        void OnContentPresenterPointerMoved(object sender, PointerRoutedEventArgs args)
        {
```

```
        uint id = args.Pointer.PointerId;
        Point point = args.GetCurrentPoint(contentPresenter).Position;

        if (pointerDictionary.ContainsKey(id))
        {
            pointerDictionary[id] = point;
            RecalculateValue();
            args.Handled = true;
        }
    }

    void OnContentPresenterPointerReleased(object sender, PointerRoutedEventArgs args)
    {
        uint id = args.Pointer.PointerId;

        if (pointerDictionary.ContainsKey(id))
        {
            pointerDictionary.Remove(id);
            RecalculateValue();
            args.Handled = true;
        }
    }

    void OnContentPresenterSizeChanged(object sender, SizeChangedEventArgs args)
    {
        SetCrossHair();
    }

    void RecalculateValue()
    {
        if (pointerDictionary.Values.Count > 0)
        {
            Point accumPoint = new Point();

            // Average all the current touch points
            foreach (Point point in pointerDictionary.Values)
            {
                accumPoint.X += point.X;
                accumPoint.Y += point.Y;
            }
            accumPoint.X /= pointerDictionary.Values.Count;
            accumPoint.Y /= pointerDictionary.Values.Count;

            RecalculateValue(accumPoint);
        }
    }

    void RecalculateValue(Point absolutePoint)
    {
        double x = Math.Max(0, Math.Min(1, absolutePoint.X / contentPresenter.ActualWidth));
        double y = Math.Max(0, Math.Min(1, absolutePoint.Y / contentPresenter.ActualHeight));
        this.Value = new Point(x, y);
    }

    void SetCrossHair()
    {
        if (contentPresenter != null && crossHairPart != null)
        {
            Canvas.SetLeft(crossHairPart, this.Value.X * contentPresenter.ActualWidth);
            Canvas.SetTop(crossHairPart, this.Value.Y * contentPresenter.ActualHeight);
        }
    }

    static void OnValueChanged(DependencyObject obj, DependencyPropertyChangedEventArgs args)
    {
        (obj as XYSlider).SetCrossHair();
        (obj as XYSlider).OnValueChanged((Point)args.NewValue);
    }
```

```
        protected void OnValueChanged(Point value)
        {
            if (ValueChanged != null)
                ValueChanged(this, value);
        }
    }
```

接下来建立 HslColorSelector 控件。该控件派生自 UserControl 并在 XAML 文件中实例化 XYSlider、Slider 和 TextBlock。Resources 部分为 XYSlider 和 Slider 定义 ControlTemplate 对象。通过我在第 13 章讲到的对应控件，会大大简化 XYSlider 模板，因为我清楚地知道想要怎样的视觉效果，不添加任何其他东西。

```
项目: FingerPaint | 文件: HslColorSelector.xaml(片段)
<UserControl ... >
    <UserControl.Resources>
        <ControlTemplate x:Key="xySliderTemplate" TargetType="local:XYSlider">
            <Border>
                <Grid>
                    <ContentPresenter Name="ContentPresenterPart"
                                      Content="{TemplateBinding Content}" />
                    <Canvas>
                        <Path Name="CrossHairPart"
                              Fill="{TemplateBinding Foreground}"
                              Data="M 0 6 L -3 24 3 24 Z
                                    M 0 -6 L -3 -24 3 -24 Z
                                    M 6 0 L 24 -3 24 3 Z
                                    M -6 0 L -24 -2 -24 3 Z" />
                    </Canvas>
                </Grid>
            </Border>
        </ControlTemplate>

        <ControlTemplate x:Key="sliderTemplate" TargetType="Slider">
            <Grid>
                <Grid Name="HorizontalTemplate"
                      Background="Transparent"
                      Height="48">
                    <Grid.ColumnDefinitions>
                        <ColumnDefinition Width="*" />
                        <ColumnDefinition Width="Auto" />
                        <ColumnDefinition Width="Auto" />
                    </Grid.ColumnDefinitions>

                    <Rectangle Name="HorizontalTrackRect"
                               Grid.Column="0"
                               Grid.ColumnSpan="3"
                               Fill="{TemplateBinding Background}"
                               Height="12"
                               VerticalAlignment="Top" />

                    <Thumb Name="HorizontalThumb"
                           Grid.Column="1"
                           DataContext="{TemplateBinding Value}">
                        <Thumb.Template>
                            <ControlTemplate TargetType="Thumb">
                                <Path Fill="{TemplateBinding Foreground}"
                                      Data="M 0 24 L -3 48 3 48 Z" />
                            </ControlTemplate>
                        </Thumb.Template>
                    </Thumb>

                    <Rectangle Name="HorizontalDecreaseRect"
                               Grid.Column="2"
                               Fill="Transparent" />
                </Grid>
```

```
                <Grid Name="VerticalTemplate"
                      Background="Transparent"
                      Width="48">
                     <Grid.RowDefinitions>
                          <RowDefinition Height="*" />
                          <RowDefinition Height="Auto" />
                          <RowDefinition Height="Auto" />
                     </Grid.RowDefinitions>

                     <Rectangle Name="VerticalTrackRect"
                                Grid.Row="0"
                                Grid.RowSpan="3"
                                Fill="{TemplateBinding Background}"
                                Width="12"
                                HorizontalAlignment="Left" />

                     <Thumb Name="VerticalThumb"
                            Grid.Row="1"
                            DataContext="{TemplateBinding Value}">
                          <Thumb.Template>
                               <ControlTemplate TargetType="Thumb">
                                    <Path Fill="{TemplateBinding Foreground}"
                                          Data="M 24 0 L 48 -3 48 3 Z" />
                               </ControlTemplate>
                          </Thumb.Template>
                     </Thumb>

                     <Rectangle Name="VerticalDecreaseRect"
                                Grid.Row="2"
                                Fill="Transparent" />
                </Grid>
           </Grid>
      </ControlTemplate>
</UserControl.Resources>

<Grid>
     <Grid.RowDefinitions>
          <RowDefinition Height="Auto" />
          <RowDefinition Height="Auto" />
          <RowDefinition Height="Auto" />
     </Grid.RowDefinitions>

     <local:XYSlider x:Name="xySlider"
                     Grid.Row="0"
                     Template="{StaticResource xySliderTemplate}"
                     ValueChanged="OnXYSliderValueChanged">
          <Image Name="hsImage"
                 Stretch="None" />
     </local:XYSlider>

     <Slider Name="slider"
             Grid.Row="1"
             Orientation="Horizontal"
             Template="{StaticResource sliderTemplate}"
             Width="256"
             Margin="0 12"
             ValueChanged="OnSliderValueChanged">
          <Slider.Background>
               <LinearGradientBrush StartPoint="0 0" EndPoint="1 0">
                    <GradientStop Offset="0" Color="Black" />
                    <GradientStop x:Name="sliderGradientStop" Offset="0.5" />
                    <GradientStop Offset="1" Color="White" />
               </LinearGradientBrush>
          </Slider.Background>
     </Slider>

     <TextBlock Name="txtblk"
                Grid.Row="2"
```

```
                    HorizontalAlignment="Center"
                    TextAlignment="Center"
                    FontSize="24" />
    </Grid>
</UserControl>
```

请注意，用于 Slider 的 ControlTemplate 基本用其 Background 属性对控件进行着色。Background 属性在 XAML 文件结尾的 Slider 上进行。Background 为 LinearGradientBrush，范围从黑到白，在代码隐藏文件中间进行设置颜色设置。该颜色基于用户从 XYSlider 所选择的色调与饱和度组合。

代码隐藏文件定义了 Color 类型的 DependencyProperty，名为 Color。很显然，作为用户绑定的公共属性，Color 属性比 HSL 类型的公共属性更合理。Loaded 处理程序负责为主色调-饱和度网格创建位图。它采用 HSL 结构(及平均亮度值)进行转化，以获得位图像素的 RGB 值。

```
项目: FingerPaint | 文件: HslColorSelector.xaml.cs(片段)
public partial class HslColorSelector : UserControl
{
    bool doNotSetSliders = false;

    static readonly DependencyProperty colorProperty =
        DependencyProperty.Register("Color",
            typeof(Color),
            typeof(HslColorSelector),
            new PropertyMetadata(new Color(), OnColorChanged));

    public event EventHandler<Color> ColorChanged;

    public HslColorSelector()
    {
        this.InitializeComponent();
        Loaded += OnLoaded;
    }

    async void OnLoaded(object sender, RoutedEventArgs args)
    {
        // Build bitmap for hue/saturation grid
        WriteableBitmap bitmap = new WriteableBitmap(256, 256);
        byte[] pixels = new byte[4 * 256 * 256];
        int index = 0;

        for (int y = 0; y < 256; y++)
            for (int x = 0; x < 256; x++)
            {
                HSL hsl = new HSL((byte)x, (byte)(255 - y), (byte)128);
                Color clr = hsl.Color;

                pixels[index++] = clr.B;
                pixels[index++] = clr.G;
                pixels[index++] = clr.R;
                pixels[index++] = clr.A;
            }

        using (Stream pixelStream = bitmap.PixelBuffer.AsStream())
        {
            await pixelStream.WriteAsync(pixels, 0, pixels.Length);
        }
        bitmap.Invalidate();
        hsImage.Source = bitmap;
    }

    public static DependencyProperty ColorProperty
    {
```

```
            get { return colorProperty; }
        }
        public Color Color
        {
            set { SetValue(ColorProperty, value); }
            get { return (Color)GetValue(ColorProperty); }
        }

        // Event handlers for sliders
        void OnXYSliderValueChanged(object sender, Point point)
        {
            HSL hsl = new HSL(360 * point.X, 1 - point.Y, 0.5);
            sliderGradientStop.Color = hsl.Color;
            SetColorFromSliders();
        }

        void OnSliderValueChanged(object sender, RangeBaseValueChangedEventArgs args)
        {
            SetColorFromSliders();
        }

        void SetColorFromSliders()
        {
            Point point = xySlider.Value;
            double value = slider.Value;
            HSL hsl = new HSL(360 * point.X, 1 - point.Y, value / 100);

            doNotSetSliders = true;
            this.Color = hsl.Color;
            doNotSetSliders = false;
        }

        // Color property-changed handlers
        static void OnColorChanged(DependencyObject obj, DependencyPropertyChangedEventArgs args)
        {
             (obj as HslColorSelector).OnColorChanged((Color)args.NewValue);
        }

        protected void OnColorChanged(Color color)
        {
            HSL hsl = new HSL(color);

            if (!doNotSetSliders)
            {
                xySlider.Value = new Point(hsl.Hue / 360, 1 - hsl.Saturation);
                slider.Value = 100 * hsl.Lightness;
            }

            txtblk.Text = String.Format("RGB = ({0}, {1}, {2})",
                                this.Color.R, this.Color.G, this.Color.B);

            if (ColorChanged != null)
                ColorChanged(this, color);
        }
    }
```

如果从文件之外设置新的 Color 属性，OnColorChanged 会进行响应，设置 XYSlider 和 Slider 值，还会用 TextBlock 来显示 RGB 颜色值。如果用户操作 XYSlider 和 Slider 值，则设置新的 Color 属性并且调用 OnColorChanged。正常情况下，可以递归调用属性已改变的处理程序，但在这种情况下却不行，因为往返转换——从 RGB 到 HSL 再到 RGB——不会产生完全相同的值。这就是为什么用户输入改变 Color 属性会有布尔字段 doNotSetSliders 的原因。

最后，ColorSettingDialog 合并 HslColorSelector。

项目：FingerPaint | 文件：ColorSettingDialog.xaml(片段)

```
<Grid>
    <Border Background="White"
            BorderBrush="Black"
            BorderThickness="3"
            Padding="32">
        <StackPanel>
            <Path Data="M 0 50 C 80 0 160 100 256 0"
                  StrokeStartLineCap="Round"
                  StrokeEndLineCap="Round"
                  StrokeThickness="{Binding Thickness}"
                  Margin="0 12">
                <Path.Stroke>
                    <SolidColorBrush Color="{Binding Color}" />
                </Path.Stroke>
            </Path>

            <local:HslColorSelector x:Name="hslColorSelector"
                                    Foreground="Black"
                                    Color="{Binding Path=Color, Mode=TwoWay}" />
        </StackPanel>
    </Border>
</Grid>
```

对于预览，控件会显示一个基于宽度和颜色的波形曲线(见下图)。

如果不是在分辨率非常高的显示器上使用 FingerPaint，你可能会注意到上图比之前第 13 章的 FingerPaint 程序绘出来的要粗糙一点。有一个正当理由：如果用 Line、Polyline 和 Path 渲染图形对象，就会得到反锯齿。边界线是填充颜色和背景颜色的组合，线条给人流畅的感觉。但在 Petzold.Windows8.VectorDrawing 库中没有实现抗锯齿。像素可以着色，也可以不着色。

14.8 反向绘画

我曾经看过一部短片，有个人用白色颜料和滚筒画一幅大型的彩色壁画，但片子是倒着放的，壁画好像奇迹般地用滚筒画在白色表面上，像创造出来一样。

ReversePaint 程序阐述了类似技巧。XAML 文件会访问我网站上的一张位图并定义位于顶部的另一个 Image 元素。

```
项目: ReversePaint | 文件: MainPage.xaml(片段)
<Grid Background="{StaticResource ApplicationPageBackgroundThemeBrush}">
    <Image Source="http://www.charlespetzold.com/pw6/PetzoldJersey.jpg" />
    <Image Name="whiteImage" />
</Grid>
```

第二个 Image 元素的位图在 Loaded 处理程序中创建，它与下载的位图大小相同，并且全白。像 FingerPaint 一样，有 CalculateImageScaleAndOffset 方法计算用于缩放指针输入该位图的因子。为了简化代码演示，我删除了很多指针事件处理程序的注释，但你之前看见过该逻辑。OnPointeMoved 方法有两个点、一个固定线宽以及一个表示透明度的 Color 值，用来调用简化形式的 RenderOnBitmap。

```
项目: ReversePaint | 文件: MainPage.xaml.cs(片段)
public sealed partial class MainPage : Page
{
    Dictionary<uint, Point> pointerDictionary = new Dictionary<uint, Point>();
    List<double> xCollection = new List<double>();

    WriteableBitmap bitmap;
    byte[] pixels;
    Stream pixelStream;

    Point imageOffset = new Point();
    double imageScale = 1;

    public MainPage()
    {
        this.InitializeComponent();

        SizeChanged += OnMainPageSizeChanged;
        Loaded += OnMainPageLoaded;
    }

    void OnMainPageSizeChanged(object sender, SizeChangedEventArgs args)
    {
        if (bitmap != null)
            CalculateImageScaleAndOffset();
    }

    async void OnMainPageLoaded(object sender, RoutedEventArgs args)
    {
        bitmap = new WriteableBitmap(320, 400);
        pixels = new byte[4 * bitmap.PixelWidth * bitmap.PixelHeight];

        // Initialize pixels to white
        for (int index = 0; index < pixels.Length; index++)
            pixels[index] = 0xFF;

        pixelStream = bitmap.PixelBuffer.AsStream();
        await pixelStream.WriteAsync(pixels, 0, pixels.Length);
        bitmap.Invalidate();

        // Set to Image element
        whiteImage.Source = bitmap;
        CalculateImageScaleAndOffset();
    }

    void CalculateImageScaleAndOffset()
```

```
    {
        imageScale = Math.Min(this.ActualWidth / bitmap.PixelWidth,
                              this.ActualHeight / bitmap.PixelHeight);

        imageOffset = new Point((this.ActualWidth - imageScale * bitmap.PixelWidth) / 2,
                                (this.ActualHeight - imageScale * bitmap.PixelHeight) / 2);
    }

    protected override void OnPointerPressed(PointerRoutedEventArgs args)
    {
        uint id = args.Pointer.PointerId;
        Point point = args.GetCurrentPoint(this).Position;
        pointerDictionary.Add(id, point);
        CapturePointer(args.Pointer);
        base.OnPointerPressed(args);
    }

    protected override void OnPointerMoved(PointerRoutedEventArgs args)
    {
        uint id = args.Pointer.PointerId;
        Point point = args.GetCurrentPoint(this).Position;

        if (pointerDictionary.ContainsKey(id))
        {
            Point previousPoint = pointerDictionary[id];

            // Render the line
            RenderOnBitmap(previousPoint, point, 12, new Color());

            pointerDictionary[id] = point;
        }
        base.OnPointerMoved(args);
    }

    protected override void OnPointerReleased(PointerRoutedEventArgs args)
    {
        uint id = args.Pointer.PointerId;

        if (pointerDictionary.ContainsKey(id))
            pointerDictionary.Remove(id);

        base.OnPointerReleased(args);
    }

    protected override void OnPointerCaptureLost(PointerRoutedEventArgs args)
    {
        uint id = args.Pointer.PointerId;

        if (pointerDictionary.ContainsKey(id))
            pointerDictionary.Remove(id);

        base.OnPointerCaptureLost(args);
    }

    void RenderOnBitmap(Point point1, Point point2, double radius, Color color)
    {
        bool bitmapNeedsUpdate = false;

        // Adjust the points for any bitmap scaling
        Point center1 = ScaleToBitmap(point1);
        Point center2 = ScaleToBitmap(point2);

        // Create object to render the line
        RoundCappedLine line = new RoundCappedLine(center1, center2, radius);
```

```
        // Find the minimum and maximum vertical coordinates
        int yMin = (int)Math.Min(center1.Y - radius, center2.Y - radius);
        int yMax = (int)Math.Max(center1.Y + radius, center2.Y + radius);

        yMin = Math.Max(0, Math.Min(bitmap.PixelHeight, yMin));
        yMax = Math.Max(0, Math.Min(bitmap.PixelHeight, yMax));

        // Loop through all the y coordinates that contain part of the segment
        for  (int y = yMin; y < yMax; y++)
        {
            // Get the range of x coordinates in the segment
            xCollection.Clear();
            line.GetAllX(y, xCollection);

            if  (xCollection.Count == 2)
            {
                // Find the minimum and maximum horizontal coordinates
                int xMin = (int)(Math.Min(xCollection[0], xCollection[1]) + 0.5f);
                int xMax = (int)(Math.Max(xCollection[0], xCollection[1]) + 0.5f);

                xMin = Math.Max(0, Math.Min(bitmap.PixelWidth, xMin));
                xMax = Math.Max(0, Math.Min(bitmap.PixelWidth, xMax));

                // Loop through the X values
                for (int x = xMin; x < xMax; x++)
                {
                    {
                        // Set the pixel
                        int index = 4 * (y * bitmap.PixelWidth + x);
                        pixels[index + 0] = color.B;
                        pixels[index + 1] = color.G;
                        pixels[index + 2] = color.R;
                        pixels[index + 3] = color.A;
                        bitmapNeedsUpdate = true;
                    }
                }
            }
        }
        // Update bitmap
        if  (bitmapNeedsUpdate)
        {
            // Find the starting index and number of pixels
            int start = 4 * yMin * bitmap.PixelWidth;
            int count = 4 * (yMax - yMin) * bitmap.PixelWidth;

            pixelStream.Seek(start, SeekOrigin.Begin);
            pixelStream.Write(pixels, start, count);
            bitmap.Invalidate();
        }
    }

    Point ScaleToBitmap(Point pt)
    {
        return new Point((pt.X - imageOffset.X) / imageScale,
                         (pt.Y - imageOffset.Y) / imageScale);
    }
}
```

该 RenderOnBitmap 方法比 FingerPaint 中的简单，因为只需要处理恒定宽度，而且一致使用 RoundCappedLine。用几笔透明像素“喷涂”白色位图，效果如下图所示。

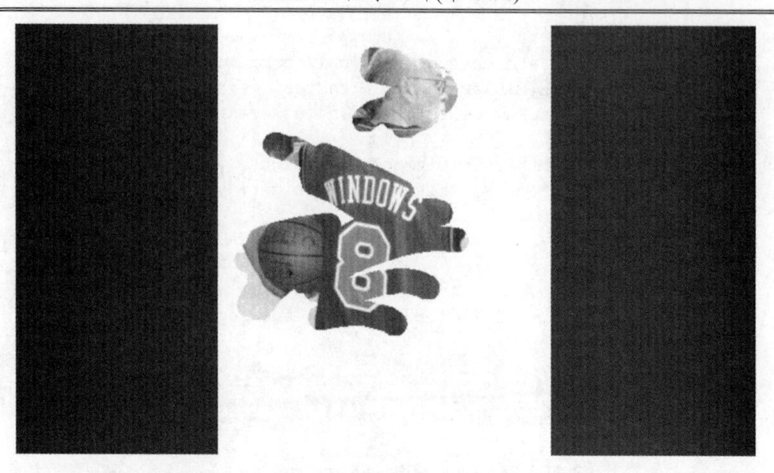

注意，PointerMoved 方法会如下调用 RenderOnBitmap。

```
RenderOnBitmap(previousPoint, point, 12, new Color());
```

Color 构造函数用设置均为零的 A、R、G 和 B 属性来创建 Color 值，该颜色有时也称为“透明黑”。如果要把颜色放入 WriteableBitmap，该 Color 构造函数比静态 Color.Transparent 属性更好。Colors.Transparent 返回 Color 值，其中 A 属性等于零，但 R、G、B 设置为 255。这种颜色有时也称为“透明白”，但不是预乘-α透明色！你需要预乘-α颜色用于 WriteableBitmap，也就是说，R、G 和 B 的属性都不大于 A。

14.9 访问照片库

应用可能直接访问照片库并枚举所有子文件夹及文件夹中的文件。该程序可以最高效地把文件以缩略图形式显示在屏幕上，以便此后再访问实际位图。

PhotoScatter 程序演示了这种情况。程序在显示照片库目录结构的页面左侧构造 ListBox 框。选择一个文件夹，程序以缩略图显示文件夹内容。可以用手指来移动、缩放和旋转图片，此时加载真正的文件，能用更高分辨率来放大图片。

下图为是程序运行效果。我的 Screenshots 文件夹目前存储了 200 多张图片，你可能会认出其中一些。

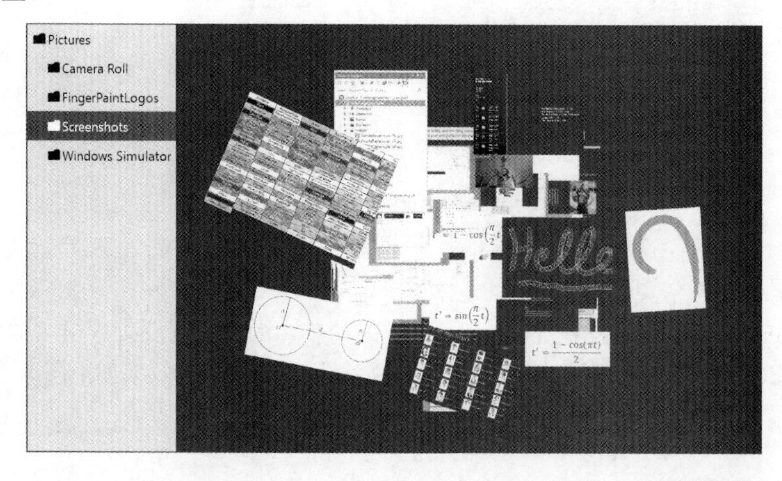

我想让显示在这里的每张图片都可以独立操作，而且能自己处理操作。为此，我创建了一个通用的 ContentControl 派生类，名为 ManipulableContentControl。该控件使用的是我在第 13 章提出的 ManipulationManager 类的奇妙版。

```
项目: PhotoScatter | 文件: ManipulationManager.cs
public class ManipulationManager
{
    TransformGroup xformGroup;
    MatrixTransform matrixXform;
    CompositeTransform compositeXform;

    public ManipulationManager() : this(new CompositeTransform())
    {
    }

    public ManipulationManager(CompositeTransform initialTransform)
    {
        xformGroup = new TransformGroup();
        matrixXform = new MatrixTransform();
        xformGroup.Children.Add(matrixXform);
        compositeXform = initialTransform;
        xfor mGroup.Children.Add(compositeXform);
        this.Matrix = xformGroup.Value;
    }

    public Matrix Matrix { private set; get; }

    public void AccumulateDelta(Point position, ManipulationDelta delta)
    {
        matrixXform.Matrix = xformGroup.Value;
        Point center = matrixXform.TransformPoint(position);
        compositeXform.CenterX = center.X;
        compositeXform.CenterY = center.Y;
        compositeXform.TranslateX = delta.Translation.X;
        compositeXform.TranslateY = delta.Translation.Y;
        compositeXform.ScaleX = delta.Scale;
        compositeXform.ScaleY = delta.Scale;
        compositeXform.Rotation = delta.Rotation;
        this.Matrix = xformGroup.Value;
    }
}
```

唯一的附加功能是一个构造函数，该函数允许用该类内使用的 CompositeTransform 对象来初始化项目方位。

为了创建 ManipulableContentControl 类，我在 Visual Studio 中创建了 UserControl 类型的新项。在 XAML 文件和代码隐藏文件中，我把 UserControl 改变为 ContentControl。正常情况下，在 UserControl 派生类中，XAML 文件定义控件内容。而在该 XAML 文件中没有定义内容，但 RenderTransform 设置为 MatrixTransform，MatrixTransform 根据 ManipulationManager 实例从代码隐藏文件进行设置。

```
项目: PhotoScatter | 文件: ManipulableContentControl.xaml
<ContentControl
    x:Class="PhotoScatter.ManipulableContentControl"
    xmlns="http://schemas.microsoft.com/winfx/2006/xaml/presentation"
    xmlns:x="http://schemas.microsoft.com/winfx/2006/xaml"
    xmlns:local="using:PhotoScatter">

    <ContentControl.RenderTransform>
        <MatrixTransform x:Name="matrixXform" />
    </ContentControl.RenderTransform>
</ContentControl>
```

以下为代码隐藏文件。注意构造函数，用 CompositeTransform 来创建 ManipulationManager 对象。

```
项目: PhotoScatter | 文件: ManipulableContentControl.xaml.cs(片段)
public sealed partial class ManipulableContentControl : ContentControl
{
    static int zIndex;
    ManipulationManager manipulationManager;

    public ManipulableContentControl(CompositeTransform initialTransform)
    {
        this.InitializeComponent();

        // Create the ManipulationManager and set MatrixTransform from it
        manipulationManager = new ManipulationManager(initialTransform);
        matrixXform.Matrix = manipulationManager.Matrix;

        this.ManipulationMode = ManipulationModes.All &
                                ~ManipulationModes.TranslateRailsX &
                                ~ManipulationModes.TranslateRailsY;
    }

    protected override void OnManipulationStarting(ManipulationStartingRoutedEventArgs args)
    {
        Canvas.SetZIndex(this, zIndex += 1);
        base.OnManipulationStarting(args);
    }

    protected override void OnManipulationDelta(ManipulationDeltaRoutedEventArgs args)
    {
        manipulationManager.AccumulateDelta(args.Position, args.Delta);
        matrixXform.Matrix = manipulationManager.Matrix;
        base.OnManipulationDelta(args);
    }
}
```

还要注意，该类维护静态的递增 zIndex 属性，用于操作一开始就把被触碰元素置于顶部。

在正常情况下，使用 TreeView 或类似名称的控件来显示目录结构，该控件提供可视化缩进和界面以展开和关闭树节点。Windows Runtime(还)没有 TreeView，因此我决定使用朴素的老式 ListBox。不展开和关闭节点，但有缩进。

ListBox 中的项目为 FolderItem 类型。

```
项目: PhotoScatter | 文件: FolderItem.cs
public class FolderItem
{
    public StorageFolder StorageFolder { set; get; }

    public int Level { set; get; }

    public string Indent
    {
        get { return new string('\x00A0', this.Level * 4); }
    }

    public Grid DisplayGrid { set; get; }
}
```

每个 FolderItem 对象代表一个文件夹。从 StorageFolder 对象获得文件夹名称，嵌套层级由代码设置为 Level 属性，Indent 属性使用该值来为每个层级构建四个空格的字符串。

FolderItem 还定义 Grid 类型的 DisplayGrid 属性。该 Grid 对象在用户第一次选择特定文件夹时进行设置并用一组对应于该文件夹中图片的 ManipulableContentControl 对象进行填充。如果用户反复浏览文件夹，保留该 Grid 以及其里面的所有内容，会避免程序重新枚举每个文件夹内容。(然而，程序不安装文件查看器，因此，如果有项目稍后添加到文件夹，程序就会不知道。)

ListBox 的 ItemTemplate 在 MainPage.xaml 中进行定义，并且引用 FolderItem 中的属性。

项目: PhotoScatter | 文件: MainPage.xaml(片段)

```
<Grid Background="{StaticResource ApplicationPageBackgroundThemeBrush}">
     <Grid.ColumnDefinitions>
          <ColumnDefinition Width="Auto" />
          <ColumnDefinition Width="*" />
     </Grid.ColumnDefinitions>

     <ListBox Name="folderListBox"
              Grid.Column="0"
              SelectionChanged="OnFolderListBoxSelectionChanged">
          <ListBox.ItemTemplate>
               <DataTemplate>
                    <ContentControl FontSize="24">
                         <StackPanel Orientation="Horizontal">
                              <TextBlock Text="{Binding Indent}" />
                              <TextBlock Text="&#xE188;"
                                         FontFamily="Segoe UI Symbol" />
                              <TextBlock Text="{Binding StorageFolder.Name}" />
                         </StackPanel>
                    </ContentControl>
               </DataTemplate>
          </ListBox.ItemTemplate>
     </ListBox>

     <Border Name="displayBorder"
          Grid.Column="1" />
</Grid>
```

请注意用于显示小文件夹图标的 Segoe UI Symbol 字型中的 0xE188 代码点。以 Indent 字符串开头，后跟 FolderItem 中 StorageFolder 对象的 Name 属性。

PhotoScatter 程序在 Package.appxmanifest 的 Capabilities 部分要求有访问照片库的权限，因为在 Loaded 事件期间，程序会递归调用 StorageFolder 的 GetFoldersAsync 方法，并同时为 ListBox 创建 FolderItem 对象，以获得完整目录树。

项目: PhotoScatter | 文件: MainPage.xaml.cs(片段)

```
public sealed partial class MainPage : Page
{
     ...

     public MainPage()
     {
          this.InitializeComponent();
          Loaded += OnMainPageLoaded;
     }

     void OnMainPageLoaded(object sender, RoutedEventArgs args)
     {
          StorageFolder storageFolder = KnownFolders.PicturesLibrary;
          BuildFolderListBox(storageFolder, 0);
          folderListBox.SelectedIndex = 0;
     }

     async void BuildFolderListBox(StorageFolder parentStorageFolder, int level)
     {
          FolderItem folderItem = new FolderItem
```

```
            {
                StorageFolder = parentStorageFolder,
                Level = level
            };
            folderListBox.Items.Add(folderItem);

            IReadOnlyList<StorageFolder> storageFolders =
                            await parentStorageFolder.GetFoldersAsync();
            foreach (StorageFolder storageFolder in storageFolders)
                BuildFolderListBox(storageFolder, level + 1);
        }
        ...
    }
```

Loaded 处理程序最后把 ListBox 的 SelectedIndex 设置为 0，并选择第一项，即图片文件夹本身。这样会触发调用 SelectionChanged 处理程序，SelectionChanged 使用 StorageFolder 的 GetFilesAsync 方法来获取文件夹中的所有文件。但对于每个 StorageFile，GetFilesAsync 方法调用 GetThumbnailAsync 获取该文件的缩略图图片。(加载缩略图比加载实际图片更好，后者可能需要相当长的时间，并消耗大量内存。)调用在 MainPage 中名为 LoadBitmapAsync 的方法(我会马上介绍)创建 Image 元素和 ManipulableContentControl，用于显示缩略图。

项目: PhotoScatter | 文件: MainPage.xaml.cs(片段)

```
public sealed partial class MainPage : Page
{
    Random rand = new Random();
    ...
    async void OnFolderListBoxSelectionChanged(object sender, SelectionChangedEventArgs args)
    {
        FolderItem folderItem = (sender as ListBox).SelectedItem as FolderItem;

        if (folderItem == null)
        {
            displayBorder.Child = null;
            return;
        }

        if (folderItem.DisplayGrid != null)
        {
            displayBorder.Child = folderItem.DisplayGrid;
            return;
        }

        Grid displayGrid = new Grid();
        folderItem.DisplayGrid = displayGrid;
        displayBorder.Child = displayGrid;

        StorageFolder storageFolder = folderItem.StorageFolder;
        IReadOnlyList<StorageFile> storageFiles = await storageFolder.GetFilesAsync();

        foreach (StorageFile storageFile in storageFiles)
        {
            StorageItemThumbnail thumbnail =
                    await storageFile.GetThumbnailAsync(ThumbnailMode.SingleItem);
            BitmapSource bitmap = await LoadBitmapAsync(thumbnail);

            if (bitmap == null)
                continue;

            // Create new Image element to display the thumbnail
            Image image = new Image
                {
                    Source = bitmap,
                    Stretch = Stretch.None,
                    Tag = ImageType.Thumbnail
```

```
            };

            // Create an initial CompositeTransform for the item
            CompositeTransform xform = new CompositeTransform();
            xform.TranslateX = (displayBorder.ActualWidth - bitmap.PixelWidth) / 2;
            xform.TranslateY = (displayBorder.ActualHeight - bitmap.PixelHeight) / 2;
            xform.TranslateX += 256 * (0.5 - rand.NextDouble());
            xform.TranslateY += 256 * (0.5 - rand.NextDouble());

            // Create the ManipulableContentControl for the Image
            ManipulableContentControl manipulableControl = new ManipulableContentControl(xform)
            {
                Content = image,
                Tag = storageFile
            };
            manipulableControl.ManipulationStarted += OnManipulableControlManipulationStarted;

            // Put it in the Grid
            displayGrid.Children.Add(manipulableControl);
        }
    }
    ...
}
```

由于 GetThumbnailAsync 和 LoadBitmapAsync 上的 await 操作符，BitmapSource 对象、Image 元素和 ManipulableContentControl 实例会依次创建，并且创建一个显示一个，从而提供了一场娱乐表演，就像图片逐渐堆积成一个有点儿随机的大堆。另一种选择是同时创建这些实例，但在大多数情况下，这样产生的线程多得会超出处理器的处理能力。

ListBox 的 SelectionChanged 处理程序对每个文件夹只执行一次。ManipulableContentControl 的 Tag 属性设置为与每项有关的 StorageFile 对象。后面用来加载实际位图(如有必要)。还要注意，每个 Image 元素的 Tag 属性设置为 ImageType.Thumbnail。ImageType.Thumbnail 为下列枚举项。

项目：PhotoScatter | 文件：ImageType.cs(片段)

```
public enum ImageType
{
    Thumbnail,
    Full,
    Transitioning
}
```

用户一开始操作某一特定项，Tag 属性就会发生变化。虽然 ManipulableContentControl 会处理 Manipulation 事件得允许移动、缩放和旋转项目，但仍然要通过 SelectionChanged 处理程序附加 ManipulationStarted 事件的处理程序。该处理程序负责用实际位图替换缩略图。

项目：PhotoScatter | 文件：MainPage.xaml.cs(片段)

```
public sealed partial class MainPage : Page
{
    ...
    async void OnManipulableControlManipulationStarted(object sender,
                                        ManipulationStartedRoutedEventArgs args)
    {
        ManipulableContentControl manipulableControl = sender as ManipulableContentControl;
        Image image = manipulableControl.Content as Image;

        If ((ImageType)image.Tag == ImageType.Thumbnail)
        {
            // Set tag to transitioning
            image.Tag = ImageType.Transitioning;
```

```
                // Load the actual bitmap file
                StorageFile storageFile = manipulableControl.Tag as StorageFile;
                BitmapSource newBitmap = await LoadBitmapAsync(storageFile);

                // This is the case for a file that BitmapDecoder can't handle
                if (newBitmap != null)
                {
                    // Get the thumbnail from the Image element
                    BitmapSource oldBitmap = image.Source as BitmapSource;

                    // Find a ScaleTransform between old and new
                    double scale = 1;

                    if (oldBitmap.PixelWidth > oldBitmap.PixelHeight)
                        scale = (double)oldBitmap.PixelWidth / newBitmap.PixelWidth;
                    else
                        scale = (double)oldBitmap.PixelHeight / newBitmap.PixelHeight;

                    // Set properties on the Image element
                    image.Source = newBitmap;
                    image.RenderTransform = new ScaleTransform
                    {
                        ScaleX = scale,
                        ScaleY = scale,
                    };
                }
                image.Tag = ImageType.Full;
            }
        }
        ...
    }
```

用位图替换缩略图可能是程序中最棘手的部分。ManipulationStarted 处理程序包含异步调用，因此，如果用户同时操作多项，则可能要处理多项的重叠事件。只有当 Image 的 Tag 属性为 ImageType.Thumbnail 时，主逻辑才会发生。Tag 会设置为 ImageType.Transitioning(不是绝对必要，但有助于调试)，并调用 LoadBitmapAsync 获得该图片。如果终于替换了缩略图，Image 元素的 Tag 属性会设置为 ImageType.Full。

我尽可能想让过程看起来很流畅，因此，常规任务计算实际位图转换成缩略图的缩放因子。每项的尺寸和方位不改变，但分辨率有所提高。

最后是 LoadBitmapAsync 的三个重载，每个均返回 BitmapSource。有些不同的方法用来获得图片及其缩略图的 IRandomAccessStream 对象，然后常规任务用你看过的代码加载文件。

```
项目: PhotoScatter | 文件: MainPage.xaml.cs(片段)
public sealed partial class MainPage : Page
{
    ...
    async Task<BitmapSource> LoadBitmapAsync(StorageFile storageFile)
    {
        BitmapSource bitmapSource = null;

        // Open the StorageFile for reading
        using (IRandomAccessStreamWithContentType stream = await storageFile.OpenReadAsync())
        {
            bitmapSource = await LoadBitmapAsync(stream);
        }

        return bitmapSource;
    }
```

```
    async Task<BitmapSource> LoadBitmapAsync(StorageItemThumbnail thumbnail)
    {
        return await LoadBitmapAsync(thumbnail as IRandomAccessStream);
    }

    async Task<BitmapSource> LoadBitmapAsync(IRandomAccessStream stream)
    {
        WriteableBitmap bitmap = null;

        // Create a BitmapDecoder from the stream
        BitmapDecoder decoder = null;

        try
        {
            decoder = await BitmapDecoder.CreateAsync(stream);
        }
        catch
        {
            // Just skip ones that aren't valid
            return null;
        }

        // Get the first frame
        BitmapFrame bitmapFrame = await decoder.GetFrameAsync(0);

        // Get the pixels
        PixelDataProvider dataProvider =
                await bitmapFrame.GetPixelDataAsync(BitmapPixelFormat.Bgra8,
                                        BitmapAlphaMode.Premultiplied,
                                        new BitmapTransform(),
                                        ExifOrientationMode.RespectExifOrientation,
                                        ColorManagementMode.ColorManageToSRgb);

        byte[] pixels = dataProvider.DetachPixelData();

        // Create WriteableBitmap and set the pixels
        bitmap = new WriteableBitmap((int)bitmapFrame.PixelWidth,
                                     (int)bitmapFrame.PixelHeight);

        using (Stream pixelStream = bitmap.PixelBuffer.AsStream())
        {
            pixelStream.Write(pixels, 0, pixels.Length);
        }

        bitmap.Invalidate();
        return bitmap;
    }
}
```

14.10　捕捉相机照片

你已经看过 Windows Runtime 应用如何从头创建 WriteableBitmap 对象或从文件加载现有位图。程序获取位图还有一些其他方式。比如，在第 17 章中，你会看到程序如何直接或通过剪贴板从其他应用获取位图。

应用也可以从电脑的内置摄像头获得位图。有两种方法，而且如果你愿意服从 Windows 8 的控制，Windows 8 就可以显示其正常拍照界面，过程非常简单。

为了使应用能使用电脑摄像头，必须在 Package.appxmanifest 文件中声明。在 Visual Studio 中打开该文件，选择 Capabilities 标签，再单击 Webcam。

我给 EasyCameraCapture 程序做好了这些事情。以下为 MainPage.xaml 文件。

项目: EasyCameraCapture | 文件: MainPage.xaml(片段)

```
<Grid Background="{StaticResource ApplicationPageBackgroundThemeBrush}">
    <Image Name="image" />

    <Button Content="Capture Photo!"
            FontSize="48"
            HorizontalAlignment="Left"
            VerticalAlignment="Top"
            Click="OnButtonClick" />
</Grid>
```

Button 的 Click 处理程序实例化 Windows.Media.Capture 命名空间所定义的 CameraCaptureUI 类，并调用 CaptureFileAsync。

项目: EasyCameraCapture | 文件: MainPage.xaml.cs(片段)

```
async void OnButtonClick(object sender, RoutedEventArgs args)
{
    CameraCaptureUI cameraCap = new CameraCaptureUI();
    StorageFile storageFile = await cameraCap.CaptureFileAsync(CameraCaptureUIMode.Photo);

    if (storageFile != null)
    {
        IRandomAccessStreamWithContentType stream = await storageFile.OpenReadAsync();
        BitmapImage bitmap = new BitmapImage();
        await bitmap.SetSourceAsync(stream);
        image.Source = bitmap;
    }
}
```

在调用 CaptureFileAsync 之前，程序可以设置 CameraCaptureUI 的各种属性来选择文件格式、选择像素尺寸、启用裁剪等。

应用调用 CaptureFileAsync 时，Windows 8 会切换到非常像普通 Windows 8 相机应用的界面。唯一重要的区别在于视频模式按钮被禁用(但可以通过把 CameraCaptureUIMode.PhotoOrVideo 传递到 CaptureFileAsync 方法而启用)，而且左上角有一个圆形的左箭头。

如果要返回 EasyCameraCapture 应用，可以按下该圆形左箭头，此时所返回的 StorageFile 为 null。或者轻击或点击屏幕，并按下底部的圆形复选标记，这样就可以拍摄照片啦。

返回到程序，StorageFile 对象就会引用存储在应用本地存储的 TempState 目录中的文件。EasyCameraCapture 代码会直接显示文件(见下图)。

应用可能要启用 FileSavePicker，让用户保存图片或者自动保存到照片库中。或许应用会用拍摄到的图片做一些特定的事情，如果照片库中有特殊目录会很方便。(标准 Windows 8 相机应用会把照片存储到 Pictures 的 Camera Roll 目录。)为此，需要设置应用允许访问照片库，就像常规 Windows 8 相机应用一样。

可以降低相机接口等级，编写自己的完整摄像头应用，包括视频预览、摄像头选择(如果有多个摄像头)、启动照片拍摄等。

HarderCameraCapture 项目展示了基础功能。XAML 文件包含你以前没看过的东西(用于预览视频的 CaptureElement)以及一个老朋友。

```
项目: HarderCameraCapture | 文件: MainPage.xaml(片段)
<Grid Background="{StaticResource ApplicationPageBackgroundThemeBrush}">
    <CaptureElement Name="captureElement" />
    <Image Name="image" />
</Grid>
```

代码隐藏文件使用 Loaded 处理程序来执行初始化。静态 DeviceInformation.FindAllAsync 方法允许获取视频采集设备集合。DeviceInformation 对象包含一个字符串 ID 和 EnclosureLocation 属性，EnclosureLocation 属性让程序能确定计算机上每个摄像机的位置。代码试图找到正面摄像头，但如果找不到，就使用集合中的第一个(或可能是唯一一个)相机。

```
项目: HarderCameraCapture | 文件: MainPage.xaml.cs(片段)
public sealed partial class MainPage : Page
{
    MediaCapture mediaCapture = new MediaCapture();
    ...

    public MainPage()
    {
        this.InitializeComponent();
        Loaded += OnMainPageLoaded;
    }

    async void OnMainPageLoaded(object sender, RoutedEventArgs args)
    {
        DeviceInformationCollection devInfos =
            await DeviceInformation.FindAllAsync(DeviceClass.VideoCapture);

        if (devInfos.Count == 0)
        {
            await new MessageDialog("No video capture devices found").ShowAsync();
            return;
        }

        string id = null;

        // Try to find the front webcam
        foreach (DeviceInformation devInfo in devInfos)
        {
            if (devInfo.EnclosureLocation != null &&
                devInfo.EnclosureLocation.Panel == Windows.Devices.Enumeration.Panel.Front)
            id = devInfo.Id;
        }

        // If not available, just pick the first one
        if (id == null)
            id = devInfos[0].Id;

        // Create initialization settings
```

```
            MediaCaptureInitializationSettings settings = new MediaCaptureInitializationSettings();
            settings.VideoDeviceId = id;
            settings.StreamingCaptureMode = StreamingCaptureMode.Video;

            // Initialize the MediaCapture device
            await mediaCapture.InitializeAsync(settings);

            // Associate with the CaptureElement
            captureElement.Source = mediaCapture;

            // Start the preview
            await mediaCapture.StartPreviewAsync();
        }
        ...
}
```

一旦获取到设备 ID，Loaded 处理程序就会继续创建 MediaCaptureInitializationSettings 对象并用来初始化定义为字段的 MediaCapture 对象。MediaCapture 对象在 XAML 文件成为 CaptureElement 实例化来源。

Loaded 处理程序结束后，开始预览工作。如果相机在计算机正面，则应该出现你自己的实况视频。

我还为拍照实现了 Tapped 处理程序。MediaCapture 类有 CapturePhotoToStorageFileAsync 和 CapturePhotoToStreamAsync 两种方法。我选择了使用流方式，并拍摄照片到内存流。此时，BitmapDecoder 可以获取像素位。程序借用 FingerPaint 程序的 HSL 结构以增加所有像素的饱和度，并创建 WriteableBitmap。

项目：HarderCameraCapture | 文件：MainPage.xaml.cs(片段)

```
public sealed partial class MainPage : Page
{
    ...
    bool ignoreTaps = false;
    ...
    async protected override void OnTapped(TappedRoutedEventArgs args)
    {
        if (ignoreTaps)
            return;

        // Capture photo to memory stream
        ImageEncodingProperties imageEncodingProps = ImageEncodingProperties.CreateJpeg();
        InMemoryRandomAccessStream memoryStream = new InMemoryRandomAccessStream();
        await mediaCapture.CapturePhotoToStreamAsync(imageEncodingProps, memoryStream);

        // Use BitmapDecoder to get pixels array
        BitmapDecoder decoder = await BitmapDecoder.CreateAsync(memoryStream);
        PixelDataProvider pixelProvider = await decoder.GetPixelDataAsync();
        byte[] pixels = pixelProvider.DetachPixelData();

        // Saturate the colors
        for (int index = 0; index < pixels.Length; index += 4)
        {
            Color color = Color.FromArgb(pixels[index + 3],
                                         pixels[index + 2],
                                         pixels[index + 1],
                                         pixels[index + 0]);
            HSL hsl = new HSL(color);
            hsl = new HSL(hsl.Hue, 1.0, hsl.Lightness);
            color = hsl.Color;

            pixels[index + 0] = color.B;
            pixels[index + 1] = color.G;
            pixels[index + 2] = color.R;
            pixels[index + 3] = color.A;
```

```
        }

        // Create a WriteableBitmap and initialize it
        WriteableBitmap bitmap = new WriteableBitmap((int)decoder.PixelWidth,
                                                     (int)decoder.PixelHeight);
        Stream pixelStream = bitmap.PixelBuffer.AsStream();
        await pixelStream.WriteAsync(pixels, 0, pixels.Length);
        bitmap.Invalidate();

        // Display the bitmap
        image.Source = bitmap;

        // Set a timer for the image
        DispatcherTimer timer = new DispatcherTimer
        {
            Interval = TimeSpan.FromSeconds(2.5)
        };
        timer.Tick += OnTimerTick;
        timer.Start();
        ignoreTaps = true;
        base.OnTapped(args);
    }

    void OnTimerTick(object sender, object args)
    {
        // Disable the timer
        DispatcherTimer timer = sender as DispatcherTimer;
        timer.Stop();
        timer.Tick -= OnTimerTick;

        // Get rid of the bitmap
        image.Source = null;
        ignoreTaps = false;
    }
}
```

我不想一直留着照片，因此，程序把 DispatcherTimer 设置为 20.5 秒。在此期间的触屏都会被忽略，但过了这段时间，就直接从屏幕上移除照片，我们又回到实况视频。

当然，高饱和度颜色可能有点吓人(见下图)。

第15章　原　　生

在 Windows 8 编程中，并非所有语言都生而平等，这是一个悲惨的事实。从理论上讲，任何编程语言都可以访问可用于 Windows Store 应用的任何类或功能，只不过整个 API 建立在组件对象模型(COM)之上。在理智编程的现实世界中，要想轻松访问 Windows 8 API 的特定区域则取决于你正在使用的编程语言。

例如，只有使用 C#和 Visual Basic 托管语言的程序员才可以直接访问 Windows 8 应用的.NET API，即以单词 System 为开头的命名空间。而对于同等功能，C++程序员则要用 Platform 命名空间中的 C++运行库和类。

另一方面，Windows 8 应用可以访问 Win32 和 COM API 的子集，但这些函数和类只方便提供给 C++程序员。要获得相同 API，C#程序员则需要费尽力气。

本章讲解如何完成这些繁琐工作。我会讨论两种基本方法。第一种方法称为“平台调用”(也称为 PInvoke 或 P/Invoke)，平台调用从.NET 编程开始就存在，用来访问 Win32 函数或其他动态链接库(DLL)函数。平台调用特别适合访问“扁平”API，也就是说，其中一项功能是独立的(或者由其他函数提供的引用句柄)，而不是合并成类。

第二种方法涉及用 C++写的“封装”DLL，并从 C#程序中访问该 DLL。这种方法更适合面向对象的 API，特别适合统称为 DirectX 的高性能图形和音频类。

在 Windows 8 应用中，用一种语言写而供另一种语言访问的 DLL 必须为特殊格式，称为 Windows Runtime Component。Visual Studio 可以创建 Windows Runtime Component，但这些库的功能有一堆规则和限制。

请记住，不能使用这些方法来赋予程序访问 Windows Store 应用不允许的函数。不能使用这些方法来访问任意 Win32 函数。仅限于 Windows 8 应用允许的子集的函数。也不能调用 DLL 中的函数，DLL 会调用不在该子集中的 Win32 函数。

15.1　P/Invoke 简介

假设你正在浏览可用于新的 Windows 8 应用的 Win32 函数子集，看到了一个你想要的。在文档中其形式如下：

```
void WINAPI GetNativeSystemInfo(__out LPSYSTEM_INFO lpSystemInfo);
```

如果你完全不熟悉 Win32 API，肯定觉得这些东西无用。大写字母标识符一般是在各种 Windows 头文件中通过 C#define 或 typedef 语句进行定义。在安装了 Visual Studio 的计算机上，可以在 C:/Program Files(x86)/Windows Kits/8.0 目录的子目录中找到这些头文件。最基本的是 Windows.h、WinDef.h、WinBase.h 和 winnt.h。WINAPI 标识符和_stdcall 是相同的，_stdcall 是 C 程序调用 Win32 函数的标准调用约定。LPSYSTEM_INFO 在 SYSTEM_INFO 结构中是 SYSTEM_INFO 的长指针，这里的“长”指的是比 Windows 刚刚出现时的 16 位指针要宽。SYSTEM_INFO 结构定义如下：

```
typedef struct _SYSTEM_INFO {
    union {
        DWORD  dwOemId;
        Struct  {
            WORD wProcessorArchitecture;
            WORD wReserved;
        };
    };
    DWORD       dwPageSize;
    LPVOID      lpMinimumApplicationAddress;
    LPVOID      lpMaximumApplicationAddress;
    DWORD_PTR dwActiveProcessorMask;
    DWORD       dwNumberOfProcessors;
    DWORD       dwProcessorType;
    DWORD       dwAllocationGranularity;
    WORD        wProcessorLevel;
    WORD        wProcessorRevision;
} SYSTEM_INFO;
```

带数据类型为小写字母的字段引语是一种简单的匈牙利命名法，匈牙利出生的查尔斯・西蒙尼(Charles Simonyi)发明了这种命名，因此而得名。匈牙利命名法由 Windows API 和一些关于 Windows 编程的古老书籍而得到普及，但对于应用编程不再广泛使用。

在 Windows 语法中，一个 WORD 为一个 16 位无符号值，C#程序员将其作为 ushort。DWORD 是一个双 WORD 或一个 32 位无符号值或 uint。注意 long 型引用，不是 64 位 C#long，而是 C++ long，和 int 相同或 32 位。

LPVOID 翻译为“没有类型的长指针”，在标准 C 中为 vid*，DWORD_PTR 要么为无符号的 32 位，要么为 64 位整数，取决于是在 32 位还是 64 位处理器上运行 Windows。这些等同于 C# IntPtr。

需要知道这些 Windows API 数据类型如何对应于 C#数据类型，因为要从 C#程序使用这种结构，就需要在 C#中进行重新定义。幸运的是，SYSTEM_INFO 文档表明 dwOemId 字段过时了，也就是说可以忽略 union，可利用公共字段直接创建 C#结构，也许在此过程中可以给结构指定更具有 C#风格的名称：

```
struct SystemInfo
{
    public ushort wProcessorArchitecture;
    public ushort wReserved;
    public uint dwPageSize;
    public IntPtr lpMinimumApplicationAddress;
    public IntPtr lpMaximumApplicationAddress;
    public IntPtr dwActiveProcessorMask;
    public uint dwNumberOfProcessors;
    public uint dwProcessorType;
    public uint dwAllocationGranularity;
    public ushort wProcessorLevel;
    public ushort wProcessorRevision;
}
```

在 C#中，如果想从结构外部访问字段，就必须把字段定义为 public。当然，也可以重命名所有字段(例如，ProcessorArchitecture 和 PageSize)。

还可以指定大小相同的不同数据类型(例如，short 而非 ushort，int 而非 uint)，如果已经知道实际值不会溢出符号类型的话。对于 Windows API，你在做的事情就是提供内存块。完整结构在 32 位 Windows 中占 36 字节内存，而在 64 位 Windows 中占 48 字节内存。

很多时候，在 P/Invoke 代码中，你会看到前面再加如下属性的结构：

```
[StructLayout(LayoutKind.Sequential)]
struct SystemInfo
{
    ...
}
```

StructLayoutAttribute 类和 LayoutKind 枚举都在 System.Runtime.InteropServices 命名空间中进行定义，该命名空间有很多与 P/Invoke 相关的其他类。属性明确表明这些字段应视为连续并按字节边界对齐。

现在要把结构传递给 GetNativeSystemInfo 函数，必须声明函数本身。为此，可以用也在 System.Runtime.InteropServices 中进行定义的 DllImportAttribute。至少必须注明动态链接库，而在链接库中可以找到该函数。文档表明 GetNativeSystemInfo 在 kernel32.DLL 中进行定义。以下为函数声明：

```
[DllImport("kernel32.dll")]
static extern void GetNativeSystemInfo(out SystemInfo systemInfo);
```

该声明必须出现 C#类定义中，并且和其他方法同一级别。把函数声明为 static(常见于正则 C#类)，也可以为 extern(并不常见但表示实际实施函数是在类之外)。如果想让函数在类之外可见，则可以赋予其 public 关键字。

除了 extern，函数声明似乎是 C#方法。该方法返回 void，并且一个参数应用 SystemInfo 对象。许多 Windows API 调用需要或返回使用指针参数结构的信息，并且可以通过 out 或 ref 来定义这些参数。它们在功能上相同，但在调用函数之前 C#编译器会用 ref 检查初始化的值类型。

在该类的其他一些方法中，可以定义 SystemInfo 类型的值，并将其作为正常静态方法来调用该函数：

```
SystemInfo systemInfo;
GetNativeSystemInfo(out systemInfo);
```

我们来看看一个完整的程序。用于 SystemInfoPInvoke 的 XAML 文件在表中使用 Grid 来格式化从 GetNativeSystemInfo 得到的信息。

项目：SystemInfoPInvoke | 文件：MainPage.xaml(片段)

```
<Page ...
    FontSize="24">

    <Page.Resources>
        <Style x:Key="rightJustifiedText" TargetType="TextBlock">
            <Setter Property="TextAlignment" Value="Right" />
            <Setter Property="Margin" Value="12 0 0 0" />
        </Style>
    </Page.Resources>

    <Grid Background="{StaticResource ApplicationPageBackgroundThemeBrush}">
        <Grid HorizontalAlignment="Center"
              VerticalAlignment="Center">
            <Grid.RowDefinitions>
                <RowDefinition Height="Auto" />
                <RowDefinition Height="Auto" />
                <RowDefinition Height="Auto" />
                <RowDefinition Height="Auto" />
                <RowDefinition Height="Auto" />
                <RowDefinition Height="Auto" />
                <RowDefinition Height="Auto" />
                <RowDefinition Height="Auto" />
                <RowDefinition Height="Auto" />
```

```
            </Grid.RowDefinitions>

            <Grid.ColumnDefinitions>
                <ColumnDefinition Width="Auto" />
                <ColumnDefinition Width="Auto" />
            </Grid.ColumnDefinitions>

            <TextBlock Text="Processor Architecture: " Grid.Row="0" Grid.Column="0" />
            <TextBlock Name="processorArchitecture" Grid.Row="0" Grid.Column="1"
                       Style="{StaticResource rightJustifiedText}" />

            <TextBlock Text="Page Size: " Grid.Row="1" Grid.Column="0" />
            <TextBlock Name="pageSize" Grid.Row="1" Grid.Column="1"
                       Style="{StaticResource rightJustifiedText}" />

            <TextBlock Text="Minimum Application Addresss: " Grid.Row="2" Grid.Column="0" />
            <TextBlock Name="minAppAddr" Grid.Row="2" Grid.Column="1"
                       Style="{StaticResource rightJustifiedText}" />

            <TextBlock Text="Maximum Application Addresss: " Grid.Row="3" Grid.Column="0" />
            <TextBlock Name="maxAppAddr" Grid.Row="3" Grid.Column="1"
                       Style="{StaticResource rightJustifiedText}" />

            <TextBlock Text="Active Processor Mask: " Grid.Row="4" Grid.Column="0" />
            <TextBlock Name="activeProcessorMask" Grid.Row="4" Grid.Column="1"
                       Style="{StaticResource rightJustifiedText}" />

            <TextBlock Text="Number of Processors: " Grid.Row="5" Grid.Column="0" />
            <TextBlock Name="numberProcessors" Grid.Row="5" Grid.Column="1"
                       Style="{StaticResource rightJustifiedText}" />

            <TextBlock Text="Allocation Granularity: " Grid.Row="6" Grid.Column="0" />
            <TextBlock Name="allocationGranularity" Grid.Row="6" Grid.Column="1"
                       Style="{StaticResource rightJustifiedText}" />

            <TextBlock Text="Processor Level: " Grid.Row="7" Grid.Column="0" />
            <TextBlock Name="processorLevel" Grid.Row="7" Grid.Column="1"
                       Style="{StaticResource rightJustifiedText}" />

            <TextBlock Text="Processor Revision: " Grid.Row="8" Grid.Column="0" />
            <TextBlock Name="processorRevision" Grid.Row="8" Grid.Column="1"
                       Style="{StaticResource rightJustifiedText}" />
        </Grid>
    </Grid>
</Page>
```

在代码隐藏文件中，结构和外部函数声明都在MainPage类中进行定义。外部函数必须在类定义中进行声明，但结构却不需要，就像其他普通C#结构一样可以在一个完全不同的文件中进行声明。以下是完整的代码隐藏文件。

```
项目: SystemInfoPInvoke | 文件: MainPage.xaml.cs
using System;
using System.Runtime.InteropServices;
using Windows.UI.Xaml.Controls;
namespace SystemInfoPInvoke
{
    public sealed partial class MainPage : Page
    {
        [StructLayout(LayoutKind.Sequential)]
        struct SystemInfo
        {
            public ushort wProcessorArchitecture;
            public byte wReserved;
            public uint dwPageSize;
            public IntPtr lpMinimumApplicationAddress;
            public IntPtr lpMaximumApplicationAddress;
            public IntPtr dwActiveProcessorMask;
            public uint dwNumberOfProcessors;
```

```
            public uint dwProcessorType;
            public uint dwAllocationGranularity;
            public ushort wProcessorLevel;
            public ushort wProcessorRevision;
        }
        [DllImport("kernel32.dll")]
        static extern void GetNativeSystemInfo(out SystemInfo systemInfo);
        enum ProcessorType
        {
            x86 = 0,
            ARM = 5,
            ia64 = 6,
            x64 = 9,
            Unknown = 65535
        };
        public MainPage()
        {
            this.InitializeComponent();
            SystemInfo systemInfo = new SystemInfo();
            GetNativeSystemInfo(out systemInfo);
            processorArchitecture.Text =
                ((ProcessorType)systemInfo.wProcessorArchitecture).ToString();
            pageSize.Text = systemInfo.dwPageSize.ToString();
            minAppAddr.Text = ((ulong)systemInfo.lpMinimumApplicationAddress).ToString("X");
            maxAppAddr.Text = ((ulong)systemInfo.lpMaximumApplicationAddress).ToString("X");
            activeProcessorMask.Text = ((ulong)systemInfo.dwActiveProcessorMask).ToString("X");
            numberProcessors.Text = systemInfo.dwNumberOfProcessors.ToString("X");
            allocationGranularity.Text = systemInfo.dwAllocationGranularity.ToString();
            processorLevel.Text = systemInfo.wProcessorLevel.ToString();
            processorRevision.Text = systemInfo.wProcessorRevision.ToString("X");
        }
    }
}
```

文档表明 wProcessorArchitecture 字段可以采取 0 值(用于 x86 架构)、6 值(用于 Intel 安腾)、9 值(用于 x64 架构)以及适用于“未知”的 0xFFFF。对于 ARM 处理器(如微软 Surface 的首个发布)的值在文档中并没有表明，但所有可能的值均为 winnt.h 中定义的以 PROCESSOR_ARCHITECTURE 开头的常量，PROCESSOR_ARCHITECTURE_ARM 定义为 5。

为了简化格式化 wProcessorArchitecture 值，我定义了一个小 enum，称为 ProcessorType 并将 wProcessorArchitecture 值转换为该枚举。对于 IntPtr 字段，我转换为 ulong 并用十六进制显示。在我写本书所用的平板电脑上运行此程序。

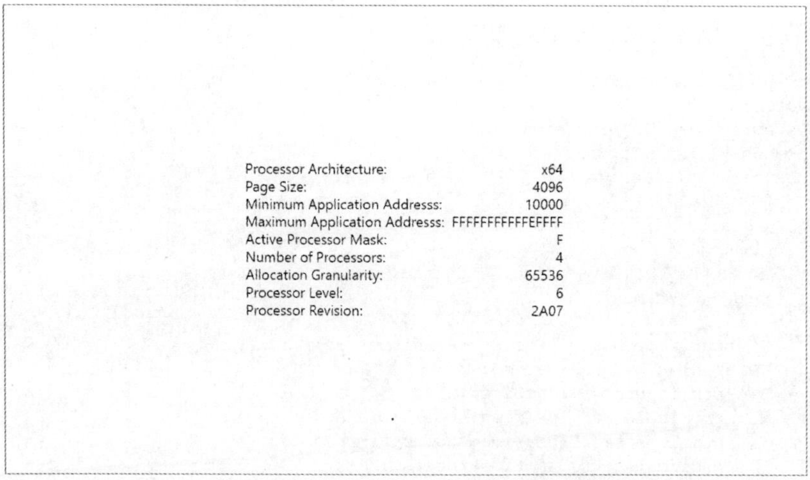

Processor Architecture:	x64
Page Size:	4096
Minimum Application Addresss:	10000
Maximum Application Addresss:	FFFFFFFFFFFEFFFF
Active Processor Mask:	F
Number of Processors:	4
Allocation Granularity:	65536
Processor Level:	6
Processor Revision:	2A07

该平板有 64 位处理器。程序在微软 Surface 上运行时，如下图所示。

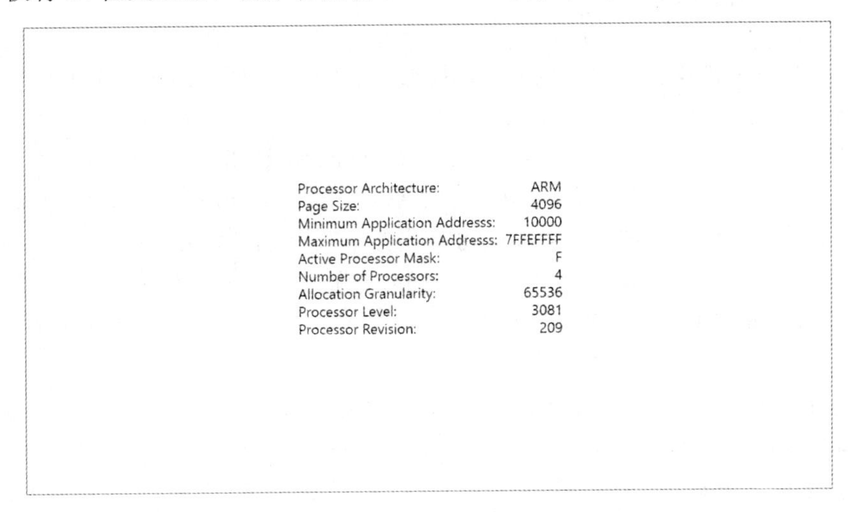

15.2 一 些 帮 助

在使用 P/Invoke 来定义结构和声明函数时，务必要正确使用。例如，必须提供该函数所在的 DLL 的正确文件名。(而在我开发的本章第一个项目中，我输入的是“kernel32.lib”而不是“kernel32.DLL”，我没搞明白为什么不行。)如果访问的不是系统 DLL，则必须确保该 DLL 被应用所引用。还必须正确拼写函数名并正确声明所有参数。P/Invoke 可没有 IntelliSense!

结构和函数声明往往比较复杂。有一个维基网站 www.pinvoke.net 能够提供一些帮助，很多人都贡献了结构定义和函数声明，你可以直接复制并粘贴到自己的代码中。也允许你贡献自己写的一些代码!

15.3 时 区 信 息

假设你想写一个 Windows Store 应用来显示世界各地不同地点的时钟(我在 2000 年为 PC Magazine 写过 ClockRack 程序)。其 Windows 8 系统版本看起来可能如下图所示。

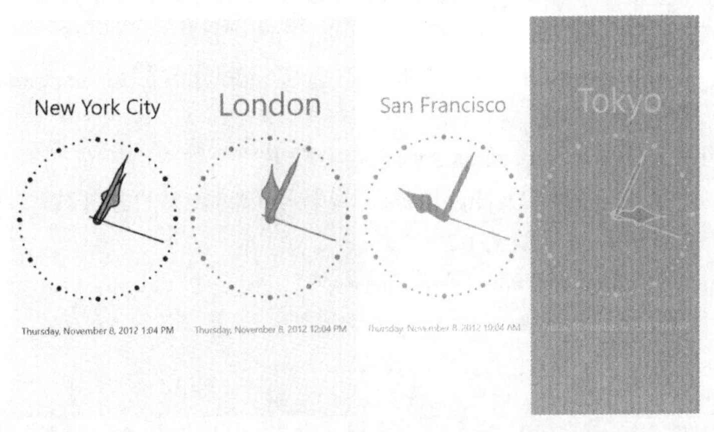

这种程序允许你添加新的时钟、设置位置和时区，指定独特颜色，并在应用设置中保留这些信息。

Windows 内置功能支持计算不同时区的时间，尤其能处理日光节约时间(在一些地方称为“夏令时”)问题，如果能利用该功能，就太好了。

你会非常有热情在 System 命名空间发现 TimeZoneInfo 类，而且会注意到静态 GetSystemTimeZones 方法返回世界各地所有时区的 TimeZoneInfo 对象集合。然而，如果尝试使用这种方法时，你会发现对 Windows 8 应用并不可用。在 Windows 8 应用中唯一可获得的 TimeZoneInfo 对象，是适合当前系统时区设置的对象或不常用的协调世界时对象(Universal Coordinated Time，UTC)，(更通常但不准确)的说法是格林尼治标准时间。

然而，Windows 8 应用的确能访问向你提供信息的若干 Win32 函数。Win32 EnumDynamicTimeZoneInformation 函数以 DYNAMIC_TIME_ZONE_INFORMATION 结构形式枚举世界各地时区：

```
typedef struct _TIME_DYNAMIC_ZONE_INFORMATION{
    LONG Bias;
    WCHAR StandardName[32];
    SYSTEMTIME StandardDate;
    LONG StandardBias;
    WCHAR DaylightName[32];
    SYSTEMTIME DaylightDate;
    LONG DaylightBias;
    WCHAR TimeZoneKeyName[128];
    BOOLEAN DynamicDaylightTimeDisabled;
}   DYNAMIC_TIME_ZONE_INFORMATION, *PDYNAMIC_TIME_ZONE_INFORMATION;
```

以下为扩展版 TIME_ZONE_INFORMATION 结构：

```
typedef struct _TIME_ZONE_INFORMATION {
    LONG Bias;
    WCHAR StandardName[32];
    SYSTEMTIME StandardDate;
    LONG StandardBias;
    WCHAR DaylightName[32];
    SYSTEMTIME DaylightDate;
    LONG DaylightBias;
}   TIME_ZONE_INFORMATION, *PTIME_ZONE_INFORMATION;
```

WCHAR 为宽 16 位 Unicode 字符，而其字符数组基本上是零结尾的字符串。StandardName 就像是 Eastern Standard Time 字符串，DaylightName 就像是 Eastern Daylight Time 字符串。DYNAMIC_TIME_ ZONE_INFORMATION 结构中的 TimeZoneKeyName 是在 Windows 注册表中使用的键值。在 Windows 8 中，这些注册表项可以在 HKEY_LOCAL_MACHINE/SOFTWARE/Microsoft/Windows NT/CurrentVersion/Time Zones 找到，而且匹配 StandardName。

Bias 字段为从 Universal Coordinated Time 中减去的分钟数后所获得的本地时间。对于美国东部时区，即为 300 分钟。DaylightBias 是从标准时间减去的分钟数，以转换为夏令时，通常为-60，而 StandardBias 始终为 0。

DaylightDate 和 StandardDate 字段表示何时切换到夏令时间并回到标准时，它为 SYSTEMTIME 类型：

```
typedef struct _SYSTEMTIME {
    WORD wYear;
    WORD wMonth;
```

```
    WORD wDayOfWeek;
    WORD wDay;
    WORD wHour;
    WORD wMinute;
    WORD wSecond;
    WORD wMilliseconds;
}   SYSTEMTIME, *PSYSTEMTIME;
```

SYSTEMTIME 结构最主要用于和 Win32 GetLocalTime 和 GETSYSTEMTIME 函数一起分别获取当前时间和 UTC。TIME_ZONE_INFORMATION 结构中的 SYSTEMTIME 值特别编码，用于表示转换日期：wHour 和 wMinute 字段表示转换时间，wMonth 字段表示转换月份(例如，3 代表 3 月)，wDayOfWeek 字段表示转换星期(例如，1 代表为星期日)，wDay 字段表示在指定月份内某周某天所发生的特定事件(例如，2 代表该月的第二个星期日，5 代表最后一个星期日)。

Windows 能区分在标准时间和夏令时之间进行切换的地点，还能区分每年动态更改日期的地点。后者称为使用动态 DST，但年份信息不能直接可用。

GetTimeZoneInformationForYear 函数接受年份参数与指向 DYNAMIC_TIME_ZONE_INFORMATION 结构的指针，并返回指向 TIME_ZONE_INFORMATION 结构带适合年份信息的指针。SystemTimeToTzSpecificLocalTime 有指向 TIME_ZONE_INFORMATION 结构的指针，以及指向可能从 GETSYSTEMTIME 函数获得的 SYSTEMTIME 结构的指针，并返回表示当地时区时间的 SYSTEMTIME 结构。因此，程序没有必要执行自身时间转换。

类似 ClockRack 的程序需要工具使用户为特定地点选择时区。最好和 Windows 8 为用户提供的时区选择功能保持一致。

看一看 Windows 8 为用户提供的功能：启用 Windows 8 超级按钮、选择 Settings、点击底部的 Change PC Settings 标签。这样启用一个程序，标题为 PC Settings，并带有一个列表。选择 General，会在顶部看到有时区选择的组合框。每个时区通过和 UTC 的差值来进行识别，有时通过时区名称，经常也有一些示范城市。例如，对于 Romance Standard Time(欧洲中部时间)，组合框显示如下。

```
(UTC +01:00)Brussles, Copenhagen, Madrid, Paris
```

在 Windows 的时区注册表部分，你会发现这些标签用 Display 名称进行识别，但 Win32 函数并没有提供这一信息。你需要访问注册表来获取，但对 Windows 8 应用而言，没有现在的 Win32 函数能够访问注册表。

当然没什么能够阻止你写一个小的桌面.NET 程序来访问完整 TimeZoneInfo 类并格式化所得到的字符串，以便定义能包含到 Windows 8 程序的 Dictionary 的对象。我用来生成该列表的.NET 程序代码显示在 Dictionary 定义上面的注释中。

```
项目: ClockRack | 文件: TimeZoneManager.Display.cs(片段)
namespace ClockRack
{
    public partial class TimeZoneManager
    {
        // Generated from tiny .NET program:
        // foreach (TimeZoneInfo info in TimeZoneInfo.GetSystemTimeZones())
        // Console.WriteLine("{{ \"{0}\", \"{1}\" }},", info.StandardName, info.DisplayName);
        static Dictionary<string, string> displayStrings = new Dictionary<string, string>
        {
```

```
            { "Dateline Standard Time", "(UTC-12:00) International Date Line West" },
            { "UTC-11", "(UTC-11:00) Coordinated Universal Time-11" },
            { "Hawaiian Standard Time", "(UTC-10:00) Hawaii" },
            { "Alaskan Standard Time", "(UTC-09:00) Alaska" },
            { "Pacific Standard Time (Mexico)", "(UTC-08:00) Baja California" },

            ...

            { "Kamchatka Standard Time", "(UTC+12:00) Petropavlovsk-Kamchatsky - Old" },
            { "Tonga Standard Time", "(UTC+13:00) Nuku'alofa" },
            { "Samoa Standard Time", "(UTC+13:00) Samoa" }
        };
    }
}
```

字典在 ClockRack 项目中我称为 TimeZoneManager 类。我用该类来加强 P/Invoke 逻辑。TimeZoneManager 外部没有代码可以访问 Win32 函数或结构。

TimeZoneManager 类实例化一次，就可以用于应用的整个运行过程。该类作为以下值的集合使时区数据可用于程序的其余部分。

```
项目: ClockRack | 文件: TimeZoneDisplayInfo.cs
namespace ClockRack
{
    public struct TimeZoneDisplayInfo
    {
        public int Bias { set; get; }
        public string TimeZoneKey { set; get; }
        public string Display { set; get; }
    }
}
```

Bias 属性仅用于排序。TimeZoneKey 和 DYNAMIC_TIME_ZONE_INFORMATION 结构中的 TimeZoneKeyName 相同，而 Display 属性从 displayStrings 字典中获得。

TimeZoneManager.cs 文件中的 TimeZoneManager 类的一部分先定义必要的 Win32 结构，并声明该类所要求的三个 Win32 函数。

```
项目: ClockRack | 文件: TimeZoneManager.cs(片段)
public partial class TimeZoneManager
{
    [StructLayout(LayoutKind.Sequential)]
    struct SYSTEMTIME
    {
        public ushort wYear;
        public ushort wMonth;
        public ushort wDayOfWeek;
        public ushort wDay;
        public ushort wHour;
        public ushort wMinute;
        public ushort wSecond;
        public ushort wMilliseconds;
    }

    [StructLayout(LayoutKind.Sequential, CharSet = CharSet.Unicode)]
    struct TIME_ZONE_INFORMATION
    {
        public int Bias;
        [MarshalAs(UnmanagedType.ByValTStr, SizeConst = 32)]
        public string StandardName;
        public SYSTEMTIME StandardDate;
        public int StandardBias;
        [MarshalAs(UnmanagedType.ByValTStr, SizeConst = 32)]
        public string DaylightName;
        public SYSTEMTIME DaylightDate;
```

```
        public int DaylightBias;
    }

    [StructLayout(LayoutKind.Sequential, CharSet = CharSet.Unicode)]
    struct DYNAMIC_TIME_ZONE_INFORMATION
    {
        public int Bias;
        [MarshalAs(UnmanagedType.ByValTStr, SizeConst = 32)]
        public string StandardName;
        public SYSTEMTIME StandardDate;
        public int StandardBias;
        [MarshalAs(UnmanagedType.ByValTStr, SizeConst = 32)]
        public string DaylightName;
        public SYSTEMTIME DaylightDate;
        public int DaylightBias;
        [MarshalAs(UnmanagedType.ByValTStr, SizeConst = 128)]
        public string TimeZoneKeyName;
        public byte DynamicDaylightTimeDisabled;
    }

    [DllImport("Advapi32.dll")]
    static extern uint EnumDynamicTimeZoneInformation(uint index,
                                   ref DYNAMIC_TIME_ZONE_INFORMATION dynamicTzi);

    [DllImport("kernel32.dll")]
    static extern byte GetTimeZoneInformationForYear(ushort year,
                                   ref DYNAMIC_TIME_ZONE_INFORMATION dtzi,
                                   out TIME_ZONE_INFORMATION tzi);

    [DllImport("kernel32.dll")]
    static extern byte SystemTimeToTzSpecificLocalTime(ref TIME_ZONE_INFORMATION tzi,
                                   ref SYSTEMTIME utc, out SYSTEMTIME local);
    ...
}
```

回想一下 DYNAMIC_TIME_ZONE_INFORMATION 和 TIME_ZONE_INFORMATION 结构中的若干字段定义为 WCHAR 数组。这在 C#不可行，因为 C#数组始终为从堆中指向内存的指针。MarshalAs 属性则允许表明这些字段应该作为特定最大长度的 C#字符串。

TimeZoneManager 类的构造函数重复调用 EnumDynamicTimeZoneInformation，直到返回非零值(表明到达列表末尾)。在不同版本的 Windows 中，列表项数量可能略有变化，我看一共有 101 项。每一项都存储在私有字典中，并且每一项都变成 TimeZoneDisplayInfo 值，并添加到名称为 DisplayInformation 的公开可用集合。

```
项目: ClockRack | 文件: TimeZoneManager.cs(片段)
public partial class TimeZoneManager
{
    ...
    // Internal dictionary for looking up DYNAMIC_TIME_ZONE_INFORMATION values from keys
    Dictionary<string, DYNAMIC_TIME_ZONE_INFORMATION> dynamicTzis =
                         New Dictionary<string, DYNAMIC_TIME_ZONE_INFORMATION>();

    public TimeZoneManager()
    {
        uint index = 0;
        DYNAMIC_TIME_ZONE_INFORMATION tzi = new DYNAMIC_TIME_ZONE_INFORMATION();
        List<TimeZoneDisplayInfo> displayInformation = new List<TimeZoneDisplayInfo>();

        // Enumerate through time zones
        while (0 == EnumDynamicTimeZoneInformation(index, ref tzi))
        {
            dynamicTzis.Add(tzi.TimeZoneKeyName, tzi);

            // Create TimeZoneDisplayInfo for public property
```

```
            TimeZoneDisplayInfo displayInfo = new TimeZoneDisplayInfo
            {
                Bias = tzi.Bias,
                TimeZoneKey = tzi.TimeZoneKeyName
            };

            // Look up the display string
            if (displayStrings.ContainsKey(tzi.TimeZoneKeyName))
            {
                displayInfo.Display = displayStrings[tzi.TimeZoneKeyName];
            }
            else if (displayStrings.ContainsKey(tzi.StandardName))
            {
                displayInfo.Display = displayStrings[tzi.StandardName];
            }
            // Or calculate one
            else
            {
                if (tzi.Bias == 0)
                    displayInfo.Display = "(UTC) ";
                else
                    displayInfo.Display = String.Format("(UTC{0}{1:D2}:{2:D2}) ",
                                                tzi.Bias > 0 ? '-' : '+',
                                                Math.Abs(tzi.Bias) / 60,
                                                Math.Abs(tzi.Bias) % 60);
                displayInfo.Display += tzi.TimeZoneKeyName;
            }

            // Add to collection
            displayInformation.Add(displayInfo);

            // Prepare for next iteration
            index += 1;
            tzi = new DYNAMIC_TIME_ZONE_INFORMATION();
        }

        // Sort the display information items
        displayInformation.Sort((TimeZoneDisplayInfo info1, TimeZoneDisplayInfo info2) =>
            {
                return info2.Bias.CompareTo(info1.Bias);
            });

        // Set to the publicly available property
        this.DisplayInformation = displayInformation;
    }

    // Public interface
    public IList<TimeZoneDisplayInfo> DisplayInformation { protected set; get; }
    ...
}
```

很快就能看到 DisplayInformation 属性用作 ComboBox 的 ItemsSource。

TimeZoneManager 的最后一个方法是根据时区键值把 UTC 时间转换为本地时间。这和 DYNAMIC_TIME_ZONE_INFORMATION 结构的 TimeZoneKeyName 字段以及 TimeZoneDisplayInfo 结构的 TimeZoneKey 属性是相同的。

项目: ClockRack | 文件: TimeZoneManager.cs(片段)

```
public partial class TimeZoneManager
{
    ...
    public DateTime GetLocalTime(string timeZoneKey, DateTime utc)
    {
        // Convert to Win32 SYSTEMTIME
        SYSTEMTIME utcSysTime = new SYSTEMTIME
        {
```

```
                wYear = (ushort)utc.Year,
                wMonth = (ushort)utc.Month,
                wDay = (ushort)utc.Day,
                wHour = (ushort)utc.Hour,
                wMinute = (ushort)utc.Minute,
                wSecond = (ushort)utc.Second,
                wMilliseconds = (ushort)utc.Millisecond
            };

            // Convert to local time
            DYNAMIC_TIME_ZONE_INFORMATION dtzi = dynamicTzis[timeZoneKey];
            TIME_ZONE_INFORMATION tzi = new TIME_ZONE_INFORMATION();
            GetTimeZoneInformationForYear((ushort)utc.Year, ref dtzi, out tzi);

            SYSTEMTIME localSysTime = new SYSTEMTIME();
            SystemTimeToTzSpecificLocalTime(ref tzi, ref utcSysTime, out localSysTime);

            // Convert SYSTEMTIME to DateTime
            return new DateTime(localSysTime.wYear, localSysTime.wMonth, localSysTime.wDay,
                                localSysTime.wHour, localSysTime.wMinute,
                                localSysTime.wSecond, localSysTime.wMilliseconds);
        }
    }
```

该方法把.NET DateTime 转换成 Win32 SYSTEMTIME，从私有字典中获得 DYNAMIC_TIME_ZONE_INFORMATION，然后调用 GetTimeZoneInformationForYear，返回 TIME_ZONE_INFORMATION 结构形式的信息，再传递给 SystemTimeToTzSpecificLocalTime 函数。所得 SYSTEMTIME 再转换回.Net DateTime。

我对该方法并不完全满意，让我来告诉你为什么。ClockRack 程序显示多个时钟，并使用 CompositionTarget.Rendering 方法来获取更新后的 DateTime.UtcNow 值，用于所有时钟。(我想这可能比 GetLocalTime 方法调用 Win32 GETSYSTEMTIME 函数来给每个时钟 UTC 获取 SYSTEMTIME 值效率高。)我不满意的是要反复调用 GetTimeZoneInformationForYear 方法。对每个时区实际只需要调用一次该功能，而 TIME_ZONE_INFORMATION 在随后调用中可以重复使用。然而，如果程序从 12 月 31 日运行至 1 月 1 日，由于新年需要重新调用。所以我决定不用这种逻辑把类搞乱。

传递给 GetTimeZoneInformationForYear 的年份应该是本地年份，而不是 UTC 年份，还有一些别的事情做得也不完全正确。只有在 UTC 新年临界 24 小时内，两年才可能不同，在类似程序中，实际上不应该有什么问题，因为标准时间和夏令时之间的转换发生在这年下半年。

然而，如果在南半球的一个特定地点决定在一个公历年中采用夏令时，而下一年并不采用，或者反过来，则该地点在 12 月 31 日午夜会经历本地时间变化，计算两年之间新年过渡期间的时间是否正确。

但我们继续往下看。

你可以从第 10 章中的 AnalogClock 程序识别出实际时钟(称为 TimeZoneClock 的 UserControl 派生类)，但我对其进行转换以通过数据绑定来使用视图模式。我用因子 10 减少所有坐标和尺寸。因为可能会在非常狭小的空间内(例如辅屏模式)显示很多时钟，因此，时钟尺寸必须能够大幅度变化。我采用的方法是涉及 Viewbox 和自定义面板，最适合时钟定义的尺寸小于可能的尺寸。

模拟时钟表面有两个 TextBlock 元素包围，这是另一个区别。最上面的 TextBlock 元素显示位置，而底部的则显示当前日期和时间。如果底部没有文字时间，你可能会困惑时间是中午之前还是之后。

每个 TextBlock 元素都有 RowDefinition 对 Grid 设置的固定高度，但两者也各有 Viewbox，(如果文本太长，就压缩到适合长度)。总体而言，整个时钟是一个 Grid，并有设置 e 绑定的 Background 属性。Grid 会占据整个可用区域。其内部是 Viewbox，会调整其内容大小。而内容是另一个固定大小为 30*20 的 Grid，但实际大小由 Viewbox 根据可用空间进行控制。

项目: ClockRack | 文件: TimeZoneClock.xaml(片段)

```
<UserControl ...
    Name="ctrl">

    <UserControl.Resources>
        <Style TargetType="TextBlock">
            <Setter Property="Margin" Value="12 0" />
            <Setter Property="TextAlignment" Value="Center" />
        </Style>

        <Style TargetType="Path">
            <Setter Property="StrokeThickness" Value="0.2" />
            <Setter Property="StrokeStartLineCap" Value="Round" />
            <Setter Property="StrokeEndLineCap" Value="Round" />
            <Setter Property="StrokeLineJoin" Value="Round" />
            <Setter Property="StrokeDashCap" Value="Round" />
            <Setter Property="Fill" Value="Gray" />
        </Style>
    </UserControl.Resources>

    <UserControl.Foreground>
        <SolidColorBrush Color="{Binding Foreground}" />
    </UserControl.Foreground>

    <Grid>
        <Grid.Background>
            <SolidColorBrush Color="{Binding Background}" />
        </Grid.Background>

        <Viewbox>
            <Grid Width="20"
                  Height="30"
                  HorizontalAlignment="Center"
                  VerticalAlignment="Center">

            <Grid.RowDefinitions>
                <RowDefinition Height="5" />
                <RowDefinition Height="20" />
                <RowDefinition Height="5" />
            </Grid.RowDefinitions>

            <Viewbox Grid.Row="0">
                <TextBlock Text="{Binding Location}" />
            </Viewbox>

            <Grid Grid.Row="1">

                    <!-- Transform for entire clock -->
                    <Grid.RenderTransform>
                        <TranslateTransform X="10" Y="10" />
                    </Grid.RenderTransform>

                    <!-- Small tick marks -->
                    <Path Fill="{x:Null}"
                          Stroke="{Binding ElementName=ctrl, Path=Foreground}"
```

```
                    StrokeThickness="0.3"
                    StrokeDashArray="0 3.14159">
                <Path.Data>
                    <EllipseGeometry RadiusX="90" RadiusY="90" />
                </Path.Data>
            </Path>

            <!-- Large tick marks -->
            <Path Fill="{x:Null}"
                    Stroke="{Binding ElementName=ctrl, Path=Foreground}"
                    StrokeThickness="0.6"
                    StrokeDashArray="0 7.854">
                <Path.Data>
                    <EllipseGeometry RadiusX="90" RadiusY="90" />
                </Path.Data>
            </Path>

            <!-- Hour hand pointing straight up -->
            <Path Data="M 0 -6 C 0 -3, 2 -3, 0.5 -2 L 0.5 0
                               C 0.5 0.75, --/5 0.75, -0.5 0 L -0.5 -2
                               C -2 -3, 0 -3, 0 -6"
                    Stroke="{Binding ElementName=ctrl, Path=Foreground}">
                <Path.RenderTransform>
                    <RotateTransform Angle="{Binding HourAngle}" />
                </Path.RenderTransform>
            </Path>

            <!-- Minute hand pointing straight up -->
            <Path Data="M 0 -8 C 0 -7.5, 0 -7, 0.25 -6 L 0.25 0
                               C 0.25 0.5, -0.25 0.5, -0.25 0 L -0.255 -6
                               C 0 -7, 0 -7.5, 0 -8.0"
                    Stroke="{Binding ElementName=ctrl, Path=Foreground}">
                <Path.RenderTransform>
                    <RotateTransform Angle="{Binding MinuteAngle}" />
                </Path.RenderTransform>
            </Path>

            <!-- Second hand pointing straight up -->
            <Path Data="M 0 1 L 0 -8"
                    Stroke="{Binding ElementName=ctrl, Path=Foreground}">
                <Path.RenderTransform>
                    <RotateTransform Angle="{Binding SecondAngle}" />
                </Path.RenderTransform>
            </Path>
        </Grid>

        <Viewbox Grid.Row="2">
            <TextBlock Text="{Binding FormattedDateTime}" />
        </Viewbox>
    </Grid>
  </Viewbox>
 </Grid>
</UserControl>
```

TextBlock 元素和所有 RotateTransform 元素都绑定到视图模式中的一个属性。在 TimeZoneClock.xaml 文件开头，你会看到视图模式还包括 Color 类型的属性，分别称为 Foreground 和 Background。代码隐藏文件只有一个 InitializeComponent 调用。

为了让程序相对简单，我决定把每个时钟的背景和前景颜色定为 140 种颜色，这些颜色有名称，对应于静态 Colors 类项。TimeZoneClock 类的视图模式会如你期望的定义 Color 类型的 Foreground 和 Background 属性，但也定义 ForegroundName 和 BackgroundName 属性，只要其中一个属性发生改变，另一个也会有一些反射性逻辑变化。

项目: ClockRack | 文件: TimeZoneClockViewModel.cs(片段)

```
public class TimeZoneClockViewModel : INotifyPropertyChanged
{
    string location = "New York City", timeZoneKey = "Eastern Standard Time";
    Color background = Colors.Yellow, foreground = Colors.Blue;
    string backgroundName = "Yellow", foregroundName = "Blue";
    DateTime dateTime;
    string formattedDateTime;
    double hourAngle, minuteAngle, secondAngle;
    TypeInfo colorsTypeInfo = typeof(Colors).GetTypeInfo();

    public event PropertyChangedEventHandler PropertyChanged;

    public string Location
    {
        set { SetProperty<string>(ref location, value); }
        get { return location; }
    }

    public string TimeZoneKey
    {
        set { SetProperty<string>(ref timeZoneKey, value); }
        get { return timeZoneKey; }
    }

    public string BackgroundName
    {
        set
        {
            if (SetProperty<string>(ref backgroundName, value))
                this.Background = NameToColor(value);
        }
        get { return backgroundName; }
    }

    public Color Background
    {
        set
        {
            if (SetProperty<Color>(ref background, value))
                this.BackgroundName = ColorToName(value);
        }
        get { return background; }
    }

    public string ForegroundName
    {
        set
        {
            if (SetProperty<string>(ref foregroundName, value))
                this.Foreground = NameToColor(value);
        }
        get { return foregroundName; }
    }

    public Color Foreground
    {
        set
        {
            if (SetProperty<Color>(ref foreground, value))
                this.ForegroundName = ColorToName(value);
        }
        get { return foreground; }
    }

    public DateTime DateTime
    {
```

```
            set
            {
                if (SetProperty<DateTime>(ref dateTime, value))
                {
                    this.FormattedDateTime = String.Format("{0:D} {1:t}", value, value);
                    this.SecondAngle = 6 * (dateTime.Second + dateTime.Millisecond / 1000.0);
                    this.MinuteAngle = 6 * dateTime.Minute + this.SecondAngle / 60;
                    this.HourAngle = 30 * (dateTime.Hour % 12) + this.MinuteAngle / 12;
                }
            }
            get { return dateTime; }
        }

        public string FormattedDateTime
        {
            set { SetProperty<string>(ref formattedDateTime, value); }
            get { return formattedDateTime; }
        }

        public double HourAngle
        {
            set { SetProperty<double>(ref hourAngle, value); }
            get { return hourAngle; }
        }

        public double MinuteAngle
        {
            set { SetProperty<double>(ref minuteAngle, value); }
            get { return minuteAngle; }
        }

        public double SecondAngle
        {
            set { SetProperty<double>(ref secondAngle, value); }
            get { return secondAngle; }
        }

        Color NameToColor(string name)
        {
            return(Color)colorsTypeInfo.GetDeclaredProperty(name).GetValue(null);
        }

        string ColorToName(Color color)
        {
            foreach (PropertyInfo property in colorsTypeInfo.DeclaredProperties)
                if (color.Equals((Color)property.GetValue(null)))
                    return property.Name;

            return "";
        }

        protected bool SetProperty<T>(ref T storage, T value,
                                      [CallerMemberName] string propertyName = null)
        {
            if (object.Equals(storage, value))
                return false;

                storage = value;
                OnPropertyChanged(propertyName);
                return true;
        }

        protected void OnPropertyChanged(string propertyName)
        {
            if (PropertyChanged != null)
                PropertyChanged(this, new PropertyChangedEventArgs(propertyName));
        }
    }
```

该视图模式还包括 DateTime 属性，只要 DateTime 发生变化，HourAngle、MinuteAngle 和 SecondAngle 属性也会随之变化，并且驱动 TimeZoneClock.xaml 中的三个 RotateTransform 对象。

为了显示多个时钟，我需要一个面板把所有时钟都显示在页面范围内，并给每个时钟分配最佳空间。第 11 章中提到的 UniformGrid 面板似乎接近我的要求，但又不完全是。例如，假设有七个时钟，UniformGrid 把它们分成两行来显示。UniformGrid 会把五个时钟放在第一行，其余两个放在第二行。如果第一行放四个，而第二行放三个，这样更均匀，摆放起来更美观。

ClockRack 程序包含对第 11 章的 Petzold.ProgrammingWindows6.Chapter11 库的引用，但从 UniformGrid 派生了名为 DistributedUniformGrid 的面板。该新类的逻辑是给每行分配大致相当的项目。每一行项目间隔相同。

项目：ClockRack | 文件：DistributedUniformGrid.cs

```
using System;
using Windows.Foundation;
using Windows.UI.Xaml.Controls;
using Petzold.ProgrammingWindows6.Chapter11;
namespace ClockRack

{
    public class DistributedUniformGrid : UniformGrid
    {
        protected override Size ArrangeOverride(Size finalSize)
        {
            int index = 0;
            double cellWidth = finalSize.Width / cols;
            double cellHeight = finalSize.Height / rows;
            int displayed = 0;

            if (this.Orientation == Orientation.Vertical)
            {
                for (int row = 0; row < rows; row++)
                {
                    double y = row * cellHeight;
                    int accumDisplay = (int)Math.Ceiling((row + 1.0) *
                        this.Children.Count / rows);
                    int display = accumDisplay - displayed;
                    cellWidth = Math.Round(finalSize.Width / display);
                    double x = 0;

                    for (int col = 0; col < display; col++)
                    {
                        if (index < this.Children.Count)
                            this.Children[index].Arrange(new Rect(x, y, cellWidth, cellHeight));

                        x += cellWidth;
                        index++;
                    }
                    displayed += display;
                }
            }
            else
            {
                for (int col = 0; col < cols; col++)
                {
                    double x = col * cellWidth;
                    int accumDisplay =
                        (int)Math.Ceiling((col + 1.0) * this.Children.Count / cols);
                    int display = accumDisplay - displayed;
                    cellHeight = Math.Round(finalSize.Height / display);
```

```
                        double y = 0;

                        for (int row = 0; row < display; row++)
                        {
                            if (index < this.Children.Count)
                                this.Children[index].Arrange(new Rect(x, y, cellWidth, cellHeight));

                            y += cellHeight;
                            index++;
                        }
                        displayed += display;
                    }
                }
                return finalSize;
            }
        }
    }
```

MainPage.xaml 文件包含 DistributedUniformGrid 来控制时钟控件。

项目：ClockRack | 文件：MainPage.xaml(片段)

```
<Page ... >
    <Grid Background="{StaticResource ApplicationPageBackgroundThemeBrush}">
        <Grid Name="contentGrid"
              Background="Transparent">

            <DistributedUniformGrid Name="uniformGrid"
                                    Orientation="Vertical" />
        </Grid>
    </Grid>
</Page>
```

MainPage 类的构造函数负责从应用设置中填入 DistributedUniformGrid。程序用 Windows.Storage 命名空间中的的 ApplicationData 类来存储每个时钟的四个文本项：位置名称(由用户选择)、用来识别时区的时区键、前景色名称和背景色名称。对于第一个时钟，存储四个文本项目使用的键值为“0Location”、“0TimeZoneKey”、“0Foreground”和“0Background”，第二时钟的键值以数字 1 开头，以此类推。如果检索每组设置，则创建并初始化 TimeZoneClock 和 TimeZoneClockViewModel。

项目：ClockRack | 文件：MainPage.xaml.cs(片段)

```
public sealed partial class MainPage : Page
{
    ...
    IPropertySet appSettings = ApplicationData.Current.LocalSettings.Values;

    public MainPage()
    {
        this.InitializeComponent();

        // Load application settings for clocks
        int index = 0;

        while (appSettings.ContainsKey(index.ToString() + "Location"))
        {
            string preface = index.ToString();

            TimeZoneClock clock = new TimeZoneClock
            {
                DataContext = new TimeZoneClockViewModel
                {
                    Location = appSettings[preface + "Location"] as string,
                    TimeZoneKey = appSettings[preface + "TimeZoneKey"] as string,
                    ForegroundName = appSettings[preface + "Foreground"] as string,
                    BackgroundName = appSettings[preface + "Background"] as string
```

```
                },
            };
            uniformGrid.Children.Add(clock);
            index += 1;
        }

        // If there are no settings, make a default Clock
        if (uniformGrid.Children.Count == 0)
        {
            TimeZoneClock clock = new TimeZoneClock
            {
                DataContext = new TimeZoneClockViewModel()
            };
            uniformGrid.Children.Add(clock);
        }

        // Set the Suspending handler
        Application.Current.Suspending += OnApplicationSuspending;

        // Start the Rendering event
        CompositionTarget.Rendering += OnCompositionTargetRendering;
    }

    void OnApplicationSuspending(object sender, SuspendingEventArgs args)
    {
        appSettings.Clear();

        for (int index = 0; index < uniformGrid.Children.Count; index++)
        {
            TimeZoneClock timeZoneClock = uniformGrid.Children[index] as TimeZoneClock;
            TimeZoneClockViewModel viewModel =
                    timeZoneClock.DataContext as TimeZoneClockViewModel;
            string preface = index.ToString();

            appSettings[preface + "Location"] = viewModel.Location;
            appSettings[preface + "TimeZoneKey"] = viewModel.TimeZoneKey;
            appSettings[preface + "Foreground"] = viewModel.ForegroundName;
            appSettings[preface + "Background"] = viewModel.BackgroundName;
        }
    }
    ...
}
```

像往常一样，这些设置在 Suspending 事件中进行保存。

构造函数最后触发 CompositionTarget.Rendering 事件。它负责使用 TimeZoneManager 实例，基于当前 UTC 时间和时区键值来获取每个时钟的本地时间。

项目：ClockRack | 文件：MainPage.xaml.cs(片段)

```
public sealed partial class MainPage : Page
{
    TimeZoneManager timeZoneManager = new TimeZoneManager();
    ...
    void OnCompositionTargetRendering(object sender, object args)
    {
        // Get the time once
        DateTime utc = DateTime.UtcNow;

        foreach (UIElement child in uniformGrid.Children)
        {
            TimeZoneClockViewModel viewModel =
                 (child as FrameworkElement).DataContext as TimeZoneClockViewModel;
            string timeZoneKey = viewModel.TimeZoneKey;

            // Set the local time from the TimeZoneManager
            viewModel.DateTime = timeZoneManager.GetLocalTime(timeZoneKey, utc);
        }
```

```
        }
        ...
}
```

右击显示 PopupMenu，其中有三项：Add、Edit 和 Delete。Edit 和 Delete 从属于被点击的特定时钟，因此 OnRightTapped 覆写会先发现这个时钟。对象传递到这三项的处理程序。即使对于 Add，也需要点击时钟，因为在点击时钟后逻辑会插入新时钟。只要有多个时钟，Delete 就就会出现在菜单中。

项目：ClockRack | 文件：MainPage.xaml.cs(片段)

```
async protected override void OnRightTapped(RightTappedRoutedEventArgs args)
{
    // Check if the parent of the click element is a TimeZoneClock
    FrameworkElement element = args.OriginalSource as FrameworkElement;

    while (element != null)
    {
        if (element is TimeZoneClock)
            break;

        element = element.Parent as FrameworkElement;
    }

    if (element == null)
        return;

    // Create a PopupMenu
    PopupMenu popupMenu = new PopupMenu();
    popupMenu.Commands.Add(new UICommand("Add...", OnAddMenuItem, element));
    popupMenu.Commands.Add(new UICommand("Edit...", OnEditMenuItem, element));

    if (uniformGrid.Children.Count > 1)
        popupMenu.Commands.Add(new UICommand("Delete", OnDeleteMenuItem, element));

    args.Handled = true;
    base.OnRightTapped(args);

    // Display the menu
    await popupMenu.ShowAsync(args.GetPosition(this));
}
```

对于 Add 菜单项，必须创建新的 TimeZoneClock(带有对应 TimeZoneClockViewModel)并将其插入集合中。点击时钟后，总是要插入新时钟。

项目：ClockRack | 文件：MainPage.xaml.cs(片段)

```
void OnAddMenuItem(IUICommand command)
{
    // Create new TimeZoneClock
    TimeZoneClock timeZoneClock = new TimeZoneClock
    {
        DataContext = new TimeZoneClockViewModel()
    };

    // Insert after the tapped clock
    TimeZoneClock clickedClock = command.Id as TimeZoneClock;
    int index = uniformGrid.Children.IndexOf(clickedClock);
    uniformGrid.Children.Insert(index + 1, timeZoneClock);
}
```

Delete 也相当容易，但程序会坚持用 MessageDialog 来得到删除确认。

项目：ClockRack | 文件：MainPage.xaml.cs(片段)

```
async void OnDeleteMenuItem(IUICommand command)
{
```

```
        TimeZoneClock timeZoneClock = command.Id as TimeZoneClock;
        TimeZoneClockViewModel viewModel = timeZoneClock.DataContext as TimeZoneClockViewModel;

        MessageDialog msgdlg = new MessageDialog("Delete clock from collection?",
                                                 viewModel.Location);
        msgdlg.Commands.Add(new UICommand("OK"));
        msgdlg.Commands.Add(new UICommand("Cancel"));
        msgdlg.DefaultCommandIndex = 0;
        msgdlg.CancelCommandIndex = 1;

        IUICommand msgDlgCommand = await msgdlg.ShowAsync();

        If  (msgDlgCommand.Label == "OK")
            uniformGrid.Children.Remove(command.Id as TimeZoneClock);
    }
```

当然，Edit 比较复杂，除非你完全满意纽约市时钟的蓝色前景和黄色背景，否则就得在增加新时钟之后启用 Edit 菜单。Edit 菜单项实例化 SettingsDialog(很快就会看到)作为 Popup 子对象。SettingsDialog 需要访问 TimeZoneManager 实例，因此，TimeZoneManager 对象就会向 SettingsDialog 提供构造函数。该方法主要负责定位 Popup，以便从视觉上和点击时钟相关联，但又不至于伸出屏幕边缘。

项目：ClockRack | 文件：MainPage.xaml.cs(片段)

```
    void OnEditMenuItem(IUICommand command)
    {
        TimeZoneClock timeZoneClock = command.Id as TimeZoneClock;
        SettingsDialog settingsDialog = new SettingsDialog(timeZoneManager);
        settingsDialog.DataContext = timeZoneClock.DataContext;

        // Create Popup with SettingsDialog child
        Popup popup = new Popup
        {
            Child = settingsDialog,
            IsLightDismissEnabled = true
        };

        settingsDialog.SizeChanged += (sender, args) =>
            {
                // Get clock center
                Point position = new Point(timeZoneClock.ActualWidth / 2,
                                           timeZoneClock.ActualHeight / 2);

                // Convert to Page coordinates
                position = timeZoneClock.TransformToVisual(this).TransformPoint(position);

                // Position popup so lower-left or lower-right corner
                // aligns with center of edited clock
                if  (position.X > this.ActualWidth / 2)
                    position.X -= settingsDialog.ActualWidth;
                position.Y -= settingsDialog.ActualHeight;

                // Adjust for size of page
                if  (position.X + settingsDialog.ActualWidth > this.ActualWidth)
                    position.X = this.ActualWidth - settingsDialog.ActualWidth;

                if  (position.X < 0)
                    position.X = 0;

                if  (position.Y < 0)
                    position.Y = 0;

                // Set the Popup position
                popup.HorizontalOffset = position.X;
                popup.VerticalOffset = position.Y;
            };
```

```
        popup.IsOpen = true;
    }
```

以下为 SettingsDialog 效果。第一个字段为 EditBox，允许你在标签进行输入；其他三个字段使用 ComboBox 来显示当前选中的菜单项。如果控件接收到输入焦点，ComboBox 就会打开并显示项目列表：

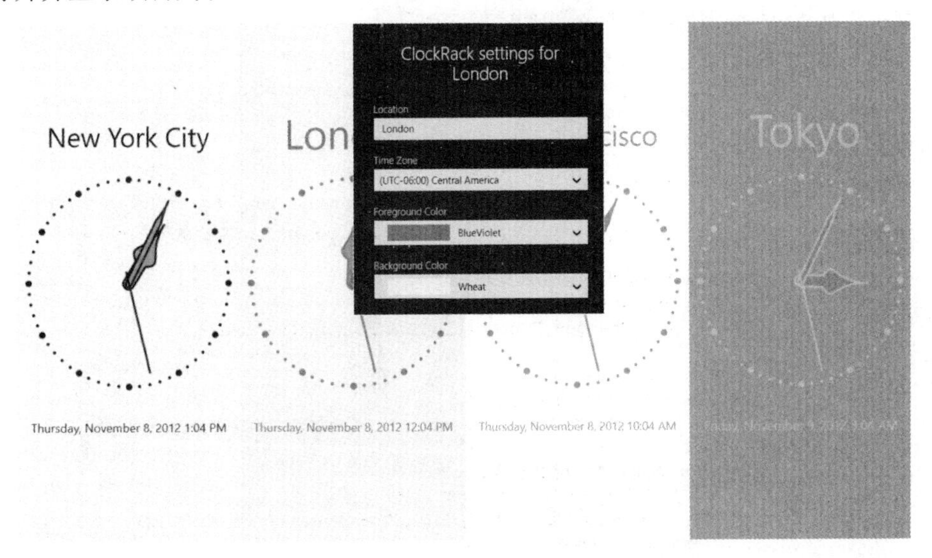

SettingsDialog 对象的 DataContext 设定为正在编辑的 TimeZoneClock 的 DataContext。DataContext 属性是 TimeZoneClockViewModel 的对象，而 XAML 文件已经绑定到该类的 Location、TimeZoneKey、ForegroundName 和 BackgroundName 属性。请注意，绑定到 Location 属性的 TextBox 会设置其 TextChanged 事件，可以使代码隐藏文件“手动”更新 TimeZoneClockViewModel 的 Location 属性，然后更新弹窗的顶部内容。

项目：ClockRack | 文件：SettingsDialog.xaml(片段)

```
<UserControl ... >
    <UserControl.Resources>
        <Style x:Key="DialogCaptionTextStyle"
               TargetType="TextBlock"
               BasedOn="{StaticResource CaptionTextStyle}">
            <Setter Property="FontSize" Value="14.67" />
            <Setter Property="FontWeight" Value="SemiLight" />
            <Setter Property="Margin" Value="0 16 0 8" />
        </Style>

        <DataTemplate x:Key="colorItemTemplate">
            <!-- Item is SettingsDialog.ColorItem -->
            <StackPanel Orientation="Horizontal">
                <Rectangle Width="96" Height="24" Margin="12 6">
                    <Rectangle.Fill>
                        <SolidColorBrush Color="{Binding Color}" />
                    </Rectangle.Fill>
                </Rectangle>

                <TextBlock Text="{Binding Name}"
                           VerticalAlignment="Center" />
            </StackPanel>
        </DataTemplate>
    </UserControl.Resources>
```

```
        <!-- DataContext is TimeZoneClockViewModel -->
        <Border Background="{StaticResource ApplicationPageBackgroundThemeBrush}"
                BorderBrush="{StaticResource ApplicationForegroundThemeBrush}"
                BorderThickness="1"
                Padding="7 0 0 0"
                Width="384">
            <StackPanel Margin="24">
                <TextBlock Text="ClockRack settings for"
                           Style="{StaticResource SubheaderTextStyle}"
                           TextAlignment="Center" />

                <TextBlock Text="{Binding Location}"
                           Style="{StaticResource SubheaderTextStyle}"
                           TextAlignment="Center"
                           Margin="0 0 0 12" />

                <!-- Location -->
                <TextBlock Text="Location"
                            Style="{StaticResource DialogCaptionTextStyle}" />

                <TextBox Name="locationTextBox"
                         Text="{Binding Location}"
                         TextChanged="OnLocationTextBoxTextChanged" />

                <!-- Time Zone -->
                <TextBlock Text="Time Zone"
                           Style="{StaticResource DialogCaptionTextStyle}" />

                <ComboBox Name="timeZoneComboBox"
                          SelectedValuePath="TimeZoneKey"
                          SelectedValue="{Binding TimeZoneKey, Mode=TwoWay}">
                    <ComboBox.ItemTemplate>
                        <!-- Data is TimeZoneDisplayInfo -->
                        <DataTemplate>
                            <TextBlock Text="{Binding Display}" />
                        </DataTemplate>
                    </ComboBox.ItemTemplate>
                </ComboBox>

                <!-- Foreground and Background Colors -->
                <TextBlock Text="Foreground Color"
                            Style="{StaticResource DialogCaptionTextStyle}" />

                <ComboBox  Name="foregroundComboBox"
                          ItemTemplate="{StaticResource colorItemTemplate}"
                          SelectedValuePath="Name"
                          SelectedValue="{Binding ForegroundName, Mode=TwoWay}" />

                <TextBlock Text="Background Color"
                           Style="{StaticResource DialogCaptionTextStyle}" />

                <ComboBox  Name="backgroundComboBox"
                          ItemTemplate="{StaticResource colorItemTemplate}"
                          SelectedValuePath="Name"
                          SelectedValue="{Binding BackgroundName, Mode=TwoWay}" />
            </StackPanel>
        </Border>
    </UserControl>
```

代码隐藏文件(如下所示)为三个 ComboBox 控件提供集合。时区的 ComboBox 由 TimeZoneManager 的 DisplayInformation 属性填充，而第一个 ComboBox 的标记引用了 TimeZoneKey 和 Display 属性。

我已经对 UniformGrid 使用了 Petzold.ProgrammingWindows6.Chapter11 库，因此，我决定使用 NamedColor 类来获得 NamedColor 对象集合。正如 XAML 文件所示，用于两个

ComboBox 控件的 ItemTemplat 引用 NamedColor 的 Color 和 Name 属性，而每个 ComboBox 则表明 SelectedValuePath 为 Name 属性。

```
项目: ClockRack | 文件: SettingsDialog.xaml.cs(片段)
public sealed partial class SettingsDialog : UserControl
{
    public SettingsDialog(TimeZoneManager timeZoneManager)
    {
        this.InitializeComponent();

        // Set ItemsSource for time zone ComboBox
        timeZoneComboBox.ItemsSource = timeZoneManager.DisplayInformation;

        // Set ItemsSource for foreground and background ComboBoxes
        foregroundComboBox.ItemsSource = NamedColor.All;
        backgroundComboBox.ItemsSource = NamedColor.All;
    }

    void OnLocationTextBoxTextChanged(object sender, TextChangedEventArgs args)
    {
        (this.DataContext as TimeZoneClockViewModel).Location = (sender as TextBox).Text;
    }
}
```

ClockRack 就这么多代码。

15.4　DirectX 的 Windows Runtime Component 封装器

P/Invoke 能够很好地用于调用 Win32 API 中的各种函数，而处理 DirectX 却是另一回事。DirectX 对 P/Invoke 有点不合适，最好通过 C++代码来访问。如果想从 C#程序使用 DirectX，可以用 C++写一个包含所有 DirectX 代码的 Windows Runtime Component 并从它访问。对于 DirectX 的一些小领域，你自己就可以搞定，也可以寻求更广泛的解决方案，例如 http://code.google.com/p/sharpdx 所提供的开源 SharpDX 库。

然而，你可能会觉得通过封装库用 C#程序来访问 DirectX 是在自欺欺人。使用 DirectX 的一个原因是出于性能，并且常常涉及 DirectX 库本身的性能(独立于使用 DirectX 库的语言)和应用代码的性能。如果用 C++而不是用 C#写代码，应用代码通常运行得更快(即使用 C#代码可能会写得更快，错误更少)。因此，你可能想用 C++来写一些或全部 DirectX 应用代码。

我下面要展示的 DirectXWrapper 库绝对少见。我刻意将其限定义为三项具体工作：获取安装在系统上的字型列表、获取特定字型的字型规格以及在 SurfaceImageSource 对象上画线条。SurfaceImageSource 实际上是一个位图，只不过没有我在第 14 章中实现的画线算法。

为了在 Visual Studio 中创建该库，我做了一个新的解决方案及项目，名为 DirectXWrapper。在 New Project 对话框的左侧列表中，我将其名称指定为 C++ Windows Store 项目。我在对话框中心区所选的模板是 Windows Runtime Component。这是一个 Windows 8 库，可以用一种语言进行编码(本例中为 C++)，也可以通过其他 Windows Store 应用进行访问，包括用 C#、Visual Basic 和 JavaScript 程序。由于这种灵活性，Windows Runtime Component 有非常严格的限制。Windows Runtime Component 无法执行语言之外的事情。

Windows Runtime Component 最显著的限制如下。

- 公共类必须密封或非实例化。
- 公共方法的参数和返回值必须为 Windows Runtime 类型。
- 公共 C++类和结构必须定义为 ref(意为引用计数)。
- 结构的公共成员限于字段。

Windows 8 帮助文档描述了其他限制。搜索“Creating Windows Runtime Components”可以了解更多。

DirectXWrapper 项目需要引用一些默认不包含的 C++库。在 Solution Explorer 中，我用鼠标右键点击项目名称，并选择 Properties，随后出现一个标题为 DirectXWrapper Property Pages 的对话框。对话框顶部是一个 Platform 组合框。我选择 All Platforms。在对话框左侧，我选择 Configuration Properties、Linker 和 Input.。在所得项列表的顶部是一个 Additional Dependencies 的字段。单击该字段，然后选择 Edit。你随后会看到一个 Additional Dependencies 对话框。我在列表中加入 3 个 DirectX 库：

- d2d1 .lib
- d3d11 .lib
- dwrite .lib

前两个库用于 2D 和 3D 图形，在 SurfaceImageSource 上绘图是要求两者都有的。第三个库用于 DirectWrite。

DirectXWrapper 库还要求访问和这些库相关的一些头文件。在 pch.h(“预编译头文件”)文件中，我包括了必要的几个头文件。

```
项目: DirectXWrapper | 文件: pch.h
#pragma once

#include <wrl.h>
#include <d2d1_1.h>
#include <d3d11_1.h>
#include <dwrite.h>
#include <windows.ui.xaml.media.dxinterop.h>
```

wrl.h 头文件代表 Windows Runtime Library 并包含对处理 Windows 8 应用 COM 有用的定义。windows.ui.xaml.media.dxinterop.h 头文件有一个声明，声明使用 SurfaceImageSource 类所需的 ISurfaceImageSourceNative 接口。

15.5 DirectWrite 和字型

DirectWrite 是专用于文本的高性能显示的 DirectX 子集。即使你不需要这种高性能显示，DirectWrite 也提供了 Windows Runtime 缺少的一些工具，特别值得一提的是能获取已安装的字型列表和字型规格。

为了访问 DirectWrite，我决定在 DirectXWrapper 库中通过 DirectWrite 接口一一对应的方法来定义类。这些接口都以 IDWrite 开头：I 代表 interface，DWRITE 代表 DirectWrite。我所对应的类直接以 Write 开头。

一开始会有一点乱，但我要依次讨论下表所示的对应关系。

DirectWrite 接口	DirectXWrapper 类
IDWriteFactory	WriteFactory
IDWriteFontCollection	WriteFontCollection
IDWriteFontFamily	WriteFontFamily
IDWriteFont	WriteFont
IDWriteLocalizedString	WriteLocalizedStrings

在很多情况下，DirectWrite 接口中的方法名称(例如，IDWriteFont 中的 GetMetrics 方法)都可以直接进行复制：WriteFont 类也有 GetMetrics 方法。我并不想复制接口中的所有方法。

为了使用 DirectWrite，程序一开始会调用 DWriteCreateFactory 函数来获取 IDWriteFactory 类型的对象。在许多其他方法中，IDWriteFactory 接口会定义 GetSystemFontCollection 来获取系统当前所安装的字型。

我在我自己的 WriteFactory 类中封装了 IDWriteFactory。如下为 C++头文件。

项目: DirectXWrapper | 文件: WriteFactory.h

```
#pragma once

#include "WriteFontCollection.h"

namespace DirectXWrapper
{
    public ref class WriteFactory sealed
    {
    private:
        Microsoft::WRL::ComPtr<IDWriteFactory> pFactory;

    public:
        WriteFactory();
        WriteFontCollection^ GetSystemFontCollection();
        WriteFontCollection^ GetSystemFontCollection(bool checkForUpdates);
    };
}
```

WriteFactory 类用 ref 和 sealed 定义，在 Windows Runtime Component 中需要用于公共 C++类。ref 表示该类必须用 ref new 初始化，而不仅仅是用 new，构造函数返回引用计数句柄，而不是返回指针。

从 DWriteCreateFactory 获得的 IDWriteFactory 对象存储为 ComPtr 私有字段，ComPtr 在 Microsoft.Wrl 命名空间中进行定义(或在使 C++语法的 Microsoft::WRL 命名空间)。ComPtr 是“Common Object Model pointer”的简称(把指针指向 COM 对象)，比如 IDWriteFactory 变成“智能指针”，也就是引用计数，并且适当释放自己的资源。在 Windows 8 的 DirectX 代码中，推荐采用这种方式来维持指向 COM 对象的指针。

这三个公共方法也在头文件中进行定义：一个构造函数和两个版本的 GetSystemFontCollection 方法。这些方法返回 WriteFontCollection 对象。返回的对象不是 DirectWrite 类型。不能是 DirectWrite 类型，因为 Windows Runtime Component 中的公共方法只能返回 Windows Runtime 类型。返回的对象是 DirectXWrapper 库的另一个类。(^)表示 WriteFontCollection 是句柄，而不是指针，也就是说，也用 ref 关键字进行定义，在 C++中用 ref new，而不只是用 new 进行初始化。

头文件中提到的 WriteFontCollection 类要求在开头加入 WriteFontCollection.h 头文件。

实现 WriteFactory 类是在 WriteFactory.cpp 文件中进行的。

项目: DirectXWrapper | 文件: WriteFactory.cpp

```
#include "pch.h"
#include "WriteFactory.h"

using namespace DirectXWrapper;
using namespace Platform;
using namespace Microsoft::WRL;

WriteFactory::WriteFactory()
{
    HRESULT hr = DWriteCreateFactory(DWRITE_FACTORY_TYPE_SHARED,
                                     __uuidof(IDWriteFactory),
                                     &pFactory);

    if (!SUCCEEDED(hr))
        throw ref new COMException(hr);
}

WriteFontCollection^ WriteFactory::GetSystemFontCollection()
{
    return GetSystemFontCollection(false);
}

WriteFontCollection^ WriteFactory::GetSystemFontCollection(bool checkForUpdates)
{
    ComPtr<IDWriteFontCollection> pFontCollection;

    HRESULT hr = pFactory->GetSystemFontCollection(&pFontCollection, checkForUpdates);

    If (!SUCCEEDED(hr))
        throw ref new COMException(hr);

    return ref new WriteFontCollection(pFontCollection);
}
```

构造函数调用 DWriteCreateFactory 函数以获取 IDWriteFactory 对象。uuidof 操作符获得标识此对象的 GUID。DirectX 函数和方法很多时候返回的都是 HRESULT 类型的值。该值就是一个表示成功或失败的数字，但很重要，不可忽略。如果发生错误，Windows 8 程序的标准方法会抛出 COMException 类型的异常。请注意用于初始化 COMException 类的 ref new，它其为 Windows Runtime 类型。

WriteFactory 类中的 GetSystemFontCollection 方法使用 IDWriteFactory 对象来调用该接口的 GetSystemFontCollection 方法，以获得指向 DirectWrite IDWriteFontCollection 接口的指针。指针传递到 WriteFontCollection 构造函数。请再次注意 ref new。

以下为 WriteFontCollection 头文件。

项目: DirectXWrapper | 文件: WriteFontCollection.h

```
#pragma once

#include "WriteFontFamily.h"

namespace DirectXWrapper
{
    public ref class WriteFontCollection sealed
    {
    private:
        Microsoft::WRL::ComPtr<IDWriteFontCollection> pFontCollection;
```

```
    internal:
        WriteFontCollection(Microsoft::WRL::ComPtr<IDWriteFontCollection> pFontCollection);

    public:
        bool FindFamilyName(Platform::String^ familyName, int * index);
        int GetFontFamilyCount();
        WriteFontFamily^ GetFontFamily(int index);
    };
}
```

构造函数定义为只限于库内部。不能为 private，因为不能从类的外部访问该构造函数(WriteFontFactory 类显然需要调用该函数)。但也不能为 public，因为构造函数的参数并不是 Windows Runtime 类型。还要注意对 Platform 命名空间定义的 String 类的使用。String 类是 Windows Runtime 类型，等同于在 System 命名空间定义的 C# String 类。

WriteFontCollection 的实现如下所示。

```
项目: DirectXWrapper | 文件: WriteFontCollection.cpp
#include "pch.h"
#include "WriteFontCollection.h"
#include "WriteFontFamily.h"

using namespace DirectXWrapper;
using namespace Platform;
using namespace Microsoft::WRL;

WriteFontCollection::WriteFontCollection(ComPtr<IDWriteFontCollection> pFontCollection)
{
    this->pFontCollection = pFontCollection;
}

bool WriteFontCollection::FindFamilyName(String^ familyName, int * index)
{
    uint32 familyIndex;
    BOOL exists;
    HRESULT hr = this->pFontCollection->FindFamilyName(familyName->Data(), &familyIndex, &exists);

    if (!SUCCEEDED(hr))
        throw ref new COMException(hr);

    *index = familyIndex;

    return exists != 0;
}

int WriteFontCollection::GetFontFamilyCount()
{
    return pFontCollection->GetFontFamilyCount();
}

WriteFontFamily^ WriteFontCollection::GetFontFamily(int index)
{
    ComPtr<IDWriteFontFamily> pfontFamily;

    HRESULT hr = pFontCollection->GetFontFamily(index, &pfontFamily);

    if (!SUCCEEDED(hr))
        throw ref new COMException(hr);

    return ref new WriteFontFamily(pfontFamily);
}
```

从该集合获取特定字型的过程分为两步。首先，必须调用有特定名称(如 Times New Roman)调用 FindFamilyName，以获得集合中的索引。索引随后传递到 GetFontFamily，以得到 IDWriteFontFamily 对象(如果使用 DirectWrite)或 WriteFontFamily 对象(如果使用 DirectXWrapper 库)。

另一种方法是使用传递给 GetFontFamily 的索引枚举集合中的所有字型，不超过 GetFontFamilyCount 的返回值。

以下为 WriteFontFamily 头文件。

```
项目: DirectXWrapper | 文件: WriteFontFamily.h
#pragma once

#include "WriteLocalizedStrings.h"
#include "WriteFont.h"

namespace DirectXWrapper
{
    public ref class WriteFontFamily sealed
    {
    private:
        Microsoft::WRL::ComPtr<IDWriteFontFamily> pFontFamily;

    internal:
        WriteFontFamily(Microsoft::WRL::ComPtr<IDWriteFontFamily> pFontFamily);

    public:
        WriteLocalizedStrings^ GetFamilyNames();
        WriteFont^ GetFirstMatchingFont(Windows::UI::Text::FontWeight fontWeight,
                                        Windows::UI::Text::FontStretch fontStretch,
                                        Windows::UI::Text::FontStyle fontStyle);
    };
}
```

看一看 GetFirstMatchingFont 参数，它均为 Windows Runtime 类型，因为都是在 Windows.UI.Text 命名空间中定义的。FontWeight 是一种结构，为 FontWeights 类中的静态属性类型，而 FontStretch 和 FontStyle 均为枚举。在 IDWriteFontFamily 接口所实施的 GetFirstMatchingFont 方法中，参数类型为 DWRITE_FONT_WEIGHT、DWRITE_FONT_STRETCH 和 DWRITE_FONT_STYLE，均为枚举。有趣的是，FontStretch 和 FontStyle 值可以直接转换，两个枚举有相同的值，表明 DirectWrite 是 Windows Runtime 的文本输出基础。

```
项目: DirectXWrapper | 文件: WriteFontFamily.cpp
#include "pch.h"
#include "WriteFontFamily.h"

using namespace DirectXWrapper;
using namespace Platform;
using namespace Microsoft::WRL;
using namespace Windows::UI::Text;

WriteFontFamily::WriteFontFamily(ComPtr<IDWriteFontFamily> pFontFamily)
{
    this->pFontFamily = pFontFamily;
}

WriteLocalizedStrings^ WriteFontFamily::GetFamilyNames()
{
    ComPtr<IDWriteLocalizedStrings> pFamilyNames;

    HRESULT hr = pFontFamily->GetFamilyNames(&pFamilyNames);

    if (!SUCCEEDED(hr))
        throw ref new COMException(hr);
```

```
    return ref new WriteLocalizedStrings(pFamilyNames);
}

WriteFont^ WriteFontFamily::GetFirstMatchingFont(FontWeight fontWeight,
                                                  FontStretch fontStretch,
                                                  FontStyle fontStyle)
{
    // Convert font weight from Windows Runtime to DirectX
    DWRITE_FONT_WEIGHT writeFontWeight = DWRITE_FONT_WEIGHT_NORMAL;

    if (fontWeight.Equals(FontWeights::Black))
        writeFontWeight = DWRITE_FONT_WEIGHT_BLACK;

    else if (fontWeight.Equals(FontWeights::Bold))
        writeFontWeight = DWRITE_FONT_WEIGHT_BOLD;

    else if (fontWeight.Equals(FontWeights::ExtraBlack))
        writeFontWeight = DWRITE_FONT_WEIGHT_EXTRA_BLACK;

    else if (fontWeight.Equals(FontWeights::ExtraBold))
        writeFontWeight = DWRITE_FONT_WEIGHT_EXTRA_BOLD;

    else if (fontWeight.Equals(FontWeights::ExtraLight))
        writeFontWeight = DWRITE_FONT_WEIGHT_EXTRA_LIGHT;

    else if (fontWeight.Equals(FontWeights::Light))
        writeFontWeight = DWRITE_FONT_WEIGHT_LIGHT;

    else if (fontWeight.Equals(FontWeights::Medium))
        writeFontWeight = DWRITE_FONT_WEIGHT_MEDIUM;

    else if (fontWeight.Equals(FontWeights::Normal))
        writeFontWeight = DWRITE_FONT_WEIGHT_NORMAL;

    else if (fontWeight.Equals(FontWeights::SemiBold))
        writeFontWeight = DWRITE_FONT_WEIGHT_SEMI_BOLD;

    else if (fontWeight.Equals(FontWeights::SemiLight))
        writeFontWeight = DWRITE_FONT_WEIGHT_SEMI_LIGHT;

    else if (fontWeight.Equals(FontWeights::Thin))
        writeFontWeight = DWRITE_FONT_WEIGHT_THIN;

    // Convert font stretch from Windows Runtime to DirectX
    DWRITE_FONT_STRETCH writeFontStretch = (DWRITE_FONT_STRETCH)fontStretch;

    // Convert font style from Windows Runtime to DirectX
    DWRITE_FONT_STYLE writeFontStyle = (DWRITE_FONT_STYLE)fontStyle;

    ComPtr<IDWriteFont> pWriteFont = nullptr;
    HRESULT hr = pFontFamily->GetFirstMatchingFont(writeFontWeight,
                                                    writeFontStretch,
                                                    writeFontStyle,
                                                    &pWriteFont);

    if (!SUCCEEDED(hr))
        throw ref new COMException(hr);

    return ref new WriteFont(pWriteFont);
}
```

字型家族通常有一个名称，比如 Times New Roman，但在 DirectWrite 中，字型名称则有针对不同地点和语言的若干名称。GetFamilyNames 方法返回的不是一个名称，而是存储在 IDWriteLocalizedStrings 中的名称集合。这些字符串是由标准地点名称进行识别，例如，EN-US 代表美式英语。

项目: DirectXWrapper | 文件: WriteLocalizedStrings.h

```
#pragma once

namespace DirectXWrapper
{
    public ref class WriteLocalizedStrings sealed
    {
    private:
        Microsoft::WRL::ComPtr<IDWriteLocalizedStrings> pLocalizedStrings;

    internal:
        WriteLocalizedStrings(Microsoft::WRL::ComPtr<IDWriteLocalizedStrings>
                                                                    pLocalizedStrings);

    public:
        int GetCount();
        Platform::String^ GetLocaleName(int index);
        Platform::String^ GetString(int index);
        bool FindLocaleName(Platform::String^ localeName, int * index);
    };
}
```

实现如下所示。

项目: DirectXWrapper | 文件: WriteLocalizedStrings.cpp

```
#include "pch.h"
#include "WriteLocalizedStrings.h"

using namespace DirectXWrapper;
using namespace Platform;
using namespace Microsoft::WRL;

WriteLocalizedStrings::WriteLocalizedStrings(ComPtr<IDWriteLocalizedStrings> pLocalizedStrings)
{
    this->pLocalizedStrings = pLocalizedStrings;
}

int WriteLocalizedStrings::GetCount()
{
    return this->pLocalizedStrings->GetCount();
}

String^ WriteLocalizedStrings::GetLocaleName(int index)
{
    UINT32 length = 0;
    HRESULT hr = this->pLocalizedStrings->GetLocaleNameLength(index, &length);

    if (!SUCCEEDED(hr))
        throw ref new COMException(hr);

    wchar_t* str = new (std::nothrow) wchar_t[length + 1];

    if (str == nullptr)
        throw ref new COMException(E_OUTOFMEMORY);

    hr = this->pLocalizedStrings->GetLocaleName(index, str, length + 1);

    if (!SUCCEEDED(hr))
        throw ref new COMException(hr);

    String^ string = ref new String(str);
    delete[] str;
    return string;
}

String^ WriteLocalizedStrings::GetString(int index)
{
```

```
    UINT32 length = 0;
    HRESULT hr = this->pLocalizedStrings->GetStringLength(index, &length);

    if (!SUCCEEDED(hr))
        throw ref new COMException(hr);

    wchar_t* str = new (std::nothrow) wchar_t[length + 1];

    if (str == nullptr)
        throw ref new COMException(E_OUTOFMEMORY);

    hr = this->pLocalizedStrings->GetString(index, str, length + 1);

    if (!SUCCEEDED(hr))
        throw ref new COMException(hr);

    String^ string = ref new String(str);
    delete[] str;
    return string;
}

bool WriteLocalizedStrings::FindLocaleName(String^ localeName, int * index)
{
    uint32 localeIndex = 0;
    BOOL exists = false;
    RESULT hr = this->pLocalizedStrings->FindLocaleName(localeName->Data(),
                                                  &localeIndex, &exists);

    if (!SUCCEEDED(hr))
        throw ref new COMException(hr);

    *index = localeIndex;

    return exists != 0;
}
```

代码中的大部分混乱涉及分配 C++字符串(真正的字符数组)来调用 DirectWrite 方法，并将其转换为从 DirectXWrapper 实现返回的 Windows Runtime String 对象。

以下为 WriteFont 头文件。

项目：DirectXWrapper | 文件：WriteFont.h

```
#pragma once

#include "WriteFontMetrics.h"

namespace DirectXWrapper
{
    public ref class WriteFont sealed
    {
    private:
        Microsoft::WRL::ComPtr<IDWriteFont> pWriteFont;

    internal:
        WriteFont(Microsoft::WRL::ComPtr<IDWriteFont> pWriteFont);

    public:
        bool HasCharacter(UINT32 unicodeValue);
        bool IsSymbolFont();
        WriteFontMetrics GetMetrics();
    };
}
```

实现如下所示。

项目：DirectXWrapper | 文件：WriteFont.cpp

```
#include "pch.h"
```

```
#include "WriteFont.h"

using namespace DirectXWrapper;
using namespace Platform;
using namespace Microsoft::WRL;

WriteFont::WriteFont(ComPtr<IDWriteFont> pWriteFont)
{
    this->pWriteFont = pWriteFont;
}

WriteFontMetrics WriteFont::GetMetrics()
{
    DWRITE_FONT_METRICS fontMetrics;
    this->pWriteFont->GetMetrics(&fontMetrics);

    WriteFontMetrics writeFontMetrics =
    {
        fontMetrics.designUnitsPerEm,
        fontMetrics.ascent,
        fontMetrics.descent,
        fontMetrics.lineGap,
        fontMetrics.capHeight,
        fontMetrics.xHeight,
        fontMetrics.underlinePosition,
        fontMetrics.underlineThickness,
        fontMetrics.strikethroughPosition,
        fontMetrics.strikethroughThickness
    };

    return writeFontMetrics;
}

bool WriteFont::HasCharacter(UINT32 unicodeValue)
{
    BOOL exists = 0;
    HRESULT hr = this->pWriteFont->HasCharacter(unicodeValue, &exists);

    if (!SUCCEEDED(hr))
        throw ref new COMException(hr);

    return exists != 0;
}

bool WriteFont::IsSymbolFont()
{
    return this->pWriteFont->IsSymbolFont() != 0;
}
```

GetMetrics 方法的 DirectWrite 版本填充 DWRITE_FONT_METRICS 类型结构。当然，Windows Runtime Component 不能直接返回该结构，因此我自己定义了这个结构。

项目：DirectXWrapper | 文件：WriteFontMetrics.h

```
#pragma once

namespace DirectXWrapper
{
    public value struct WriteFontMetrics
    {
        UINT16 DesignUnitsPerEm;
        UINT16 Ascent;
        UINT16 Descent;
        INT16 LineGap;
        UINT16 CapHeight;
        UINT16 XHeight;
        INT16 UnderlinePosition;
        UINT16 UnderlineThickness;
```

```
        INT16 StrikethroughPosition;
        UINT16 StrikethroughThickness;
    };
}
```

前面就是 DirectXWrapper 库中实现 DirectWrite 的所有代码。显然，可供 DirectWrite 使用的多于我试图提供给 C#程序的，但我现在有了两项基本工作所需要的东西。

我们来枚举已安装字型。EnumerateFonts 项目是一个普通的 Windows 8 C#项目，只不过在解决方案资源管理器中，我是在用鼠标右键点击解决方案名称，并选择 Add an Existing Project。我添加的项目为 DirectXWrapper。和通常一样，当调用库项目的时候，我也右键单击 EnumerateFonts 中的 References 部分，选择 Add Reference，并在 Add Reference 对话框中选择左侧的 Projects 和 DirectXWrapper。

EnumerateFonts 的 XAML 文件包含 ListBox。

```
项目: EnumerateFonts | 文件: MainPage.xaml(片段)
<Grid Background="{StaticResource ApplicationPageBackgroundThemeBrush}">
    <ListBox Name="lstbox">
        <ListBox.ItemTemplate>
            <DataTemplate>
                <TextBlock Text="{Binding}"
                           FontFamily="{Binding}"
                           FontSize="24" />
            </DataTemplate>
        </ListBox.ItemTemplate>
    </ListBox>
</Grid>
```

显然，ItemTemplate 预计 ListBox 会填满字型家族名称。每个名称都用基于字型库里的字型进行显示。ListBox 通过代码隐藏文件的构造函数进行填充。

```
项目: EnumerateFonts | 文件: MainPage.xaml.cs
using Windows.UI.Xaml.Controls;
using DirectXWrapper;

namespace EnumerateFonts
{
    public sealed partial class MainPage : Page
    {
        public MainPage()
        {
            this.InitializeComponent();

            WriteFactory writeFactory = new WriteFactory();
            WriteFontCollection writeFontCollection =
                        writeFactory.GetSystemFontCollection();

            int count = writeFontCollection.GetFontFamilyCount();
            string[] fonts = new string[count];

            for (int i = 0; i < count; i++)
            {
                WriteFontFamily writeFontFamily =
                        writeFontCollection.GetFontFamily(i);

                WriteLocalizedStrings writeLocalizedStrings =
                        writeFontFamily.GetFamilyNames();
                int index;

                if (writeLocalizedStrings.FindLocaleName("en-us", out index))
                {
                    fonts[i] = writeLocalizedStrings.GetString(index);
                }
```

```
                    else
                    {
                        fonts[i] = writeLocalizedStrings.GetString(0);
                    }
                }
                lstbox.ItemsSource = fonts;
            }
        }
    }
```

正如你所看到的，DirectXWrapper 类及方法就像普通 Windows Runtime 类一样访问及使用。程序试图找到名为“EN-US”地域的字型；如果没有，则获取集合中的第一个。而实际上，许多 Windows 8 字型都只有一个名称，但一些为远东语言设计的字型则有另外的中文、韩文和日文名称。

下图所示的这些字体可能你自己的系统上也有。

SimSun-ExtB
KodchiangUPC
Kokila
Shonar Bangla
Mangal
BrowalliaUPC
Sakkal Majalla
LilyUPC
Palatino Linotype
MoolBoran
Franklin Gothic
Cordia New
Arial
AngsanaUPC
JasmineUPC

15.6 配置和平台

Visual Studio 的标准工具栏包括两个下拉组合框，分别名为 Solution Configurations 和 Solution Platforms。

Solution Configurations 有以下三个选项：

- Debug
- Release
- Configuration Manager

前两项允许你以两种不同的方式来编译程序，而在正常情况下，你会在程序开发过程中使用 Debug 配置。但如果已经完成大部分调试工作，就可以切换到 Release 配置以获得更好的代码优化和性能。

在本书中，EnumerateFonts 之前所有项目的 Solution Platforms 均显示五个选项：

- Any CPU
- ARM

- x64
- x86
- Configuration Manager

在本书中，EnumerateFonts 之前的所有项目，Solution Platforms 可能已经为了 C#项目显示了默认选项，即 Any CPU。

本来就应该是这样。在 Visual Studio 中编译 C#程序时，源代码编译成中间语言(Intermediate Language 或称 IL)。在程序运行时，IL 编译成适合相应处理器的原生代码。这是使用 C#类似托管语言的一大优势：分布式可执行文件由独立于相应处理器的中间语言组成。即使程序使用 P/Invoke 来访问 Win32 函数也这样。

通过 C#项目，可以使用 Solution Platforms 组合框切换到 ARM、x64 或 x86 平台。编译器仍然会产生中间语言，但可执行文件只能在特定处理器上运行。如果指定 ARM，程序只能在用 ARM 处理器的机器上运行。如果指定 X64，程序只能在 64 位 Intel 处理器上运行。如果指定 X86，程序可以在 32 位和 64 位 Intel 处理器上运行。

一般情况下，都不想限制 C#程序在特定处理器上运行，除非心须这样做。你想让 Solution Platforms 读取 Any CPU。(如果想让程序在 Intel 和 ARM 处理器中有一些不同，可以使用本章前面介绍的 GetNativeSystemInfo 函数。)

然而，一旦 C++代码开始引入应用，一切都会发生改变。Visual Studio 不会把 C++代码编译成中间语言，而是将其编译为特定处理器的原生机器代码。可执行文件只在该类处理器上运行，为 32 位 Intel 处理器编译的代码也在 64 位 Intel 处理器上运行。

此外，如果有一个多项目应用包括 C#代码和 C++代码(如 EnumerateFonts 解决方案)，则多个项目的平台必须相同，并且必须匹配运行应用的平台。至于 Any CPU，C#项目似乎可以引用 X64 C++项目，但事实并非如此。

要查看单个项目的平台，可以从任一组合框中选择 Configuration Manager 并启用 Configuration Manager 对话框。对于该解决方案中的 C#项目，有以下平台选项：

- Any CPU
- ARM
- x64
- x86

而对于 C++项目，则有以下平台选项：

- ARM
- Win32
- x64

C++ Win32 平台选项等同于 C# x86 平台。

平台选项唯一可能的组合分别如下：

- 在 ARM 处理器上运行的 C# ARM 和 C++ ARM
- 在 Intel 64 位处理器上运行的的 C# x64 和 C++ x64
- 在 Intel 32 位或 64 位处理器上运行的 C# x86 和 C++ Win32

如果从 Solution Platforms 组合框中选择 ARM、X64、X86 或者 Win32，就会得到以上三种组合之一。

对于本书的任何程序，试一试从 Solution Platforms 组合框中选择 ARM，然后按功能键 F5。没有在基于 ARM 的设备上运行 Visual Studio，因此项目生成成功，但不会进行部署，因为 Visual Studio 并没有在 ARM 处理器上运行。

如果在基于 ARM 的机器上运行 Windows 8(比如微软 Surface 初始版本)则可能要在该设备上测试程序。不能在 Surface 上运行 Visual Studio，因此必须通过另一种方式在 Surface 上运行应用。

为了调试和测试，最简单的方式就是远程部署，可以通过 WiFi 网络进行，详情可参见 Tim Heuer 的一篇博客文章 http://timheuer.com/blog/archive/2012/10/26/remote-debugging-windows-store-apps-on-surface-arm-devices.aspx。一旦设置好，Surface 就运行远程调试器，也不是休眠或显示锁屏状态，做出以下两个选择：

- 从 Solution Platform 组合框选择远程机器平台
- 从 Solution Platform 组合框左侧的下拉列表选择 Remote Machine

因为目标机器与平台相关，因此按以上顺序选择两项会有用。如果先选择 Remote Machine，再选择平台，Visual Studio 就会切换到本地计算机。

如果解决方案只包括 C#代码，Solution Platform 组合框则应该为 Any CPU，即不考虑要部署的机器。如果解决方案有一些 C++代码并要部署到基于 ARM 的设备(比如微软 Surface)，则应该从 Solution Platform 组合框选择 ARM。

对于分布式，无论是上传应用到 Windows Store，还是把程序部署到其他机器，都要涉及另一种方法。需要通过从 Store 菜单中选择 Create App Packages，并在 Visual Studio 中创建应用包。

在此过程中，你会看到标题为 Select and Configure Packages 架构表的对话框。如果项目只有 C#代码，可以选择 Neutral 架构，等同于 Any CPU。然而，如果项目有 C++代码，则不能选择 Neutral。必须选择一个或多个其他选项：x64、x86 和 ARM。你可能要选择全部三项，以便部署到任何类型的机器上。

如果不向 Windows Store 上传程序包，而是要安装到另一台计算机上，Visual Studio 就会为你选的各种架构生成目录。每个目录都包含 Windows PowerShell 脚本(扩展名为 ps1 的文件)可以运行用来部署应用。一个方法是把目录复制到 U 盘(或者让 Visual Studio 创建)，将 U 盘插到其他机器(如微软 Surface)上，然后运行脚本在这台机器上安装应用。

15.7 解读字型规格

字型规格指特定字型的字符和字符串大小。在大多数情况下，处理 Windows 8 程序中的文本往往不需要字型规格信息。TextBlock 元素决定特定文本及字型大小，通常这就够了。然而，如果你打算进行复杂文本布局，则有必要有字型规格，有一些不常见任务偶尔会要求字型规格。

我要把讨论限定于垂直规格，讨论高度而非宽度。垂直规格随字型、字型风格(斜体)、字型粗细(粗体)和字型大小而变化，但又独立于任何特定字符或字符串。

LookAtFontMetrics 程序以可视化方式演示了 TextBlock 元素计算所得文本字符串的大小和 DirectWrite 所提供字型规格之间的关系。该项目和 EnumerateFonts 一样都引用了

DirectXWrapper 项目。XAML 文件不仅有相似 ListBox，在 Border 中还包括一些半定义 Line 元素的 TextBlock。

```
项目: LookAtFontMetrics | 文件: MainPage.xaml(片段)
<Grid Background="{StaticResource ApplicationPageBackgroundThemeBrush}">
    <Grid.ColumnDefinitions>
        <ColumnDefinition Width="Auto" />
        <ColumnDefinition Width="*" />
    </Grid.ColumnDefinitions>

    <ListBox Name="lstbox"
             Grid.Column="0"
             Width="300"
             SelectionChanged="OnListBoxSelectionChanged">
        <ListBox.ItemTemplate>
            <DataTemplate>
                <TextBlock Text="{Binding}"
                           FontFamily="{Binding}"
                           FontSize="24" />
            </DataTemplate>
        </ListBox.ItemTemplate>
    </ListBox>

    <Grid Grid.Column="1"
          HorizontalAlignment="Center"
          VerticalAlignment="Center">
        <Border BorderBrush="{StaticResource ApplicationForegroundThemeBrush}"
                BorderThickness="1">
            <Grid>
                <Grid.Resources>
                    <Style TargetType="Line">
                        <Setter Property="Stroke" Value="Red" />
                        <Setter Property="StrokeThickness" Value="2" />
                        <Setter Property="X1" Value="0" />
                    </Style>
                </Grid.Resources>

                <TextBlock Name="txtblk"
                           Text="Texting"
                           FontSize="192"
                           SizeChanged="OnTextBlockSizeChanged" />

                <Line x:Name="ascenderLine" Y1="0" Y2="0" />
                <Line x:Name="capsHeightLine" />
                <Line x:Name="xHeightLine" />
                <Line x:Name="baselineLine" Stroke="Blue" />
                <Line x:Name="descenderLine" />
                <Line x:Name="lineGapLine" />
            </Grid>
        </Border>
    </Grid>
</Grid>
```

程序的构造函数和之前的程序非常相似，两者都用可用字型填充 ListBox。程序还设置了 TextBlock 的 FontFamily 属性并从 DirectXWrapper 获取 WriteFontMetrics 值来处理 ListBox 中的 SelectionChanged 事件。程序使用这些字型规格来设置不同 Line 元素的 Y1 和 Y2 属性。

```
项目: LookAtFontMetrics | 文件: MainPage.xaml.cs(片段)
public sealed partial class MainPage : Page
{
    WriteFactory writeFactory;
    WriteFontCollection writeFontCollection;

    public MainPage()
    {
```

```
        this.InitializeComponent();

        writeFactory = new WriteFactory();
        writeFontCollection = writeFactory.GetSystemFontCollection();
        int count = writeFontCollection.GetFontFamilyCount();
        string[] fonts = new string[count];

        for (int i = 0; i < count; i++)
        {
            WriteFontFamily writeFontFamily = writeFontCollection.GetFontFamily(i);
            WriteLocalizedStrings writeLocalizedStrings = writeFontFamily.GetFamilyNames();
            int nameCount = writeLocalizedStrings.GetCount();
            int index;

            if (writeLocalizedStrings.FindLocaleName("en-us", out index))
            {
                fonts[i] = writeLocalizedStrings.GetString(index);
            }
        }

        lstbox.ItemsSource = fonts;

        Loaded += (sender, args) =>
            {
                lstbox.SelectedIndex = 0;
            };
    }

    void OnListBoxSelectionChanged(object sender, SelectionChangedEventArgs args)
    {
        string fontFamily = (sender as ListBox).SelectedItem as string;

        if (fontFamily == null)
            return;

        txtblk.FontFamily = new FontFamily(fontFamily);
        int index;
        if (writeFontCollection.FindFamilyName(fontFamily, out index))
        {
            WriteFontFamily writeFontFamily = writeFontCollection.GetFontFamily(index);
            WriteFont writeFont = writeFontFamily.GetFirstMatchingFont(FontWeights.Normal,
                                                                  FontStretch.Normal,
                                                                  FontStyle.Normal);

            WriteFontMetrics fontMetrics = writeFont.GetMetrics();
            double fontSize = txtblk.FontSize;
            double ascent = fontSize * fontMetrics.Ascent / fontMetrics.DesignUnitsPerEm;
            double capsHeight = fontSize * fontMetrics.CapHeight / fontMetrics.DesignUnitsPerEm;
            double xHeight = fontSize * fontMetrics.XHeight / fontMetrics.DesignUnitsPerEm;
            double descent = fontSize * fontMetrics.Descent / fontMetrics.DesignUnitsPerEm;
            double lineGap = fontSize * fontMetrics.LineGap / fontMetrics.DesignUnitsPerEm;

            baselineLine.Y1 = baselineLine.Y2 = ascent;
            capsHeightLine.Y1 = capsHeightLine.Y2 = ascent - capsHeight;
            xHeightLine.Y1 = xHeightLine.Y2 = ascent - xHeight;
            descenderLine.Y1 = descenderLine.Y2 = ascent + descent;
            lineGapLine.Y1 = lineGapLine.Y2 = ascent + descent + lineGap;
        }
    }

    void OnTextBlockSizeChanged(object sender, SizeChangedEventArgs args)
    {
        double width = txtblk.ActualWidth;
        ascenderLine.X2 = width;
        capsHeightLine.X2 = width;
        xHeightLine.X2 = width;
        baselineLine.X2 = width;
        descenderLine.X2 = width;
```

```
            lineGapLine.X2 = width;
        }
    }
```

DirectWrite 中的 DWRITE_FONT_METRICS 有一个字段名为 designUnitsPerEm，大多数字型都有一个不错的约整数值，比如 256、1024、2048 或 4096，只不过偶尔会出现 1000 这样的独特值。正如其名，它表示用于定义字型特点的网格高度。结构中的所有其他高度都是相对于该设计高度的。这就是为什么该结构所有字段都是整数的原因。为了能够获得特定字型和字型大小的像素高值，这种结构的字段必须乘以 FontSize 再被 designUnitsPerEm 除。

这就是 LookAtFontMetrics 程序在设置所有 Line 元素的 Y1 和 Y2 属性时所做的事情。对于一些字型，结果看起来显示不完全正确，因为这些字型主要用于非拉丁字母。但对用于拉丁字母语言的标准字型，根据字型规格计算所得的线条十分精确，如下图所示。

字符在其之上的线条(即蓝色线条，如果你正在阅读本书的电子版)为基线。在很多字型中，圆形字符(如“e”)会超过基线下方一点点。从基线往上到下一条线为 x-高度线，也就是小写字母的高度。同样，一些圆体字符略高于 x-高度线。接下来是大写高度线，表明大写字母高度。上行高度线甚至更高(在 TextBlock 自身计算所得的矩形最顶端)并且表明一些字母可能出现的变音符，如元音变音(Ü)。低于基线的区域用于有些字母会下降到基线之下。最底层有一间距，对于许多字型而言都是零。下图使用了原始 DWRITE_FONT_METRICS 结构所定义的名称。

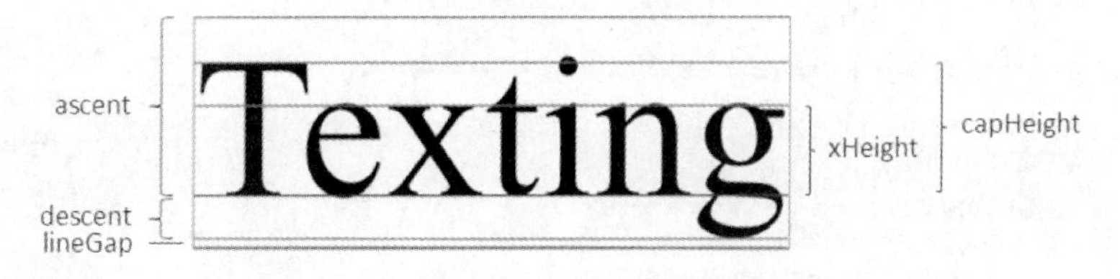

TextBlock 计算自身的高度是基于 ascent、descent 和 lineGap 字段的总和。

第 10 章展示过一个程序，用于显示文本字符串的倾斜阴影(见下图)。

我当时说过，不可能像这样将任意字型做成阴影从基线向后倾斜的效果。你需要知道字型规格，而现在来试一试。

以下是 BaselineTiltedShadow 的 XAML 文件。它有另一个用于系统字型的 ListBox，还有一些文字和阴影。

项目：BaselineTiltedShadow | 文件：MainPage.xaml(片段)

```
<Grid Background="{StaticResource ApplicationPageBackgroundThemeBrush}">
    <Grid.ColumnDefinitions>
        <ColumnDefinition Width="Auto" />
        <ColumnDefinition Width="*" />
    </Grid.ColumnDefinitions>

    <ListBox Name="lstbox"
             Grid.Column="0"
             Width="300"
             SelectionChanged="OnListBoxSelectionChanged">
        <ListBox.ItemTemplate>
            <DataTemplate>
                <TextBlock Text="{Binding}"
                           FontFamily="{Binding}"
                           FontSize="24" />
            </DataTemplate>
        </ListBox.ItemTemplate>
    </ListBox>

    <Grid Grid.Column="1"
          HorizontalAlignment="Center"
          VerticalAlignment="Center">

        <TextBlock Name="shadowTextBlock"
                   Text="shadow"
                   FontSize="192"
                   Foreground="Gray">
            <TextBlock.RenderTransform>
                <CompositeTransform ScaleY="1.5" SkewX="-60" />
            </TextBlock.RenderTransform>
        </TextBlock>

        <TextBlock Name="foregroundTextBlock"
                   Text="shadow"
                   FontSize="192" />
```

```
    </Grid>
</Grid>
```

两个 TextBlock 元素都缺少 fontFamily 属性，从代码隐藏文件中设置为 ListBox 中所选择的字型。阴影部分的 TextBlock 也缺少 RenderTransformOrigin 属性。初始化 ListBox 的代码隐藏文件的构造函数和之前程序的一样；SelectionChanged 属性的重要部分根据基线以上字型高度部分的比例来计算用于阴影的 RenderTransformOrigin。

项目：BaselineTiltedShadow | 文件：MainPage.xaml.cs(片段)

```
void OnListBoxSelectionChanged(object sender, SelectionChangedEventArgs args)
{
    string fontFamily = (sender as ListBox).SelectedItem as string;

    if (fontFamily == null)
        return;

    foregroundTextBlock.FontFamily = new FontFamily(fontFamily);
    shadowTextBlock.FontFamily = foregroundTextBlock.FontFamily;

    int index;
    if (writeFontCollection.FindFamilyName(fontFamily, out index))
    {
        WriteFontFamily writeFontFamily = writeFontCollection.GetFontFamily(index);
        WriteFont writeFont = writeFontFamily.GetFirstMatchingFont(FontWeights.Normal,
                                                                   FontStretch.Normal,
                                                                   FontStyle.Normal);

        WriteFontMetrics fontMetrics = writeFont.GetMetrics();

        double fractionAboveBaseline = (double)fontMetrics.Ascent /
            (fontMetrics.Ascent + fontMetrics.Descent + fontMetrics.LineGap);

        shadowTextBlock.RenderTransformOrigin = new Point(0, fractionAboveBaseline);
    }
}
```

结果如下图所示。

15.8 用 SurfaceImageSource 绘画

DirectXWrapper 另外还有一个类。我称之为 SurfaceImageSourceRenderer，它在结构上不同于封装 DirectWrite 类的库中其他的类。SurfaceImageSourceRenderer 类实例化所有 DirectX 对象并用来提供高层次接口，以在 SurfaceImageSource 类型的对象上绘制线条。

SurfaceImageSource 派生自 ImageSource，因此可以设置为 Image 的 Source 属性或设置为 ImageBrush 的 ImageSource 属性。SurfaceImageSourceRenderer 基本上是位图。但是，可以使用 DirectX 在该位图上绘制图形(或文字)。SurfaceImageSourceRenderer 类执行所有必要系统开销，并公开三个公共方法：Clear、DrawLine 和 Update。很明显，该类可以进一步扩展，包括更多东西。

以下为头文件。

项目：DirectXWrapper | 文件：SurfaceImageSourceRenderer.h

```
#pragma once

namespace DirectXWrapper
{
    public ref class SurfaceImageSourceRenderer sealed
    {
    private:
        int width, height;
        Microsoft::WRL::ComPtr<ID2D1Factory> pFactory;
        Microsoft::WRL::ComPtr<ID3D11Device> pd3dDevice;
        Microsoft::WRL::ComPtr<ID3D11DeviceContext> pd3dContext;
        Microsoft::WRL::ComPtr<ISurfaceImageSourceNative> sisNative;
        Microsoft::WRL::ComPtr<IDXGIDevice> pDxgiDevice;
        Microsoft::WRL::ComPtr<ID2D1BitmapRenderTarget> bitmapRenderTarget;
        Microsoft::WRL::ComPtr<ID2D1Bitmap> bitmap;
        Microsoft::WRL::ComPtr<ID2D1SolidColorBrush> solidColorBrush;
        Microsoft::WRL::ComPtr<ID2D1StrokeStyle> strokeStyle;
        bool needsUpdate;

    public:
        SurfaceImageSourceRenderer(
                Windows::UI::Xaml::Media::Imaging::SurfaceImageSource^ surfaceImageSource,
                int width, int height);
        void Clear(Windows::UI::Color color);
        void DrawLine(Windows::Foundation::Point pt1, Windows::Foundation::Point pt2,
                    Windows::UI::Color color, double thickness);
        void Update();

    private:
        ID2D1RenderTarget * CreateRenderTarget(Microsoft::WRL::ComPtr<IDXGISurface> pSurface);
        D2D1::ColorF ConvertColor(Windows::UI::Color color);
    };
}
```

公共构造函数要求有一个 SurfaceImageSource 类的实例。这在公共构造函数中是允许的，因为它是在 Windows.UI.Xaml.Media.Imaging 命名空间中定义的 Windows Runtime 类型。该构造函数包含非 DirectX 专家也能从头写出来的代码。我自然是写不出来的，因此，我从其他样例项目中提取大部分代码来说明如何使用 SurfaceImageSource。

项目：DirectXWrapper | 文件：SurfaceImageSourceRenderer.cpp(片段)

```
SurfaceImageSourceRenderer::SurfaceImageSourceRenderer(SurfaceImageSource^surfaceImageSource,
                                                        int width, int height)
```

```
{
    // Save the image width and height
    this->width = width;
    this->height = height;

    // Create Factory
    D2D1_FACTORY_OPTIONS options = { D2D1_DEBUG_LEVEL_NONE };

    HRESULT hr = D2D1CreateFactory(D2D1_FACTORY_TYPE_SINGLE_THREADED,
                                   __uuidof(ID2D1Factory),
                                   &options,
                                   &pFactory);

    if (!SUCCEEDED(hr))
        throw ref new COMException(hr);

    // Create ISurfaceImageSourceNative object
    IInspectable* sisInspectable = (IInspectable*)
                        reinterpret_cast<IInspectable*>(surfaceImageSource);
    sisInspectable->QueryInterface(__uuidof(ISurfaceImageSourceNative), (void **)&sisNative);

    // Create Device and Device Context
    D3D_FEATURE_LEVEL featureLevels[] =
    {
        D3D_FEATURE_LEVEL_11_1,
        D3D_FEATURE_LEVEL_11_0,
        D3D_FEATURE_LEVEL_10_1,
        D3D_FEATURE_LEVEL_10_0,
        D3D_FEATURE_LEVEL_9_3,
        D3D_FEATURE_LEVEL_9_2,
        D3D_FEATURE_LEVEL_9_1,
    };

    hr = D3D11CreateDevice (nullptr,
                            D3D_DRIVER_TYPE_HARDWARE,
                            0,
                            D3D11_CREATE_DEVICE_SINGLETHREADED |
                            D3D11_CREATE_DEVICE_BGRA_SUPPORT,
                            featureLevels,
                            ARRAYSIZE(featureLevels),
                            D3D11_SDK_VERSION,
                            &pd3dDevice,
                            nullptr,
                            &pd3dContext);

    if (!SUCCEEDED(hr))
        throw ref new COMException(hr);

    // Get DXGIDevice
    hr = pd3dDevice.As(&pDxgiDevice);

    if (!SUCCEEDED(hr))
        throw ref new COMException(hr);

    sisNative->SetDevice(pDxgiDevice.Get());

    // Begin drawing
    RECT update = { 0, 0, width, height };
    POINT offset;
    IDXGISurface * dxgiSurface;
    hr = sisNative->BeginDraw(update, &dxgiSurface, &offset);

    if (!SUCCEEDED(hr))
        throw ref new COMException(hr);

    ID2D1RenderTarget * pRenderTarget = CreateRenderTarget(dxgiSurface);

    // But only go far enough to create compatible BitmapRenderTarget
```

```
    // and get the Bitmap for updating the surface
    pRenderTarget->CreateCompatibleRenderTarget(&bitmapRenderTarget);
    bitmapRenderTarget->GetBitmap(&bitmap);

    // End drawing
    sisNative->EndDraw();
    pRenderTarget->Release();
    dxgiSurface->Release();

    // Create a SolidColorBrush for drawing lines
    bitmapRenderTarget->CreateSolidColorBrush(D2D1::ColorF(0, 0, 0, 0), &solidColorBrush);

    // Create StrokeStyle for drawing lines
    D2D1_STROKE_STYLE_PROPERTIES strokeStyleProperties =

    {
        D2D1_CAP_STYLE_ROUND,
        D2D1_CAP_STYLE_ROUND,
        D2D1_CAP_STYLE_ROUND,
        D2D1_LINE_JOIN_ROUND,
        10,
        D2D1_DASH_STYLE_SOLID,
        0
    };

    hr = pFactory->CreateStrokeStyle(&strokeStyleProperties, nullptr, 0, &strokeStyle);

    if (!SUCCEEDED(hr))
        throw ref new COMException(hr);
}
```

该构造函数使用了在绘画操作时发挥作用的一个私有方法。该方法在实际绘的画地方创建 ID2D1RenderTarget 类型的对象。

```
项目: DirectXWrapper | 文件: SurfaceImageSourceRenderer.cpp(片段)
ID2D1RenderTarget* SurfaceImageSourceRenderer::CreateRenderTarget(ComPtr<IDXGISurface>pSurface)
{
    D2D1_PIXEL_FORMAT format =
    {
        DXGI_FORMAT_UNKNOWN,
        D2D1_ALPHA_MODE_PREMULTIPLIED
    };

    float dpiX, dpiY;
    pFactory->GetDesktopDpi(&dpiX, &dpiY);

    D2D1_RENDER_TARGET_PROPERTIES properties =
    {
        D2D1_RENDER_TARGET_TYPE_DEFAULT,
        format,
        dpiX,
        dpiY,
        D2D1_RENDER_TARGET_USAGE_NONE,
        D2D1_FEATURE_LEVEL_DEFAULT
    };

    ID2D1RenderTarget * pRenderTarget;
    HRESULT hr = pFactory->CreateDxgiSurfaceRenderTarget(pSurface.Get(),
                                                         &properties, &pRenderTarget);

    if (!SUCCEEDED(hr))
        throw ref new COMException(hr);

    return pRenderTarget;
}
```

请注意，在构造函数和该方法中，我定义了两个指针指向 DirectX 对象(命名为

dxgiSurface 和 pRenderTarget)，而没有使用 ComPtr 封装器。这是因为我只在很短时间内使用这些对象，然后就通过调用 Release 方法手动释放。

Clear 方法主要调用存储为字段的 ID2D1BitmapRenderTarget 对象的 Clear 方法。

```
项目: DirectXWrapper | 文件: SurfaceImageSourceRenderer.cpp(片段)
void SurfaceImageSourceRenderer::Clear(Color color)
{
    bitmapRenderTarget->BeginDraw();
    bitmapRenderTarget->Clear(ConvertColor(color));
    bitmapRenderTarget->EndDraw();
    needsUpdate = true;
}
```

Clear 是公共方法，而且参数必须为 Windows Runtime 类型，因此也正是如此。然而，Windows Runtime 所定义的 Color 结构和 DirectX 中的各种颜色结构并不相同，也就是说，颜色必须通过以下私有方法进行转化。

```
项目: DirectXWrapper | 文件: SurfaceImageSourceRenderer.cpp(片段)
D2D1::ColorF SurfaceImageSourceRenderer::ConvertColor(Color color)
{
    D2D1::ColorF colorf(color.R / 255.0f,
                        color.G / 255.0f,
                        color.B / 255.0f,
                        color.A / 255.0f);
    return colorf;
}
```

点也必须进行转换。我已经定义的公共 DrawLine 方法会渲染两点之间的线条，但该方法会首先把这些 Windows Runtime Point 值转换成 DirectX D2D1_POINT_2F 值，再传递给 ID2D1BitmapRenderTarget 对象的 DrawLine 方法。

```
项目: DirectXWrapper | 文件: SurfaceImageSourceRenderer.cpp(片段)
void SurfaceImageSourceRenderer::DrawLine(Point point1, Point point2,
                                          Color color, double thickness)
{
    // Convert the points
    D2D1_POINT_2F pt1 = { (float)point1.X, (float)point1.Y };
    D2D1_POINT_2F pt2 = { (float)point2.X, (float)point2.Y };

    // Convert the color for the SolidColorBrush
    solidColorBrush->SetColor(ConvertColor(color));

    // Draw the line
    bitmapRenderTarget->BeginDraw();
    bitmapRenderTarget->DrawLine(pt1, pt2, solidColorBrush.Get(),
                                 (float)thickness,
                                 strokeStyle.Get());
    bitmapRenderTarget->EndDraw();
    needsUpdate = true;
}
```

显然，ID2D1BitmapRenderTarget 接口还定义了除 DrawLine 之外的许多其他方法，但如果你要在应用中大量运用这些方法，则更应该至少把应用的一部分移植成 C++。

Clear 和 DrawLine 在 ID2D1BitmapRenderTarget 上进行绘画，而 SurfaceImageSource 对象必须从 ID2D1BitmapRenderTarget 上进行更新。而这发生在 Update 方法中。

```
项目: DirectXWrapper | 文件: SurfaceImageSourceRenderer.cpp(片段)
void SurfaceImageSourceRenderer::Update()
{
    // Check if needs update
```

```
    if (!needsUpdate)
        return;

    needsUpdate = false;

    // Begin drawing
    RECT update = { 0, 0, width, height };
    POINT offset;
    IDXGISurface * dxgiSurface;
    HRESULT hr = sisNative->BeginDraw(update, &dxgiSurface, &offset);

    if (!SUCCEEDED(hr))
        throw ref new COMException(hr);

    ID2D1RenderTarget * renderTarget = CreateRenderTarget(dxgiSurface);
    renderTarget->BeginDraw();

    // Draw the bitmap to the surface
    D2D1_RECT_F rect = { 0, 0, (float)width, (float)height };
    renderTarget->DrawBitmap(bitmap.Get(), &rect);

    // End drawing
    renderTarget->EndDraw();
    sisNative->EndDraw();

    // Release update resources
    renderTarget->Release();
    dxgiSurface->Release();
}
```

SurfaceImageSourceRenderer 源代码也就这么多。SpinPaint 项目还演示了该类。这个程序显示一个旋转磁盘(见下图)，可以在屏幕上按住或移动手指就可以在磁盘上进行绘画。但所画的东西还会作为镜像再画三次，这样一来，便用最少的精力就创建了一个有趣的模式。

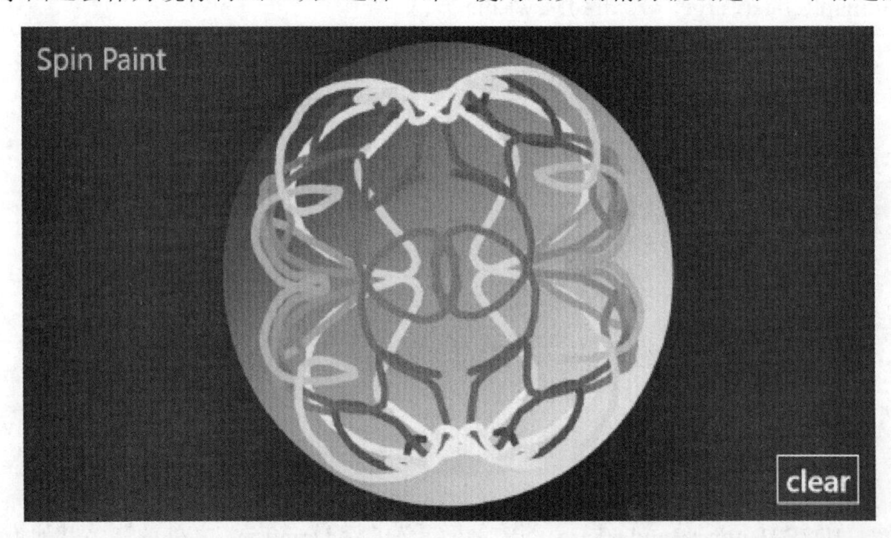

我写过第一个 SpinPaint 版本，用于咖啡桌大小的电脑，现在称为 Microsoft PixelSense。我使用 WriteableBitmap 把程序移植到 Silverlight，又使用 XNA 把程序移植到 Windows Phone 7。

在 Windows 8 版本的 SpinPaint 中，XAML 文件定义了一个名为 referencePanel 的 Grid，位于页面中心。在 Loaded 事件中，Grid 有和 SurfaceImageSource 尺寸相同的矩形，SurfaceImageSource 也是由 Grid 生成。在该 Grid 中，有另一个名为 rotatingPanel 的(顾名思

义)旋转 Grid。Grid 内部包含一些背景底纹，只是让之前画好的东西更加明显。Grid 最顶部是 Image 元素，用于显示 SurfaceImageSource 位图和剪切圈。

项目: SpinPaint | 文件: MainPage.xaml(片段)

```
<Grid Background="{StaticResource ApplicationPageBackgroundThemeBrush}">
    <Grid Name="referencePanel"
          Margin="24"
          HorizontalAlignment="Center"
          VerticalAlignment="Center">

        <Grid Name="rotatingPanel">
            <Grid.RenderTransform>
                <RotateTransform x:Name="rotate" />
            </Grid.RenderTransform>

            <Ellipse>
                <Ellipse.Fill>
                    <LinearGradientBrush>
                        <GradientStop Offset="0" Color="Black" />
                        <GradientStop Offset="1" Color="White" />
                    </LinearGradientBrush>
                </Ellipse.Fill>
            </Ellipse>

            <Image Name="image"
                   Stretch="None" />

            <!-- Cover all but a circle (poor man's clipping) -->
            <Path Fill="{StaticResource ApplicationPageBackgroundThemeBrush}"
                  Stretch="Uniform">
                <Path.Data>
                    <GeometryGroup>
                        <RectangleGeometry Rect="0 0 100 100" />
                        <EllipseGeometry Center="50 50" RadiusX="50" RadiusY="50" />
                    </GeometryGroup>
                </Path.Data>
            </Path>
        </Grid>
    </Grid>

    <TextBlock x:Name="pageTitle"
               Text="Spin Paint"
               FontSize="48"
               Margin="24">
        <TextBlock.Foreground>
            <SolidColorBrush />
        </TextBlock.Foreground>
    </TextBlock>

    <Button Content="clear"
            HorizontalAlignment="Right"
            VerticalAlignment="Bottom"
            FontSize="48"
            Margin="24"
            Click="OnClearButtonClick" />
</Grid>
```

Loaded 处理程序基于屏幕尺寸和视图状态来决定适合 SurfaceImageSource 对象的尺寸。Loaded 处理程序还从 DirectXWrapper 库创建 SurfaceImageSourceRenderer。

项目: SpinPaint | 文件: MainPage.xaml.cs(片段)

```
public sealed partial class MainPage : Page
{
    ...

    int dimension;
```

```
    SurfaceImageSourceRenderer surfaceImageSourceRenderer;
    RotateTransform inverseRotate = new RotateTransform();

    public MainPage()
    {
        InitializeComponent();
        Loaded +=OnMainPageLoaded;
    }

    void OnMainPageLoaded(object sender, RoutedEventArgs args)
    {
        // Find the dimension of the square bitmap
        if  (ApplicationView.Value == ApplicationViewState.FullScreenPortrait)
        {
            dimension = (int)(this.ActualWidth - referencePanel.Margin.Left
                                               - referencePanel.Margin.Right);
        }
        else
        {
            dimension = (int)(this.ActualHeight - referencePanel.Margin.Top
                                                - referencePanel.Margin.Bottom);
        }

        // Set this size to the reference panel so it doesn't get distorted in Snapped view
        referencePanel.Width = dimension;
        referencePanel.Height = dimension;

        // Create the SurfaceImageSource and renderer
        SurfaceImageSource surfaceImageSource = new SurfaceImageSource(dimension, dimension);
        surfaceImageSourceRenderer = new SurfaceImageSourceRenderer(surfaceImageSource,
                                                                    dimension, dimension);

        image.Source = surfaceImageSource;

        // Set rotation centers
        rotate.CenterX = dimension / 2;
        rotate.CenterY = dimension / 2;

        inverseRotate.CenterX = dimension / 2;
        inverseRotate.CenterY = dimension / 2;

        // Start the event
        CompositionTarget.Rendering += OnCompositionTargetRendering;
    }

    ...

    void OnClearButtonClick(object sender, RoutedEventArgs e)
    {
        SurfaceImageSource surfaceImageSource = new SurfaceImageSource(dimension, dimension);
        surfaceImageSourceRenderer = new SurfaceImageSourceRenderer(surfaceImageSource,
                                                                    dimension, dimension);

        image.Source = surfaceImageSource;
    }
}
```

Clear 按钮直接创建预定尺寸的新 SurfaceImageSource 和 SurfaceImageSourceRenderer。

类似于 FingerPaint 系列程序，也可以用多个手指在 SpinPaint 上绘画。然而，从概念上来说是在在一个旋转磁盘上绘画，可以把手指保持在屏幕上并进行喷绘。换句话说，手指不动或不产生任何 PointerMoved 事件的情况下，仍然可以绘画！

这就要求一个稍微不同的方法来处理 Pointer 事件。当然，仍然要维护字典，而其包含的 FingerInfo 值有 LastPosition 和 ThisPosition 字段。然而，在 OnPointerPressed 覆写中，

LastPosition 初始化为无穷大坐标，而在 OnPointerPressed 和 OnPointerMoved 中，ThisPosition 字段设置为当前手指位置。除了 LastPosition 字段初始化以外，这些覆写不会把该字段设置其他东西。

项目：SpinPaint | 文件：MainPage.xaml.cs(片段)

```
public sealed partial class MainPage : Page
{
    class FingerInfo
    {
        public Point LastPosition;
        public Point ThisPosition;
    }

    Dictionary<uint, FingerInfo> fingerTouches = new Dictionary<uint, FingerInfo>();
    ...
    protected override void OnPointerPressed(PointerRoutedEventArgs args)
    {
        uint id = args.Pointer.PointerId;
        Point pt = args.GetCurrentPoint(referencePanel).Position;
        if (fingerTouches.ContainsKey(id))
            fingerTouches.Remove(id);
        FingerInfo fingerInfo = new FingerInfo
        {
            LastPosition = new Point(Double.PositiveInfinity, Double.PositiveInfinity),
            ThisPosition = pt
        };
        fingerTouches.Add(id, fingerInfo);
        CapturePointer(args.Pointer);
        base.OnPointerPressed(args);
    }

    protected override void OnPointerMoved(PointerRoutedEventArgs args)
    {
        uint id = args.Pointer.PointerId;
        Point pt = args.GetCurrentPoint(referencePanel).Position;
        if (fingerTouches.ContainsKey(id))
            fingerTouches[id].ThisPosition = pt;
        base.OnPointerMoved(args);
    }

    protected override void OnPointerReleased(PointerRoutedEventArgs args)
    {
        uint id = args.Pointer.PointerId;
        Point pt = args.GetCurrentPoint(referencePanel).Position;
        if (fingerTouches.ContainsKey(id))
            fingerTouches[id].ThisPosition = pt;
        base.OnPointerMoved(args);
    }

    protected override void OnPointerReleased(PointerRoutedEventArgs args)
    {
        uint id = args.Pointer.PointerId;
        if (fingerTouches.ContainsKey(id))
            fingerTouches.Remove(id);
        base.OnPointerReleased(args);
    }

    protected override void OnPointerCaptureLost(PointerRoutedEventArgs args)
    {
        uint id = args.Pointer.PointerId;
        if (fingerTouches.ContainsKey(id))
        fingerTouches.Remove(id);
        base.OnPointerCaptureLost(args);
    }
    ...
}
```

所有真正有趣的事情是发生在 CompositionTarget.Rendering 事件处理程序中。根据应用当前运行时长可以计算旋转角度来旋转名为 rotatingPanel 的 Grid，并用于计算喷绘颜色。该颜色也应用到在页面左上角显示应用名称的 TextBlock。

项目：SpinPaint | 文件：MainPage.xaml.cs(片段)
```
public sealed partial class MainPage : Page
{
void OnCompositionTargetRendering(object sender, object args)
{
    // Get elapsed seconds since app began
    TimeSpan timeSpan = (args as RenderingEventArgs).RenderingTime;
    double seconds = timeSpan.TotalSeconds;

    // Calculate rotation angle
    rotate.Angle = (360 * seconds / 5) % 360;

    // Calculate color and brush
    Color clr;
    double fraction = 6 * (seconds % 10) / 10;

    if (fraction < 1)
        clr = Color.FromArgb(255, 255, (byte)(fraction * 255), 0);
    else if (fraction < 2)
        clr = Color.FromArgb(255, (byte)(255 - (fraction - 1) * 255), 255, 0);
    else if (fraction < 3)
        clr = Color.FromArgb(255, 0, 255, (byte)((fraction - 2) * 255));
    else if (fraction < 4)
        clr = Color.FromArgb(255, 0, (byte)(255 - (fraction - 3) * 255), 255);
    else if (fraction < 5)
        clr = Color.FromArgb(255, (byte)((fraction - 4) * 255), 0, 255);
    else
        clr = Color.FromArgb(255, 255, 0, (byte)(255 - (fraction - 5) * 255));

    (pageTitle.Foreground as SolidColorBrush).Color = clr;

    // All done if nobody's touching
    if (fingerTouches.Count == 0)
        return;

    ...
}
```

对于正在触摸屏幕的每个手指，都会通过旋转 FingerInfo 的 ThisPosition 字段，该点不再处在屏幕坐标系中，而是相对于旋转后的 Image 元素。用于绘画的正是该点和 FingerInfo 的 LastPosition 字段。四次调用 SurfaceImageSourceRenderer 的 DrawLine 方法会在四张位图的象限中绘制四条单独的线：

项目：SpinPaint | 文件：MainPage.xaml.cs(片段)
```
public sealed partial class MainPage : Page
{
void OnCompositionTargetRendering(object sender, object args)
{
    ...

    bool bitmapNeedsUpdate = false;

    foreach (FingerInfo fingerInfo in fingerTouches.Values)
    {
        // Find point relative to rotated bitmap
        inverseRotate.Angle = -rotate.Angle;
        Point point1 = inverseRotate.TransformPoint(fingerInfo.ThisPosition);

        if (!Double.IsPositiveInfinity(fingerInfo.LastPosition.X))
```

```
        {
            Point point2 = fingerInfo.LastPosition;
            float thickness = 12;

            // Draw the lines
            surfaceImageSourceRenderer.DrawLine(point1, point2, clr, thickness);
            surfaceImageSourceRenderer.DrawLine(new Point(dimension - point1.X, point1.Y),
                                                new Point(dimension - point2.X, point2.Y),
                                                clr, thickness);
            surfaceImageSourceRenderer.DrawLine(new Point(point1.X, dimension - point1.Y),
                                                new Point(point2.X, dimension - point2.Y),
                                                clr, thickness);
            surfaceImageSourceRenderer.DrawLine(new Point(dimension - point1.X,
                                                          dimension - point1.Y),
                                                new Point(dimension - point2.X,
                                                          dimension - point2.Y),
                                                clr, thickness);
            bitmapNeedsUpdate = true;
        }
        fingerInfo.LastPosition = point1;
    }

    // Update bitmap
    if (bitmapNeedsUpdate)
    {
        surfaceImageSourceRenderer.Update();
    }
}
```

也正是该旋转手指的位置被作为 LastPosition 存储回到 FingerInfo 对象。不动手指也可以绘画，过程如下：即使手指没有移动，OnPointerPressed 重写也可以得到手指当前位置并且一直存储在 FingerInfo 的 ThisPosition 字段中。在每次调用 CompositionTarget.Rendering 期间，就会用一个新角度旋转 ThisPosition 字段，并且从 LastPosition 字段绘制线条。转动位置的值作为 LastPosition 存储回 FingerInfo，用于准备下一次迭代。

有趣的是，我第一次考虑 Micorsoft PixelSense 的 SpinPaint 程序的时候，如果用静态 Contacts 类，即使不处理任何触摸事件也可以获得所有触摸屏幕的手指当前位置。我可以把手指触摸处理状态而不是处理为事件，因此，处理 CompositionTarget.Rendering 处理程序的中的手指事件显得很自然。

如果把 SpinPaint 程序移植到只有处理触摸事件的环境中，我得模仿 Contact 类的触摸状态。FingerInfo 的 ThisPosition 字段精确如下：在任何时候，字典中的 FingerInfo 对象都表示屏幕上所有手指的当前位置。但如果没有将触摸处理为状态而不是事件的环境，我不确定自己会这么考虑 SpinPaint 程序。

这让我更相信懂得得越多，就越能跳出固有思维。

第 16 章 富 文 本

“富文本”这个术语曾指通过不同字型、大小和风格来显示文本，但现在，这些特征都很常见，富文本的含义就显得比较模糊，指不同寻常的东西。本章大部分内容集中在 RichTextBlock 元素和 RichEditBox 控件，两者(正如其名)是 TextBlock 和 TextBox 的增强版。不过，本章还会强调一些注意事项，可以帮助处理更广泛的文本处理任务。

围绕字型的专业术语已经随着数字印刷的转换而有所改变。“字体”(typeface)一词在传统上表示字符字形的特定设计风格。常见字体是 Times New Roman 和 Helvetica。这些字体设计经常有变体，最常见的是斜体(italic)和粗体(boldface)，因此，字体家族会包括 Times New Roman、Times New Roman Italic 和 Times New Roman Bold。

字型(font)是用特定风格对特定字体的具体实现：例如，在数字印刷术之前，10 磅的 Helvetica Bold。在过去，特定字型中的每个字符都是一个金属制的活字。

随着人们开始用计算机处理文本，有两个趋势导致这些术语变得日益模糊。首先，用户首先把斜体或粗体当成字体的属性，而不是其本性。例如，从 Times New Roman 到 Times New Roman Italic，不是要改变特定字，而把 Italic 属性应用到文字上更方便，也不需要考虑基本字形。其次，随着数字轮廓字型技术的出现，例如 TrueType，字型字符的大小变成了非常简单的缩放过程，因此，人们不再认为大小是构成字型规范的重要组成部分。

为了帮助满足这种不同思维方式，术语“字型家族”(font family)开始普遍使用。字型库很像传统字体。有诸如 Times New Roman 或 Helvetica 的名称。字型库在 Windows Runtime 中用 fontFamily 类和 TextBlock 以及 Control 类的 fontFamily 属性进行实现。对于一个完整字型规范，Windows 8 程序结合其他字型相关属性((FontSize、FontStyle、FontWeight 和 FontStretch)来使用 fontFamily。

然而，底层技术则采用比较传统的方法。在 Windows 中，字型采用字型文件来实现，扩展名通常为.TTF(“TrueType font”)。可以在/Windows/Fonts 目录找到这些文件。许多这些文件可能随 Windows 一起安装；而其他一些则可能通过各种应用添加到集合中。Windows Explorer 管理该目录的方式和传统目录有一些区别，因此无法直接看到文件名。(该目录中还有特定大小的位图字型，但用于命令行窗口。)

在/Windows/Fonts 目录中显示的是字型家族名，不是文件名，比如下图所示的 Georgia。

Georgia

请注意，它看上去像是若干个文档文件。如果双击，就会看到下图，在另一个窗口中显示字型家族成员的单个字型文件。

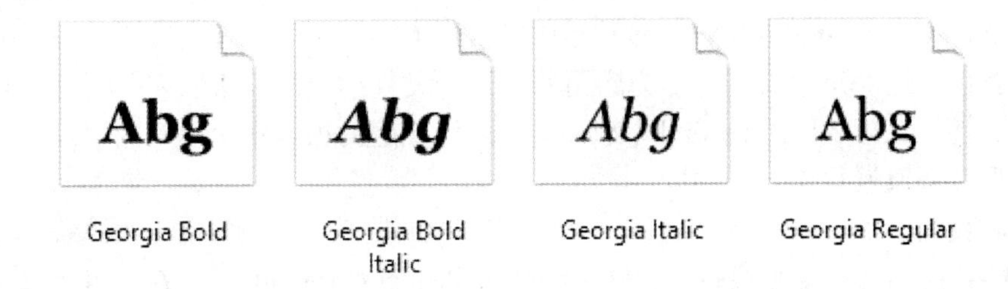

如果用鼠标右键单击其中一个，随后弹出属性窗口，会发现每个都是一个单独文件，依次为 georgiab.ttf、georgiaz.ttf、georgiai.ttf 和 georgia.ttf。每个文件都包含对很多字符的可缩放轮廓，并不是所有 Unicode 字符，而是较大的字符集。

有些字型家族不仅仅包含 Italic 和 Bold 变体，例如，还有 Oblique 、Light 或 Demibold，有些字型家族包含 Compressed 或 Expanded 变体。如果指定 FontStyle、FontWeight 和 FontStretch 属性的特定组合，Windows 就会引用一个合适的字型文件。

有些字型家族不包含 Italic 和 Bold 变体。这些字型家族是可以模拟的，即把字符向右倾斜或让字符笔画变宽一点。

16.1　专 用 字 体

Windows Store app 可以使用/Windows/Fonts 目录下的任何空心字型，但正如第 15 章所讨论的，需要 DirectWrite 来实际枚举可用字库。

有时，程序需要使用没有安装在 Windows 中的字型。一种传统解决方案是随应用提供字体并让用户安装，但在某些情况下，程序可能需要字体是专用的。这些字型可能已经由字型厂商授权严格用于该特定应用。在这种情况下，字型应该保持为该应用专用。

在这种情况下，这些专用字型文件可以处理为应用内容并有效嵌入可执行应用。PrivateFonts 项目演示了如何做到这一点。我为该项目创建了一个名为 Fonts 的文件夹，并添加 8 个 TrueType 文件，如下图所示。

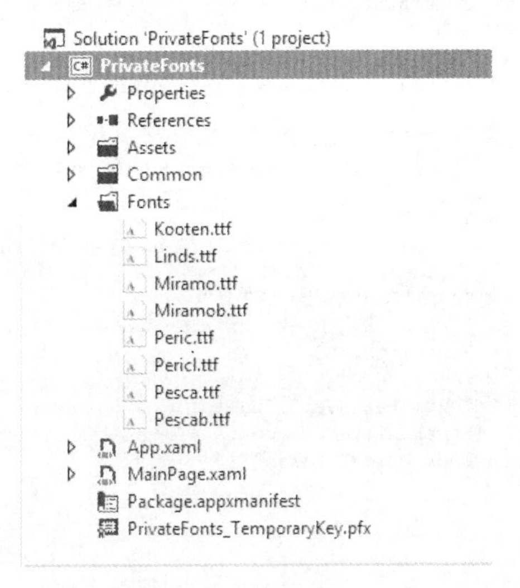

每一个字型文件都有其 Build Action，并设置为 Content 的默认值。

如果要在 Visual Studio 中创建这种程序给其他人运行，则不能直接把任意字型文件添加到项目。由于这些字型文件会作为应用的一部分发行给用户，所以你必须有权限。对于很多字型文件，包括随 Windows 和 Windows 应用一起的很多字型，随应用发行字型文件要求从字型商获取发行许可。

然而，我加到 PrivateFonts 项目中的特定字型文件并没有这种限制。如果你是 XNA 程序员，可能认得出这些字型文件，微软已经从 Ascender Corporation 获得许可，可以将其随应用一起发行。

XAML 文件中的 TextBlock 元素可以访问这些字型，与 fontFamily 属性的格式略有不同。通常情况下，可以把 fontFamily 设置为字型家族名，比如 Times New Roman 或 Segoe UI。要使用专用字体，必须指定表示字型文件位置的 URI，后面跟上哈希(或数字)标符，再跟上字型家族名称，如下所示。

```
项目: PrivateFonts | 文件: MainPage.xaml(片段)
<Page ... FontSize="36">
    <Grid Background="{StaticResource ApplicationPageBackgroundThemeBrush}">
        <Grid.ColumnDefinitions>
            <ColumnDefinition Width="*" />
            <ColumnDefinition Width="*" />
            <ColumnDefinition Width="*" />
            <ColumnDefinition Width="*" />
        </Grid.ColumnDefinitions>

        <StackPanel Grid.Column="0">
            <TextBlock Text="Kootenay"
                       FontFamily="ms-appx:///Fonts/Kooten.ttf#Kootenay" />

            <TextBlock Text="Lindsey"
                       FontFamily="ms-appx:///Fonts/Linds.ttf#Lindsey" />

            <TextBlock Text="Miramonte"
                       FontFamily="ms-appx:///Fonts/Miramo.ttf#Miramonte" />

            <TextBlock Text="Miramonte Bold"
                       FontFamily="ms-appx:///Fonts/Miramob.ttf#Miramonte" />

            <TextBlock Text="Pericles"
                       FontFamily="ms-appx:///Fonts/Peric.ttf#Pericles" />

            <TextBlock Text="Pericles Light"
                       FontFamily="ms-appx:///Fonts/Pericl.ttf#Pericles" />

            <TextBlock Text="Pescadero"
                       FontFamily="ms-appx:///Fonts/Pesca.ttf#Pescadero" />

            <TextBlock Text="Pescadero Bold"
                       FontFamily="ms-appx:///Fonts/Pescab.ttf#Pescadero" />

            <TextBlock Text="Pescadero Bold*"
                       FontFamily="ms-appx:///Fonts/Pesca.ttf#Pescadero"
                       FontWeight="Bold" />

            <TextBlock Text="Pescadero Italic*"
                       FontFamily="ms-appx:///Fonts/Pesca.ttf#Pescadero"
                       FontStyle="Italic" />
        </StackPanel>
        ...
    </Grid>
</Page>
```

FontFamily 字符串包括用于引用应用文件嵌入内容的 MS-APPX 前缀，后跟 Fonts 文件夹以及 Fonts 文件夹中的文件名。以下是 Kooten.ttf 文件的 URI：

```
ms-appx:///Fonts/Kooten.ttf
```

移除 MS-APPX:///前缀也可以。

该 URI 之后是一个哈希符号和在这个字体文件中字体的字库名称：FontFamily="ms-appx:///Fonts/Kooten.ttf#Kootenay"。

该字型家族名不是文件名(但可以是)。为了获得没有存储在/Windows/Fonts 目录下的任意 TrueType 字型文件的字型家族名称，可以在 Windows Explorer 中右键单击字型文件，然后从弹出的快捷菜单中选择 Properties 或 Preview。

Miramo.ttf 文件是 Miramonte 的普通版；粗体版在 Miramob.ttf 文件中。请注意，在这两种情况下，标记中所指定字型家族名是 Miramonte。如果 Windows 安装了这两种字型文件，则可以把 fontFamily 设置为 Miramonte，以此来引用其中的一种字型，还能把 FontWeight 设置为 Bold 而得到粗体版本。如果使用包含字型文件的语法，字型家族名相同，不需要设置 FontWeight。

同样，Peric.ttf 文件有普通 Pericles 字体，Pericl.ttf 文件包含浅色版；普通 Pescadero 字体在 Pesca .ttf 中，而粗体版是在 Pescab.TTF 中。

请注意，最后两个 TextBlock 元素均引用包含 Pescadero 普通版的文件，但 FontWeight 和 FontStyle 属性分别设置为 Bold 和 Italic。这些属性应用于常规字体，因此会模拟这些样式，我用星号和脚注做了标注。

PrivateFonts 程序实际显示四列文字。下图所示为从前面 XAML 文件产生的第一列。

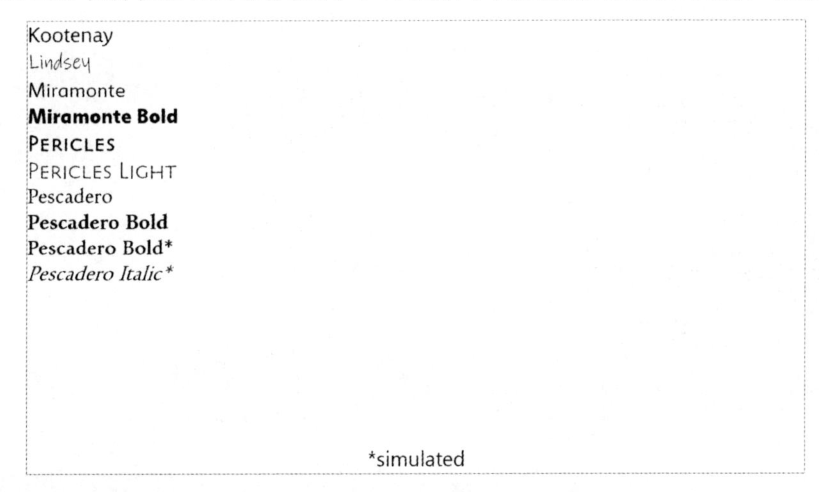

请注意，实际粗体字型不同于模拟的粗体。另外，还要注意 Pericles 字型如何显示大小写字母！

如果要在代码中引用一种专用字体，可以使用 XAML 中使用的相同字符串来创建 fontFamily 对象：

```
txtblk.FontFamily = new FontFamily("ms-appx:///Fonts/Linds.ttf#Lindsey");
```

PrivateFonts 程序显示了四列非常相似的文字，这还只是第一列，所以此时的程序还不完整。

16.2 初试 Glyphs

有一种可以代替 TextBlock 的元素称为 Glyphs。Glyphs 比 TextBlock 难用一些，但能分隔单个字符。

PrivateFonts 中第二列文本用以下标记进行显示。

项目: PrivateFonts | 文件: MainPage.xaml(片段)

```
<Grid Grid.Column="1">
    <Glyphs UnicodeString="Kootenay"
            FontUri="ms-appx:///Fonts/Kooten.ttf"
            FontRenderingEmSize="36"
            Fill="Black"
            OriginX="0"
            OriginY="45" />

    <Glyphs UnicodeString="Lindsey"
            FontUri="ms-appx:///Fonts/Linds.ttf"
            FontRenderingEmSize="36"
            Fill="Black"
            OriginX="0"
            OriginY="90" />

    <Glyphs UnicodeString="Miramonte"
            FontUri="ms-appx:///Fonts/Miramo.ttf"
            FontRenderingEmSize="36"
            Fill="Black"
            OriginX="0"
            OriginY="135" />

    <Glyphs UnicodeString="Miramonte Bold"
            FontUri="ms-appx:///Fonts/Miramob.ttf"
            FontRenderingEmSize="36"
            Fill="Black"
            OriginX="0"
            OriginY="180" />

    <Glyphs UnicodeString="Pericles"
            FontUri="ms-appx:///Fonts/Peric.ttf"
            FontRenderingEmSize="36"
            Fill="Black"
            OriginX="0"
            OriginY="225" />

    <Glyphs UnicodeString="Pericles Light"
            FontUri="ms-appx:///Fonts/Pericl.ttf"
            FontRenderingEmSize="36"
            Fill="Black"
            OriginX="0"
            OriginY="270" />

    <Glyphs UnicodeString="Pescadero"
            FontUri="ms-appx:///Fonts/Pesca.ttf"
            FontRenderingEmSize="36"
            Fill="Black"
            OriginX="0"
            OriginY="315" />

    <Glyphs UnicodeString="Pescadero Bold"
            FontUri="ms-appx:///Fonts/Pescab.ttf"
            FontRenderingEmSize="36"
            Fill="Black"
            OriginX="0"
```

```
            OriginY="360" />

    <Glyphs UnicodeString="Pescadero Bold*"
            FontUri="ms-appx:///Fonts/Pesca.ttf"
            StyleSimulations="BoldSimulation"
            FontRenderingEmSize="36"
            Fill="Black"
            OriginX="0"
            OriginY="405" />

    <Glyphs UnicodeString="Pescadero Italic*"
            FontUri="ms-appx:///Fonts/Pesca.ttf"
            StyleSimulations="ItalicSimulation"
            FontRenderingEmSize="36"
            Fill="Black"
            OriginX="0"
            OriginY="450" />
</Grid>
```

Glyphs 定义的是 UnicodeString 属性，而不是 Text 属性。其他三个属性均为必需：FontUri 属性(顾名思义)是字型文件的 URI。请注意，只有 URI；而不需要提供字型家族名称。Glyphs 在低层处理字型文件，它不知道字型家族名。FontRenderingEmSize 等同于 FontSize 属性，但没有默认值。fill 属性也没有默认值。

请注意，最后两项引用了普通 Pescadero 字型文件，但把 StyleSimulations 分别设置为 BoldSimulation 和 ItalicSimulation。StyleSimulations 枚举还包括 BoldItalicSimulation。

OriginX 和 OriginY 属性表明文本相对于父级的位置，或者更准确地说，相对于父级放置元素的地方。此处的父级仅仅是一个单格 Grid，而不是在第一列所使用的 StackPanel。(Glyphs 元素集合的父级通常为 Canvas)。源点指定 baseline 的左侧，未指定文本顶端。我直接把第一个 OriginY 设置为 45，并且每次以 45 增加。从下图可以看出 PrivateFonts 显示的前两列对比效果。

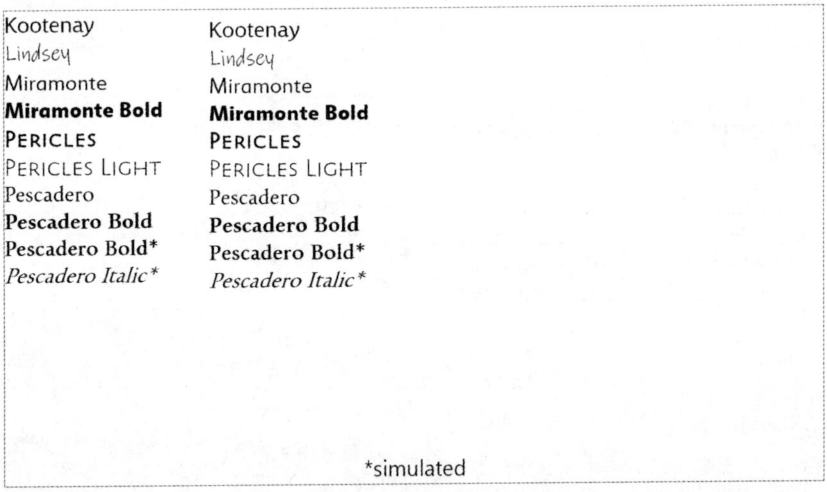

Kootenay　Kootenay
Lindsey　Lindsey
Miramonte　Miramonte
Miramonte Bold　**Miramonte Bold**
PERICLES　PERICLES
PERICLES LIGHT　PERICLES LIGHT
Pescadero　Pescadero
Pescadero Bold　**Pescadero Bold**
Pescadero Bold*　Pescadero Bold*
*Pescadero Italic**　*Pescadero Italic**

*simulated

在实际程序中，会使用字型规格定位每个 Glyphs 元素。

如果需要在代码中设置 FontUri 属性，只需创建 URI 对象并使用在 XAML 文件看到的相同字符串：

```
glyphs.FontUri = new Uri("ms-appx:///Fonts/Linds.ttf");
```

Glyph 元素并不能对多行文本自动换行。然而，该例并没有展示一个叫 Indices 的属性，这需要提供额外的系统开销来高精度分隔单个字符。也可以用 Indices 字符串表示替代字符，例如连字，连字是两个或两个以上图像字符的风格组合。

16.3 本地存储的字型文件

Glyphs 元素最常见于 XPS(XML Paper Specification)创建的文档中，XPS 是微软结合 WPF 开发的文档。XPS 是一种固定页文档格式。也就是说，该文档所有页面的大小和布局均为固定，很像 Adobe PDF。

包含 XPS 文档的文件是一个“包”，本质上是一个 ZIP 文件，包含字型文件、位图和文档每页的单独文件。文档每页是一个包含 FixedPage(Windows Runtime 中无定义)根元素的 XAML 文件，一般包含一些以 ImageBrush 对象形式显示图形和位图的 Path 元素，还包含用于显示文本的 Glyphs 元素。Glyphs 元素有自己的 FontUri 属性并设置为引用 XPS 包中字体文件的 URI。这些 Glyphs 元素已经设置好所有属性，以正确定位页中的文本。

WPF 程序可以渲染 XPS 文件，并没有太多麻烦。Windows 8 程序则要做很多事情。程序需要打开 XPS 包并解析单个 FixedPage 文件。对于每一页，程序需要实例化代码中的各种 Path、ImageBrush 和 Glyph 对象，所有时间都在补偿 Windows Runtime 不支持的任何 XPS 功能。

在 XPS 包中，ImageBrush 和 Glyphs 元素的 URI 引用包内的位图文件和字型文件。这些位图和字型文件需要复制到应用的本地存储，URI 则需要修改以指向该存储。

PrivateFonts 程序演示了该过程的可行性。第一次运行程序时，第三列显示使用 Windows 8 默认字体的文本，而不是使用任何专用字体，还没有第四列(见下图)。

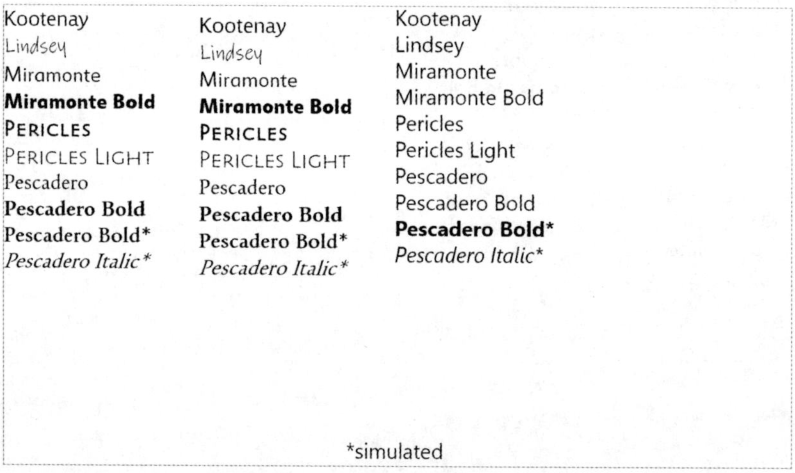

Kootenay	Kootenay	Kootenay
Lindsey	Lindsey	Lindsey
Miramonte	Miramonte	Miramonte
Miramonte Bold	**Miramonte Bold**	Miramonte Bold
PERICLES	PERICLES	Pericles
PERICLES LIGHT	PERICLES LIGHT	Pericles Light
Pescadero	Pescadero	Pescadero
Pescadero Bold	**Pescadero Bold**	Pescadero Bold
Pescadero Bold*	Pescadero Bold*	**Pescadero Bold***
*Pescadero Italic**	*Pescadero Italic**	*Pescadero Italic**

*simulated

之后再运行程序，第三列和第四列匹配前两列(见下图)。

Kootenay	Kootenay	Kootenay	Kootenay
Lindsey	Lindsey	Lindsey	Lindsey
Miramonte	Miramonte	Miramonte	Miramonte
Miramonte Bold	**Miramonte Bold**	**Miramonte Bold**	**Miramonte Bold**
PERICLES	PERICLES	PERICLES	PERICLES
PERICLES LIGHT	PERICLES LIGHT	PERICLES LIGHT	PERICLES LIGHT
Pescadero	Pescadero	Pescadero	Pescadero
Pescadero Bold	**Pescadero Bold**	**Pescadero Bold**	**Pescadero Bold**
Pescadero Bold*	Pescadero Bold*	Pescadero Bold*	Pescadero Bold*
*Pescadero Italic**	*Pescadero Italic**	*Pescadero Italic**	*Pescadero Italic**

*simulated

这种差异是程序结构所造成的偶然事件。第一次运行程序时，字体文件复制到应用的本地存储。

项目：PrivateFonts | 文件：MainPage.xaml.cs（片段）

```
public sealed partial class MainPage : Page
{
    public MainPage()
    {
        this.InitializeComponent();
        Loaded += OnLoaded;
    }

    async void OnLoaded(object sender, RoutedEventArgs args)
    {
        StorageFolder localFolder = ApplicationData.Current.LocalFolder;
        bool folderExists = false;

        try
        {
            StorageFolder fontsFolder = await localFolder.GetFolderAsync("Fonts");
            folderExists = true;
        }
        catch (Exception)
        {
        }

        if (!folderExists)
        {
            StorageFolder fontsFolder = await localFolder.CreateFolderAsync("Fonts");

            string[] fonts = { "Kooten.ttf", "Linds.ttf", "Miramo.ttf", "Miramob.ttf",
                               "Peric.ttf", "pericl.ttf", "Pesca.ttf", "Pescab.ttf" };

            foreach (string font in fonts)
            {
                // Copy from application content to IBuffer
                string uri = "ms-appx:///Fonts/" + font;
                IBuffer buffer = await PathIO.ReadBufferAsync(uri);

                // Copy from IBuffer to local storage
                StorageFile fontFile = await fontsFolder.CreateFileAsync(font);
                await FileIO.WriteBufferAsync(fontFile, buffer);
            }
        }
    }
}
```

Loaded 处理程序检查本地存储是否存在名为 Fonts 的目录。如果不存在，则创建，然后把所有字型文件以相同名称复制到该目录。(如果可以同步进行，我会在构造函数之前的 InitializeComponent 中完成，以便在程序第一次运行时，XAML 分析器即可使用该文件。)

引用本地存储文件的标记和引用程序内容的标记非常相同，只不过前缀为 ms-appdata，且需要在 Fonts 目录前面加上 local。我敢肯定，不需要看也能明白总体思路。

项目：PrivateFonts | 文件：MainPage.xaml(片段)

```
<Page ... FontSize="36">
    <Grid Background="{StaticResource ApplicationPageBackgroundThemeBrush}">
        <Grid.ColumnDefinitions>
            <ColumnDefinition Width="*" />
            <ColumnDefinition Width="*" />
            <ColumnDefinition Width="*" />
            <ColumnDefinition Width="*" />
        </Grid.ColumnDefinitions>

        ...

        <StackPanel Grid.Column="2">
            <TextBlock Text="Kootenay"
                       FontFamily="ms-appdata:///local/Fonts/Kooten.ttf#Kootenay" />

            ...

            <TextBlock Text="Pescadero Italic*"
                       FontFamily="ms-appdata:///local/Fonts/Pesca.ttf#Pescadero"
                       FontStyle="Italic" />
        </StackPanel>

        <Grid Grid.Column="3">
            <Glyphs UnicodeString="Kootenay"
                    FontUri="ms-appdata:///local/Fonts/Kooten.ttf"
                    FontRenderingEmSize="36"
                    Fill="Black"
                    OriginX="0"
                    OriginY="45" />

            ...

            <Glyphs UnicodeString="Pescadero Italic*"
                    FontUri="ms-appdata:///local/Fonts/Pesca.ttf"
                    StyleSimulations="ItalicSimulation"
                    FontRenderingEmSize="36"
                    Fill="Black"
                    OriginX="0"
                    OriginY="450" />
        </Grid>

        <TextBlock Text="*simulated"
                   Grid.ColumnSpan="4"
                   VerticalAlignment="Bottom"
                   HorizontalAlignment="Center" />
    </Grid>
</Page>
```

如果需要在代码中做这些，就使用在 XAML 文件中看到的相同字符串：

```
txtblk.FontFamily = new FontFamily("ms-appdata:///local/Fonts/Linds.ttf#Lindsey");

glyphs.FontUri = new Uri("ms-appdata:///local/Fonts/Linds.ttf");
```

在 Windows 8 可以渲染的所有现有复杂文件格式中，XPS 几乎肯定是最简单的，因为其内容与在 Windows Runtime 中发现的元素类似，所有页面已经构建，所有图形和 Glyphs

元素已精确定位在页面中。更有挑战性的是回流页格式，比如 EPUB，而程序主要负责在页面中定位文字。这项工作需要更加熟悉字型规格。

对字型规格最简单的方法涉及 TextBlock 调用 Measure，并获取宽度高度。这样可以确定段落中单个字的位置，段落中何处转行、页面之间何处分段。然而，如果需要基线之上的不同字号或字型家族的单个 TextBlock 元素，则需要更多信息。

此时，需要检查字型文件内部来提取字型规格或开始使用 DirectWrite。一旦开始对字型规格使用 DirectWrite，很有可能发现 DirectWrite 是布局页面的最佳工具。

16.4　排版功能增强

Windows.UI.Xaml.Documents 命名空间中的 Typography 类只包括用来增强文本附加属性的集合。可以将这些附加属性插入到页的根元素、TextBlock 元素或 Run 元素中，以控制文本显示的方方面面。但是会有一个陷阱，即不能保证这些功适用于所有字体。事实上，你会发现为了找到能响应这些附加属性的字体，要花很长时间！

在以下例子中，我在很大程度上依赖于 Typography 类的 WPF 版文档，能匹配一些含有特定字型的附加属性。其中一些实例涉及到 Lindsey、Miramonte、Pescadero 和 Pericles 字型，这些字型均作为程序内容。

项目：TypographyDemo | 文件：MainPage.xaml(片段)

```
<Page ... >
    <Page.Resources>
        <Style TargetType="TextBlock">
            <Setter Property="FontSize" Value="48" />
            <Setter Property="Margin" Value="6 6 6 0" />
        </Style>
    </Page.Resources>

    <Grid Background="{StaticResource ApplicationPageBackgroundThemeBrush}">
        <StackPanel>
            <TextBlock Text="Small Caps are Nice for Titles"
                       Typography.Capitals="SmallCaps" />

            <TextBlock Text="Some random contextual alternates make script look more natural"
                       FontFamily="ms-appx:///Fonts/Linds.ttf#Lindsey"
                       Typography.ContextualAlternates="True" />

            <TextBlock Text="Stacked fractions: 1/2 1/4 1/8 1/3 2/3"
                       FontFamily="Palatino Linotype"
                       Typography.Fraction="Stacked" />

            <TextBlock Text="Historical forms: Four score and seven years ago"
                       FontFamily="Palatino Linotype"
                       Typography.HistoricalForms="True" />

            <TextBlock Text="Numeral alignment for tables: 0123456789"
                       FontFamily="ms-appx:///Fonts/Miramo.ttf#Miramonte"
                       Typography.NumeralAlignment="Tabular" />

            <TextBlock Text="Old-style numbers: 0123456789"
                       FontFamily="Palatino Linotype"
                       Typography.NumeralStyle="OldStyle" />

            <TextBlock Text="Standard Swashes With The Pescadero Font"
                       FontFamily="ms-appx:///Fonts/Pesca.ttf#Pescadero"
                       Typography.StandardSwashes="1" />
```

```
            <TextBlock Text="Slashed Zero: 0"
                       FontFamily="ms-appx:///Fonts/Miramo.ttf#Miramonte"
                       Typography.SlashedZero="True" />

            <TextBlock Text="STYLISTIC ALTERNATES WITH THE PERICLES FONT"
                       FontFamily="ms-appx:///Fonts/Peric.ttf#Pericles"
                       Typography.StylisticAlternates="1" />

            <TextBlock FontFamily="Palatino Linotype">
                       Sucrose is C<Run Typography.Variants="Inferior">12</Run
                                >H<Run Typography.Variants="Inferior">22</Run
                                >O<Run Typography.Variants="Inferior">11</Run>
            </TextBlock>
        </StackPanel>
    </Grid>
</Page>
```

显示效果如下图所示。

SMALL CAPS ARE NICE FOR TITLES

Some random contextual alternates make script look more natural

Stacked fractions: $\frac{1}{2}$ $\frac{1}{4}$ $\frac{1}{8}$ $\frac{1}{3}$ $\frac{2}{3}$

Hiſtorical formſ: Four ſcore and ſeven yearſ ago

Numeral alignment for tables: 0123456789

Old-style numbers: 0123456789

Standard Swashes With The Pescadero Font

Slashed Zero: 0

STYLISTIC ALTERNATES WITH THE PERICLES FONT

Sucrose is $C_{12}H_{22}O_{11}$

16.5 RichTextBlock 和段落

虽然我们会继续把 TextBlock 首选用于最多一段长度的文本，但 RichTextBlock 能提供若干增强功能。RichTextBlock 没有 Text 属性，也没有以 Inlines 派生形式指定文本的 Inlines 属性。RichTextBlock 定义的是名为 Blocks 的属性，Blocks 属性是 Block 派生集合。像 Inline 一样，Block 派生自 TextElement，并由此获取许多和文本相关的属性。此外，Block 还定义如下属性：

- LineHeight
- LineStackingStrategy
- Margin
- TextAlignment

此外，也像 Inline 一样，Block 本身没有实例化。目前源自 Block 的唯一类是 Paragraph，定义两个属性：

- Inlines，Inlines 派生类集合
- Textindent，用于设置段落的第一行缩进

因此， RichTextBlock 基本上是段落集合。Margin 属性用于定于段落之间的空间，而 TextIndent 可缩进第一行。

MadTeaParty 项目在 RichTextBlock 内使用 ScrollViewer，允许你细读刘易斯·卡罗尔《爱丽思漫游奇境》的第 7 章，包括三幅约翰·坦尼尔所画的插图。以下为 XAML 文件片段。

项目: MadTeaParty | 文件: MainPage.xaml(片段)

```
<Page ... >
    <Grid Background="{StaticResource ApplicationPageBackgroundThemeBrush}">
        <ScrollViewer Width="720"
                      Padding="40 20">

            <!-- Text and images from http://ebooks.adelaide.edu.au/c/carroll/lewis/alice/ -->

            <RichTextBlock FontFamily="Cambria"
                           FontSize="24">
                <Paragraph Margin="0 12" TextAlignment="Center" FontSize="40">
                    <Italic>Alice's Adventures in Wonderland</Italic>
                    <LineBreak/>
                    by
                    <LineBreak/>
                    Lewis Carroll
                </Paragraph>

                <Paragraph Margin="0 24 0 36" TextAlignment="Center" FontSize="30">
                    Chapter VII
                    <LineBreak />
                    A Mad Tea-Party
                </Paragraph>

                <Paragraph Margin="0 6">
                    There was a table set out under a tree in front of the
                    house, and the March Hare and the Hatter were having tea at
                    it: a Dormouse was sitting between them, fast asleep, and
                    the other two were using it as a cushion, resting their
                    elbows on it, and talking over its head. 'Very uncomfortable
                    for the Dormouse,' thought Alice; 'only, as it's asleep, I
                    suppose it doesn't mind.'
                </Paragraph>

                <Paragraph Margin="0 6" TextIndent="48">
                    The table was a large one, but the three were all crowded
                    together at one corner of it: 'No room! No room!' they
                    cried out when they saw Alice coming. 'There's
                    <Italic>plenty</Italic> of room!' said Alice indignantly,
                    and she sat down in a large arm-chair at one end of the table.
                </Paragraph>

                <Paragraph Margin="0 6" TextIndent="48">
                    'Have some wine,' the March Hare said in an encouraging tone.
                </Paragraph>

                <Paragraph Margin="0 6" TextIndent="48">
                    Alice looked all round the table, but there was nothing on it
                    but tea. 'I don't see any wine,' she remarked.
                </Paragraph>

                <Paragraph Margin="0 6" TextIndent="48">
                    'There isn't any,' said the March Hare.
                </Paragraph>

...

                <Paragraph Margin="0 6" TextIndent="48">
                    'It
                    <Italic>is</Italic> the same thing with you,' said the
```

```
                Hatter, and here the conversation dropped, and the party sat
                silent for a minute, while Alice thought over all she could
                remember about ravens and writing-desks, which wasn't much.
            </Paragraph>

            <Paragraph Margin="0 6" TextAlignment="Center">
                <InlineUIContainer>
                    <Image Source="Images/ChapterVII-1.png" Stretch="None" />
                </InlineUIContainer>
            </Paragraph>

            <Paragraph Margin="0 6" TextIndent="48">
                The Hatter was the first to break the silence. 'What day of
                the month is it?' he said, turning to Alice: he had taken
                his watch out of his pocket, and was looking at it uneasily,
                shaking it every now and then, and holding it to his ear.
            </Paragraph>

            ...

            <Paragraph Margin="0 6" TextIndent="48">
                Just as she said this, she noticed that one of the trees
                had a door leading right into it. 'That's very curious!'
                she thought. 'But everything's curious today. I think I
                may as well go in at once.' And in she went.
            </Paragraph>

            <Paragraph Margin="0 6" TextIndent="48">
                Once more she found herself in the long hall, and close to
                the little glass table. 'Now, I'll manage better this time,'
                she said to herself, and began by taking the little golden
                key, and unlocking the door that led into the garden. Then
                she went to work nibbling at the mushroom (she had kept a
                piece of it in her pocket) till she was about a foot high:
                then she walked down the little passage: and
                <Italic>then</Italic> — she found herself at last in the
                beautiful garden, among the bright flower-beds and the cool
                fountains.
            </Paragraph>
        </RichTextBlock>
    </ScrollViewer>
  </Grid>
</Page>
```

Paragraph 不是派生自 FrameworkElement，因此没有 Style 属性。如果要对很多 Paragraph 对象中设置相同属性，则需要明确。“A Mad Tea-Party”中大多数段落都有 12 像素段落间距的 Margin 属性，以及缩进第一行 48 个像素的 TextIndent 属性。

InlineUIContainer 不与 TextBlock 一起用，但与 RichTextBlock 一起用。因此，可以在文本中嵌入 UIElement 派生类。可能包括 TextBlock，因此该功能提供了在段落中嵌入文本的方法，而段落对 Text 属性包含了绑定。然而，嵌入的 TextBlock 元素本身不能换行。

在 MadTeaParty 程序中，Image 元素成为 RichTextBlock 的一部分。这要求 Image 元素放入 InlineUIContainer 对象，而后者需要放入 Paragraph。现在没有办法处理围绕图片的段落文本。如果想要用 C#和 XAML 进行处理，则需要测量文本的单个单词，并且自己进行定位。

把第 7 章向下滚动到看到三张图片中的第三张，如下图所示。

she got up in great disgust, and walked off; the Dormouse fell asleep instantly, and neither of the others took the least notice of her going, though she looked back once or twice, half hoping that they would call after her: the last time she saw them, they were trying to put the Dormouse into the teapot.

'At any rate I'll never go *there* again!' said Alice as she picked her way through the wood. 'It's the stupidest tea-party I ever was at in all my life!'

16.6　RichTextBlock 选择

MadTeaParty 程序在运行的时候，用手指轻按一个单词。所选单词两端会出现有两个圆形把手。可以抓住这些把手并扩展选择。点击选择，会出现一个小菜单，并把所选内容复制到剪贴板。

或者，如果没有触摸屏，可以按下用鼠标按钮来选择文本。然后用鼠标右键单击，弹出包含 Copy 选项的快捷菜单。

RichTextBlock 实施 SelectionChanged 事件、获取所选文本的 SelectedText 属性(但不是取代或删除)以及 SelectionStart 和 SelectionEnd 属性。后两个属性为 TextPointer 类型，TextPointer 不仅提供抵消 TextBlock 内所选文本，也表明选择相对于 TextBlock 的像素位置。

快捷菜单显示之前还有 ContextMenuOpening 事件。如果把事件参数的 Handled 设置属性为 true，则不会显示菜单，也就是说，你可以显示定制快捷菜单。

如果想让 RichTextBlock 不允许文本被选中，则可以把 IsTextSelectionEnabled 设置为 false。

16.7　RichTextBlock 和超限

图书和阅读历史上有两种显示扩展文字的基本方式：卷轴(常见于古埃及、中国、希腊和罗马的地中海文化)或单页集合(最先形成于公元后最初几个世纪的欧洲)。

电脑上也有两种常见格式：大部分使用滚动的网页，但把文本分成多页的大多数电子书阅读器。

我刚刚演示了 RichTextBlock 可呈现滚动文本，但 RichTextBlock 则能给文档分页，大大超越了之前的文本显示元素。这些页可以按顺序显示(比如在电子书阅读器中)或以相邻列显示。

过程如下：把所有想要显示的文本放入 RichTextBlock，然后赋予它一个有限尺寸，也就是说，无论手动还是作为正常网页布局的一部分，使其受限于 Measure 传递。如果 RichTextBlock 包含的文本多于其在所分配空间里显示的文本，则 HasOverflowContent 属性变为 true。为了显示 RichTextBlock 以后的第二页文本，可以创建 RichTextBlockOverflow 类实例，并把该实例设置为 RichTextBlock 的 OverflowContentTarget 属性。

RichTextBlockOverflow 还定义了 HasOverflowContent 和 OverflowContentTarget 属性，因此可以为每个页创建额外的 RichTextBlockOverflow，并把它们串成一条链。RichTextBlockOverflow 元素从父级 RichTextBlock 继承了所有文字相关属性(FontFamily、FontSize 等等)。

如果能估计出需要显示文档的最大页数，则完全可以在 XAML 中使用数据绑定来执行链接工作。YoungGoodmanBrown 项目展示了具体实现。文本来自纳撒尼尔·霍桑(Nathaniel Hawthorne)的短篇小说《好小伙布朗》(*Young Goodman Brown*)，是我从古登堡计划拿来的。正如在 A Mad Tea-Party 一章把整个文本放入一个单独 RichTextBlock，我把 RichTextBlock 的 OverflowContentTarget 属性赋予给 RichTextBlockOverflow 元素，因而形成链条。

项目：YoungGoodmanBrown | 文件：MainPage.xaml(片段)

```
<Page ... >
    <Page.Resources>
        <local:BooleanToVisibilityConverter x:Key="booleanToVisibility" />

        <Style TargetType="RichTextBlock">
            <Setter Property="Width" Value="480" />
            <Setter Property="Margin" Value="24 0 24 0" />
            <Setter Property="FontSize" Value="18" />
            <Setter Property="TextAlignment" Value="Justify" />
        </Style>

        <Style TargetType="RichTextBlockOverflow">
            <Setter Property="Width" Value="480" />
            <Setter Property="Margin" Value="24 0 24 0" />
        </Style>
    </Page.Resources>

    <Grid Background="{StaticResource ApplicationPageBackgroundThemeBrush}">
        <ScrollViewer HorizontalScrollBarVisibility="Hidden"
                      VerticalScrollBarVisibility="Disabled">
            <StackPanel Orientation="Horizontal">

                <!-- Text from http://www.gutenberg.org/files/512/512-h/512-h.htm -->

                <RichTextBlock Name="richTextBlock"
                               OverflowContentTarget="{Binding ElementName=overflow1}">
                    <Paragraph TextAlignment="Center">
                        YOUNG GOODMAN BROWN
                    </Paragraph>

                    <Paragraph TextAlignment="Center" Margin="0 12">
                        by
                        <LineBreak />
                        Nathaniel Hawthorne
                    </Paragraph>

                    <Paragraph Margin="0 6" TextIndent="48">
                        Young Goodman Brown came forth at sunset into the street at Salem
                        village; but put his head back, after crossing the threshold, to
                        exchange a parting kiss with his young wife. And Faith, as the wife was
                        aptly named, thrust her own pretty head into the street, letting the
                        wind play with the pink ribbons of her cap while she called to Goodman
```

```
        Brown.
    </Paragraph>

    <Paragraph Margin="0 6" TextIndent="48">
        "Dearest heart," whispered she, softly and rather sadly, when her lips
        were close to his ear, "prithee put off your journey until sunrise and
        sleep in your own bed to-night. A lone woman is troubled with such
        dreams and such thoughts that she's afeard of herself sometimes. Pray
        tarry with me this night, dear husband, of all nights in the year."
    </Paragraph>
    ...
</RichTextBlock>

<RichTextBlockOverflow Name="overflow1"
          Visibility="{Binding ElementName=richTextBlock,
                       Path=HasOverflowContent,
                       Converter={StaticResource booleanToVisibility}}"
          OverflowContentTarget="{Binding ElementName=overflow2}" />

<RichTextBlockOverflow Name="overflow2"
          Visibility="{Binding ElementName=overflow1,
                       Path=HasOverflowContent,
                       Converter={StaticResource booleanToVisibility}}"
          OverflowContentTarget="{Binding ElementName=overflow3}" />

<RichTextBlockOverflow Name="overflow3"
          Visibility="{Binding ElementName=overflow2,
                       Path=HasOverflowContent,
                       Converter={StaticResource booleanToVisibility}}"
          OverflowContentTarget="{Binding ElementName=overflow4}" />

<RichTextBlockOverflow Name="overflow4"
          Visibility="{Binding ElementName=overflow3,
                       Path=HasOverflowContent,
                       Converter={StaticResource booleanToVisibility}}"
          OverflowContentTarget="{Binding ElementName=overflow5}" />

<RichTextBlockOverflow Name="overflow5"
          Visibility="{Binding ElementName=overflow4,
                       Path=HasOverflowContent,
                       Converter={StaticResource booleanToVisibility}}"
          OverflowContentTarget="{Binding ElementName=overflow6}" />

<RichTextBlockOverflow Name="overflow6"
          Visibility="{Binding ElementName=overflow5,
                       Path=HasOverflowContent,
                       Converter={StaticResource booleanToVisibility}}"
          OverflowContentTarget="{Binding ElementName=overflow7}" />

<RichTextBlockOverflow Name="overflow7"
          Visibility="{Binding ElementName=overflow6,
                       Path=HasOverflowContent,
                       Converter={StaticResource booleanToVisibility}}"
          OverflowContentTarget="{Binding ElementName=overflow8}" />

<RichTextBlockOverflow Name="overflow8"
          Visibility="{Binding ElementName=overflow7,
                       Path=HasOverflowContent,
                       Converter={StaticResource booleanToVisibility}}"
          OverflowContentTarget="{Binding ElementName=overflow9}" />

<RichTextBlockOverflow Name="overflow9"
          Visibility="{Binding ElementName=overflow8,
                       Path=HasOverflowContent,
                       Converter={StaticResource booleanToVisibility}}"
          OverflowContentTarget="{Binding ElementName=overflow10}" />

<RichTextBlockOverflow Name="overflow10"
```

```
          Visibility="{Binding ElementName=overflow9,
                               Path=HasOverflowContent,
                               Converter={StaticResource booleanToVisibility}}"
          OverflowContentTarget="{Binding ElementName=overflow11}" />

<RichTextBlockOverflow Name="overflow11"
          Visibility="{Binding ElementName=overflow10,
                               Path=HasOverflowContent,
                               Converter={StaticResource booleanToVisibility}}"
          OverflowContentTarget="{Binding ElementName=overflow12}" />

<RichTextBlockOverflow Name="overflow12"
          Visibility="{Binding ElementName=overflow11,
                               Path=HasOverflowContent,
                               Converter={StaticResource booleanToVisibility}}"
          OverflowContentTarget="{Binding ElementName=overflow13}" />

<RichTextBlockOverflow Name="overflow13"
          Visibility="{Binding ElementName=overflow12,
                               Path=HasOverflowContent,
                               Converter={StaticResource booleanToVisibility}}"
          OverflowContentTarget="{Binding ElementName=overflow14}" />

<RichTextBlockOverflow Name="overflow14"
          Visibility="{Binding ElementName=overflow13,
                               Path=HasOverflowContent,
                               Converter={StaticResource booleanToVisibility}}"
          OverflowContentTarget="{Binding ElementName=overflow15}" />

<RichTextBlockOverflow Name="overflow15"
          Visibility="{Binding ElementName=overflow14,
                               Path=HasOverflowContent,
                               Converter={StaticResource booleanToVisibility}}"
          OverflowContentTarget="{Binding ElementName=overflow16}" />

<RichTextBlockOverflow Name="overflow16"
          Visibility="{Binding ElementName=overflow15,
                               Path=HasOverflowContent,
                               Converter={StaticResource booleanToVisibility}}"
          OverflowContentTarget="{Binding ElementName=overflow17}" />

<RichTextBlockOverflow Name="overflow17"
          Visibility="{Binding ElementName=overflow16,
                               Path=HasOverflowContent,
                               Converter={StaticResource booleanToVisibility}}"
          OverflowContentTarget="{Binding ElementName=overflow18}" />

<RichTextBlockOverflow Name="overflow18"
          Visibility="{Binding ElementName=overflow17,
                               Path=HasOverflowContent,
                               Converter={StaticResource booleanToVisibility}}"
          OverflowContentTarget="{Binding ElementName=overflow19}" />

<RichTextBlockOverflow Name="overflow19"
          Visibility="{Binding ElementName=overflow18,
                               Path=HasOverflowContent,
                               Converter={StaticResource booleanToVisibility}}"
          OverflowContentTarget="{Binding ElementName=overflow20}" />

<RichTextBlockOverflow Name="overflow20"
          Visibility="{Binding ElementName=overflow19,
                               Path=HasOverflowContent,
                               Converter={StaticResource booleanToVisibility}}"
          OverflowContentTarget="{Binding ElementName=overflow21}" />

<RichTextBlockOverflow Name="overflow21"
          Visibility="{Binding ElementName=overflow20,
                               Path=HasOverflowContent,
```

```
                                Converter={StaticResource booleanToVisibility}}" />
            </StackPanel>
        </ScrollViewer>
    </Grid>
</Page>
```

如果前一个 RichTextBlockOverflow 没有任何超限内容，则隐藏下一个 RichTextBlockOverflow，并且除了最后一个外，每个 RichTextBlockOverflow 通过绑定把溢出超限内容放入下一个。所有 RichTextBlockOverflow 元素用原始 RichTextBlock 共享 ScrollViewer 的水平 StackPanel，因此，文本会形成水平滚动的列，如下图所示。

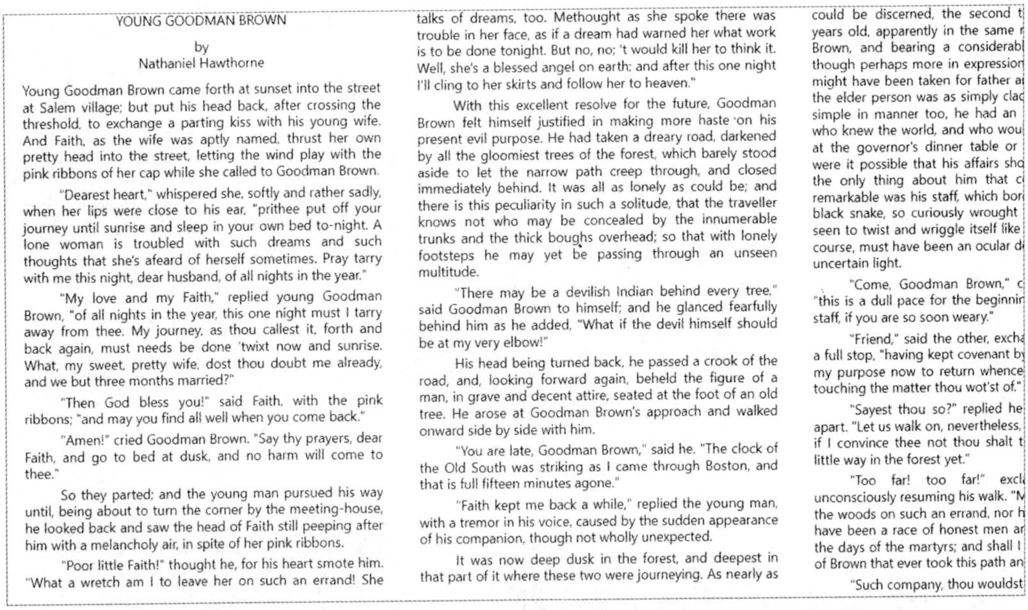

YOUNG GOODMAN BROWN

by

Nathaniel Hawthorne

Young Goodman Brown came forth at sunset into the street at Salem village; but put his head back, after crossing the threshold, to exchange a parting kiss with his young wife. And Faith, as the wife was aptly named, thrust her own pretty head into the street, letting the wind play with the pink ribbons of her cap while she called to Goodman Brown.

"Dearest heart," whispered she, softly and rather sadly, when her lips were close to his ear, "prithee put off your journey until sunrise and sleep in your own bed to-night. A lone woman is troubled with such dreams and such thoughts that she's afeard of herself sometimes. Pray tarry with me this night, dear husband, of all nights in the year."

"My love and my Faith," replied young Goodman Brown, "of all nights in the year, this one night must I tarry away from thee. My journey, as thou callest it, forth and back again, must needs be done 'twixt now and sunrise. What, my sweet, pretty wife, dost thou doubt me already, and we but three months married?"

"Then God bless you!" said Faith, with the pink ribbons; "and may you find all well when you come back."

"Amen!" cried Goodman Brown. "Say thy prayers, dear Faith, and go to bed at dusk, and no harm will come to thee."

So they parted; and the young man pursued his way until, being about to turn the corner by the meeting-house, he looked back and saw the head of Faith still peeping after him with a melancholy air, in spite of her pink ribbons.

"Poor little Faith!" thought he, for his heart smote him. "What a wretch am I to leave her on such an errand! She talks of dreams, too. Methought as she spoke there was trouble in her face, as if a dream had warned her what work is to be done tonight. But no, no; 't would kill her to think it. Well, she's a blessed angel on earth; and after this one night I'll cling to her skirts and follow her to heaven."

With this excellent resolve for the future, Goodman Brown felt himself justified in making more haste on his present evil purpose. He had taken a dreary road, darkened by all the gloomiest trees of the forest, which barely stood aside to let the narrow path creep through, and closed immediately behind. It was all as lonely as could be; and there is this peculiarity in such a solitude, that the traveller knows not who may be concealed by the innumerable trunks and the thick boughs overhead; so that with lonely footsteps he may yet be passing through an unseen multitude.

"There may be a devilish Indian behind every tree," said Goodman Brown to himself; and he glanced fearfully behind him as he added, "What if the devil himself should be at my very elbow!"

His head being turned back, he passed a crook of the road, and, looking forward again, beheld the figure of a man, in grave and decent attire, seated at the foot of an old tree. He arose at Goodman Brown's approach and walked onward side by side with him.

"You are late, Goodman Brown," said he. "The clock of the Old South was striking as I came through Boston, and that is full fifteen minutes agone."

"Faith kept me back a while," replied the young man, with a tremor in his voice, caused by the sudden appearance of his companion, though not wholly unexpected.

It was now deep dusk in the forest, and deepest in that part of it where these two were journeying. As nearly as could be discerned, the second t
years old, apparently in the same r
Brown, and bearing a considerabl
though perhaps more in expressior
might have been taken for father a
the elder person was as simply cla
simple in manner too, he had an
who knew the world, and who wou
at the governor's dinner table or
were it possible that his affairs sho
the only thing about him that c
remarkable was his staff, which bor
black snake, so curiously wrought
seen to twist and wriggle itself like
course, must have been an ocular d
uncertain light.

"Come, Goodman Brown," c
"this is a dull pace for the beginnir
staff, if you are so soon weary."

"Friend," said the other, excha
a full stop, "having kept covenant b
my purpose now to return whence
touching the matter thou wot'st of."

"Sayest thou so?" replied he
apart. "Let us walk on, nevertheless,
if I convince thee not thou shalt t
little way in the forest yet."

"Too far! too far!" excl
unconsciously resuming his walk. "M
the woods on such an errand, nor h
have been a race of honest men ar
the days of the martyrs; and shall I
of Brown that ever took this path an

"Such company, thou wouldst

这些列比延伸到屏幕整个宽度的文本可读性更强。

当然，如果不提供足够数量的 RichTextBlockOverflow 元素，文本会被截断，会永远无法读到故事结尾。因此，最好用代码生成 RichTextBlockOverflow 元素。

如果在 Visual Studio 中创建 Grid App 或 Split App 类型的项目，会在 Common 文件夹得到 RichTextColumns 类。该类派生自 Panel 并在其 MeasureOverride 方法中产生 RichTextBlockOverflow 元素。

下面来讲另一种方法。像前两个项目一样，在接下来的项目中，MainPage.xaml 文件包含 RichTextBlock 元素的完整文本。这一次是 F.司各特·菲茨杰拉德(F. Scott Fitzgerald)所著《伯尼丝剪头发》(*Bernice Bobs Her Hair*)中爵士时代的青少年故事。

项目：BerniceBobsHerHair | 文件：MainPage.xaml(片段)

```
<Page ... >
    <Page.Resources>
        <Style TargetType="RichTextBlock">
            <Setter Property="Width" Value="480" />
            <Setter Property="Margin" Value="24 0 24 0" />
            <Setter Property="FontSize" Value="18" />
            <Setter Property="TextAlignment" Value="Justify" />
        </Style>

        <Style TargetType="RichTextBlockOverflow">
            <Setter Property="Width" Value="480" />
            <Setter Property="Margin" Value="24 0 24 0" />
```

```
        </Style>
    </Page.Resources>

    <Grid Background="{StaticResource ApplicationPageBackgroundThemeBrush}">
        <ScrollViewer HorizontalScrollBarVisibility="Hidden"
                      VerticalScrollBarVisibility="Disabled">
            <StackPanel Name="stackPanel"
                        Orientation="Horizontal">
                <RichTextBlock SizeChanged="OnRichTextBlockSizeChanged">
                    <Paragraph TextAlignment="Center" FontSize="36" Margin="0 0 0 12">
                        "Bernice Bobs Her Hair"
                        <LineBreak />
                        by
                        <LineBreak />
                        F. Scott Fitzgerald
                    </Paragraph>

                    <Paragraph Margin="0 6">
                        After dark on Saturday night one could stand on the first tee of
                        the golf-course and see the country-club windows as a yellow
                        expanse over a very black and wavy ocean. The waves of this
                        ocean, so to speak, were the heads of many curious caddies, a few
                        of the more ingenious chauffeurs, the golf professional's deaf
                        sister&#x2014;and there were usually several stray, diffident waves who
                        might have rolled inside had they so desired. This was the
                        gallery.
                    </Paragraph>
                    ...
                    <Paragraph TextIndent="48" Margin="0 6">
                        Then picking up her staircase she set off at a half-run down the
                        moonlit street.
                    </Paragraph>
                </RichTextBlock>
            </StackPanel>
        </ScrollViewer>
    </Grid>
</Page>
```

注意 SizeChanged 处理程序对 RichTextBlock 的设置。该处理程序会去除之前大小变化期间所创建的所有 RichTextBlockOverflow 元素，然后会重新创建一批。

项目: BerniceBobsHerHair | 文件: MainPage.xaml.cs(片段)

```
void OnRichTextBlockSizeChanged(object sender, SizeChangedEventArgs args)
{
    RichTextBlock richTextBlock = sender as RichTextBlock;

    if (richTextBlock.ActualHeight == 0)
        return;

    // Get rid of all previous RichTextBlockOverflow objects
    while (stackPanel.Children.Count > 1)
        stackPanel.Children.RemoveAt(1);

    if (!richTextBlock.HasOverflowContent)
        return;

    // Create first RichTextBlockOverflow
    RichTextBlockOverflow richTextBlockOverflow = new RichTextBlockOverflow();
    richTextBlock.OverflowContentTarget = richTextBlockOverflow;
    stackPanel.Children.Add(richTextBlockOverflow);

    // Measure it
    richTextBlockOverflow.Measure(new Size(richTextBlockOverflow.Width, this.ActualHeight));

    // If it has overflow content, repeat the process
    while (richTextBlockOverflow.HasOverflowContent)
    {
```

```
            RichTextBlockOverflow newRichTextBlockOverflow = new RichTextBlockOverflow();
            richTextBlockOverflow.OverflowContentTarget = newRichTextBlockOverflow;
            richTextBlockOverflow = newRichTextBlockOverflow;
            stackPanel.Children.Add(richTextBlockOverflow);
            richTextBlockOverflow.Measure(new Size(richTextBlockOverflow.Width, this.ActualHeight)
        }
    }
```

请注意对 RichTextBlock 的 Measure 和 RichTextBlockOverflow 方法的调用。有必要强制该元素确定有多少文字可以在矩形内，并恰当设定 HasOverflowContent 属性。结果如下图所示。

"Bernice Bobs Her Hair"
by
F. Scott Fitzgerald

After dark on Saturday night one could stand on the first tee of the golf-course and see the country-club windows as a yellow expanse over a very black and wavy ocean. The waves of this ocean, so to speak, were the heads of many curious caddies, a few of the more ingenious chauffeurs, the golf professional's deaf sister—and there were usually several stray, diffident waves who might have rolled inside had they so desired. This was the gallery.

The balcony was inside. It consisted of the circle of wicker chairs that lined the wall of the combination clubroom and ballroom. At these Saturday-night dances it was largely feminine; a great babel of middle-aged ladies with sharp eyes and icy hearts behind lorgnettes and large bosoms. The main function of the balcony was critical. It occasionally showed grudging admiration, but never approval, for it is well known among ladies over thirty-five that when the younger set dance in the summer-time it is with the very worst intentions in the world, and if they are not bombarded with stony eyes stray couples will dance weird barbaric interludes in the corners, and the more popular, more dangerous, girls will sometimes be kissed in the parked limousines of unsuspecting dowagers.

But, after all, this critical circle is not close enough to the stage to see the actors' faces and catch the subtler byplay. It can only frown and lean, ask questions and make satisfactory deductions from its set of postulates, such as the one which states that every young man with a large income leads the life of a hunted partridge. It never really appreciates the drama of the shifting, semi-cruel world of adolescence. No; boxes, orchestra-circle, principals, and chorus be represented by the medley of faces and voices that sway to the plaintive African rhythm of Dyer's dance orchestra.

From sixteen-year-old Otis Ormonde, who has two more years at Hill School, to G. Reece Stoddard, over whose bureau at home hangs a Harvard law diploma; from little Madeleine Hogue, whose hair still feels strange and uncomfortable on top of her head, to Bessie MacRae, who has been the life of the party a little too long—more than ten years—the medley is not only the centre of the stage but contains the only people capable of getting an unobstructed view of it.

With a flourish and a bang the music stops. The couples exchange artificial, effortless smiles, facetiously repeat "*la-de-da-dadum-dum*," and then the clatter of young feminine voices soars over the burst of clapping.

A few disappointed stags caught in midfloor as they had been about to cut in subsided listlessly back to the walls, because this was not like the riotous Christmas dances—these summer hops were considered just pleasantly warm and exciting, where even the younger marrieds rose and performed ancient waltzes and terrifying fox trots to the tolerant amusement of their younger brothers and sisters.

Warren McIntyre, who casually attended Yale, being one of the unfortunate stags, felt in his dinner-coat pocket for a cigarette and strolled out onto the wide, semidark veranda, where couples were scattered at tables, filling the lantern-hung night with vague words and hazy laughter. He nodded here and there at the less absorbed and as he passed each couple some half-forgotten fragment of a story played in his mind, for it was not a large city and every one was Who's Who to every one else's past. There, for example, were Jim Strain and Ethel Demorest, engaged for three years. Every one managed to hold a job for more would marry him. Yet how bored th wearily Ethel regarded Jim sometin why she had trained the vines of wind-shaken poplar.

Warren was nineteen and ratl his friends who hadn't gone East t boys, he bragged tremendously ab when he was away from it. There w who regularly made the rounds o and football games at Princeton, Ya there was black-eyed Roberta Dil famous to her own generation a Cobb; and, of course, there was besides having a fairylike face and tongue was already justly celebrate cart-wheels in succession during th dance at New Haven.

Warren, who had grown up Marjorie, had long been "crazy ab seemed to reciprocate his feeling w she had tried him by her infallible gravely that she did not love him. she was away from him she forgot other boys. Warren found this dis Marjorie had been making little trip first two or three days after each ar heaps of mail on the Harveys' hall t various masculine handwritings. To during the month of August she h cousin Bernice from Eau Claire, and see her alone. It was always neces

16.8　分页的危险

我们已经看到 RichTextBlock 和 RichTextBlockOverflow 对几个短篇小说的应用，问题自然归结于一点：能否对整部小说使用同样的方法？

我们来试一试。但我们先尝试一部短篇小说，例如，乔治·艾略特(George Eliot)的著作《织工马南》(*Silas Marner*)。SilasMarner 程序会显示实际页面，而不是显示文本列，并且使用 FlipView 来显示 RichTextBlock 和 RichTextBlockOverflow 元素。

FlipView 能非常好地赋予程序真实电子书阅读器的界面外观。每页都可以占据整个屏幕(或接近全屏幕)，可以滑动手指来前后翻页。另外，我把 Slider 放到页面底部，提供阅读进度的视觉展现，并能非常快速导航到任何一页。

与本书其他程序相比，SilasMarner 更适合于竖屏模式而不是横屏模式。在横屏模式中，视线长度太宽而无法舒适地阅读，如下图所示。

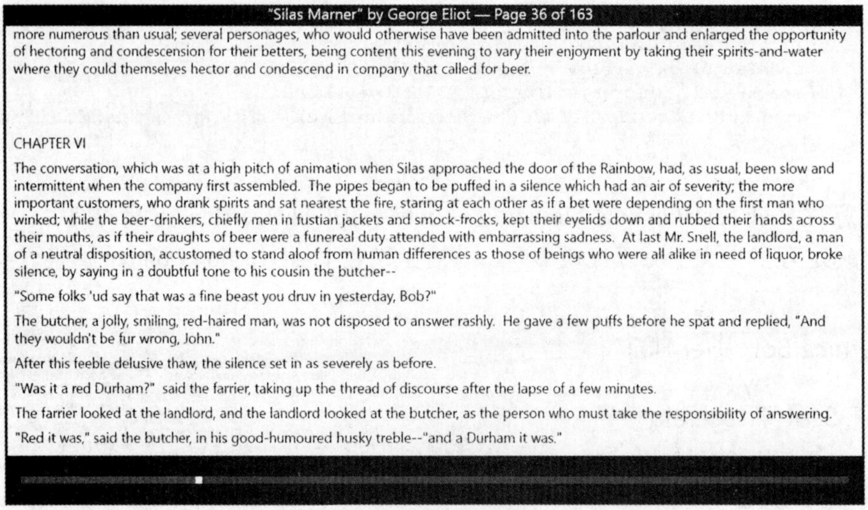

"Silas Marner" by George Eliot — Page 36 of 163

more numerous than usual; several personages, who would otherwise have been admitted into the parlour and enlarged the opportunity of hectoring and condescension for their betters, being content this evening to vary their enjoyment by taking their spirits-and-water where they could themselves hector and condescend in company that called for beer.

CHAPTER VI

The conversation, which was at a high pitch of animation when Silas approached the door of the Rainbow, had, as usual, been slow and intermittent when the company first assembled. The pipes began to be puffed in a silence which had an air of severity; the more important customers, who drank spirits and sat nearest the fire, staring at each other as if a bet were depending on the first man who winked; while the beer-drinkers, chiefly men in fustian jackets and smock-frocks, kept their eyelids down and rubbed their hands across their mouths, as if their draughts of beer were a funereal duty attended with embarrassing sadness. At last Mr. Snell, the landlord, a man of a neutral disposition, accustomed to stand aloof from human differences as those of beings who were all alike in need of liquor, broke silence, by saying in a doubtful tone to his cousin the butcher--

"Some folks 'ud say that was a fine beast you druv in yesterday, Bob?"

The butcher, a jolly, smiling, red-haired man, was not disposed to answer rashly. He gave a few puffs before he spat and replied, "And they wouldn't be fur wrong, John."

After this feeble delusive thaw, the silence set in as severely as before.

"Was it a red Durham?" said the farrier, taking up the thread of discourse after the lapse of a few minutes.

The farrier looked at the landlord, and the landlord looked at the butcher, as the person who must take the responsibility of answering.

"Red it was," said the butcher, in his good-humoured husky treble--"and a Durham it was."

如果横过来，阅读体验会好很多，如下图所示。

"Silas Marner" by George Eliot — Page 39 of 137

leave off skulking i' the dark and i' lone places--let 'em come where there's company and candles."

"As if ghos'es 'ud want to be believed in by anybody so ignirant!" said Mr. Macey, in deep disgust at the farrier's crass incompetence to apprehend the conditions of ghostly phenomena.

CHAPTER VII

Yet the next moment there seemed to be some evidence that ghosts had a more condescending disposition than Mr. Macey attributed to them; for the pale thin figure of Silas Marner was suddenly seen standing in the warm light, uttering no word, but looking round at the company with his strange unearthly eyes. The long pipes gave a simultaneous movement, like the antennae of startled insects, and every man present, not excepting even the sceptical farrier, had an impression that he saw, not Silas Marner in the flesh, but an apparition; for the door by which Silas had entered was hidden by the high-screened seats, and no one had noticed his approach. Mr. Macey, sitting a long way off the ghost, might be supposed to have felt an argumentative triumph, which would tend to neutralize his share of the general alarm. Had he not always said that when Silas Marner was in that strange trance of his, his soul went loose from his body? Here was the demonstration: nevertheless, on the whole, he would have been as well contented without it. For a few moments there was a dead silence, Marner's want of breath and agitation not allowing him to speak. The landlord, under the habitual sense that he was bound to keep his house open to all company, and confident in the protection of his unbroken neutrality, at last took on himself the task of adjuring the ghost.

"Master Marner," he said, in a conciliatory tone, "what's lacking to you? What's your business here?"

"Robbed!" said Silas, gaspingly. "I've been robbed! I want the constable--and the Justice--and Squire Cass--and Mr. Crackenthorp."

"Lay hold on him, Jem Rodney," said the landlord, the idea of a ghost subsiding; "he's off his head, I doubt. He's wet through."

Jem Rodney was the outermost man, and sat conveniently near Marner's standing-place; but he declined to give his services.

"Come and lay hold on him yourself, Mr. Snell, if you've a mind," said Jem, rather sullenly. "He's been robbed, and murdered too, for what I know," he added, in a muttering tone.

"Jem Rodney!" said Silas, turning and fixing his strange eyes on the suspected man.

"Aye, Master Marner, what do you want wi' me?" said Jem, trembling a little, and seizing his drinking-can as a defensive weapon.

然而，我坚持程序竖屏编程这个偏好。在实际电子书阅读器上(我广泛使用该术语来描述程序中读取大块文本的任何功能)能有效分页至关重要。即使用户不能改变屏幕尺寸，大多数电子书阅读器仍然允许用户改变字型或字型大小，而这会影响到书的分页。

在平板电脑上运行 SilasMarner，可以旋转，尝试辅屏视图，并直接观察 RichTextBlock

和 RichTextBlockOverflow 元素对文档重新分页需要多长时间。

电子书阅读器有一个臭名昭著的问题，涉及到维护和显示有意义的页码。任何时候文档重新分页，就会有不同页数。SilasMarner 程序在本地存储中保存信息，可以让你从上次离开时的地方继续阅读，但程序没有为此目的保存页码。相反，程序用当前页除以页面总数计算得到。这是程序启动之间需要维护的的唯一值。

项目：SilasMarner | 文件：MainPage.xaml.cs(片段)

```
public sealed partial class MainPage : Page
{
    double fractionRead;

    public MainPage()
    {
        this.InitializeComponent();

        // Save and reload fraction read
        IPropertySet propertySet = ApplicationData.Current.LocalSettings.Values;

        Application.Current.Suspending += (sender, args) =>
            {
                propertySet["FractionRead"] = fractionRead;
            };

        if (propertySet.ContainsKey("FractionRead"))
            fractionRead = (double)propertySet["FractionRead"];
    }
    ...
}
```

如果新的页面大小和程序最后保存的值不完全相同，则看不到完全相同的一页，但至少很接近。

真正的电子书阅读器可能会下载书籍。对于该演示程序，我把古登堡计划文件作为程序资源。把本书分段是代码隐藏文件的责任。

因此，XAML 文件中没有书的文本。而 XAML 文件在中心有一个 FlipView，还会显示书的标题、当前页和页数。头部当然不是绝对必需，但我想清楚表示页数和当前页码。Slider 位于底部。

项目：SilasMarner | 文件：MainPage.xaml(片段)

```
<Page ... >
    <Grid Background="{StaticResource ApplicationPageBackgroundThemeBrush}">
        <Grid.RowDefinitions>
            <RowDefinition Height="Auto" />
            <RowDefinition Height="*" />
            <RowDefinition Height="Auto" />
        </Grid.RowDefinitions>

        <StackPanel Grid.Row="0"
                    Orientation="Horizontal"
                    HorizontalAlignment="Center">
            <StackPanel.Resources>
                <Style TargetType="TextBlock">
                    <Setter Property="FontSize" Value="24" />
                </Style>
            </StackPanel.Resources>

            <TextBlock Text="&#x201C;Silas Marner&#x201D; by George Eliot" />
            <TextBlock Text=" &#x2014; Page " />
            <TextBlock Name="pageNumber" />
            <TextBlock Text=" of " />
            <TextBlock Name="pageCount" />
```

```
            </StackPanel>

            <FlipView Name="flipView"
                      Grid.Row="1"
                      Background="White"
                      SizeChanged="OnFlipViewSizeChanged"
                      SelectionChanged="OnFlipViewSelectionChanged">
                <FlipView.ItemsPanel>
                    <ItemsPanelTemplate>
                        <StackPanel Orientation="Horizontal" />
                    </ItemsPanelTemplate>
                </FlipView.ItemsPanel>
            </FlipView>

            <Slider Name="pageSlider"
                    Grid.Row="2"
                    Margin="24 12 24 0"
                    ValueChanged="OnPageSliderValueChanged" />
        </Grid>
    </Page>
```

程序的关键部分在 FlipView 的 SizeChanged 处理程序。基于 FlipView 的大小，程序必须生成适当数量的 RichTextBlockOverflow 元素。

程序启动后，SizeChanged 事件第一次触发，处理程序必须访问书籍文件，并分成段落。在古登堡计划纯文本文件中，每段都包含硬回车行序列。段落用空白行来分隔。处理程序必须创建 Paragraph 对象，并添加到 RichTextBlock 元素，以处理该文本。

项目: SilasMarner | 文件: MainPage.xaml.cs(片段)

```
async void OnFlipViewSizeChanged(object sender, SizeChangedEventArgs args)
{
    // Get the size of the FlipView
    Size containerSize = args.NewSize;

    // Actual value gets modified during processing here, so save it
    double saveFractionRead = fractionRead;

    // First time through after program is launched
    if (flipView.Items.Count == 0)
    {
        // Load book resource
        IList<string> bookLines =
            await PathIO.ReadLinesAsync("ms-appx:///Books/pg550.txt",
                                        UnicodeEncoding.Utf8);

        // Create RichTextBlock
        RichTextBlock richTextBlock = new RichTextBlock
        {
            FontSize = 22,
            Foreground = new SolidColorBrush(Colors.Black)
        };

        // Create paragraphs
        Paragraph paragraph = new Paragraph();
        paragraph.Margin = new Thickness(12);
        richTextBlock.Blocks.Add(paragraph);

        foreach (string line in bookLines)
        {
            // End of paragraph, make new Paragraph
            if (line.Length == 0)
            {
                paragraph = new Paragraph();
                paragraph.Margin = new Thickness(12);
                richTextBlock.Blocks.Add(paragraph);
            }
```

```
            // Continue the paragraph
            else
            {
                string textLine = line;
                char lastChar = line[line.Length - 1];

                if (lastChar != ' ')
                    textLine += ' ';

                if (line[0] == ' ')
                    paragraph.Inlines.Add(new LineBreak());

                paragraph.Inlines.Add(new Run { Text = textLine });
            }
        }

        // Make RichTextBlock the same size as the FlipView
        flipView.Items.Add(richTextBlock);
        richTextBlock.Measure(containerSize);

        // Generate RichTextBlockOverflow elements
        if (richTextBlock.HasOverflowContent)
        {
            // Add the first one
            RichTextBlockOverflow richTextBlockOverflow = new RichTextBlockOverflow();
            richTextBlock.OverflowContentTarget = richTextBlockOverflow;
            flipView.Items.Add(richTextBlockOverflow);
            richTextBlockOverflow.Measure(containerSize);

            // Add subsequent ones
            while (richTextBlockOverflow.HasOverflowContent)
            {
                RichTextBlockOverflow newRichTextBlockOverflow = new RichTextBlockOverflow();
                richTextBlockOverflow.OverflowContentTarget = newRichTextBlockOverflow;
                richTextBlockOverflow = newRichTextBlockOverflow;
                flipView.Items.Add(richTextBlockOverflow);
                richTextBlockOverflow.Measure(containerSize);
            }
        }
    }
    ...
}
```

在随后触发的 SizeChanged 处理程序中，程序可以清除 FlipView 并重新开始，但我决定如果需要则添加新的 RichTextBlockOverflow 元素或者移除不再需要的，试一下更高效率。

项目：SilasMarner | 文件：MainPage.xaml.cs(片段)

```
async void OnFlipViewSizeChanged(object sender, SizeChangedEventArgs args)
{
    ...
    // Subsequent SizeChanged events
    else
    {
        // Resize all the items in the FlipView
        foreach (object obj in flipView.Items)
        {
            (obj as FrameworkElement).Measure(containerSize);
        }

        // Generate new RichTextBlockOverflow elements if needed
        while ((flipView.Items[flipView.Items.Count - 1]
                    as RichTextBlockOverflow).HasOverflowContent)
        {
            RichTextBlockOverflow richTextBlockOverflow =
                    flipView.Items[flipView.Items.Count - 1] as RichTextBlockOverflow;
```

```
                RichTextBlockOverflow newRichTextBlockOverflow = new RichTextBlockOverflow();
                richTextBlockOverflow.OverflowContentTarget = newRichTextBlockOverflow;
                richTextBlockOverflow = newRichTextBlockOverflow;
                flipView.Items.Add(richTextBlockOverflow);
                richTextBlockOverflow.Measure(args.NewSize);
            }
            // Remove superfluous RichTextBlockOverflow elements
            while (!(flipView.Items[flipView.Items.Count - 2]
                                as RichTextBlockOverflow).HasOverflowContent)
            {
                flipView.Items.RemoveAt(flipView.Items.Count - 1);
            }
        }
        ...
    }
```

然而，我发现(你可能也会发现)，这种逻辑似乎计算了数量不足的 RichTextBlockOverflow 元素。整部小说虽然保留下来了，但文件末尾的一些许可信息被截断。我不知道为什么会这样。

SizeChanged 最后初始化标题文本和 Slider，把 FlipView 的 SelectedIndex 属性设置为基于 fractionRead 的值。

项目: SilasMarner | 文件: MainPage.xaml.cs(片段)

```
async void OnFlipViewSizeChanged(object sender, SizeChangedEventArgs args)
{
    ...
    // Initialize the header and Slider
    int count = flipView.Items.Count;
    pageNumber.Text = "1";              // probably modified soon
    pageCount.Text = count.ToString();
    pageSlider.Minimum = 1;
    pageSlider.Maximum = flipView.Items.Count;
    pageSlider.Value = 1;                   // probably modified soon

    // Go to approximate page
    fractionRead = saveFractionRead;
    flipView.SelectedIndex = (int)Math.Min(count - 1, fractionRead * count);
}
```

这是程序的大部分。用于 FlipView 的 SelectionChanged 处理程序改变标题和 Slider，而用于 Slider 的 ValueChanged 处理程序改变 FlipView 的 SelectedIndex 属性。

项目: SilasMarner | 文件: MainPage.xaml.cs(片段)

```
public sealed partial class MainPage : Page
{
    ...
    void OnFlipViewSelectionChanged(object sender, SelectionChangedEventArgs args)
    {
        int pageNum = flipView.SelectedIndex + 1;
        pageNumber.Text = pageNum.ToString();
        fractionRead = (pageNum - 1.0) / flipView.Items.Count;
        pageSlider.Value = pageNum;
    }

    void OnPageSliderValueChanged(object sender, RangeBaseValueChangedEventArgs args)
    {
        flipView.SelectedIndex = Math.Min(flipView.Items.Count, (int)args.NewValue) - 1;
    }
}
```

现在，程序完成了，如何发挥作用？

我认为程序并不是很好。每次 SizeChanged 处理程序执行时都需要几秒钟，代码不能移动到辅助线程，因为几乎所有代码都涉及用户界面对象。另外，我发现在元素内容周围有一些偏移，表明我在挑战 RichTextBlock 分页。

这些问题意味着 RichTextBlock 必须放弃用于大型文档分页。程序本身必须承担这些任务，最有效的方式是基于文本规格信息。如果有兴趣探讨这些问题，以及可能的解决方案，可以参考我写的一系列文章，发表于 2011 年的 6～11 月 *MSDN Magazine* (http://msdn.microsoft.com/en-us/magazine，选择期号并下载)。这些文章介绍 Windows Phone 7 代码，但原理与处理 Windows 8 非常类似。

把大文件分成章节尤其绝对有用。在传统排版以及电子书阅读器中，章节代表分页重新开始的位置点。在特定章节内，可以随每个新页面而按需执行分页。

16.9　使用 RichEditBox 富文本编辑

就像 TextBlock 有增强版称为 RichTextBlock 一样，TextBox 也有增强版，但并不称为 RichTextBox，实际上称为 RichEditBox。

如果把 TextBox 当成传统 Windows Notepad 程序的“引擎”，则可以把 RichEditBox 视为 Windows WordPad 程序的引擎。RichEditBox 可以用程序调整选择文本范围或者(更常见的)允许用户选择文本范围，并将特殊字符和段落格式应用到该选择。RichEditBox 有内置文件加载及保存选项，但该选项只支持古怪的 RTF。

以下讨论只简单触及 RichEditBox。你想要通过 Document 属性来探讨该类的独特功能。Document 属性在内部设置为实施 ITextDocument 接口的对象，并在 Windows.UI.Text 命名空间中进行定义。接口支持加载并保存到数据流、用来设置和获取默认字符和段落格式的方法以及在文档范围内设置文本格式的方法。

ITextDocument 还支持 Selection 属性，Selection 指用户所选文档区域。Selection 属性为实施 ITextRange 接口的 ITextSelection 类型。ITextRange 接口支持剪贴板复制和粘贴，并规定 CharacterFormat 和 PragraphFormat 属性，后两者分别引用实施 ITextCharacterFormat 和 ITextParagraphFormat 接口的对象。

我们来用 RichEditBox 构建一个基本的富文本编辑器，命名为 RichTextEditor。程序在顶部有应用栏，以应用字符格式(左侧)和段落格式(右侧)，底部应用栏用于加载和保存文件，如下图所示。

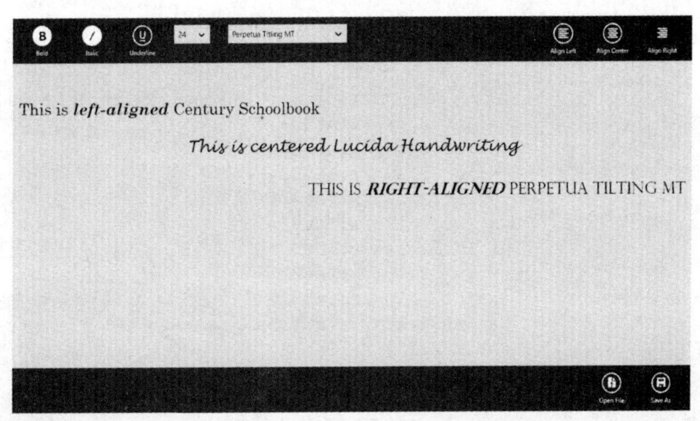

一旦开始组合这类程序，你会意识到难点并不在于 RichEditBox 编程接口，该接口计算如何组织用户界面。我用了 StandardStyles.xaml 中的 8 种按钮样式，但 StandardStyles.xaml 包含字型、字型颜色和字型大小的按钮样式，但如果要使用这些按钮，则需要调用弹出对话框，而这又是我在该程序中想避免的。因此，大小和字型家族库实现为应用栏中的 ComboBox 控件，而且没有颜色选择。就像我刚刚说的，我只是蜻蜓点水一样谈谈 RichEditBox。

以下为 XAML 文件。可以看出，RichEditBox 的标记由于只有两个 AppBar 定义而显得很少。

项目: RichTextEditor | 文件: MainPage.xaml(片段)

```
<Page
    x:Class="RichTextEditor.MainPage"
    xmlns="http://schemas.microsoft.com/winfx/2006/xaml/presentation"
    xmlns:x="http://schemas.microsoft.com/winfx/2006/xaml"
    xmlns:local="using:RichTextEditor"
    xmlns:d="http://schemas.microsoft.com/expression/blend/2008"
    xmlns:mc="http://schemas.openxmlformats.org/markup-compatibility/2006"
    mc:Ignorable="d">

    <Grid Background="{StaticResource ApplicationPageBackgroundThemeBrush}">
        <RichEditBox Name="richEditBox" />
    </Grid>

    <Page.TopAppBar>
        <AppBar Opened="OnTopAppBarOpened">
            <Grid>
                <StackPanel Orientation="Horizontal"
                            HorizontalAlignment="Left">
                    <!-- For CheckBox's, need to comment out BackgroundCheckedGlyph in
                         AppBarButtonStyle in StandardStyles.xaml -->
                    <CheckBox Name="boldAppBarCheckBox"
                              Style="{StaticResource BoldAppBarButtonStyle}"
                              Checked="OnBoldAppBarCheckBoxChecked"
                              Unchecked="OnBoldAppBarCheckBoxChecked" />

                    <CheckBox Name="italicAppBarCheckBox"
                              Style="{StaticResource ItalicAppBarButtonStyle}"
                              Checked="OnItalicAppBarCheckBoxChecked"
                              Unchecked="OnItalicAppBarCheckBoxChecked" />

                    <CheckBox Name="underlineAppBarCheckBox"
                              Style="{StaticResource UnderlineAppBarButtonStyle}"
                              Checked="OnUnderlineAppBarCheckBoxChecked"
                              Unchecked="OnUnderlineAppBarCheckBoxChecked" />

                    <ComboBox Name="fontSizeComboBox"
                              Width="72"
                              Margin="12 12 24 36"
                              SelectionChanged="OnFontSizeComboBoxSelectionChanged">
                        <x:Int32>8</x:Int32>
                        <x:Int32>9</x:Int32>
                        <x:Int32>10</x:Int32>
                        <x:Int32>11</x:Int32>
                        <x:Int32>12</x:Int32>
                        <x:Int32>14</x:Int32>
                        <x:Int32>16</x:Int32>
                        <x:Int32>18</x:Int32>
                        <x:Int32>20</x:Int32>
                        <x:Int32>22</x:Int32>
                        <x:Int32>24</x:Int32>
                        <x:Int32>26</x:Int32>
                        <x:Int32>28</x:Int32>
```

```
                        <x:Int32>36</x:Int32>
                        <x:Int32>48</x:Int32>
                        <x:Int32>72</x:Int32>
                    </ComboBox>

                    <ComboBox Name="fontFamilyComboBox"
                              Width="240"
                              Margin="12 12 24 36"
                              SelectionChanged="OnFontFamilyComboBoxSelectionChanged" />
                </StackPanel>
                <StackPanel Orientation="Horizontal"
                            HorizontalAlignment="Right">
                    <StackPanel Name="alignmentPanel"
                                Orientation="Horizontal">
                        <RadioButton Name="alignLeftAppBarRadioButton"
                                     Style="{StaticResource AlignLeftAppBarButtonStyle}"
                                     Checked="OnAlignAppBarRadioButtonChecked" />

                        <RadioButton Name="alignCenterAppBarRadioButton"
                                     Style="{StaticResource AlignCenterAppBarButtonStyle}"
                                     Checked="OnAlignAppBarRadioButtonChecked" />

                        <RadioButton Name="alignRightAppBarRadioButton"
                                     Style="{StaticResource AlignRightAppBarButtonStyle}"
                                     Checked="OnAlignAppBarRadioButtonChecked" />
                    </StackPanel>
                </StackPanel>
            </Grid>
        </AppBar>
    </Page.TopAppBar>
    <Page.BottomAppBar>
        <AppBar>
            <Grid>
                <StackPanel Orientation="Horizontal"
                            HorizontalAlignment="Left" />

                <StackPanel Orientation="Horizontal"
                            HorizontalAlignment="Right">

                    <Button Style="{StaticResource OpenFileAppBarButtonStyle}"
                            Click="OnOpenAppBarButtonClick" />

                    <Button Style="{StaticResource SaveAppBarButtonStyle}"
                            AutomationProperties.Name="Save As"
                            Click="OnSaveAsAppBarButtonClick" />
                </StackPanel>
            </Grid>
        </AppBar>
    </Page.BottomAppBar>
</Page>
```

CheckBox 用于表示标签为 Bold、Italic 和 Underline 的三个按钮。这三项要么开启要么关闭。用于字型大小的 CheckBox 采用明确值进行初始化。文本编辑器通常提供输入自定义值功能，但由于简洁原因，我去除该功能。字体家族的 CheckBox 在代码隐藏文件中进行初始化。用于文本对齐的三个按钮是一组 RadioButton 控件。

因为该项目要通过安装在系统中的字型列表填写第二个 CheckBox，因此包括第 15 章的 DirectXWrapper 项目。ComboBox 在 Loaded 处理程序中进行初始化。Loaded 处理程序还从本地存储加载进行中的文档，以及涉及用户最后所选文档的两个设置。文档和这些设置保存到 Suspending 处理程序。

```
项目: RichTextEditor | 文件: MainPage.xaml.cs(片段)
public sealed partial class MainPage : Page
{
```

```
public MainPage()
{
    this.InitializeComponent();
    Loaded += OnLoaded;
    Application.Current.Suspending += OnAppSuspending;
}

async void OnLoaded(object sender, RoutedEventArgs args)
{
    // Get fonts from DirectXWrapper library
    WriteFactory writeFactory = new WriteFactory();
    WriteFontCollection writeFontCollection =
            writeFactory.GetSystemFontCollection();

    int count = writeFontCollection.GetFontFamilyCount();
    string[] fonts = new string[count];

    for (int i = 0; i < count; i++)
    {
        WriteFontFamily writeFontFamily =
                            writeFontCollection.GetFontFamily(i);

        WriteLocalizedStrings writeLocalizedStrings =
                            writeFontFamily.GetFamilyNames();

        int index;

        if (writeLocalizedStrings.FindLocaleName("en-us", out index))
            fonts[i] = writeLocalizedStrings.GetString(index);
        else
            fonts[i] = writeLocalizedStrings.GetString(0);
    }

    Array.Sort<string>(fonts);
    fontFamilyComboBox.ItemsSource = fonts;

    // Load current document
    StorageFolder localFolder = ApplicationData.Current.LocalFolder;

    try
    {
        StorageFile storageFile = await localFolder.CreateFileAsync("RichTextEditor.rtf","
                                        CreationCollisionOption.OpenIfExists);
        IRandomAccessStream stream = await storageFile.OpenAsync(FileAccessMode.Read);
        richEditBox.Document.LoadFromStream(TextSetOptions.FormatRtf, stream);
    }
    catch
    {
        // Ignore exceptions here
    }

    // Load selection settings
    IPropertySet propertySet = ApplicationData.Current.LocalSettings.Values;

    if (propertySet.ContainsKey("SelectionStart"))
        richEditBox.Document.Selection.StartPosition = (int)propertySet["SelectionStart"];

    if (propertySet.ContainsKey("SelectionEnd"))
        richEditBox.Document.Selection.EndPosition = (int)propertySet["SelectionEnd"];
}

async void OnAppSuspending(object sender, SuspendingEventArgs args)
{
    SuspendingDeferral deferral = args.SuspendingOperation.GetDeferral();

    // Save current document
    StorageFolder localFolder = ApplicationData.Current.LocalFolder;
```

```
        try
        {
            StorageFile storageFile = await localFolder.CreateFileAsync("RichTextEditor.rtf",
                                        CreationCollisionOption.ReplaceExisting);
            IRandomAccessStream stream = await storageFile.OpenAsync(FileAccessMode.ReadWrite);
            richEditBox.Document.SaveToStream(TextGetOptions.FormatRtf, stream);
        }
        catch
        {
            // Ignore exceptions here
        }

        // Save selection settings
        IPropertySet propertySet = ApplicationData.Current.LocalSettings.Values;
        propertySet["SelectionStart"] = richEditBox.Document.Selection.StartPosition;
        propertySet["SelectionEnd"] = richEditBox.Document.Selection.EndPosition;

        deferral.Complete();
    }
    ...
}
```

ITextDocument 接口所定义的 LoadFromStream 和 SaveToStream 方法需要 FormatRtf 枚举项加载并保存 RTF 文件。否则，该方法只加载并保存纯文本。

Suspending 处理程序只保存 ITextSelection 对象的 StartPosition 和 EndPosition 属性，而 ITextSelection 显示为 ITextDocument 的 Selection 属性。如果实际没有选择文字，则这些值相同，同时表明光标在文档的当前位置，即插入键入文本的当前插入点。

程序不保存任何格式信息，因为程序不维护适用于新文档或纯文本文件的任何默认格式。应用栏上的格式项仅适用于文档(或插入点)的特定选择。所有格式规范均属于 RichEditBox 所维护的文档内部。当然，程序可能允许用户为整个文档选择默认格式，但这是另一个用户界面的问题。

程序不维护任何格式信息，因此应用栏出现时，程序必须初始化顶端应用栏中的所有文本格式项，以 Opened 事件为标志。这些项均基于文档中的当前选择或插入点而进行初始化。可从 ITextSelection 对象的 CharacterFormat 和 ParagraphFormat 属性获得当前设置，ITextSelection 显示为 ITextDocument 的 Selection 属性。

项目: RichTextEditor | 文件: MainPage.xaml.cs(片段)

```
void OnTopAppBarOpened(object sender, object args)
{
    // Get the character formatting at the current selection
    ITextCharacterFormat charFormat = richEditBox.Document.Selection.CharacterFormat;

    // Set the CheckBox app bar buttons
    boldAppBarCheckBox.IsChecked = charFormat.Bold == FormatEffect.On;
    italicAppBarCheckBox.IsChecked = charFormat.Italic == FormatEffect.On;
    underlineAppBarCheckBox.IsChecked = charFormat.Underline == UnderlineType.Single;

    // Set the two ComboBox's
    fontSizeComboBox.SelectedItem = (int)charFormat.Size;
    fontFamilyComboBox.SelectedItem = charFormat.Name;

    // Get the paragraph alignment and set the RadioButton's
    ParagraphAlignment paragraphAlign =
                        richEditBox.Document.Selection.ParagraphFormat.Alignment;
    alignLeftAppBarRadioButton.IsChecked = paragraphAlign == ParagraphAlignment.Left;
    alignCenterAppBarRadioButton.IsChecked = paragraphAlign == ParagraphAlignment.Center;
    alignRightAppBarRadioButton.IsChecked = paragraphAlign == ParagraphAlignment.Right;
}
```

ITextCharacterFormat 对象定义 Bold、Italic 和 Underline 属性(如你所见)，而且还用熟悉的 FontStyle 属性和 Weight 属性进行补充，Weight 属性是对应于 FontWeights 类的属性的数值。

FormatEffect 取值为 ON、OFF、Toggle 和 Undefined 枚举。如果当前选择含有一些斜体和非斜体文本，则 Italic 属性的值为 FormatEffect.Undefined，而且相应应用栏按钮应该可能设置为不确定状态，但通过标准应用栏的风格，这种状态看起来与未选中状态相同，因此我没有理会。

请注意，供选择的字型家族由 ITextCharacterFormat 对象的字符串 Name 属性所提供。该属性名称非常常见，所以很容易被忽视。

Bold、Italic 和 Underline 按钮处理方式与此类似。ITextCharacterFormat 对象的 Bold、Italic 和 Underline 属性根据 CheckBox 状态进行设置，因此，这些设置应用于当前选择或插入点。

项目：RichTextEditor | 文件：MainPage.xaml.cs(片段)

```
void OnBoldAppBarCheckBoxChecked(object sender, RoutedEventArgs args)
{
    richEditBox.Document.Selection.CharacterFormat.Bold =
        (sender as CheckBox).IsChecked.Value ? FormatEffect.On : FormatEffect.Off;
}

void OnItalicAppBarCheckBoxChecked(object sender, RoutedEventArgs args)
{
    richEditBox.Document.Selection.CharacterFormat.Italic =
        (sender as CheckBox).IsChecked.Value ? FormatEffect.On : FormatEffect.Off;
}

void OnUnderlineAppBarCheckBoxChecked(object sender, RoutedEventArgs args)
{
    richEditBox.Document.Selection.CharacterFormat.Underline =
        (sender as CheckBox).IsChecked.Value ? UnderlineType.Single : UnderlineType.None;
}
```

两个 ComboBox 控件的处理程序也一样简单。

项目：RichTextEditor | 文件：MainPage.xaml.cs(片段)

```
void OnFontSizeComboBoxSelectionChanged(object sender, SelectionChangedEventArgs args)
{
    ComboBox comboBox = sender as ComboBox;
    if (comboBox.SelectedItem != null)
    {
        richEditBox.Document.Selection.CharacterFormat.Size = (int)comboBox.SelectedItem;
    }
}

void OnFontFamilyComboBoxSelectionChanged(object sender, SelectionChangedEventArgs args)
{
    ComboBox comboBox = sender as ComboBox;
    if (comboBox.SelectedItem != null)
    {
        richEditBox.Document.Selection.CharacterFormat.Name = (string)comboBox.SelectedItem;
    }
}
```

应用栏上的最后一个格式项适用于段落。基于所选 RadioButton，ITextParagraphFormat 对象的 Alignment 属性设置为 ParagraphAlignment 枚举项之一。

项目：RichTextEditor | 文件：MainPage.xaml.cs(片段)

```
void OnAlignAppBarRadioButtonChecked(object sender, RoutedEventArgs args)
{
```

```
        ParagraphAlignment paragraphAlign = ParagraphAlignment.Undefined;

        if (sender == alignLeftAppBarRadioButton)
            paragraphAlign = ParagraphAlignment.Left;

        else if (sender == alignCenterAppBarRadioButton)
                paragraphAlign = ParagraphAlignment.Center;

        else if (sender == alignRightAppBarRadioButton)
                paragraphAlign = ParagraphAlignment.Right;

        richEditBox.Document.Selection.ParagraphFormat.Alignment = paragraphAlign;
    }
```

MainPage 中唯一剩余的代码处理底部应用栏的 Open File 和 Save As 按钮。程序允许加载和保存扩展名为.txt 和.rtf 的文件。在适合当前文档的 Loaded 和 Suspending 处理程序的代码之后，这段代码相当明确。

项目：RichTextEditor | 文件：MainPage.xaml.cs(片段)

```
async void OnOpenAppBarButtonClick(object sender, RoutedEventArgs args)
{
    FileOpenPicker picker = new FileOpenPicker();
    picker.FileTypeFilter.Add(".txt");
    picker.FileTypeFilter.Add(".rtf");
    StorageFile storageFile = await picker.PickSingleFileAsync();

    // If user presses Cancel, result is null
    if (storageFile == null)
        return;

    TextSetOptions textOptions = TextSetOptions.None;

    if (storageFile.ContentType != "text/plain")
        textOptions = TextSetOptions.FormatRtf;

    string message = null;

    try
    {
        IRandomAccessStream stream = await storageFile.OpenAsync(FileAccessMode.Read);
        richEditBox.Document.LoadFromStream(textOptions, stream);
    }
    catch (Exception exc)
    {
        message = exc.Message;
    }

    if (message != null)
    {
        MessageDialog msgdlg = new MessageDialog("The file could not be opened. " +
                                                "Windows reports the following error: " +
                                                message, "RichTextEditor");
        await msgdlg.ShowAsync();
    }
}

async void OnSaveAsAppBarButtonClick(object sender, RoutedEventArgs args)
{
    FileSavePicker picker = new FileSavePicker();
    picker.DefaultFileExtension = ".rtf";
    picker.FileTypeChoices.Add("Rich Text Document", new List<string> { ".rtf" });
    picker.FileTypeChoices.Add("Text Document", new List<string> { ".txt" });
    StorageFile storageFile = await picker.PickSaveFileAsync();

    // If user presses Cancel, result is null
    if (storageFile == null)
```

```
        return;

    TextGetOptions textOptions = TextGetOptions.None;

    if (storageFile.ContentType != "text/plain")
        textOptions = TextGetOptions.FormatRtf;

    string message = null;

    try
    {
        IRandomAccessStream stream = await storageFile.OpenAsync(FileAccessMode.ReadWrite);
        richEditBox.Document.SaveToStream(textOptions, stream);
    }
    catch (Exception exc)
    {
        message = exc.Message;
    }

    if (message != null)
    {
        MessageDialog msgdlg = new MessageDialog("The file could not be saved. " +
                                                 "Windows reports the following error: " +
                                                 message, "RichTextEditor");
        await msgdlg.ShowAsync();
    }
}
```

基于文件选择器所返回的 StorageFile 的 MIME 类型，这两种方法决定是否使用 TextSetOptions.FormatRtf 和 TextGetOptions.FormatRtf 标志。我的经验表明，文件选择器表明所选文件 MIME 类型要么是扩展名为.txt 的 text/plain 文件，要么是扩展名为.rtf 的 application/msword 文件，但如果 text/rtf 和 application/rtf 的 MIME 类型也与 RTF 文件相关联，我还是会小心硬编码进入程序的后一个 MIME 类型。

如果没有指定 FormatRtf 标志，RichEditBox 方法会保存并加载纯文本文件。然而，SaveToStream 方法使用 Unicode(或 UTF-16)编码保存纯文本，而文件中的每个字符均占用两个字节。这种编码并不常用于纯文本文件，而且文件开头不包含表示编码的字节顺序标记(BOM)。Windows Notepad 可以加载这些文件，显然会从文件内容检查而决定编码，但第 7 章的 PrimitivePad 程序却不能。在遇到数据流的第一个零时就停止读取。

保存自 PrimitivePad 的保存文件为 UTF-8 编码，但没有 BOM，因此，RichEditBox 的 LoadFromStream 方法假设编码为 UTF-16。也就是说，RichTextEditor 无法正确加载保存自 PrimitivePad 的文件。文件中的每两个字节将视为包括单个 Unicode 字符，因此拉丁字母的成对字符大都显示为中国汉字。

用 RichEditBox 保存和加载纯文本文件更好的解决方案可能是利用 GetText 和 SetText 方法以及常规 Windows Runtime 文件 I / O 功能。

16.10 自行文本输入

当然，TextBox 和 RichEditBox 控件为程序提供了从计算机键盘获得文字输入的最佳方式。但如果你想实现实现自行文字输入，该怎么办？

UIElement 类定义 KeyDown 和 KeyUp 事件，并且 Control 类用 OnKeyDown 和 OnKeyUp 虚拟方法进行补充。然而，信息以 VirtualKey 值的形式传递给程序。VirtualKey 是大枚举，

其枚举项可用于键盘上的所有键。该信息适用于获取包括功能键或光标移动键的活动，但并非适用于字母数字输入。很难以一种与语言无关、设备无关的方式从键盘获取字符。

获取字符输入有一个更好的事件，称为 CharacterReceived，但该事件不由 UIElement 定义。而是由 CoreWindow 定义，可以从与应用关联的 Window 对象轻松获得 CoreWindow。

GettingCharacterInput 项目简单演示了该方法。XAML 文件中包含显示输入字符的 TextBlock。

项目：GettingCharacterInput | 文件：MainPage.xaml(片段)

```
<Page ... >
    <Grid Background="{StaticResource ApplicationPageBackgroundThemeBrush}">
        <TextBlock Name="txtblk"
                   FontSize="24"
                   TextWrapping="Wrap" />
    </Grid>
</Page>
```

代码隐藏文件附加一个处理程序用于 CoreWindow 定义的 CharacterReceived 事件，并获取该窗口的所有输入字符。字符是计算为 char 值的无符号整数。只需要对 Backspace 键进行特殊处理。

项目：GettingCharacterInput | 文件：MainPage.xaml.cs

```
using Windows.UI.Core;
using Windows.UI.Xaml;
using Windows.UI.Xaml.Controls;

namespace GettingCharacterInput
{
    public sealed partial class MainPage : Page
    {
        public MainPage()
        {
            this.InitializeComponent();

            Window.Current.CoreWindow.CharacterReceived += OnCoreWindowCharacterReceived;
        }

        void OnCoreWindowCharacterReceived(CoreWindow sender, CharacterReceivedEventArgs args)
        {
            // Process Backspace key
            if (args.KeyCode == 8 && txtblk.Text.Length > 0)
            {
                txtblk.Text = txtblk.Text.Substring(0, txtblk.Text.Length - 1);
}
            // All other keys
            else
            {
                txtblk.Text += (char)args.KeyCode;
            }
        }
    }
}
```

Backspace 键是我唯一提供的“编辑”功能。事件需要执行 KeyUp 和 KeyDown 来实现在键入的文本字符串内通过光标移动键也来回移动。你可能还想添加方法来选择包括键盘或指针的文本。对于更专业的实现，则需要画一个光标以及独立的彩色文本字符和背景来表示选择。也就是说，你可能会对每个字符使用单个 TextBlock 元素来显示所输入的文本(我肯定 TextBox 和 RichEditBox 看起来非常适合)。

GettingCharacterInput 项目最大的问题在于仅从物理键盘获得输入。如果需要从

TextBox 和 RichEditBox 屏幕弹出的触控键盘输入，过程更复杂。以下为基础版本。

BetterCharacterInput 项目中的 MainPage.xaml 实例化自定义控件，名为 RudimentaryTextBox。

```
项目: BetterCharacterInput | 文件: MainPage.xaml(片段)
<Grid Background="{StaticResource ApplicationPageBackgroundThemeBrush}">
     <local:RudimentaryTextBox Background="DarkBlue"
                               Width="320"
                               Height="320" />
</Grid>
```

RudimentaryTextBox 类派生自 UserControl，而视觉元素主要包含显示键入文本的 TextBlock。

```
项目: BetterCharacterInput | 文件: RudimentaryTextBox.xaml
<UserControl
     x:Class="BetterCharacterInput.RudimentaryTextBox"
     xmlns="http://schemas.microsoft.com/winfx/2006/xaml/presentation"
     xmlns:x="http://schemas.microsoft.com/winfx/2006/xaml">
     <Grid Background="DarkBlue">
          <TextBlock Name="txtblk"
                     Foreground="Yellow"
                     FontSize="24"
                     TextWrapping="Wrap" />
     </Grid>
</UserControl>
```

RudimentaryTextBox 代码隐藏文件中的 CharacterReceived 事件处理程序与先前项目中的相同，除了仅在控件有输入焦点时才连接处理程序。该类定义了一个用于键入输入的简单 Text 属性。

```
项目: BetterCharacterInput | 文件: RudimentaryTextBox.xaml.cs
using Windows.UI.Core;
using Windows.UI.Xaml;
using Windows.UI.Xaml.Automation.Peers;
using Windows.UI.Xaml.Controls;
using Windows.UI.Xaml.Input;

namespace BetterCharacterInput
{
     public sealed partial class RudimentaryTextBox : UserControl
     {
          public RudimentaryTextBox()
          {
               this.InitializeComponent();
               this.IsTabStop = true;
               this.Text = "";
          }

          public string Text { set; get; }

          protected override void OnTapped(TappedRoutedEventArgs args)
          {
               this.Focus(FocusState.Programmatic);
               base.OnTapped(args);
          }

          protected override void OnGotFocus(RoutedEventArgs args)
          {
               Window.Current.CoreWindow.CharacterReceived += OnCoreWindowCharacterReceived;
               base.OnGotFocus(args);
          }
```

```
        protected override void OnLostFocus(RoutedEventArgs args)
        {
            Window.Current.CoreWindow.CharacterReceived -= OnCoreWindowCharacterReceived;
            base.OnLostFocus(args);
        }

        protected override AutomationPeer OnCreateAutomationPeer()
        {
            return new RudimentaryTextBoxPeer(this);
        }

        void OnCoreWindowCharacterReceived(CoreWindow sender, CharacterReceivedEventArgs args)
        {
            // Process Backspace key
            if (args.KeyCode == 8 && txtblk.Text.Length > 0)
            {
                txtblk.Text = txtblk.Text.Substring(0, txtblk.Text.Length - 1);
            }
            // All other keys
            else
            {
                txtblk.Text += (char)args.KeyCode;
            }
        }
    }
}
```

在有多个用于获取键盘输入的自定义控件的实际应用中，你可能希望由页面而非控件本身来决定何时获取键盘输入。

RudimentaryTextBox 唯一真正独特的部分是 OnCreateAutomationPeer 方法覆写。自动化同级提供编程控件来控制用户输入的功能，通常用来实现辅助技术和应用测试。为了能使控件在获得输入焦点时调用屏幕上的触控键盘，必须有一个自定义的自动化同级，它派生自 FrameworkElementAutomationPeer，并实施 IValueProvider 和 ITextProvider 接口。

该自定义自动化同级类也必须覆写 FrameworkElementAutomationPeer 构造函数和 GetPatternCore 方法。实现 IValueProvider 需要两个属性和一个方法。而实现 ITextProvider 还需要两个属性和四种方法，但如果这样做只是为了用触控键盘输入提供自定义空间，则可以用非常简单的方式来定义这些方法和属性。

示例代码不简单，但很接近。

```
项目: BetterCharacterInput | 文件: RudimentaryTextBoxPeer.cs
using Windows.Foundation;
using Windows.UI.Xaml.Automation;
using Windows.UI.Xaml.Automation.Peers;
using Windows.UI.Xaml.Automation.Provider;

namespace BetterCharacterInput
{
    public sealed class RudimentaryTextBoxPeer : FrameworkElementAutomationPeer,
                                                 IValueProvider, ITextProvider
    {
        RudimentaryTextBox rudimentaryTextBox;

        public RudimentaryTextBoxPeer(RudimentaryTextBox owner)
            : base(owner)
        {
            this.rudimentaryTextBox = owner;
        }

        // Override
        protected override object GetPatternCore(PatternInterface patternInterface)
```

```
        {
            if (patternInterface == PatternInterface.Value ||
                patternInterface == PatternInterface.Text)
            {
                return this;
            }
            return base.GetPatternCore(patternInterface);
        }

        // Required for IValueProvider
        public string Value
        {
            get { return rudimentaryTextBox.Text; }
        }

        public bool IsReadOnly
        {
            get { return false; }
        }

        public void SetValue(string value)
        {
            rudimentaryTextBox.Text = value;
        }

        // Required for ITextProvider
        public SupportedTextSelection SupportedTextSelection
        {
            get { return SupportedTextSelection.None; }
        }

        public ITextRangeProvider DocumentRange
        {
            get { return null; }
        }

        public ITextRangeProvider RangeFromPoint(Point pt)
        {
            return null;
        }

        public ITextRangeProvider RangeFromChild(IRawElementProviderSimple child)
        {
            return null;
        }

        public ITextRangeProvider[] GetVisibleRanges()
        {
            return null;
        }

        public ITextRangeProvider[] GetSelection()
        {
            return null;
        }
    }
}
```

从 Value 属性返回 null，并从 SetValue 属性去除正文，可以进一步简化。

点击深蓝色 RudimentaryTextBox 控件时，会弹出虚拟键盘(不过要想得到重大事件的屏幕截图，完全是另外一回事)。

第 17 章　共享和打印

将手指扫过 Windows 8 屏幕的右边缘(或按快捷键 Windows+C)，就会弹出当前日期和时间，以及五个称为 charm(超级按钮)的图标栏(见下图)。

中心的超级按钮直接进入开始屏幕，但其他按钮也可以为应用提供服务。如果用户点击按钮，每个按钮都会弹出一个窗格。如果应用在屏幕上(辅屏状态除外)，就都支持和其他四个按钮有关联的功能。

本章首先探讨如何处理 Settings 和 Share 两个图标，再关注 Devices 按钮，主要是想让程序访问打印机。

17.1　设置和弹窗

点击本书目前任何程序中的Settings按钮，都只会看到一项。该Permission项由Windows提供，它列出了你所写的应用在 Package.appxmanifest 文件 Capabilities 部分中已申请的权限。运行时，应用可以将项目添加到 Settings 列表，并且所添加的项会将 Permissions 项目推到列表底部。通常情况下，这些附加项提供的是有关程序的信息，可能有 About、Credits、Terms of Use 或 Privacy Statement 等标签。其他项可以从用户获得输入，可能会贴上 Options 或 Feedback 标签。

通常以一种非常熟悉的方式来实现已经添加到 Settings 列表中的每一项，即带 UserControl 的 Popup。按照惯例，Popup 定位在应用右边边缘并延伸到整个屏幕高度。

让我来演示一下，把传统 About 对话框添加到 FingerPaint 项目。我想把该程序提交到 Windows Store，想在 About 对话框展示本书封面和一个购买图书链接(链接到微软出版社的经销商网站)。我首先给 FingerPaint 项目添加一个新的 UserControl 项，并称之为 AboutBox 类。以下为 XAML 文件。

项目：FingerPaint | 文件：AboutBox.xaml(片段)

```
<UserControl ... Width="400">
    <UserControl.Transitions>
        <TransitionCollection>
            <EntranceThemeTransition FromHorizontalOffset="400" />
        </TransitionCollection>
    </UserControl.Transitions>

    <Grid>
        <Border BorderBrush="Black"
                BorderThickness="1"
                Background="#404040"
                Margin="0 12"
                Padding="0 24">
            <StackPanel>
                <StackPanel Orientation="Horizontal">
                    <Button Style="{StaticResource PortraitBackButtonStyle}"
                            Foreground="Black"
                            Click="OnBackButtonClick" />

                    <TextBlock Text="About"
                               Style="{StaticResource HeaderTextStyle}" />
                </StackPanel>

                <TextBlock FontSize="24"
                           FontWeight="Light"
                           TextWrapping="Wrap"
                           Margin="24">
                    This program was written by Charles Petzold
                    and is just one of many example programs in
                    his book
                    <Italic>Programming Windows</Italic>,
                    6th edition.
                    <LineBreak />
                    <LineBreak />
                    You can purchase a copy at many bookstores
                    or directly from the O'Reilly website.
                </TextBlock>

                <HyperlinkButton HorizontalAlignment="Center"
                    NavigateUri="http://shop.oreilly.com/product/0790145369079.do">
                    <StackPanel>
                        <Image Source="Assets/BookCover.gif"
                               Stretch="None" />
                        <TextBlock TextAlignment="Center">
                               <Italic>Programming Windows</Italic>,
                               6th edition
                        </TextBlock>
                    </StackPanel>
                </HyperlinkButton>
            </StackPanel>
```

```
        </Border>
    </Grid>
</UserControl>
```

我已经为给控件指定了特定宽度，但没有指定特定高度，因为它会拉伸到显示窗口的高度。我在顶部和底部预留了一点空间，并且提供过渡，空间看上去似乎是从右侧滑入。控件有 Back 按钮，包含 Click 处理程序，还有链接到 O' Reilly 网站的本书目录页 URL 的 HyperlinkButton。HyperlinkButton 的内容包括 Image 元素，Image 元素引用了我添加到 Assets 文件夹中的一张位图。

用于 Back 按钮的 Click 处理程序十分确定该控件的上层结构是 Popup，因此，将 Popup 的 IsOpen 属性设置为 false。

```
项目:FingerPaint | 文件: AboutBox.xaml.cs(片段)
void OnBackButtonClick(object sender, RoutedEventArgs args)
{
    // Dismiss Popup
    Popup popup = this.Parent as Popup;

    if (popup != null)
        popup.IsOpen = false;
}
```

这是解除弹窗的一种方式。

实现 About 对话框对 FingerPaint 程序的影响非常小。MainPage 构造函数获取 SettingsPane 对象并给 CommandsRequested 事件附加处理程序。

```
项目:FingerPaint | 文件: MainPage.xaml.cs(片段)
public MainPage()
{
    ...
    // Install a handler for the Settings pane
    SettingsPane settingsPane = SettingsPane.GetForCurrentView();
    settingsPane.CommandsRequested += OnSettingsPaneCommandsRequested;
    ...
}
```

SettingsPane 及相关类和枚举都占用 Windows.UI.ApplicationSettings 命名空间。从概念上讲，SettingsPane 对象指用户按下 Settings 按钮时 Windows 所显示的窗格。这就是为什么是获取而不是创建 SettingsPane 对象的原因。如果窗格自身显示，就会请求应用添加额外项。这就是 CommandsRequested 处理程序所做的事情。

挂接到其他超级按钮也相类似。SettingsPane 还有 Show 方法，能用编程来显示设置窗格，但对于大多数目的，你可能只想为 CommandsRequested 事件安装处理程序。不要保留 SettingsPane 对象，可以把 MainPage 结构中的两个语句组合成一个：

```
SettingsPane.GetForCurrentView().CommandsRequested += OnSettingsPaneCommandsRequested;
```

如果程序运行，用户按下 Settings 按钮，就会调用 CommandsRequested 事件。这时可以给 Settings 窗格增加额外命令。每次按下 Settings 按钮的时候都可以增加额外命令，因此，可以根据这些额外命令来判断当前应用状态。

FingerPaint 通过给列表添加 SettingsCommand 对象来处理 CommandsRequested 事件。

```
项目:FingerPaint | 文件: MainPage.xaml.cs(片段)
void OnSettingsPaneCommandsRequested(SettingsPane sender,
                                     SettingsPaneCommandsRequestedEventArgs args)
{
```

```
        SettingsCommand aboutCommand = new SettingsCommand(0, "About", OnAboutInvoked);
        args.Request.ApplicationCommands.Add(aboutCommand);
    }
```

该命令有一个 ID(我没有使用，因此设置为 0)、一个文本标签以及一个用户选择该命令时会调用的方法。程序从 CommandsRequested 处理程序返回之后，窗格就显示新的 About 信息。

以下为处理该命令的方法。

```
项目:FingerPaint | 文件: MainPage.xaml.cs(片段)
void OnAboutInvoked(IUICommand command)
{
    AboutBox aboutBox = new AboutBox();
    aboutBox.Height = this.ActualHeight;
    Popup popup = new Popup
    {
        IsLightDismissEnabled = true,
        Child = aboutBox,
        IsOpen = true,
        HorizontalOffset = this.ActualWidth - aboutBox.Width
    };
}
```

Popup 出现在页面右侧的固定位置，因此用于设置 AboutBox 的 Height 以及设置和 Popup 的 HorizontalOffset 的代码很简单，效果如下图所示。

IsLightDismissEnabled 属性设置可以确保如果用户按住 Popup 以外的任何地方，Popup 就会消除，AboutBox 中的 Back 按钮也提供消除。如果用户按下 Hyperlink 按钮，Popup 就会消除，同时启动 IE 浏览器。

17.2 通过剪贴板共享

通过 Share 超级图标共享数据涉及 Windows.ApplicationModel.DataTransfer 和 Windows.ApplicationModel.DataTransfer.ShareTarget 命名空间中的类。第一个命名空间还包括一个非常传统的机制来传输 Windows 应用数据，即剪贴板。

我想先添加剪贴板来支持 FingerPaint，再处理 Share 超级图标。给处理位图的程序添加剪切板有些复杂。你可能想实现选择 API 功能以便用户能刻出绘画的矩形区域，然后复制到剪贴板。如果还想实现粘贴 API 功能，允许要进入的位图定位到当前图片的某个位置。

但我要采用一个简单的方法即 Copy 命令将整个作品复制到剪贴板，Paste 命令把输入的位图作为新图片，就像从文件中载入，只不过没有文件名。

第一件事情是把 Copy 和 Paste 按钮添加到应用栏。

```
项目:FingerPaint | 文件: MainPage.xaml(片段)
<Page.BottomAppBar>
    <AppBar>
        <Grid>
            <StackPanel Orientation="Horizontal"
                        HorizontalAlignment="Left">
                ...
                <Button Style="{StaticResource CopyAppBarButtonStyle}"
                        Click="OnCopyAppBarButtonClick" />

                <Button Name="pasteAppBarButton"
                        Style="{StaticResource PasteAppBarButtonStyle}"
                        Click="OnPasteAppBarButtonClick" />
            </StackPanel>
            ...
        </Grid>
    </AppBar>
</Page.BottomAppBar>
```

Paste 按钮需要名称，因为必须通过基于剪贴板上实际位图数据情况的代码来启用和禁用 Paste 按钮。

我决定在 MainPage 的另一个部分类实现文件中实现所有共享数据代码，该文件名为 MainPage.Share.cs。Mainpage 构造函数会调用该类中的一个方法。

```
项目:FingerPaint | 文件: MainPage.xaml.cs(片段)
public MainPage()
{
    ...
    // Call a method in MainPage.Share.cs
    InitializeSharing();
    ...
}
```

在某种程度上，方法会检查 Paste 按钮是否已启用并设置一个事件处理程序，而当剪贴板内容发生改变时，就调用该事件处理程序。

```
项目:FingerPaint | 文件: MainPage.Share.cs(片段)
public sealed partial class MainPage : Page
{
    void InitializeSharing()
    {
        // Initialize the Paste button and provide for updates
        CheckForPasteEnable();
        Clipboard.ContentChanged += OnClipboardContentChanged;
        ...
    }
    ...
    void OnClipboardContentChanged(object sender, object args)
    {
        CheckForPasteEnable();
    }
    void CheckForPasteEnable()
    {
        pasteAppBarButton.IsEnabled = CheckClipboardForBitmap();
    }

    bool CheckClipboardForBitmap()
```

```
    {
        DataPackageView dataView = Clipboard.GetContent();
        return dataView.Contains(StandardDataFormats.Bitmap);
    }
    ...
}
```

Clipboard 是一个只有四个方法和一个事件的静态类。其中两个最重要的方法是 getContent 和 setContent。getContent 返回 DataPackageView 对象，该对象提供检查当前剪贴板内容的简便方法并确定位图是否存在。

SetContent 需要 DataPackage 对象，该对象有许多方法可以把各种形式的数据复制到剪贴板，其中包括一个名为 SetBitmap 的方法。剪贴板 Copy 按钮的处理程序会创建 DataPackage 并把操作设置为 Move，也就是说，项目不会进一步处理放入剪贴板的位图。

```
项目:FingerPaint | 文件: MainPage.Share.cs(片段)
async void OnCopyAppBarButtonClick(object sender, RoutedEventArgs args)
{
    DataPackage dataPackage = new DataPackage
    {
        RequestedOperation = DataPackageOperation.Move,
    };
    dataPackage.SetBitmap(await GetBitmapStream(bitmap));

    Clipboard.SetContent(dataPackage);
    this.BottomAppBar.IsOpen = false;
}
```

然而，SetBitmap 方法并不想让东西和 BitmapSource 一样平凡。相反，SetBitmap 想要会引用编码位图图片的 RandomAccessStreamReference。可以从 InMemoryRandomAccessStream 创建 RandomAccessStreamReference。这工作是调用 SetBitmap 时所引用的 GetBitmapStream 方法来完成的。

注意，GetBitmapStream 调用的参数是作为字段存储在 Mainpage 中的 WriteableBitmap。我对 GetBitmapStream 进行了泛化处理，它可以从该参数创建自身的 pixels 数组，但应该可以访问同样以字段形式存储在 Mainpage 中的 pixels 数组。

```
项目:FingerPaint | 文件: MainPage.Share.cs(片段)
async Task<RandomAccessStreamReference> GetBitmapStream(WriteableBitmap bitmap)
{
    // Get a pixels array for this bitmap
    byte[] pixels = new byte[4 * bitmap.PixelWidth * bitmap.PixelHeight];
    Stream stream = bitmap.PixelBuffer.AsStream();
    await stream.ReadAsync(pixels, 0, pixels.Length);

    // Create a BitmapEncoder associated with a memory stream
    InMemoryRandomAccessStream memoryStream = new InMemoryRandomAccessStream();
    BitmapEncoder encoder = await BitmapEncoder.CreateAsync(BitmapEncoder.PngEncoderId,memoryStream);
    // Set the pixels into that encoder
    encoder.SetPixelData(BitmapPixelFormat.Bgra8, BitmapAlphaMode.Premultiplied,
            (uint)bitmap.PixelWidth, (uint)bitmap.PixelHeight, 96, 96, pixels);
    await encoder.FlushAsync();

    // Return a RandomAccessStreamReference
    return RandomAccessStreamReference.CreateFromStream(memoryStream);
}
```

Paste 逻辑比较复杂，不仅是因为要根据剪贴板中的位图内容来启用 Paste 按钮。如果没有保存当前绘画，而且用户按下 Paste 按钮，程序就会要求应该保存或放弃绘画，好像用户已经选择加载新文件一样。

也就是说，Paste 按钮的处理程序应该调用 Mainpage.File.cs 中的 CheckIfOkToTrashFile 方法，并且向其传递的方法应该在执行 Paste 时执行。在调用 CheckIfOkToTrashFile 之前，我并不清楚应该怎么处理传入的位图。我担心用户可能选择保存现有图片，而在此期间剪贴板内容就会改变。通过立刻获取像素矩阵，我避开了这个问题。但代码尚未创建 WriteableBitmap。在此之前要求新位图的几个关联项必须以字段形式存储。

项目:FingerPaint | 文件: MainPage.Share.cs(片段)

```
public sealed partial class MainPage : Page
{
    int pastedPixelWidth, pastedPixelHeight;
    byte[] pastedPixels;
    ...

    async void OnPasteAppBarButtonClick(object sender, RoutedEventArgs args)
    {
        // Temporarily disable the Paste button
        Button button = sender as Button;
        button.IsEnabled = false;

        // Get the Clipboard contents and check for a bitmap
        DataPackageView dataView = Clipboard.GetContent();

        if  (dataView.Contains(StandardDataFormats.Bitmap))
        {
            // Get the stream reference and a stream
            RandomAccessStreamReference streamRef = await dataView.GetBitmapAsync();
            IRandomAccessStreamWithContentType stream = await streamRef.OpenReadAsync();

            // Create a BitmapDecoder for reading the bitmap
            BitmapDecoder decoder = await BitmapDecoder.CreateAsync(stream);
            BitmapFrame bitmapFrame = await decoder.GetFrameAsync(0);
            PixelDataProvider pixelProvider =
                await bitmapFrame.GetPixelDataAsync(BitmapPixelFormat.Bgra8,
                                                    BitmapAlphaMode.Premultiplied,
                                                    new BitmapTransform(),
                                                    ExifOrientationMode.RespectExifOrientation,
                                                    ColorManagementMode.ColorManageToSRgb);

            // Save information sufficient for creating WriteableBitmap
            pastedPixelWidth = (int)bitmapFrame.PixelWidth;
            pastedPixelHeight = (int)bitmapFrame.PixelHeight;
            pastedPixels = pixelProvider.DetachPixelData();

            // Check if it's OK to replace the current painting
            await CheckIfOkToTrashFile(FinishPasteBitmapAndPixelArray);
        }

        // Re-enable the button and close the app bar
        button.IsEnabled = true;
        this.BottomAppBar.IsOpen = false;
    }

    async Task FinishPasteBitmapAndPixelArray()
    {
        bitmap = new WriteableBitmap(pastedPixelWidth, pastedPixelHeight);
        pixels = pastedPixels;

        // Transfer pixels to bitmap, among other chores
        await InitializeBitmap();

        // Set AppSettings properties for new image
        appSettings.LoadedFilePath = null;
        appSettings.LoadedFilename = null;
        appSettings.IsImageModified = false;
```

```
    }
    ...
}
```

要实现剪贴板支持，还需要一项工作。许多用户都熟悉通过快捷键 Ctrl+C 和 Ctrl+V 进行复制和粘贴操作，因此，我给 Mainpage.Share.cs 也加入了该项支持，同时还利用了现有按钮的处理程序。

```
项目:FingerPaint | 文件: MainPage.Share.cs(片段)
public sealed partial class MainPage : Page
{
    ...
    void InitializeSharing()
    {
        ...
        // Watch for accelerator keys for Copy and Paste
        Window.Current.CoreWindow.Dispatcher.AcceleratorKeyActivated +=
                                OnAcceleratorKeyActivated;
        ...
    }
    ...
    void OnAcceleratorKeyActivated(CoreDispatcher sender, AcceleratorKeyEventArgs args)
    {
        if ((args.EventType == CoreAcceleratorKeyEventType.SystemKeyDown ||
             args.EventType == CoreAcceleratorKeyEventType.KeyDown) &&
             (args.VirtualKey == VirtualKey.C || args.VirtualKey == VirtualKey.V))
        {
            CoreWindow window = Window.Current.CoreWindow;
            CoreVirtualKeyStates down = CoreVirtualKeyStates.Down;

            // Only want case where Ctrl is down
            if ((window.GetKeyState(VirtualKey.Shift) & down) == down ||
                (window.GetKeyState(VirtualKey.Control) & down) != down ||
                (window.GetKeyState(VirtualKey.Menu) & down) == down)
            {
                return;
            }

            if (args.VirtualKey == VirtualKey.C)
            {
                OnCopyAppBarButtonClick(null, null);
            }
            else if (args.VirtualKey == VirtualKey.V)
            {
                OnPasteAppBarButtonClick(pasteAppBarButton, null);
            }
        }
    }
}
```

17.3 Share 超级按钮

程序可以通过两种方式使用 Share 超级按钮。我要展示一个程序如何把数据提供给其他程序。把一个应用作为其他应用的数据接收对象，相当困难。第二项工作需要应用在 Package.appxmanifest 的 Declarations 部分中把自己声明为共享目标，以一种独特状态进行激活，并为该目的提供特殊的用户界面。

程序可以通过给 DataTransferManager 实例设置事件处理程序而成为 Share 提供者，而给其他应用提供位图的程序也通过复制位图到剪贴板的同一个

RandomAccessStreamReference 而成为 Share 提供者。在 MainPage.Share.cs 中已经定义了 GetBitmapStream 方法，因此用来支持 Share 按钮的额外代码微不足道。

项目:FingerPaint | 文件: MainPage.Share.cs(片段)

```
public sealed partial class MainPage : Page
{
    ...
    void InitializeSharing()
    {
        ...
        // Hook into the Share pane for providing data
        DataTransferManager.GetForCurrentView().DataRequested += OnDataTransferDataRequested;
    }

    async void OnDataTransferDataRequested(DataTransferManager sender,
                                           DataRequestedEventArgs args)
    {
        DataRequestDeferral deferral = args.Request.GetDeferral();

        // Get a stream reference and hand it over
        RandomAccessStreamReference reference = await GetBitmapStream(bitmap);
        args.Request.Data.SetBitmap(reference);
        args.Request.Data.Properties.Title = "Finger Paint";
        args.Request.Data.Properties.Description = "Share this painting with another app";

        deferral.Complete();
    }
}
```

现在，FingerPaint 正在运行，如果用户选择 Share 按钮，窗格不会报告“This app can’t share”，而会说“Finger Paint”和“Share this painting with another app”。显然，已经调用了 DataRequested 事件处理程序，并且 Windows 有 RandomAccessStreamReference，因此，在接着 Share 窗格中出现是一个可以接受位图数据的应用列表。程序提供了位图，因此不需要程序的进一步交互。

17.4 基本打印

在本书目前显示的任何程序中，如果激活超级按钮并按下 Device 按钮，会得到 Device 窗格，该窗格不会提及关于打印机的任何情况。应用需要先注册到 Windows 8，说明它可以打印。

有三个命名空间在打印中发挥作用。

- Windows.UI.Xaml.Printing 命名空间有 PrintDocument 类并支持其事件。正如名称所示，PrintDocument 代表程序用户希望打印的东西。
- Windows.Graphics.Printing 命名空间有 PrintManager，是 Windows 8 列出的打印机和打印机选项的窗格接口；以及 PrintTask、PrintTaskRequest 和 PrintTaskOptions 类。打印“任务”(task)和打印“工作”(job)是一回事，即打印特定文档的打印机特定用途。
- Windows.Graphics.Printing.OptionDetails 命名空间包含用来自定义打印选项的类。

打印机 API 的很多内容涉及系统开销，而不涉及给打印页面实际定义文本和图形的过程。事实上，Windows 8 应用在打印纸上进行打印的方式和在屏幕上绘图的方式一样：通过派生自 UIElement 类的实例的可视树。一般情况下，根元素为带子类的 Border 或 Panel。

可以在 XAML 中定义该视觉树，但可能更多时候是在代码中创建。

定义显示在屏幕上的元素，有一种有用的指导，即把视频显示当成每英寸有 96 个像素的分辨率。对打印机也这么做，只不过更精确。不管打印机的实际分辨率如何，总是可以把它当作 96 DPI 的设备。

用户点击 Device 按钮图时，为了让 Windows 8 列出打印机设备，首要任务就是设置事件处理程序：

```
PrintManager printManager = PrintManager.GetForCurrentView();
printManager.PrintTaskRequested += OnPrintManagerPrintTaskRequested
```

两行可以合并为一行：

```
PrintManager.GetForCurrentView().PrintTaskRequested += OnPrintManagerPrintTaskRequested;
```

静态 GetForCurrentView 方法获得与程序窗口相关的 PrintManager 实例。为 PrintTaskRequested 事件设置一个处理程序，程序就声称它是可打印的。处理程序如下：

```
void OnPrintManagerPrintTaskRequested(PrintManager sender, PrintTaskRequestedEventArgs args)
{
    ...
}
```

如果用户点击 Device 按钮(或按下 Windows+K)，就会调用处理程序，但(你很快就会看到)它需要调用另一个带回调函数的方法，以便 Windows 可以显示打印机列表。

只有当应用准备好要实际打印东西的时候，才会附加 PrintTaskRequested 处理程序。如果应用在可以进行打印之前要求从用户获取一些初步信息或需要加载文档，则不应该把处理程序附加到 PrintTaskRequested 事件。如果程序又发现自身处于不可打印的情况，则应该解除处理程序：

```
PrintManager.GetForCurrentView().PrintTaskRequested -= OnPrintManagerPrintTaskRequested;
```

对于本章的示例程序，我主要在表示整个过程的 OnNavigatedTo 和 OnNavigatedFrom 覆写中附加和解除该事件处理程序。

PrintTaskRequested 事件处理程序是执行简单打印的程序所需 5 种回调方法和事件处理程序中的一种。五个方法均为必需。此外，甚至在 PrintTaskRequested 事件激活之前，程序就需要创建 PrintDocument 对象，并附加三个事件处理程序，以做好打印准备。

我们来看一个完整的程序，该程序会打印一个一页文档，包含一个 TextBlock(上面写着“Hello, Printer!”)。HelloPrinter 项目中的 XAML 文件在程序逻辑中并没有起作用，而只是通知新用户如何打印内容。

项目:HelloPrinter | 文件: MainPage.xaml(片段)

```
<Grid Background="{StaticResource ApplicationPageBackgroundThemeBrush}">
    <TextBlock FontSize="48"
               HorizontalAlignment="Center"
               VerticalAlignment="Center"
               TextAlignment="Center">
        Hello, Printer!
        <LineBreak />
        <Run FontSize="24">
            (Invoke charms, select Devices and a printer)
        </Run>
    </TextBlock>
</Grid>
```

代码隐藏文件定义了三个字段，其中之一就是该程序要打印的 TextBlock。

```
项目:HelloPrinter | 文件: MainPage.xaml.cs(片段)
public sealed partial class MainPage : Page
{
    PrintDocument printDocument;
    IPrintDocumentSource printDocumentSource;
    // UIElement to print
    TextBlock txtblk = new TextBlock
    {
        Text = "Hello, Printer!",
        FontFamily = new FontFamily("Times New Roman"),
        FontSize = 48,
        Foreground = new SolidColorBrush(Colors.Black)
    };
    ...
}
```

PrintDocument 对象表示应用要打印的东西。一般情况下，程序只创建一个 PrintDocument 对象，并用于一个打印任务。在某些情况下，程序可以维护多个 PrintDocument 对象，可能一个打印整个文档，另一个打印文档大纲，第三个打印缩略图，但不要为每个打印任务创造新的 PrintDocument 对象。(正如你会看到的，请求打印任务的时候，根本来不及创建 PrintDocument！)如果方便，可以从 PrintDocument 派生类，用来封装一些打印逻辑，但在 PrintDocument 中没有要覆写的东西。

对于处理单一文档类型的程序，你可能会像我一样把 PrintDocument 和 IPrintDocumentSource 定义为字段类型并在程序初始化过程中创建 PrintDocument 对象。

```
项目:HelloPrinter | 文件: MainPage.xaml.cs(片段)
public sealed partial class MainPage : Page
{
    ...
    public MainPage()
    {
        this.InitializeComponent();

        // Create PrintDocument and attach handlers
        printDocument = new PrintDocument();
        printDocumentSource = printDocument.DocumentSource;
        printDocument.Paginate += OnPrintDocumentPaginate;
        printDocument.GetPreviewPage += OnPrintDocumentGetPreviewPage;
        printDocument.AddPages += OnPrintDocumentAddPages;
    }
    ...
}
```

第二个字段 IPrintDocumentSource 类型的对象从 PrintDocument 对象获得。此外，PrintDocument 定义的三个事件都需要处理程序。这些事件处理程序负责提供页面计数以及在打印预览和打印过程中的实际页数。

HelloPrinter 程序在 OnNavigatedTo 期间为 PrintManager 的 PrintTaskRequested 事件附加事件处理程序，并且在 OnNavigatedFrom 期间将其解除，在第一种情况下使用两个语句，而在第二个情况下的使用一个带一点变化的语句。

```
项目:HelloPrinter | 文件: MainPage.xaml.cs(片段)
public sealed partial class MainPage : Page
{
    ...
    protected override void OnNavigatedTo(NavigationEventArgs args)
    {
```

```
        // Attach PrintManager handler
        PrintManager printManager = PrintManager.GetForCurrentView();
        printManager.PrintTaskRequested += OnPrintManagerPrintTaskRequested;

        base.OnNavigatedTo(args);
    }

    protected override void OnNavigatedFrom(NavigationEventArgs e)
    {
        // Detach PrintManager handler
        PrintManager.GetForCurrentView().PrintTaskRequested -= OnPrintManagerPrintTaskRequested;

        base.OnNavigatedFrom(e);
    }
    ...
}
```

在真实的程序中，如果程序能够打印，就附加此处理程序。如果不需要打印，就解除处理程序。

如果程序已经附加该处理程序并且用户手指从屏幕右侧滑过，选择 Device，则调用 PrintTaskRequested 事件处理程序。以下是响应该事件的的标准方式。

项目:HelloPrinter | 文件: MainPage.xaml.cs(片段)

```
void OnPrintManagerPrintTaskRequested(PrintManager sender, PrintTaskRequestedEventArgs args)
{
    args.Request.CreatePrintTask("Hello Printer", OnPrintTaskSourceRequested);
}
```

PrintTaskRequested 事件的事件参数包括 Request 类型属性，程序通常调用 Request 对象的 CreatePrintTask 方法来做出响应，并向其传递打印任务名称(可能会是应用名称或由应用要打印的文档名称)以及回调函数。CreatePrintTask 方法返回 PrintTask 对象，但通常没有必要保留该对象。

Windows 8 显示打印机列表。下图为屏幕上的内容(不同的屏幕上可能有所不同)。

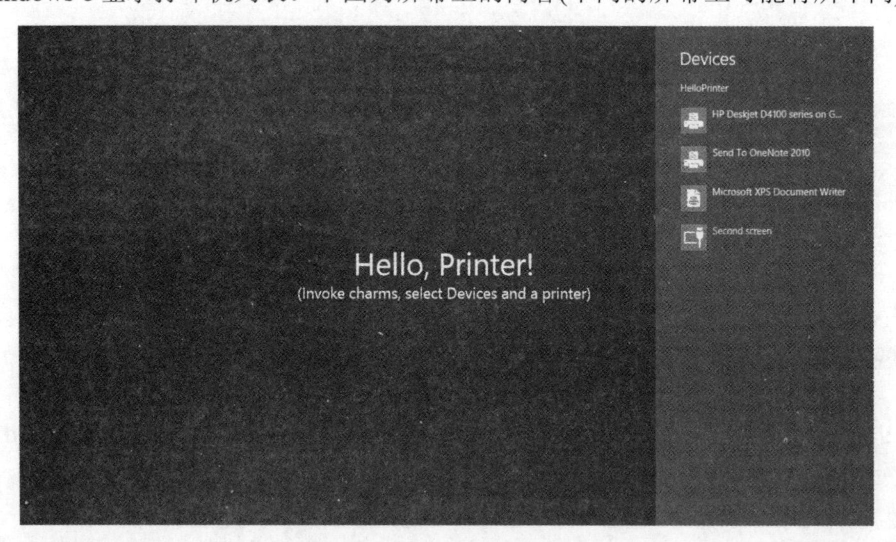

唯一的、真正的打印机在列表中位列第一。后两个将打印输出保存为文件。最后一个不是打印机，配置的是连接到平板电脑上的第二个显示器。

如果用户选择列表中的任意一项，则调用我命名为 OnPrintTaskSourceRequested 的回调函数。在最简单的情况下，处理程序可以通过调用事件参数的 SetSource 做出响应，并将早先从 PrintDocument 对象获得的 IPrintDocumentSource 对象中传递给 SetSource。

```
项目:HelloPrinter | 文件: MainPage.xaml.cs(片段)
void OnPrintTaskSourceRequested(PrintTaskSourceRequestedArgs args)
{
    args.SetSource(printDocumentSource);
}
```

该方法将控制返回给 Windows，并显示一个打印机特定窗格(见下图)。

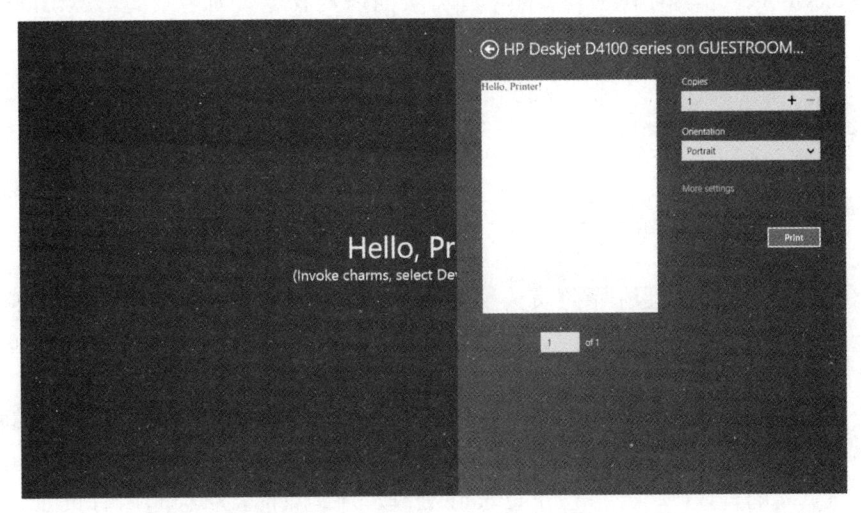

屏幕顶部的打印机名称看起来可能有些奇怪。这台特定打印机实际并没有连接在我工作的平板电脑上，而是连接到我家客房里的电脑，我在客房写本章节的内容，而你所看到的就是这台计算机的一部分名称。

右上框用于指定打印份数，下拉框可以选择打印纸张(Portrait 或 Landscape)。这些都是打印机的标准设置。(对于 Send To OneNote 2010 和 Microsoft XPS Document Writer 选项，该区域只显示 Orientation 选项。)按下 More settings 按钮，可以选择页面大小以及一些打印机特定选项。

左侧窗格是打印页面预览。如果文档有多个页面，则可以通过下方的选择框来选择需要查看的页面。页面预览为 FlipView 控件，从一边扫到另一边最简单。

总页数来自 Paginate 事件的处理程序，Paginate 事件是 PrintDocument 定义的三个事件之一。所有三个事件的处理程序都附加在 MainPage 构造中。在 HelloPrinter 中，可以直接如下实现 Paginate 处理程序。

```
项目:HelloPrinter | 文件: MainPage.xaml.cs(片段)
void OnPrintDocumentPaginate(object sender, PaginateEventArgs args)
{
    printDocument.SetPreviewPageCount(1, PreviewPageCountType.Final);
}
```

通过 Paginate 处理程序，应用可以准备好所有需要打印的页面，然后调用 PrintDocument 方法，以显示有多少页并表明这是初步计数还是最终计数。(如果不可能或者不方便一次性进行所有工作，情况会更复杂一些。)

PrintDocument 定义的 GetPreviewPage 事件的处理程序提供打印页面预览而且也在 Mainpage 构造函数中设置较早。

```
项目:HelloPrinter | 文件: MainPage.xaml.cs(片段)
void OnPrintDocumentGetPreviewPage(object sender, GetPreviewPageEventArgs args)
{
    printDocument.SetPreviewPage(args.PageNumber, txtblk);
}
```

事件参数的 PageNumber 属性为基于 1，并且其范围从 1 到 SetPreviewPageCount 调用所指定的数量。对于本特定程序，PageNumber 总是等于 1。程序通过调用 PrintDocument 的 SetPreviewPage 方法来响应该事件，并将页面以及我定义为字段的 TextBlock 传递给 SetPreviewPage。这就是在打印预览中显示的东西。

按下 Print 按钮，调用最后一个事件处理程序。

```
项目:HelloPrinter | 文件: MainPage.xaml.cs(片段)
void OnPrintDocumentAddPages(object sender, AddPagesEventArgs args)
{
    printDocument.AddPage(txtblk);
    printDocument.AddPagesComplete();
}
```

AddPages 事件的处理程序负责为文档每一页调用 AddPage。通常情况下，传递给 SetPreviewPage 方法的对象都相同，但如果你希望，也可以让这些对象有所不同。程序最后调用 AddPagesComplete。打印窗格消失，(运气好的话)你会听到熟悉的打印声。

注意！Paginate 处理程序可以多次调用，尤其是用户开始处理各种打印机选项的时候。如果程序实际上做了大量工作来分页文件，而且页面的实际布局不会改变，你可能不想重复调用。在真实的程序中，一般在 Paginate 处理程序运行时，你会将所有页面集中到 List 对象中，然后交付给 GetPreviewPage 和 addPages 处理程序。

HelloPrinter 打印的 TextBlock 被指定 48 的 FontSize。如果显示在不同尺寸和分辨率的视频显示器上，TextBlock 的大小可能会有稍微不同。但打印的时候，48 的 FontSize 是一种精确测量，意味着 48/96 英寸，即半英寸或 36 点。

注意，我把可打印 TextBlock 的 Foreground 属性指定为黑色。由于程序采用黑色主题，默认 Foreground 属性为白色，如果没有明确的 Foreground 设置，TextBlock 会获取默认值，而在白纸上就显示不出来。这种事情会让你困惑好几天！处理打印机代码的时候，养成习惯(使用红色和蓝色等颜色)，会帮助你减少打印白色字的机会。

如果检查 HelloPrinter 中的代码，你可能会觉得有一些方法可以简化。例如，你可能认为不需要一开始就创建 PrintDocument 并将其保存为字段。可以直接在 OnPrintTaskSourceRequested 方法中创建 PrintDocument，设置三个事件处理程序，然后提取 IPrintDocumentSource 对象。不同 PrintDocument 事件处理程序可以从 sender 参数访问 PrintDocument。

但这是行不通的。需要在用户界面线程创建并访问创建 PrintDocument，而 PrintTaskRequested 处理器和我命名为 OnPrintTaskSourceRequested 的回调函数并不在用户界面线程运行。想调用 PrintTaskRequested 事件来创建 PrintDocument，为时已太晚了。

17.5 可打印边距和不可打印边距

即使我谨慎设置了用黑色打印 TextBlock，但我的打印机还是无法正确打印，而且可能你的打印机也无法正确打印。TextBlock 和页面左上角直接对齐，而大多数打印机根本无法达到纸边缘，也就是说，一部分文本会被切掉。

如果尝试通过把 TextBlock 的 HorizontalAlignment 和 VerticalAlignment 属性设置为 Center 来解决该问题，你会发现此时这些属性根本没用。对齐值是相对于父元素的，而 TextBlock 却没有父元素，因为它是打印机页面上的顶层元素。由于同样原因，Margin 也没用。然而，设置 TextBlock 的 Padding 属性有用，因为这是 TextBlock 处理程序本身的事情。

一个更好的通用解决方案是把每个打印机页面变成一个可视树，该可视树以顶层 Border 元素开头。打印时，Border 占据整张纸，但 Border 可以包括非零 Padding 属性，该属性能够有效地为整个页面提供边距。

PaginateEventArgs 和 AddPagesEventArgs 都包括 PrintTaskOptions 类型的属性，名为 PrintTaskOptions。该对象的大多数属性都对应打印机属性，用户可以手动设置。这些属性的名称像 Collation、NumberOfCopies、Orientation 和 PrintQuality。程序可以访问这些属性以定制打印，但通常并不是必须。本章稍后展示程序如何初始化这些属性，并添加一些自定义选项。

PrintTaskOptions 还有一个名为 GetPageDescription 的方法。假设每个页面均为不同大小，参数就是基于零的页码。该方法返回的 PrintPageDescription 结构有 DpiX 和 DpiY 属性，会报告打印机的实际分辨率(值通常为 600 或 1200)以及以 1/96 英寸为单位的 Size 类型的 Pagesize。对于大小为 8 1/2×11 的美国标准竖排信纸，PageSize 属性为 816 和 1056。

PrintPageDescription 结构还包括 Rect 类型的 ImageableRect 属性，该属性表示打印机能够实际打印的页面矩形区域。对于我的打印机上的信件尺寸纸张，该矩形的左上角为(120.48110.35748)，尺寸为(791.04,9880.1575)，均已 1/96 英寸为单位。与 816×1056 的 PageSize 比较一下。执行一些减法，你会得出结论，即打印机无法打印在左右边缘的 12.48 个单位、顶部 11.35748 个单位以及底部的 56.48502 个单位。在横排模式中，Pagesize 和 ImageableRect 反映出页面的特定方位。

我们来看看这些数字有多么精确。PrintPrintableArea 程序在 XAML 文件声明了其名称。

```
项目:PrintPrintableArea | 文件: MainPage.xaml(片段)
<Grid Background="{StaticResource ApplicationPageBackgroundThemeBrush}">
    <TextBlock Text="Print Printable Area"
               FontSize="24"
               HorizontalAlignment="Center"
               VerticalAlignment="Center" />
</Grid>
```

代码隐藏文件的结构非常像 HelloPrinter，只不过程序要打印的颜色较多，包括红色背景的 Border、白色背景和黑色轮廓嵌套的 Border 以及居中的 TextBlock。

你还会注意到，我没有定义传递给 CreatePrintTask 方法的单独回调方法，而将其定义为匿名 lambda 函数。这是一种常见做法，但对于更复杂的打印逻辑，我不会固执地采用这种做法。

```
项目:PrintPrintableArea | 文件: MainPage.xaml.cs(片段)
public sealed partial class MainPage : Page
{
    PrintDocument printDocument;
    IPrintDocumentSource printDocumentSource;

    // Element to print
```

```
Border border = new Border
{
    Background = new SolidColorBrush(Colors.Red),
    Child = new Border
    {
        Background = new SolidColorBrush(Colors.White),
        BorderBrush = new SolidColorBrush(Colors.Black),
        BorderThickness = new Thickness(1),
        Child = new TextBlock
        {
            Text = "Print Printable Area",
            FontFamily = new FontFamily("Times New Roman"),
            FontSize = 24,
            Foreground = new SolidColorBrush(Colors.Black),
            HorizontalAlignment = HorizontalAlignment.Center,
            VerticalAlignment = VerticalAlignment.Center
        }
    }
};

public MainPage()
{
    this.InitializeComponent();
    // Create PrintDocument and attach handlers
    printDocument = new PrintDocument();
    printDocumentSource = printDocument.DocumentSource;
    printDocument.Paginate += OnPrintDocumentPaginate;
    printDocument.GetPreviewPage += OnPrintDocumentGetPreviewPage;
    printDocument.AddPages += OnPrintDocumentAddPages;
}

protected override void OnNavigatedTo(NavigationEventArgs args)
{
    // Attach PrintManager handler
    PrintManager.GetForCurrentView().PrintTaskRequested += OnPrintManagerPrintTaskRequested;

    base.OnNavigatedTo(args);
}

protected override void OnNavigatedFrom(NavigationEventArgs e)
{
    // Detach PrintManager handler
    PrintManager.GetForCurrentView().PrintTaskRequested -= OnPrintManagerPrintTaskRequested;

    base.OnNavigatedFrom(e);
}

void OnPrintManagerPrintTaskRequested(PrintManager sender, PrintTaskRequestedEventArgs args)
{
    args.Request.CreatePrintTask("Print Printable Area", (requestArgs) =>
    {
        requestArgs.SetSource(printDocumentSource);
    });
}

void OnPrintDocumentPaginate(object sender, PaginateEventArgs args)
{
    PrintPageDescription printPageDescription = args.PrintTaskOptions.GetPageDescription(0);
    // Set Padding on outer Border
    double left = printPageDescription.ImageableRect.Left;
    double top = printPageDescription.ImageableRect.Top;
    double right = printPageDescription.PageSize.Width
                    - left - printPageDescription.ImageableRect.Width;
    double bottom = printPageDescription.PageSize.Height
                    - top - printPageDescription.ImageableRect.Height;
    border.Padding = new Thickness(left, top, right, bottom);

    printDocument.SetPreviewPageCount(1, PreviewPageCountType.Final);
```

```
    }

    void OnPrintDocumentGetPreviewPage(object sender, GetPreviewPageEventArgs args)
    {
        printDocument.SetPreviewPage(args.PageNumber, border);
    }

    void OnPrintDocumentAddPages(object sender, AddPagesEventArgs args)
    {
        printDocument.AddPage(border);
        printDocument.AddPagesComplete();
    }
}
```

另一个显著区别是 PrintDocument 的 Paginate 事件处理程序。处理程序获得 PrintPageDescription 结构并计算出 Padding 值，并应用于所打印元素的外 Border。正如你所看到的，打印预览显示一直到纸张边缘的外边框红色背景(见下图)。

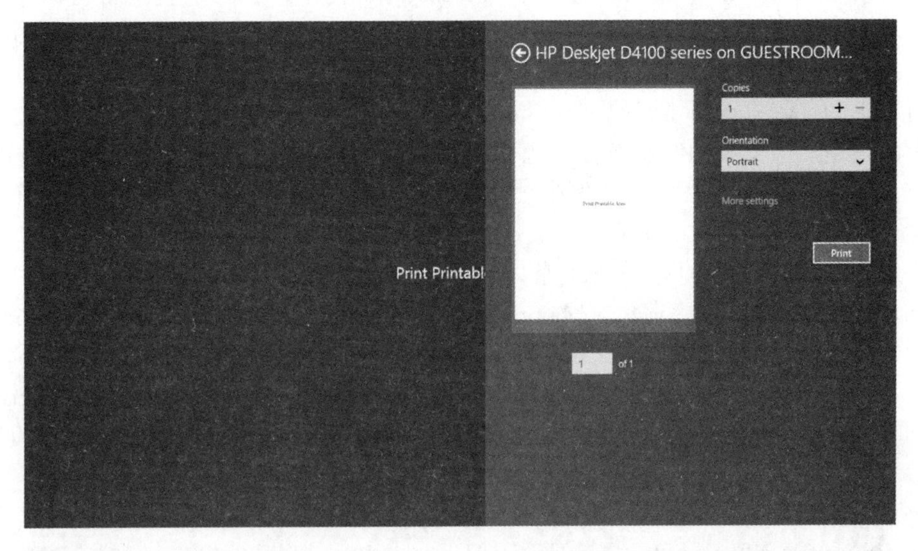

显示预览时，试试在 Portrait 和 Landscape 之间进行切换。每次变化都会引起再次调用 Paginate 处理程序并重新计算外边框的 Padding 值。

打印预览显然没有反映出打印到纸张边缘时的打印机限制。否则，红色区域将不可见。我很高兴地发现实际打印出来的的页面显示黑色内边框没有问题，只是带了一点点外 Border 的红色背景痕迹，这表明打印机的 ImageableRect 值是精确的。

虽然打印图片和其他位图的程序可能想适应页面打印，但大多数打印程序都会设置较大的页边距(或许一英寸左右)，或者为固定尺寸，或者由用户自定义。在这些情况下，通常都没有必要访问 ImageableRect 属性。

接下来我会举一个用户自定义页边距的例子。

17.6　分页过程

一般情况下，Windows 应用能够打印很多页，页面数量取决于许多因素，例如文档长度字号、纸张尺寸、页边距以及页面竖排或横排模式。

Paginate 事件处理程序的目的，不仅要准备好页面预览和打印，而且还要统计文档页数。某些情况下，分页可能需要一段时间，有一些办法能避免立刻完成分页，但如果能立刻完成分页，就是最简单的。

我们重新运行第 4 章的 DependencyObjectClassHierarchy 程序，并对其增加打印选项，以此来检查一个相当短的分页任务。你可能还记得，DependencyObjectClassHierarchy 为每个 DependencyObject 派生类都创建 TextBlock，并都放入 ScrollViewer 的 StackPanel。PrintableClassHierarchy 程序的 XAML 文件和以前的版本一样。

```
项目:PrintableClassHierarchy | 文件: MainPage.xaml(片段)
<Grid Background="{StaticResource ApplicationPageBackgroundThemeBrush}">
    <ScrollViewer>
        <StackPanel Name="stackPanel" />
    </ScrollViewer>
</Grid>
```

我还决定采用含 TextBlock 子元素的 StackPanel 来用于打印，但有一个本质的区别：对于屏幕，只有一个 StackPanel，因为它在 ScrollViewer 中。而对于打印，每页都必须有 StackPanel，而且页面只包含 TextBlock 元素。

对屏幕和打印机使用相同的 TextBlock 元素很简单。理论上讲，可以打印已经显示在屏幕上的元素，但就我的经验而言，这种方法并没有如期真正实现。有一个主要的限制，即特定元素不能有两个父类。在本例中，所打印的 TextBlock 必须是打印机页面的 StackPane 子元素，因此，不能同时是屏幕上 StackPanel 的子元素。

因此，类层次结构的程序的修订版本会创建 TextBlock 元素的单独整体集合并存储在名为 printerTextBlocks 的字段中。Mainpage 类的这部分与 DependencyObjectClassHierarchy 中的代码隐藏文件十分相似，只不过 TextBlock 代码被分成单独的各个方法，以利于创建 TextBlock 元素的两个平行集合。注意，打印机 TextBlock 元素在 DisplayAndPrinterPrep 方法中(重命名自先前的 Display 方法)已指定了明确的黑色 Foreground。程序片段中没有显示大多数打印支持。

```
项目:PrintableClassHierarchy | 文件: MainPage.xaml.cs(片段)
public sealed partial class MainPage : Page
{
    Type rootType = typeof(DependencyObject);
    TypeInfo rootTypeInfo = typeof(DependencyObject).GetTypeInfo();
    List<Type> classes = new List<Type>();
    Brush highlightBrush;

    // Printing support
    List <TextBlock> printerTextBlocks = new List<TextBlock>();
    Brush blackBrush = new SolidColorBrush(Colors.Black);
    ...
    public MainPage()
    {
        this.InitializeComponent();
        highlightBrush =
            new SolidColorBrush(new UISettings().UIElementColor(UIElementType.Highlight));

        // Accumulate all the classes that derive from DependencyObject
        AddToClassList(typeof(Windows.UI.Xaml.DependencyObject));

        // Sort them alphabetically by name
        classes.Sort((t1, t2) =>
        {
            return String.Compare(t1.GetTypeInfo().Name, t2.GetTypeInfo().Name);
        });

        // Put all these sorted classes into a tree structure
```

```
        ClassAndSubclasses rootClass = new ClassAndSubclasses(rootType);
        AddToTree(rootClass, classes);

        // Display the tree using TextBlocks added to StackPanel
        DisplayAndPrinterPrep(rootClass, 0);
        ...
    }
    ...
    void AddToClassList(Type sampleType)
    {
        Assembly assembly = sampleType.GetTypeInfo().Assembly;

        foreach (Type type in assembly.ExportedTypes)
        {
            TypeInfo typeInfo = type.GetTypeInfo();

            if (typeInfo.IsPublic && rootTypeInfo.IsAssignableFrom(typeInfo))
                classes.Add(type);
        }
    }

    void AddToTree(ClassAndSubclasses parentClass, List<Type> classes)
    {
        foreach (Type type in classes)
        {
            Type baseType = type.GetTypeInfo().BaseType;

            if (baseType == parentClass.Type)
            {
                ClassAndSubclasses subClass = new ClassAndSubclasses(type);
                parentClass.Subclasses.Add(subClass);
                AddToTree(subClass, classes);
            }
        }
    }

    void DisplayAndPrinterPrep(ClassAndSubclasses parentClass, int indent)
    {
        TypeInfo typeInfo = parentClass.Type.GetTypeInfo();

        // Create TextBlock and add to StackPanel
        TextBlock txtblk = CreateTextBlock(typeInfo, indent);
        stackPanel.Children.Add(txtblk);

        // Create TextBlock and add to printer list
        txtblk = CreateTextBlock(typeInfo, indent);
        txtblk.Foreground = blackBrush;
        printerTextBlocks.Add(txtblk);

        // Call this method recursively for all subclasses
        foreach (ClassAndSubclasses subclass in parentClass.Subclasses)
            DisplayAndPrinterPrep(subclass, indent + 1);
    }

    TextBlock CreateTextBlock(TypeInfo typeInfo, int indent)
    {
        // Create TextBlock with type name
        TextBlock txtblk = new TextBlock();
        txtblk.Inlines.Add(new Run { Text = new string(' ', 8 * indent) });
        txtblk.Inlines.Add(new Run { Text = typeInfo.Name });

        // Indicate if the class is sealed
        if (typeInfo.IsSealed)
            txtblk.Inlines.Add(new Run
            {
                Text = " (sealed)",
                Foreground = highlightBrush
            });

        // Indicate if the class can't be instantiated
```

```
        IEnumerable<ConstructorInfo> constructorInfos = typeInfo.DeclaredConstructors;
        int publicConstructorCount = 0;

        foreach (ConstructorInfo constructorInfo in constructorInfos)
            if (constructorInfo.IsPublic)
                publicConstructorCount += 1;

        if (publicConstructorCount == 0)
            txtblk.Inlines.Add(new Run
            {
                Text = " (non-instantiable)",
                Foreground = highlightBrush
            });

        return txtblk;
    }
    ...
}
```

打印支持的剩余部分与你以前看过的很相似，除非需要打印很多页。Paginate 方法发挥作用，将格式化页面存储到 printerPages 字段。每个对象均为 Padding 值设置为 96(1 英寸的)的 Border 以及带页面等于早先创建的 TextBlock 元素的子 StackPanel。

请记住，如果用户切换 Portrait 或 Landscape 模式、信件或 Legal 页大小，Paginate 处理程序可能会多次调用。由于程序要处理 TextBlock 元素固定集合，而且禁止元素有多重父级，因此，要确保所有 TextBlock 元素都不是以前创建的 StackPanel 的子集，它对 Paginate 方法非常重要。

项目:PrintableClassHierarchy | 文件: MainPage.xaml.cs(片段)

```
public sealed partial class MainPage : Page
{
    ...
    PrintDocument printDocument;
    IPrintDocumentSource printDocumentSource;
    List<UIElement> printerPages = new List<UIElement>();

    public MainPage()
    {
        ...
        // Create PrintDocument and attach handlers
        printDocument = new PrintDocument();
        printDocumentSource = printDocument.DocumentSource;
        printDocument.Paginate += OnPrintDocumentPaginate;
        printDocument.GetPreviewPage += OnPrintDocumentGetPreviewPage;
        printDocument.AddPages += OnPrintDocumentAddPages;
    }

    protected override void OnNavigatedTo(NavigationEventArgs args)
    {
        // Attach PrintManager handler
        PrintManager.GetForCurrentView().PrintTaskRequested += OnPrintManagerPrintTaskRequested;

        base.OnNavigatedTo(args);
    }

    protected override void OnNavigatedFrom(NavigationEventArgs e)
    {
        // Detach PrintManager handler
        PrintManager.GetForCurrentView().PrintTaskRequested -= OnPrintManagerPrintTaskRequested;

        base.OnNavigatedFrom(e);
    }
    ...
    void OnPrintManagerPrintTaskRequested(PrintManager sender, PrintTaskRequestedEventArgs args)
    {
        args.Request.CreatePrintTask("Dependency Property Class Hierarchy", (requestArgs) =>
        {
```

```
            requestArgs.SetSource(printDocumentSource);
        });
    }

    void OnPrintDocumentPaginate(object sender, PaginateEventArgs args)
    {
        // Verbosely set some variables for the page margin
        double leftMargin = 96;
        double topMargin = 96;
        double rightMargin = 96;
        double bottomMargin = 96;

        // Clear out previous printerPage collection
        foreach (UIElement printerPage in printerPages)
            ((printerPage as Border).Child as Panel).Children.Clear();

        printerPages.Clear();

        // Initialize page construction
        Border border = null;
        StackPanel stackPanel = null;
        double maxPageHeight = 0;
        double pageHeight = 0;

        // Look through the list of TextBlocks
        for (int index = 0; index < printerTextBlocks.Count; index++)
        {
            // A null Border object signals a new page
            if (border == null)
            {
                // Calculate the height available for text
                uint pageNumber = (uint)printerPages.Count;
                maxPageHeight =
                args.PrintTaskOptions.GetPageDescription(pageNumber).PageSize.Height;
                maxPageHeight -= topMargin + bottomMargin;
                pageHeight = 0;

                // Create StackPanel and Border
                stackPanel = new StackPanel();
                border = new Border
                {
                    Padding = new Thickness(leftMargin, topMargin, rightMargin, bottomMargin),
                    Child = stackPanel
                };

                // Add to the list of pages
                printerPages.Add(border);
            }

            // Get the TextBlock and measure it
            TextBlock txtblk = printerTextBlocks[index];
            txtblk.Measure(Size.Empty);

            // Check if OK to add TextBlock to this page
            if (stackPanel.Children.Count == 0 ||
                pageHeight + txtblk.ActualHeight < maxPageHeight)
            {
                stackPanel.Children.Add(txtblk);
                pageHeight += Math.Ceiling(txtblk.ActualHeight);
            }

            // Otherwise, it's the end of the page
            else
            {
                // No longer working with this Border object
                border = null;

                // Reprocess this TextBlock
                index--;
            }
        }
```

```
            // Notify about the final page count
            printDocument.SetPreviewPageCount(printerPages.Count, PreviewPageCountType.Final);
        }

        void OnPrintDocumentGetPreviewPage(object sender, GetPreviewPageEventArgs args)
        {
            printDocument.SetPreviewPage(args.PageNumber, printerPages[args.PageNumber - 1]);
        }

        void OnPrintDocumentAddPages(object sender, AddPagesEventArgs args)
        {
            foreach (UIElement printerPage in printerPages)
                printDocument.AddPage(printerPage);

            printDocument.AddPagesComplete();
        }
    }
```

分页策略涉及把纸页高度各减去顶部和底部 1 英寸页边距，计算得出 maxPageHeight 值。每个 TextBlock 都添加到该页面的 StackPanel，而另一个名为 pageHeight 的变量也会增加。方法对每个 TextBlock 都调用 Measure 方法并计算其大小，如果 TextBlock 高度加上 pageHeight 会超过 maxPageHeight 值，则需要一个新页面。

GetPreviewPage 处理程序使用事件参数中基于 1 的 PageNumber 属性来访问 printerPages 列表中的相应元素。AddPages 处理程序对所有页面都调用 AddPage。

在下图所示的预览中，可以在打印整张列表之前检查不同页面。

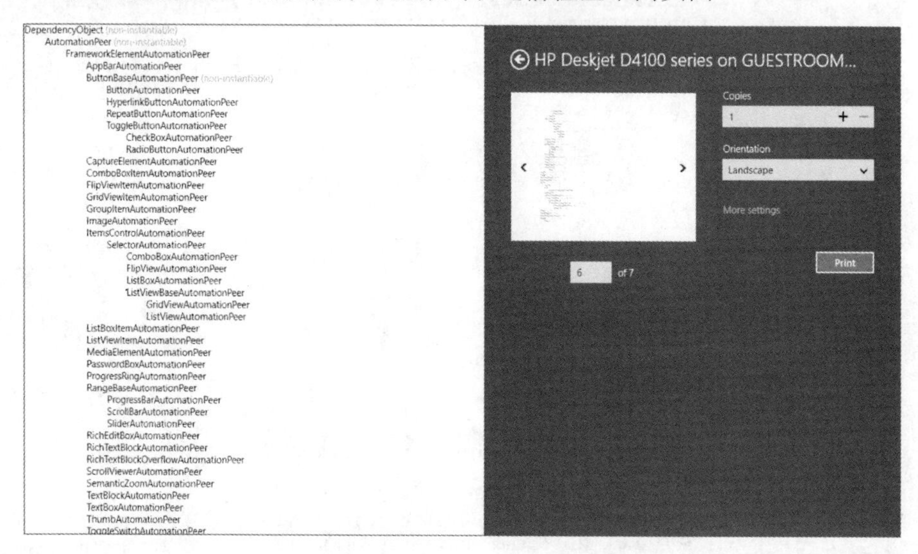

你可能已经注意到，分页逻辑基于每个 TextBlock 高度来增加 pageHeight，如下所示：

```
pageHeight += Math.Ceiling(txtblk.ActualHeight);
```

我一开始没有用 Math.Ceiling 调用。默认 FontSize 为 11，ActualHeight 值为 13.2，有了这两个值，程序会赋予每个 StackPanel 65 行文本，使其显示在竖排模式的 9 英寸内。然而，预览并打印出来后，只有 62 行可见。所得 StackPanel 比分配给它的空间要大，用于堆叠 StackPanel 中文本的行间距明显大于 13.2，因此导致每页都会剪掉 3 个 TextBlock 元素。

此时使用 Math.Ceiling 会导致每页有 61 行文字，有点朝另外方向偏离，但至少文本不会消失。

不过，这确实有点奇怪。当然，在视频显示器中将像素边界与文本对齐，对可读性而言具有十分重要的意义，这就是为什么坐标会四舍五入到下一个最高像素。然而在打印机上，每英寸也有 600 像素(或左右)，因此，四舍五入不需要基于 96 DPI 的设备。

分页非常复杂，尤其是涉及文本时。元素没有准确出现在你希望它在打印机页上的位置，如果长期遇到该问题，你可能想切换到使用 Canvas。对于复杂文本布局，Glyphs 比 TextBlock 更流行，而如果使用 Glyphs 遇到困难，你可能会想探讨 DirectWrite 并渲染到屏幕和打印机页上。

如果你喜欢另一个方向(通过 Windows Runtime 来确定如何显示文本)，使用 RichTextBlock(第 16 章的主题)可能有用。

17.7　自定义打印属性

PrintableClassHierarchy 中的一英寸边距为硬编码。假设想让用户可以选择页边距。我们来给用户设置打印字体大小的选项。

自定义打印机设置面板并不复杂，还可以让 Windows Runtime 负责创建合适的控件和管理输入的大部分工作。

自定义是在 PrintManager 的 PrintTaskRequested 事件处理程序中完成的。该处理程序目前如下所示：

```
void OnPrintManagerPrintTaskRequested(PrintManager sender, PrintTaskRequestedEventArgs args)
{
    args.Request.CreatePrintTask("My Print Task Title", OnPrintTaskSourceRequested);
}
```

或者也可以作为匿名 lambda 函数用于回调：

```
void OnPrintManagerPrintTaskRequested(PrintManager sender, PrintTaskRequestedEventArgs args)
{
    args.Request.CreatePrintTask("My Print Task Title", (requestArgs) =>
    {
        requestArgs.SetSource(printDocumentSource);
    });
}
```

无论采用哪种方式，CreatePrintTask 调用都会实际返回 PrintTask 类型的对象，使其可以保存到局部变量：

```
void OnPrintManagerPrintTaskRequested(PrintManager sender, PrintTaskRequestedEventArgs args)
{
    PrintTask printTask = args.Request.CreatePrintTask("My Print Task Title", ... );
}
```

通过可能习以为常的循环静态调用，从 PrintTask 对象可以得到 PrintTaskOptionDetails 类型的对象：

```
PrintTaskOptionDetails optionDetails =
        PrintTaskOptionDetails.GetFromPrintTaskOptions(printTask.Options);
```

PrintTaskOptionDetails 和相关类均在 Windows.Graphics.Printing.OptionDetails 命名空间中进行定义。

如果愿意，你可以删除打印机设置面板中的第一页内的所有选项：

```
optionDetails.DisplayedOptions.Clear();
```

现在看不到改变拷贝数量或横竖排的选项。当然也可以选择把这些设置还原，也许通过相反的顺序：

```
optionDetails.DisplayedOptions.Add(StandardPrintTaskOptions.Orientation);
optionDetails.DisplayedOptions.Add(StandardPrintTaskOptions.Copies);
```

StandardPrintTaskOptions 是一个静态类，并且属性代表用字符串 ID 进行识别的标准打印机选项。StandardPrintTaskOptions.Orientation 实际上是字符串 PageOrientation，StandardPrintTaskOptions.Copies 是字符串“JobCopiesAllDocuments”。如果适合，你可以初始化这些选项：

```
optionDetails.Options[StandardPrintTaskOptions.Orientation].TrySetValue(
                PrintOrientation.Landscape);
```

PrintOrientation 是 Windows.Graphics.Printing 中的 11 个类似枚举之一。

如果认为适合，则可以添加一个不太常见的选项：

```
optionDetails.DisplayedOptions.Add(StandardPrintTaskOptions.Collation);
```

也可以添加你自己的选项。仅限于两种类型的自定义选项：文本字段或类似 Orientation 选项的选项扩展列表。

我们来创建一个新项目为 CustomizableClassHierarchy。程序和 PrintableClassHierarchy 大同小异，但一些自定义值被定义为字段，这些字段初始化为程序认为合适的值。

```
项目:CustomizableClassHierarchy | 文件: MainPage.xaml.cs(片段)
public sealed partial class MainPage : Page
{
    ...
    // Initial values of custom printer settings
    double fontSize = new TextBlock().FontSize;
    double leftMargin = 96; // 1 inch
    double topMargin = 72; // 3/4 inch
    double rightMargin = 96;
    double bottomMargin = 72;
    ...
}
```

这些字段通过 PrintManager 的 PrintTaskRequested 事件处理程序进行访问。用户点击 Devices 按钮(可能是在选择打印机的过程中)，则触发该事件。

```
项目:CustomizableClassHierarchy | 文件: MainPage.xaml.cs(片段)
void OnPrintManagerPrintTaskRequested(PrintManager sender, PrintTaskRequestedEventArgs args)
{
    PrintTask printTask = args.Request.CreatePrintTask("Dependency Property Class Hierarchy",
                                                       (requestArgs) =>
    {
        requestArgs.SetSource(printDocumentSource);
    });

    PrintTaskOptionDetails optionDetails =
                PrintTaskOptionDetails.GetFromPrintTaskOptions(printTask.Options);

    // Add item for font size
    optionDetails.CreateTextOption("idFontSize", "Font size (in points)");
    optionDetails.DisplayedOptions.Add("idFontSize");
    optionDetails.Options["idFontSize"].TrySetValue((72 * fontSize / 96).ToString());

    // Add items for page margins
    optionDetails.CreateTextOption("idLeftMargin", "Left margin (in inches)");
    optionDetails.DisplayedOptions.Add("idLeftMargin");
    optionDetails.Options["idLeftMargin"].TrySetValue((leftMargin / 96).ToString());
```

```
        optionDetails.CreateTextOption("idTopMargin", "Top margin (in inches)");
        optionDetails.DisplayedOptions.Add("idTopMargin");
        optionDetails.Options["idTopMargin"].TrySetValue((topMargin / 96).ToString());

        optionDetails.CreateTextOption("idRightMargin", "Right margin (in inches)");
        optionDetails.DisplayedOptions.Add("idRightMargin");
        optionDetails.Options["idRightMargin"].TrySetValue((rightMargin / 96).ToString());

        optionDetails.CreateTextOption("idBottomMargin", "Bottom margin (in inches)");
        optionDetails.DisplayedOptions.Add("idBottomMargin");
        optionDetails.Options["idBottomMargin"].TrySetValue((bottomMargin / 96).ToString());

        // Set handler for the option changing
        optionDetails.OptionChanged += OnOptionDetailsOptionChanged;
    }
```

每个自定义选项至少需要两个步骤，也可能是三个步骤。首先必须创建选项，在此过程中，指定其 ID 字符串以及出现在打印机设置面板中的标签。然后把自定义项添加到 DisplayedOptions 集合。第三个步骤可选，但要设置初始值。在我的代码中，存储这些值的字段由像素转换成点(用于字体大小)和英寸(用于边界值)。

该方法最后为 OptionChanged 事件设置事件处理程序。该事件触发用于更改所有打印机的选项，而不仅是自定义选项。对于如下图所示的的文本项，不会每个按键都触发该事件，而只有按下 Enter 键、丢失输入焦点或按下 Press 按钮的时候才触发。下图所示为自定义设置面板。

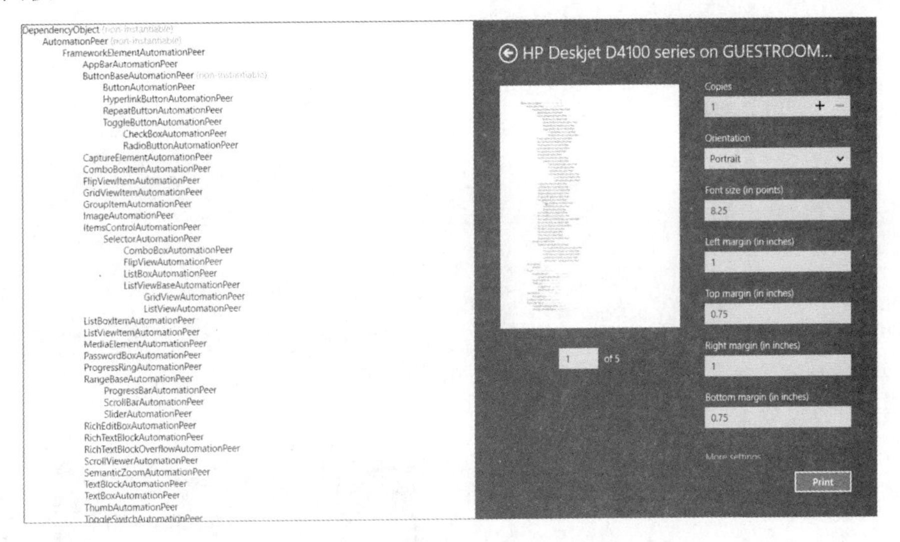

有没有看到五个新项目？我知道这看起来已经超越自定义选项的可用大小限制，但列表可滚动。

以下为 OptionChanged 事件处理程序的实现。这里进行验证并发出信号，表明需要用新值刷新打印预览，再次调用 Paginate 处理程序。PrintTaskOptionChangedEventArgs 类只定义一种属性，即 object 类型，名为 OptionId(但的确是字符串)，表明所改变选项，但也需要用到 sender 参数。也就是在 PrintTaskRequested 处理程序自定义过程中所使用的 PrintTaskOptionDetails 对象。

项目:CustomizableClassHierarchy | 文件: MainPage.xaml.cs(片段)

```
async void OnOptionDetailsOptionChanged(PrintTaskOptionDetails sender,
                                         PrintTaskOptionChangedEventArgs args)
```

```
{
    if (args.OptionId == null)
        return;

    string optionId = args.OptionId.ToString();
    string strValue = sender.Options[optionId].Value.ToString();
    string errorText = String.Empty;
    double value = 0;

    switch (optionId)
    {
        case "idFontSize":
            if (!Double.TryParse(strValue, out value))
                errorText = "Value must be numeric";

            else if (value < 4 || value > 36)
                errorText = "Value must be between 4 and 36";
            break;

        case "idLeftMargin":
        case "idTopMargin":
        case "idRightMargin":
        case "idBottomMargin":
            if (!Double.TryParse(strValue, out value))
                errorText = "Value must be numeric";

            else if (value < 0 || value > 2)
                errorText = "Value must be between 0 and 2";
            break;
    }

    sender.Options[optionId].ErrorText = errorText;

    // If there's no error, then invalidate the preview
    if (String.IsNullOrEmpty(errorText))
    {
        await this.Dispatcher.RunAsync(CoreDispatcherPriority.Normal, () =>
            {
                printDocument.InvalidatePreview();
            });
    }
}
```

如果其中一个选项的输入出了问题，就需要为该选项把 ErrorText 属性设置为简短但有用的文本字符串。该字符串向用户显示为红色。如果设置了任何选项的 ErrorText，则禁用 Print 按钮。效果如下图所示。

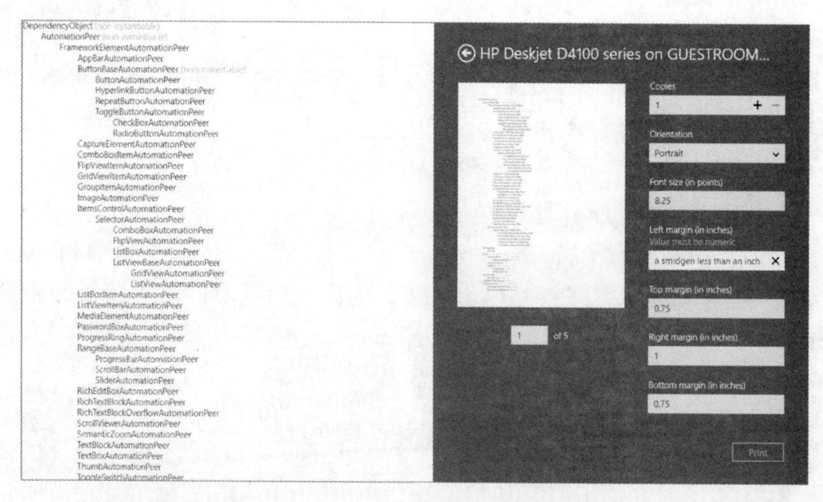

请注意，错误消息下面的事件进行了下移。如果你提供的错误信息多于一行，则折行。

如果没有错误，则调用 PrintDocument 对象的 InvalidatePreview 方法。请注意，需要 CoreDispatcher 以强制该调用发生在用户界面线程。OptionChanged 处理程序运行在辅助线程。

InvalidatePreview 调用导致对 Paginate 触发新的 Paginate 事件。新版本的 Paginate 处理程序通过获取所有自定义值并将其转换成可用数字。字体大小应用于所有要打印的已存储 TextBlock 元素，使用该方法旧版本中的页边距值。

```
项目:CustomizableClassHierarchy | 文件: MainPage.xaml.cs(片段)
void OnPrintDocumentPaginate(object sender, PaginateEventArgs args)
{
    // Get values of custom settings
    PrintTaskOptionDetails optionDetails =
                PrintTaskOptionDetails.GetFromPrintTaskOptions(args.PrintTaskOptions);
    fontSize = 96 * Double.Parse(optionDetails.Options["idFontSize"].Value.ToString()) / 72;
    leftMargin = 96 * Double.Parse(optionDetails.Options["idLeftMargin"].Value.ToString());
    topMargin = 96 * Double.Parse(optionDetails.Options["idTopMargin"].Value.ToString());
    rightMargin = 96 * Double.Parse(optionDetails.Options["idRightMargin"].Value.ToString());
    bottomMargin = 96 * Double.Parse(optionDetails.Options["idBottomMargin"].Value.ToString());

    // Set FontSize of stored TextBlocks
    foreach (TextBlock txtblk in printerTextBlocks)
        txtblk.FontSize = fontSize;
    ...
}
```

如果再多做一点工作，就可以检查用户输入的页边距值足够高，以避免文字出现在不可打印区域。在 OptionChanged 处理程序中，很容易从 PrintTaskOptionDetails 对象访问页面描述：

```
Rect imageableRect = sender.GetPageDescription(0).ImageableRect;
```

17.8　打印每月计划

有时，如果要进行一个长期项目，我喜欢打印月历并贴到墙上。这些日历不需要任何花哨功能，它只需要有大量空白能够用来写每天的事情。

PrintMonthlyPlanner 程序的唯一目的是打印用户所指定范围的月历。主页如下图所示。

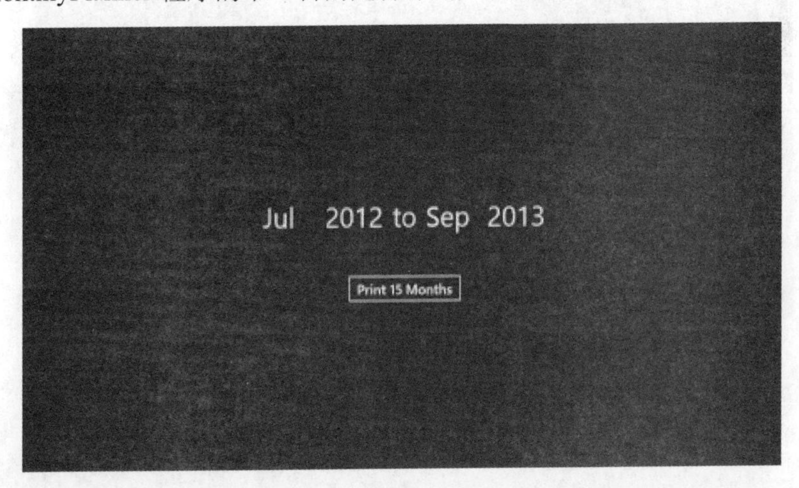

月份和年份通过 FlipView 控件来选择。只有开始月份小于或等于结束月份时，才会启用该按钮。按钮的 Click 处理程序只需要一行代码：

```
await PrintManager.ShowPrintUIAsync();
```

尽管用户一般都会调用超级按钮和 Devices 窗格，程序也可以这样做。一般情况下，此选项会保留给只在特殊场合打印的程序，例如 Print Ticket Confirmation。有趣的是，调用 ShowPrintUIAsync 带来的窗格与 Devices 按钮略有不同(见下图)。

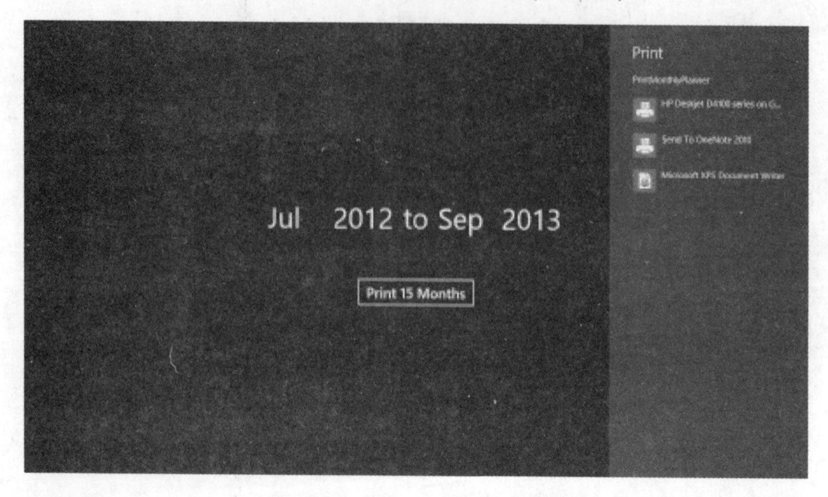

PrintMonthlyPlanner 程序专门用于打印，因此，要打印的页面并不由程序显示，只在屏幕上的打印窗格中可见，如下图所示。

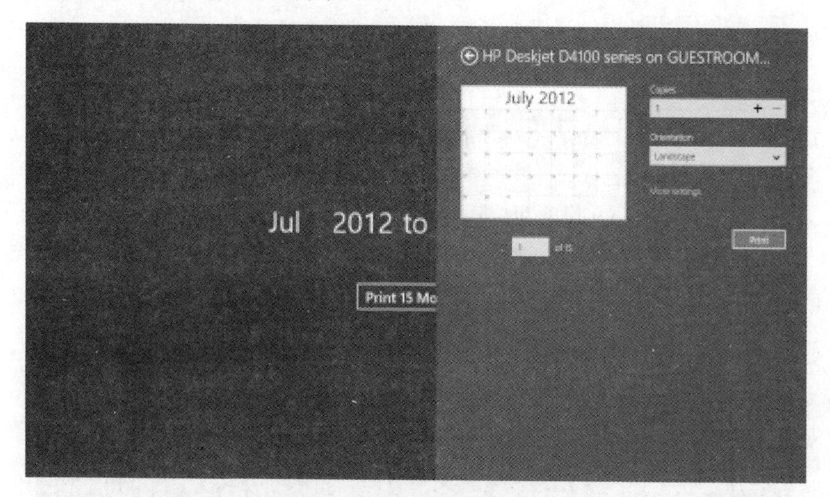

请注意，Orientation 设置为 Landscape。假设横排页面的日历效果更好，程序已经设置了这个初始值。每页都打印到可打印边距非常靠边的地方。

我创建了一个自定义控件，用户可以选择月份和年份。该空间称为 MonthYearSelect，XAML 文件揭示了两个模板化的 FlipView 控件，二者都有作为 ItemsPanel 的 StackPanel。

```
项目:PrintMonthlyPlanner | 文件: MonthYearSelect.xaml(片段)
<UserControl ... >
    <UserControl.Resources>
        <Style TargetType="FlipView">
```

```
            <Setter Property="ItemsPanel">
                <Setter.Value>
                    <ItemsPanelTemplate>
                        <StackPanel Orientation="Vertical" />
                    </ItemsPanelTemplate>
                </Setter.Value>
            </Setter>

            <Setter Property="ItemTemplate">
                <Setter.Value>
                    <DataTemplate>
                        <TextBlock Text="{Binding}" VerticalAlignment="Center" />
                    </DataTemplate>
                </Setter.Value>
            </Setter>
        </Style>
    </UserControl.Resources>

    <Grid>
        <StackPanel Orientation="Horizontal">
            <FlipView x:Name="monthFlipView"
                      SelectionChanged="OnMonthYearSelectionChanged" />

            <TextBlock Text=" " />

            <FlipView x:Name="yearFlipView"
                      SelectionChanged="OnMonthYearSelectionChanged" />
        </StackPanel>
    </Grid>
</UserControl>
```

为了能部分利用 Windows Runtime 的新功能，我决定把对该类的公共接口做成 Calendar 对象，而不是传统的.NET DateTime。我希望该程序适用于任何类型的日历，但 Calendar 类似乎并没有超越于标准格里高利历法。我甚至找不到一种方法来确定每星期的第一天应该是星期天(大多数地方的标准)还是星期一(例如在法国)。

我还发现，Calendar 是一个类而不是一个结构，这会困扰我随 FlipView 每次旋转而创建新的 Calendar 对象。我决定，控件只创建一个 Calendar 对象并可以更改该单个对象的 Month 和 Year 属性。但在这种情况下，不能把 Calendar 显示为一个从属属性，从属属性是带控件的，所以 Calendar 类型的属性就是一个普通的老属性，名为 MonthYear，并由 MonthYearChanged 事件进行补充(表明 Month 或 Year 新值)。

项目:PrintMonthlyPlanner | 文件: MonthYearSelect.xaml.cs(片段)

```
public sealed partial class MonthYearSelect : UserControl
{
    public event EventHandler MonthYearChanged;

    public MonthYearSelect()
    {
        this.InitializeComponent();

        // Create Calendar with current date
        Calendar calendar = new Calendar();
        calendar.SetToNow();

        // Fill the first FlipView with the abbreviated month names
        DateTimeFormatter monthFormatter =
            new DateTimeFormatter(YearFormat.None, MonthFormat.Abbreviated,
                                  DayFormat.None, DayOfWeekFormat.None);

        for (int month = 1; month <= 12; month++)
        {
            string strMonth = monthFormatter.Format(
```

```
                    new DateTimeOffset(2000, month, 15, 0, 0, 0, TimeSpan.Zero));
            monthFlipView.Items.Add(strMonth);
        }

        // Fill the second FlipView with years (5 years before current, 25 after)
        for (int year = calendar.Year - 5; year <= calendar.Year + 25; year++)
        {
            yearFlipView.Items.Add(year);
        }

        // Set the FlipViews to the current month and year
        monthFlipView.SelectedIndex = calendar.Month - 1;
        yearFlipView.SelectedItem = calendar.Year;
        this.MonthYear = calendar;
    }

    public Calendar MonthYear { private set; get; }

    void OnMonthYearSelectionChanged(object sender, SelectionChangedEventArgs args)
    {
        if (this.MonthYear == null)
            return;

        if (monthFlipView.SelectedIndex != -1)
            this.MonthYear.Month = (int)monthFlipView.SelectedIndex + 1;

        if (yearFlipView.SelectedIndex != -1)
            this.MonthYear.Year = (int)yearFlipView.SelectedItem;

        // Fire the event
        if (MonthYearChanged != null)
            MonthYearChanged(this, EventArgs.Empty);
    }
}
```

MainPage.xaml 文件实例化两个 MonthYearSelect 控件。

项目:PrintMonthlyPlanner | 文件: MainPage.xaml(片段)

```
<Page ...
    FontSize="48">

    <Grid Background="{StaticResource ApplicationPageBackgroundThemeBrush}">
        <Grid HorizontalAlignment="Center"
              VerticalAlignment="Center">
            <Grid.ColumnDefinitions>
                <ColumnDefinition Width="Auto" />
                <ColumnDefinition Width="*" />
                <ColumnDefinition Width="Auto" />
            </Grid.ColumnDefinitions>

            <Grid.RowDefinitions>
                <RowDefinition Height="Auto" />
                <RowDefinition Height="Auto" />
            </Grid.RowDefinitions>

            <local:MonthYearSelect x:Name="monthYearSelect1"
                                   Grid.Row="0" Grid.Column="0"
                                   Height="144"
                                   VerticalAlignment="Center"
                                   MonthYearChanged="OnMonthYearChanged" />

            <TextBlock Text=" to "
                       Grid.Row="0" Grid.Column="1"
                       VerticalAlignment="Center" />

            <local:MonthYearSelect x:Name="monthYearSelect2"
                                   Grid.Row="0" Grid.Column="2"
                                   Height="144"
```

```
                                          VerticalAlignment="Center"
                                          MonthYearChanged="OnMonthYearChanged" />

            <Button Name="printButton"
                    Content="Print 1 Month"
                    Grid.Row="1" Grid.Column="0" Grid.ColumnSpan="3"
                    FontSize="24"
                    HorizontalAlignment="Center"
                    Margin="0 24"
                    Click="OnPrintButtonClick" />
        </Grid>
    </Grid>
</Page>
```

该程序与本章其他程序有一些不同，打印功能在应用运行过程中不可用。只有两个 MonthYearSelect 控件达到月份的有效范围，打印功能才启用。随着这两个控件的每一次变化，程序需要为 Button 创建新标签，决定是应该启用还是禁用按钮，并确定是要附加还是解除 PrintTaskRequested 事件。该逻辑是 Mainpage 类的初始化部分的主要内容。

```
项目:PrintMonthlyPlanner | 文件: MainPage.xaml.cs(片段)
public sealed partial class MainPage : Page
{
    PrintDocument printDocument;
    IPrintDocumentSource printDocumentSource;
    List<UIElement> calendarPages = new List<UIElement>();
    bool printingEnabled;

    public MainPage()
    {
        this.InitializeComponent();

        // Create PrintDocument and attach handlers
        printDocument = new PrintDocument();
        printDocumentSource = printDocument.DocumentSource;
        printDocument.Paginate += OnPrintDocumentPaginate;
        printDocument.GetPreviewPage += OnPrintDocumentGetPreviewPage;
        printDocument.AddPages += OnPrintDocumentAddPages;
    }

    void OnMonthYearChanged(object sender, EventArgs args)
    {
        // Calculate number of months and check if it's non-negative
        int printableMonths = GetPrintableMonthCount();
        printButton.Content = String.Format("Print {0} Month{1}", printableMonths,
                                            printableMonths > 1 ? "s" : "");
        printButton.IsEnabled = printableMonths > 0;

        // Attach or detach PrintManager handler
        if (printingEnabled != printableMonths > 0)
        {
            PrintManager printManager = PrintManager.GetForCurrentView();

            if (printableMonths > 0)
                printManager.PrintTaskRequested += OnPrintManagerPrintTaskRequested;
            else
                printManager.PrintTaskRequested -= OnPrintManagerPrintTaskRequested;

            printingEnabled = printableMonths > 0;
        }
    }

    int GetPrintableMonthCount()
    {
        Calendar cal1 = monthYearSelect1.MonthYear;
        Calendar cal2 = monthYearSelect2.MonthYear;
```

```
        return cal2.Month - cal1.Month + 1 + 12 * (cal2.Year - cal1.Year);
    }

    async void OnPrintButtonClick(object sender, RoutedEventArgs args)
    {
        await PrintManager.ShowPrintUIAsync();
    }

    void OnPrintManagerPrintTaskRequested(PrintManager sender, PrintTaskRequestedEventArgs args)
    {
        // Create PrintTask
        PrintTask printTask = args.Request.CreatePrintTask("Monthly Planner",
                                        OnPrintTaskSourceRequested);

        // Set orientation to landscape
        PrintTaskOptionDetails optionDetails =
            PrintTaskOptionDetails.GetFromPrintTaskOptions(printTask.Options);

        PrintOrientationOptionDetails orientation =
            optionDetails.Options[StandardPrintTaskOptions.Orientation] as
                                        PrintOrientationOptionDetails;

        orientation.TrySetValue(PrintOrientation.Landscape);
    }

    void OnPrintTaskSourceRequested(PrintTaskSourceRequestedArgs args)
    {
        args.SetSource(printDocumentSource);
    }
    ...
}
```

注意，PrintTaskRequested 处理程序访问 Orientation 选项，并将其初始化为 Landscape。每次用户打开打印机窗格时都会发生。有可能用户实际不希望在横排模式下打印月历。你可能想跟踪用户最终采用的设置，获取设置，在 Paginate 处理程序中将其保存为字段，并在下一次打印机窗格出现时使用该设置。用户偏好甚至可以保存在用户设置中，以供下一次程序运行时使用。

创建页面是 Paginate 处理程序的任务，Paginate 处理程序把页面保存在字段中用于 GetPreviewPage 和 addPages 处理程序。这些页面围绕 Grid 建立，该 Grid 有对应一周七天的七列、取决于本月周数(范围可从二月的四个到其他月份的六个)的行数，还有一排在顶部显示月份和年份标题。

```
项目:PrintMonthlyPlanner | 文件: MainPage.xaml.cs(片段)
public sealed partial class MainPage : Page
{
    ...
    void OnPrintDocumentPaginate(object sender, PaginateEventArgs args)
    {
        // Prepare to generate pages
        uint pageNumbers = 0;
        calendarPages.Clear();
        Calendar calendar = monthYearSelect1.MonthYear.Clone();
        calendar.Day = 1;
        Brush black = new SolidColorBrush(Colors.Black);

        // For each month
        do
        {
            PrintPageDescription printPageDescription =
                        args.PrintTaskOptions.GetPageDescription(pageNumber);
```

```
// Set Padding on outer Border
double left = printPageDescription.ImageableRect.Left;
double top = printPageDescription.ImageableRect.Top;
double right = printPageDescription.PageSize.Width
            - left - printPageDescription.ImageableRect.Width;
double bottom = printPageDescription.PageSize.Height
            - top - printPageDescription.ImageableRect.Height;
Border border = new Border { Padding = new Thickness(left, top, right, bottom) };

// Use Grid for calendar cells
Grid grid = new Grid();
border.Child = grid;
int numberOfWeeks = (6 + (int)calendar.DayOfWeek + calendar.LastDayInThisMonth) / 7;

for (int row = 0; row < numberOfWeeks + 1; row++)
    grid.RowDefinitions.Add(new RowDefinition
    {
        Height = new GridLength(1, GridUnitType.Star)
    });

for (int col = 0; col < 7; col++)
    grid.ColumnDefinitions.Add(new ColumnDefinition
    {
        Width = new GridLength(1, GridUnitType.Star)
    });

// Month and year display at top
Viewbox viewbox = new Viewbox
{
    Child = new TextBlock
    {
        Text = calendar.MonthAsSoloString() + " " + calendar.YearAsString(),
        Foreground = black,
        FontSize = 96,
        HorizontalAlignment = HorizontalAlignment.Center
    }
};
Grid.SetRow(viewbox, 0);
Grid.SetColumn(viewbox, 0);
Grid.SetColumnSpan(viewbox, 7);
grid.Children.Add(viewbox);

// Now loop through the days of the month
for (int day = 1, row = 1, col = (int)calendar.DayOfWeek;
    day <= calendar.LastDayInThisMonth; day++)
{
    Border dayBorder = new Border
    {
        BorderBrush = black,

        // Avoid double line drawing
        BorderThickness = new Thickness
        {
            Left = day == 1 || col == 0 ? 1 : 0,
            Top = day - 7 < 1 ? 1 : 0,
            Right = 1,
            Bottom = 1
        },

        // Put day of month in upper-left corner
        Child = new TextBlock
        {
            Text = day.ToString(),
            Foreground = black,
            FontSize = 24,
            HorizontalAlignment = HorizontalAlignment.Left,
```

```
                        VerticalAlignment = VerticalAlignment.Top
                    }
                };
                Grid.SetRow(dayBorder, row);
                Grid.SetColumn(dayBorder, col);
                grid.Children.Add(dayBorder);

                if (0 == (col = (col + 1) % 7))
                    row += 1;
            }
            calendarPages.Add(border);
            calendar.AddMonths(1);
            pageNumber += 1;
        }
        while (calendar.Year < monthYearSelect2.MonthYear.Year ||
               calendar.Month <= monthYearSelect2.MonthYear.Month);

        printDocument.SetPreviewPageCount(calendarPages.Count, PreviewPageCountType.Final);
    }

    void OnPrintDocumentGetPreviewPage(object sender, GetPreviewPageEventArgs args)
    {
        printDocument.SetPreviewPage(args.PageNumber, calendarPages[args.PageNumber - 1]);
    }

    void OnPrintDocumentAddPages(object sender, AddPagesEventArgs args)
    {
        foreach (UIElement calendarPage in calendarPages)
            printDocument.AddPage(calendarPage);

        printDocument.AddPagesComplete();
    }
}
```

17.9　打印可选范围页

本章的下一个程序是试验完全失控的情景。我想演示如何把选项添加到印刷窗格，以允许用户可以选择打印页面范围。同时，我还想展示如何在屏幕与打印机之间共享 UIElement 实例。

对于该演示，我选择更改第 4 章中比阿特丽克斯·波特(Beatrix Potter)的作品《汤姆小猫》。为了方便打印成书页，我决定每页都是单独 UserControl 派生类。而对于在屏幕显示的内容，通过一个可滚动 StackPanel 就可以直接集合这些 UserControl 页。

该机制没有问题，只不过在末尾我用了 57 个 UserControl 派生类，名称从 TomKitten03 到 TomKitten59，其中的数字表示原书页。但事实证明，我真的无法对屏幕和打印机使用相同的控件实例，除非我想把书中的每一页文字或图片都打印在打印页的左上角(这当然是无法接受的)。

显示在屏幕上的元素随定义其和父级关系的布局过程而变化，而且在一般情况下无法将这些元素从可视树上解除并期望可以在打印机上很好地呈现。你也不能搞乱这些元素。也不能仅仅为了打印机就设置新属性，因为这些属性会影响到屏幕显示结果。也不能把它们放入另一个容器，因为会违反一个元素只能有一个父级的规则。

最后，我意识到可以对屏幕和打印机都使用相同的 57 个 UserControl 派生类，但只有在它们是单独实例的情况下，也就是说，每个控件都会实例化两次：一次是用于屏幕的

Mainpage.xaml，第二次是用于打印机的 Mainpage.xaml.cs。

因此，从某种意义而言，该试验失败了，因为我无法不能直接重复使用实例，但它说明了包含 57 个 UserControl 派生类的 Visual Studio 项目的不足！Visual Studio 尽力加载和编译所有 XAML 文件，而我们程序员也应该很细心。这种方式不可用于制作电子书！

另一方面，程序的确演示了如何把选择要打印的页面范围功能添加到打印机选项。

为了把一组统一风格应用到 MainPage.xaml 的 UserControl 的派生类以及应用到 Mainpage.xaml.cs 中的 UserControl 派生类实例，我把所有 Style 定义都移到 App.xaml。这样一来，在整个应用中都可用。

```
项目:PrintableTomKitten | 文件: App.xaml
<Application
    x:Class="PrintableTomKitten.App"
    xmlns="http://schemas.microsoft.com/winfx/2006/xaml/presentation"
    xmlns:x="http://schemas.microsoft.com/winfx/2006/xaml"
    xmlns:local="using:PrintableTomKitten">

    <Application.Resources>
        <ResourceDictionary>
            <ResourceDictionary.MergedDictionaries>
                <ResourceDictionary Source="Common/StandardStyles.xaml"/>
            </ResourceDictionary.MergedDictionaries>

            <Style x:Key="commonTextStyle" TargetType="TextBlock">
                <Setter Property="FontFamily" Value="Century Schoolbook" />
                <Setter Property="FontSize" Value="36" />
                <Setter Property="Foreground" Value="Black" />
                <Setter Property="Margin" Value="0 12" />
            </Style>

            <Style x:Key="paragraphTextStyle" TargetType="TextBlock"
             BasedOn="{StaticResource commonTextStyle}">
                <Setter Property="TextWrapping" Value="Wrap" />
            </Style>

            <Style x:Key="frontMatterTextStyle" TargetType="TextBlock"
             BasedOn="{StaticResource commonTextStyle}">
                <Setter Property="TextAlignment" Value="Center" />
            </Style>

            <Style x:Key="imageStyle" TargetType="Image">
                <Setter Property="Stretch" Value="None" />
                <Setter Property="HorizontalAlignment" Value="Center" />
            </Style>
        </ResourceDictionary>
    </Application.Resources>
</Application>
```

MainPage.xaml 文件把书中的所有单个页面都列在 StackPanel 上。以下代码省去了中间部分。

```
项目:PrintableTomKitten | 文件: MainPage.xaml(片段)
<Page
    x:Class="PrintableTomKitten.MainPage"
    xmlns="http://schemas.microsoft.com/winfx/2006/xaml/presentation"
    xmlns:x="http://schemas.microsoft.com/winfx/2006/xaml"
    xmlns:local="using:PrintableTomKitten">

    <Grid Background="White">
        <ScrollViewer>
            <StackPanel Name="bookPageStackPanel"
                        MaxWidth="640"
                        HorizontalAlignment="Center">
```

```
                <local:TomKitten03 />
                <local:TomKitten04 />
                <local:TomKitten05 />
                <local:TomKitten06 />
                <local:TomKitten07 />
                <local:TomKitten08 />
                <local:TomKitten09 />

                <local:TomKitten10 />
                <local:TomKitten11 />
                <local:TomKitten13 />
                <local:TomKitten12 />

                <local:TomKitten14 />
                <local:TomKitten15 />
                <local:TomKitten17 />
                <local:TomKitten16 />
                ...
                <local:TomKitten50 />
                <local:TomKitten51 />
                <local:TomKitten53 />
                <local:TomKitten52 />

                <local:TomKitten54 />
                <local:TomKitten55 />
                <local:TomKitten56 />
                <local:TomKitten57 />

                <local:TomKitten59 />
                <local:TomKitten58 />
            </StackPanel>
        </ScrollViewer>
    </Grid>
</Page>
```

其中一些看似混乱。像第 4 章中讨论的，我发现有必要对一些文字和图片页进行交换，以提供更流畅的阅读体验。

如下所示，只包含一张图片的页相当小。

```
项目:PrintableTomKitten | 文件: TomKitten20.xaml
<UserControl
    x:Class="PrintableTomKitten.TomKitten20"
    xmlns="http://schemas.microsoft.com/winfx/2006/xaml/presentation"
    xmlns:x="http://schemas.microsoft.com/winfx/2006/xaml">

    <Image Source="Images/tom20.jpg" Style="{StaticResource imageStyle}" />
</UserControl>
```

许多页面都只有一段文字，如下所示。

```
项目:PrintableTomKitten | 文件: TomKitten21.xaml
<UserControl
    x:Class="PrintableTomKitten.TomKitten21"
    xmlns="http://schemas.microsoft.com/winfx/2006/xaml/presentation"
    xmlns:x="http://schemas.microsoft.com/winfx/2006/xaml">

    <Grid VerticalAlignment="Center"
          MaxWidth="640">
        <TextBlock Style="{StaticResource paragraphTextStyle}">
              Tom Kitten was very fat, and he had grown;
            several buttons burst off. His mother sewed them on again.
        </TextBlock>
    </Grid>
</UserControl>
```

注意 Grid 的 VerticalAlignment 和 maxWidth 设置。这些设置可优化打印机。如果显示在屏幕上，VerticalAlignment 设置没有用，因为它是竖排 StackPanel 的子类，而 StackPanel 本身有 640 的 maxWidth 设置。

如下所示，超过一段文字的页需要 StackPanel。

```
项目:PrintableTomKitten | 文件: TomKitten21.xaml
<UserControl
     x:Class="PrintableTomKitten.TomKitten22"
     xmlns="http://schemas.microsoft.com/winfx/2006/xaml/presentation"
     xmlns:x="http://schemas.microsoft.com/winfx/2006/xaml">

     <Grid VerticalAlignment="Center"
           MaxWidth="640">
          <StackPanel>
               <TextBlock Style="{StaticResource paragraphTextStyle}">
                      When the three kittens were ready, Mrs.
                    Tabitha unwisely turned them out into the garden, to be
                    out of the way while she made hot buttered toast.
               </TextBlock>

               <TextBlock Style="{StaticResource paragraphTextStyle}">
                      "Now keep your frocks clean, children! You
                    must walk on your hind legs. Keep away from the dirty
                    ash-pit, and from Sally Henny Penny, and from the
                    pig-stye and the Puddle-Ducks."
               </TextBlock>
          </StackPanel>
     </Grid>
</UserControl>
```

这就是该项目所有的 XAML。

如你所知，现在的程序都提供打印全部或部分文档的选项，这很常见。这些选项通常标有类似 All、Selection 和 Custom Range 标签。PrintableTomKitten 项目中没有选择概念，因此，我的选项只限于 Print all pages 和 Print custom range。

自定义范围同时包含单独网面以及逗号分隔的连续页面的范围，如 2-4，7，9-11，这也很常见。以下 CustomPageRange 类的构造函数接受带自定义页面范围的字符串并将信息转换成连续页面列表。对于字符串 2-4，7，9-11，PageMapping 属性会设置为整数 2、3、4、7、9、10、11 列表。如果在某种方式上字符串无效，PageMapping 则为 null，IsValid 返回 false。

```
项目:PrintableTomKitten | 文件: CustomPrintRange.cs
using System;
using System.Collections.Generic;

namespace PrintableTomKitten
{
     public class CustomPageRange
     {
          // Structure used internally
          struct PageRange
          {
               public PageRange(int from, int to) : this()
               {
                    this.From = from;
                    this.To = to;
               }

               public int From { private set; get; }
               public int To { private set; get; }
          }
```

```
        public CustomPageRange(string str, int maxPageNumber)
        {
            List<PageRange> pageRanges = new List<PageRange>();
            string[] strRanges = str.Split(',');

            foreach (string strRange in strRanges)
            {
                int dashIndex = strRange.IndexOf('-');

                // Just one page number
                if (dashIndex == -1)
                {
                    int page;

                    if (Int32.TryParse(strRange.Trim(), out page) &&
                        page > 0 && page <= maxPageNumber)
                    {
                        pageRanges.Add(new PageRange(page, page));
                    }
                    else
                    {
                        return;
                    }
                }
                // Two page numbers separated by a dash
                else
                {
                    string strFrom = strRange.Substring(0, dashIndex);
                    string strTo = strRange.Substring(dashIndex + 1);
                    int from, to;

                    if (Int32.TryParse(strFrom.Trim(), out from) &&
                        Int32.TryParse(strTo.Trim(), out to) &&
                        from > 0 && from <= maxPageNumber &&
                        to > 0 && to <= maxPageNumber &&
                        from <= to)
                    {
                        pageRanges.Add(new PageRange(from, to));
                    }
                    else
                    {
                        return;
                    }
                }
            }

            // If we made it to this, the input string is valid
            this.PageMapping = new List<int>();

            // Define a mapping to page numbers
            foreach (PageRange pageRange in pageRanges)
                for (int page = pageRange.From; page <= pageRange.To; page++)
                    this.PageMapping.Add(page);
        }

        // Zero-based in, one-based out
        public IList<int> PageMapping { private set; get; }

        public bool IsValid
        {
            get { return this.PageMapping != null; }
        }
    }
}
```

PrintableTomKitten 程序在两个地方使用 CustomPageRange 类：验证用户在打印机选项面板进行输入的时候以及稍后在 Paginate 事件处理程序中。第二种情况下，

CustomPageRange 对象存储为字段，供 GetPreviewPage 和 addPages 处理程序使用。

MainPage.xaml.cs 文件如下所示。请注意开头的大数组，包含只用于打印的所有书页的额外实例。

项目:PrintableTomKitten | 文件: MainPage.xaml.cs(片段)

```
public sealed partial class MainPage : Page
{
    PrintDocument printDocument;
    IPrintDocumentSource printDocumentSource;
    CustomPageRange customPageRange;
    UIElement[] bookPages =
    {
        new TomKitten03(), new TomKitten04(), new TomKitten05(), new TomKitten06(),
        new TomKitten07(), new TomKitten08(), new TomKitten09(), new TomKitten10(),
        new TomKitten11(), new TomKitten12(), new TomKitten13(), new TomKitten14(),
        new TomKitten15(), new TomKitten16(), new TomKitten17(), new TomKitten18(),
        new TomKitten19(), new TomKitten20(), new TomKitten21(), new TomKitten22(),
        new TomKitten23(), new TomKitten24(), new TomKitten25(), new TomKitten26(),
        new TomKitten27(), new TomKitten28(), new TomKitten29(), new TomKitten30(),
        new TomKitten31(), new TomKitten32(), new TomKitten33(), new TomKitten34(),
        new TomKitten35(), new TomKitten36(), new TomKitten37(), new TomKitten38(),
        new TomKitten39(), new TomKitten40(), new TomKitten41(), new TomKitten42(),
        new TomKitten43(), new TomKitten44(), new TomKitten45(), new TomKitten46(),
        new TomKitten47(), new TomKitten48(), new TomKitten49(), new TomKitten50(),
        new TomKitten51(), new TomKitten52(), new TomKitten53(), new TomKitten54(),
        new TomKitten55(), new TomKitten56(), new TomKitten57(), new TomKitten58(),
        new TomKitten59()
    };

    public MainPage()
    {
        this.InitializeComponent();

        // Create PrintDocument and attach handlers
        printDocument = new PrintDocument();
        printDocumentSource = printDocument.DocumentSource;
        printDocument.Paginate += OnPrintDocumentPaginate;
        printDocument.GetPreviewPage += OnPrintDocumentGetPreviewPage;
        printDocument.AddPages += OnPrintDocumentAddPages;
    }

    protected override void OnNavigatedTo(NavigationEventArgs args)
    {
        // Attach PrintManager handler
        PrintManager.GetForCurrentView().PrintTaskRequested +=
        OnPrintManagerPrintTaskRequested;

        base.OnNavigatedTo(args);
    }

    protected override void OnNavigatedFrom(NavigationEventArgs e)
    {
        // Detach PrintManager handler
        PrintManager.GetForCurrentView().PrintTaskRequested -= OnPrintManagerPrintTaskRequested;
        base.OnNavigatedFrom(e);
    }

    void OnPrintManagerPrintTaskRequested(PrintManager sender, PrintTaskRequestedEventArgs args)
    {
        PrintTask printTask = args.Request.CreatePrintTask("The Tale of Tom Kitten",
                                                          OnPrintTaskSourceRequested);

        // Get PrintTaskOptionDetails for making changes to options
        PrintTaskOptionDetails optionDetails =
            PrintTaskOptionDetails.GetFromPrintTaskOptions(printTask.Options);
```

```
        // Create the custom item
        PrintCustomItemListOptionDetails pageRange =
                    optionDetails.CreateItemListOption("idPrintRange", "Print range");
        pageRange.AddItem("idPrintAll", "Print all pages");
        pageRange.AddItem("idPrintCustom", "Print custom range");

        // Add it to the options
        optionDetails.DisplayedOptions.Add("idPrintRange");

        // Create a page-range edit item also, but this only
        //    comes into play when user selects "Print custom range"
        optionDetails.CreateTextOption("idCustomRangeEdit", "Custom Range");

        // Set a handler for the OptionChanged event
        optionDetails.OptionChanged += OnOptionDetailsOptionChanged;
    }

    void OnPrintTaskSourceRequested(PrintTaskSourceRequestedArgs args)
    {
        args.SetSource(printDocumentSource);
    }
    ...
}
```

此前讲过如何使用 PrintTaskOptionDetails 的 CreateTextOption 方法来创建自定义文本输入字段。唯一可以替代文本输入字段的方法涉及这里显示的 CreateItemListOption 方法。这会产生类似 Orientation 选项的一个互斥选项列表。指定字符串 ID 和标签。该方法返回 PrintCustomItemListOptionDetails 类型的对象。为此，需要用 ID 字符串和标签添加列表项，然后把相同的 ID 添加到 DisplayedOptions 集合。下图为初始效果。

但还要注意，PrintTaskRequested 处理程序还调用 CreateTextOption 为自定义页面范围创建文本输入字段：

```
optionDetails.CreateTextOption("idCustomRangeEdit", "Custom Range");
```

创建好字段，但还不会添加到 DisplayedOptions 集合。你想要只有用户选择 Print custom range 才显示该条目。

该逻辑发生在 OptionChanged 处理程序中。如果选项 ID 字符串为 idPrintCustom，则把由字符串 idCustomRangeEdit 标识的文本输入字段添加到 DisplayedOptions 集合，而如果 ID 字符串为 idPrintAll，则必须从 DisplayedOptions 中移除。

```
项目:PrintableTomKitten | 文件: MainPage.xaml.cs(片段)
public sealed partial class MainPage : Page
{
    ...
    async void OnOptionDetailsOptionChanged(PrintTaskOptionDetails sender,
                                    PrintTaskOptionChangedEventArgs args)
    {
        if (args.OptionId == null)
            return;

        string optionId = args.OptionId.ToString();
        string strValue = sender.Options[optionId].Value.ToString();
        string errorText = String.Empty;

        switch (optionId)
        {
            case "idPrintRange":
                switch (strValue)
                {
                    case "idPrintAll":
                        if (sender.DisplayedOptions.Contains("idCustomRangeEdit"))
                            sender.DisplayedOptions.Remove("idCustomRangeEdit");
                        break;

                    case "idPrintCustom":
                        sender.DisplayedOptions.Add("idCustomRangeEdit");
                        break;
                }
                break;

            case "idCustomRangeEdit":
                // Check to see if CustomPageRange accepts this
                if (!new CustomPageRange(strValue, bookPages.Length).IsValid)
                {
                    errorText = "Use the form 2-4, 7, 9-11";
                }
                break;
        }

        sender.Options[optionId].ErrorText = errorText;

        // If no error, then invalidate the preview
        if (String.IsNullOrEmpty(errorText))
        {
            await this.Dispatcher.RunAsync(CoreDispatcherPriority.Normal, () =>
            {
                printDocument.InvalidatePreview();
            });
        }
    }
    ...
}
```

如果 idCustomRangeEdit 控件可见，也可以接收来自该控件的通知。为了确定范围是否有效，需要调用 CustomPageRange 构造函数并设置可能的错误文本。下图为成功解析的页面范围。

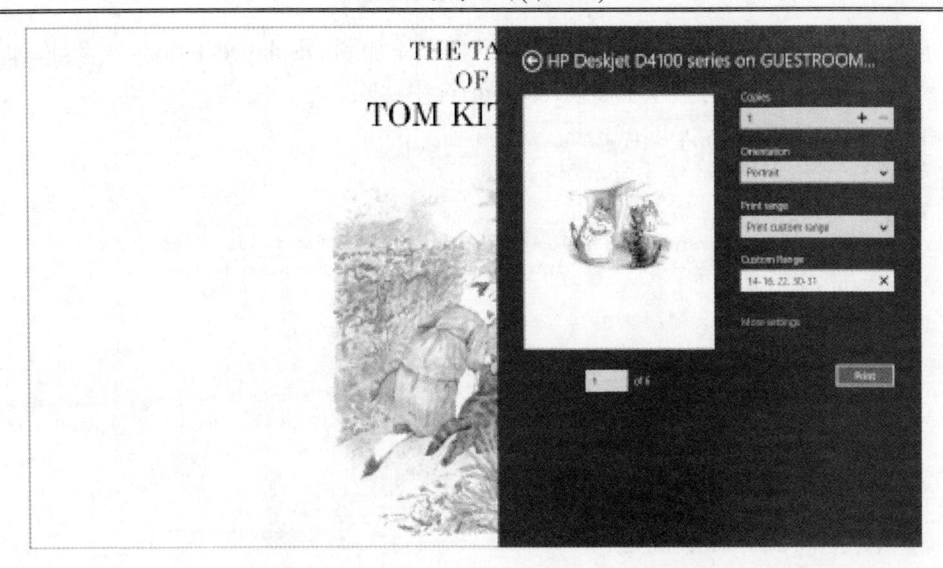

注意，预览下方的页面数字表示应该打印的页数，但并不表示用户选择的实际页数。我不肯定这是否能够真正解决该问题，除非把页面范围选择移动到标准选项中，而这一点，我们就无法控制了。

另外请注意，OptionChanged 处理程序并没有将 CustomPageRange 对象保存为字段。并不需要将其保存在该处理程序中，而且应该避免这样做。如果用户在来回选择选项，很难跟踪哪些是实际选定可见的，哪些不是。

相反，可以在 PrintDocument 事件的三个处理程序中获取选项的最终设置。在本例中，Paginate 处理程序获取该设置并将 CustomPageRange 对象存储为字段，而其他两种方法可以访问 CustomPageRange。

项目:PrintableTomKitten | 文件: MainPage.xaml.cs(片段)

```
public sealed partial class MainPage : Page
{
    ...
    void OnPrintDocumentPaginate(object sender, PaginateEventArgs args)
    {
        // Obtain the print range option
        PrintTaskOptionDetails optionDetails =
                PrintTaskOptionDetails.GetFromPrintTaskOptions(args.PrintTaskOptions);

        string strValue = optionDetails.Options["idPrintRange"].Value as string;

        if (strValue == "idPrintCustom")
        {
            // Parse the print range for GetPreviewPage and AddPages
            string strPageRange = optionDetails.Options["idCustomRangeEdit"].Value as string;
            customPageRange = new CustomPageRange(strPageRange, bookPages.Length);
        }
        else
        {
            // Make sure field is null if printing all pages
            customPageRange = null;
        }

        int pageCount = bookPages.Length;

        if (customPageRange != null && customPageRange.IsValid)
            pageCount = customPageRange.PageMapping.Count;
```

```
            printDocument.SetPreviewPageCount(pageCount, PreviewPageCountType.Final);
        }

        void OnPrintDocumentGetPreviewPage(object sender, GetPreviewPageEventArgs args)
        {
            int oneBasedIndex = args.PageNumber;

            if (customPageRange != null && customPageRange.IsValid)
                oneBasedIndex = customPageRange.PageMapping[args.PageNumber - 1];

            printDocument.SetPreviewPage(args.PageNumber, bookPages[oneBasedIndex - 1]);
        }

        void OnPrintDocumentAddPages(object sender, AddPagesEventArgs args)
        {
            if (customPageRange != null && customPageRange.IsValid)
            {
                foreach (int oneBasedIndex in customPageRange.PageMapping)
                    printDocument.AddPage(bookPages[oneBasedIndex - 1]);
            }
            else
            {
                foreach (UIElement bookPage in bookPages)
                    printDocument.AddPage(bookPage);
            }

            printDocument.AddPagesComplete();
        }
    }
```

我前面提过开始做《汤姆小猫》可打印版本时，是想确定是否可以在屏幕与打印机之间共享元素。如果想看看打印从 MainPage 显示的 UserControl 派生类实例时会发生什么，可以直接在 OnPrintDocumentGetPreviewPage 和 OnPrintDocumentAddPages 方法中用 bookPageStackPanel.Children 更换 bookPages。

17.10　关　　键

可能打印很多页的程序会遇到分页问题。也许程序需要一些时间来确定到底有多少页要打印。

在传递给 CreatePrintTask 的回调方法中(即我调用 OnPrintTaskSourceRequested 的方法)，可以先调用事件参数中的 SetSource，再使用事件参数来延迟执行异步任务：

```
PrintTaskSourceRequestedDeferral deferral = args.GetDeferral();
await BigJobInvolvingPrintingAsync();
deferral.Complete();
```

在这种情况下，显示带有所选打印机名称的打印窗格，但在打印机名称下方有一个配有文字“App preparing to print”的旋转进度环。用户可能无法享受该体验，但这是应用无须挂在用户界面线程上就可以获取一点点时间的有效途径。

也请记住，PrintDocument 的 SetPreviewPageCount 方法的第二个参数是 PreviewPageCountType 枚举项，可以为 Intermediate，也可以为 Final。不需要限制对 Paginate 处理程序调用该方法。一开始可以用初步页数进行调用，然后用分页继续后台任务。针对用户界面线程的 Dispatcher，可以额外调用 SetPreviewPageCount 来更新计数。

为了帮助应用告知用户长打印任务的进度，PrintTask 定义了三个事件，分别名为 Previewing、Submitting、Progressing 和 Completed。

17.11 打印 FingerPaint 艺术画

从你第一次开始使用这本书中的各种 FingerPaint 程序，我相信你就渴望打印自己的作品，并将它贴在冰箱门上。当然，总是可以对 FingerPaint 屏幕进行截屏并打印出来，但我们把打印支持集成到 FingerPaint 内部。

为了尽量不影响现有 FingerPaint 代码，我决定从 PrintDocument 中产生派生类，称为 BitmapPrintDocument。我提到过可以这么做，虽然所得类没有覆写方法，但确实使引用某些对象更容易了。

BitmapPrintDocument 类在 MainPage 构造函数中进行实例化。

项目:FingerPaint | 文件: MainPage.xaml.cs(片段)

```
public MainPage()
{
    ...
    // Create a PrintDocument derivative for handling printing
    new BitmapPrintDocument(() => { return bitmap; });
}
```

注意构造函数 BitmapPrintDocument 的参数，它太怪了！问题在于如果打印该位图，BitmapPrintDocument 类肯定需要引用名为 bitmap 的 WriteableBitmap 字段，但不能直接传递到 BitmapPrintDocument 构造函数。无论什么时候程序从文件或剪贴板加载图片或者创建新画布的时候，bitmap 字段都会发生变化。因此，我用 Func<BitmapSource>类型的参数定义了 BitmapPrintDocument 构造函数，这样一来，只要 BitmapPrintDocument 需要当前位图，就可以直接回调 MainPage。

BitmapPrintDocument 将该参数保存为字段并执行标准初始化。

项目:FingerPaint | 文件: BitmapPrintDocument.cs(片段)

```
public class BitmapPrintDocument : PrintDocument
{
    Func<BitmapSource> getBitmap;
    IPrintDocumentSource printDocumentSource;

    // Element to print
    Border border = new Border
    {
        Child = new Image()
    };

    public BitmapPrintDocument(Func<BitmapSource> getBitmap)
    {
        this.getBitmap = getBitmap;

        // Get IPrintDocumentSource and attach event handlers
        printDocumentSource = this.DocumentSource;
        this.Paginate += OnPaginate;
        this.GetPreviewPage += OnGetPreviewPage;
        this.AddPages += OnAddPages;

        // Attach PrintManager handler
        PrintManager.GetForCurrentView().PrintTaskRequested +=
                                OnPrintDocumentPrintTaskRequested;
```

```
    }
    ...
}
```

PrintTaskRequested 处理程序是第一个需要位图的，因为它设置的是打印页面的初始方向与位图方向相一致。

```
项目:FingerPaint | 文件: BitmapPrintDocument.cs(片段)
async void OnPrintDocumentPrintTaskRequested(PrintManager sender,
                                             PrintTaskRequestedEventArgs args)
{
    PrintTaskRequestedDeferral deferral = args.Request.GetDeferral();

    // Obtain PrintTask
    PrintTask printTask = args.Request.CreatePrintTask("Finger Paint",
                                                       OnPrintTaskSourceRequested);

    // Probably set orientation to landscape
    PrintTaskOptionDetails optionDetails =
        PrintTaskOptionDetails.GetFromPrintTaskOptions(printTask.Options);

    PrintOrientationOptionDetails orientation =
        optionDetails.Options[StandardPrintTaskOptions.Orientation] as
                                                 PrintOrientationOptionDetails;

    bool bitmapIsLandscape = false;

    await border.Dispatcher.RunAsync(CoreDispatcherPriority.Normal, () =>
    {
        BitmapSource bitmapSource = getBitmap();
        bitmapIsLandscape = bitmapSource.PixelWidth > bitmapSource.PixelHeight;
    });

    orientation.TrySetValue(bitmapIsLandscape ? PrintOrientation.Landscape :
                                                PrintOrientation.Portrait);
    deferral.Complete();
}
```

请注意，CoreDispatcher 对象必须用来访问用户界面线程中的位图。此时，另一些事件处理程序应该会很熟悉，不同之处在于引用 PrintDocument 方法是直接引用 this 方法。

```
项目:FingerPaint | 文件: BitmapPrintDocument.cs(片段)
void OnPrintTaskSourceRequested(PrintTaskSourceRequestedArgs args)
{
    args.SetSource(printDocumentSource);
}

void OnPaginate(object sender, PaginateEventArgs args)
{
    PrintPageDescription pageDesc = args.PrintTaskOptions.GetPageDescription(0);

    // Get the Bitmap
     (border.Child as Image).Source = getBitmap();

    // Set Padding on the Border
    double left = pageDesc.ImageableRect.Left;
    double top = pageDesc.ImageableRect.Top;
    double right = pageDesc.PageSize.Width - left - pageDesc.ImageableRect.Width;
    double bottom = pageDesc.PageSize.Height - top - pageDesc.ImageableRect.Height;
    border.Padding = new Thickness(left, top, right, bottom);

    this.SetPreviewPageCount(1, PreviewPageCountType.Final);
}

void OnGetPreviewPage(object sender, GetPreviewPageEventArgs args)
{
```

```
        this.SetPreviewPage(args.PageNumber, border);
    }

    void OnAddPages(object sender, AddPagesEventArgs args)
    {
        this.AddPage(border);
        this.AddPagesComplete();
    }
```

Image 元素有 Uniform 的默认 Stretch 模式，会使位图显示最大化，同时保持满屏长宽比。此外，OnPaginate 方法设置 Border 的 Padding 属性，以免页面中不可打印区域，因为你当然想把手绘画打印到打印机允许的区域。

第 18 章　传感器与 GPS

近些年来，计算机已经演化为开发新的传感器功能。这可不是新片中的情节！许多电脑(尤其是平板电脑和其他移动设备)都包括一些硬件能让机器知道它在三维空间的方位、在地球表面的位置、附近光的亮度甚至计算机在用户手中旋转的速度。

这些硬件统称为“传感器”，其软件接口能在 Windows.Devices.Sensors 命名空间中找到，帮助程序确定其地理位置的类则在 Windows.Devices.Geolocation 命名空间中。承担后一项工作的硬件经常称为 GPS(卫星全球定位系统)，但计算机也常常通过网络连接来确定其地理位置。

本章着重讲 SimpleOrientationSensor、Accelerometer、Compass、Inclinometer、OrientationSensor 和 Geolocator 类所提供的信息，但恐怕我会跳过不常用的 LightSensor 和 Gyrometer 类，后者测量计算机的角速度。

为了完全利用本章的内容，你需要让电脑运行示例程序并在空中移动电脑，甚至把电脑举过头顶。如果你的 Windows 8 开发机器像我的一样是固定在桌子上的，则需要一台平板电脑(比如微软 Surface)，并远程部署程序，就像 Tim Heuer 在其博客文章中讨论的一样，其网址为 http://timheuer.com/blog/archive/2012/10/26/remote-debugging-windows-store-apps-on-surface-arm-devices.aspx。

本章的一些示例程序均改编自我 2012 年 6 月到 12 月发表于 *MSDN Magazine* 上的有关 Windows Phone 7.5 传感器的文章。

18.1　方位和定位

正如其名字所表示的，我讨论的最简单传感器就是 SimpleOrientationSensor，让程序大致了解计算机如何在三维空间中定位，但没有细节。如果要实例化 SimpleOrientationSensor 类，可以调用一个静态方法：

```
SimpleOrientationSensor simpleOrientationSensor = SimpleOrientationSensor.GetDefault();
```

在应用中只需要调用一次，因此，这段代码可以显示为一个字段定义，允许在整个类的范围内访问该对象。如果 SimpleOrientationSensor.GetDefault 方法返回 null，则说明电脑没有定位功能。

任何时候都可以从 SimpleOrientationSensor 对象获取表明当前方位的值：

```
SimpleOrientation simpleOrientation = simpleOrientationSensor.GetCurrentOrientation();
```

SimpleOrientation 枚举类型包含以下 6 项：

- NotRotated
- Rotated90DegreesCounterclockwise
- Rotated180DegreesCounterclockwise
- Rotated270DegreesCounterclockwise

- Faceup
- Facedown

对这 6 项的信息限制是 SimpleOrientationSensor 中的 Simple 那部分。

如果方位发生改变，也可以通过事件获得通知。可以对 OrientationChanged 事件设置处理程序，如下所示：

```
simpleOrientationSensor.OrientationChanged+= OnSimpleOrientationChanged;
```

该事件只有在方向发生变化时才会触发，如果电脑保持相对静止，该事件不会发生。如果需要初始值，可以额外调用 GetCurrentOrientation 方法来设置事件处理程序。

该事件处理程序在其自身线程中运行，这样可以和用户界面线程进行交互，而你需要通过户界面线程来使用 CoreDispatcher 对象：

```
async void OnSimpleOrientationChanged(SimpleOrientationSensor sender,
                              SimpleOrientationSensorOrientationChangedEventArgs args)
{
    await this.Dispatcher.RunAsync(CoreDispatcherPriority.Normal, () =>
        {
            ...
    });
}
```

名字超长的事件参数有 SimpleOrientation 枚举类型的 Orientation 属性以及 DateTimeOffset 类型的 TimeStamp 属性。

你可能会问：我不是已经有了定位信息吗？Windows.Graphics.Display 命名空间不是已经提供了吗？我不是用了 DisplayProperties 类及其 NativeOrientation 和 CurrentOrientation 静态属性，还有用于定位信息的 OrientationChanged 事件吗？你应该记得这两个静态属性会返回 DisplayOrientations 枚举类型项：

- Landscape
- Portrait
- LandscapeFlipped
- PortraitFlippe

SimpleOrientationSensor 和 DisplayProperties 类当然相关，但重点是要知道如何相关：SimpleOrientationSensor 类表明计算机在三维空间如何定位。DisplayProperties.CurrentOrientation 属性则表明 Windows 如何自动重新定位程序窗口来补偿计算机。换句话说，SimpleOrientationSensor 报告硬件定位，DisplayProperties.CurrentOrientation 则报告能响应硬件定位的软件定位。

OrientationAndOrientation 项目试图区分这两种定位。XAML 文件只定义了几个 TextBlock 元素标签并显示一些信息。

```
项目：OrientationAndOrientation | 项目：MainPage.xaml(片段)
<Page ... FontSize="24">
    <Grid Background="{StaticResource ApplicationPageBackgroundThemeBrush}">
        <Grid HorizontalAlignment="Center"
              VerticalAlignment="Center">
            <Grid.RowDefinitions>
                <RowDefinition Height="Auto" />
                <RowDefinition Height="Auto" />
            </Grid.RowDefinitions>
```

```
            <Grid.ColumnDefinitions>
                <ColumnDefinition Width="Auto" />
                <ColumnDefinition Width="Auto" />
            </Grid.ColumnDefinitions>

            <TextBlock Text="SimpleOrientationSensor: "
                       Grid.Row="0"
                       Grid.Column="0" />

            <TextBlock Name="orientationSensorTextBlock"
                       Grid.Row="0"
                       Grid.Column="1"
                       TextAlignment="Right" />

            <TextBlock Text="DisplayProperties.CurrentOrientation: "
                        Grid.Row="1"
                        Grid.Column="0" />

            <TextBlock Name="displayOrientationTextBlock"
                        Grid.Row="1"
                        Grid.Column="1"
                        TextAlignment="Right" />
        </Grid>
    </Grid>
</Page>
```

代码隐藏文件定义两个方法，只用于设置 Grid 第二列的两个 TextBlock 元素。既从构造函数中调用两种方法设置初始值，同时也从两个事件的处理程序调用。

```
项目: OrientationAndOrientation | 文件: MainPage.xaml.cs(片段)
public sealed partial class MainPage : Page
{
    SimpleOrientationSensor simpleOrientationSensor = SimpleOrientationSensor.GetDefault();

    public MainPage()
    {
    vthis.InitializeComponent();

        // SimpleOrientationSensor initialization
        if  (simpleOrientationSensor != null)
        {
            SetOrientationSensorText(simpleOrientationSensor.GetCurrentOrientation());
            simpleOrientationSensor.OrientationChanged += OnSimpleOrientationChanged;
        }

        // DisplayProperties initialization
        SetDisplayOrientationText(DisplayProperties.CurrentOrientation);
        DisplayProperties.OrientationChanged += OnDisplayPropertiesOrientationChanged;
    }

    // SimpleOrientationSensor handler
    async void OnSimpleOrientationChanged(SimpleOrientationSensor sender,
                            SimpleOrientationSensorOrientationChangedEventArgs args)
    {
        await this.Dispatcher.RunAsync(CoreDispatcherPriority.Normal, () =>
            {
                SetOrientationSensorText(args.Orientation);
            });
    }

    void SetOrientationSensorText(SimpleOrientation simpleOrientation)
    {
        orientationSensorTextBlock.Text = simpleOrientation.ToString();
    }

    // DisplayProperties handler
    void OnDisplayPropertiesOrientationChanged(object sender)
```

```
        {
            SetDisplayOrientationText(DisplayProperties.CurrentOrientation);
        }

        void SetDisplayOrientationText(DisplayOrientations displayOrientation)
        {
            displayOrientationTextBlock.Text = displayOrientation.ToString();
        }
    }
```

注意，SimpleOrientationSensor 是作为字段进行实例化的，但构造函数在访问对象之前会检查非 null 值。

在一台有原生横向定位的平板电脑上运行该程序(DisplayProperties.NativeOrientation 属性会返回 DisplayOrientations.Landscape)，如果没有采取措施来阻止 Windows 8 变化方位(比如将平板放在扩展坞上)，那么随着顺时针方向逐渐旋转平板电脑，你一般会发现两个方位指示器有下表所示的对应关系。

SimpleOrientationSensor	DisplayProperties.CurrentOrientation
NotRotated	Landscape
Rotated270DegreesCounterClockwise	Portrait
Rotated180DegreesCounterClockwise	LandscapeFlipped
Rotated90DegreesCounterClockwise	PortraitFlipped

SimpleOrientationSensor 还报告 Faceup 和 Facedown 值，两者和 DisplayOrientations 枚举没有对应关系。

上表大致适用于能原生横向定位的平板，而具有原生纵向定位的移动设备则有下表所示的对应关系。

SimpleOrientationSensor	DisplayProperties.CurrentOrientation
NotRotated	Portrait
Rotated270DegreesCounterClockwise	LandscapeFlipped
Rotated180DegreesCounterClockwise	PortraitFlipped
Rotated90DegreesCounterClockwise	Landscape

此外，应用可以要求 Windows 补偿计算机定位，要么通过 Package.appxmanifest 文件，要么在软件中设置 DisplayProperties.AutoRotationPreferences 属性。而在这种情况下，应用运行时，DisplayProperties.CurrentOrientation 不会改变。有些平板也有一个硬件开关，用户可以切换以停止 Windows 自动旋转屏幕。而此时，你会看到下表所示的对应关系。

SimpleOrientationSensor	DisplayProperties.CurrentOrientation
NotRotated	PortraitFlipped
Rotated270DegreesCounterClockwise	PortraitFlipped
Rotated180DegreesCounterClockwise	PortraitFlipped
Rotated90DegreesCounterClockwise	PortraitFlipped

如果想自己补偿方位，也是可以的。可以指令 Windows 不执行任何方位变化，然后用 SimpleOrientationSensor 来决定电脑如何真正定位。然而，请记住 Package.appxmanifest 文

件和 DisplayProperties.AutoRotationPreferences 所做的只是你偏好的，并不是 Windows 实际会做的事情，因此，如果 Windows 调整显示器的方位与偏好相反，可能需要做进一步调整。

防止自动旋转最安全的方法是把 DisplayProperties.AutoRotationPreferences 设定为 DisplayProperties.NativeOrientation，本章稍后将进行讨论。

18.2　加速度、力、重力和矢量

毫无疑问，SimpleOrientationSensor 在内部可以访问称为“加速计”的硬件。加速计是测量加速度的装置，知道电脑的加速度初看好像没有什么用。然而，根据基本物理知识可以知道，特别是牛顿第二运动定律：

$$F=ma$$

力等于质量乘以加速度，而有一种力几乎无法摆脱，即重力。计算机的加速计通常用来测量重力，并回答基本问题“哪条路向下？”

可以通过 Accelerometer 类更直接地访问加速硬件。要实例化 Accelerometer 类，可以使用和 SimpleOrientationSensor 方法名的静态方法：

```
Accelerometer accelerometer = Accelerometer.GetDefault();
```

如果 Accelerometer.GetDefault 方法返回 null，则表示电脑没有加速计或 Windows 8 没有识别出来。如果没有加速计，应用就不能运行，需要通知用户缺少加速计。

任何时候都可以获得加速度的当前值：

```
AccelerometerReading accelerometerReading = accelerometer.GetCurrentReading();
```

SimpleOrientationSensor 中的类似方法称为 GetCurrentOrientation。

最好检查 GetCurrentReading 返回的值是否为 null。AccelerometerReading 定义以下四个属性：

- double 类型的 AccelerationX
- double 类型的 AccelerationY
- double 类型的 AccelerationZ
- DateTimeOffset 类型的 Timestamp

三个 double 值共同构成三维矢量，指向相对于设备的地球。

还可以为 Accelerometer 对象附加事件处理程序：

```
accelerometer.ReadingChanged += OnAccelerometerReadingChanged;
```

SimpleOrientationSensor 中类似的事件称为 OrientationChanged。和 OrientationChanged 一样，ReadingChanged 处理程序在单独线程中运行，因此可以进行如下处理：

```
async void OnAccelerometerReadingChanged(Accelerometer sender,
                                    AccelerometerReadingChangedEventArgs args)
{
    await this.Dispatcher.RunAsync(CoreDispatcherPriority.Normal, () =>
        {
            ...
        });
}
```

AccelerometerReadingChangedEventArgs 定义了 AccelerometerReading 类型的属性，名为 Reading，与从 GetCurrentReading 返回的对象相同。

你预计 ReadingChanged 处理程序多久调用一次？如果电脑静止，则可能根本不调用！因此，如果需要初始 Accelerometer 读数，则应该在一开始就调用 GetCurrentReading。

如果计算机正在移动并在空间中改变方位，参数变化(在一定标准范围内)但不超过从 Accelerometer 的 ReportInterval 属性获得的以毫秒为单位的时间间隔时，则调用 ReadingChanged 处理程序。我看到默认值为 112，也就是说，以不快于于每秒 9 次的频率调用 ReadingChanged 方法。

可以把 ReportInterval 设置为其他值，但不能低于 MinimumReportInterval 属性的返回值，我发现返回值为 16 毫秒，即每秒约 60 次。把 ReportInterval 设置为 MinimumReportInterval 获取数据最大量；把 ReportInterval 设置为零，返回默认设置。

Windows.Devices.Sensors 中的所有其他传感器类都和 Accelerometer 的软件界面相同。这些类都有以下几项：

- 静态 GetDefault 方法
- GetCurrentReading 实例方法
- ReportInterval 属性
- MinimumReportInterval 属性
- ReadingChanged 事件

只有 SimpleOrientationSensor 与其他的不同。

如果电脑为静止，AccelerometerReading 类的 AccelerationX、AccelerationY 和 AccelerationZ 属性会定义一个指向地球中心的矢量。矢量通常用黑体字坐标标记，例如(**x**，**y**，**z**)，以区别于三维空间中的点(x，y，z)。点是在空间中的位置；而矢量是方向和量值。矢量和点当然相关：矢量(X，Y，Z)的方向是从点(0，0，0)到点(x，y，z)，矢量的量值是该线的长度。但矢量并不是线本身，也没有位置。

矢量量值，可以用勾股定理的三维形式进行计算：

$$Magnitude = \sqrt{x^2 + y^2 + z^2}$$

任何三维矢量都必须相对于一个特定三维坐标系统，从 AccelerometerReading 对象获得的矢量也不例外。对于原生横向定位的平板，加在设备硬件上的坐标系统如下图所示。

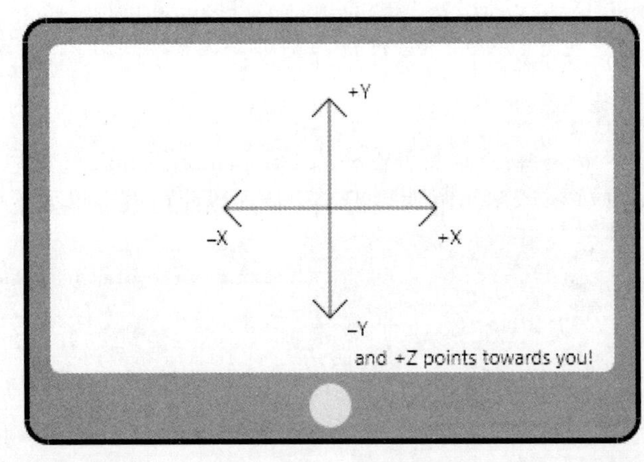

注意，Y 增加方向为向上，这和常规二维图形相反。正 Z 轴指向屏幕外。这一惯例通常称为“右手”坐标系统。如果右手食指在正 X 轴方向，并且中指在正 Y 轴方向，则拇指指向正 Z 轴。

或者，如果把右手手指蜷缩起来，把正 X 轴旋转到正 Y 轴，则拇指指向正 Z 轴。这种方法适用于顺序 X、Y、Z 轴的任何两个轴：蜷缩右手手指把正 Y 轴旋转正 Z 轴，则拇指指向正 X 方向。或蜷缩右手手指把正 Z 轴旋转为正 X 轴，则拇指指向正 Y 轴。

右手规则也可以用于确定绕轴旋转的方向。对于绕 X 轴旋转(例如)，让右手拇指指向正 X，而手指蜷缩在围绕该轴正向旋转角度的方向。

对于原生纵向定位的设备，坐标系统与用户角度相同(见下图)。

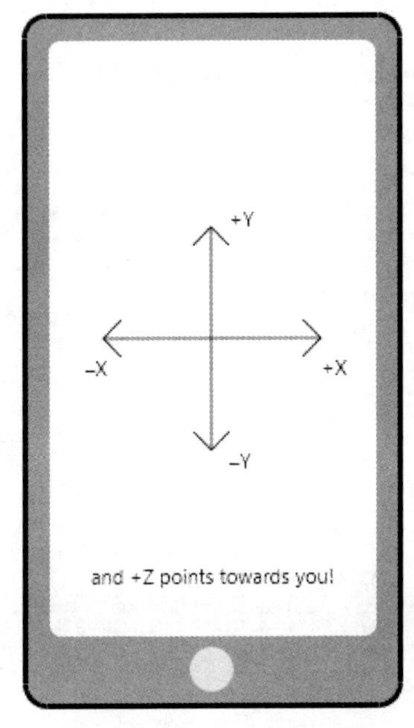

虽然我一直没能证实这一点，但传统笔记本电脑的坐标系统是基于键盘进行记录，而不是基于屏幕。X 轴沿键盘宽度，Y 轴沿键盘高度，而 Z 轴位于键盘之外。

坐标系统固化在设备硬件上，Accelerometer 矢量点指向相对于该坐标系统的地球中心。例如，平板电脑在原生方向上保持竖直，加速度矢量点在-Y 方向。该矢量大小约为 1，因此，该矢量位于区域(0, -1, 0)的某处。设备放在平整表面且屏幕朝上，则矢量位于区域(0, 0, -1)的某处。

量值 1 表示矢量以 g 为单位，即由地球表面重力作用所造成的的加速度或约 32 英尺每平方秒。如果带平板去月球，量值就约为 0.17。如果让平板自由落体(如果你敢的话)，加速度矢量的量值将下降到零，直到击中地面。

以下是 AccelerometerAndSimpleOrientation 程序，用来显示来自 Accelerometer 和 SimpleOrientationSensor 的值。XAML 文件包含一些 TextBlock 元素标签以及来自代码隐藏文件的等待值。

项目：AccelerometerAndSimpleOrientation | 文件：MainPage.xaml(片段)

```
<Page ... >
    <Page.Resources>
        <Style TargetType="TextBlock">
```

```
            <Setter Property="FontSize" Value="24" />
            <Setter Property="Margin" Value="24 12 24 12" />
        </Style>
    </Page.Resources>

    <Grid Background="{StaticResource ApplicationPageBackgroundThemeBrush}">

        <Grid HorizontalAlignment="Center"
              VerticalAlignment="Center">
            <Grid.RowDefinitions>
                <RowDefinition Height="Auto" />
                <RowDefinition Height="Auto" />
                <RowDefinition Height="Auto" />
                <RowDefinition Height="Auto" />
                <RowDefinition Height="Auto" />
            </Grid.RowDefinitions>

            <Grid.ColumnDefinitions>
                <ColumnDefinition Width="Auto" />
                <ColumnDefinition Width="Auto" />
            </Grid.ColumnDefinitions>

            <TextBlock Grid.Row="0" Grid.Column="0" Text="Accelerometer X:" />
            <TextBlock Grid.Row="1" Grid.Column="0" Text="Accelerometer Y:" />
            <TextBlock Grid.Row="2" Grid.Column="0" Text="Accelerometer Z:" />
            <TextBlock Grid.Row="3" Grid.Column="0" Text="Magnitude:"
                        Margin="24 24" />
            <TextBlock Grid.Row="4" Grid.Column="0" Text="Simple Orientation:" />

            <TextBlock Grid.Row="0" Grid.Column="1" Name="accelerometerX"
                       TextAlignment="Right" />
            <TextBlock Grid.Row="1" Grid.Column="1" Name="accelerometerY"
                       TextAlignment="Right"/>
            <TextBlock Grid.Row="2" Grid.Column="1" Name="accelerometerZ"
                       TextAlignment="Right"/>
            <TextBlock Grid.Row="3" Grid.Column="1" Name="magnitude"
                       TextAlignment="Right"
                       VerticalAlignment="Center" />
            <TextBlock Grid.Row="4" Grid.Column="1" Name="simpleOrientation"
                       TextAlignment="Right" />
        </Grid>
    </Grid>
</Page>
```

代码隐藏文件比之前的程序有更多功能。如果 Accelerometer 或 SimpleOrientationSensor 不能实例化，程序会报告给用户。此外，还有一件好事情，如果程序不使用就不运行 Accelerometer，因为会消耗电量。为了表示程序规范，程序在 OnNavigatedTo 覆写中附加了处理程序并在 OnNavigatedFrom 中分离处理程序。除此之外，程序结构和之前的程序非常类似。

```
项目: AccelerometerAndSimpleOrientation | 文件: MainPage.xaml.cs(片段)
public sealed partial class MainPage : Page
{
    Accelerometer accelerometer = Accelerometer.GetDefault();
    SimpleOrientationSensor simpleOrientationSensor = SimpleOrientationSensor.GetDefault();

    public MainPage()
    {
        this.InitializeComponent();
        this.Loaded += OnMainPageLoaded;
    }

    async void OnMainPageLoaded(object sender, RoutedEventArgs args)
    {
        if (accelerometer == null)
```

```
            await new MessageDialog("Cannot start Accelerometer").ShowAsync();

        if (simpleOrientationSensor == null)
            await new MessageDialog("Cannot start SimpleOrientationSensor").ShowAsync();
    }

    // Attach event handlers
    protected override void OnNavigatedTo(NavigationEventArgs args)
    {
        if (accelerometer != null)
        {
            SetAccelerometerText(accelerometer.GetCurrentReading());
            accelerometer.ReadingChanged += OnAccelerometerReadingChanged;
        }

        if (simpleOrientationSensor != null)
        {
            SetSimpleOrientationText(simpleOrientationSensor.GetCurrentOrientation());
            simpleOrientationSensor.OrientationChanged += OnSimpleOrientationChanged;
        }
        base.OnNavigatedTo(args);
    }

    // Detach event handlers
    protected override void OnNavigatedFrom(NavigationEventArgs args)
    {
        if (accelerometer != null)
            accelerometer.ReadingChanged -= OnAccelerometerReadingChanged;

        if (simpleOrientationSensor != null)
            simpleOrientationSensor.OrientationChanged -= OnSimpleOrientationChanged;

        base.OnNavigatedFrom(args);
    }

    // Accelerometer handler
    async void OnAccelerometerReadingChanged(Accelerometer sender,
                              AccelerometerReadingChangedEventArgs args)
    {
        await this.Dispatcher.RunAsync(CoreDispatcherPriority.Normal, () =>
            {
                SetAccelerometerText(args.Reading);
            });
    }

    void SetAccelerometerText(AccelerometerReading accelerometerReading)
    {
        if (accelerometerReading == null)
            return;

        accelerometerX.Text = accelerometerReading.AccelerationX.ToString("F2");
        accelerometerY.Text = accelerometerReading.AccelerationY.ToString("F2");
        accelerometerZ.Text = accelerometerReading.AccelerationZ.ToString("F2");
        magnitude.Text =
            Math.Sqrt(Math.Pow(accelerometerReading.AccelerationX, 2) +
            Math.Pow(accelerometerReading.AccelerationY, 2) +
            Math.Pow(accelerometerReading.AccelerationZ, 2)).ToString("F2");
    }

    // SimpleOrientationSensor handler
    async void OnSimpleOrientationChanged(SimpleOrientationSensor sender,
                              SimpleOrientationSensorOrientationChangedEventArgs args)
    {
        await this.Dispatcher.RunAsync(CoreDispatcherPriority.Normal, () =>
            {
                SetSimpleOrientationText(args.Orientation);
            });
    }
```

```
        void SetSimpleOrientationText(SimpleOrientation simpleOrientation)
        {
            this.simpleOrientation.Text = simpleOrientation.ToString();
        }
    }
```

下图所示为是这个程序在我本书所用平板电脑上运行后所得到的效果，平板电脑放在扩展坞上。

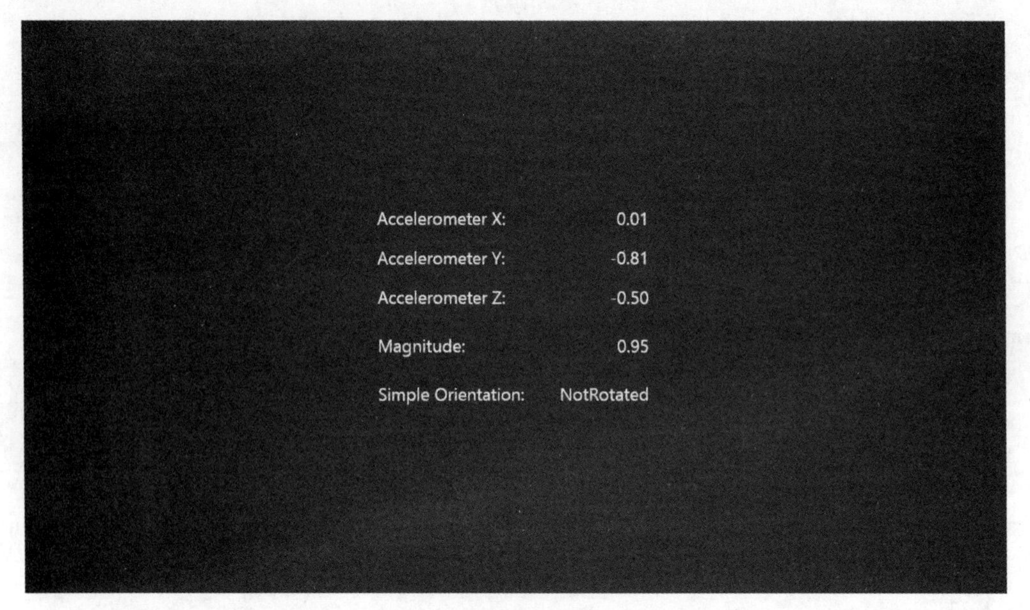

不要看到量值不精确等于 1 就感到心慌。这并不意味着你已经不知不觉离开地球表面，只是加速硬件并不总是如我们想得那么精确罢了。

Y 和 Z 分量均为负，表明平板向后倾斜。前面已经提到过，如果平板直线上升，矢量理论上为(0，-1，0)，如果放在桌子上并且屏幕竖直向上，则矢量理论上为**(0, 0, -1)**。而在两个位置之间，平板绕着 X 轴旋转。将 Y 和 Z 值传递给 Math.Atan2 方法，则可以得到旋转角度。

如果在手持设备上运行该程序，可以在不同方向上转动设备来看效果。一般会看出 SimpleOrientationSensor 和 Accelerometer 的如下对应关系。

SimpleOrientationSensor	Accelerometer 矢量
NotRotated	~ (0, –1, 0)
Rotated90DegreesCounterClockwise	~ (–1, 0, 0)
Rotated180DegreesCounterClockwise	~ (0, 1, 0)
Rotated270DegreesCounterClockwise	~ (1, 0, 0)
Faceup	~ (0, 0, –1)
Facedown	~ (0, 0, 1)

约等于符号(~)可以随便解释。Accelerometer 矢量显然表明在达到某个值之前有很多变化，而该值会促使改变 SimpleOrientationSensor。

AccelerometerAndSimpleOrientation 程序没有表明任何偏好方向，因此，在空中四处移动平板电脑时，Windows 会假设你不想看到数据倒过来，因此会自动改变显示方向。应该看到 SimpleOrientationSensor 值和屏幕方向之间的对应关系，但这只是因为 Windows 会基于这些值改变显示方向！如果禁止 Windows(以任何方式)更改屏幕方向，并不影响通过该程序显示的信息。

事实上，显示方向不断发生变化是非常恼人的。每一次显示方向变化时，屏幕更新都会暂停一会儿，内容也会伸缩以响应这种变化。你可能认为通过 Accelerometer 改变屏幕内容的程序应该也可以禁止自动显示方向改变。

出于该原因，本章剩余所有程序都在程序构造函数中包含一个简单语句，把偏好定位设置为原生定位：

```
DisplayProperties.AutoRotationPreferences = DisplayProperties.NativeOrientation;
```

你会发现，如果在手持设备上运行 AccelerometerAndSimpleOrientation 程序并快速移动，加速度矢量的方向和量值也会随之变化，不再显示来自地球中心的重力。例如，如果向左猛移设备，加速度矢量会指向右边(但只有当装置正在加速时)。如果以稳定速度移动，加速度矢量会停下来并重新指向地球。如果在移动时突然停止设备，加速度矢量也会显示速度变化。

Accelerometer 类还定义了一个名为 Shaken 的事件，没有其他信息。Shaken 事件对需要扔一对骰子、推荐另一家餐厅、擦除绘图或撤消意外擦除等操作的程序非常有用。

Accelerometer 的一个常见应用是气泡水平仪。XAML 文件实例化 4 个 Ellipse 元素。其中 3 个为同心轮廓，而第 4 个是气泡本身。

```
项目: BubbleLevel | 文件: MainPage.xaml(片段)
<Grid Background="{StaticResource ApplicationPageBackgroundThemeBrush}">
    <Grid Name="centeredGrid"
          HorizontalAlignment="Center"
          VerticalAlignment="Center">
        <Ellipse Name="outerCircle"
                 Stroke="{StaticResource ApplicationForegroundThemeBrush}" />

        <Ellipse Name="halfCircle"
                 Stroke="{StaticResource ApplicationForegroundThemeBrush}" />

        <Ellipse Width="24"
                 Height="24"
                 Stroke="{StaticResource ApplicationForegroundThemeBrush}" />

        <Ellipse Fill="Red"
                 Width="24"
                 Height="24"
                 HorizontalAlignment="Center"
                 VerticalAlignment="Center">
            <Ellipse.RenderTransform>
                <TranslateTransform x:Name="bubbleTranslate" />
            </Ellipse.RenderTransform>
        </Ellipse>
    </Grid>
</Grid>
```

代码隐藏文件把 DisplayProperties.AutoRotationPreferences 设置为 DisplayProperties.NativeOrientation。Windows 不会自动改变程序的显示方向。程序还使用 SizeChanged 处理程序来设置 outerCircle 和 halfCircle 的尺寸。

项目: BubbleLevel | 文件: MainPage.xaml.cs(片段)
```
public sealed partial class MainPage : Page
{
    Accelerometer accelerometer = Accelerometer.GetDefault();

    public MainPage()
    {
        this.InitializeComponent();
        DisplayProperties.AutoRotationPreferences = DisplayProperties.NativeOrientation;
        Loaded += OnMainPageLoaded;
        SizeChanged += OnMainPageSizeChanged;
    }

    async void OnMainPageLoaded(object sender, RoutedEventArgs args)
    {
        if (accelerometer != null)
        {
            accelerometer.ReportInterval = accelerometer.MinimumReportInterval;
            SetBubble(accelerometer.GetCurrentReading());
            Accelerometer .ReadingChanged += OnAccelerometerReadingChanged;
        }
        else
        {
            await new MessageDialog("Accelerometer is not available").ShowAsync();
        }
    }

    void OnMainPageSizeChanged(object sender, SizeChangedEventArgs args)
    {
        double size = Math.Min(args.NewSize.Width, args.NewSize.Height);
        outerCircle.Width = size;
        outerCircle.Height = size;
        halfCircle.Width = size / 2;
        halfCircle.Height = size / 2;
    }

    async void OnAccelerometerReadingChanged(Accelerometer sender,
                                    AccelerometerReadingChangedEventArgs args)
    {
        await this.Dispatcher.RunAsync(CoreDispatcherPriority.Normal, () =>
        {
            SetBubble(args.Reading);
        });
    }

    void SetBubble(AccelerometerReading accelerometerReading)
    {
        if (accelerometerReading == null)
            return;

        double x = accelerometerReading.AccelerationX;
        double y = accelerometerReading.AccelerationY;

        bubbleTranslate.X = -x * centeredGrid.ActualWidth / 2;
        bubbleTranslate.Y = y * centeredGrid.ActualHeight / 2;
    }
}
```

SetBubble 方法看起来很简单：只需要加速度矢量的 X 和 Y 分量，并使用两者来设置中心气泡的 X 和 Y 坐标，缩放到外圆周的半径。但考虑一下平板电脑面朝上或面朝下放在桌子上的情况。加速度矢量的 Z 分量为 1 或-1，而 X 和 Y 分量均为 0，也就是说气泡位于屏幕中心。这是正确的。

现在拿好平板电脑，屏幕垂直于地球。Z 分量变为 0。也就是说加速度矢量的量值完全来源于 X 和 Y 分量。换而言之：

$$X^2+y^2=1^2$$

这就是在两个维度上圆方程，因此，气泡位于外围圆的某个地方。气泡的确切位置基于平板电脑绕着 Z 轴的当前旋转。

加速度矢量向下指向地球中心，气泡会动起来，也就是说需要将加速度矢量的 x 和 Y 分量的符号转换为二维屏幕坐标。但记住，加速度矢量的 Y 轴是从屏幕坐标反转，因此只有 X 分量需要符号交换，程序的最后两行代码显示了这种情况。

下图所示为是程序在 Microsoft Surface 平板电脑上运行的效果。

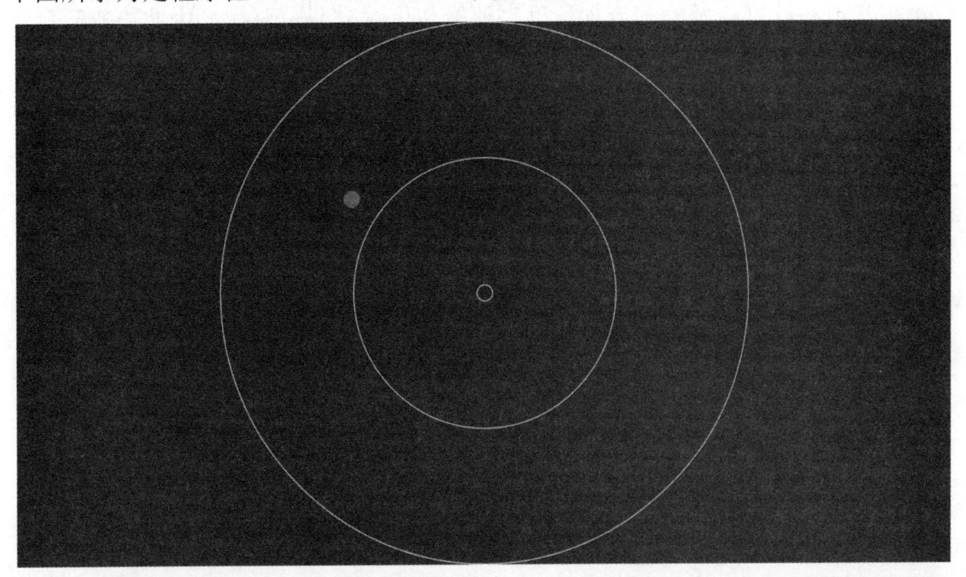

当然，屏幕截图并没有完全捕捉气泡在现实生活中如何抖动。Accelerometer 值还相当原始，不过在现实应用中，你会想让气泡顺滑一点，这是接下来两个程序要完成的。

18.3 跟随滚球

加速计常见于手持设备游戏中。例如，如果有一个模拟驾驶汽车的游戏，用户可以通过向左或向右倾斜电脑来实现汽车转向。

以下两个程序模拟球在屏幕上的滚动。如果平板电脑与地平行并在电脑上平衡放置一个真实的球，可以通过倾斜屏幕使球左右滚动。越倾斜，球的加速度越大。接下来的两个程序以类似方式移动虚拟球。和气泡水平仪程序一样，两个程序忽略 Accelerometer 矢量的 Z 分量，只用 X 和 Y 分量来管理屏幕二维表面的加速度。

在 TiltAndRoll 程序中，如果球撞击到边缘，其垂直于边上的速度将全部丢失，如果在该方向仍有速度，则继续沿着边缘滚动。XAML 文件定义该球。通过设置 Center 属性，EllipseGeometry 允许球定位在特定坐标。

```
项目: TiltAndRoll | 文件: MainPage.xaml(片段)
<Grid Background="{StaticResource ApplicationPageBackgroundThemeBrush}">
    <Path Fill="Red">
        <Path.Data>
```

```
            <EllipseGeometry x:Name="ball" />
        </Path.Data>
    </Path>
</Grid>
```

代码隐藏文件开始为 GRAVITY 定义常量，以像素/平方秒为单位。理论上讲，无摩擦力的滑动球完全受重力控制，但滚动球的加速度是重力加速度的 2/3。(细节可参考 A. P. French, Newtonian Mechanics, W. W. Norton, 1971, 652-3)。也就是说，每平方秒 32 英尺乘以每英尺 12 英寸，乘以每英寸 96 像素，再乘以 2/3，结果约为 25000，可以计算出 GRAVITY，但我大大降低了速度。

二维矢量值在计算涉及到二维加速度、速度和位置时非常有用，因此第 13 章包含了 Vector2 结构。

球需要保持滚动而不考虑 Accelerometer 的 ReadingChanged 事件触发，因此程序没有为该事件安装处理程序，而是通过 CompositionTarget.Rendering 获取当前值，应用于球。注意，Accelerometer 读数的 X 和 Y 分量用来创建 Vector2 值，然后将其与上一次的值进行平均，而上一次的值本身就是其上一个值的平均值，并依此类推。这是极简的平滑形式。

项目: TiltAndRoll | 文件: MainPage.xaml.cs(片段)

```
public sealed partial class MainPage : Page
{
    const double GRAVITY = 5000; // pixels per second squared
    const double BALL_RADIUS = 32;

    Accelerometer accelerometer = Accelerometer.GetDefault();
    TimeSpan timeSpan;
    Vector2 acceleration;
    Vector2 ballPosition;
    Vector2 ballVelocity;
    public MainPage()
    {
        this.InitializeComponent();
        DisplayProperties.AutoRotationPreferences = DisplayProperties.NativeOrientation;

        ball.RadiusX = BALL_RADIUS;
        ball.RadiusY = BALL_RADIUS;

        Loaded += OnMainPageLoaded;
    }

        async void OnMainPageLoaded(object sender, RoutedEventArgs args)
    {
        if (accelerometer == null)
        {
            await new MessageDialog("Accelerometer is not available").ShowAsync();
        }
        else
        {
            CompositionTarget.Rendering += OnCompositionTargetRendering;
        }
    }

    void OnCompositionTargetRendering(object sender, object args)
    {
        AccelerometerReading reading = accelerometer.GetCurrentReading();

        if (reading == null)
            return;

        // Get elapsed time since last event
        TimeSpan timeSpan = (args as RenderingEventArgs).RenderingTime;
```

```
        double elapsedSeconds = (timeSpan - this.timeSpan).TotalSeconds;
        this.timeSpan = timeSpan;

        // Convert accelerometer reading to display coordinates
        double x = reading.AccelerationX;
        double y = -reading.AccelerationY;

        // Get current X-Y acceleration and smooth it
        acceleration = 0.5 * (acceleration + new Vector2(x, y));

        // Calculate new velocity and position
        ballVelocity += GRAVITY * acceleration * elapsedSeconds;
        ballPosition += ballVelocity * elapsedSeconds;

        // Check for hitting edge
        if (ballPosition.X - BALL_RADIUS < 0)
        {
            ballPosition = new Vector2(BALL_RADIUS, ballPosition.Y);
            ballVelocity = new Vector2(0, ballVelocity.Y);
        }
        if (ballPosition.X + BALL_RADIUS > this.ActualWidth)
        {
            ballPosition = new Vector2(this.ActualWidth - BALL_RADIUS, ballPosition.Y);
            ballVelocity = new Vector2(0, ballVelocity.Y);
        }
        if (ballPosition.Y - BALL_RADIUS < 0)
        {
            ballPosition = new Vector2(ballPosition.X, BALL_RADIUS);
            ballVelocity = new Vector2(ballVelocity.X, 0);
        }
        if (ballPosition.Y + BALL_RADIUS > this.ActualHeight)
        {
            ballPosition = new Vector2(ballPosition.X, this.ActualHeight - BALL_RADIUS);
            ballVelocity = new Vector2(ballVelocity.X, 0);
        }
        ball.Center = new Point(ballPosition.X, ballPosition.Y);
    }
}
```

两项重要计算为：

```
ballVelocity += GRAVITY * acceleration * elapsedSeconds;
ballPosition += ballVelocity * elapsedSeconds;
```

请记住，acceleration、ballVelocity 和 ballPosition 都是 Vector2 值，因此都具有 X 和 Y 分量。速度增加了 acceleration 乘以运行时间，位置上升了(速度乘以运行时间)。然后只需要做检查新位置是否在页面边界之外。如果是，移回到页面内且速度的一个分量设置为零。

物理效果很真实。如果增加和减小倾斜，球的加速度大小也相应增加和减少。此外，程序对速度和位置处理实际公式，因此容易增加一些反弹。有一个简单方法是当球击中边缘时，不把速度分量设置为零，而是设置为负值或之前的值。然而，这意味着球反弹后具有相同速度的大小，而这不现实。设置一个称为 BOUNCE 的衰减因子。TiltAndBounce 程序和 TiltAndRoll 相同，只不过 BOUNCE 常量和 CompositionTarget.Rendering 处理程序中的球移动逻辑不同。

```
项目: TiltAndBounce | 文件: MainPage.xaml.cs(片段)
public sealed partial class MainPage : Page
{
    ...
    const double BOUNCE = -2.0 / 3; // fraction of velocity
    ...
    void OnCompositionTargetRendering(object sender, object args)
    {
```

```
            AccelerometerReading reading = accelerometer.GetCurrentReading();

            if (reading == null)
                return;

            // Get elapsed time
            TimeSpan timeSpan = (args as RenderingEventArgs).RenderingTime;
            double elapsedSeconds = (timeSpan - this.timeSpan).TotalSeconds;
            this.timeSpan = timeSpan;

            // Convert accelerometer reading to display coordinates
            double x = reading.AccelerationX;
            double y = -reading.AccelerationY;

            // Get current X-Y acceleration and smooth it
            acceleration = 0.5 * (acceleration + new Vector2(x, y));

            // Calculate new velocity and position
            ballVelocity += GRAVITY * acceleration * elapsedSeconds;
            ballPosition += ballVelocity * elapsedSeconds;

            // Check for bouncing off edge
            bool needAnotherLoop = true;

            while (needAnotherLoop)
            {
                needAnotherLoop = false;

                if (ballPosition.X - BALL_RADIUS < 0)
                {
                    ballPosition = new Vector2(-ballPosition.X + 2 * BALL_RADIUS, ballPosition.Y);
                    ballVelocity = new Vector2(BOUNCE * ballVelocity.X, ballVelocity.Y);
                    needAnotherLoop = true;
                }
                else if (ballPosition.X + BALL_RADIUS > this.ActualWidth)
                {
                    ballPosition = new Vector2(-ballPosition.X + 2 *
                                               (this.ActualWidth - BALL_RADIUS),
                                               ballPosition.Y);
                    ballVelocity = new Vector2(BOUNCE * ballVelocity.X, ballVelocity.Y);
                    needAnotherLoop = true;
                }
                else if (ballPosition.Y - BALL_RADIUS < 0)
                {
                    ballPosition = new Vector2(ballPosition.X, -ballPosition.Y + 2 * BALL_RADIUS);
                    ballVelocity = new Vector2(ballVelocity.X, BOUNCE * ballVelocity.Y);
                    needAnotherLoop = true;
                }
                else if (ballPosition.Y + BALL_RADIUS > this.ActualHeight)
                {
                    ballPosition = new Vector2(ballPosition.X,
                                               -ballPosition.Y + 2 *
                                               (this.ActualHeight - BALL_RADIUS));
                    ballVelocity = new Vector2(ballVelocity.X, BOUNCE * ballVelocity.Y);
                    needAnotherLoop = true;
                }
            }
            ball.Center = new Point(ballPosition.X, ballPosition.Y);
        }
    }
```

在 TiltAndRoll 程序中，球可能在同一事件中超越两个相邻边缘，但这种情况需要用一系列 if 语句进行处理。在该程序中，球从一个边缘反弹可能会跨过另一边，也就是说，循环需要反复测试球的位置，直到没有反弹。

18.4　两 个 北 极

虽然加速计能告诉你哪条路向下跌，但并没有揭示设备在三维空间的完整方向。要搞明白我的意思，需要在手持设备上运行 AccelerometerAndSimpleOrientation 程序。站起来用一些奇怪的方式拿着设备。加速计会告诉你哪个方向向下。现在，整个身体绕一圈。平板电脑在空间旋转 360 度，但加速计报告的值几乎相同，因为方向向下对于装置而言仍然不变。

如果把平板电脑以加速度矢量转一圈，有什么变化？答案之一是：相对于平板电脑的北方。这就是为什么罗盘传感器如此重要的原因：罗盘提供决定平板电脑方向的缺失因素。由罗盘和加速计相结合，就可以推导出平板电脑在三维空间的完整方向。或者也可以让 Windows 来完成。

Compass 类的结构非常类似于 Accelerometer，并且 CompassReading 类有两个属性，分别为：HeadingMagneticNorth 和 HeadingTrueNorth。这些都是以度为单位的角度，并且显示计算机相对于北方的角度。如果平板电脑平行于地球放置，则角度应该接近于零，指向屏幕的顶部为北方。(我指的“顶”是指本章前面所示图中正 Y 轴方向。)如果让平板电脑屏幕朝向东，则角度增大。

当然，这些角度在世界各地的某些位置并不相同。平板电脑中包含磁力计，会响应磁北极(与地球磁场对齐)，所以正北指地球围绕旋转的轴线。注意，HeadingMagneticNorth 属性是 double 类型，但 HeadingTrueNorth 是可空 double 类型，暗示该值不可用。

我们来试试。SimpleCompass 项目的 XAML 文件定义了两个图形箭头，其源点在屏幕中心，并且竖直向上。

```
项目: SimpleCompass | 文件: MainPage.xaml(片段)
<Grid Background="{StaticResource ApplicationPageBackgroundThemeBrush}">
    <Canvas HorizontalAlignment="Center"
            VerticalAlignment="Center">
        <Path Fill="Magenta"
              Data="M -10 0 L 10 0, 10 -300, 0 -350, -10 -300 Z">
            <Path.RenderTransform>
                <RotateTransform x:Name="magNorthRotate" />
            </Path.RenderTransform>
        </Path>

        <Path Name="trueNorthPath"
              Fill="Blue"
              Data="M -10 0 L 10 0, 10 -300, 0 -350, -10 -300 Z">
            <Path.RenderTransform>
                <RotateTransform x:Name="trueNorthRotate" />
            </Path.RenderTransform>
        </Path>
    </Canvas>
</Grid>
```

有两个帮助记忆的颜色，Magenta(洋红)代表地磁北极，Blue(蓝)代表地理北极。

如果 HeadingTrueNorth 的值为 null，代码隐藏文件会隐藏第二个 Path。

```
项目: SimpleCompass | 文件: MainPage.xaml.cs(片段)
public sealed partial class MainPage : Page
{
```

```
    Compass compass = Compass.GetDefault();

    public MainPage()
    {
        this.InitializeComponent();
        DisplayProperties.AutoRotationPreferences = DisplayProperties.NativeOrientation;
        Loaded += OnMainPageLoaded;
    }

    async void OnMainPageLoaded(object sender, RoutedEventArgs args)
    {
        if (compass != null)
        {
            ShowCompassValues(compass.GetCurrentReading());
            compass.ReportInterval = compass.MinimumReportInterval;
            compass.ReadingChanged += OnCompassReadingChanged;
        }
        else
        {
            await new MessageDialog("Compass is not available").ShowAsync();
        }
    }

    async void OnCompassReadingChanged(Compass sender, CompassReadingChangedEventArgs args)
    {
        await this.Dispatcher.RunAsync(CoreDispatcherPriority.Normal, () =>
        {
            ShowCompassValues(args.Reading);
        });
    }

    void ShowCompassValues(CompassReading compassReading)
    {
        if (compassReading == null)
            return;

        magNorthRotate.Angle = -compassReading.HeadingMagneticNorth;

        if (compassReading.HeadingTrueNorth.HasValue)
        {
            trueNorthPath.Visibility = Visibility.Visible;
            trueNorthRotate.Angle = -compassReading.HeadingTrueNorth.Value;
        }
        else
        {
            trueNorthPath.Visibility = Visibility.Collapsed;
        }
    }
}
```

请注意，两个箭头的旋转角度设定为 HeadingMagneticNorth 和 HeadingTrueNorth 属性的负数。这些值表示计算机相对于北极的转动，因此箭头需要相对转动，并显示相对于计算机的北极方向。

在我一直为本书使用的两台平板电脑上(包括微软 Surface)，结果令人失望。两台机器上 HeadingTrueNorth 值始终为 null。在微软 Surface 上，地磁北极价值相当不稳定。在三星平板电脑上，只有范围从 0 到 180 度的值！在我的 Technical Editor 的平板电脑上，HeadingMagneticNorth 始终为 0。

从理论上讲，地磁北极如果知道计算机位置，则可以计算地理北极，但 Package.appxmanifest 的定位能力没有起到帮助作用。

不过，我们可以祈求好运，希望罗盘硬件能够结合加速计数据好好工作并提供完整的方位信息。

18.5　陀螺仪 = 加速计 + 罗盘

有两个类从内部结合加速计和罗盘数据并平滑结果，Inclinometer(陀螺仪)传感器是两者之一。Inclinometer 类提供 yaw(偏航角)、pitch(俯仰角)和 roll(滚转角)信息，这些都是飞行动力学所使用的术语。

偏航角、俯仰角和滚转角经常称为欧拉，得名于 18 世纪探索三维旋转数学的数学家莱昂哈德 • 欧拉。如果你正在驾驶飞机，偏航角表示飞机机头所面对的罗盘方向。如果飞机向左或向右改变方向，偏航角则随之改变。俯仰角指飞机机头的角度。由于机头往上爬升和下降俯冲，俯仰角也随之改变。滚转角由向左或向右倾斜实现。

要理解如何应用于计算机，你可能想让平板电脑像魔毯一样自己能“飞”。假设你坐在屏幕顶端的位置起飞(当然是以平板电脑的原生方位)。在之前章节所示的坐标系统中，偏航角绕 Z 轴旋转，俯仰角绕 X 轴转动，而滚转角绕 Y 轴旋转。

YawPitchRoll 程序还可以帮助你以视觉方式表达这些角度。XAML 文件包含一些用于线条的 Rectangle 元素、一些用于滚动球的 Ellipse 元素以及一些文字。

```
项目: YawPitchRoll | 文件: MainPage.xaml(片段)
<Grid Background="{StaticResource ApplicationPageBackgroundThemeBrush}">
    <!-- Pitch -->
    <Rectangle Fill="Blue"
               Width="3"
               HorizontalAlignment="Center"
               VerticalAlignment="Stretch" />

    <Path Name="pitchPath"
          Stroke="Blue">
        <Path.Data>
            <EllipseGeometry x:Name="pitchEllipse" RadiusX="20" RadiusY="20" />
        </Path.Data>
    </Path>

    <!-- Roll -->
    <Rectangle Fill="Red"
               Height="3"
               HorizontalAlignment="Stretch"
               VerticalAlignment="Center" />

    <Path Name="rollPath"
          Stroke="Red"
          Fill="Red">
        <Path.Data>
            <EllipseGeometry x:Name="rollEllipse" RadiusX="20" RadiusY="20" />
        </Path.Data>
    </Path>

    <Grid>
        <Grid.RowDefinitions>
            <RowDefinition Height="*" />
            <RowDefinition Height="*" />
        </Grid.RowDefinitions>
        <Grid.ColumnDefinitions>
            <ColumnDefinition Width="*" />
            <ColumnDefinition Width="*" />
        </Grid.ColumnDefinitions>
```

```
            <!-- Pitch -->
            <TextBlock Text="PITCH"
                       Grid.Row="0"
                       Grid.Column="0"
                       Foreground="Blue"
                       HorizontalAlignment="Right"
                       Margin="0 0 24 0" />

            <TextBlock Name="pitchValue"
                       Grid.Row="0"
                       Grid.Column="1"
                       Foreground="Blue"
                       HorizontalAlignment="Left"
                       Margin="24 0 0 0" />

            <!-- Roll -->
            <TextBlock Text="ROLL"
                       Grid.Row="1"
                       Grid.Column="0"
                       Foreground="Red"
                       HorizontalAlignment="Left"
                       VerticalAlignment="Top"
                       Margin="0 108 0 0">
                <TextBlock.RenderTransform>
                    <RotateTransform Angle="-90" />
                </TextBlock.RenderTransform>
            </TextBlock>

            <TextBlock Name="rollValue"
                       Grid.Row="0"
                       Grid.Column="0"
                       Foreground="Red"
                       HorizontalAlignment="Left"
                       VerticalAlignment="Bottom">
                <TextBlock.RenderTransform>
                    <RotateTransform Angle="-90" />
                </TextBlock.RenderTransform>
            </TextBlock>

            <!-- Yaw -->
            <Grid Grid.Row="0"
                       Grid.Column="1"
                       HorizontalAlignment="Stretch"
                       VerticalAlignment="Bottom"
                       RenderTransformOrigin="0 1">
                <StackPanel Orientation="Horizontal"
                            HorizontalAlignment="Center">
                    <TextBlock Text="YAW = " Foreground="Green" />
                    <TextBlock Name="yawValue" Foreground="Green" />
                </StackPanel>

                <Rectangle Fill="Green"
                           Height="3"
                           HorizontalAlignment="Stretch"
                           VerticalAlignment="Bottom" />

                <Grid.RenderTransform>
                    <TransformGroup>
                        <RotateTransform Angle="-90" />
                        <RotateTransform x:Name="yawRotate" />
                    </TransformGroup>
                </Grid.RenderTransform>
            </Grid>
        </Grid>
    </Grid>
```

从代码隐藏文件中可以看到，Inclinometer 类实例化，而且用得与 Accelerometer 及

Compass 一样多。

```
项目: YawPitchRoll | 文件: MainPage.xaml.cs(片段)
public sealed partial class MainPage : Page
{
     Inclinometer inclinometer = Inclinometer.GetDefault();

     public MainPage()
     {
          this.InitializeComponent();
          DisplayProperties.AutoRotationPreferences = DisplayProperties.NativeOrientation;
          Loaded += OnMainPageLoaded;
     }

     async void OnMainPageLoaded(object sender, RoutedEventArgs args)
     {
          if  (inclinometer == null)
          {
               await new MessageDialog("Cannot obtain Inclinometer").ShowAsync();
          }
          else
          {
               ShowYawPitchRoll(inclinometer.GetCurrentReading());
               inclinometer.ReportInterval = inclinometer.MinimumReportInterval;
               inclinometer.ReadingChanged += OnInclinometerReadingChanged;
          }
     }

     async void OnInclinometerReadingChanged(Inclinometer sender,
                                        InclinometerReadingChangedEventArgs args)
     {
          await this.Dispatcher.RunAsync(CoreDispatcherPriority.Normal, () =>
               {
                    ShowYawPitchRoll(args.Reading);
               });
     }

     void ShowYawPitchRoll(InclinometerReading inclinometerReading)
     {
          if  (inclinometerReading == null)
               return;

          double yaw = inclinometerReading.YawDegrees;
          double pitch = inclinometerReading.PitchDegrees;
          double roll = inclinometerReading.RollDegrees;

          yawValue.Text = yaw.ToString("F0") + "°";
          pitchValue.Text = pitch.ToString("F0") + "°";
          rollValue.Text = roll.ToString("F0") + "°";

          yawRotate.Angle = yaw;

          if  (pitch <= 90 && pitch >= -90)
          {
               pitchPath.Fill = pitchPath.Stroke;
               pitchEllipse.Center = new Point(this.ActualWidth / 2,
                                        this.ActualHeight * (pitch + 90) / 180);
          }
          else
          {
               pitchPath.Fill = null;

               if  (pitch > 90)
                    pitchEllipse.Center = new Point(this.ActualWidth / 2,
                                        this.ActualHeight * (270 - pitch) / 180);
               else // pitch < -90
                    pitchEllipse.Center = new Point(this.ActualWidth / 2,
```

```
                                  this.ActualHeight * (-90 - pitch) / 180);
        }
        rollEllipse.Center = new Point(this.ActualWidth * (roll + 90) / 180,
                                       this.ActualHeight / 2);
    }
}
```

没有罗盘数据秘密来源为陀螺仪提供信息。作为 Compass 读数，YawDegrees 属性不稳定(或有限)，但能补足，即 YawDegrees 与 Compass 读数之和总是约等于 360。如果平板屏幕上升到一个水平面，偏航线会指向北(或在附近)，且俯仰球和滚转球的均位于中心。如果向上或向下倾斜平板电脑顶部，PitchDegrees 范围从平板电脑竖直时的 90 度到顶点向下时的-90 度。如果平板电脑向右或向左倾斜，RollDegrees 范围从 90 度到-90 度。下图所示为平板电脑的顶部和左侧都提升时的视图。

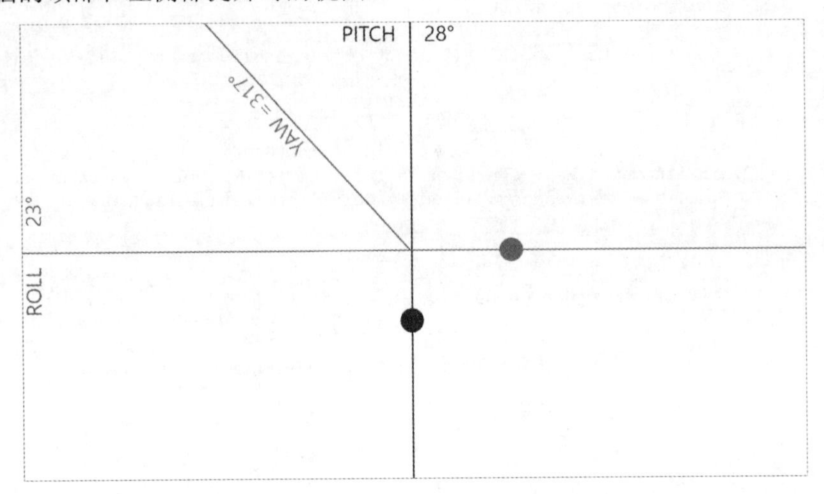

如果屏幕朝下，YawDegrees 则指向南，PitchDegrees 值范围从 90 度至 180 度、从-90 度到-180 度。程序用空心红球代表这些值。

如果正在处理的程序中会有某些东西绕着屏幕飞来飞去，这些欧拉角可能就是你需要的。然而，你可能想要更数学化的东西。这就是下一个类的目的。

18.6 OrientationSensor(方向传感器) = 加速计+罗盘

有几种三维空间方式代表旋转，这些方式彼此之间都可以进行转换。OrientationSensor 类与 Inclinometer 从某种意义上非常相似，因为 OrientationSensor 类结合了来自加速计和罗盘的信息并在三维空间提供完整方位。OrientationSensor 通过两个类的实例提供定位：

- SensorQuaternion
- SensorRotationMatrix

四元数在数学上非常有趣。就像虚数就可以表示在两维空间的旋转一样，四元数表示在三维空间的旋转。游戏程序员尤其喜欢代表旋转的四元数，因为可以平滑插入四元数。

旋转矩阵是一个没有最后一排及最后一列的规则变换矩阵。规则三维变换矩阵为 4 行 4 列。SensorRotationMatrix 类定义了 3 行 3 列。这种矩阵不能代表平移或视角，并且按照惯例没有缩放和倾斜。但 SensorRotationMatrix 类在三维空间很容易用于旋转对象。

在我写本书时所用的三星平板电脑上，SensorRotationMatrix 包含所有零值，因此本书中使用该矩阵的程序都不起作用。Microsoft Surface 的结果更好。

处理该旋转矩阵的时候，换个角度看可能有所帮助。我描述过 Accelerometer 和 Compass 的值如何相对于本章前面已经讲过的 3D 坐标系统。处理来自 OrientationSensor 类的旋转矩阵工作时，这样有助于可视化两个 3D 坐标系统，一个用于设备，而另一个用于地球：

- 在计算机 3D 坐标系统中，正 Y 指向屏幕顶部，正 X 正指向右边，而正 Z 轴指向屏幕之外，正如我之前展示的
- 在地球坐标系统中，正 Y 指向北，正 X 指向东，而正 Z 轴从地面垂直向上

如果电脑屏幕朝上、顶端向北，放在水平表面，两个坐标系统则一致。(理论上)SensorRotationMatrix 变为恒等矩阵：1s 在对角线，其他为 0s。否则，该矩阵会描述地球如何相对于计算机旋转，而计算机是由欧拉角所描述的旋转对立面。

AxisAngleRotation 程序演示了这种差异，程序还会计算代表三维空间旋转另一种方法，即围绕三维矢量的旋转。XAML 文件由一些无趣的 TextBlock 元素、功能标签和其他等待文本组合而成。

项目：AxisAngleRotation | 文件：MainPage.xaml(片段)

```
<Page ... >
    <Page.Resources>
        <Style x:Key="DefaultTextBlockStyle" TargetType="TextBlock">
            <Setter Property="FontFamily" Value="Lucida Sans Unicode" />
            <Setter Property="FontSize" Value="36" />
            <Setter Property="Margin" Value="0 0 48 0" />
        </Style>

        <Style x:Key="rightText" TargetType="TextBlock"
               BasedOn="{StaticResource DefaultTextBlockStyle}">
            <Setter Property="TextAlignment" Value="Right" />
            <Setter Property="Margin" Value="48 0 0 0" />
        </Style>
    </Page.Resources>

    <Grid Background="{StaticResource ApplicationPageBackgroundThemeBrush}">
        <StackPanel HorizontalAlignment="Center"
                    VerticalAlignment="Center">

            <!-- Grid showing Pitch, Roll, and Yaw -->
            <Grid HorizontalAlignment="Center">
                <Grid.RowDefinitions>
                    <RowDefinition Height="Auto" />
                    <RowDefinition Height="Auto" />
                    <RowDefinition Height="Auto" />
                </Grid.RowDefinitions>

                <Grid.ColumnDefinitions>
                    <ColumnDefinition Width="Auto" />
                    <ColumnDefinition Width="Auto" />
                </Grid.ColumnDefinitions>

                <Grid.Resources>
                    <Style TargetType="TextBlock"
                           BasedOn="{StaticResource DefaultTextBlockStyle}" />
                </Grid.Resources>

                <TextBlock Text="Pitch: " Grid.Row="0" Grid.Column="0" />
                <TextBlock Name="pitchText" Grid.Row="0" Grid.Column="1"
                           Style="{StaticResource rightText}" />
```

```
          <TextBlock Text="Roll: " Grid.Row="1" Grid.Column="0" />
          <TextBlock Name="rollText" Grid.Row="1" Grid.Column="1"
                     Style="{StaticResource rightText}" />

          <TextBlock Text="Yaw: " Grid.Row="2" Grid.Column="0" />
          <TextBlock Name="yawText" Grid.Row="2" Grid.Column="1"
                      Style="{StaticResource rightText}" />
      </Grid>

      <!-- Grid for RotationMatrix -->
      <Grid HorizontalAlignment="Center"
            Margin="0 48">
          <Grid.RowDefinitions>
              <RowDefinition Height="Auto" />
              <RowDefinition Height="Auto" />
              <RowDefinition Height="Auto" />
          </Grid.RowDefinitions>

          <Grid.ColumnDefinitions>
              <ColumnDefinition Width="Auto" />
              <ColumnDefinition Width="Auto" />
              <ColumnDefinition Width="Auto" />
          </Grid.ColumnDefinitions>

          <Grid.Resources>
              <Style TargetType="TextBlock" BasedOn="{StaticResource rightText}" />
          </Grid.Resources>

          <TextBlock Name="m11Text" Grid.Row="0" Grid.Column="0" />
          <TextBlock Name="m12Text" Grid.Row="0" Grid.Column="1" />
          <TextBlock Name="m13Text" Grid.Row="0" Grid.Column="2" />

          <TextBlock Name="m21Text" Grid.Row="1" Grid.Column="0" />
          <TextBlock Name="m22Text" Grid.Row="1" Grid.Column="1" />
          <TextBlock Name="m23Text" Grid.Row="1" Grid.Column="2" />

          <TextBlock Name="m31Text" Grid.Row="2" Grid.Column="0" />
          <TextBlock Name="m32Text" Grid.Row="2" Grid.Column="1" />
          <TextBlock Name="m33Text" Grid.Row="2" Grid.Column="2" />
      </Grid>

      <!-- Axis/Angle rotation display -->
      <Grid HorizontalAlignment="Center">
          <Grid.RowDefinitions>
              <RowDefinition Height="Auto" />
              <RowDefinition Height="Auto" />
          </Grid.RowDefinitions>

          <Grid.ColumnDefinitions>
              <ColumnDefinition Width="Auto" />
              <ColumnDefinition Width="Auto" />
          </Grid.ColumnDefinitions>

          <Grid.Resources>
              <Style TargetType="TextBlock"
                     BasedOn="{StaticResource DefaultTextBlockStyle}" />
          </Grid.Resources>

          <TextBlock Text="Angle:" Grid.Row="0" Grid.Column="0" />
          <TextBlock Name="angleText" Grid.Row="0" Grid.Column="1"
                     TextAlignment="Center"/>
          <TextBlock Text="Axis:" Grid.Row="1" Grid.Column="0" />
          <TextBlock Name="axisText" Grid.Row="1" Grid.Column="1"
                     TextAlignment="Center" />
      </Grid>
    </StackPanel>
  </Grid>
</Page>
```

代码隐藏文件实例化 Inclinometer，获得偏航角、俯仰角和滚转角，还实例化 OrientationSensor，获得(并显示)旋转矩阵，并将其转换为轴/角旋转。

```
项目: AxisAngleRotation | 文件: MainPage.xaml.cs(片段)
public sealed partial class MainPage : Page
{
    Inclinometer inclinometer = Inclinometer.GetDefault();
    OrientationSensor orientationSensor = OrientationSensor.GetDefault();

    public MainPage()
    {
        this.InitializeComponent();
        DisplayProperties.AutoRotationPreferences = DisplayProperties.NativeOrientation;
        Loaded += OnMainPageLoaded;
    }

    async void OnMainPageLoaded(object sender, RoutedEventArgs args)
    {
        if (inclinometer == null)
        {
            await new MessageDialog("Inclinometer is not available").ShowAsync();
        }
        else
        {
            // Start the Inclinometer events
            ShowYawPitchRoll(inclinometer.GetCurrentReading());
            inclinometer.ReadingChanged += OnInclinometerReadingChanged;
        }

        if (orientationSensor == null)
        {
            await new MessageDialog("OrientationSensor is not available").ShowAsync();
        }
        else
        {
            // Start the OrientationSensor events
            ShowOrientation(orientationSensor.GetCurrentReading());
            orientate onSensor.ReadingChanged += OrientationSensorChanged;
        }
    }

    async void OnInclinometerReadingChanged(Inclinometer sender,
                                    InclinometerReadingChangedEventArgs args)
    {
        await this.Dispatcher.RunAsync(CoreDispatcherPriority.Normal, () =>
        {
            ShowYawPitchRoll(args.Reading);
        });
    }

    void ShowYawPitchRoll(InclinometerReading inclinometerReading)
    {
        if (inclinometerReading == null)
            return;

        yawText.Text = inclinometerReading.YawDegrees.ToString("F0") + "°";
        pitchText.Text = inclinometerReading.PitchDegrees.ToString("F0") + "°";
        rollText.Text = inclinometerReading.RollDegrees.ToString("F0") + "°";
    }

    async void OrientationSensorChanged(OrientationSensor sender,
                                OrientationSensorReadingChangedEventArgs args)
    {
        await this.Dispatcher.RunAsync(CoreDispatcherPriority.Normal, () =>
        {
            ShowOrientation(args.Reading);
        });
```

```
        }

        void ShowOrientation(OrientationSensorReading orientationReading)
        {
            if (orientationReading == null)
                return;

            SensorRotationMatrix matrix = orientationReading.RotationMatrix;

            If (matrix == null)
                return;

            m11Text.Text = matrix.M11.ToString("F3");
            m12Text.Text = matrix.M12.ToString("F3");
            m13Text.Text = matrix.M13.ToString("F3");

            m21Text.Text = matrix.M21.ToString("F3");
            m22Text.Text = matrix.M22.ToString("F3");
            m23Text.Text = matrix.M23.ToString("F3");

            m31Text.Text = matrix.M31.ToString("F3");
            m32Text.Text = matrix.M32.ToString("F3");
            m33Text.Text = matrix.M33.ToString("F3");

            // Convert rotation matrix to axis and angle
            double angle = Math.Acos((matrix.M11 + matrix.M22 + matrix.M33 - 1) / 2);
            angleText.Text = (180 * angle / Math.PI).ToString("F0");

            if (angle != 0)
            {
                double twoSine = 2 * Math.Sin(angle);
                double x = (matrix.M23 - matrix.M32) / twoSine;
                double y = (matrix.M31 - matrix.M13) / twoSine;
                double z = (matrix.M12 - matrix.M21) / twoSine;

                axisText.Text = String.Format("({0:F2} {1:F2} {2:F2})", x, y, z);
            }
        }
    }
```

下图所示屏幕截图来自 Microsoft Surface，三个欧拉角在顶部，中间为旋转矩阵，底部为派生轴/角旋转。

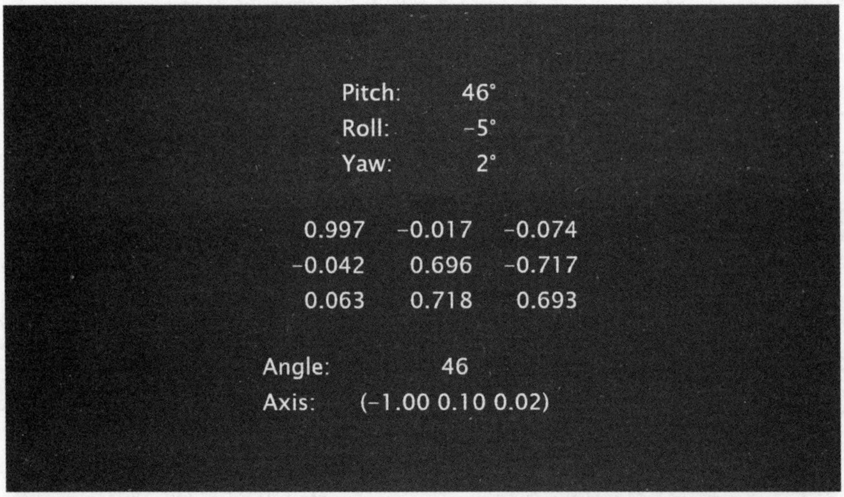

Pitch: 46°
Roll: –5°
Yaw: 2°

0.997 –0.017 –0.074
–0.042 0.696 –0.717
0.063 0.718 0.693

Angle: 46
Axis: (–1.00 0.10 0.02)

对于该屏幕截图，我保持平板电脑大致朝北，因此偏航角几乎为零。平板电脑向左倾斜一点点，滚转角会变负。但我已 46 度提高平板电脑顶部。屏幕底部显示了来自旋转矩阵的相同角度。但看一看轴，非常接近矢量(-1，0，0)，为负 X 轴。使用右手法则，把拇指指向负X轴方向。手指蜷缩表示正角度旋转的方向，因此证实了我前面讲的：旋转矩阵描述了地球相对于计算机的旋转。

也就是说，如果想用旋转矩阵来表示电脑相对于地球的旋转，则需要反转矩阵。SensorRotationMatrix 类本身不会反转，但 Matrix3D 结构可以反转。(注意，Matrix3D 在 Windows.UI.Xaml.Media.Media3D 命名空间中进行定义，并结合 Matrix3DProjection 使用。)非常容易从 SensorRotationMatrix 对象创建 Matrix3D 值，然后再翻转。

接下来我要使用该方法在 3D 空间中创建另一种方位代表形式。

18.7　方位角和海拔

从概念上讲，我们生活在一个天体内。如果需要描述在 3D 空间中相对于自己位置的对象，距离并不重要，相对于这个地球的点非常方便。地球尤其适合让你通过电脑屏幕看到虚拟现实及增强现实。在这类程序中，手拿平板电脑，就好像用平板背面的摄像头进行拍照，但在屏幕上所看到的东西是基于屏幕方位(全部或部分)由程序而生成的。如果以弧线平移屏幕，则可以看到世界的不同部分。

地球在陆地范围有熟悉的模拟。如果需要确定地球上的位置，我们会使用经纬度，两者都是与地球中心顶点的夹角。从概念上讲，我们通过赤道把地球分为两半。纬度线与赤道平行，以正角测量赤道以北(最大为 90 度，在北极)，以负角测量赤道以南(到南极的－90 度)。经度角度基于通过以本初子午线测量两极的大圆，本初子午线是通过英格兰格林尼治的经线。

我们可以以大致相同的方式描述地球上的点，但如果我们在球体中心往外看，就不同了。

向任意方向伸出手臂。我们怎样才能确定位置？第一，向上或向下摆动手臂，让手臂保持水平，让手臂与地球表面平行。在运动期间在手臂摆动产生的角度称为海拔。

正海拔值高于地平线；而负值则低于地平线。竖直向上称为最高点，即海拔为 90 度。竖直向下成为最底点，即海拔为-90 度。

此时手臂仍然指向地平线，对吗？现在，摆动手臂指向北方。刚刚手臂第二次摆动形成的角称为“方位角”。

总之，海拔和方位角构成地平坐标，如此命名是因为地平线将地球分为两半，类似于地面坐标中的赤道。

地平坐标没有距离信息。在日食过程中，太阳和月亮有相同地平坐标。地平坐标不是在 3D 空间中的位置，而是从观看者的角度在 3D 空间中的方向。就此意义而言，地平坐标像 3D 矢量，而不同的是，矢量用直角坐标表达，而地平坐标为球形。

为了让推导水平坐标的工作容易一点，我们先定义一个 Vector3 结构来封装 3D 矢量。

```
项目: EarthlyDelights | 文件: Vector3.cs
using System;
using Windows.Foundation;
using Windows.UI.Xaml.Media;
using Windows.UI.Xaml.Media.Media3D;
```

```
namespace Petzold.Windows8.VectorDrawing
{
    public struct Vector3
    {
        // Constructors
        public Vector3(double x, double y, double z)
            : this()
        {
            X = x;
            Y = y;
            Z = z;
        }

        // Properties
        public double X { private set; get; }
        public double Y { private set; get; }
        public double Z { private set; get; }

        public double LengthSquared
        {
            get { return X * X + Y * Y + Z * Z; }
        }

        public double Length
        {
            get { return Math.Sqrt(LengthSquared); }
        }

        public Vector3 Normalized
        {
            get
            {
                double length = this.Length;

                if  (length != 0)
                {
                    return new Vector3(this.X / length,
                                       this.Y / length,
                                       this.Z / length);
                }
                return new Vector3();
            }
        }

        // Static properties
        public static Vector3 UnitX
        {
            get { return new Vector3(1, 0, 0); }
        }

        public static Vector3 UnitY
        {
            get { return new Vector3(0, 1, 0); }
        }

        public static Vector3 UnitZ
        {
            get { return new Vector3(0, 0, 1); }
        }

        // Static methods
        public static Vector3 Cross(Vector3 v1, Vector3 v2)
        {
            return new Vector3(v1.Y * v2.Z - v1.Z * v2.Y,
                               v1.Z * v2.X - v1.X * v2.Z,
                               v1.X * v2.Y - v1.Y * v2.X);
        }
```

```
        public static double Dot(Vector3 v1, Vector3 v2)
        {
            return v1.X * v2.X + v1.Y * v2.Y + v1.Z * v2.Z;
        }

        public static double AngleBetween(Vector3 v1, Vector3 v2)
        {
            return 180 / Math.PI * Math.Acos(Vector3.Dot(v1, v2) /
                                    v1.Length * v2.Length);
        }

        public static Vector3 Transform(Vector3 v, Matrix3D m)
        {
            double x = m.M11 * v.X + m.M21 * v.Y + m.M31 * v.Z + m.OffsetX;
            double y = m.M12 * v.X + m.M22 * v.Y + m.M32 * v.Z + m.OffsetY;
            double z = m.M13 * v.X + m.M23 * v.Y + m.M33 * v.Z + m.OffsetZ;
            double w = m.M14 * v.X + m.M24 * v.Y + m.M34 * v.Z + m.M44;
            return new Vector3(x / w, y / w, z / w);
        }

        // Operators
        public static Vector3 operator +(Vector3 v1, Vector3 v2)
        {
            return new Vector3(v1.X + v2.X, v1.Y + v2.Y, v1.Z + v2.Z);
        }

        public static Vector3 operator -(Vector3 v1, Vector3 v2)
        {
            return new Vector3(v1.X - v2.X, v1.Y - v2.Y, v1.Z - v2.Z);
        }

        public static Vector3 operator *(Vector3 v, double d)
        {
            return new Vector3(d * v.X, d * v.Y, d * v.Z);
        }

        public static Vector3 operator *(double d, Vector3 v)
        {
            return new Vector3(d * v.X, d * v.Y, d * v.Z);
        }

        public static Vector3 operator /(Vector3 v, double d)
        {
            return new Vector3(v.X / d, v.Y / d, v.Z / d);
        }

        public static Vector3 operator -(Vector3 v)
        {
            return new Vector3(-v.X, -v.Y, -v.Z);
        }

        // Overrides
        public override string ToString()
        {
            return String.Format("({0} {1} {2})", X, Y, Z);
        }
    }
}
```

这种结构有很多优点，包括传统点积和叉积，还有通过 Matrix3D 值乘以 Vector3 值的 Transform 方法。在实践中，这个 Matrix3D 值可能代表的是旋转，乘法可以有效旋转 3D 空间中的矢量。

如果把平板电脑放在空中，看着屏幕，我们就在寻找相对于计算机坐标系统的方向，尤其是穿过屏幕背面的矢量方向，即负 Z 轴或**(0, 0, -1)**。我们需要将其转换为地平坐标。

基于 OrientationSensor 所提供的 SensorRotationMatrix 对象，我们创建一个名为 matrix 的 Matrix3D 值。该值可以反转，表示从计算机坐标系到地球坐标系统的变换：

```
matrix.Invert();
```

利用该矩阵把**(0, 0, - 1)**矢量(由 Vector3 结构提供的静态 unitz 属性的负值)变换为地球直角坐标系：

```
Vector3 vector = Vector3.Transform(-Vector3.UnitZ, matrix);
```

该矢量在直角坐标系中，需要转换为地平坐标。回想一下在地球坐标系中，Z 坐标指向地球之外。如果平板电脑保持竖直，穿过设备背面并转化为地球坐标的轴有 Z 分量，为零。这意味着，可以通过从二维笛卡儿坐标到极坐标的众所周知转换而计算方位角，我们把它从弧度转换为角度：

```
double azimuth = 180 * Math.Atan2(vector.X, vector.Y) / Math.PI;
```

不管怎么变换矢量的 Z 分量，该公式实际上都有效。海拔范围只在负 90 度和正 90 度之间，因此，可以用反正弦函数计算：

```
double altitude = 180 * Math.Asin(vector.Z) / Math.PI;
```

但我们少了一些东西。我们已经把三维旋转矩阵转换为只有两个组件的坐标，因为它只限于地球表面。如果把平板电脑指向地球中的东西，然后绕轴旋转，会发生什么？海拔和方位一样，但电脑屏幕上的画面会通过旋转而改变。该缺失部分有时也称为倾斜(tilt)。计算倾斜有点难，但 HorizontalCoordinate 结构显示了数学方法。

```
项目: EarthlyDelights | 文件: HorizontalCoordinate.cs
using System;
using Windows.UI.Xaml.Media.Media3D;

namespace Petzold.Windows8.VectorDrawing
{
    public struct HorizontalCoordinate
    {
        public HorizontalCoordinate(double azimuth, double altitude, double tilt)
            : this()
        {
            this.Azimuth = azimuth;
            this.Altitude = altitude;
            this.Tilt = tilt;
        }

        public HorizontalCoordinate(double azimuth, double altitude)
            : this(azimuth, altitude, 0)
        {
        }

        // Eastward from north
        public double Azimuth { private set; get; }

        public double Altitude { private set; get; }

        public double Tilt { private set; get; }

        public static HorizontalCoordinate FromVector(Vector3 vector)
        {
            double altitude = 180 * Math.Asin(vector.Z) / Math.PI;
            double azimuth = 180 * Math.Atan2(vector.X, vector.Y) / Math.PI;
```

```
        return new HorizontalCoordinate(azimuth, altitude);
    }

    public static HorizontalCoordinate FromMotionMatrix(Matrix3D matrix)
    {
        // Invert the matrix
        matrix.Invert();

        // Transform (0, 0, -1) -- the vector extending from the lens
        Vector3 zAxisTransformed = Vector3.Transform(-Vector3.UnitZ, matrix);

        // Get the horizontal coordinates
        HorizontalCoordinate horzCoord = FromVector(zAxisTransformed);

        // Find the theoretical HorizontalCoordinate for the transformed +Y vector
        //    if the device is upright
        double yUprightAltitude = 0;
        double yUprightAzimuth = 0;

        if (horzCoord.Altitude > 0)
        {
            yUprightAltitude = 90 - horzCoord.Altitude;
            yUprightAzimuth = 180 + horzCoord.Azimuth;
        }
        else
        {
            yUprightAltitude = 90 + horzCoord.Altitude;
            yUprightAzimuth = horzCoord.Azimuth;
        }
        Vector3 yUprightVector =
            new HorizontalCoordinate(yUprightAzimuth, yUprightAltitude).ToVector();

        // Find the real transformed +Y vector
        Vector3 yAxisTransformed = Vector3.Transform(Vector3.UnitY, matrix);

        // Get the angle between the upright +Y vector and the real transformed +Y vector
        double dotProduct = Vector3.Dot(yUprightVector, yAxisTransformed);
        Vector3 crossProduct = Vector3.Cross(yUprightVector, yAxisTransformed);
        crossProduct = crossProduct.Normalized;

        // Sometimes dotProduct is slightly greater than 1, which
        // raises an exception in the angleBetween calculation, so....
        dotProduct = Math.Min(dotProduct, 1);
        double angleBetween = 180 * Vector3.Dot(zAxisTransformed, crossProduct)
                                  * Math.Acos(dotProduct) / Math.PI;
        horzCoord.Tilt = angleBetween;

        return horzCoord;
    }

    public Vector3 ToVector()
    {
        double x = Math.Cos(Math.PI * this.Altitude / 180) *
                   Math.Sin(Math.PI * this.Azimuth / 180);
        double y = Math.Cos(Math.PI * this.Altitude / 180) *
                   Math.Cos(Math.PI * this.Azimuth / 180);
        double z = Math.Sin(Math.PI * this.Altitude / 180);

        return new Vector3((float)x, (float)y, (float)z);
    }

    public override string ToString()
    {
        return String.Format("Azi: {0} Alt: {1} Tilt: {2}",
                             this.Azimuth, this.Altitude, this.Tilt);
    }
  }
}
```

有了转换，就可以写一个天文程序来根据如何定位屏幕显示夜空的特定部分，很像我 2012 年 10 月发表于 MSDN Magazine 上的一个 Windows Phone 7.5 程序。但我们现在要做的东西仍然有一点模糊。

如果你想的位图比电脑屏幕大，但你不想缩小位图来适合屏幕，该怎么办？传统解决方案是使用滚动条。但有一个更现代的解决方案，让你能用手指来移动位图。

但另一种方法是将位图放在地球内饰。把平板电脑举向空中，改变屏幕方向来查看该图片。当然，我们并不希望真正拉伸位图，使位图适合符合地球内饰！可以直接使用水平滚动的方位角和垂直滚动的海拔。

EarthlyDelights 程序允许你查看一张大位图(7793*4409 像素)，该位图是 500 年前希罗尼穆斯·博斯(Hieronymus Bosch))所绘的油画《人间乐园》(*The Garden of Earthly Delights*)。程序从维基百科下载图片。下图是程序运行在 Surface 上所显示的一部分图片。

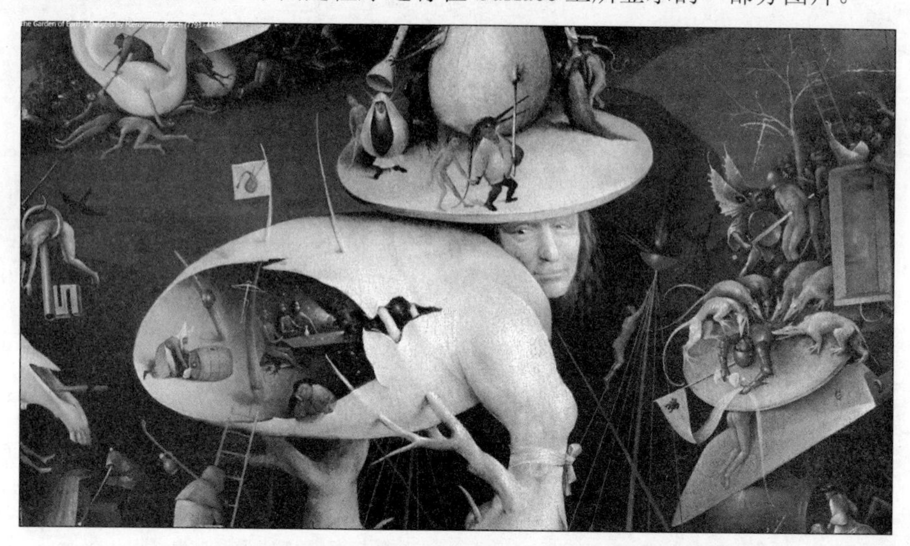

程序没有扫描或缩放图片的交互界面。而是完全基于改变屏幕方向。然而，如果点击屏幕，程序会应用缩放，把整张图片用在正常模式下可查看部分视图的矩形来显示(见下图)。

虽然这个功能会使程序复杂化，但我发现它非常重要。

XAML 文件中最重要的部分很明显是 Image 元素。注意，Image 的 Stretch 属性设置为 None，并包含不带 URI 源的 BitmapImage 对象。包含 Image 的 Grid 在 Canvas 中，以便大于屏幕时(而且会大于)不至于被剪切掉。

```
项目: EarthlyDelights | 文件: MainPage.xaml(片段)
<Grid Background="{StaticResource ApplicationPageBackgroundThemeBrush}">
    <!-- Two items displayed only during downloading -->
    <ProgressBar Name="progressBar"
                 VerticalAlignment="Center"
                 Margin="96 0" />

    <TextBlock Name="statusText"
               Text="downloading image..."
               HorizontalAlignment="Center"
               VerticalAlignment="Center" />

    <Canvas>
        <Grid>
            <Image Stretch="None">
                <Image.Source>
                    <BitmapImage x:Name="bitmapImage"
                                 DownloadProgress="OnBitmapImageDownloadProgress"
                                 ImageFailed="OnBitmapImageFailed"
                                 ImageOpened="OnBitmapImageOpened" />
                </Image.Source>
            </Image>

            <Border Name="outlineBorder"
                    BorderBrush="White"
                    HorizontalAlignment="Left"
                    VerticalAlignment="Top">

                <Rectangle Name="outlineRectangle"
                           Stroke="Black" />

                <Border.RenderTransform>
                    <CompositeTransform x:Name="borderTransform" />
                </Border.RenderTransform>
            </Border>

            <Grid.RenderTransform>
                <CompositeTransform x:Name="imageTransform" />
            </Grid.RenderTransform>
        </Grid>
    </Canvas>

    <TextBlock Name="titleText"
               Margin="2 "/>
</Grid>
```

内嵌 Rectangle 的 Border 用于缩小视图，显示通常占整个屏幕图片的一部分，但在普通视图下也可以看到这个矩形。外部的 CompositeTransform 适用于 Image 和 Border。在普通视图中，什么也不做。内部 CompositeTransform 会把 Border 定位到普通模式可见的相同图片区域。

Loaded 处理程序检查 OrientationSensor 是否可用，如果可用，则直接设置 BitmapImage 对象的 UriSource 属性并开始下载。如果位图下载成功，则可以获取像素尺寸，同时把页面尺寸保存为字段。

```
项目: EarthlyDelights | 文件: MainPage.xaml.cs(片段)
public sealed partial class MainPage: Page
```

```
{
    ...
    OrientationSensor orientationSensor = OrientationSensor.GetDefault();
    double pageWidth, pageHeight, maxDimension;
    int imageWidth, imageHeight;
    string title = "The Garden of Earthly Delights by Hieronymus Bosch";
    double zoomInScale;
    double rotation;
    bool isZoomView;

    public MainPage()
    {
        this.InitializeComponent();
        DisplayProperties.AutoRotationPreferences = DisplayProperties.NativeOrientation;
        Loaded += OnMainPageLoaded;
        SizeChanged += OnMainPageSizeChanged;
    }
    ...
    async void OnMainPageLoaded(object sender, RoutedEventArgs args)
    {
        if (orientationSensor == null)
        {
            await new MessageDialog("OrientationSensor is not available",
                                    "Earthly Delights").ShowAsync();

            progressBar.Visibility = Visibility.Collapsed;
            statusText.Visibility = Visibility.Collapsed;
        }
        else
        {
            bitmapImage.UriSource =
                new Uri("http://upload.wikimedia.org/ ... Bosch_High_Resolution_2.jpg");
        }
    }

    void OnMainPageSizeChanged(object sender, SizeChangedEventArgs args)
    {
        // Save the page dimensions
        pageWidth = this.ActualWidth;
        pageHeight = this.ActualHeight;
        maxDimension = Math.Max(pageWidth, pageHeight);

        // Initialize some values
        outlineBorder.Width = pageWidth;
        outlineBorder.Height = pageHeight;
        borderTransform.CenterX = pageWidth / 2;
        borderTransform.CenterY = pageHeight / 2;
    }

    void OnBitmapImageDownloadProgress(object sender, DownloadProgressEventArgs args)
    {
        progressBar.Value = args.Progress;
    }

    async void OnBitmapImageFailed(object sender, ExceptionRoutedEventArgs args)
    {
        progressBar.Visibility = Visibility.Collapsed;
        statusText.Visibility = Visibility.Collapsed;

        await new MessageDialog("Could not download image: " + args.ErrorMessage,
                                "Earthly Delights").ShowAsync();
    }

    void OnBitmapImageOpened(object sender, RoutedEventArgs args)
    {
        progressBar.Visibility = Visibility.Collapsed;
        statusText.Visibility = Visibility.Collapsed;
```

```
            // Save image dimensions
            imageWidth = bitmapImage.PixelWidth;
            imageHeight = bitmapImage.PixelHeight;
            titleText.Text = String.Format("{0} ({1}\x00D7{2})", title, imageWidth, imageHeight);

            // Initialize image transforms
            zoomInScale = Math.Min(pageWidth / imageWidth, pageHeight / imageHeight);

            // Start OrientationSensor going
            if (orientationSensor != null)
            {
                ProcessNewOrientationReading(orientationSensor.GetCurrentReading());
                orientationSensor.ReportInterval = orientationSensor.MinimumReportInterval;
                orientationSensor.ReadingChanged += OnOrientationSensorReadingChanged;
            }
        }

        async void OnOrientationSensorReadingChanged(OrientationSensor sender,
                                            OrientationSensorReadingChangedEventArgs args)
        {
            await this.Dispatcher.RunAsync(CoreDispatcherPriority.Normal, () =>
            {
                ProcessNewOrientationReading(args.Reading);
            });
        }
        ...
    }
```

ProcessNewOrientationReading 方法创建来自 SensorRotationMatrix 的 Matrix3D 对象，并通过它推导 HorizontalCoordinate 值。

项目: EarthlyDelights | 文件: MainPage.xaml.cs(片段)

```
void ProcessNewOrientationReading(OrientationSensorReading orientationReading)
{
    if (orientationReading == null)
        return;

    // Get the rotation matrix & convert to horizontal coordinates
    SensorRotationMatrix m = orientationReading.RotationMatrix;

    if (m == null)
        return;

    Matrix3D matrix3d = new Matrix3D(m.M11, m.M12, m.M13, 0,
                                     m.M21, m.M22, m.M23, 0,
                                     m.M31, m.M32, m.M33, 0,
                                     0, 0, 0, 1);

    if (!matrix3d.HasInverse)
        return;

    HorizontalCoordinate horzCoord = HorizontalCoordinate.FromMotionMatrix(matrix3d);

    // Set the transform center on the Image element
    imageTransform.CenterX = (imageWidth + maxDimension) *
                             (180 + horzCoord.Azimuth) / 360 - maxDimension / 2;
    imageTransform.CenterY = (imageHeight + maxDimension) *
                             (90 - horzCoord.Altitude) / 180 - maxDimension / 2;

    // Set the translation on the Border element
    borderTransform.TranslateX = imageTransform.CenterX - pageWidth / 2;
    borderTransform.TranslateY = imageTransform.CenterY - pageHeight / 2;

    // Get rotation from Tilt
    rotation = -horzCoord.Tilt;
    UpdateImageTransforms();
}
```

该方法负责设置一些变换，其他变化都在 UpdateImageTransforms 方法中(在方法末尾会看到调用 UpdateImageTransforms)设置。如果方位角为 0(平板电脑指向北极时)、海拔为 0(平板电脑竖直时)，centerX 和 CenterY 属性设置为位图中心。否则，设置为整个宽度和高度，并包括不显示图片的环绕区域边缘(否则，方案会需要显示屏幕左侧显示位图的右边缘以及在屏幕右侧显示位图的左边缘。)

我想让缩放操作采用动画形式，因此指定了 MainPage 依赖属性，点击屏幕时，该依赖属性就成为动画目标。

```
项目: EarthlyDelights | 文件: MainPage.xaml.cs(片段)
public sealed partial class MainPage : Page
{
    // Dependency property for zoom-in transition animation
    static readonly DependencyProperty interpolationFactorProperty =
        DependencyProperty.Register("InterpolationFactor",
                                    typeof(double),
                                    typeof(MainPage),
                                    new PropertyMetadata(0.0, OnInterpolationFactorChanged));

    ...
    // Interpolation Factor property
    public static DependencyProperty InterpolationFactorProperty
    {
        get { return interpolationFactorProperty; }
    }

    public double InterpolationFactor
    {
        set { SetValue(InterpolationFactorProperty, value); }
        get { return (double)GetValue(InterpolationFactorProperty); }
    }
    ...
    protected override void OnTapped(TappedRoutedEventArgs e)
    {
        // Animate the InterpolationFactor property
        DoubleAnimation doubleAnimation = new DoubleAnimation
        {
            EnableDependentAnimation = true,
            To = isZoomView ? 0 : 1,
            Duration = new Duration(TimeSpan.FromSeconds(1))
        };
        Storyboard.SetTarget(doubleAnimation, this);
        Storyboard.SetTargetProperty(doubleAnimation, "InterpolationFactor");
        Storyboard storyboard = new Storyboard();
        storyboard.Children.Add(doubleAnimation);
        storyboard.Begin();
        isZoomView ^= true;
        base.OnTapped(e);
    }

    static void OnInterpolationFactorChanged(DependencyObject obj,
                                            DependencyPropertyChangedEventArgs args)
    {
        (obj as MainPage).UpdateImageTransforms();
    }
    ...
}
```

OnInterpolationFactorChanged 方法还会调用 UpdateImageTransforms 来完成大部分繁重工作。

```
项目: EarthlyDelights | 文件: MainPage.xaml.cs(片段)
void UpdateImageTransforms()
{
    // If being zoomed out, set scaling
    double interpolatedScale = 1 + InterpolationFactor * (zoomInScale - 1);
    imageTransform.ScaleX =
    imageTransform.ScaleY = interpolatedScale;

    // Move transform center to screen center
    imageTransform.TranslateX = pageWidth / 2 - imageTransform.CenterX;
    imageTransform.TranslateY = pageHeight / 2 - imageTransform.CenterY;

    // If being zoomed out, adjust for scaling
    imageTransform.TranslateX -= InterpolationFactor *
                     (pageWidth / 2 - zoomInScale * imageTransform.CenterX);
    imageTransform.TranslateY -= InterpolationFactor *
                     (pageHeight / 2 - zoomInScale * imageTransform.CenterY);

    // If being zoomed out, center image in screen
    imageTransform.TranslateX += InterpolationFactor *
                     (pageWidth - zoomInScale * imageWidth) / 2;
    imageTransform.TranslateY += InterpolationFactor *
                     (pageHeight - zoomInScale * imageHeight) / 2;

    // Set border thickness
    outlineBorder.BorderThickness = new Thickness(2 / interpolatedScale);
    outlineRectangle.StrokeThickness = 2 / interpolatedScale;

    // Set rotation on image and border
    imageTransform.Rotation = (1 - InterpolationFactor) * rotation;
    borderTransform.Rotation = -rotation;
}
```

如果有新的 OrientationSensor 值或者缩放操作的 InterpolationFactor 属性改变，就调用该方法。如果你有兴趣了解该方法如何工作，则需要消除所有插值代码来进行清理。把 InterpolationFactor 设置为 0，再设置为 1，就会发现方法非常直观。

18.8　必应地图和必应地图图块

Geolocator 类没有被当成是传感器，它完全位于另一个命名空间 Windows.Devices.Geolocation。然而，Geolocator 类启用相类似，会告诉你电脑什么时候改变了地理位置以及新位置是什么。

你需要在 Package.appxmanifest 文件的 Capabilities 部分特别指出应用要求位置信息。用户第一次运行程序时，Windows 8 会得到确认。

一般情况下，要结合地图来使用 Geolocator 位置。Windows 8 没有内置必应地图控件，但可以下载工具包把它添加到应用。访问 www.bingmapsportal.com 可以获取所需要的证书密钥。

但我要为本章最后一个程序采用一种稍微不同的方法。我要展示基于平板电脑定位来旋转的地图。这种旋转允许地图上的北方与实际北方(或平板电脑所认为的北方)保持一致。为了做到这一点，我不使用必应地图控件，而是通过 Bing Maps SOAP 服务下载单个图块，然后拼接成一张复合地图。当然，还是需要证书密钥的。

运行 RotatingMap 程序时，你会想用手指浏览和缩放地图。但行不通。程序没有触控界面！为简单起见，程序直接以当前位置为地图中心，如果位置变化，则重新定位地图。程序提供应用栏按钮进行放大和缩小并可以在道路和鸟瞰视图之间进行切换，但仅此而已。

以下为 XAML 文件。所有构成地图的图块都进入名为 imageCanvas 的 Canvas。注意，RotateTransform 围绕其中心旋转 Canvas。

项目：RotatingMap | 文件：MainPage.xaml(片段)

```
<Page ... >
    <Grid Background="{StaticResource ApplicationPageBackgroundThemeBrush}">
        <Canvas Name="imageCanvas"
                HorizontalAlignment="Center"
                VerticalAlignment="Center">
            <Canvas.RenderTransform>
                <RotateTransform x:Name="imageCanvasRotate" />
            </Canvas.RenderTransform>
        </Canvas>

        <!-- Circle to show location -->
        <Ellipse Name="locationDisplay"
                 Width="24"
                 Height="24"
                 Stroke="Red"
                 StrokeThickness="6"
                 HorizontalAlignment="Center"
                 VerticalAlignment="Center"
                 Visibility="Collapsed" />

        <!-- Arrow to show north -->
        <Border HorizontalAlignment="Left"
                VerticalAlignment="Top"
                Margin="12"
                Background="Black"
                Width="36"
                Height="36"
                CornerRadius="18">
            <Path Stroke="White"
                  StrokeThickness="3"
                  Data="M 18 4 L 18 24 M 12 12 L 18 4 24 12">
                <Path.RenderTransform>
                    <RotateTransform x:Name="northArrowRotate"
                                     CenterX="18"
                                     CenterY="18" />
                </Path.RenderTransform>
            </Path>
        </Border>

        <!-- "powered by bing" display -->
        <Border Background="Black"
                HorizontalAlignment="Center"
                VerticalAlignment="Bottom"
                Margin="12"
                CornerRadius="12"
                Padding="3">

            <StackPanel Name="poweredByDisplay"
                        Orientation="Horizontal"
                        Visibility="Collapsed">
                <TextBlock Text=" powered by "
                           Foreground="White"
                           VerticalAlignment="Center" />
                <Image Stretch="None">
                    <Image.Source>
                        <BitmapImage x:Name="poweredByBitmap" />
                    </Image.Source>
                </Image>
            </StackPanel>
```

```
        </Border>
    </Grid>

    <Page.BottomAppBar>
        <AppBar Name="bottomAppBar"
                IsEnabled="False">
            <StackPanel Orientation="Horizontal"
                        HorizontalAlignment="Right">

                <!-- Must remove reference to BackgroundCheckedGlyph in
                     AppBarButtonStyle to use it for a CheckBox -->

                <CheckBox Name="streetViewAppBarButton"
                          Style="{StaticResource StreetAppBarButtonStyle}"
                          AutomationProperties.Name="Street View"
                          Checked="OnStreetViewAppBarButtonChecked"
                          Unchecked="OnStreetViewAppBarButtonChecked" />

                <Button Name="zoomInAppBarButton"
                        Style="{StaticResource ZoomInAppBarButtonStyle}"
                        Click="OnZoomInAppBarButtonClick" />

                <Button Name="zoomOutAppBarButton"
                        Style="{StaticResource ZoomOutAppBarButtonStyle}"
                        Click="OnZoomOutAppBarButtonClick" />
            </StackPanel>
        </AppBar>
    </Page.BottomAppBar>
</Page>
```

可以来回传递大量 XML 文件，“手动”使用 Bing Maps SOAP 服务，但更好的方法是通过 Visual Studio 生成的代理类来使用 Web 服务。该代理类让 Web 服务可作为一些结构、枚举和异步方法调用。为了将代理服务器添加到 RotatingMap 程序，我用鼠标右键点击 Visual Studio 中解决方案资源管理器的项目名称，并从菜单中选择“添加服务引用”。对话框请求地址时，我在 Imagery Service 中粘贴 URL(在以下网址可以找到与 Bing 地图相关联的其他三个 Web 服务 URL，http://msdn.microsoft.com/en-us/library/cc966738.aspx)。我将其命名为 ImageryService，也就是说，Visual Studio 生成了使用 RotatingMap.ImageryService 命名空间的代码。

该服务有两类请求，分别是 GetMapUriAsync 和 GetImageryMetadataAsync。第一个请求允许获得特定位置的静态地图，但我喜欢选择另一个请求，获得下载单个地图图块和拼成完整地图所需要的信息。

我们先看包含 MainPage 构造函数的 RotatingMap 代码。代码只保存两个值作为应用设置：地图样式(MapStyle 枚举项，在为 Web 服务所生成的代码中，表示道路或鸟瞰视图)和整数缩放级别。

```
项目: RotatingMap | 文件: MainPage.xaml.cs(片段)
public sealed partial class MainPage : Page
{
    ...
    // Saved as application settings
    MapStyle mapStyle = MapStyle.Aerial;
    int zoomLevel = 12;

    public MainPage()
    {
        this.InitializeComponent();
        DisplayProperties.AutoRotationPreferences = DisplayProperties.NativeOrientation;
        Loaded += OnMainPageLoaded;
```

```
            SizeChanged += OnMainPageSizeChanged;

            // Get application settings (and later save them)
            IPropertySet propertySet = ApplicationData.Current.LocalSettings.Values;

            if (propertySet.ContainsKey("ZoomLevel"))
                zoomLevel = (int)propertySet["ZoomLevel"];

            if (propertySet.ContainsKey("MapStyle"))
                mapStyle = (MapStyle)(int)propertySet["MapStyle"];

            Application.Current.Suspending += (sender, args) =>
            {
                propertySet["ZoomLevel"] = zoomLevel;
                propertySet["MapStyle"] = (int)mapStyle;
            };
        }
        ...
    }
```

在 Loaded 处理程序访问 Web 服务。两个调用必须进行：一个获得道路视图的地图元数据，另一个用于鸟瞰图。信息会保存在名为 ViewParams 局部类的两个实例中。元数据最重要的部分是下载单个地图图块的 URI 模板。ViewParams 类也有最小和最大缩放级别字段，缩放比例范围为 1～21，假设上限为 21，以下是代码的其他部分。

项目：RotatingMap | 文件：MainPage.xaml.cs(片段)

```
public sealed partial class MainPage : Page
{
    ...
    // Storage of parameters for two views
    class ViewParams
    {
        public string UriTemplate;
        public int MinimumLevel;
        public int MaximumLevel;
    }
    ViewParams aerialParams;
    ViewParams roadParams;

    Geolocator geolocator = new Geolocator();
    Inclinometer inclinometer = Inclinometer.GetDefault();
    ...
    async void OnMainPageLoaded(object sender, RoutedEventArgs args)
    {
        // Initialize the Bing Maps imagery service
        ImageryServiceClient imageryServiceClient =
            new ImageryServiceClient(
                ImageryServiceClient.EndpointConfiguration.BasicHttpBinding_IImageryService);

        // Make two requests for road and aerial views
        ImageryMetadataRequest request = new ImageryMetadataRequest
        {
            Credentials = new Credentials
            {
                ApplicationId = "put your own credentials string here"
            },
            Style = MapStyle.Road
        };
        Task<ImageryMetadataResponse> roadStyleTask =
                imageryServiceClient.GetImageryMetadataAsync(request);

        request = new ImageryMetadataRequest
        {
            Credentials = new Credentials
            {
```

```
            ApplicationId = "put your own credentials string here"
        },
        Style = MapStyle.Aerial
    };
    Task<ImageryMetadataResponse> aerialStyleTask =
            imageryServiceClient.GetImageryMetadataAsync(request);

    // Wait for both tasks to complete
    Task.WaitAll(roadStyleTask, aerialStyleTask);

    // Check if everything is OK
    if (!roadStyleTask.IsCanceled && !roadStyleTask.IsFaulted &&
        !aerialStyleTask.IsCanceled && !aerialStyleTask.IsCanceled)
        {
            // Get the "powered by" bitmap
            poweredByBitmap.UriSource = roadStyleTask.Result.BrandLogoUri;
            poweredByDisplay.Visibility = Visibility.Visible;

            // Get the URIs and min/max zoom levels
            roadParams = CreateViewParams(roadStyleTask.Result.Results[0]);
            aerialParams = CreateViewParams(aerialStyleTask.Result.Results[0]);

            // Get the current location
            Geoposition geoPosition = await geolocator.GetGeopositionAsync();
            GetLongitudeAndLatitude(geoPosition.Coordinate);
            RefreshDisplay();

            // Get updated locations
            geolocator.PositionChanged += OnGeolocatorPositionChanged;

            // Enable the application bar
            bottomAppBar.IsEnabled = true;
            streetViewAppBarButton.IsChecked = mapStyle == MapStyle.Road;

            // Get the current yaw
            if (inclinometer != null)
            {
                SetRotation(inclinometer.GetCurrentReading());
                inclinometer.ReadingChanged += OnInclinometerReadingChanged;
            }
        }
    }

    ViewParams CreateViewParams(ImageryMetadataResult result)
    {
        string uri = result.ImageUri;
        uri = uri.Replace("{subdomain}", result.ImageUriSubdomains[0]);
        uri = uri.Replace("&token={token}", "");
        uri = uri.Replace("{culture}", "en-us");

        return new ViewParams
        {
            UriTemplate = uri,
            MinimumLevel = result.ZoomRange.From,
            MaximumLevel = result.ZoomRange.To
        };
    }
    ...
}
```

两个异步调用需要获得两个视图的元数据，但两个调用并不彼此依赖，因此可以同时运行。这似乎是 Task.WaitAll 方法的完美应用，该方法会一直等待，直到多个任务项目都完成。

两个 Web 服务调用成功完成后，启动 Geolocator 和 Inclinometer。Inclinometer 仅用于获取偏航值，该值可用于旋转地图和旋转指示北方的箭头。

项目: RotatingMap | 文件: MainPage.xaml.cs(片段)
```
public sealed partial class MainPage : Page
{
    ...
    async void OnInclinometerReadingChanged(Inclinometer sender,
                                            InclinometerReadingChangedEventArgs args)
    {
        await this.Dispatcher.RunAsync(CoreDispatcherPriority.Normal, () =>
        {
            SetRotation(args.Reading);
        });
    }

    void SetRotation(InclinometerReading inclinometerReading)
    {
        if (inclinometerReading == null)
            return;

        imageCanvasRotate.Angle = inclinometerReading.YawDegrees;
        northArrowRotate.Angle = inclinometerReading.YawDegrees;
    }
    ...
}
```

Loaded 处理程序完成后，程序会有两个 URI 模板，可以用来下载单独地图图块。形成必应地图的图块是大小为 256 像素的正方形位图。每个图块与特定经度、纬度及缩放级别相关联，并且包含一部分通过墨卡托投影的扁平世界的图片。

在 1 级中，整个地球，或者说，地球上纬度位于+85.08 度和-85.05 度之间的那部分，由四个图块覆盖(见下图)。

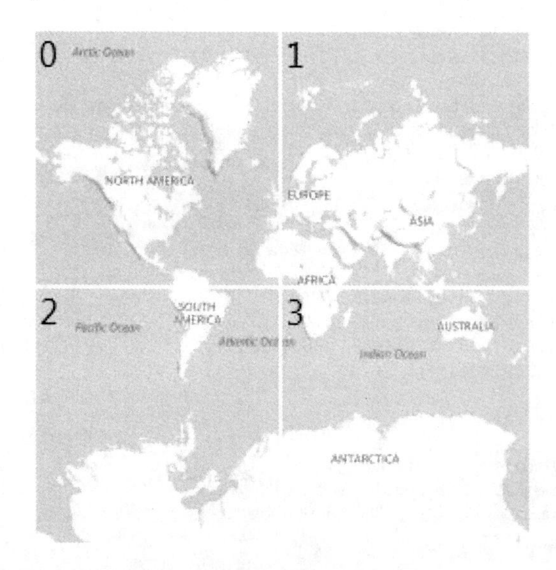

下面要讨论图块中的这些数字。图块是 256 像素的正方形，因此赤道上的每个像素大约覆盖 49 英里。

在 2 级中，有 16 个图块覆盖地球(见下图)。

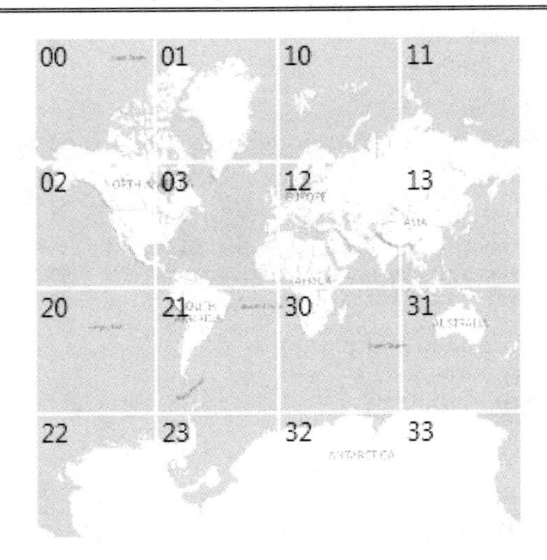

这些图块也是 256 像素的正方形，因此赤道上的每个像素约为 24 英里。

1 级中每个图块的覆盖面积等于 2 级中 4 个图块的覆盖面积，以此类推：3 级有 64 个图块，4 级有 256 个图块，并一直上升到 21 级，(原则上)覆盖地球需要超过 4 万亿个图块，每像素 3 英寸的赤道上有 2 百万水平和 2 百万垂直的分辨率。

如何用一致方式组织这么多图块？注意，为了使通过 Web 服务最有效地提供这些图块，牵涉到三个维度(经度、纬度和缩放级别)，而覆盖相同区域的图块在服务器上临近存储。

显然，可以采用巧妙的编号方案 quadkey。每个图块都有一个唯一的 quadkey。从 Bing Maps Web 服务获得的 URI 模板包含一个占位符“{quadkey }”，可以引用实际图块代替。两个图表示在左上角特定图块的 quadkeys。前导零很重要！Quadkey 中的数字数量等于缩放级别。21 级中的的图块均标识为 21 位 quadkeys。

在 quadkey 的数字总是 0、1、2 或 3，表明 quadkeys 实际上是基 4 数字。在二进制中，数字 0、1、2 和 3 为 00、01、10 和 11。第一位是垂直坐标，第二位是地平坐标。因此，位会对应于交错的经纬度。

正如你所见，1 级中的每个图块相当于 2 级的 4 个图块，这样就可以把图块认为具有父子关系。子图块的 quadkey 总是以与其父开始的相同数字作为开头，但会增加一个数字，取决于相对于其父的位置。可以直接从子 quadkey 截掉最后一位数字而获得父 quadkey。

要使用 Bing Maps Web 服务，需要根据经纬度推导出 quadkey。以下 GetLongitudeAndLatitude 方法的代码显示了第一步。为了把来自 Geolocator 的经纬度转换成相对 double 值，范围从 0 到 1，然后再转换为整数值。

```
项目: RotatingMap | 文件: MainPage.xaml.cs (片段)
public sealed partial class MainPage : Page
{
    const int BITRES = 29;
    ...
    int integerLongitude = -1;
    int integerLatitude = -1;
    ...
    async void OnGeolocatorPositionChanged(Geolocator sender, PositionChangedEventArgs args)
    {
        await this.Dispatcher.RunAsync(CoreDispatcherPriority.Normal, () =>
            {
                GetLongitudeAndLatitude(args.Position.Coordinate);
```

```
                RefreshDisplay();
            });
        }

        void GetLongitudeAndLatitude(Geocoordinate geoCoordinate)
        {
            locationDisplay.Visibility = Visibility.Visible;

            // Calculate integer longitude and latitude
            double relativeLongitude = (180 + geoCoordinate.Longitude) / 360;
            integerLongitude = (int)(relativeLongitude * (1 << BITRES));

            double sinTerm = Math.Sin(Math.PI * geoCoordinate.Latitude / 180);
            double relativeLatitude = 0.5 - Math.Log((1 + sinTerm) / (1 - sinTerm)) / (4 * Math.PI);
            integerLatitude = (int)(relativeLatitude * (1 << BITRES));
        }
        ...
    }
```

BITRES 值为 29，是 21 级 quadkey 中的 21 位数字，加上图块像素尺寸的 8 位，也就是说，这些整数值识别了一个经度和一个纬度，精确到最高缩放等级图块上最接近的像素。integerLongitude 的计算并不重要，但 integerLatitude 更复杂，因为离赤道越远，墨卡托地图投影就越会压缩纬度。

例如：纽约市中央公园的中心，经度为-73.965368，纬度为 40.783271。相对的 double 值(只有若干小数位)为 0.29454 和 0.37572。29 位的整数值如下(以二进制表示，4 位一组，便于阅读)：

```
0 1001 0110 1100 1110 0000 1000 0000
0 1100 0000 0101 1110 1011 0000 0000
```

假设你想要该经度和纬度在 12 级缩放的图块。你需要整数经度和纬度的前 12 位(注意，所得数字分组方式略有不同)：

```
0100 1011 0110
0110 0000 0010
```

有两个二进制数，但要形成 quadkey，需要将其组合成基 4 数字。没有经过实际位循环的代码，不能这样做，但为了便于说明，可以直接把纬度上的所有位增加一倍，如果基 4 的值，则进行相加：

```
  0100 1011 0110
+ 0220 0000 0020
--------------------------      -
  0320 1011 0130
```

这就是在从 Web 服务所得 URI 模板中需要替换为“{quadkey }”占位符的 quadkey。所得 URI 引用一个 256 像素的正方形位图。

下面是 RotatingMap 中的例程，RotatingMap 从截断的整数经度和纬度构建 quadkey。为清楚起见，该逻辑被分离，首先显示一个长整数推导，再显示字符串。

项目：RotatingMap | 文件：MainPage.xaml.cs(片段)

```
public sealed partial class MainPage : Page
{
    ...
    StringBuilder strBuilder = new StringBuilder();
    ...
    string ToQuadKey(int longitude, int latitude, int level)
    {
        long quadkey = 0;
```

```
            int mask = 1 << (level - 1);

            for (int i = 0; i < level; i++)
            {
                quadkey <<= 2;

                if ((longitude & mask) != 0)
                    quadkey |= 1;

                if ((latitude & mask) != 0)
                    quadkey |= 2;

                mask >>= 1;
            }

            strBuilder.Clear();

            for (int i = 0; i < level; i++)
            {
                strBuilder.Insert(0, (quadkey & 3).ToString());
                quadkey >>= 2;
            }

            return strBuilder.ToString();
        }
        ...
    }
```

该 quadkey 引用包含所需经纬度的图块，但精确经纬度的位置实际上位于图块内某处。通过所需 quadkey 位数之后的整数经纬度的下一个 8 位，可以确定图块内的像素位置。

现在我们回到主页。整页必须用 256 个像素的正方形图块覆盖，图块矩阵可旋转，并且用户当前位置被定位在屏幕中心的中心图块某处， 因此，SizeChanged 处理程序决定需要多少图块，需要创建多少 Image 元素。名为 sqrtNumTiles 的字段指“图块数量的平方根。”对于 1366×768 像素的显示屏，值为 9。图块(和 Image 元素)的总数是其平方，或 81。

```
项目: RotatingMap | 文件: MainPage.xaml.cs(片段)
public sealed partial class MainPage : Page
{
    ...
    int sqrtNumTiles;   // always an odd number
    ...
    void OnMainPageSizeChanged(object sender, SizeChangedEventArgs args)
    {
        // Clear out the existing Image elements
        imageCanvas.Children.Clear();

        // Determine how many Image elements are needed
        double diagonal = Math.Sqrt(Math.Pow(args.NewSize.Width, 2) +
                          Math.Pow(args.NewSize.Height, 2));

        sqrtNumTiles = 1 + 2 * (int)Math.Ceiling((diagonal / 2) / 256);

        // Create Image elements for a sqrtNumTiles x sqrtNumTiles array
        for (int i = 0; i < sqrtNumTiles * sqrtNumTiles; i++)
        {
            Image image = new Image
            {
                Source = new BitmapImage(),
                Stretch = Stretch.None
            };
            imageCanvas.Children.Add(image);
        }
        RefreshDisplay();
```

```
        }
        ...
}
```

RefreshDisplay 方法执行真正的工作。RefreshDisplay 循环所有 Image 元素，并给每个 Image 确定 quadkey(以及 URI)。

```
项目: RotatingMap | 文件: MainPage.xaml.cs(片段)
public sealed partial class MainPage : Page
{
    ...
    void RefreshDisplay()
    {
        if (roadParams == null || aerialParams == null)
            return;

        if (integerLongitude == -1 || integerLatitude == -1)
            return;

        // Get coordinates and pixel offsets based on current zoom level
        int croppedLongitude = integerLongitude >> BITRES - zoomLevel;
        int croppedLatitude = integerLatitude >> BITRES - zoomLevel;
        int xPixelOffset = (integerLongitude >> BITRES - zoomLevel - 8) % 256;
        int yPixelOffset = (integerLatitude >> BITRES - zoomLevel - 8) % 256;

        // Prepare for the loop
        string uriTemplate = (mapStyle == MapStyle.Road ? roadParams : aerialParams).
            UriTemplate;
        int index = 0;
        int maxValue = (1 << zoomLevel) - 1;

        // Loop through the array of Image elements
        for (int row = -sqrtNumTiles / 2; row <= sqrtNumTiles / 2; row++)
            for (int col = -sqrtNumTiles / 2; col <= sqrtNumTiles / 2; col++)
            {
                // Get the Image and BitmapImage
                Image image = imageCanvas.Children[index] as Image;
                BitmapImage bitmap = image.Source as BitmapImage;
                index++;

                // Check if we've gone beyond the bounds
                if (croppedLongitude + col < 0 ||
                    croppedLongitude + col > maxValue ||
                    croppedLatitude + row < 0 ||
                    croppedLatitude + row > maxValue)
                {
                    bitmap.UriSource = null;
                }
                else
                {
                    // Calculate a quadkey and set URI to bitmap
                    int longitude = croppedLongitude + col;
                    int latitude = croppedLatitude + row;
                    string strQuadkey = ToQuadKey(longitude, latitude, zoomLevel);
                    string uri = uriTemplate.Replace("{quadkey}", strQuadkey);
                    bitmap.UriSource = new Uri(uri);
                }

                // Position the Image element
                Canvas.SetLeft(image, col * 256 - xPixelOffset);
                Canvas.SetTop(image, row * 256 - yPixelOffset);
            }
    }
    ...
}
```

剩下的部分只牵涉到应用栏按钮。放大/缩小按钮基于当前选择视图的最小/最大缩放级别而仔细设置启用和禁用，尽管(正如我已经讲过的)程序的其他部分显然相当肯定最大缩放等级为 21。

项目：RotatingMap | 文件：MainPage.xaml.cs(片段)

```
public sealed partial class MainPage : Page
{
    ...
    void OnStreetViewAppBarButtonChecked(object sender, RoutedEventArgs args)
    {
        ToggleButton btn = sender as ToggleButton;
        ViewParams viewParams = null;

        if (btn.IsChecked.Value)
        {
            mapStyle = MapStyle.Road;
            viewParams = roadParams;
        }
        else
        {
            mapStyle = MapStyle.Aerial;
            viewParams = aerialParams;
        }

        zoomLevel = Math.Max(viewParams.MinimumLevel,
                    Math.Min(viewParams.MaximumLevel, zoomLevel));

        RefreshDisplay();
        RefreshButtons();
    }

    void OnZoomInAppBarButtonClick(object sender, RoutedEventArgs args)
    {
        zoomLevel += 1;
        RefreshDisplay();
        RefreshButtons();
    }

    void OnZoomOutAppBarButtonClick(object sender, RoutedEventArgs args)
    {
        zoomLevel -= 1;
        RefreshDisplay();
        RefreshButtons();
    }

    void RefreshButtons()
    {
        ViewParams viewParams =
                streetViewAppBarButton.IsChecked.Value ? roadParams : aerialParams;
        zoomInAppBarButton.IsEnabled = zoomLevel < viewParams.MaximumLevel;
        zoomOutAppBarButton.IsEnabled = zoomLevel > viewParams.MinimumLevel;
    }
}
```

我们还不太习惯看到已旋转地图的熟悉区域，因此曼哈顿岛在该视图中看上去有点怪(见下图)。

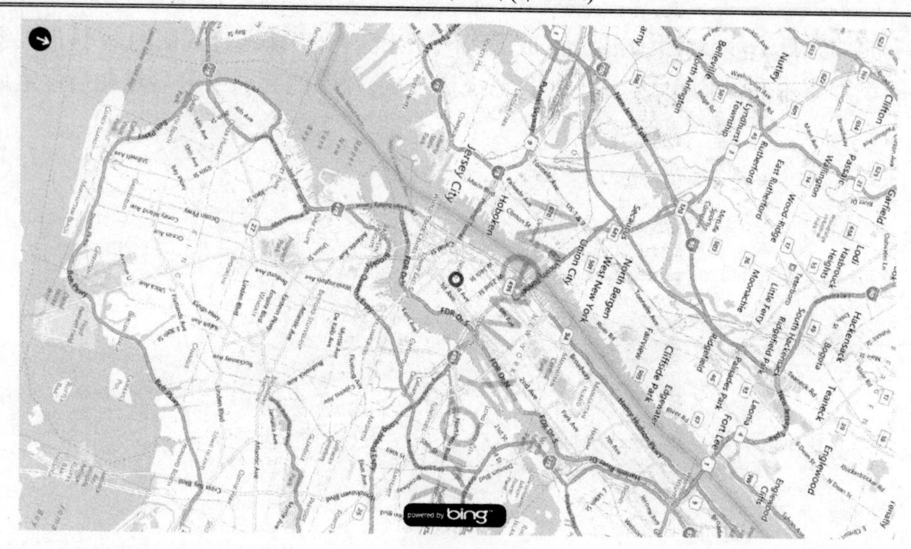

但如果你站在一个陌生地方，想用平板电脑找到你在哪里，根据现实旋转地图就非常有用。也许有一天，显示城市和街道的标签也可以旋转。

第 19 章　手　写　笔

本章的内容充满争议。争议主题是在计算领域未来不确定的输入设备，争论双方都有强烈的热情。2010 年，史蒂夫·乔布斯讨论了其他平板电脑和 iPad 竞争的可能性，并宣称“如果你看到了触控笔，事情就搞砸了。”[①]

然而，试过在触摸屏上使用基于传统鼠标的应用的人当然不同意他的观点。手写笔不如手指方便，也不如多个手指那么万能，但它像鼠标一样精确，而且通常比手指更适合从菜单中选中条目、选择以及删除输入。我在本书中大部分时候使用的三星 700T 有一支触控笔，我认为，如果在没有鼠标的机器上使用 Visual Studio，触控笔则是必须的。

因此，我曾万分焦急地等待第一款 Surface 设备的面世。会包含手写笔吗？我甚至认为有没有包含决定了本书是否包含有关触控笔的章节！

根据史蒂夫·乔布斯的标准，Surface 当然没有“搞砸”。第一款 Surface 电脑不包括触控笔。我很失望，但无论如何，我还是选择在本书中包含本章节。

我个人更喜欢用触控笔一词来指这些输入设备，但从本章这里术语开始将与 Windows Runtime 编程接口保持一致，在接口中称之为“手写笔”。而手写笔输入和经渲染的图形输出称为“墨迹”。

读过第 13 章，你知道了如何处理和渲染手写笔输入。然而，处理手写笔时，Windows.UI.Input.Inking 命名空间会提供额外的以下功能：

- 在涂墨之外的擦除和选择模式
- 为了更平滑过渡，把折线输入转换为贝塞尔曲线
- 手写识别
- 保存墨迹到文件，并从文件中加载墨迹

只不过，本章不探讨手写识别。

有趣的是，这些功能实际上并不需要手写笔！理论上讲，可以通过触摸或鼠标输入做本章中的任何事。然而，触摸或鼠标输入在手写方面很笨拙，因为用手指画出的文本往往太大，用鼠标绘制的文本往往又太抖。用笔就刚刚好。人们预计大小和形状适合书写的设备备至少存在了两千年。

之前我引用《纽约时报》的文章，内容都是与电容笔相关。电容笔设计用来补充手指在电容触屏上的操作。除了精度和可操作性之外，与手指相比较，电容笔并没有提供任何真正的优势。

有更多功能的是电子笔，有时称为“数字化仪”或“数字触控笔”，但这些笔要求屏幕能响应这种笔输入。本书中使用的三星 700T 平板就是这种情况。这种笔有一个小尖头(直

① David Pogue, “On Touch Screens, Rest Your Finger by Using a Stylus,” http://www.nytimes.com/2012/08/02/technology/personaltech/on-touch-screens-rest-your-finger-by-using-a-stylus -state-of-the-art.html。

径约 1 毫米)，而另一端有一个“橡皮擦”，笔管上有一个按键。PointerPointProperties 类定义了两个属性 IsEraser 和 IsInverted，如果“橡皮擦”而不是笔尖触碰屏幕，则两个属性均为 true。这一般用来擦除之前的输入。如果使用了笔尖且按下按键，则 IsBarrelButtonPressed 属性为 true。这通常用于选择。

除非专门为电磁笔用户编写程序，否则一般都会用软件选项来补充电磁笔的擦除和选择功能，但本章为了精简示例项目，会跳过该功能。

19.1 InkManager 集合

Windows.UI.Input.Inking 命名空间里有 InkManager 类的成员。InkManager 类是应用和使用笔相关功能的入口。

InkManager 实例维护某个特定输入页面的所有墨迹。如果程序实现的是便笺本，就像本章最后一个程序一样，便笺本上的每个页面有自己的 InkManager。

InkManager 对象维护 InkStroke 类型对象的集合。每个 InkStroke 是一条连续曲线，一般都用笔触碰屏幕、移动并抬起而创建的。InkStroke 和特定 InkDrawingAttributes 对象相关，后者的主要目的是显示笔画颜色以及笔尖形状和大小，尽管(正如你会看到的)InkManager 和 InkStroke 并不真的处理这些绘画属性。

每个 InkStroke 都包含一个 InkStrokeRenderingSegment 对象集合。InkStrokeRenderingSegment 是特定笔压力、倾斜和扭曲的单条贝塞尔曲线。压力通常用于渲染笔画时计算线的宽度。值范围从 0 到 1，就像 PointerPointProperties 的 Pressure 属性一样。支持倾斜和扭曲的笔很少见。

在程序的帮助下，InkManager 可以积累笔输入并将输入变成贝塞尔曲线，但其自身无法进行渲染。渲染完全是你的责任，一般需要两步操作。

- 随着用户用笔绘画或书写，用 Polyline、Line 或 Path 进行渲染线条。
- 每个笔画完成后，用贝塞尔曲线取代这些元素。

你已经看过与 FingerPaint 程序有关的第一步渲染。为了阐明使用 InkManager 的基本知识，我专注于第一个示例项目 SimpleInking 中的第二步渲染。SimpleInking 程序非常简单，以至于要把笔从屏幕上抬起来才能看到实际上画的内容。

以下为 XAML 文件。注意，我把 Grid 涂成白色，这通常是笔输入的惯例。

项目：SimpleInking | 文件：MainPage.xaml(片段)

```
<Page ... >

    <Grid Name="contentGrid"
          Background="White" />
</Page>
```

在默认情况下，笔输入为黑色。

我采用了单个 InkManager 对象。

项目：SimpleInking | 文件：MainPage.xaml(片段)

```
public sealed partial class MainPage : Page
{
    InkManager inkManager = new InkManager();
    bool hasPen;
```

```
    public MainPage()
    {
        this.InitializeComponent();

        // Check if there's a pen among the pointer input devices
        foreach (PointerDevice device in PointerDevice.GetPointerDevices())
            hasPen |= device.PointerDeviceType == PointerDeviceType.Pen;
    }
    ...
}
```

构造函数决定机器是否支持手写笔，如果支持，hasPen 字段设置为 true。对该项目，我决定忽略所有支持手写笔的电脑的无笔指针输入，但会接受所有不支持手写笔的电脑的指针输入。这样一来，就允许程序用于 Microsoft Surface。

InkManager 定义以下三个方法，可以结合 Pointer 事件来使用，它们允许 InkManager 对象积累指针输入：

- ProcessPointerDown，可从 PointerPressed 事件处理程序进行调用
- ProcessPointerUpdate，可从 PointerMoved 事件处理程序进行多次调用
- ProcessPointerUp，可从 PointerReleased 事件处理程序进行调用

每个方法的参数是均为 PointerPoint 对象，可以通过调用 GetCurrentPoint 或 GetIntermediatePoints 从 PointerRoutedEventArgs 获得。PointerPoint 对象不仅包括指针位置，而且包括指针 ID(允许 InkManager 跟踪多个指针)和 PointerPointProperties，包括压力和倾斜。

以下为 SimpleInking 中的 OnPointerPressed 覆写。

```
项目: SimpleInking | 文件: MainPage.xaml.cs(片段)
protected override void OnPointerPressed(PointerRoutedEventArgs args)
{
    if (args.Pointer.PointerDeviceType == PointerDeviceType.Pen || !hasPen)
    {
        PointerPoint pointerPoint = args.GetCurrentPoint(this);
        inkManager.ProcessPointerDown(pointerPoint);
    }
    base.OnPointerPressed(args);
}
```

if 语句检查 Pen 的设备类型，但如果计算机不支持手写笔也允许其他指针设备。如果想支持所有指针输入设备，则可以删除整个 if 语句。在任何情况下，只要把 PointerPoint 对象传送给 InkManager 的 ProcessPointerDown 方法。处理 OnPointerMoved 稍微复杂一点。

```
项目: SimpleInking | 文件: MainPage.xaml.cs(片段)
protected override void OnPointerMoved(PointerRoutedEventArgs args)
{
    if ((args.Pointer.PointerDeviceType == PointerDeviceType.Pen || !hasPen) &&
        args.Pointer.IsInContact)
    {
        IEnumerable<PointerPoint> points = args.GetIntermediatePoints(this).Reverse();

        foreach (PointerPoint point in points)
            inkManager.ProcessPointerUpdate(point);
    }
    base.OnPointerMoved(args);
}
```

调用 ProcessPointerUpdate 允许 InkManager 积累所有墨迹笔画片段。为了尽可能保真

笔输入，代码使用了 GetIntermediatePoints，而没有使用 GetCurrentPoint，并采用 LINQ Reverse 操作符进行逆转。

注意，OnPointerMoved 的 if 语句包括对 IsInContact 属性的检查。你应该记得，在屏幕附近实际触碰到屏幕之前，手写笔就开始生成 OnPointerMoved 事件。如果 if 语句不检查 IsInContact，则调用 InkManager 的 ProcessPointerUpdate，再调用 ProcessPointerDown，而这会引发异常。

到目前为止，程序还没有进行任何绘画。任何合理的程序都应该能够顺利绘画。程序为 OnPointerReleased 方法保存所有绘画，我们先来看看 InkManager 的开销。

项目：SimpleInking | 文件：MainPage.xaml.cs(片段)

```
protected override void OnPointerReleased(PointerRoutedEventArgs args)
{
    if (args.Pointer.PointerDeviceType != PointerDeviceType.Pen && hasPen)
        return;

    inkManager.ProcessPointerUp(args.GetCurrentPoint(this));

    // Render the most recent InkStroke
    ...
    base.OnPointerReleased(args);
}
```

在渲染新的 InkStroke 对象之前，需要先了解 InkDrawingAttributes。

19.2 墨迹绘画属性

尽管 InkManager 自身并不进行渲染，但它维护着与渲染墨迹有关的属性信息。这些信息封装在 InkDrawingAttributes 类中，该类具有以下属性和默认值。

属 性	值
Color	Black (0xFF000000)
PenTip	Circle
Size	(2, 2)
FitToCurve	true
IgnorePressure	false

如果自己要创建新的 InkDrawingAttributes 实例，就可以用这些 InkManager 内部为 InkDrawingAttributes 创建并保留的默认值。

这些属性确实有助于需要渲染墨迹笔画的应用程序！它们不会影响 InkManager 的操作，因为 InkManager 自身不进行渲染。

PenTip 的唯一另一个选项是 Rectangle，本例中(Size 类型的)Size 属性描述笔尖大小。对于默认 Circle 笔尖，可以使用 Size 值的 Width 属性来决定渲染线条的宽度。

FitToCurve 属性表明墨迹是否应该渲染为贝塞尔曲线，而不管如何设置，InkManager 都会把指针输入转换为贝塞尔曲线，IgnorePressure 属性表明渲染墨迹无需考虑压力信息，但无论怎样设置，InkManager 都会包括压力信息。

InkManager 通过这些默认属性创建 InkDrawingAttributes 对象并在内部进行维护。然而，程序不能访问该对象。如果想给 InkManager 对象设置不同的默认属性，必须采用如下方式：

```
InkDrawingAttributes inkDrawingAttributes = new InkDrawingAttributes();
inkDrawingAttributes.Color = Colors.Red;
inkDrawingAttributes.Size = new Size(6, 6);
inkManager.SetDefaultDrawingAttributes(inkDrawingAttributes);
```

创建新的 InkDrawingAttributes 对象并传递给 InkManager，不要假设程序和 InkManager 共享该对象，应用对对象进行的任何更改都反映到 InkManager 维护的内部对象上。如果进一步更改 InkDrawingAttributes，必须再次调用 SetDefaultDrawingAttributes 方法，才能使更改生效。

正如你看到的，使用 InkManager 的程序通过调用 InkManager 方法 ProcessPointerDown、ProcessPointerUpdate(多次)和 ProcessPointerUp，来处理正常序列的 PointerPressed、多个 PointerMoved 和 PointerReleased 事件。该序列调用完成后，InkManager 创建新的 InkStroke 并将其添加到集合中。

InkStroke 对象代表从指针触碰屏幕开始到指针抬起为止的连续墨迹笔画。InkStroke 有 InkDrawingAttributes 类型的 DrawingAttributes 属性，DrawingAttributes 属性由 InkManager 基于其内部默认 InkDrawingAttributes 对象创建。

例如，假设要通过调用 ProcessPointerDown、ProcessPointerUpdate(多次)和 ProcessPointerUp 来创建新的 InkManager 并处理指针输入。得到的 InkStroke 对象有一个 InkDrawingAttributes 对象，表示黑色笔，大小为(2, 2)。现在，程序创建新的 InkDrawingAttributes 对象、设置两个属性并通过之前展示过的代码调用 SetDefaultDrawingAttributes。ProcessPointerDown、ProcessPointerUpdate 和 ProcessPointerUp 的下一个序列调用产生第二个 InkStroke 对象，但其 DrawingAttributes 属性表示红色笔，大小为(6, 6)。

但这并不是一成不变的。可以创建新的 InkDrawingAttributes 对象，设置给单个 InkStroke 对象，也可以对引用自 InkStroke 对象 DrawingAttributes 属性的现有 InkDrawingAttributes 对象，改变其任何属性值。

调用 ProcessPointerUp 方法后，InkManager 把内部积累的新笔画所有点转换为一条或多条贝塞尔曲线，构成新的 InkStroke 对象。通过 GetStrokes 方法，这个新的 InkStroke 添加到内部集合。

笔画完成后，SimpleInking 程序渲染每个笔画，因此只关注对集合里最近的 InkStroke，可以通过如下方法获得：

```
IReadOnlyList<InkStroke> inkStrokes = inkManager.GetStrokes();
InkStroke inkStroke = inkStrokes[inkStrokes.Count - 1];
```

该 InkStroke 有 DrawingAttributes 属性和 InkStrokeRenderingSegment 对象集合，代表一系列相连的贝塞尔曲线。通过调用 GetRenderingSegments，程序可以获得区段集合。

每个 InkStrokeRenderingSegment 包含三个 Point 类型属性：

- BezierControlPoint1
- BezierControlPoint2

- Position

集合中的第一个 InkStrokeRenderingSegment 对象中，三个点都相同。这是完整曲线的第一个点。每个后续 InkStrokeRenderingSegment 都从该点继续，同时有两个控制点和一个结束点。

此外，每个 InkStrokeRenderingSegment 还包含四个 float 类型属性：

- Pressure
- TiltX
- TiltY
- Twist

这显然适用于我一直在用的更漂亮的手写笔系统！我在三星 700 T 上使用手写笔时，看到 Pressure 值从 0 到 1，但其他三个属性的默认值为 0.5。理论上，TiltX 和 TiltY 属性表明笔杆相对于屏幕如何倾斜，Twist 属性仅适用于矩形笔尖，表明矩形笔尖相对于屏幕轴如何旋转。

本章一直都会考虑 Pressure 值。正如你所记得的，在 FingerPaint 程序中，如果忽略压力，Polyline 就适合渲染连接曲线，但考虑压力就需要单个 Line 元素或单个 Path 元素，而每个 Line 元素都可能有不同的线宽度模拟不同宽度的一条直线。

SimpleInking 中的代码把每个 InkStrokeRenderingSegment(第一个除外)绘制成带成 PathGeometry 的 Path 元素，并包含带单个 BezierSegment 的单个 PathFigure。Path 的 Stroke 属性为 SolidColorBrush，创建自 InkStroke 的 DrawingAttributes 属性的 Color 属性。StrokeThickness 属性是 InkStroke 的 DrawingAttributes 的 Size 属性及特定 InkStrokeRenderingSegment 的 Pressure 属性的产物。

项目：SimpleInking | 文件：MainPage.xaml.cs(片段)

```
protected override void OnPointerReleased(PointerRoutedEventArgs args)
{
    ...
    // Render the most recent InkStroke
    IReadOnlyList<InkStroke> inkStrokes = inkManager.GetStrokes();
    InkStroke inkStroke = inkStrokes[inkStrokes.Count - 1];

    // Create SolidColorBrush used for all segments in the stroke
    Brush brush = new SolidColorBrush(inkStroke.DrawingAttributes.Color);

    // Get the segments
    IReadOnlyList<InkStrokeRenderingSegment> inkSegments = inkStroke.GetRenderingSegments();

    // Notice loop starts at 1
    for  (int i = 1; i < inkSegments.Count; i++)
    {
        InkStrokeRenderingSegment inkSegment = inkSegments[i];

        // Create a BezierSegment from the points
        BezierSegment bezierSegment = new BezierSegment
        {
            Point1 = inkSegment.BezierControlPoint1,
            Point2 = inkSegment.BezierControlPoint2,
            Point3 = inkSegment.Position
        };

        // Create a PathFigure that begins at the preceding Position
        PathFigure pathFigure = new PathFigure
        {
```

```
                StartPoint = inkSegments[i - 1].Position,
                IsClosed = false,
                IsFilled = false
            };
            pathFigure.Segments.Add(bezierSegment);

            // Create a PathGeometry with that PathFigure
            PathGeometry pathGeometry = new PathGeometry();
            pathGeometry.Figures.Add(pathFigure);

            // Create a Path with that PathGeometry
            Path path = new Path
            {
                Stroke = brush,
                StrokeThickness = inkStroke.DrawingAttributes.Size.Width *
                                      inkSegment.Pressure,
                StrokeStartLineCap = PenLineCap.Round,
                StrokeEndLineCap = PenLineCap.Round,
                Data = pathGeometry
            };

            // Add it to the Grid
            contentGrid.Children.Add(path);
        }
        ...
    }
```

for 循环从 InkStrokeRenderingSegment 对象的集合中的 1 开始，因为第一个对象仅仅表示起点，而每个后续 InkStrokeRenderingSegment 均为单一贝塞尔曲线。在每个 PathGeometry 中，PathFigure 里的 StartPoint 均为之前 InkStrokeRenderingSegment 的 Position 属性。

尽管在抬起笔之前无法明白我画的是什么，我还是画出了如下图所示的消息。

InkManager 把折线转换为一系列贝塞尔曲线，不仅是为了渲染一条更平滑的线，而且是为了减少数据数量。本特定例子包括下表所示的五个 InkStroke 对象：两个用于 H，一个用于 Hello 的剩余部分，一个用于 P，最后一个用于 Pen 的剩余部分。你可能很想知道指定 InkManager 原始折线片段的数量以及其所创建的 InkStrokeRenderingSegment 数量。

笔 画	折线片段	贝塞尔片段
0	44	9
1	75	14
2	213	29
3	96	16
4	105	16

即使考虑到每个折线片段是一个点、每个贝塞尔曲线片段(第一段之后)是三个点，仍然会有效减少数据量。

如果想忽略压力，可以为整个 InkStroke 采用单个 Path 元素，并通过 PolyBezierSegment 单个实例来创建几何图形，以通过来自 InkStrokeRenderingSegment 的每个点填充 Points 集合，但第一个点仅用于设置 PathFigure 的 StartPoint 属性。这种替代方法如下所示。

```
// Render the most recent InkStroke
IReadOnlyList<InkStroke> inkStrokes = inkManager.GetStrokes();
InkStroke inkStroke = inkStrokes[inkStrokes.Count - 1];

// Create a PolyBezierSegment for all the segments in that stroke
IReadOnlyList<InkStrokeRenderingSegment> inkSegments = inkStroke.GetRenderingSegments();
PolyBezierSegment polyBezierSegment = new PolyBezierSegment();

for (int i = 1; i < inkSegments.Count; i++)
{
    InkStrokeRenderingSegment inkSegment = inkSegments[i];

    polyBezierSegment.Points.Add(inkSegment.BezierControlPoint1);
    polyBezierSegment.Points.Add(inkSegment.BezierControlPoint2);
    polyBezierSegment.Points.Add(inkSegment.Position);
}

// Create a PathFigure that begins at first point
PathFigure pathFigure = new PathFigure
{
    StartPoint = inkSegments[0].Position,
    IsClosed = false,
    IsFilled = false
};
pathFigure.Segments.Add(polyBezierSegment);

// Create a PathGeometry with that PathFigure
PathGeometry pathGeometry = new PathGeometry();
pathGeometry.Figures.Add(pathFigure);

// Create a Path with that PathGeometry
Path path = new Path
{
    Stroke = new SolidColorBrush(inkStroke.DrawingAttributes.Color),
    StrokeThickness = inkStroke.DrawingAttributes.Size.Width,
    StrokeStartLineCap = PenLineCap.Round,
    StrokeEndLineCap = PenLineCap.Round,
    Data = pathGeometry
};

// Add it to the Grid
contentGrid.Children.Add(path);
```

显然，InkManager 不会试图捕捉指针。这是你自己必须要做的事。然而，如果积累指针输入抬起笔的时候，一旦笔飘出控制范围，InkManager 就会优雅地进行恢复。如果没有全部调用 ProcessPointerUp，调用 ProcessPointerDown 就不会抛出异常。

19.3　擦除和其他增强功能

SimpleInking 程序有一项明显的增强功能，即在实际用笔绘画的同时渲染折线。(这实际上是最低标准，而不是增强！)如果忽略笔的压力，折线还算是一个实际的 Polyline 元素，如果不忽略，则折线是 Line 或 Path 元素的集合。但逻辑一旦实现，可以选择：笔画完成的时候，可以用贝塞尔曲线替换折线或者在屏幕上保留折线。

如果用 SimpleInking 程序尝试不同笔压力，则可能倾向于在屏幕上保留折线。如果仔细观察经渲染的贝塞尔曲线，你可能不喜欢看到的东西，我也不怪你。有时候容易看到一条贝塞尔曲线结束的地方以及下一条开始的地方，因为线的宽度不连续。在笔画开始和结束的地方特别明显。(在之前的屏幕截图中，看看单词“Pen”两个笔画结束的地方。)

一旦开始思考这个问题，这些贝塞尔曲线就不应该有统一的宽度。例如，如果一个笔画由四个 InkStrokeRenderingSegment 对象组成，压力值为 0.25、0.5、0.6、0.4，则第一条贝塞尔曲线应该有一个可变宽度，开始的宽度基于 0.25 并增加到基于 0.5 的宽度，第二条贝塞尔曲线的宽度应该基于从 0.5 到 0.6 增加，最后一条贝塞尔曲线的宽度应该基于 0.6 到 0.4 减少。

也就是说，不能只从 InkStrokeRenderingSegment 对象来设置 BezierSegment 属性。你可能想用贝塞尔曲线的点和压力值合成笔画的 outline，然后通过 Path 填补，就像我在 FingerPaint5 里处理线条那样。但很显然，这项工作对贝塞尔曲线比直线在算法上复杂得多。

还有一个问题。SimpleInking 程序在每一个新笔画完成时都进行渲染并把 Path 元素添加到名为 contentGrid 的 Grid 里。如果在画折线的同时创建折线，并且用 Bezier 曲线取代，则需要从 Grid 中删除早期元素。如果擦除，在某些时候则可能需要从 Grid 中删除一些贝塞尔曲线，但你不确定哪些需要挑出来，除非用某种方法对其进行标记。

这两个问题暗示非常容易在 OnPointerReleased 期间清理 Grid 的 Children 集合以及从一开始就渲染内容。如果擦除某些内容，通常就需要这么做，但不需要一直都这么做，尤其是定义单独的 Grid 元素用于渲染初始线条和最终的贝塞尔曲线的。

InkManager 有 InkManipulationMode 类型的 Mode 属性，有以下三个枚举项：

- Inking
- Erasing
- Selecting

显然，默认值为 Inking。为了能够擦除，可以在 OnPointerPressed 期间把属性设置为 OnPointerPressed，然后正常处理，但不做任何渲染。在随后调用 InkManager 的 ProcessPointerUpdate 方法中，只要笔的移动跨越了现有笔画，InkManager 就从其集合中移除该笔画。

虽然可以在 OnPointerReleased 期间重新渲染所有剩余的 InkStroke 对象，但在 OnPointerMoved 期间笔画实际就已经删除，因此不需要等待 OnPointerReleased 向用户反馈笔画删除。

新项目为 InkAndErase。为了简化移除在绘制新笔画过程中所创建的初始 Line 元素，

XAML 文件包含两个同级的 Grid 元素。

```
项目: InkAndErase | 文件: MainPage.xaml(片段)
<Page ... >

    <Grid Background="White">
        <Grid Name="contentGrid" />
        <Grid Name="newLineGrid" />
    </Grid>
</Page>
```

contentGrid 用于通过贝塞尔曲线渲染的所有笔画，newLineGrid 用于在笔画过程中所创建的 Line 元素。这种分离容易通过清理 newLineGrid 的 Children 集合来去掉 Line 部分。

代码隐藏文件创建 InkDrawingAttributes 对象，将它设置为非默认值(为了不同)的 InkManager，但把 InkDrawingAttributes 对象作为字段进行维护，有助于 OnPointerMoved 覆写中线条绘画代码。因为程序自身执行一些指针输入处理，因此会定义 Dictionary 用于存储每个指针的相关信息。

```
项目: InkAndErase | 文件: MainPage.xaml.cs(片段)
public sealed partial class MainPage : Page
{
    InkManager inkManager = new InkManager();
    InkDrawingAttributes inkDrawingAttributes = new InkDrawingAttributes();
    bool hasPen;

    Dictionary<uint, Point> pointerDictionary = new Dictionary<uint, Point>();

    public MainPage()
    {
        this.InitializeComponent();

        // Check if there's a pen among the pointer input devices
        foreach (PointerDevice device in PointerDevice.GetPointerDevices())
            hasPen |= device.PointerDeviceType == PointerDeviceType.Pen;

        // Default drawing attributes
        inkDrawingAttributes.Color = Colors.Blue;
        inkDrawingAttributes.Size = new Size(6, 6);
        inkManager.SetDefaultDrawingAttributes(inkDrawingAttributes);
    }
    ...
}
```

在 InkAndErase 中，贝塞尔曲线的渲染代码和之前的程序几乎相同，但我将其分离成两个方法，以允许整个墨迹集合可以重绘或可用于单个 InkStroke 对象的绘制。

```
项目: InkAndErase | 文件: MainPage.xaml.cs(片段)
public sealed partial class MainPage : Page
{
    ...
    void RenderAll()
    {
        contentGrid.Children.Clear();

        foreach (InkStroke inkStroke in inkManager.GetStrokes())
            RenderStroke(inkStroke);
    }

    void RenderStroke(InkStroke inkStroke)
    {
        Brush brush = new SolidColorBrush(inkStroke.DrawingAttributes.Color);
        IReadOnlyList<InkStrokeRenderingSegment> inkSegments = inkStroke.GetRenderingSegments();
```

```
            for (int i = 1; i < inkSegments.Count; i++)
            {
                InkStrokeRenderingSegment inkSegment = inkSegments[i];

                BezierSegment bezierSegment = new BezierSegment
                {
                    Point1 = inkSegment.BezierControlPoint1,
                    Point2 = inkSegment.BezierControlPoint2,
                    Point3 = inkSegment.Position
                };

                PathFigure pathFigure = new PathFigure
                {
                    StartPoint = inkSegments[i - 1].Position,
                    IsClosed = false,
                    IsFilled = false
                };
                pathFigure.Segments.Add(bezierSegment);

                PathGeometry pathGeometry = new PathGeometry();
                pathGeometry.Figures.Add(pathFigure);

                Path path = new Path
                {
                    Stroke = brush,
                    StrokeThickness = inkStroke.DrawingAttributes.Size.Width *
                                                inkSegment.Pressure,
                    StrokeStartLineCap = PenLineCap.Round,
                    StrokeEndLineCap = PenLineCap.Round,
                    Data = pathGeometry
                };
                contentGrid.Children.Add(path);
            }
        }
    }
```

除了与 InkManager 对象交互的代码外，Pointer 事件处理的大部分内容非常类似于第 13 章的压力敏感 FingerPaint4 程序。OnPointerPressed 覆写是程序检查指针输入设备是笔的唯一地方。随后的 Pointer 覆写通过 pointerDictionary 里的指针 ID 键来确定绘画操作是否正在进行中。

OnPointerPressed 覆写基于 IsEraser 属性将 InkManager 进入擦除模式，也就是说，用户用笔的橡皮擦端来触碰屏幕。真正的程序可能有应用栏按钮为没有使用高档笔的用户把 InkManager 放入擦除模式。

```
项目: InkAndErase | 文件: MainPage.xaml.cs(片段)
protected override void OnPointerPressed(PointerRoutedEventArgs args)
{
    if (args.Pointer.PointerDeviceType == PointerDeviceType.Pen || !hasPen)
    {
        // Get information
        PointerPoint pointerPoint = args.GetCurrentPoint(this);
        uint id = pointerPoint.PointerId;

        // Initialize for inking or erasing
        if (!pointerPoint.Properties.IsEraser)
        {
            inkManager.Mode = InkManipulationMode.Inking;
        }
        else
        {
            inkManager.Mode = InkManipulationMode.Erasing;
        }
```

```
            // Give PointerPoint to InkManager
            inkManager.ProcessPointerDown(pointerPoint);

            // Add an entry to the dictionary
            pointerDictionary.Add(args.Pointer.PointerId, pointerPoint.Position);

            // Capture the pointer
            CapturePointer(args.Pointer);
        }
        base.OnPointerPressed(args);
    }
```

OnPointerPressed 覆写以捕获指针结束。

OnPointerMoved 重写就像 FingerPaint4 程序一样创建和渲染 Line 元素，但只是不擦除。擦除的时候，检查 ProcessPointerUpdate 的返回值。如果笔画已经从集合中删除，则返回值为非空 Rect 对象，以显示必须重绘的屏幕区域。该方法通过重新渲染所有笔画集合进行响应，现在失去了已删除的笔画。

项目：InkAndErase | 文件：MainPage.xaml.cs(片段)

```
protected override void OnPointerMoved(PointerRoutedEventArgs args)
{
    // Get information
    PointerPoint pointerPoint = args.GetCurrentPoint(this);
    uint id = pointerPoint.PointerId;

    if (pointerDictionary.ContainsKey(id))
    {
        foreach (PointerPoint point in args.GetIntermediatePoints(this).Reverse())
        {
            // Give PointerPoint to InkManager
            object obj = inkManager.ProcessPointerUpdate(point);

            if (inkManager.Mode == InkManipulationMode.Erasing)
            {
                // See if something has actually been removed
                Rect rect = (Rect)obj;

                if (rect.Width != 0 && rect.Height != 0)
                {
                    RenderAll();
                }
            }
            else
            {
                // Render the line
                Point point1 = pointerDictionary[id];
                Point point2 = pointerPoint.Position;

                Line line = new Line
                {
                    X1 = point1.X,
                    Y1 = point1.Y,
                    X2 = point2.X,
                    Y2 = point2.Y,
                    Stroke = new SolidColorBrush(inkDrawingAttributes.Color),
                    StrokeThickness = inkDrawingAttributes.Size.Width *
                                  pointerPoint.Properties.Pressure,
                    StrokeStartLineCap = PenLineCap.Round,
                    StrokeEndLineCap = PenLineCap.Round
                };
                newLineGrid.Children.Add(line);
                pointerDictionary[id] = point2;
            }
        }
```

```
        }
        base.OnPointerMoved(args);
    }
```

注意，Line 元素放入 newLineGrid，但渲染贝塞尔曲线笔画涉及 contentGrid。

调用 OnPointerReleased 的时候，所有擦除应该都已经完成。然而，任何墨迹操作都需要通过渲染 contentGrid 中的新笔画并从 newLineGrid 移除初始的 Line 元素来完成。

```
项目: InkAndErase | 文件: MainPage.xaml.cs(片段)
protected override void OnPointerReleased(PointerRoutedEventArgs args)
{
    // Get information
    PointerPoint pointerPoint = args.GetCurrentPoint(this);
    uint id = pointerPoint.PointerId;

    if (pointerDictionary.ContainsKey(id))
    {
        // Give PointerPoint to InkManager
        inkManager.ProcessPointerUp(pointerPoint);

        if (inkManager.Mode == InkManipulationMode.Inking)
        {
            // Get rid of the little Line segments
            newLineGrid.Children.Clear();

            // Render the new stroke
            IReadOnlyList<InkStroke> inkStrokes = inkManager.GetStrokes();
            InkStroke inkStroke = inkStrokes[inkStrokes.Count - 1];
            RenderStroke(inkStroke);
        }
        pointerDictionary.Remove(id);
    }
    base.OnPointerReleased(args);
}
```

程序已经捕获了指针，因此应该有对 PointerCaptureLost 事件的处理程序。程序通过从 newLineGrid 删除初始线条以及重新渲染其余一切来处理 PointerCaptureLost 事件。

```
项目: InkAndErase | 文件: MainPage.xaml.cs(片段)
protected override void OnPointerCaptureLost(PointerRoutedEventArgs args)
{
    uint id = args.Pointer.PointerId;

    if (pointerDictionary.ContainsKey(id))
    {
        pointerDictionary.Remove(id);
        newLineGrid.Children.Clear();
        RenderAll();
    }
    base.OnPointerCaptureLost(args);
}
```

19.4　选择笔画

InkManipulationMode 的第三项是 Selecting。对于电磁笔，当按下笔管按键时，你想在 PointerPressed 事件期间让 InkManager 进入选择模式。下一个示例程序就要做这些，但真正的应用还应该有程序选项，使用户可以手动让 InkManager 进入选择模式。

在选择模式中，传递给 ProcessPointerUpdate 方法的点被视为定义封闭区域。你可能想渲染线条，但要用能区分墨迹的方式。该封闭线完成后，ProcessPointerUp 返回非空 Rect 值，表示所选笔画的边界矩形。如果没有选中笔画，则 Rect 为空。如果选中笔画，则集合中被选中的 InkStroke 对象把其 Selected 属性设置为 true。

在实际使用中，选择使用该封闭线我觉得有点古怪。我通常必须试几次才能生效。

也可以通过编程来选择笔画，即使用 InkManager 的 SelectWithLine 或 SelectWithPolyLine 方法，手动切换 InkStroke 的 Selected 属性，但我不展示这些方法。这些方法允许你实现自己选择的协议，而不依赖于 InkManager，而此时，在选择操作正在进行时，不会调用 InkManager 方法。

如果用户选择一个或多个 InkStroke 对象，就用某种方式进行高亮显示。还需要提供程序选项来处理这些选中项目。InkManager 类自身定义了名为 DeleteSelected、CopySelectedToClipboard 和 MoveSelected 的方法。最后一个方法把笔画从当前位置平移指定偏移量。还可以将笔画从剪贴板粘贴到 InkManager。

你可能还想定义应用栏控件来改变所选笔画的颜色或笔画宽度。没有选中笔画时，你可能想通过同一应用栏控件来设置默认颜色和笔画宽度。这里用不上 InkManager，但用户界面设计是用 InkManager 来完成的。

以下项目名为 InkEraseSelect，演示了所有三种模式。就像 InkAndErase 一样，InkEraseSelect 包含两个 Grid 元素用于渲染墨迹和初始线条。

项目: InkEraseSelect | 文件: MainPage.xaml(片段)

```
<Page ... >
    <Grid Background="White">
        <Grid Name="contentGrid" />
        <Grid Name="newLineGrid" />
    </Grid>

    <Page.BottomAppBar>
        <AppBar Name="bottomAppBar"
                Opened="OnAppBarOpened">
            <StackPanel Orientation="Horizontal"
                        HorizontalAlignment="Left">

                <Button Name="copyAppBarButton"
                        Style="{StaticResource CopyAppBarButtonStyle}"
                        Click="OnCopyAppBarButtonClick" />

                <Button Name="cutAppBarButton"
                        Style="{StaticResource CutAppBarButtonStyle}"
                        Click="OnCutAppBarButtonClick" />

                <Button Name="pasteAppBarButton"
                        Style="{StaticResource PasteAppBarButtonStyle}"
                        Click="OnPasteAppBarButtonClick" />

                <Button Name="deleteAppBarButton"
                        Style="{StaticResource DeleteAppBarButtonStyle}"
                        Click="OnDeleteAppBarButtonClick" />
            </StackPanel>
        </AppBar>
    </Page.BottomAppBar>
</Page>
```

XAML 文件还有一组应用程序栏按钮用于标准选项 Copy、Cut、Paste 和 Delete。

代码隐藏文件和之前的程序一样，只不过定义了用于对所选项目封闭线进行着色的 Brush。

```
项目: InkEraseSelect | 文件: MainPage.xaml.cs(片段)
public sealed partial class MainPage : Page
{
    InkManager inkManager = new InkManager();
    InkDrawingAttributes inkDrawingAttributes = new InkDrawingAttributes();
    bool hasPen;

    Dictionary<uint, Point> pointerDictionary = new Dictionary<uint, Point>();
    Brush selectionBrush = new SolidColorBrush(Colors.Red);

    public MainPage()
    {
        this.InitializeComponent();

        // Check if there's a pen among the pointer input devices
        foreach (PointerDevice device in PointerDevice.GetPointerDevices())
            hasPen |= device.PointerDeviceType == PointerDeviceType.Pen;

        // Default drawing attributes
        inkDrawingAttributes.Color = Colors.Blue;
        inkDrawingAttributes.Size = new Size(6, 6);
        inkManager.SetDefaultDrawingAttributes(inkDrawingAttributes);
    }
    ...
}
```

OnPointerPressed 覆写还会检查笔杆按键。如果按下按键，则设置选择模式。在真实的程序中，你要有选项能够用于为没有笔杆按键的设置选择模式。如果进入选择模式，程序就会画一个简单外壳，宽度均匀，因此为此目的创建 Polyline，并将其添加到 newLineGrid。

```
项目: InkEraseSelect | 文件: MainPage.xaml.cs(片段)
protected override void OnPointerPressed(PointerRoutedEventArgs args)
{
    if (args.Pointer.PointerDeviceType == PointerDeviceType.Pen || !hasPen)
    {
        // Get information
        PointerPoint pointerPoint = args.GetCurrentPoint(this);
        uint id = pointerPoint.PointerId;

        // Initialize for erasing, selecting, or inking
        if (pointerPoint.Properties.IsEraser)
        {
            inkManager.Mode = InkManipulationMode.Erasing;
        }
        else if (pointerPoint.Properties.IsBarrelButtonPressed)
        {
            inkManager.Mode = InkManipulationMode.Selecting;

            // Create Polyline for showing enclosure
            Polyline polyline = new Polyline
            {
                Stroke = selectionBrush,
                StrokeThickness = 1
            };
            polyline.Points.Add(pointerPoint.Position);
            newLineGrid.Children.Add(polyline);
        }
        else
        {
            inkManager.Mode = InkManipulationMode.Inking;
        }

        // Give PointerPoint to InkManager
        inkManager.ProcessPointerDown(pointerPoint);
```

```
            // Add an entry to the dictionary
            pointerDictionary.Add(args.Pointer.PointerId, pointerPoint.Position);

            // Capture the pointer
            CapturePointer(args.Pointer);
        }
        base.OnPointerPressed(args);
    }
```

在 OnPointerMoved 覆写中，擦除和墨迹模式和之前的程序一样。对于选择模式，Polyline 直接继续第 13 章中的 FingerPaint1 程序。

```
项目: InkEraseSelect | 文件: MainPage.xaml.cs(片段)
protected override void OnPointerMoved(PointerRoutedEventArgs args)
{
    // Get information
    PointerPoint pointerPoint = args.GetCurrentPoint(this);
    uint id = pointerPoint.PointerId;

    if (pointerDictionary.ContainsKey(id))
    {
        foreach (PointerPoint point in args.GetIntermediatePoints(this).Reverse())
        {
            Point point1 = pointerDictionary[id];
            Point point2 = pointerPoint.Position;

            // Give PointerPoint to InkManager
            object obj = inkManager.ProcessPointerUpdate(point);

            if (inkManager.Mode == InkManipulationMode.Erasing)
            {
                // See if something has actually been removed
                Rect rect = (Rect)obj;

                if (rect.Width != 0 && rect.Height != 0)
                {
                    RenderAll();
                }
            }
            else if (inkManager.Mode == InkManipulationMode.Selecting)
            {
                Polyline polyline = newLineGrid.Children[0] as Polyline;
                polyline.Points.Add(point2);
            }
            else // inkManager.Mode == InkManipulationMode.Inking
            {
                // Render the line
                Line line = new Line
                {
                    X1 = point1.X,
                    Y1 = point1.Y,
                    X2 = point2.X,
                    Y2 = point2.Y,
                    Stroke = new SolidColorBrush(inkDrawingAttributes.Color),
                        StrokeThickness = inkDrawingAttributes.Size.Width *
                                          pointerPoint.Properties.Pressure,
                        StrokeStartLineCap = PenLineCap.Round,
                        StrokeEndLineCap = PenLineCap.Round
                };
                newLineGrid.Children.Add(line);
            }
            pointerDictionary[id] = point2;
        }
    }
    base.OnPointerMoved(args);
}
```

当然，FingerPoint1 程序中可能有多个 Polyline 元素和触碰屏幕的多个手指相关，多个 Polyline 元素存储在字典中。虽然 InkManager 可以处理多个手指，但不能处理多支笔，而且由于在本程序中选择只限定用于一支笔，因此不可能存在多个 Polyline 元素。允许包括触控输入的替代选择方式的程序需要处理多个定义的同时选择区域！

对于选择模式，OnPointerReleased 覆写删除定义封闭的 Polyline 并调用 RenderAll。渲染逻辑负责不同渲染所选择的笔画。

```
项目: InkEraseSelect | 文件: MainPage.xaml.cs(片段)
public sealed partial class MainPage : Page
{
    ...
    protected override void OnPointerReleased(PointerRoutedEventArgs args)
    {
        // Get information
        PointerPoint pointerPoint = args.GetCurrentPoint(this);
        uint id = pointerPoint.PointerId;

        if (pointerDictionary.ContainsKey(id))
        {
            // Give PointerPoint to InkManager
            inkManager.ProcessPointerUp(pointerPoint);

            if (inkManager.Mode == InkManipulationMode.Inking)
            {
                // Get rid of the little line segments
                newLineGrid.Children.Clear();

                // Render the new stroke
                IReadOnlyList<InkStroke> inkStrokes = inkManager.GetStrokes();
                InkStroke inkStroke = inkStrokes[inkStrokes.Count - 1];
                RenderStroke(inkStroke);
            }
            else if (inkManager.Mode == InkManipulationMode.Selecting)
            {
                // Get rid of the encircling line
                newLineGrid.Children.Clear();

                // Render everything so selected items are identified
                RenderAll();
            }
            pointerDictionary.Remove(id);
        }
        base.OnPointerReleased(args);
    }

    protected override void OnPointerCaptureLost(PointerRoutedEventArgs args)
    {
        uint id = args.Pointer.PointerId;

        if (pointerDictionary.ContainsKey(id))
        {
            pointerDictionary.Remove(id);
            newLineGrid.Children.Clear();
            RenderAll();
        }
        base.OnPointerCaptureLost(args);
    }
    ...
}
```

选择封闭完成以及抬起笔之前看的样子，如下图所示，此时封闭线从屏幕上移除。

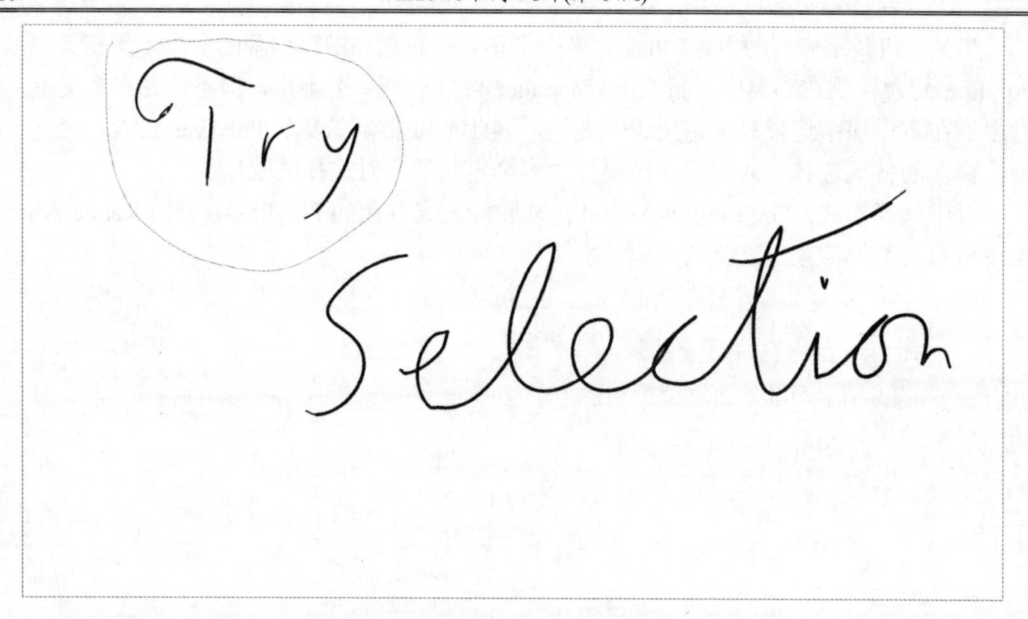

对于该程序，我把渲染逻辑分成三种方法。RenderStroke 方法调用 RenderBeziers，但对于所选笔画调用两次，第一次调用采用银色和更宽的笔画。

项目：InkEraseSelect | 文件：MainPage.xaml.cs(片段)

```
public sealed partial class MainPage : Page
{
    ...
    void RenderAll()
    {
        contentGrid.Children.Clear();

        foreach (InkStroke inkStroke in inkManager.GetStrokes())
            RenderStroke(inkStroke);
    }

    public void RenderStroke(InkStroke inkStroke)
    {
        Color color = inkStroke.DrawingAttributes.Color;
        double penSize = inkStroke.DrawingAttributes.Size.Width;

        if (inkStroke.Selected)
            RenderBeziers(contentGrid, inkStroke, Colors.Silver, penSize + 24);

        RenderBeziers(contentGrid, inkStroke, color, penSize);
    }

    static void RenderBeziers(Panel panel, InkStroke inkStroke, Color color, double penSize)
    {
        Brush brush = new SolidColorBrush(color);
        IReadOnlyList<InkStrokeRenderingSegment> inkSegments = inkStroke.GetRenderingSegments();

        for (int i = 1; i < inkSegments.Count; i++)
        {
            InkStrokeRenderingSegment inkSegment = inkSegments[i];

            BezierSegment bezierSegment = new BezierSegment
            {
                Point1 = inkSegment.BezierControlPoint1,
                Point2 = inkSegment.BezierControlPoint2,
                Point3 = inkSegment.Position
            };
```

```
                PathFigure pathFigure = new PathFigure
                {
                    StartPoint = inkSegments[i - 1].Position,
                    IsClosed = false,
                    IsFilled = false
                };
                pathFigure.Segments.Add(bezierSegment);

                PathGeometry pathGeometry = new PathGeometry();
                pathGeometry.Figures.Add(pathFigure);

                Path path = new Path
                {
                    Stroke = brush,
                    StrokeThickness = penSize * inkSegment.Pressure,
                    StrokeStartLineCap = PenLineCap.Round,
                    StrokeEndLineCap = PenLineCap.Round,
                    Data = pathGeometry
                };
                panel.Children.Add(path);
            }
        }
        ...
    }
```

我之所以把 RenderBeziers 设置为静态，只是想准确演示该方法需要什么参数来渲染单个笔画。

通过这种方法来识别的所选笔画，如下图所示。

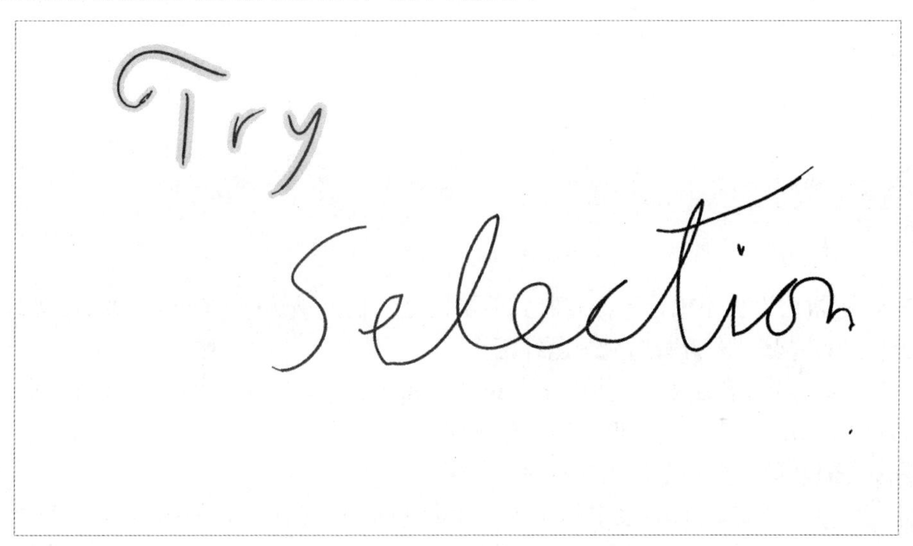

这当然是显示所选笔画的一种方法，但也可以考虑其他方法。

如果打开应用栏，Opened 处理程序会启用或禁用四个按钮。其中三个按钮的启用基于 InkManager 中所选笔画的出现；Paste 按钮的启用基于 InkManager 的 CanPasteFromClipboard 属性。

```
项目: InkEraseSelect | 文件: MainPage.xaml.cs(片段)
public sealed partial class MainPage : Page
{
    ...
    void OnAppBarOpened(object sender, object args)
    {
        bool isAnythingSelected = false;
```

```
        foreach (InkStroke inkStroke in inkManager.GetStrokes())
            isAnythingSelected |= inkStroke.Selected;
            copyAppBarButton.IsEnabled = isAnythingSelected;
            cutAppBarButton.IsEnabled = isAnythingSelected;
            pasteAppBarButton.IsEnabled = inkManager.CanPasteFromClipboard();
            deleteAppBarButton.IsEnabled = isAnythingSelected;
    }

    void OnCopyAppBarButtonClick(object sender, RoutedEventArgs args)
    {
        inkManager.CopySelectedToClipboard();

        foreach (InkStroke inkStroke in inkManager.GetStrokes())
            inkStroke.Selected = false;
        RenderAll();
        bottomAppBar.IsOpen = false;
    }

    void OnCutAppBarButtonClick(object sender, RoutedEventArgs args)
    {
        inkManager.CopySelectedToClipboard();
        inkManager.DeleteSelected();
        RenderAll();
        bottomAppBar.IsOpen = false;
    }

    void OnPasteAppBarButtonClick(object sender, RoutedEventArgs args)
    {
        inkManager.PasteFromClipboard(new Point());
        RenderAll();
        bottomAppBar.IsOpen = false;
    }

    void OnDeleteAppBarButtonClick(object sender, RoutedEventArgs args)
    {
        inkManager.DeleteSelected();
        RenderAll();
        bottomAppBar.IsOpen = false;
    }
}
```

Copy 逻辑假设你不想让笔画复制到剪贴板之后仍然被选中。Cut 和 Delete 处理程序不需要处理类似情况，因为所选笔画已经消失。

有趣的是，如果复制墨迹到剪贴板，InkManager 还会将墨迹转换为位图和增强元文件，因此，这些剪贴板格式对粘贴也可用。有些程序(尤其是 Microsoft Word)可以直接从剪贴板读取墨迹，但从剪贴板粘贴位图的程序更常见。

所有复制到剪贴板的墨迹中的坐标标准化为最小渲染坐标为(0，0)，这就是为什么 PasteFromClipboard 方法需要 Point 参数的原因。如果没有指定 Point(就是我采用的方法)，所粘贴的墨迹会出现在左上角。而实现粘贴功能的真正程序，需要给用户某种方式来指定在页面的什么地方粘贴墨迹。类似逻辑也可以用于实现 InkManager 所支持的 MoveSelected 方法。

19.5 黄色拍纸簿

我写本书的时候，用了很多窄横隔线的黄色标准拍纸簿。这是我最喜欢的媒介，可以用来做笔记、写想法、画出代码的相互关系以及解决数学问题。我不确定会为写书转而使用电子黄色拍纸簿，但我给电子黄色拍纸簿一个公平机会，我要写一个应用来试试。

YellowPad 程序非商业级，但确实比本章目前为止出现的程序有更多的功能。

YellowPad 支持多个页面在 FlipView 控件可见。因此，可以通过手指扫屏进行翻页。程序确保 FlipView 永远不会到达集合末尾，而如果达到末尾，总是创建一个新页面。

YellowPad 还会演示 InkManager 所定义的 LoadAsync 和 SaveAsync 方法，在 Suspending 事件期间把所有页面内容保存在本地应用存储，并在下次程序运行时进行加载。

YellowPad 有你刚刚看到的应用栏条目并加入条目，可以设置笔宽度并对当前页或所选笔画进行着色。

为了从用户界面加强分离这种逻辑，我定义了一个名为 InkFileManager 的类。如果 InkManager 不封闭，则 InkFileManager 派生自 InkManager，而 InkFileManager 会实例化 InkManager 以及默认 InkDrawingAttributes 对象并公开为公共属性。InkFileManager 还包括一个方法，用新的绘图属性值来更新 InkManager。

项目：YellowPad | 文件：InkFileManager.cs(片段)

```
public class InkFileManager
{
    string id;
    ...

    public InkFileManager(string id)
    {
        this.id = id;
        this.InkManager = new InkManager();
        this.InkDrawingAttributes = new InkDrawingAttributes();
    }

    public InkManager InkManager
    {
        private set;
        get;
    }

    public InkDrawingAttributes InkDrawingAttributes
    {
        private set;
        get;
    }

    ...

    public void UpdateAttributes()
    {
        this.InkManager.SetDefaultDrawingAttributes(this.InkDrawingAttributes);
    }
    ...
}
```

InkFileManager 还包含和选择有关的几个常规任务。

项目：YellowPad | 文件：InkFileManager.cs(片段)

```
public class InkFileManager
{
    ...
    public bool IsAnythingSelected
    {
        get
        {
            bool isAnythingSelected = false;
```

```
            foreach (InkStroke inkStroke in this.InkManager.GetStrokes())
                isAnythingSelected |= inkStroke.Selected;

            return isAnythingSelected;
        }
    }

    public void UnselectAll()
    {
        if (IsAnythingSelected)
        {
            foreach (InkStroke inkStroke in this.InkManager.GetStrokes())
                inkStroke.Selected = false;

            RenderAll();
        }
    }
    ...
}
```

我还把所有贝塞尔曲线渲染逻辑都移到该文件中。除了 InkManager 对象本身，渲染逻辑唯一需要的就是用于添加 Path 元素的 Panel。该重要信息通过一个名为 RenderTarget 的公共属性来提供。所选笔画的渲染和之前程序的一样。

项目：YellowPad | 文件：InkFileManager.cs(片段)

```
public class InkFileManager
{
    ...
    public Panel RenderTarget
    {
        set;
        get;
    }
    ...

    public void RenderAll()
    {
        this.RenderTarget.Children.Clear();

        foreach (InkStroke inkStroke in this.InkManager.GetStrokes())
            RenderStroke(inkStroke);
    }

    public void RenderStroke(InkStroke inkStroke)
    {
        Color color = inkStroke.DrawingAttributes.Color;
        double penSize = inkStroke.DrawingAttributes.Size.Width;

        if (inkStroke.Selected)
            RenderBeziers(this.RenderTarget, inkStroke, Colors.Silver, penSize + 24);

        RenderBeziers(this.RenderTarget, inkStroke, color, penSize);
    }

    static void RenderBeziers(Panel panel, InkStroke inkStroke, Color color, double penSize)
    {
        Brush brush = new SolidColorBrush(color);
        IReadOnlyList<InkStrokeRenderingSegment> inkSegments = inkStroke.GetRenderingSegments();

        for (int i = 1; i < inkSegments.Count; i++)
        {
            InkStrokeRenderingSegment inkSegment = inkSegments[i];
            BezierSegment bezierSegment = new BezierSegment
            {
                Point1 = inkSegment.BezierControlPoint1,
                Point2 = inkSegment.BezierControlPoint2,
```

```
                    Point3 = inkSegment.Position
                };

                PathFigure pathFigure = new PathFigure
                {
                    StartPoint = inkSegments[i - 1].Position,
                    IsClosed = false,
                    IsFilled = false
                };
                pathFigure.Segments.Add(bezierSegment);
                    PathGeometry pathGeometry = new PathGeometry();
                    pathGeometry.Figures.Add(pathFigure);

                Path path = new Path
                {
                    Stroke = brush,
                    StrokeThickness = penSize * inkSegment.Pressure,
                    StrokeStartLineCap = PenLineCap.Round,
                    StrokeEndLineCap = PenLineCap.Round,
                    Data = pathGeometry
                };
                panel.Children.Add(path);
            }
        }
        ...
    }
```

最后，InkFileManager 有两个公共方法，帮助证明其名称中的“文件”(File)部分。LoadAsync 方法加载之前保存的墨迹和设置或如果是新创建页面，则设置默认值。SaveAsync 方法把 InkManager 的当前内容以及与 InkManager 相关的笔宽度和颜色保存到本地应用。两者都利用原始传递给构造函数并保存为 ID 字符串。该 ID 字符串对程序维护的每个 InkFileManager 对象均为唯一。正如你将看到的，仅仅为转为字符串的索引(0、1、2 等)。

```
项目: YellowPad | 文件: InkFileManager.cs(片段)
public class InkFileManager
{
    ...
    bool isLoaded;
    ...

    public async Task LoadAsync()
    {
        if (isLoaded)
            return;

        // Load previously saved ink
        StorageFolder storageFolder = ApplicationData.Current.LocalFolder;

        try
        {
            StorageFile storageFile =
                    await storageFolder.GetFileAsync("Page" + id + ".ink");

            using (IRandomAccessStream stream =
                        await storageFile.OpenAsync(FileAccessMode.Read))
            {
                await this.InkManager.LoadAsync(stream.GetInputStreamAt(0));
            }
        }
        catch
        {
            // Do nothing if an exception occurs
        }
```

```
        // Load saved settings
        IPropertySet appData = ApplicationData.Current.LocalSettings.Values;

        // Pen size setting
        double penSize = 4;

        if (appData.ContainsKey("PenSize" + id))
            penSize = (double)appData["PenSize" + id];

        this.InkDrawingAttributes.Size = new Size(penSize, penSize);

        // Color setting
        if (appData.ContainsKey("Color + id"))
        {
            byte[] argb = (byte[])appData["Color + id"];
            this.InkDrawingAttributes.Color =
                Color.FromArgb(argb[0], argb[1], argb[2], argb[3]);
        }

        // Set default drawing attributes
        UpdateAttributes();
        isLoaded = true;
    }

    public async Task SaveAsync()
    {
        if (!isLoaded)
            return;

        // Save the ink
        StorageFolder storageFolder = ApplicationData.Current.LocalFolder;

        try
        {
            StorageFile storageFile =
                    await storageFolder.CreateFileAsync("Page" + id + ".ink",
                                    CreationCollisionOption.ReplaceExisting);

            using (IRandomAccessStream stream =
                        await storageFile.OpenAsync(FileAccessMode.ReadWrite))
            {
                await this.InkManager.SaveAsync(stream.GetOutputStreamAt(0));
            }
        }
        catch
        {
            // Do nothing if an exception occurs
        }

        // Save settings
        IPropertySet appData = ApplicationData.Current.LocalSettings.Values;

        // Save pen size
        appData["PenSize" + id] = this.InkDrawingAttributes.Size.Width;

        // Save color
        Color color = this.InkDrawingAttributes.Color;
        byte[] argb = { color.A, color.R, color.G, color.B };
        appData["Color" + id] = argb;
    }
}
```

在 YellowPad 程序中，每个 InkFileManager 都和一个名为 YellowPadPage 的 UserControl 派生类相关。以下是该类的 XAML 文件，包括在视觉上用黄色背景模仿标准黄色拍纸簿以及朝向页面左边的两条红色竖线。

项目：YellowPad | 文件：YellowPadPage.xaml

```
<UserControl
    x:Class="YellowPad.YellowPadPage"
    xmlns="http://schemas.microsoft.com/winfx/2006/xaml/presentation"
    xmlns:x="http://schemas.microsoft.com/winfx/2006/xaml"
    xmlns:local="using:YellowPad">

    <Grid>
        <Viewbox>
            <Grid Name="sheetPanel"
                  Width="816" Height="1056"
                  Background="#FFFF80">

                <Line Stroke="Red" X1="132" Y1="0" X2="132" Y2="1056" />
                <Line Stroke="Red" X1="138" Y1="0" X2="138" Y2="1056" />

                <Grid Name="contentGrid" />
                <Grid Name="newLineGrid" />
            </Grid>
        </Viewbox>
    </Grid>
</UserControl>
```

控件集成了一个 Viewbox，可以适应任何大小的窗口。

从两个内部 Grid 元素的名称可以猜到，代码隐藏文件负责处理所有指针输入。然而，我发现在一台没有笔的设备上运行该程序是有问题的。记住，YellowPadPage 实例在 FlipView 里，而 FlipView 想用其自己的触控输入来改变所选项。我决定去除允许在没有笔的情况下运行程序的逻辑。结果，YellowPad 程序还是认为是在使用的真正的笔。

YellowPadPage 构造函数负责在页面上绘制蓝色分隔线。

项目：YellowPad | 文件：YellowPadPage.xaml.cs(片段)

```
public YellowPadPage()
{
    this.InitializeComponent();

    // Draw horizontal lines in blue
    Brush blueBrush = new SolidColorBrush(Colors.Blue);

    For  (int y = 120; y < sheetPanel.Height; y += 24)
        sheetPanel.Children.Add(new Line
        {
            X1 = 0,
            Y1 = y,
            X2 = sheetPanel.Width,
            Y2 = y,
            Stroke = blueBrush
        });
}
```

YellowPadPage 控件还定义了一个新的 InkFileManager 类型依赖属性。以下为系统开销。

项目：YellowPad | 文件：YellowPadPage.xaml.cs(片段)

```
public sealed partial class YellowPadPage : UserControl
{
    static readonly DependencyProperty inkFileManagerProperty =
        DependencyProperty.Register("InkFileManager",
                                    typeof(InkFileManager),
                                    typeof(YellowPadPage),
                                    new PropertyMetadata(null, OnInkFileManagerChanged));
    ...
```

```
    // Overhead for InkFileManager dependency property
    public static DependencyProperty InkFileManagerProperty
    {
        get { return inkFileManagerProperty; }
    }

    public InkFileManager InkFileManager
    {
        set { SetValue(InkFileManagerProperty, value); }
        get { return (InkFileManager)GetValue(InkFileManagerProperty); }
    }

    static void OnInkFileManagerChanged(DependencyObject obj,
                                        DependencyPropertyChangedEventArgs args)
    {
        (obj as YellowPadPage).OnInkFileManagerChanged(args);
    }

    async void OnInkFileManagerChanged(DependencyPropertyChangedEventArgs args)
    {
        contentGrid.Children.Clear();
        newLineGrid.Children.Clear();

        if (args.NewValue != null)
        {
            await this.InkFileManager.LoadAsync();
            this.InkFileManager.RenderTarget = contentGrid;
            this.InkFileManager.RenderAll();
        }
    }
    ...
}
```

如果把该 InkFileManager 属性设置为一个新的 InkFileManager 实例，属性更改处理程序就调用 LoadAsync 加载任何现有的墨迹和设置，把 RenderTarget 设置为自身的 contentGrid，然后将 InkFileManager 渲染之前就有的所有墨迹。

YellowPadPage 的其余部分假定该 InkFileManager 属性已经设置并致力于处理 Pointer 事件。除了通过 InkFileManager 属性获得 InkManager 和与该页面相关的 InkDrawingAttributes 对象以及渲染笔画，逻辑和之前看到的程序几乎相同。

项目：YellowPad | 文件：YellowPadPage.xaml.cs(片段)

```
public sealed partial class YellowPadPage : UserControl
{
    ...
    Dictionary<uint, Point> pointerDictionary = new Dictionary<uint, Point>();
    Brush selectionBrush = new SolidColorBrush(Colors.Red);
    ...
    protected override void OnPointerPressed(PointerRoutedEventArgs args)
    {
        if (args.Pointer.PointerDeviceType == PointerDeviceType.Pen)
        {
            // Get information
            PointerPoint pointerPoint = args.GetCurrentPoint(sheetPanel);
            uint id = pointerPoint.PointerId;
            InkManager inkManager = this.InkFileManager.InkManager;

            // Initialize for inking, erasing, or selecting
            if (pointerPoint.Properties.IsEraser)
            {
                inkManager.Mode = InkManipulationMode.Erasing;
                this.InkFileManager.UnselectAll();
            }
            else if (pointerPoint.Properties.IsBarrelButtonPressed)
            {
                inkManager.Mode = InkManipulationMode.Selecting;
```

```
                // Create Polyline for showing enclosure
                Polyline polyline = new Polyline
                {
                    Stroke = selectionBrush,
                    StrokeThickness = 1
                };
                polyline.Points.Add(pointerPoint.Position);
                newLineGrid.Children.Add(polyline);
            }
            else
            {
                inkManager.Mode = InkManipulationMode.Inking;
                this.InkFileManager.UnselectAll();
            }

            // Give PointerPoint to InkManager
            inkManager.ProcessPointerDown(pointerPoint);

            // Add an entry to the dictionary
            pointerDictionary.Add(args.Pointer.PointerId, pointerPoint.Position);

            // Capture the pointer
            this.CapturePointer(args.Pointer);
        }
        base.OnPointerPressed(args);
    }

    protected override void OnPointerMoved(PointerRoutedEventArgs args)
    {
        // Get information
        PointerPoint pointerPoint = args.GetCurrentPoint(sheetPanel);
        uint id = pointerPoint.PointerId;
        InkManager inkManager = this.InkFileManager.InkManager;
        InkDrawingAttributes inkDrawingAttributes =
                            this.InkFileManager.InkDrawingAttributes;

        if (pointerDictionary.ContainsKey(id))
        {
            foreach (PointerPoint point in args.GetIntermediatePoints(sheetPanel).Reverse())
            {
                Point point1 = pointerDictionary[id];
                Point point2 = pointerPoint.Position;

                // Give PointerPoint to InkManager
                object obj = inkManager.ProcessPointerUpdate(point);

                if (inkManager.Mode == InkManipulationMode.Erasing)
                {
                    // See if something has actually been removed
                    Rect rect = (Rect)obj;

                    if (rect.Width != 0 && rect.Height != 0)
                    {
                        this.InkFileManager.RenderAll();
                    }
                }
                else if (inkManager.Mode == InkManipulationMode.Selecting)
                {
                    Polyline polyline = newLineGrid.Children[0] as Polyline;
                    polyline.Points.Add(point2);
                }
                else // inkManager.Mode == InkManipulationMode.Inking
                {
                    // Render the line
                    Line line = new Line
                    {
                        X1 = point1.X,
                        Y1 = point1.Y,
                        X2 = point2.X,
```

```
                                  Y2 = point2.Y,
                                  Stroke = new SolidColorBrush(inkDrawingAttributes.Color),
                                  StrokeThickness = inkDrawingAttributes.Size.Width *
                                                    pointerPoint.Properties.Pressure,
                                  StrokeStartLineCap = PenLineCap.Round,
                                  StrokeEndLineCap = PenLineCap.Round
                              };
                              newLineGrid.Children.Add(line);
                          }
                          pointerDictionary[id] = point2;
                      }
                  }
                  base.OnPointerMoved(args);
              }

              protected override void OnPointerReleased(PointerRoutedEventArgs args)
              {
                  // Get information
                  PointerPoint pointerPoint = args.GetCurrentPoint(sheetPanel);
                  uint id = pointerPoint.PointerId;
                  InkManager inkManager = this.InkFileManager.InkManager;

                  if (pointerDictionary.ContainsKey(id))
                  {
                      // Give PointerPoint to InkManager
                      inkManager.ProcessPointerUp(pointerPoint);

                      if (inkManager.Mode == InkManipulationMode.Inking)
                      {
                          // Get rid of the little line segments
                          newLineGrid.Children.Clear();

                          // Render the new stroke
                          IReadOnlyList<InkStroke> inkStrokes = inkManager.GetStrokes();
                          InkStroke inkStroke = inkStrokes[inkStrokes.Count - 1];
                          this.InkFileManager.RenderStroke(inkStroke);
                      }
                      else if (inkManager.Mode == InkManipulationMode.Selecting)
                      {
                          // Get rid of the enclosure line
                          newLineGrid.Children.Clear();

                          // Render everything so selected items are identified
                          this.InkFileManager.RenderAll();
                      }
                      pointerDictionary.Remove(id);
                  }
                  base.OnPointerReleased(args);
              }

              protected override void OnPointerCaptureLost(PointerRoutedEventArgs args)
              {
                  uint id = args.Pointer.PointerId;

                  if (pointerDictionary.ContainsKey(id))
                  {
                      pointerDictionary.Remove(id);
                      newLineGrid.Children.Clear();
                      this.InkFileManager.RenderAll();
                  }
                  base.OnPointerCaptureLost(args);
              }
          }
```

YellowPadPage 通过数据绑定获得 InkFileManager 实例。MainPage 中的 FlipView 控件包 InkFileManager 对象集合(每个页面一个)FlipView 的 ItemTemplate 受绑定到控件的 ItemsSource 条目的 YellowPadPage 支配(尽管不是用标记来显示)。

```
项目: YellowPad | 文件: MainPage.xaml(片段)
<Page ... >
    <Page.Resources>
        <local:IndexToPageNumberConverter x:Key="indexToPageNumber" />
    </Page.Resources>

    <Grid Background="{StaticResource ApplicationPageBackgroundThemeBrush}">
        <FlipView Name="flipView"
                  SelectionChanged="OnFlipViewSelectionChanged">
            <FlipView.ItemTemplate>
                <DataTemplate>
                    <Grid HorizontalAlignment="Center"
                          VerticalAlignment="Center">

                        <local:YellowPadPage InkFileManager="{Binding}" />

                        <TextBlock Name="pageNumTextBlock"
                                   HorizontalAlignment="Right"
                                   VerticalAlignment="Top"
                                   FontSize="12"
                                   Foreground="Black"
                                   Margin="6"
                                   Text="{Binding ElementName=flipView,
                                                  Path=SelectedIndex,
                                   Converter={StaticResource indexToPageNumber}}" />
                    </Grid>
                </DataTemplate>
            </FlipView.ItemTemplate>
        </FlipView>
    </Grid>
    <Page.BottomAppBar>
    ...
    </Page.BottomAppBar>
</Page>
```

DataTemplate 中定义的 TextBlock 连同 YellowPadPage 显示当前页码。Text 属性的绑定引用一个特定绑定转换器，后者把从零开始的索引转换为文本标签。

```
项目: YellowPad | 文件: IndexToPageNumberConverter.cs
using System;
using Windows.UI.Xaml.Data;

namespace YellowPad
{
    public class IndexToPageNumberConverter : IValueConverter
    {
        public object Convert(object value, Type targetType, object parameter, string language)
        {
            return String.Format("Page {0}", (int)value + 1);
        }

        public object ConvertBack(object value, Type targetType, object parameter, string lang)
        {
            return value;
        }
    }
}
```

正如你看到的，每个 InkFileManager 实例保存并恢复与该页面相关的应用设置，包括页面的墨迹内容。MainPage 代码保存并恢复与应用自身的相关的设置。两个整数项：页面数量(即 InkFileManager 对象集合的项目数量)和当前页面索引(即 FlipView 的 SelectedIndex 属性)。

```
项目: YellowPad | 文件: MainPage.xaml.cs(片段)
public sealed partial class MainPage : Page
{
```

```
    ObservableCollection<InkFileManager> inkFileManagers =
                                new ObservableCollection<InkFileManager>();
    public MainPage()
    {
        this.InitializeComponent();
        Loaded += OnMainPageLoaded;
        Application.Current.Suspending += OnApplicationSuspending;
    }

    void OnMainPageLoaded(object sender, RoutedEventArgs args)
    {
        // Load application settings
        IPropertySet appData = ApplicationData.Current.LocalSettings.Values;

        // Get the page count
        int pageCount = 1;

        if (appData.ContainsKey("PageCount"))
            pageCount = (int)appData["PageCount"];

        // Create that many InkFileManager objects
        for (int i = 0; i < pageCount; i++)
            inkFileManagers.Add(new InkFileManager(i.ToString()));

        // Set the collection to the FlipView
        flipView.ItemsSource = inkFileManagers;

        // Set the SelectedIndex of the PageView
        if (appData.ContainsKey("PageIndex"))
            flipView.SelectedIndex = (int)appData["PageIndex"];
    }

    async void OnApplicationSuspending(object sender, SuspendingEventArgs args)
    {
        SuspendingDeferral deferral = args.SuspendingOperation.GetDeferral();

        // Save all the InkFileManager contents
        foreach (InkFileManager inkFileManager in inkFileManagers)
            await inkFileManager.SaveAsync();

        // Save the page count and current page index
        IPropertySet appData = ApplicationData.Current.LocalSettings.Values;
        appData["PageCount"] = inkFileManagers.Count;
        appData["PageIndex"] = flipView.SelectedIndex;

        deferral.Complete();
    }

    void OnFlipViewSelectionChanged(object sender, SelectionChangedEventArgs args)
    {
        // If at the end of the FlipView, add another item!
        if (flipView.SelectedIndex == flipView.Items.Count - 1)
            inkFileManagers.Add(new InkFileManager(flipView.Items.Count.ToString()));
    }
    ...
}
```

注意，Loaded 处理程序为当前页面数量创建所有 InkFileManager 对象，但 InkFileManager 构造函数只创建 InkManager 和 InkDrawingAttributes 实例。没有加载任何之前保存的墨迹。InkFileManager 实例实际绑定到 YellowPadPage 才加载任何之前保存的墨迹。记住，FlipView 的条目面板为 VirtualizingStackPanel，只为其需要的条目创建可视树。也就是说，加载之前保存的墨迹需要延伸一段较长时间，并发生在用户浏览不同页面期间。一些页面可能根本不加载，而不加载的页面则不需要再保存。

程序的剩余部分致力于处理应用栏上的按钮，包括你看过的有缺陷的 Paste 逻辑。除了与四个剪贴板相关的按钮之外，应用栏还包括两个非常相似的模板化 ComboBox 控件，一个用于笔的宽度，另一个用于笔的颜色。

```
项目: YellowPad | 文件: MainPage.xaml(片段)
<AppBar Name="bottomAppBar"
    Opened="OnAppBarOpened">
<Grid>
    <StackPanel Orientation="Horizontal"
                HorizontalAlignment="Left">

        <Button Name="copyAppBarButton"
                Style="{StaticResource CopyAppBarButtonStyle}"
                Click="OnCopyAppBarButtonClick" />

            <Button Name="cutAppBarButton"
                    Style="{StaticResource CutAppBarButtonStyle}"
                    Click="OnCutAppBarButtonClick" />

            <Button Name="pasteAppBarButton"
                    Style="{StaticResource PasteAppBarButtonStyle}"
                    Click="OnPasteAppBarButtonClick" />

            <Button Name="deleteAppBarButton"
                    Style="{StaticResource DeleteAppBarButtonStyle}"
                    Click="OnDeleteAppBarButtonClick" />
        </StackPanel>

        <StackPanel Orientation="Horizontal"
                    HorizontalAlignment="Right">
            <ComboBox Name="penSizeComboBox"
                      SelectionChanged="OnPenSizeComboBoxSelectionChanged"
                      Width="200"
                      Margin="20 0">
                <x:Double>2</x:Double>
                <x:Double>3</x:Double>
                <x:Double>4</x:Double>
                <x:Double>5</x:Double>
                <x:Double>7</x:Double>
                <x:Double>10</x:Double>

                <ComboBox.ItemTemplate>
                    <DataTemplate>
                        <Path StrokeThickness="{Binding}"
                              Stroke="Black"
                              StrokeStartLineCap="Round"
                              StrokeEndLineCap="Round"
                              Data="M 0 0 C 50 20 100 0 150 20" />
                    </DataTemplate>
                </ComboBox.ItemTemplate>
            </ComboBox>

            <ComboBox Name="colorComboBox"
                      SelectionChanged="OnColorComboBoxSelectionChanged"
                      Width="200"
                      Margin="20 0">
                <Color>#FF0000</Color>
                <Color>#800000</Color>
                <Color>#FFFF00</Color>
                <Color>#808000</Color>
                <Color>#00FF00</Color>
                <Color>#008000</Color>
                <Color>#00FFFF</Color>
                <Color>#008080</Color>
                <Color>#0000FF</Color>
                <Color>#000080</Color>
```

```
                <Color>#FF00FF</Color>
                <Color>#800080</Color>
                <Color>#C0C0C0</Color>
                <Color>#808080</Color>
                <Color>#404040</Color>
                <Color>#000000</Color>

                <ComboBox.ItemTemplate>
                    <DataTemplate>
                        <Path StrokeThickness="6"
                              StrokeStartLineCap="Round"
                              StrokeEndLineCap="Round"
                              Data="M 0 0 C 50 20 100 0 150 20">
                            <Path.Stroke>
                                <SolidColorBrush Color="{Binding}" />
                            </Path.Stroke>
                        </Path>
                    </DataTemplate>
                </ComboBox.ItemTemplate>
            </ComboBox>
        </StackPanel>
    </Grid>
</AppBar>
```

为了保持程序简单，我没有实现对竖屏或辅屏模式进行调整的相应功能。这些模式会导致按钮和组合框重叠。

应用栏条目适用于显示在 FlipView 中的当前页。此外，两个 ComboBox 控件可能适用于页面，也就是说，适用于该页面与当前 InkFileManager 相关的 InkDrawingAttributes 对象，或者适用于页面上选中的条目。打开应用栏时，这些控件必须适当初始化。

项目: YellowPad | 文件: MainPage.xaml.cs(片段)

```
void OnAppBarOpened(object sender, object args)
{
    InkFileManager inkFileManager = (InkFileManager)flipView.SelectedItem;

    copyAppBarButton.IsEnabled = inkFileManager.IsAnythingSelected;
    cutAppBarButton.IsEnabled = inkFileManager.IsAnythingSelected;
    pasteAppBarButton.IsEnabled = inkFileManager.InkManager.CanPasteFromClipboard();
    deleteAppBarButton.IsEnabled = inkFileManager.IsAnythingSelected;

    if (!inkFileManager.IsAnythingSelected)
    {
        // Set initial selected item
        Size size = inkFileManager.InkDrawingAttributes.Size;
        penSizeComboBox.SelectedItem = (size.Width + size.Height) / 2;
        colorComboBox.SelectedItem = inkFileManager.InkDrawingAttributes.Color;
    }
    else
    {
        penSizeComboBox.SelectedItem = null;
        colorComboBox.SelectedItem = null;
    }
}
```

此方法更复杂的版本会循环所选笔画并确定它们是否都有相同的颜色或宽度。如果是，这些值就可以用来初始化两个 ComboBox 控件。现在，如果选择笔画，每个 ComboBox 没有赋予所选值。

除了必须通过从 FlipView 的 SelectedItem 属性提供的 InkFileManager 访问 InkManager，处理四个剪贴板项都非常类似于之前的程序。

项目: YellowPad | 文件: MainPage.xaml.cs(片段)

```
public sealed partial class MainPage : Page
```

```
{
    ...
    void OnCopyAppBarButtonClick(object sender, RoutedEventArgs args)
    {
        InkFileManager inkFileManager = (InkFileManager)flipView.SelectedItem;
        inkFileManager.InkManager.CopySelectedToClipboard();

        foreach (InkStroke inkStroke in inkFileManager.InkManager.GetStrokes())
            inkStroke.Selected = false;

        inkFileManager.RenderAll();
        bottomAppBar.IsOpen = false;
    }

    void OnCutAppBarButtonClick(object sender, RoutedEventArgs args)
    {
        InkFileManager inkFileManager = (InkFileManager)flipView.SelectedItem;
        inkFileManager.InkManager.CopySelectedToClipboard();
        inkFileManager.InkManager.DeleteSelected();
        inkFileManager.RenderAll();
        bottomAppBar.IsOpen = false;
    }

    void OnPasteAppBarButtonClick(object sender, RoutedEventArgs args)
    {
        InkFileManager inkFileManager = (InkFileManager)flipView.SelectedItem;
        inkFileManager.InkManager.PasteFromClipboard(new Point());
        inkFileManager.RenderAll();
        bottomAppBar.IsOpen = false;
    }

    void OnDeleteAppBarButtonClick(object sender, RoutedEventArgs args)
    {
        InkFileManager inkFileManager = (InkFileManager)flipView.SelectedItem;
        inkFileManager.InkManager.DeleteSelected();
        inkFileManager.RenderAll();
        bottomAppBar.IsOpen = false;
    }
    ...
}
```

处理两个 ComboBox 控件非常相似。在这两种情况下，要么赋予 InkFileManager 的 InkDrawingAttributes 对象新值用于未来的绘制，要么新值用来更新所选笔画。

```
项目: YellowPad | 文件: MainPage.xaml.cs(片段)
public sealed partial class MainPage : Page
{
    ...
    void OnPenSizeComboBoxSelectionChanged(object sender, SelectionChangedEventArgs args)
    {
        if (penSizeComboBox.SelectedItem == null)
            return;

        InkFileManager inkFileManager = (InkFileManager)flipView.SelectedItem;

        double penSize = (double)penSizeComboBox.SelectedItem;
        Size size = new Size(penSize, penSize);

        if (!inkFileManager.IsAnythingSelected)
        {
            inkFileManager.InkDrawingAttributes.Size = size;
            inkFileManager.UpdateAttributes();
        }
        else
        {
            foreach (InkStroke inkStroke in inkFileManager.InkManager.GetStrokes())
                if (inkStroke.Selected)
                {
                    InkDrawingAttributes drawingAttrs = inkStroke.DrawingAttributes;
```

```
                        drawingAttrs.Size = size;
                        inkStroke.DrawingAttributes = drawingAttrs;
                    }
                inkFileManager.RenderAll();
            }
        }

        void OnColorComboBoxSelectionChanged(object sender, SelectionChangedEventArgs args)
        {
            if (colorComboBox.SelectedItem == null)
                return;

            InkFileManager inkFileManager = (InkFileManager)flipView.SelectedItem;

            Color color = (Color)colorComboBox.SelectedItem;

            if (!inkFileManager.IsAnythingSelected)
            {
                inkFileManager.InkDrawingAttributes.Color = color;
                inkFileManager.UpdateAttributes();
            }
            else
            {
                foreach (InkStroke inkStroke in inkFileManager.InkManager.GetStrokes())
                    if (inkStroke.Selected)
                    {
                        InkDrawingAttributes drawingAttrs = inkStroke.DrawingAttributes;
                        drawingAttrs.Color = color;
                        inkStroke.DrawingAttributes = drawingAttrs;
                    }
                inkFileManager.RenderAll();
            }
        }
    }
```

程序当然还存在一些缺陷。例如，可以为当前页面或所选笔画设置颜色和笔的宽度，但不能设置值用于所有将来新建的页面。每个新页面会从 InkFileManager 类里默认的硬编码开始。

程序真正需要的是 GridView 控件，用于显示所有页面的缩略图并允许你可以在浏览页面或选择页面进行删除或打印，甚至分组页面。

但是，这就是软件的本质。软件从来不会真正完成，因为不需要完成，在这方面，它和书很不一样。

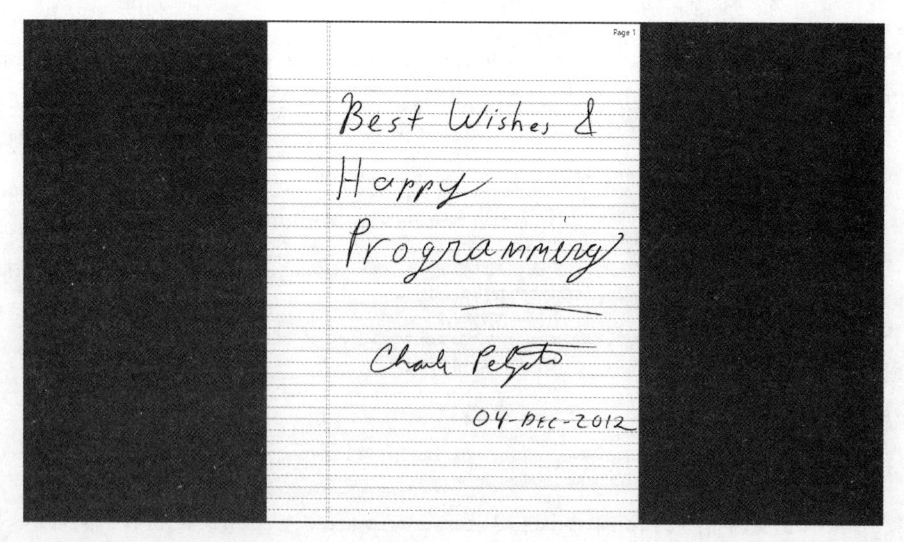